T0189254

Lecture Notes of the Institute for Computer Sciences, Social Informatics and Telecommunications Engineering 468

The LNICST series publishes ICST's conferences, symposia and workshops. It reports state-of-the-art results in areas related to the scope of the Institute.

LNICST reports state-of-the-art results in areas related to the scope of the Institute. The type of material published includes

- Proceedings (published in time for the respective event)
- Other edited monographs (such as project reports or invited volumes)

LNICST topics span the following areas:

- General Computer Science
- E-Economy
- E-Medicine
- Knowledge Management
- Multimedia
- Operations, Management and Policy
- Social Informatics
- Systems

Weina Fu · Lin Yun
Editors

Advanced Hybrid Information Processing

6th EAI International Conference, ADHIP 2022
Changsha, China, September 29–30, 2022
Proceedings, Part I

Editors
Weina Fu
Hunan Normal University
Changsha, China

Lin Yun
Harbin Engineering University
Harbin, China

ISSN 1867-8211 ISSN 1867-822X (electronic)
Lecture Notes of the Institute for Computer Sciences, Social Informatics
and Telecommunications Engineering
ISBN 978-3-031-28786-2 ISBN 978-3-031-28787-9 (eBook)
https://doi.org/10.1007/978-3-031-28787-9

This Springer imprint is published by the registered company Springer Nature Switzerland AG
The registered company address is: Gewerbestrasse 11, 6330 Cham, Switzerland

Preface

We are delighted to introduce the proceedings of the Sixth European Alliance for Innovation (EAI) International Conference on Advanced Hybrid Information Processing (ADHIP 2022). This conference brought together researchers, developers and practitioners from around the world who are leveraging and developing hybrid information processing technologies as well as related learning, training, and practice methods. The theme of ADHIP 2022 was "Hybrid Information Processing in Meta World".

The technical program of ADHIP 2022 consisted of 109 full papers, which were selected from 276 submissions by at least 3 reviewers each, including 2 invited papers in oral presentation sessions at the main conference tracks. The conference tracks were: Track 1, Information Extraction and Processing in Digital World; Track 2, Education Based Methods in Learning and Teaching; Track 3, Various Systems for Digital World. The technical program also featured two keynote speeches: "Graph Learning for Combinatorial Optimization", by Yun Peng from Guangzhou University, China, which focused on the advantages of using graph learning models for combinatorial optimization, and presented a thorough overview of recent studies of graph learning-based CO methods and several remarks on future research directions; and "Intelligent Fire Scene Analysis Using Efficient Convolutional Neural Networks", by Khan Muhammad from Sungkyunkwan University, Republic of Korea, which presented currently available approaches for early fire detection and highlighted some major drawbacks of current methods, and it also discussed a few representative vision-based fire detection, segmentation, and analysis methods along with the available fire datasets, and the major challenges in this area.

Coordination with the steering chair Imrich Chlamtac was essential for the success of the conference. We sincerely appreciate his constant support and guidance. It was also a great pleasure to work with such an excellent organizing committee team, we appreciate their hard work in organizing and supporting the conference. In particular, the Technical Program Committee, led by our TPC Chair, Khan Muhammad who completed the peer-review process of technical papers and made a high-quality technical program. We are also grateful to the Conference Managers, to Ivana Bujdakova for her support, and to all the authors who submitted their papers to the ADHIP 2022 conference.

We strongly believe that the ADHIP conference provides a good forum for researchers, developers, and practitioners to discuss all the science and technology aspects that are relevant to hybrid information processing. We also expect that future ADHIP conferences will be as successful and stimulating, as indicated by the contributions presented in this volume.

Weina Fu
Lin Yun

Organization

Steering Committee

Chair

Imrich Chlamtac — University of Trento, Italy

Organizing Committee

General Chair

Yun Lin — Harbin Engineering University, China

TPC Chair

Khan Muhammad — Sungkyunkwan University, Republic of Korea

Web Chair

Lei Chen — Georgia Southern University, USA

Publicity and Social Media Chair

Jerry Chun-Wei Lin — Western Norway University of Applied Sciences, Norway

Workshop Chair

Gautam Srivastava — Brandon University, Canada

Sponsorship and Exhibits Chair

Marcin Wozniak — Silesian University of Technology, Poland

Publications Chair

Weina Fu	Hunan Normal University, China

Posters and PhD Track Chair

Peng Gao	Hunan Normal University, China

Local Chair

Cuihong Wen	Hunan Normal University, China

Technical Program Committee

Adam Zielonka	Silesian University of Technology, Poland
Amin Taheri-Garavand	Lorestan University, Iran
Arun Kumar Sangaiah	Vellore Institute of Technology, India
Ashutosh Dhar Dwivedi	Technical University of Denmark, Denmark
Chen Cen	Institute for Infocomm Research, Singapore
Chunli Guo	Inner Mongolia University, China
Dan Sui	Changchun University of Technology, China
Danda Rawat	Howard University, USA
Dang Thanh	Hue Industrial College, Vietnam
Dongye Liu	Inner Mongolia University, China
Fanyi Meng	Harbin Institute of Technology, China
Feng Chen	Xizang Minzu University, China
Fida Hussain Memon	JEJU National University, Republic of Korea
Gautam Srivastava	Brandon University, Canada
Guanglu Sun	Harbin University of Science and Technology, China
Hari Mohan Pandey	Edge Hill University, UK
Heng Li	Henan Finance University, China
Jerry Chun-Wei Lin	Western Norway University of Applied Sciences, Norway
Jianfeng Cui	Xiamen University of Technology, China
Keming Mao	Northeastern University, China
Khan Muhammad	Sungkyunkwan University, Republic of Korea
Lei Ma	Beijing Polytechnic, China
Marcin Woźniak	Silesian University of Technology, Poland
Mu-Yen Chen	National Cheng Kung University, Taiwan
Norbert Herencsar	Brno University of Technology, Czech Republic

Contents – Part I

Contents – Part II

Practical Model of College Students' Innovation and Entrepreneurship Education Based on Social Cognitive Career Theory

Miaomiao Xu[1] and Jun Li[2,3](✉)

[1] Anhui Xinhua University, Hefei 230088, China
[2] Shenzhen Academy of Inspection and Quarantine, Shenzhen 518033, China
jyh20010@126.com
[3] Shenzhen Customs Information Center, Shenzhen 518033, China

Abstract. In the face of fierce competition in the employment environment, it is very necessary for college students to carry out innovation and entrepreneurship education. The practical model of college students' innovation and entrepreneurship education has been running for a long time. Therefore, a practical model of college students' innovation and entrepreneurship education is designed. By building a bridge between students and the society, we can extract the characteristics of the times, obtain the types of social practice, formulate a modular practice system in line with the actual situation of colleges and universities, improve the awareness of entrepreneurship, optimize the innovation and entrepreneurship education mechanism, and build a practical model of innovation and entrepreneurship education for college students. The experimental results show that the running time of the model is shorter than that of the other two models, which shows that the model is more effective when combined with social cognitive career theory.

Keywords: Social cognitive career theory · College students · Innovation and entrepreneurship · Educational practice · Teaching mode · Social environment

1 Introduction

At present, with the deepening reform of College entrepreneurship education, college students' entrepreneurship education has achieved certain results. At the same time, the new normal of economic and social development also puts forward new requirements of the times for college students' entrepreneurship education. Due to the complexity of society and the difference in interpretation of rules, college students may encounter many unpredictable difficulties and problems in the initial stage of entrepreneurship. In the face of these difficulties, some college students' entrepreneurial passion and enthusiasm are stifled, and they have the idea of shrinking. Under the multiple effects of national policy support, social coordination and University promotion, more and more college students have started their own entrepreneurial journey. Therefore, entrepreneurship education

W. Fu and L. Yun (Eds.): ADHIP 2022, LNICST 468, pp. 1–13, 2023.
https://doi.org/10.1007/978-3-031-28787-9_1

and practice of college students must become a common priority of the whole society [1, 2].

The current "entrepreneurship" is not only a means of creating wealth and making a living, but also a process of acquiring ability and value orientation that is internalized in the heart and externalized in the line. It is also a process of educational appeal that is related to the atmosphere of the times. This requires ideological and political educators to give full play to their abilities and roles, and make efforts to create a good entrepreneurial atmosphere for college students by means of publicity and coordination. College Students' entrepreneurship education is not only the cultivation of entrepreneurship knowledge and skills, but also the education of ability, quality and values in the field of Ideological and political education. How to train college students to establish entrepreneurial awareness, cultivate entrepreneurial spirit, improve entrepreneurial ability, lead them to internalize advanced entrepreneurial ideas into their own thinking patterns and motivation judgments, and externalize them into specific entrepreneurial behavior patterns is not only the logical premise for colleges and universities to carry out entrepreneurial education, but also the social realistic demand to realize entrepreneurship to drive employment and alleviate employment pressure, and is also the key to promoting the all-round development of college students. To this end, a practical model of College Students' innovation and entrepreneurship education based on social cognitive vocational theory is proposed, which builds a bridge between students and society through entrepreneurship practice activities, extracts the characteristics of the era of College Students' entrepreneurship education, obtains social practice types, formulates a modular practice system that conforms to the reality of colleges and universities, improves entrepreneurship awareness, optimizes innovation and entrepreneurship education mechanisms, and constructs a practical model of College Students' innovation and entrepreneurship education. College Students' Entrepreneurship must be closely combined with the reality of economic development, integrated into the general environment of social development, and realized self entrepreneurship and promoted social employment and economic development through a variety of ways such as sole proprietorship, joint-stock operation, relying on franchises and agents. Open up a path of entrepreneurship for college students with Chinese characteristics. Developing effective entrepreneurship education for college students is undoubtedly the best way to resolve the conflict of employment concepts and realize entrepreneurship education.

2 Practical Model of College Students' Innovation and Entrepreneurship Education Based on Social Cognitive Career Theory

2.1 Extracting the Time Characteristics of College Students' Entrepreneurship Education

The accelerated iteration of current new business forms, new technologies and new industries has brought unprecedented opportunities and challenges to the economic and social development, and the revolution of big data, artificial intelligence and Internet has led to technological innovation and progress.This new era of multi-innovation and integrated development provides entrepreneurs with a brand-new entrepreneurial living environment and unlimited space for innovation imagination, and builds a good sharing platform for the collaborative use of new resources, new projects and new information. With the continuous enrollment expansion of colleges and universities, higher education has changed from the traditional elite education model to the popular education stage, but the current higher education concept and system are still the same as the elite education, and do not adapt to the current situation of the popular education, and the curriculum system, education methods, teaching mode and other aspects are still the same as the original way. The people are the indispensable decisive factor in the process of social development. They are the historical direction and ultimate power of innovation and development. At present, the Party and the State have released the era horn of innovation-driven development, and the people need to cherish the concept of innovative developnt and actively participate in the development and reform of the times. There are three characteristics of the age of entrepreneurship education for college students, as shown in Fig. 1:

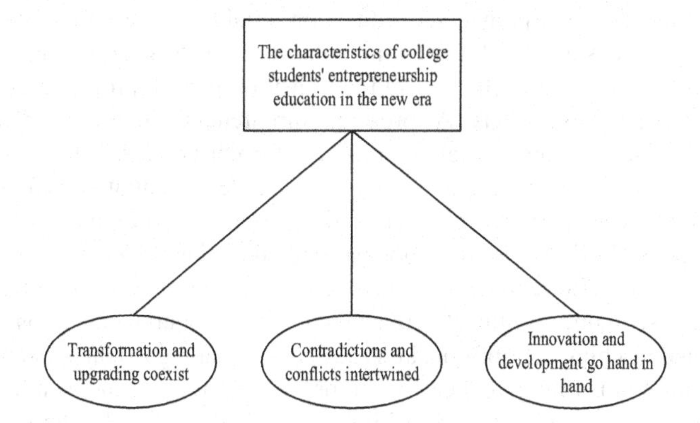

Fig. 1. Characteristics of college students in entrepreneurial education

As can be seen from Fig. 1, the characteristics of college students' entrepreneurship education mainly include: innovation and development, transformation and upgrading, contradictions and conflicts. Facing the employment problem of college students, most colleges and universities still pay attention to the increase and decrease of the employment rate of students, but ignore the essential role of entrepreneurship education. The coexistence of innovation and development can effectively stimulate the entrepreneurial enthusiasm of contemporary entrepreneurs, maximize their entrepreneurial thinking, and pursue the right entrepreneurial ideals and values. In the new historical position, the coexistence of innovation and development makes entrepreneurial activities more universal and popular, and every citizen has become a potential entrepreneur in the future. On the basis of transferring knowledge to students, entrepreneurship education should cultivate students' entrepreneurship and entrepreneurial consciousness to meet the needs of modern market economy for talents [3, 4]. Therefore, it is necessary for colleges and universities to grasp the atmosphere of the times, combine their own advantages, think seriously about the development goal of entrepreneurship education, follow the law of development, and explore scientific and reasonable standards for the goal of entrepreneurship education for college students, so as to train college students to become pioneers in entrepreneurship who undertake the mission and responsibility of the times. Traditional education mode is a subject education mode based on knowledge, which establishes a system of subject knowledge, conveys knowledge needed to students, and shows knowledge context and subject structure. But with the development of social economy, the demand of human resource market is based on ability. The traditional education model is difficult to meet the demand of the market. Transformation and upgrading are the key words of social transformation period. With the development of economic globalization, political multipolarity and cultural pluralism, the adjustment of the social system, the decentralization of government power, the integration of multiple social resources and other multiple measures have promoted the transformation and upgrading of society, and formed the era pattern of diversified distribution of resources, diversified distribution of interests and power. At present, the majority of colleges and universities in our country are still based on traditional subject education, and entrepreneurship education is also based on traditional education mode. This kind of educational model has seriously hindered the effect that entrepreneurship education should achieve, and the development of all social forces has achieved a dynamic balance. Historical experience shows that with the economic and social changes in the transition period, profound changes will take place in all areas of society. At present, entrepreneurship education in colleges and universities lacks a good social environment for entrepreneurship. Good campus environment and social environment greatly affect students' initiative and enthusiasm. Along with the transformation of economic structure, mechanism, interest distribution and concept, people's life style, behavior pattern, educational idea and value concept will be reconstructed. Entrepreneurial practice is an important part of entrepreneurial education, but also to improve the effectiveness of entrepreneurial education is the basic way. Traditional teaching methods in class are mainly teaching-oriented, students are in a passive position in learning, and can not improve the ability of independent thinking and innovation. The fierce battle for position in the ideological field, the lack of belief in the social value level and the distortion of value in the individual level caused by the erosion of the Western bad value concept and so on, these bad effects have changed the original cultural ecology, political ecology, value ecology and entrepreneurial ecology of our country. By building a bridge between the students and the society, the students can form a correct goal and value orientation of starting an undertaking in practice, strengthen the ability of organization and management, the ability of social contact and the ability of practice, and enhance the comprehensive quality of starting an undertaking in an all-round way. College students are at the stage of world outlook, outlook on life, stable and mature values, sensitive to new things, strong ability to accept, can adapt to social development and change quickly, but also vulnerable to the negative impact of multiple complex social factors, especially in the ideological field and the field of values.

2.2 Get Social Practice Types

Innovative and entrepreneurial professional social practice is a kind of social practice which is carried out in the process of professional teaching, in order to achieve the goal of students' practical application and to cultivate innovative and entrepreneurial talents.

College students' entrepreneurship education is carried out under the guidance of Marxist view of practice. First, the practical goal of college students' entrepreneurship education should be specific. This kind of social practice emphasizes the professionalism as the basis and the innovation and entrepreneurship as the aim. Internship of innovation and entrepreneurship requires students to have good basic quality, solid professional foundation, strong ability of transfer and conversion, and based on their own interests and hobbies, lay the practical needs for innovation and entrepreneurship [5, 6]. The aim of practical education is the basis and direction of college students' entrepreneurship education. Establishing a modular practical system in line with the actual situation of colleges and universities can guide the sound development of entrepreneurship education, correct the cognitive deviation in theory teaching in time, guide the entrepreneurship education not to deviate from the correct track and guide college students to establish a correct concept of entrepreneurship practice. The practice of innovative and entrepreneurial specialty emphasizes the cultivation of students' entrepreneurial skills. Compared with traditional professional practice, innovative and entrepreneurial professional practice is more inclined to the cultivation of entrepreneurial skills. Second, expand the carrier and mode of entrepreneurship practice. To some extent, rich practical carriers and diversified modes can make up for the deficiency of small audience and high similarity of cognitive behavior, which is the key factor affecting the effect of college students' entrepreneurial practice. As a result, the types of social practice are shown in Fig. 2:

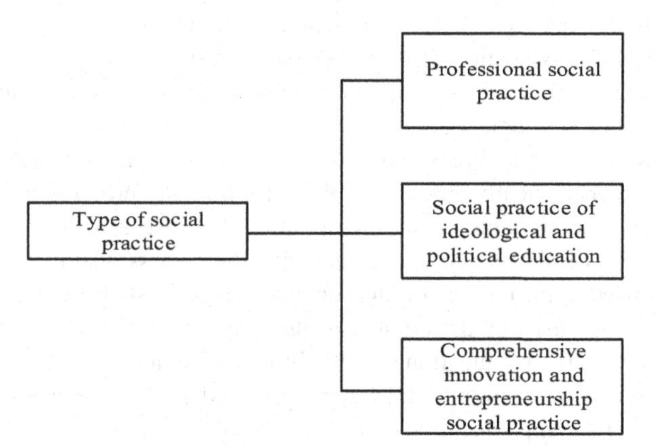

Fig. 2. Types of social practice

Through the Fig. 2, we can see that the types of social practice mainly include: professional social practice, ideological and political education social practice, comprehensive innovation and entrepreneurship social practice. Innovative and entrepreneurial professional practice brings new challenges to colleges and universities, puts forward new requirements for professional construction, sets new benchmarks for students' professional training programs, and must combine innovative and entrepreneurial education with professional education so as to ultimately create conditions that can meet the requirements of innovative and entrepreneurial professional practice. First, the school actively provide innovative and entrepreneurial professional internship opportunities.

Second, students actively and actively to find their own suitable practice units. Third, the all-round grasp of innovation and entrepreneurship education in professional education integration, serious thinking, the courage to practice. Colleges and universities should actively expand the methods of entrepreneurial practice education, explore interactive discussion, brainstorming, experience sharing and other teaching models. Professional research emphasizes on the basis of professional learning, thoughtful, problematic investigation. Compared with the traditional professional research, the innovation and entrepreneurship professional research emphasizes the innovation ability and the entrepreneurship analysis ability, and the research subject is more inclined to innovation and entrepreneurship. At the same time, a good atmosphere for entrepreneurship practice shall be created in combination with the students' entrepreneurship associations, students' entrepreneurship parks, incubators, maker studios, practice platforms for school-enterprise cooperation and other physical situations in colleges and universities, and the entrepreneurship practice shall be independently carried out with the assistance of guiding teachers, and emphasis shall be laid on the evaluation of the effectiveness of practice, so as to avoid the practice education of entrepreneurship practice becoming a mere formality. Through the professional survey of innovation and entrepreneurship, we shall understand the comprehensive application of professional contents in the society, deepen the grasp of the future development situation of this specialty, establish the social cognitive ability of students, analyze the possibility of this specialty taking the road of entrepreneurship in the future career, increase the understanding of entrepreneurship, and invest the theoretical knowledge and theoretical research learned in innovation and entrepreneurship. The education of entrepreneurial values is an important part of college students' education of entrepreneurial values, which is consistent and reasonable with the contents of ideological and political education. In order to truly carry out the professional investigation and research on innovation and entrepreneurship, the operation requirements are basically the same as those for ordinary professional investigation and research, and the contents of the investigation and research shall conform to the teaching objectives of the specialty; and secondly, the professional knowledge and the theoretical knowledge on innovation and entrepreneurship shall be combined to find the entrepreneurship analysis direction that they pay attention to. At present, college students' entrepreneurship education is still a little inadequate, even marginalized, lack of due attention and recognition, which seriously hinders the effectiveness of college students' entrepreneurship education.

2.3 Optimize the Innovation and Entrepreneurship Education Mechanism

Innovation and entrepreneurship internship is an important part of the students' complete learning system. It pays attention to stimulate students' ideas of innovation and entrepreneurship in the process of internship. Where conditions permit, teachers with practical experience in entrepreneurship shall be selected as the teaching team of entrepreneurship education for college students in priority. Compared with the traditional professional probation, the probation of innovation and entrepreneurship pays more attention to the construction of practice base. Innovative and entrepreneurial internship emphasizes that schools should actively build innovative and entrepreneurial internship bases for students to provide conditions and create a noviciate environment. At present, many colleges and universities in our country tend to select teachers of economic management, who have solid knowledge of economic management, and can understand and analyze the economic phenomena in the current social economic activities with economic management theory. In order to ensure the smooth construction of the subsequent model, the idle channel transmission mode is adopted. Therefore, when a node p transmits messages with a distance of β and a packet of δ bits to the node q, the energy consumption is as follows:

$$G = \sum \frac{|\beta - \delta|^2}{l(p, q)} \tag{1}$$

Meanwhile, when a node q receives the data sent by the node p, the energy consumed is:

$$H = \frac{1}{\phi \times \gamma} \tag{2}$$

In formula (2), ϕ represents the energy consumed by the sending circuit and the receiving circuit during data transmission, and γ represents the distance between the sending node and the receiving node. Based on formulas (1) and (2), a formula for calculating the threshold is derived:

$$L = \frac{\sigma^2}{1 - \frac{1}{\eta}} \times (\gamma - \eta) \tag{3}$$

In formula (3), σ represents the current remaining energy of the sensor node, and η represents the initial energy of the node. The level of innovation and entrepreneurship education practice base should be multi-level, the form should be rich and diverse. In addition to national innovation and entrepreneurship internship bases, innovation and entrepreneurship internship bases at the national, provincial, prefectural, school and college levels, innovation and entrepreneurship community bases, innovation and entrepreneurship job bases and innovation and entrepreneurship incubation bases shall be established in various forms. Therefore, colleges and universities should invite entrepreneurs and entrepreneurs with practical experience to the university campus, and share the entrepreneurial experience with students by giving lectures and seminars. Many colleges and universities have made beneficial attempts, such as Shanghai University of Technology has built a "Student Innovation and Entrepreneurship Center", which has many distinctive platforms for innovation and entrepreneurship. Such as virtual manufacturing technology platform, numerical control manufacturing technology platform, electrical automation technology platform, medical equipment and food safety technology platform, public business service platform, women's vocational coaching camp, etc., to become a base for innovation and entrepreneurship. At the same time, teachers teaching college students entrepreneurship education courses should create opportunities for them to go out of campus, enter the enterprise, personally experience the process of entrepreneurship practice, and improve students' entrepreneurship practice ability [7]. Colleges and universities should provide more opportunities for the teachers who are engaged in the course teaching of college students' entrepreneurship education to study and communicate, and improve their entrepreneurship analysis ability and theoretical level. And create appropriate conditions, so that they can really participate in some entrepreneurial projects to experience the practice of entrepreneurship, accumulation of entrepreneurial experience, entrepreneurship teaching theory and practice together, so that the overall level of entrepreneurship education has been improved. Therefore, schools should build a comprehensive platform for innovation and entrepreneurship education according to resources, professional characteristics and teaching needs, and establish various innovation and entrepreneurship education practice bases.

2.4 Construction of Practice Model of Social Cognitive Career Theory

Social Cognitive Career Theory introduces the evolution of career development from the perspective of social cognition, trying to explore the complex relationship between individual characteristics and occupational background, cognitive and interpersonal factors, and the complex relationship between self-orientation and the impact of external environment on career. The main theoretical source of social cognitive career theory is Bandura's social cognitive theory, which is the application of social cognitive theory in career development. The purpose of social cognitive career theory is to provide a potential unified theoretical framework. In the theory of socio-cognitive occupations, cognitive variables, personal factors and background factors are the three core variables that interact with each other in the theory, as shown in Fig. 3:

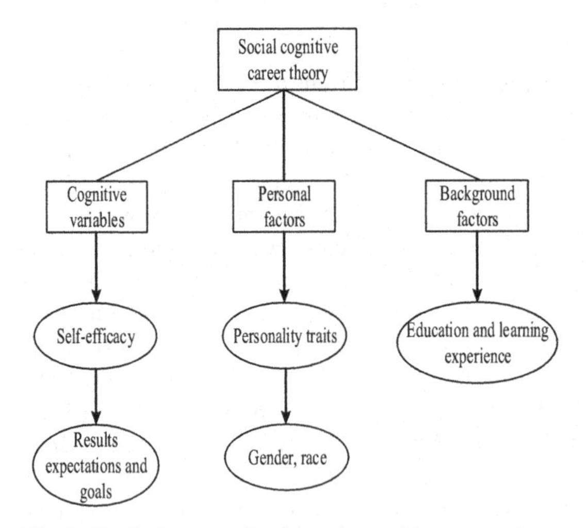

Fig. 3. Basic framework of social cognitive career theory

As can be seen from Fig. 3, in addition to cognitive variables, personal factors and background factors, it also includes: self-efficacy, expectation of results, personal goals, personality traits, gender, race, education, learning experience and other details. Expectation of results refers to an individual's belief that certain actions will lead to certain results. Personal goals are the goals that individuals expect to achieve in their specific activities. Social Cognitive Career Theory, which emphasizes the interaction between self-directed thoughts and social processes in guiding human behavior, has proved to be heuristic and applied in a wide range of psychosocial fields, such as career choice, educational achievement, organizational management and emotional response. Social cognitive career theory includes three interrelated sub-models, namely, occupational interest model, occupational choice model and job performance model, in which the three core variables play a significant role. One of the aims of social cognitive career theory is to update and expand the research by establishing closer links with the advancement of social cognitive theory and the empirical basis of occupational and non-occupational fields. Innovative and entrepreneurial school association is a combination of market-oriented operation mode and project management. Students learn the relevant skills of market operation, project development and project management and try to start a business. Entrepreneurial ability refers to the efficiency of entrepreneurial activities caused by the entrepreneur's subjective initiative, which is one of the conditions to effectively ensure the smooth progress of entrepreneurial activities [8]. It is different from the ordinary associations in colleges and universities that enterprises carry out innovative and entrepreneurial activities in the form of associations to attract students to start their own businesses. Entrepreneurial ability mainly includes people's general ability, as well as insight into market business opportunities, the ability to obtain and analyze business information, the ability to make quick decisions, the ability to coordinate teams, the ability to manage, the ability to adapt to society, the ability to communicate and coordinate, and the ability to innovate independently. Because of the support and work habits of the enterprise, the innovative and entrepreneurial school associations help students to learn different knowledge structures in terms of work concepts, operation methods and evaluation standards, and also further enhance the enterprise culture. The cultivation of entrepreneurial ability is a key part of entrepreneurship education in colleges and universities. At the beginning of entrepreneurship, entrepreneurship ability such as professional knowledge and skills is an essential and rigid condition for entrepreneurs, and has the ability to capture and analyze market information sensitively. Innovative and entrepreneurial school associations are an effective carrier for entrepreneurship practice, which speeds up the process of students' socialization and strengthens students' entrepreneurship awareness. For this kind of activities, it is necessary to have teachers with strong entrepreneurial ability as instructors. The instructors play an important role in the practical activities of the associations, which will affect the extension or cessation of a student's entrepreneurial consciousness. Only with earnest, responsible and experienced instructors can the activities of associations of innovative and entrepreneurial schools realize its role in cultivating students' entrepreneurial consciousness. Master business management ability and interpersonal skills, to have the ability to know people, but also have to withstand setbacks of the ability to resist pressure. The summary of a lot of ability, just be complete pioneering ability.

3 Experimental Analysis

In order to verify the validity of the practical model of innovation and entrepreneurship education for college students, an experiment was conducted. The practical model of undergraduates' innovation and entrepreneurship education based on cluster analysis and the practical model of undergraduates' innovation and entrepreneurship education based on deep learning are selected to carry out experimental tests with the practical model of undergraduates' innovation and entrepreneurship education in this paper. Under the conditions of different correlation coefficients, the running time of the three models is respectively tested. The experimental results are shown in Tables 1, 2, 3 and 4:

Table 1. Correlation coefficient 0.105 Model run time (ms)

Number of experiments	Practical model of innovation and entrepreneurship education for college students based on cluster analysis	Practical model of innovation and entrepreneurship education for college students based on deep learning	Practical model of innovation and entrepreneurship education for college students
1	102.33	107.31	89.64
2	105.62	106.54	91.25
3	104.71	105.28	85.71
4	115.28	104.39	84.62
5	106.37	108.14	83.55
6	105.49	110.58	82.59
7	112.88	113.12	84.10
8	105.42	106.87	85.61
9	106.31	109.45	84.63
10	106.15	110.64	82.91

According to Table 1, the average running time of the model and other two models is 85.46 ms, 107.06 ms and 108.23 ms, respectively.

From the Table 2, the average running time of the two models and other two models is 72.01 ms, 92.933 ms and 92.33 ms.

According to Table 3, the average running time of the model and other two models is 55.98 ms, 75.60 ms and 76.18 ms, respectively.

According to the Table 4, the average running time of the model and other two models is 35.84 ms, 54.577 ms and 52.60 ms respectively.

To sum up, the college students' innovation and entrepreneurship education practice model based on the social cognition career theory has a shorter average running time and better educational practice effect. The reason is that this model introduces the evolution perspective of career development from the perspective of social cognition, and tries to explore the relationship between individual characteristics and career related backgrounds, The complex relationship between cognition and interpersonal factors, as well as the complex relationship between self orientation and the impact of external environment on career, is conducive to reducing running time to a certain extent.

Table 2. Correlation coefficient 0.205 Model run time (ms)

Number of experiments	Practical model of innovation and entrepreneurship education for college students based on cluster analysis	Practical model of innovation and entrepreneurship education for college students based on deep learning	Practical model of innovation and entrepreneurship education for college students
1	94.16	89.67	72.61
2	92.21	92.37	73.48
3	93.19	94.02	71.92
4	92.17	93.41	70.48
5	93.54	92.01	71.09
6	94.32	92.25	72.61
7	92.55	91.47	73.46
8	91.34	96.33	72.51
9	92.17	91.05	71.25
10	93.66	90.74	70.69

Table 3. Correlation coefficient 0.305 Model run time (ms)

Number of experiments	Practical model of innovation and entrepreneurship education for college students based on cluster analysis	Practical model of innovation and entrepreneurship education for college students based on deep learning	Practical model of innovation and entrepreneurship education for college students
1	76.39	74.91	60.33
2	75.84	75.43	59.74
3	73.77	78.13	58.61
4	75.60	76.92	55.33
5	77.01	77.03	54.02
6	75.94	76.12	56.71
7	76.43	78.09	55.29
8	75.82	75.49	54.11
9	74.19	74.52	53.62
10	75.02	75.16	52.07

Table 4. Correlation coefficient 0.405 Model run time (ms)

Number of experiments	Practical model of innovation and entrepreneurship education for college students based on cluster analysis	Practical model of innovation and entrepreneurship education for college students based on deep learning	Practical model of Innovation and entrepreneurship education for college students
1	56.14	52.09	33.64
2	55.33	53.47	36.87
3	56.10	52.84	35.91
4	55.94	56.37	34.28
5	54.81	52.91	36.14
6	53.67	52.49	35.77
7	52.88	51.27	34.82
8	53.14	52.08	35.96
9	54.66	51.69	36.77
10	53.02	50.79	38.20

4 Conclusion

The designed innovation and entrepreneurship education practice model for college students has a short average running time, has a good educational practice effect, can help students achieve the training goal of innovation and entrepreneurship talents, and is an important guarantee for improving the overall innovation and entrepreneurship ability. At the same time, it enriches the academic research materials on the practical model of College Students' innovation and entrepreneurship education.

In the follow-up research, the countermeasures to improve the quality of innovation and entrepreneurship education in Colleges and universities are to establish the concept of national strategic development, collaborative innovation development and people-oriented development; In the course design, we should build the curriculum system of innovation and entrepreneurship education, reasonably increase the proportion of practical courses, and increase the curriculum content of innovative thinking development; In the construction of teaching staff, we should improve the ability of innovation and entrepreneurship professional teachers, improve the overall quality of part-time teachers, and strengthen the reserve of innovation and entrepreneurship teachers; In optimizing the environment of innovation and entrepreneurship education, we should pay attention to the positive guidance of public opinion, create a good campus cultural atmosphere, and influence innovation and entrepreneurship education with family environment; On the issue of resource allocation, we should encourage more social resources to enter the innovation and entrepreneurship activities of college students, build an education platform combining industry, University and research, and carry out innovation and entrepreneurship education exchanges between schools and colleges, so as to complement each other's advantages. In terms of policy support, we will introduce guiding policies at the national level, improve supporting policies for local support, and strengthen supervision over policy implementation.

References

1. Ma, X., Xia, F.: Research on the practice situation and countermeasures of college students' innovation and entrepreneurship education in provincial general universities. Theory Pract. Innov. Entrepreneurship **3**(23), 196–198 (2020)
2. Luo, Y., Yang, G., Zhang, L.: Giving full play to the practical and educational function of the communist youth league to promote college students' innovation and entrepreneurship education:taking the youth league committee of kunming medical university as an example. Educ. Teach. Forum (31), 124–125 (2020)
3. Zu, Q., Wei, Y.: Discussion on formation logic and construction significance of provincial college students' innovation and entrepreneurship practice education center. Exp. Technol. Manag. **37**(6), 15–18, 23 (2020)
4. Liu, H., Zhu, J.: Practice of innovation and entrepreneurship education for students in eco-environmental protection in local colleges: a case from yangtze normal university. Educ. Teach. Forum (42), 212–214 (2020)
5. Fang, Y.: Practice and exploration of integrating ideological and political elements into college students' innovation and entrepreneurship education under the background of ideological and political curriculum. J. ZhaoTong Univ. **42**(4), 18–23 (2020)
6. Jia, C., Zhang, X.: Ideological and political value connotation and implementation path of innovation and entrepreneurship practice education for higher vocational college students. Theory Pract. Innov. Entrepreneurship **3**(21), 9–10, 13 (2020)
7. Li, C., Sun, Y., Yu, Z., et al.: The research and practice of the "golden course" of innovative and entrepreneurial education for college students. Theory Pract. Innov. Entreptrneurship **3**(12), 29–30 (2020)
8. Zhu, Z., Wang, L.: Simulation of evaluation model of entrepreneurial success rate for college students based on big data analysis. Comput. Simul. **35**(4), 162–165 (2018)

3D Visualization Method of Folk Museum Collection Information Based on Virtual Reality

He Wang[1(✉)] and Zi Yang[2]

[1] Changchun Humanities and Sciences College, Changchun 130051, China
wh15091@163.com
[2] Hubei Institute of Fine Arts, Wuhan 430060, China

Abstract. Aiming at the lack of real-time simulation rendering of Folk Museum collection information, a three-dimensional visualization method of Folk Museum collection information based on virtual reality is proposed. According to the collection information classification of Folk Museum, the exhibition space of museum is divided. The museum is structured, the rendering scene of collection information is constructed by using parametric scene description language, and these data are uniformly managed by means of spatial projection. According to the attribute value of the rendered scene, a three-dimensional visualization model of the collection information of the Folk Museum is established based on virtual reality. The role agent communicates with the outside rather than directly accessing the role, which can enhance the encapsulation of data and the reusability of code, so as to reduce the response delay. The test results show that the 3D visualization method of Folk Museum collection information based on virtual reality can shorten the information response time and optimize the real-time rendering process.

Keywords: Virtual reality · Folklore Museum · Collection information · 3D visualization · Parametric scenario · Interaction design

1 Introduction

The museum collects and preserves historical relics in order to display cultural heritage in various ways. It is a place for human beings to understand history and master science and technology. In the museum, people can appreciate the beauty of history and culture through personal experience and improve their personal cultural literacy and national spirit. Folk Museum can protect the inheritance of folk culture and bring positive social significance. Through various exhibitions and activities in the Folk Museum, the public can experience the infinite value of folk culture in a specific space. As an educational place to promote people's informal learning, with the development and reform of science and technology, the construction of digital museum and virtual museum not only expands the educational function of the museum, but also provides new ideas and methods for museum learning. With the development of the times, the understanding of museum collection resources in the field of museum research is constantly refreshed

W. Fu and L. Yun (Eds.): ADHIP 2022, LNICST 468, pp. 14–27, 2023.
https://doi.org/10.1007/978-3-031-28787-9_2

in its depth and breadth. In addition to the exquisite collections of large-scale comprehensive museums, the traditional customs activities closely related to people's life and established conventions, as well as the artifacts, artists and solar terms reflecting customs activities, as well as the related cultural space should also be included in the scope of museum collection resources research. With the rapid development of science and technology and the accelerating process of social informatization, digital resources have penetrated into our daily work, study and life. In the field of culture, the application of network technology and virtual reality technology in museum education has become a new communication development trend, Relying on computer network technology and virtual reality technology, digital museum and virtual museum have gradually become new development fields [1]. The information guidance function of museum collection is an indispensable auxiliary function of museum. The museum collection information is for visitors to uniformly plan and sort out the seemingly clueless venue zoning and browsing order according to the age, society, class, plot, function and other information, so as to guide tourists to a regular and orderly browsing scheme, and prompt and guide them in key places. Digital museum digitizes the collection resources, displays the collection with the help of network media, disseminates cultural knowledge and expands the audience area of the museum. The artifacts in the collection of Folk Museum are one of the construction of customs and habits. The performance of folk artists is the transmitter of people's beautiful demands in folk activities. Folk activities are the life wisdom extracted by ancestors according to the law of natural development. They are the most essential behavioral expression of regional culture and aborigines. The museum collection information guide makes people's tour experience in the museum clear, unified and distinctive. Because the space size, information classification and plot setting of the traditional museum cannot be reconciled, the design of the guidance system of the traditional museum has to cut off and reorganize the inherent relationship between some exhibits, so that some of the connections between exhibits can not be displayed intuitively. The application of virtual reality technology and three-dimensional modeling technology in digital museum shows the cultural relics collection or historical real scene through virtual system [2]. The collection information of Folk Museum is guided by three-dimensional visualization method, which makes the traditional museum change from the site built of reinforced concrete to the Virtual Museum of data transmitted on the Internet under the new media platform. The ideas and influencing factors in the design have also changed: it is no longer limited by the size of the site and the number of exhibits The exhibits can be classified and guided from various dimensions, and the guiding forms can be diversified.

Based on virtual reality, this paper proposes a three-dimensional visualization method of collection information of Folk Museum. Based on the classification of the collection information of Folk Museum, the parametric scene description language is selected. Based on virtual reality, a three-dimensional visualization model of Folk Museum collection information is established, and the three-dimensional visualization of Folk Museum collection information is realized. This method broadens the development scope for economy and culture through the development of derivative design, broaden the development caliber for economy and culture, so as to meet the people's various needs for local culture.

2 3D Visualization Method of Collection Information of Folk Museum Based on Virtual Reality

2.1 Classification of Collection Information of Folk Museum

Based on the characteristics of traditional art forms, the three-dimensional visualization of collection information of Folk Museum also has the characteristics of interactivity, entertainment, networking and so on. It is not limited by time and space, and can interact with the audience in three-dimensional space. It is an art born in the digital information age. The collection information of folk museum can be divided into two categories: Folk images and folk portraits. Its resource structure is shown in Fig. 1.

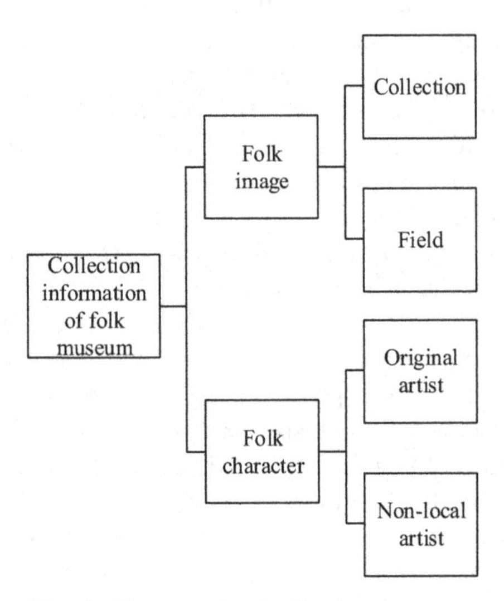

Fig. 1. Framework of collection resources

Three dimensional visualization makes the dominant static mode in the guidance system gradually change to dynamic mode. The dynamic mentioned here is not only the dynamic of visual elements, but also the dynamic of guidance data, guidance thinking and guidance path. The collection part of the Folk Museum includes artifacts, words and video, the field part includes the field area and region, the folk artist part includes native artists and non-native artists, and the group people include native people and non-native people. In this frame structure, "utensils" refer to the objects in the collection, which are used in folk activities, used by folk artists or have specific reference. Facing the full room of collections, the audience has the right to browse, choose, approach and stare, while the collections can only wait for the discovery and appreciation of the audience. Because the collection is not arranged in a narrative way, the dialogue between different collections is also quite limited. There may be an interesting story behind each collection, but there is no complete story that can be used to explain the collection of the whole room. Therefore, the display space at this time is fragmented and fragmented. The close relationship between Folk Museum and electronic equipment has laid the foundation for the dynamic display of collection information. More and more visual element symbols of virtual museum guidance system begin to transmit guidance content through dynamic images and data. The dynamic design of guidance system enriches its forms of expression and provides an opportunity for its diversified development. "Writing" refers to the written historical materials or written images with special meanings recording folk activities. "Shadow and sound" refers to the historical materials recorded in the form of video or audio or the scenes of current folk performances, practical operations and other activities. "Site area" refers to the space category of folk museum or limited local area. When the museum space is divided according to the function, it can be divided into service area, exhibition area, transition area, etc. other division methods include division according to time stage or display mode. "Region" refers to the regional spatial category divided by administrative regions. The difference between "native artist" and "non local artist" is whether the artist's folk performance or folk behavior is native. After determining the starting point and the ending point, the first-class guide moving line is formed for users according to time, material, region and other classifications. This moving line improves the visiting efficiency of visitors who are confused about the tour route and have low initiative. Considering that users with clear needs have interspersed the second-class moving lines of different paths in the first-class guide moving line, Without cutting off the connection between various lines, it is convenient for users with high initiative to visit more purposefully.

2.2 Select Parametric Scene Description Language

Folk Museum has certain particularity. The external architectural scenes and internal booths of the museum are relatively fixed, but the number of exhibits in the exhibition hall is large and scattered, the placement position of exhibits is changeable, and the information carried by exhibits is complex (including text information, audio information, video information, etc.). The variety and location of exhibits vary greatly. This requires to increase its variability and scalability when building virtual scenes. The realization of virtual scene is based on the three-dimensional simulation of the scene. Integrate the resources of the whole museum before building a virtual museum. In the three-dimensional visual management of the collection information of Folk Museum, most operations need to operate the graphics, and these operations on the graphics are based on the spatial database, so a large amount of spatial data will be used in the system. Because spatial data has a large number of spatial attributes, it is very important to manage these spatial data reasonably and effectively [3]. There are mainly two kinds of data for the collection information of Folk Museum: one is the data stored when building the system, including scene model, page and so on. The second is the data obtained in real time. For example, the location coordinate information of the first person needs to be transmitted in real time in the museum system. When a click event occurs, the click coordinates should be obtained and the corresponding text and audio data should be loaded. Due to the efficient accessibility, spatial database can store a large amount of data in a data set, can store seamless and continuous geographic data, and does not need to store geographic data in blocks [4]. Therefore, spatial database can store massive spatial data, which greatly enhances the ability of things processing. Due to the wide scope of the museum as a whole and the large amount of total model resources, there are more than ten exhibits in a theme museum, which occupy a large amount of resources after simulation. If all model resources of the whole museum are rendered when the system is loaded for the first time, the pressure on the server and memory will be great, which is bound to affect the efficiency of rendering and user experience [5]. In order to improve the efficiency of browser rendering, it is necessary to structure the museum and reduce the rendering range as much as possible. To sum up, this paper selects a parametric scene description language to build a virtual scene. XML based parametric scene description language has a good reputation for its proven architecture, its reliability, data integrity, powerful function set, scalability and the contribution of the active open source community behind its software, and has been providing users with high-performance and innovative solutions [6]. When constructing the scenario, it is implemented by XML parametric language. Considering the rendering efficiency of the whole system, the collision detection rate and the convenience of subsequent scene interaction, the scene is organized in a tree structure. For the motion model, the bounding box after motion is calculated, and the projection points in three axes are recorded. In the state of linear motion, it can be considered that the projection coordinates on the Z axis remain unchanged, so as long as the relationship between the Y axis and the X axis is considered. The projection points in motion satisfy the following relationships:

$$\begin{cases} \beta_{\min} = \varphi\alpha_{\min} + \gamma \\ \beta_{\max} = \beta_{\min} + \delta \\ \alpha_{\max} = \alpha_{\min} + \delta \end{cases} \tag{1}$$

In formula (1), $\beta_{\min}$ and $\beta_{\max}$ are the minimum and maximum values of the projection of the motion model on the Y axis respectively; $\alpha_{\min}$ and $\alpha_{\max}$ are the minimum and maximum values of the projection of the motion model on the X axis respectively; δ represents the side length of the moving bounding box; φ and γ are linear motion parameters respectively. Parameters φ and γ can be obtained according to the linear formula, and the calculation formula is as follows:

$$\begin{cases} \varphi = \dfrac{(\eta - \beta_{\min}) - (\eta' - \beta_{\min})}{(\eta - \alpha_{\min}) - (\eta' - \alpha_{\min})} \\ \gamma = \eta - \beta_{\min} - \varphi(\eta' - \alpha_{\min}) \end{cases} \tag{2}$$

In formula (2), η and η' respectively represent the coordinates of the projection points of the bounding box before and after the movement. Thus, the interactive management of attribute data, graphic image data and spatial relationship data is realized. These data are managed uniformly by means of spatial projection. XML based parametric scene description language has small volume and less space; The structure is simple, the scene can be built intuitively, and it is easy to modify; Strong cross platform and scalability; It is more flexible than data and has high access efficiency. It is suitable for 3D programs with high real-time requirements.

2.3 Establishment of 3D Visualization Model of Collection Information of Folk Museum Based on Virtual Reality

Virtual reality is generated through the interaction between learners and things in the situation, but there are no real things in the virtual situation. Therefore, interactive behavior should be set in the situation of the built virtual museum system. When learners trigger an instruction in the situation, the Folk Museum will give some feedback information, so as to achieve better experience effect [7]. In order to adapt to the complex and changeable Museum application scenarios, the three-dimensional visualization model of Folk Museum collection information needs to support dynamic application. Based on the idea of "model instance", this paper proposes a general role model (model algorithm attribute message) to form a configurable management mechanism, that is, the role algorithm and role attribute unit are granulated, and the role algorithm and attribute are dynamically bound around the specific three-dimensional application to form a role instance. The three-dimensional visualization model of Folk Museum collection information based on virtual reality should pay attention to interactive experience in the design and development, which is embodied in the model size, mapping, materials and information presentation. The design principles of these two functions correspond to the immersion, authenticity and multi perception of virtual reality technology. After exporting the 3D model in SketchUp and saving it in the "SKP" file format, it is necessary to import the exported model into lumion for real-time rendering. If you import the model directly, you may not see the effect directly after importing, which will affect the control of the overall scene. Through data-driven, enrich and improve role examples, and solve the problems of reusability and coupling of roles. Actor is the base class of the role class and has the general (basic) attributes of the role, such as role ID identification, role name, spatial location, association model, bounding box, selected status and other attributes. Therefore, make some preparations before importing, adjust lumion to the best state, and then import the model, so that there will be no phenomenon that the operation is not smooth after the model is imported or the imported model needs to be readjusted after the terrain is modified. Role agent is an encapsulation of a role and maintains all attribute and method information of the role. Communicating with the outside world through the role agent instead of directly accessing the role can enhance the encapsulation of data and the reusability of code. In the three-dimensional visual modeling of Folk Museum collection information, it is necessary to obtain the attribute value of each scene. For the data points in each relatively small area, the spatial difference method is used to solve the problem of sparse and irregular discrete distribution of museum information spatial data. After adjusting the position of the three-dimensional visualization model of the collection information of the Folk Museum and the positional relationship with the coordinate origin, the visualization models are separated and classified one by one. The core idea of spatial difference method is that the closer to the target, the greater the similarity and the greater the weight. Its weight is mainly weighted by the distance between the interpolation point and the known point. The difference point of Folk Museum collection information can be expressed as:

$$p(a,b) = \frac{\sum_{j=1}^{n} p_j \varpi_j}{\sum_{j=1}^{n} \varpi_j} \tag{3}$$

In formula (3), $p(a, b)$ represents the value of interpolation point p; p_j is the third of the sampling points j Values of points; ϖ_j is the second j The weight of a sample point in interpolation. The weight is calculated as follows:

$$\varpi_j = \left(\frac{1}{h_j}\right)^{\kappa} \tag{4}$$

In formula (4), h_j represents the distance from the j sampling point to the interpolation point; κ is the index, which significantly affects the interpolation results. The value in this paper is 2. Through the distribution of known points, data characteristics, accuracy requirements of desired results, etc., select appropriate search patterns and search ranges, as well as the number of known points to be searched, to control the way and number of their use, so as to achieve the required data modeling requirements. Import the three-dimensional models into lumion from large to small. Their positions are adjusted one by one in lumion, so that the models are rearranged and combined, thus completing the construction of the collection information scene of the Folk Museum. Virtual reality technology enables the audience to enter a newly created cultural time and space. At this time, the museum will introduce the curatorial narrative into the audience role-playing (selection), making the audience a part of the visiting situation. The three-dimensional visualization of some exhibits in the museum enables the audience to enter the original cultural time and space of the exhibits or the imagination space of artists in the virtual world, so as to further experience the objects. Considering the multi terminal application mode, the load balancer adopts a dynamic load balancing strategy based on ant colony algorithm. The load of service nodes is measured according to the load balancing factor, so as to iterate the ant colony algorithm and dynamically update the load strategy. Pheromone concentration after a task is:

$$\vartheta_t = \vartheta_{t-1}(1-\varepsilon) + \sum_{i=1}^{N} \Delta\vartheta_i \tag{5}$$

In formula (5), t represents the number of iterations; ϑ_t and ϑ_{t-1} represent the pheromone concentration after the t th and $t-1$ th iteration, respectively; ε represents Volatilization Coefficient; N represents the number of ants passing through the node; $\Delta\vartheta_i$ is the pheromone left by the i th ant at the node in this task assignment. The adaptive method is adopted to adjust the heuristic probability according to the setting threshold and the completion time of each iterative task. Therefore, the probability model of service node selection in an iteration process is obtained:

$$w_z = \frac{\mu_t \vartheta_t \sigma}{\sum_{t=1}^{m} \mu_t \vartheta_t \sigma} \tag{6}$$

In formula (6), z represents the serial number of the server node; w_z represents the probability of selecting the z th service node; μ_t represents the heuristic probability of ant routing during the t second task assignment; σ represents the load balancing measurement factor. Lumion can output still frame pictures and video animations. At this time, according to actual needs, local perspective views and lens scenes roaming from different perspectives are used to set them respectively. At the same time, in order to make the picture more vivid, the scene is given special effects such as near and far scene virtualization, time and lighting changes and transition lens, so as to enrich the display content of the lens. Make the display content of the lens richer. A complete scene role model is formed through model resource management, registration model, defining scene roles, associating role models, configuring role attributes and related behaviors (algorithms); 3D application configuration management provides scene production function, and uses role model resources to build specific 3D application scenes. By adding environment to the equation of embodied experience, the three-dimensional visualization model shifts attention to how the experience of screen and network media transcends the visual experience. Based on the understanding of place as a relationship environment, in this environment, local and global are composed of each other, which makes the Internet a potential kinship field. There are certain differences between the preview effect in lumion and the final exported image and video effect. The most obvious difference is that the final exported image and video, especially the video effect will be brighter than the preview effect, and the color scale will be whiter. In addition, once the picture is overexposed, it is difficult to adjust it in the post-processing software. If the picture is slightly dark, it can be adjusted to a certain extent in the post-processing software. Therefore, before exporting, adjust the exposure value a little smaller and the contrast slightly larger. Start the 3D application running service, load the visual application scenario and necessary data at runtime, and provide runtime services; 3D application visualization provides a three-dimensional and virtual simulation rendering environment. Users can complete specific 3D applications through human-computer interaction through the operation interface. So far, the design of 3D visualization method of collection information of Folk Museum Based on virtual reality has been completed.

3 Experimental Study

3.1 Experimental Preparation

The three-dimensional visualization of Folk Museum collection information is an important part of digital museum. As a media resource for visitors and learners to browse independently and experience learning, and also as a virtual environment for experience learning, its application and evaluation is a way to test the effectiveness and function of the system. Through surveying and mapping the site, the size parameters required for modeling are obtained, and the corresponding position data are collected by dynamic GPS. Then use CAD technology to carry out the overall plane planning and design of the Folk Museum, and then build the three-dimensional model on the basis of the plan. Based on the three-dimensional model, the real-time simulation rendering of the building scene is carried out. In the early stage, some map textures are obtained through digital cameras to facilitate the material map rendering of the building in the later stage. Finally, a three-dimensional visual dynamic display effect is output. The relevant configuration of this experiment is shown in Table 1.

Table 1. Experimental configuration

Configuration name		Parameter
Hardware configuration	CPU	i3-4160
	Main frequency	3.6 GHz
	Memory	4GB
	Graphics card	NVIDIA GeForce 9600 GT
	Video memory type	64-bit DDR3
Software environment	Windows 7 ultimate	64 bit operating system
	Visual Studio 2013	C++ language
	OSG	3.4
	OpenGL	4.3
	NVIDIA driver	174.16

3.2 Results and Analysis

In order to verify the application effect of the three-dimensional visualization method of Folk Museum collection information based on virtual reality, the response time of three-dimensional visualization of information data is selected as the test index to test the performance of the design method. The three-dimensional visualization methods of collection information of Folk Museum Based on Web GIS and data mining are compared. The response time of each three-dimensional visualization method is compared and analyzed under the condition that the collection information data of Folk Museum is 50G, 100G, 150G and 200G. The test results are shown in Tables 2, 3, 4 and 5 respectively. In order to more clearly compare the response time trends of different methods, the results in Tables 2, 3, 4 and 5 are graphically processed. As shown in Figs. 2, 3, 4 and 5. In order to more clearly compare the response time trends of different methods, the results in Tables 2, 3, 4 and 5 are graphically processed. As shown in Figs. 2, 3, 4 and 5.

Table 2. Response time of 50G collection information data (s)

Number of tests	Virtual reality	Web GIS	Data mining
1	**1.91**	2.56	2.84
2	1.84	2.63	**2.96**
3	1.65	2.86	2.53
4	1.56	2.59	2.82
5	1.63	2.65	2.75
6	1.72	2.73	2.49
7	1.65	2.44	2.85
8	1.52	**2.88**	2.65
9	1.61	2.62	2.82
10	1.82	2.51	2.86

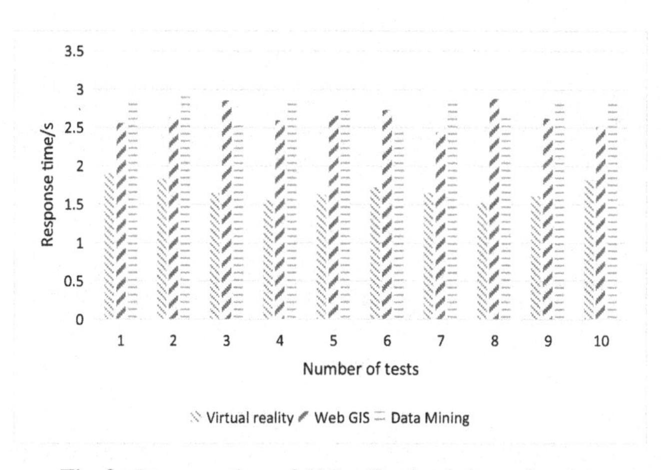

Fig. 2. Response time of 50G collection information data

In the test of 50G collection information data of Folk Museum, the response time of 3D visualization method of Folk Museum collection information based on virtual reality is 1.69 s, which is 0.95 s and 1.07 s shorter than that based on Web GIS and data mining.

In the test of 100G collection information data of Folk Museum, the response time of 3D visualization method of Folk Museum collection information based on virtual reality is 2.41 s, which is 1.00 s and 1.19 s shorter than that based on Web GIS and data mining.

Table 3. Response time of 100G collection information data (s)

Number of tests	Virtual reality	Web GIS	Data mining
1	2.23	3.16	3.57
2	2.36	3.54	3.66
3	2.47	3.48	3.25
4	**2.59**	3.66	3.42
5	2.36	3.25	3.84
6	2.55	3.32	3.51
7	2.52	3.51	3.68
8	2.43	**3.84**	3.35
9	2.31	3.25	**3.92**
10	2.25	3.12	3.83

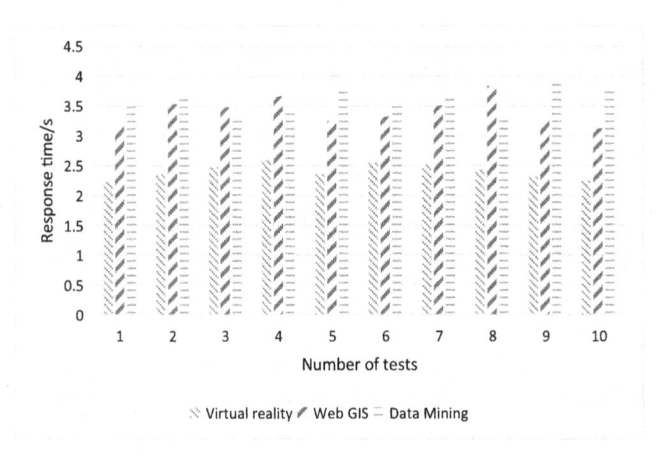

Fig. 3. Response time of 100G collection information data

In the test of 150G collection information data of Folk Museum, the response time of the three-dimensional visualization method of Folk Museum collection information based on virtual reality is 3.26 s, which is 1.43 s and 1.59 s shorter than the three-dimensional visualization method based on Web GIS and data mining.

In the test of 200G collection information data of Folk Museum, the response time of 3D visualization method of Folk Museum collection information based on virtual reality is 3.83 s, which is 2.69 s and 2.64 s shorter than that based on Web GIS and data mining. Therefore, the method proposed in this paper realizes the optimization of the real-time rendering process of the three-dimensional visualization of the collection information of the Folk Museum, and solves the problem that the collection information of the Folk

Table 4. Response time of 150G collection information data (s)

Number of tests	Virtual reality	Web GIS	Data mining
1	3.24	4.59	4.86
2	3.17	4.67	4.73
3	3.36	4.78	**4.97**
4	3.08	**4.85**	4.84
5	3.18	4.62	4.61
6	3.29	4.75	4.82
7	3.35	4.63	4.95
8	**3.42**	4.51	4.93
9	3.36	4.72	4.90
10	3.18	4.80	4.91

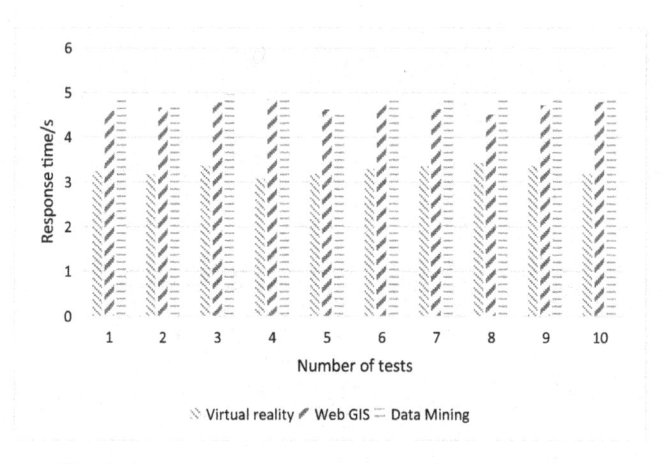

Fig. 4. Response time of 150G collection information data

Museum is difficult to respond in real time. This research provides new ideas for the construction of digital museum and has practical application value.

From the above analysis, it can be seen that the method in this paper has a good response time under different capacities of library information data. The main reason for this result is that this method uses the two formulas in part 2.2 to obtain the scene information of the Folk Museum, and combines the formula derivation in part 2.3 to obtain the weight of the collection information at different points, which improves the response efficiency.

Table 5. Response time of 200G collection information data (s)

Number of tests	Virtual reality	Web GIS	Data mining
1	3.89	6.24	**6.59**
2	3.82	6.49	6.48
3	3.76	**6.87**	6.86
4	3.83	6.54	6.63
5	3.75	6.68	6.52
6	3.78	6.76	6.25
7	3.82	6.52	6.18
8	3.81	6.35	6.46
9	**4.02**	6.21	6.54
10	3.85	6.53	6.23

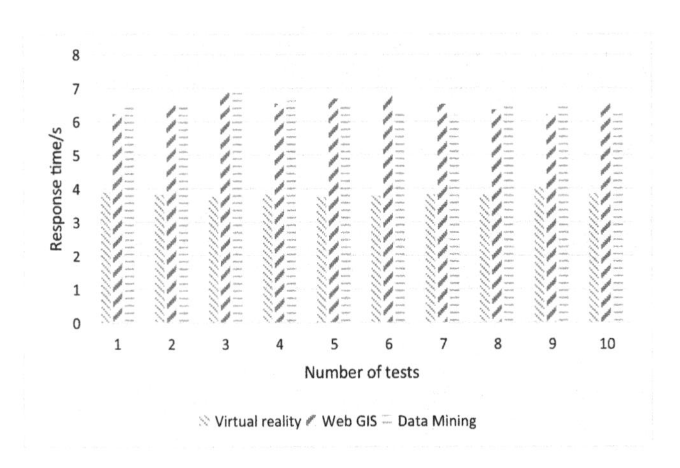

Fig. 5. Response time of 200G collection information data

4 Conclusion

The three-dimensional visualization method of Folk Museum collection information provides users with sufficient information display methods, and gives users a high sense of experience and reality. This paper mainly studies the Folk Museum in reality, and puts forward a three-dimensional visualization method of Folk Museum collection information based on virtual reality technology. According to the classification results of the collection information of Folk Museum, this paper constructs the museum. The parametric scene description language is used to construct the rendering scene of the collection information, and the virtual reality technology is used to establish the three-dimensional visualization model of the collection information of the Folk Museum. The experimental results show that this method can improve the response time of collection information and is suitable for the data transmission mode of virtual museum. Due to the limited research time and less research on caching and distribution, this paper only studies and applies it, but it also plays a certain role in improving the interaction efficiency of collection information. The functional analysis of geographic information, intelligent management of museum projects and Internet big data fusion analysis need to be improved. In the later stage, we can research and improve the data cache and distributed implementation, and constantly improve the three-dimensional visualization system of Folk Museum collection information.

Fund Project. 2022 Jilin Provincial Department of Education Scientific Research Project: Design of Interactive Picture Books for Cultural Creation of Northeast Ethnic Folk Museum in the New Media Era. Item number: JJKH20221334SK.

References

1. Zhang, S., Zhao, W., Peng, J., et al.: Augmented reality museum display system based on object 6D pose estimation. J. Northwest Univ. (Nat. Sci. Ed.) **51**(5), 816–823 (2021)
2. Li, Y.: Analysis on 3D visualization of bim in the network environment based on WebGL technology. Comput. Sci. Appl. **11**(1), 233–239 (2021)
3. Wang, Z., Peng, M., Xu, H.: Research on information visualization design of bronze ware in museums. HuNan BaoZhuang **35**(5), 57–61 (2020)
4. Yang, C.: Design of museum space multi-element interactive system based on three-dimensional representation. Mod. Electron. Tech. **44**(12), 155–158 (2021)
5. Xu, X., Tian, B.: Simulation of free handover of hybrid mobile guide information in virtual reality. Comput. Simul. **37**(1), 439–443 (2020)
6. Wang, W.: Design of 3D visualization system of geographic information based on virtual reality. J. Shenyang Inst. Eng. (Nat. Sci.) **17**(4), 74–78 (2021)
7. Shen, X., Yang, Y.: Spatial schema: immersive virtual reality promoting geospatial cognition. E-Educ. Res. **41**(5), 96–103 (2020)

Design of Pond Water Quality Monitoring System Based on Mobile UAV Technology

Shaoyong Cao(✉), Xinyan Yin, Huanqing Han, and Dongqin Li

School of Industrial Automation, Zhuhai College of Bejing Institute of Technology, Zhuhai 519085, China
suinkln45222@126.com

Abstract. At present, the monitoring range of the numerical monitoring method for fish ponds is small, which can not better protect the water quality of fish ponds and is easy to shorten the service life of the line, in order to better ensure the quality of fish pond water quality monitoring, a design method of fish pond water quality monitoring system based on mobile UAV technology is proposed, the system hardware configuration is optimized, the system software operation function is improved, and the steps of fish pond water quality monitoring with UAV technology are simplified. Finally, the experiment proves that the design system has high practicability in the practical application process and fully meets the research requirements.

Keywords: Mobile unmanned · Fishpond water quality · Water quality monitoring

1 Introduction

Traditional water quality monitoring system adopts wired transmission mode to transmit water quality related data information. Although it can accurately transmit data information, the monitoring range is small, and water wiring is also a difficult problem, which is easy to be damaged by biological eating. Long-term operation in water will reduce the service life of the line. In response to this problem, most water quality monitoring system development technicians use wireless transmission module to replace the traditional cable transmission module to solve the problem of line layout, damage [1]. At present, there are many kinds of wireless transmission modules, so how to select suitable functional modules to transmit monitoring information of aquaculture water quality has become a key research problem. At present, many water quality monitoring systems have been developed and designed by many domestic and foreign manufacturers. These systems are mainly based on the traditional SCM or ARM board technology. Aquaculture farmers will encounter the following problems when they actually use these water quality monitoring systems: The water quality monitoring equipment developed based on the traditional SCM or ARM board technology generally needs to provide stable AC power supply for it to work continuously [2]. This requires aquaculture farmers in the

W. Fu and L. Yun (Eds.): ADHIP 2022, LNICST 468, pp. 28–40, 2023.
https://doi.org/10.1007/978-3-031-28787-9_3

fish ponds near the establishment of a computer room or office, for small and medium -sized aquaculture farmers, the use of this water quality monitoring system costs too high to use. In recent years, with the adjustment of agricultural structure, China's aquaculture industry is gradually changing from the traditional artificial aquaculture to industrialization and intensive aquaculture, and water quality monitoring has become the key link of intensive aquaculture. At the same time, the developed water quality monitoring system can effectively collect water quality information, but the research on water quality control is scarce. In order to meet the requirements of sustainable development, solar power is used, but it can not work normally in rainy days [3]. In order to solve these problems, this paper selects mobile UAV technology as the core, uses temperature sensor, pH sensor and dissolved oxygen sensor to collect water quality information, transmits data information through mobile UAV technology and GSM module, and continuously supplies energy to the system through solar power module, so as to improve the function of water quality monitoring system. This paper introduces the technology system of aquaculture monitoring, which is divided into three parts: terminal layer, transport layer and management layer [4]. It points out that the collection and treatment of water quality parameters are the key link in the development of aquaculture monitoring at home and abroad. It looks forward to the information fusion of perception layer, the application of remote video transmission, the integration and intelligence of information processing technology and the combination of Internet of things and intensive aquaculture in the future.

2 Water Quality Monitoring System for Fish Ponds

The water quality monitoring system consists of 11 modules, including temperature monitoring module, pH monitoring module, dissolved oxygen monitoring module, wireless transmission module, MSP430F149 MCU, power supply module, display module, master control unit, monitoring center server, and oxygen generator. The overall system structure block diagram is shown in Fig. 1.

In Fig. 1, the whole system is divided into two parts, respectively, the water quality monitoring system information collection terminal remote monitoring terminal. Specifically, the information collection terminal of the water quality monitoring system is composed of the power supply module, temperature monitoring module, pH value monitoring module, dissolved oxygen monitoring module, display module, MSP430F149, main control unit, wireless communication module and oxygen booster; the remote monitoring terminal is composed of the server of the monitoring center and mobile phone users [5, 6]. The water quality monitoring system selects STM32 microprocessor as the core controller, uses temperature sensor, pH sensor and dissolved oxygen sensor to collect water temperature data, pH value and dissolved oxygen data, transmits them to the server of the monitoring center through wireless communication module, and continuously supplies power to the water quality monitoring system through the power supply module of the system. The monitoring center sends the received data information to the user's mobile phone, providing convenient conditions for users to view water quality information [7].

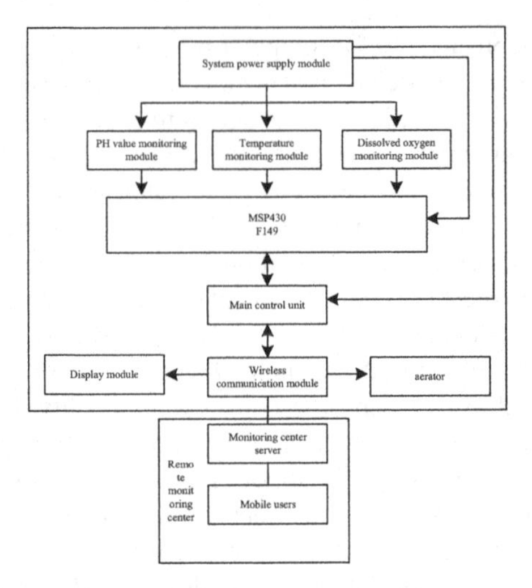

Fig. 1. Overall system structure diagram

2.1 Hardware Equipment for Water Quality Monitoring System of Fish Ponds

The technology of UAV is used to monitor and control the water quality of fishery ponds. The hardware structure of the intelligent monitoring system is shown in Fig. 2.

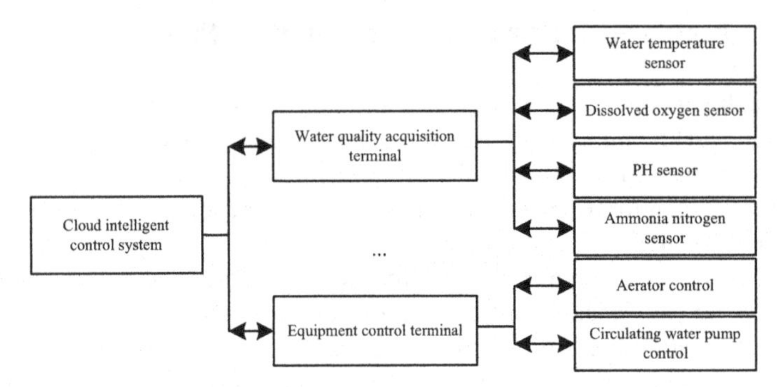

Fig. 2. System hardware structure diagram

The hardware includes two parts: water quality collecting terminal and equipment controlling terminal. Water quality collecting terminal and control terminal communicate with cloud intelligence monitoring system by mobile UAV technology. Water quality collection terminal uses RS485 interface communication to collect bottom water quality sensor data, respond to real-time data collection instructions and timing data upload. The control terminal controls the on-off of the relay by sending RS485 communication instructions, thereby controlling the operation of on-site execution equipment). The control terminal can obtain real-time water quality data through a cloud intelligent monitoring system and realize closed-loop automatic control through pre-set water quality parameter standards, or respond to user remote manual control instructions [8]. The water quality collection terminal is composed of the mobile UAV technical communication module, core controller module, water quality sensor, etc., powered by 40 W solar photovoltaic panels and storage batteries, and the control terminal is composed of the mobile UAV technical communication module, core controller module, Type 485 relay and on-site execution equipment connected thereto.

This system selects SR311A485 control relay, whose PCB is made of FR4 board. It uses industrial level chip and has lightning protection circuit. It can configure its module address by RS485 communication mode, and can use jumper lock module to set up good configuration to prevent misoperation. The product is energy-saving, long service life, stable and reliable performance. The specification of the specific parameters and control instructions are shown in Table 1, respectively:

Table 1. Mobile UAV monitoring technical parameters

Technical attributes	Specific parameters
Electric shock capacity	10 A/30 V DC
temperature range	−35 °C−−85 °C
Durability	110000 times
Rated voltage	DC10–25 V
communication interface	RS486
Baud rate	9700
Default communication format	9700, n, 8, 2

Modern monitoring and controlling system for water quality of mobile unmanned aerial vehicle is a complete system composed of water quality sensor, representative, timely and reliable water quality information, automatic control technology, computer technology and professional software, so as to realize online automatic monitoring and automatic control of various water quality parameters. Its architecture is divided into three parts: terminal layer, transport layer and management layer as shown in Fig. 3.

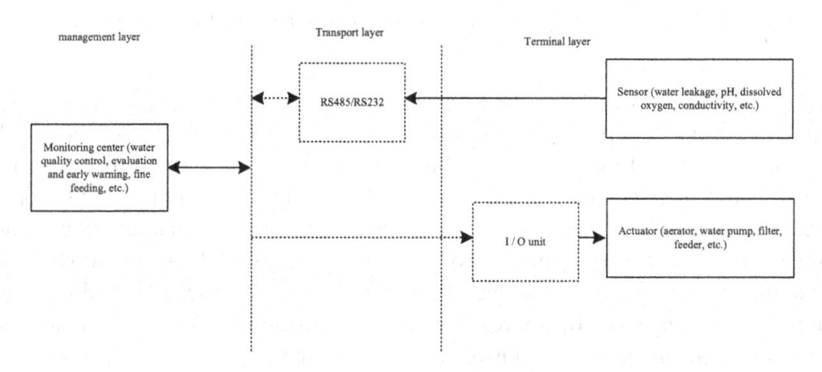

Fig. 3. Online automatic monitoring and actuator automatic control equipment

The terminal layer realizes the real-time collection, processing and automatic control of the water quality parameters. The transport layer is the key link between the terminal layer and the management layer, which is responsible for the transmission and transportation of water quality data and information. The transmission methods mainly include wired communication and wireless communication, and the field bus and mobile UAV monitoring are the main representatives of the two methods. Management is the terminal of the system, which mainly realizes real-time data acquisition, storage, analysis and processing, intelligent water quality control, evaluation and early warning, disease diagnosis and prevention, fine feed feeding and aquaculture system configuration. When designing the hardware circuit of field terminal monitoring system, the system is divided into two parts, one is the water quality terminal node hardware circuit and the other is the data display terminal node hardware circuit. Among them, the core controller of the hardware circuit of the water quality terminal node is a low power UAV monitoring SCM, and the data shows the UAV monitoring processor. The advantages of this design can reduce the workload of mobile UAV monitoring processor, improve the efficiency of water quality information collection terminal, improve real-time data, and reduce power consumption.

2.2 System Software Function Optimization

The development platform of keil u Vision 4 for C51 is selected as the software development tool of mobile UAV monitoring information collection terminal, which is developed by using the language at °C. The main program flowchart of the information acquisition terminal system is shown in Fig. 4.

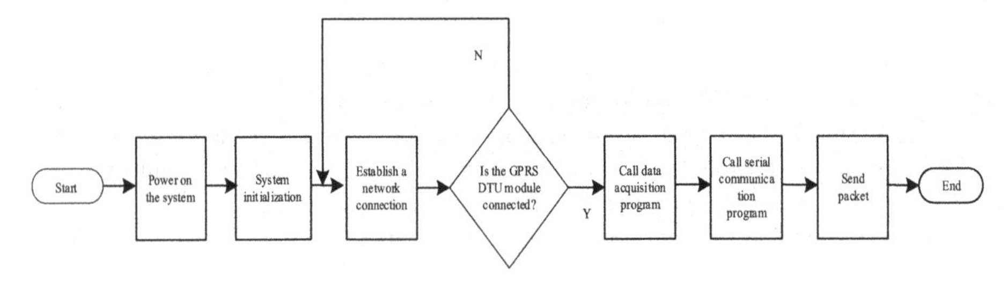

Fig. 4. Main program flow chart of information acquisition terminal system

The basic idea is: weighting the measurement results of the same kind of sensors, adjusting the weight coefficient online according to the variance of the sensors, so that the average variance of the fusion model is minimized and the optimal fusion result is obtained. The system uses the random weighted fusion algorithm of multi-sensor observation covariance, which measures the accuracy of the non-random quantity estimated from the noisy data by mean square error [9]. For the output of the fusion model, if the result of the data processing of a single sensor exceeds the set value, and the direct alarm is within the set range, the fusion result of the next fusion model to be estimated is x, and the measured values of each sensor are independent of $x_1, x_2, ..., x_n$, and the x is an unbiased estimate, then the total mean square error a2 of the n sensors is used as the data collection subroutine to collect the information about the pH value, water temperature and dissolved oxygen of aquaculture water through controlling the pH value sensor, water temperature sensor and dissolved oxygen sensor. The data collection subroutines of these sensors are the same, and the data are collected at certain time interval. The data collection subprogram design flowchart is shown in Fig. 5.

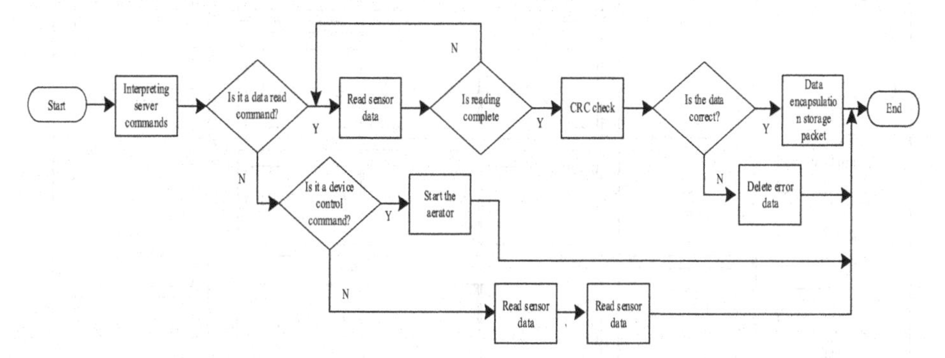

Fig. 5. Flowchart of data acquisition subroutine design

For the data interface of the server, it mainly provides the API for the embedded terminal device and the client. In the system, the data interface of the server is realized by POST request. It responds to the data upload of the water quality collecting terminal and the data query of the mobile client in real time. The operation object of the data interface is MySQL in the server, which relies on the database to respond to various requests. The persistence layer architecture chosen for the data interface design is MyBatis, whose specific structure is shown in Fig. 6. The steps for a mobile UAV to monitor MyBatis are to first write a MyBatis configuration file, create a database connection session, configure a relational mapping, complete the database, and end the session.

Breeding users can carry out remote manual control of on-site execution equipment, or turn on intelligent control. The system automatically adjusts water quality according to the pre-set water quality parameter standards and on-site water quality conditions, and adjusts water quality in advance according to dynamic prediction and early warning information.

Initialize the number of repeated EMD decompositions M, the ratio of Gaussian white noise to the standard deviation of the amplitude of the original signal t, and set the number of iterations for EMD decompositions to be $x_m(t)$. Perform and record the EMD decompositions. Perform the m decompositions as follows: Based on the relationship between the water quality data $f(t)$ collected on the spot and the Gaussian white noise sequence $n_m(t)$, the expression of the m decomposition sequence is:

$$x_m(t) = f(t) + kn_m(t)x_n \tag{1}$$

In formula (1), k represents the analysis coefficient. Decompose $x_m(t)$ to make the loop variable i 1, x1 = xn (1); make the loop variable j 1, then $y_n(t) = x_n(t) + x_m(t)$ finds all the local maxima $u_{j,m}(t) = u - a(t)$ and local minima $v_{j,m}(t)$ envelopes in the $y_n(t)$

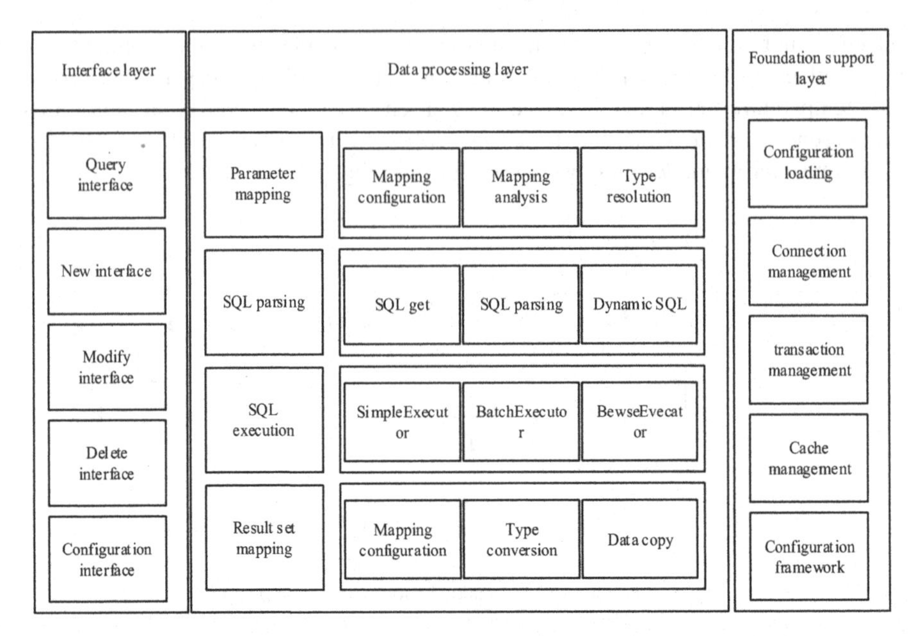

Fig. 6. My Batis persistence layer architecture for mobile UAV monitoring

sequence; where the upper and lower envelopes require all the data points enveloped, the average of which is:

$$m_{j,m}(t) = \frac{u_{j,m}(t) + 1}{2u + y_n(t)} - v_{j,m}(t) \tag{2}$$

For mobile UAV monitoring, its advantage is that it can be trained by small sample to improve the learning ability of the machine, so as to make up for the shortcomings of the machine itself and improve its operating efficiency. The complex factors of water quality change in fishery aquaculture can be solved by its generalization ability. Mobile UAV monitoring itself can control the generalization accuracy, which makes it widely used in aquaculture water quality prediction, and this is also of particular importance in small sample analysis. From the practical analysis, it can be concluded that the kernel function can not only improve the processing ability, but also further enhance the analysis dimension of the mobile UAV monitoring system. The powerful kernel function endows it with better parameter selection ability, further reduces the complexity of data processing and analysis, and greatly improves its application value.

2.3 Realization of Water Quality Monitoring of Fish Ponds

Aiming at the problem of low precision of SVR in fishery water quality parameter prediction, this paper presents a combined prediction method of UAV monitoring model based on set empirical modal decomposition, Gray Wolf optimization algorithm and SVR. Firstly, the collected water quality data are decomposed into a series of relatively stable characteristic modulus components by mobile UAV monitoring, then the mixed noise sequence is eliminated, and the optimized mobile UAV monitoring is used to establish a prediction model for the reduced noise fishery water quality data [10]. Based on the water quality environment of local fishery aquaculture, combined with the analysis of aquaculture samples and the reference of related literatures, an early warning model of fishery water quality is established. The logical flow of fishery water quality prediction and early warning model based on mobile UAV technology is shown in Fig. 7. Through the dynamic prediction and early warning of the pond aquaculture water environmental parameters, and based on the dynamic prediction and early warning information to adjust the site water quality.

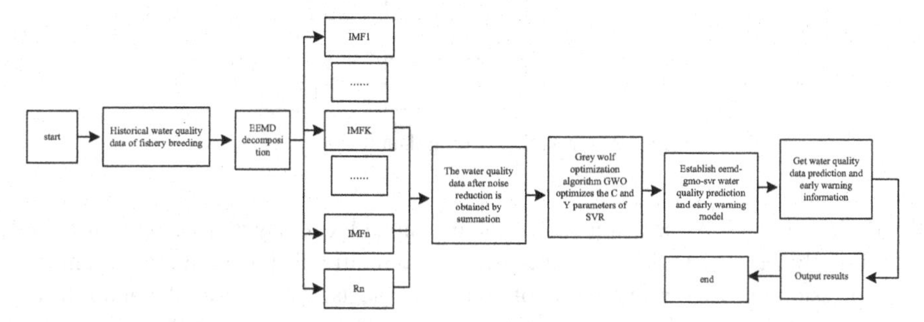

Fig. 7. Logical process of fishery water quality forecast and warning

Data fusion is the key technology of Internet of Things. Compared with Internet of Things, data fusion is equivalent to human brain. Through visual, auditory, tactile, taste access to natural information, handed over to the brain, the use of experience and reserve knowledge to perceive changes in the outside world. The monitoring system acquires water quality information through various sensors such as dissolved oxygen, temperature, pH, turbidity, ammonia nitrogen, water level and so on. Because each sensor has different measurement and characteristics, slow and fast, linear and nonlinear, real time and non-real time, data fusion technology integrates these data to get consistent description of water quality through corresponding rules, and predicts more information from them to enhance the system perception ability. Data fusion system generally includes two subsystems: multi-sensor sensing system and information processing system. Figure 8 shows a typical data fusion process.

As shown in Fig. 8, input data types may vary, for example, data sources in different representations such as text, radar, image, etc.: possible changes in the quality of the data reflected in the error, with intermittent sensor errors in some data; possible changes in the quality of the data reflected in the arrival of the data, and data from another platform that is affected by communication network overload, showing intermittent or intermittent changes, etc. Possible approaches include: the introduction of location information from external reference systems, etc., the study of targeted solutions and other data associations, the difficulty being that the number of "targets" in different event sets is inconsistent: input data is vague, inconsistent, conflicting or unreliable; input data has associated noise/errors. Generating a set of model hypotheses that may represent the real world, using methods to select the hypotheses closest to the acquired data, etc.

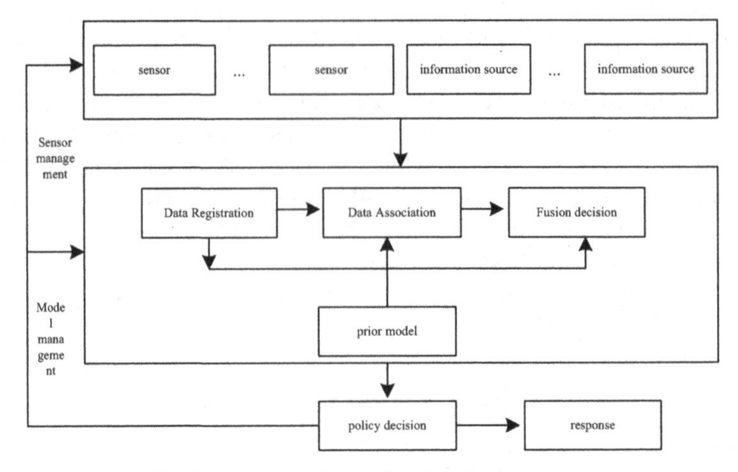

Fig. 8. Water quality testing data fusion process

The difficulty of fusion decision is that: the decision object may be more complex, and has multi-objective or multi-event, multi-level and multi-side processing requirements; the expression and processing mode of complex objects; the meta-model and relevant processing of complex dynamic situations in multiple situations; and the performance evaluation of fusion results. It is possible to use multi-level concepts, including multi-level objects, multi-level processing, multi-level meta-model, multi-side processing, and so on. In view of the different level research concrete suitable theory method and the processing structure. On the one hand, it provides output and on the other hand it is used for feedback. Feedback to the fusion part is to modify the prior model and validate the validity of the hypothetical model by continuous inspection. The function of feedback to sensor system is to control the work of the system.

3 Analysis of Experimental Results

The hex file is compiled through the USB port and written into the stm32f103zet6 single chip microcomputer. The MSP430F149 microprocessor is controlled by the single chip microcomputer to realize the control of water temperature sensor, pH sensor and dissolved oxygen sensor. The sampling frequency is 24 kHz and the time interval is 5 s. The software can automatically identify and download the COM port and support different types of microcontrollers. The information collection includes all types of sensors and their signal conditioning circuits. The specifications of each sensor are as follows (Table 2):

Table 2. Types of sensors

Sensor type	Model	Manufacturer
Temperature sensor	PE110	Guangzhou Leici Co., Ltd
Dissolved oxygen electrode	DE-110	Shanghai aiwang industry and Trade Co., Ltd
PH sensor	LQ4635T	Mettler Toledo
Water level sensor	COY12	Beijing Yiguang Sensor Technology Co., Ltd

Finally, the sensor is installed in the aquaculture base according to the layout scheme to measure the water temperature and pH dissolved oxygen, and the test results are compared and analyzed as shown in Table 3.

Table 3. Water quality monitoring accuracy test results

DS18B20 water temperature (°C)			MIK-160PH-4 Value			Kds-25b dissolved oxygen (mg / L)		
System detection	Manual monitoring	Monitoring accuracy	System detection	Manual monitoring	Monitoring accuracy	System detection	Manual monitoring	Monitoring accuracy
11.3	11.2	99.1%	7.7	7.6	98.7%	8.6	8.7	98.9%
11.9	11.7	98.3%	7.6	7.6	100%	8.5	8.5	100%
13.8	13.7	99.3%	7.6	7.5	98.7%	8.3	8.2	98.8%
18.7	18.9	99.0%	7.6	7.6	100%	8.1	8.3	97.6%
20.5	20.7	99.0%	7.7	7.6	98.7%	7.9	7.8	98.7%
23.8	24.0	99.2%	7.5	7.5	100%	7.5	7.4	98.7%
25.2	25.1	99.6%	7.6	7.6	100%	7.3	7.2	98.6%
28.5	28.5	100%	7.6	7.6	100%	6.6	6.5	98.5%

In Table 3, the deviation between the measured data of the water temperature sensor and that of the manual measurement is ± 0.2 °C and above 95%; the deviation between the measured data of the sensor and that of the manual measurement is within 0.1 °C and above 98%; the deviation between the measured data of the KDS-25B dissolved oxygen sensor and that of the manual measurement is within ± 0.2 mgL and above 95%. Therefore, the results show that the water temperature sensor, pH sensor and dissolved oxygen sensor can meet the measurement accuracy requirements, and can provide accurate data support for the remote monitoring center. Network packet loss rate test verifies the stability and reliability of system data transmission. TCP/IP protocol can guarantee the integrity of data, so the number of data packets is taken as the statistical sample. The test results from the sampling for the intervening 10 days are shown in Table 4:

Table 4. Test Packet Loss Rates and Success Rates for 10 Days

Time	Wireless communication module sends data packets (PCs.)	Data packets received by server (PCs.)	Manual control/(Times)	Response information/(piece)
2021/9/27	1445	1436	20	20
2021/9/28	1442	1442	20	20
2021/9/29	1443	1436	20	19
2021/9/30	1445	1441	20	20
2021/10/01	1439	1438	20	18
2021/10/02	1442	1438	20	20
2021/10/03	1443	1429	20	19
2021/10/04	1442	1443	20	20
2021/10/05	1437	1433	20	20
2021/10/06	1441	1432	20	18
Total	14419	13075	200	194

As can be seen from Table 4, the basic weather conditions of the aquaculture base were recorded while the experiment was carried out. The results showed that the water temperature, pH, dissolved oxygen and concentration of ammonia and nitrogen in the aquaculture pond No. 4 were directly affected by the change of the weather. Taking temperature and dissolved oxygen as examples, some of the test data are compared with those shown in Fig. 9:

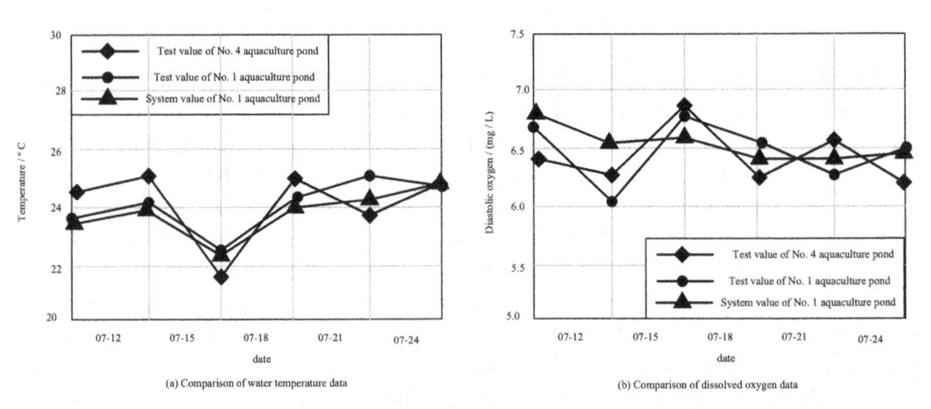

Fig. 9. Comparison chart of test data

From Fig. 9, we can see that the range of dissolved oxygen value is 6.657.31 mg/L, the precision of dissolved oxygen control is ± 0.11 mg/, and the range of dissolved oxygen value is 6.45–7.38 mg/L. In rainy days, the dissolved oxygen content in the ponds increased. On the whole, the system value of the pond is very close to the test value, and the variation range of the water quality parameters is stable and in a suitable range.

Table 5. Monitoring accuracy and error of some water quality parameters measured by the system

Accuracy and error	Temperature	PH value	Dissolved oxygen	Mass concentration of ammonia nitrogen
Monitoring accuracy	±0.16 °C	±0.08	±0.12 mg/L	±0.09 mg/L
Average relative error/%	0.18	0.19	0.17	0.49

Table 5 shows that the maximum monitoring accuracy of the parameters is ±0.2 °C, pH ±0.2, dissolved oxygen and ammonia nitrogen are ±0.2 mg/L, and the maximum average error of each parameter is 0.5%. Comparing the test results with the standard values, the maximum temperature monitoring accuracy of the system is kept at ±0.15 °C, and the maximum average relative error is 0.48%, which meets the system requirements.

To sum up, the fish pond water quality monitoring system based on mobile UAV technology can provide accurate data support for the server control of the remote monitoring center. The system value of the aquaculture pond is very close to the test value, the maximum temperature monitoring accuracy is maintained at ±0.15 °C, and the maximum average relative error is 0.48%. It has good performance and can play an important role in practical applications.

4 Conclusion

China is a large country of aquaculture, and aquaculture model is developing to intensive, factory model, and water quality monitoring has become a very important part of intensive aquaculture. Compared with developed countries, the water quality monitoring technology in our country is not mature enough, and there is still room for improvement in the research of key technologies. Through the above research, the following conclusions are obtained:

(1) The measurement accuracy of the water temperature sensor, pH sensor and dissolved oxygen sensor of the system meets the requirements, and can provide accurate data support for the server control of the remote monitoring center.
(2) The test results are analyzed in combination with the actual changes of the test environment. The overall performance is that the water temperature decreases, the pH value decreases, the dissolved oxygen increases, and the ammonia nitrogen mass concentration decreases in rainy days. The influence of weather changes on the water quality parameters of No. 4 aquaculture pond is relatively direct.
(3) When it is rainy, the dissolved oxygen of each pond tends to increase, and the system value of the pond is very close to the test value.

(4) The maximum temperature monitoring accuracy of the system is maintained at ±0.15 °C, and the maximum average relative error is 0.48%, which meets the system requirements.

For the next work, the following aspects can be studied in depth:

(1) The data acquisition and data processing technology of the terminal perception layer will be the key link to be tackled, and the research and development of theory and equipment should be strengthened.
(2) The information fusion of perception layer, the application of remote video transmission, the integration and intelligence of information processing technology, and the combination of Internet of things technology and intensive aquaculture in aquaculture water quality monitoring will be the future development trends.

Fund Project. The Special projects in Key Areas of Guangdong Province (Grant 2021ZDZX4050).

References

1. Qi, X., Yang, Z.: Joint power-trajectory-scheduling optimization in a mobile UAV-enabled network via alternating iteration. China Commun. **19**(1), 136–152 (2022)
2. Warwick, G.: HAPSMobile UAV demonstrates LTE connectivity from the stratosphere. Aeros. Daily Def. Report **274**(5), 5–15 (2020)
3. Sivakumar, S., Ramya, V.: An intuitive remote monitoring framework for water quality in fish pond using cloud computing. IOP Conf. Ser. Mater. Sci. Eng. **1085**(1), 12–17 (2021)
4. Jing, K.L., Fang, S., Zhou, Y.: Model predictive control of the fuel cell cathode system based on state quantity estimation. Comput. Simul. **37**(07), 119–122 (2020)
5. Darmalim, F., Hidayat, A.A., Cenggoro, T.W., et al.: An integrated system of mobile application and IoT solution for pond monitoring. IOP Conf. Ser.: Earth Environ. Sci. **794**(1), 12–16 (2021)
6. Tang, Q., Yu, Z., Jin, C., et al.: Completed tasks number maximization in UAV-assisted mobile relay communication system. Comput. Commun. **187**(4), 20–34 (2022)
7. Ighalo, J.O., Adeniyi, A.G., Eletta, O., et al.: Evaluation of Luffa cylindrica fibres in a biomass packed bed for the treatment of fish pond effluent before environmental release. Sustain. Water Res. Manag. **6**(6), 120–131 (2020)
8. Zhang, G., Hu, Y., Han, X., et al.: Design of distributed water quality monitoring system under circulating water aquaculture mode of freshwater pearl mussels[J]. Transactions of the Chinese Society of Agricultural Engineering **36**(07), 239–247 (2020)
9. Zuo, X., Jin, W.X., et al.: Research and application of instruments for on-line monitoring water quality of total nitrogen. Electron. Meas. Technol. **44**(14), 173–176 (2021)
10. Yu, X., Zhang, Y., Zhu, Q., et al.: Research on groundwater environment monitoring network oriented to water source protection and pollution monitoring in Guangdong Province. Environ. Monit. China **37**(05), 32–40 (2021)

Passive Positioning Method of Marine Mobile Buoy Based on Vibration Signal Extraction

Xuanqun Li[1,2,3](✉) and Yun Du[4]

[1] Institute of Oceanographic Instrumentation, Qilu University of Technology (Shandong Academy of Sciences), Qingdao 266061, China
lixxqqq@163.com

[2] Shandong Provincial Key Laboratory of Marine Monitoring Instrument Equipment Technology, Qingdao 266061, China

[3] National Engineering and Technological Research Center of Marine Monitoring Equipment, Qingdao 266061, China

[4] The University of Manchester, Manchester M13 9PL, UK

Abstract. In order to solve the problem of poor positioning effect of Marine Mobile buoys, a passive positioning method of Marine Mobile buoys based on vibration signal extraction is proposed. First of all, using the principle of signal extraction, the operational characteristic data of Marine Mobile buoys are obtained. Secondly, the interference value of the abnormal vibration signal is cleaned and removed to remove the interference signal. Finally, according to the frequency relationship between the satellite signal and the buoy transmission signal, the passive positioning of the marine mobile buoy is completed. Finally, the experiment proves that the passive location method of Marine Mobile buoy based on vibration signal extraction has high practicability and fully meets the research requirements.

Keywords: Vibration signal · Marine mobile buoy · Passive positioning

1 Introduction

Buoy is one of the important means of ocean observation. In order to achieve fine-grained observation of marine environmental parameters, it is often necessary to deploy a large number of buoys in the observation area. However, the buoy observation data without position information is meaningless. The accurate position information of the buoy observation data provides important decision support information for fishery production scheduling, search and rescue, early warning, emergency handling, behavior decision and other applications. Therefore, it is necessary to accurately locate the position of the buoy.

However, in the marine observation environment, because the buoy is relatively close to the sea surface, the radio signal is attenuated seriously in the transmission process, and its nonlinear attenuation model is more complex. Due to the influence of environmental noise, reflection, multipath, nonlinear attenuation and other negative

W. Fu and L. Yun (Eds.): ADHIP 2022, LNICST 468, pp. 41–55, 2023.
https://doi.org/10.1007/978-3-031-28787-9_4

factors in the transmission process of wireless signals, as a result, the distance estimation between the unknown buoy (the buoy whose position information is unknown and to be located) and each reference buoy (the buoy whose position information is known) has different degrees of error, resulting in low positioning accuracy.

Reference [1] proposes an ocean buoy positioning method based on asymmetric round-trip ranging, which uses the time difference between asymmetric receiving and sending ranging information between nodes to calculate the sound speed of seawater between nodes and the distance between unknown buoy nodes and their neighbor reference nodes. Convert the three-dimensional distance information into two-dimensional, and use the least square method to complete the positioning calculation. Reference [2] proposes an ocean buoy positioning method based on acoustic target motion elements, establishes an optical buoy observation mathematical model including buoy positioning error, observation time error and optical observation blur error, and uses Monte Carlo simulation method to give the positioning accuracy index of different numbers of optical buoys considering the above errors and aiming at mobile targets, and analyzes the influence of various factors on multi buoy joint positioning. Reference [3] proposes an ocean buoy positioning method based on the product season model. In view of the time series characteristics of the light buoy position data, the product season model is used to establish a mathematical model of buoy positioning, and the position information of the ocean buoy can be obtained by solving the mathematical model to complete the ocean buoy positioning.

Passive positioning is that the observation station passively receives the electromagnetic wave of the target from the vibration signal source without transmitting any vibration signal, and determines the position and motion state information of the target according to various parameters of these vibration signals. Compared with active location, passive location has the following advantages: good concealment; It can obtain a detection range much larger than that of the active radar, and can find the target in advance; Low cost. Passive positioning and guidance technology can detect and find long-range vibration signals on the sea surface and stealth targets as soon as possible, and improve the passive positioning accuracy of Marine Mobile buoys. According to the frequency of the vibration signal, the positioning of the marine mobile buoy can be divided into single station passive positioning and multi station passive positioning. In order to further improve the positioning accuracy of Marine Mobile buoys, this paper proposes a passive positioning method of Marine Mobile buoys based on vibration signal extraction.

2 Passive Positioning of Marine Mobile Buoys Based on Vibration Signal Extraction

2.1 Extraction of Vibration Signal of Marine Mobile Buoy

The marine data buoy is an ocean observation platform designed to meet the needs of marine scientific research. The development of buoy technology is closely related to marine engineering technology and sensor technology. Therefore, the development level of marine data buoys also reflects the level of marine science and technology in a country. According to different classification methods, the vibration signals of marine mobile buoys can be divided into different types, as shown in Table 1.

Table 1. Classification of vibration signals of marine mobile buoys

Classification methods	Buoy type
Application form	General purpose and special purpose
Anchoring mode	Anchoring buoy and drifting buoy
structural style	Disc, cylindrical, ship, spherical, etc

The marine data buoy includes body, communication, power supply, control, sensor and other parts involving structural design, data communication, sensor technology, energy and power technology and automation control. Due to the high cost of developing ocean data buoys, only a few major oceanic countries currently have the ability to produce, deploy and maintain large quantities of ocean buoys [4]. The buoy technology in Europe and the United States and other major marine countries started early. After years of accumulation of technology and practical experience, the buoy business has achieved a large scale. The buoy communicates between the underwater instrument and the water terminal through inductive coupling. The principle of inductive coupling communication is shown in Fig. 1.

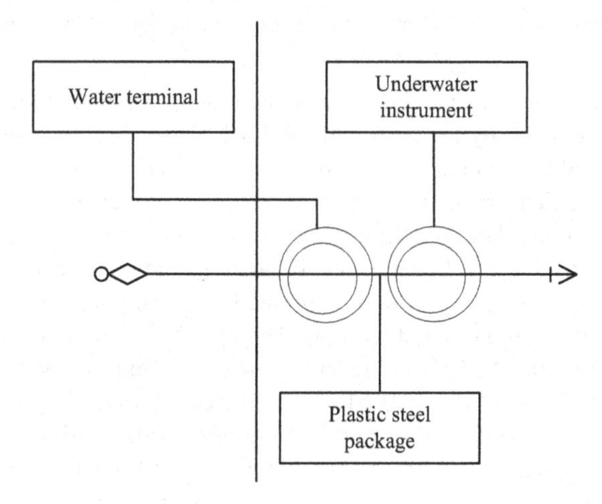

Fig. 1. Structure of coupling and transmission of vibration signal of marine mobile buoy

Both the above-water terminal and the underwater sensor are connected to the magnetic rings through coils, all the magnetic rings pass through a plastic-coated steel cable with bare ends, and the steel cable forms a loop through seawater. In the process of sending data from the water terminal to the underwater sensor, the carrier signal with the command information will be placed in the primary winding connected with the magnetic ring, and the magnetic ring can excite the induced current in the single-turn coil composed of the steel cable and the seawater, the induced current induces a voltage signal with command information in the secondary winding of the lower magnetic ring, and the sensor filters and demodulates the voltage signal to obtain the corresponding command [5]. The same process is used for the sensor to send data to the terminal on the water. The sampling period is set according to the density requirements of various meteorological factors in scientific research, and different sensors have different sampling frequencies. The underwater sensor CTD is sampled every half an hour, the meteorological sensor is collected every hour, and the GPS information is obtained every three hours, as shown in Fig. 2.

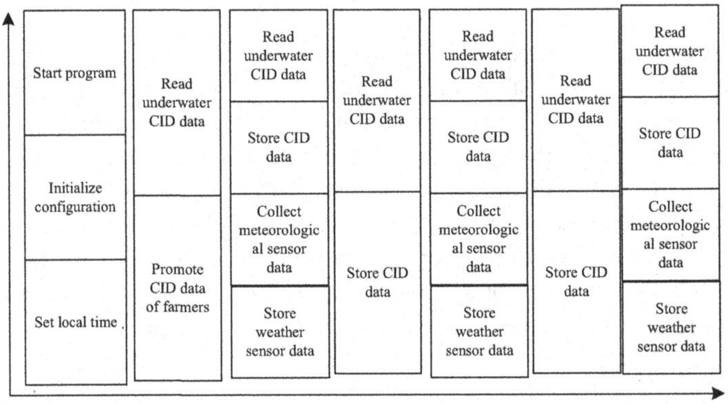

Fig. 2. Operation and management process of vibration signals of marine mobile buoys

The operation and management process of the vibration signal of the marine mobile buoys is actively initiated by the buoy control. Each sensor is either powered on regularly, or is awakened by the controller for data communication, and is not the party that initiates the communication actively. As a host, the controller must have an accurate clock to ensure accurate data sampling and communication time [6]. During the communication between the controller and the GPS module, the satellite time at the moment will be obtained, the buoy time will be calibrated, and the platform terminal and the satellite can communicate in both directions. The uplink adds a high-speed channel on the basis of retaining the ARGOSI/2 generation standard and low-power channel. The frequency band is 401.580 MHz–401.610 MHz, the center frequency is 401.595 MHz, the bandwidth is ±15 kHz, and the rate is 4800 bps, using GMSK modulation. The downlink center frequency is 4659875 MHz, the bandwidth is ±15 kHz, and the rate is 400 bps or 200 bps (Table 2).

The search and acquisition of the serial vibration signal of the marine mobile buoy refers to the use of a digital correlator to scan the dimension of the Doppler frequency shift of the specified signal in the time domain. This algorithm is relatively easy to implement. The structure of the acquisition circuit is shown in Fig. 3.

Table 2. Vibration signal system of marine mobile buoys

Passageway	Transmission mode	Rate/bps	Frequency/MHz	Modulation mode
Uplink	Standard mode	500	401.605–401.681	QPSK
	High speed mode	4900	401.96	GMSK
Downlink	Standard mode	500/300	465.98 76 MHz ± 16 kHz	PSK

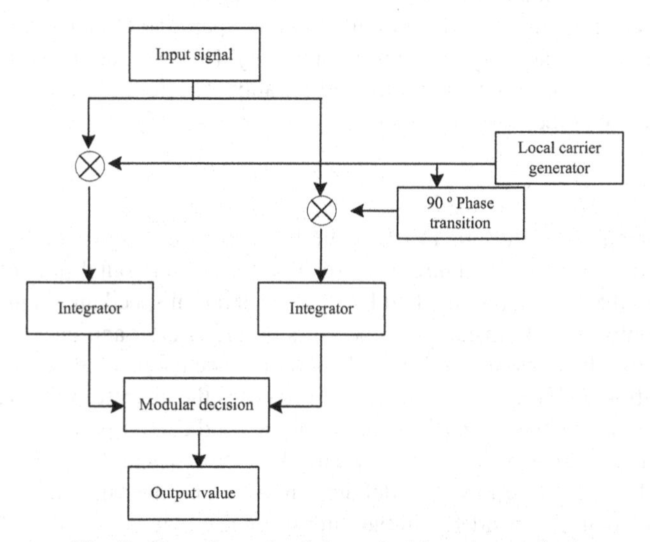

Fig. 3. Flow chart of serial sequential paving method

As can be seen from the figure, the module circuit can be completed only by a number of digital correlators, which greatly reduces the workload of the designer and saves resources. These correlators can also be multiplexed in subsequent signal tracking modules [7]. After determining the search range of the signal Doppler frequency shift, the linear search acquisition algorithm usually starts from the frequency band corresponding to the middle value of the frequency search range, and then gradually searches the frequency bands on both sides alternately, until finally detected signal or until all frequency bands are searched. Assuming that the search range of Doppler frequency shift is 2 ± 10 kHz, and the search step size is set to 500 Hz, then the receiver starts to search from the middle frequency band with Doppler frequency shift of 2kHz, and then searches for the center frequency of 1.5 kHz, 2.5 kHz, there are a total of 41 frequency bands such as 1.0 kHz and 3.0 kHz [8]. This "Christmas tree"-shaped frequency search sequence helps improve the receiver's probability of finding a satellite signal quickly. When searching for a signal in a certain frequency band, the receiver replicates a carrier signal whose frequency is the center frequency of the frequency band, and mixes it with the received signal to capture the received signal whose actual frequency is within the frequency band.

2.2 Algorithm for Interference of Vibration Signals of Marine Mobile Buoys

The marine environment is changing rapidly. As a test platform at sea, the buoy is often affected by the wind and wave environment at sea. The buoy is constantly swaying in the sea water. When the instability is too large, the marine environment measurement and communication of the drifting buoy will be affected. Therefore, stability is very important for buoys [9]. The stability of the target is not only that the electronic cabin equipment of the buoy, the temperature and depth measurement part and the environmental noise measurement part of the buoy are working normally, but also that the surface antenna and other parts can still work normally and reliably. The formula for calculating the natural frequency of the buoy signal swing is:

$$\omega_\theta = \sqrt{G * L/I} \tag{1}$$

Among them: G is the gravity; L is the height of the buoy's stable center; I is the moment of inertia of the buoy mass relative to the horizontal axis of the center of mass. The GPS positioning method has fast observation speed and high positioning accuracy. It only needs to install GPS receivers on the aircraft and buoys, and there is no restriction on the maneuverability of the aircraft. The overhead detection method is a simple positioning indication method, but it requires the aircraft to fly over the buoy, which severely limits the aircraft's maneuverability and cannot perform long-distance positioning. The estimated state is often a part of all the states in the model. If the states that do not play a big role in the model are removed, and the influence of such states on other states is fully considered in the filter design, it may be possible to reduce the filter performance. Effectively reduce the amount of computation. Mathematical model 2 of the vibration signal localization algorithm reduces the number of states for buoy positioning from 13 to 6. In this study, we try to continue to simplify the model and reduce the order to the number of states required to be estimated, so as to find a simpler calculation method [10]. The equation can be simplified to

$$X_{K+1} = X_K + W_K \tag{2}$$

By describing the statistics of the noise, there will hopefully be enough degrees of freedom in position updates to compensate for any sonobuoy drift. However, this method was not completely successful. The drift of the buoy directly affects the position of the buoy. To ensure that the state vibration signal does not diverge, the drift action of the buoy must be considered. The preset position of marine buoy measurement and control is arranged in the middle and rear of the track, which is an equipment structure with actual carrying tasks. In the marine buoy measurement and control trajectory, there are three main measurement and control units including the intermediate position node, which are named "initial preset position", "approaching preset position" and "correction preset position", and each Each type of position structure corresponds to a relatively independent marine buoy measurement and control stage.

1) Initial preset position: The initial preset position corresponds to the primary marine buoy measurement and control behavior. During this process, the marine buoy equipment always maintains the initial state, which is the first type of data information referenced by the entire preset position positioning process.
2) Approaching the preset position: Approaching the preset position corresponds to the measurement and control behavior of the intermediate marine buoys. During this process, the behavior of the marine buoy equipment changes for the first time, which is the second type of data information referenced by the entire preset position positioning process.
3) Correction of the preset position: The correction of the preset position corresponds to the monitoring and control behavior of the marine buoy at the end. During this process, the behavior of the marine buoy equipment gradually becomes stable, but the overall value level is much lower than the initial stage. This is the entire preset position. The third type of data information referenced by the positioning process. Let Q_1, Q_2, and Q_3 represent the first type of data information, the second type of data information, and the third type of data information, respectively, and the above-mentioned physical quantities can be combined to define the division of the preset stage of ocean buoy measurement and control as:

$$E = \frac{\sqrt{\left(\lambda_1 Q_1^2 + \lambda_2 Q_2^2 + \lambda_3 Q_3^2\right) - \overline{y}^2}}{|QQ'|} \tag{3}$$

Among them, λ_1, λ_2, and λ_3 respectively represent the marine buoy measurement and control behavior coefficients in the three preset stages, $\overline{y}$ represents the numerical weight of the preset position in the trajectory, and $\left|QQ'\right|$ represents the positioning vector from the starting position to the ending position.

2.3 Anchor Point Layout for Preset Positions

The layout of the preset position positioning points takes the marine buoy measurement and control nodes in the ocean as the operation target, and plans the guiding direction of the positioning vector in the behavioral coordinate system according to the structural characteristics of different preset stages. In the weighted transoceanic coordinate system, A_0 represents the initial position information of the preset position of the marine buoy measurement and control, and A_n represents the final position information of the preset position of the marine buoy measurement and control. The simultaneous formula can coordinate the positioning point of the preset position. The principle of planning Expressed as:

$$\overrightarrow{A_0 A_n} = \int_0^n \frac{\beta^2 |A_n \times A_0|^2}{E\ddot{u}} d\beta \tag{4}$$

Among them, $\overrightarrow{A_0 A_n}$ represents the marker vector of marine buoys pointing from the initial position to the final position, n represents the specific real-valued quantity of the preset position of marine buoy measurement and control from the initial position to the final position, and β represents the established curvature condition of the curve in the weighted transoceanic coordinate system, $\dot{u}$ represents the planning regression coefficient in the measurement and control trajectory. On the premise of ensuring that the overall planning position of the preset position does not change, taking the given time T as the counting condition, among all the possible curvature values in the weighted transoceanic coordinate system, a vector is randomly selected and defined as β', and let R_0 represents the lower limit layout boundary parameter, R_1 represents the upper limit layout boundary parameter, and the simultaneous formula can express the positioning point layout discriminant of the preset position of the marine buoy measurement and control as:

$$W = \frac{\left|1 - \beta' \sum_{R_0}^{R_1} T \cdot \overline{p}^2\right|}{\chi' \left|\overrightarrow{A_0 A_n}\right|} \tag{5}$$

In the above formula, $\overline{p}$ represents the average speed of the ocean buoy measurement and control device in the weighted super-ocean system, and χ' represents the layout authority parameter of the positioning point. While it was observed that the general direction of drift of the buoy was correct. When the buoy control unit starts to work, it is powered on and initialized first, and the parameters are loaded after the power is successfully powered on. Due to the variability of the sea test environment, it is impossible to configure the parameters during the test. It is necessary to configure the parameters in advance before the test, and configure the operating variables in advance. The configuration of the control unit port, the initial state of the main control chip, the parameter configuration of the data transmission radio, the serial port configuration of the temperature depth measurement node, the serial port configuration of the GPS module, the configuration of the power-on time, the configuration of the acquisition time, the configuration of the sleep time and other parameters. After these parameters are configured, it is necessary to judge whether the GPS is ready, and check whether the working status of the GPS unit can be properly timed. If not, you need to continue to wait. If the GPS module is ready, wait to start. At this time, it is necessary to judge whether the measured control unit can be powered on and collected. If it cannot be powered on and collected normally, continue to wait for the start time. If it can be powered on and collected normally, perform the normal polling acquisition mode. Finally, it is judged whether the collection of the entire unit is completed. If the collection is not completed, the polling collection is continued. If the collection is completed, the operation of the entire program is completed. The overall control flow chart of the buoy control unit is shown in the figure below (Fig. 4).

The process of configuring parameters of the buoy control unit for measuring marine environmental parameters is firstly powered on and initialized. After initialization, check whether the upper computer is connected. If it is not connected, continue to wait. If the computer is already connected, continue to the next step to check. The next step is to check the configuration of the temperature and depth measurement nodes. Through the host computer software and the buoy control panel, interactively check, check how many

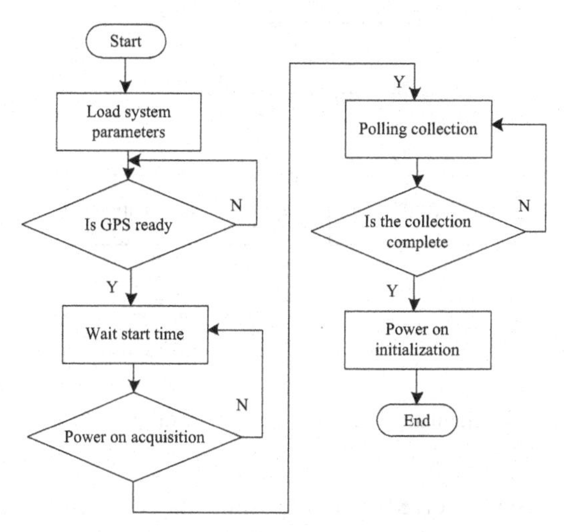

Fig. 4. Workflow of the buoy signal control unit

temperature and depth measurement probes are connected to the entire marine environment parameter measurement buoy, and check whether these temperature and depth measurement probes are all easy to use. If all the probes are available normally, proceed to the next step. If one or several probes fail, remove the failed probes before proceeding to the next step. After the probe detection is completed, the noise measurement node is detected to check whether the connected noise measurement node can work normally. The inspection method is basically the same as the detection method of the temperature and depth measurement point in the previous step.

2.4 Realization of Passive Positioning of Marine Mobile Buoy

Single-satellite frequency measurement passive positioning is to use the Doppler frequency of the signal obtained by the receiver to measure multiple times to determine the location of the radiation source (buoy). The schematic diagram of positioning is shown in Fig. 5.

When the satellite passes over the ground radiation source, the satellite receives the signal from the radiation source and measures the frequency of the received signal. Usually a low-orbit satellite passes over the radiation source for 10 to 20 min. During this time, the satellite measures the signal frequency at multiple different times, stamps it with a time stamp, and stores it. When the satellite passes over the ground station, the ground station sends an instruction to the satellite to recover the Doppler frequency of the signal measured by the satellite. Since the ephemeris of the satellite is known (obtained by means of ground measurement and control), the time to read the satellite is obtained. Press the stamp to match the time of satellite frequency measurement with the ephemeris of the satellite at that time, and then the position coordinates and speed of the satellite at the time of frequency measurement can be obtained. The above data are sent to the data processing center together, and after a certain processing, the coordinates of

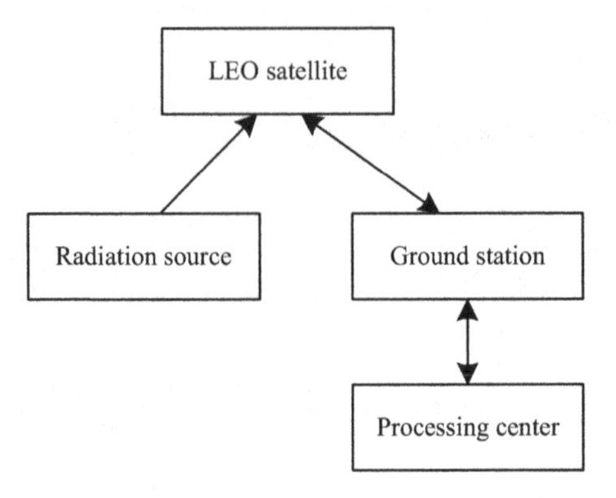

Fig. 5. Schematic diagram of passive positioning of buoys

the radiation source can be solved, and the positioning of the target can be completed. The process of signal processing is shown in the figure (Fig. 6):

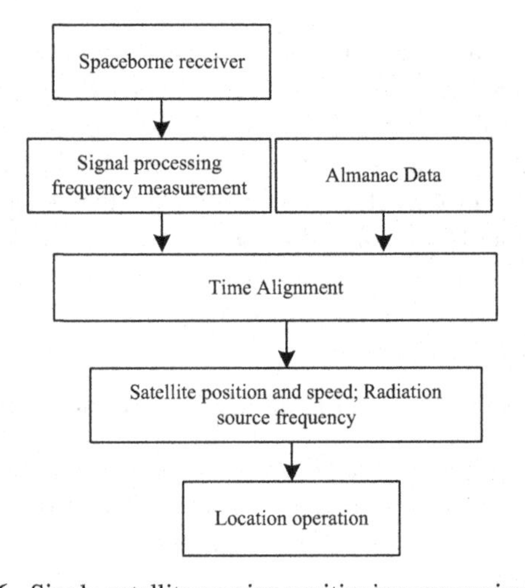

Fig. 6. Single-satellite passive positioning processing flow

According to the principle of vibration signal, the frequency f of the signal received by the satellite and the frequency f_T of the signal transmitted by the buoy have the following relationship

$$f_i = \left(1 + \frac{\left|\overrightarrow{V_i}\right| \cos\theta_i}{Zc}\right) f_T \tag{6}$$

In the formula, the speed of the buoy at time i is set to V_i, the frequency received at the viewing time is set to θ, the frequency of the buoy transmission is set to c, and the angle between the buoy connection and the direction of the satellite speed is set to

Z is the vibration frequency. Based on this, the passive positioning target of the marine mobile buoy is realized to ensure the positioning accuracy.

3 Analysis of Results

In order to effectively verify the correctness of the positioning method and the performance of the receiver, the research group built a complete set of ground test platforms. The built test platform can effectively complete the data registration of the host to the PMT and the sampling of the PMT transmitted signal, and prepares for the data processing and analysis of the software receiver in the next step. Two models of FVCOM and ROM are selected. The main purpose of this experiment is to compare and study various situations of drifting buoys at different depths, and compare the influences between them. Since the result of each experiment is the longitude and latitude of the model or drifting buoy at each time point, it is necessary to convert the difference of longitude and latitude into distance difference. The following formula is used in this experiment:

$$a = \sin\left(^{(1at2-lat1)}/2\right)^2 + \cos(lat1) \times \cos(1lat2) \times \sin(lon2 - lon1)/2\right)^2 \quad (7)$$

Increasing the measurement range to 10 V, its resolving power does not exceed 1 uV, and the accuracy exceeds 3 PPM. In this way, the calculation results of the nanovoltmeter can be used as the precise value of the measurement. The fitting and fitting relationship of Vstd-Vm of each channel is obtained by the least square method. In subsequent tests, in order to obtain more accurate measurement data, this formula must be used to correct the test data. Select several different voltage values to measure it, each with five different values, to get the average of the values. The vibration signal extraction method is fitted with origin. The vibration signal extraction method for each simulated transformed channel and its second order factor are listed in the table (Table 3).

Table 3. Fitting curve parameters

Passageway	Slope	Loading distance
Pressure	1.0013	−5.72342E–7
Precipitation	1.00185	−9.82E–4
Wind speed	1.00015	−9.16753E–7
Temperature	1.00035	−8.08158E–6
Wind direction	1.00013	−5.42656E–6
Compass	1.00035	−1.82209E–5
Long wave	1.00392	−6.46395E–5
Shortwave	1.00288	−1.47E–4

Through the VI editor, the underwater data acquisition program and the communication program of the floating buoy platform are written respectively. Through the GDB debugging tool, you can easily debug the program under the Linux of the host computer, and you can set breakpoints for debugging. Compile and generate executable files, and write Makefile files. Using the extended vibration signal algorithm, combined with the Doppler frequency shift and satellite ephemeris data generated by STK simulation, the radiation source can be better located. The positioning results are shown in the figure (Fig. 7).

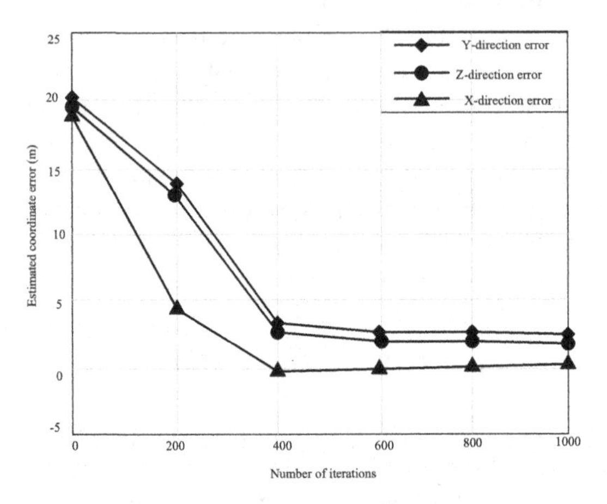

Fig. 7. Convergence diagram of buoy positioning error

In the calculation of the daily average wrong distance, the hourly distance deviation value is divided by the time-weighted value, that is, the hourly distance deviation value is divided by 24 min of work time. Every hour will get the wrong data of one kilometer per day, and then divide this value by 24, you can get the real wrong data of the next day. Horizontally, the ocean can be divided into many layers, and at each layer, the flow direction and speed of the water flow are also different, and the flow of one layer cannot be used to replace the overall flow of water. However, in this state, in order to express the data more accurately and to facilitate experiments, the ocean is often divided into several levels and divided according to different environments. The experiment used floating buoys to analyze currents at all levels of the ocean (Fig. 8).

The depth unit positioning test is carried out in the pool. The depth component of MS5837 is detected and corrected by high-precision motorized buoys. The motorized buoys have high accuracy and can be positioned within the range of $\pm 0.05\%$. The average value of the mobile buoy and the MS5837 positioning data set at each test point, and the comparison between the two is as follows (Fig. 9).

The marine mobile buoy can move according to the command of the host, and at the same time, it can also collect the information from the buoy. During the test process, the input signal can continuously excite the entire system and excite the oscillation of the entire working area, which depends on the working state of the simulation positioning platform, and requires a running curve within 0.5, so as to obtain an operating state close to the real one. Simulation results (Figs. 10 and 11).

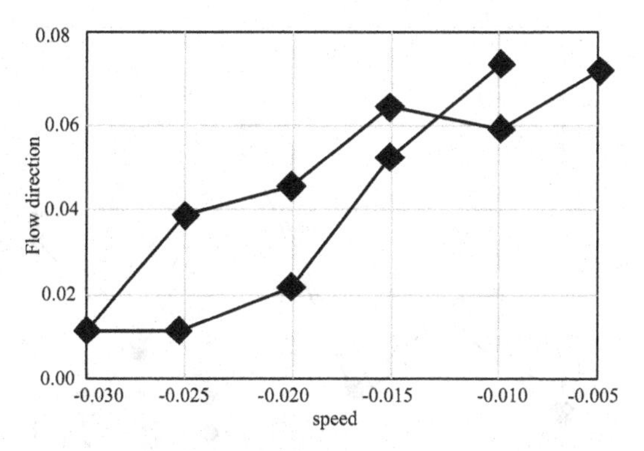

Fig. 8. The detection results of buoy positioning at different depths

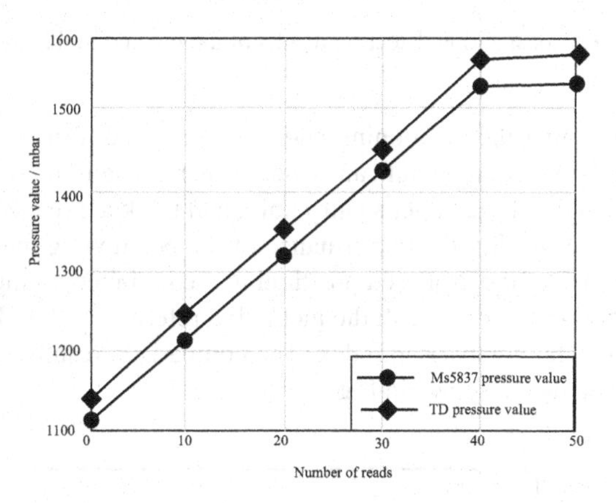

Fig. 9. Pressure values at different heights

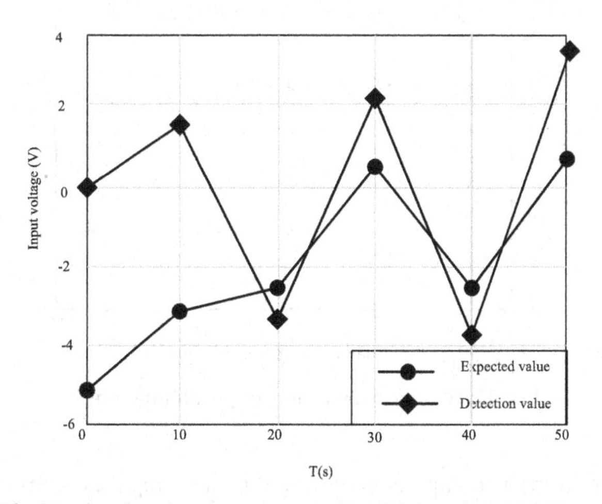

Fig. 10. The localization detection results of the method in reference [2]

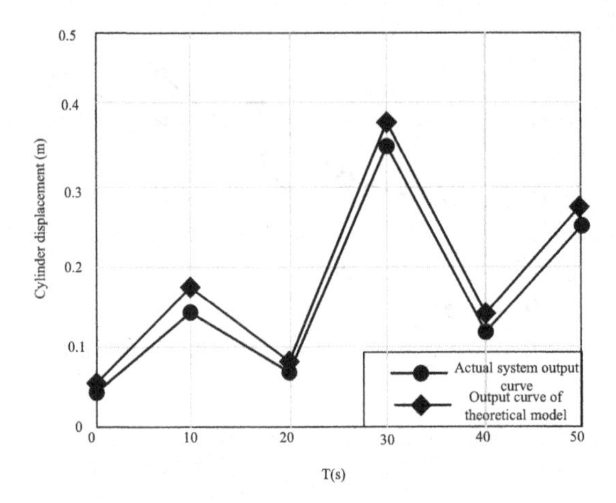

Fig. 11. Localization detection results of the method in this paper

The results show that the positioning and monitoring performance and efficiency of the method in this paper are significantly better than those in reference [2], and the tracking and positioning of the target's entry point and track are realized.

The above process verifies the performance of the positioning method. In order to verify the performance of the proposed positioning method more comprehensively, the method in this paper is compared with the method in reference [3] with the positioning time as the experimental comparison index. The comparison results of the positioning time of the two methods are shown in Fig. 12.

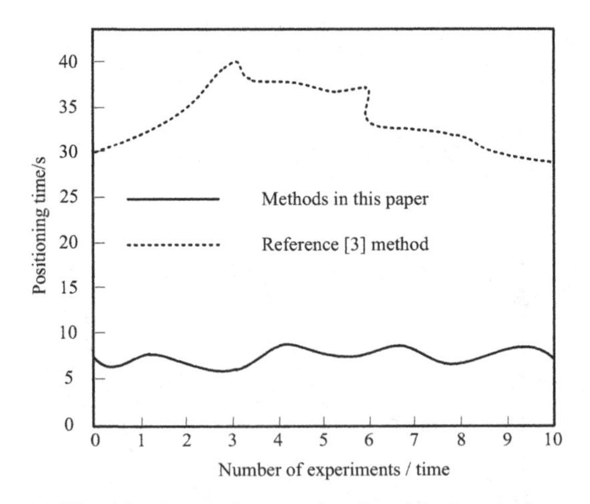

Fig. 12. Comparison results of positioning time

It can be seen from the comparison results of positioning time-consuming shown in Fig. 12 that the positioning time-consuming of the method in this paper is far lower than that of the method in reference [3]. The maximum positioning time-consuming of the method in this paper is no more than 10 s, and the maximum positioning time-consuming of the method in reference [3] is 40 s. Therefore, it shows that the method in this paper can reduce the time-consuming of passive positioning of Marine Mobile buoys.

4 Concluding Remarks

Taking the buoy positioning as the design background, the main research contents are the design and simulation of the positioning algorithm according to the requirements; the second is the software design of the display and control platform; the third is the analysis of the test data and the processing of the vector buoy array positioning. The measurement of the target water entry position and the tracking of the flight segment trajectory, according to the different signal types during target detection, the positioning algorithm is divided into a pure azimuth passive measurement algorithm based on azimuth measurement and a hyperboloid intersection positioning algorithm based on time delay measurement. The location of the target water entry point is located by using the azimuth and time delay information of the target's sound of hitting the water. According to the actual situation, two algorithms are used to complete the combination; Calculate. According to the requirements, the optimal design of the formation is carried out, and the accuracy of the two positioning algorithms in different formations is analyzed, which lays a good foundation for the design of the display and control software. The design of the display and control platform mainly uses the object-oriented design method, based on the Visual c++ 60 software development platform to complete. The experimental results show that the method in this paper is stable and has good performance, and the display and control positioning has completed the predetermined function.

References

1. Qin, Y., Liu, H., Deng, Y., et al.: A new localization method for marine wireless sensor networks based on asymmetric round-trip ranging. Acta Metrologica Sinica **41**(09), 1039–1047 (2020)
2. Guo, J., Sun, J., Wang, K.: Multi-optical buoys joint localization algorithm and simulation. J. Unmanned Undersea Syst. **29**(02), 176–182 (2021)
3. Wu, Z., Xiang, L., Xiao, H., et al.: Prediction the position of light buoy using multiplicative seasonal ARIMA model. Electron. Meas. Technol. **44**(14), 8–16 (2021)
4. Xu, L., Jin, Y., Xue, H., et al.: Buoy drifting position modeling based on telemetry data. J. Shanghai Maritime Univ. **42**(04), 26–32 (2021)
5. Liu, X., Zheng, J., Wu, H., et al.: LoRa buoy network coverage optimization algorithm based on virtual force. Appl. Res. Comput. **37**(12), 3768–3772 (2020)
6. Luo, Q., Song, Y.: Maximum likelihood localization of buoy for marnine environmental observation. Wirel. Internet Technol. **17**(22), 7–10 (2020)
7. Gao, S., Li, J., Hu, C., et al.: Design of positioning communication device for marine environmental monitoring buoy based on BDS. (09), 41–45+68 (2020)
8. Liu, X., Nan, Y., Xie, R., et al.: DDPG optimization based on dynamic inverse of aircraft attitude control. Comput. Simul. **37**(07), 37–43 (2020)
9. Li, L., Lv, T., Zhang, J.: Design of satellite communication and positioning system applied to smart float. Transd. Microsyst. Technol. **39**(01), 91–94 (2020)
10. Lu, K., Rao, X., Wang, H., et al.: The influences of different mooring systems on the hydrodynamic performance of buoy. Acta Armamentarii **43**(01), 120–130 (2022)

Optimization of Data Query Method Based on Fuzzy Theory

Yunwei Li[1](✉) and Lei Ma[2]

[1] Beijing Youth Politics College, Beijing 100102, China
liyunwei@bjypc.edu.cn
[2] Beijing Polytechnic, Beijing 100016, China

Abstract. In order to achieve the research goal of fast and accurate query of massive complex data, this study proposes a data query optimization method based on fuzzy theory. Firstly, the characteristics of the data to be queried are identified combined with the fuzzy theory, and the characteristics are classified. Then, the data management model is constructed to optimize the data query management process. The experimental results show that this method can effectively ensure the accuracy and comprehensiveness of massive data query, and prove that it can fully meet the research requirements.

Keywords: Fuzzy theory · Data query · Data management · Feature recognition · Feature classification

1 Introduction

There are a lot of fuzzy or uncertain information in relational database, and the traditional data query method has a poor effect on this kind of information, mainly manifested in the inaccurate query results. In this case, relevant scholars have proposed a probabilistic data query method based on sensor detection, and carried out research on nearest neighbor and probabilistic query problems [1]. However, it is worth noting that the probability model is only a theoretical model from the perspective of data processing, and does not give a query model according to the characteristics of the data itself and the actual needs of users [2].

Based on the above analysis, this paper studies the data storage and query distribution mechanism in fuzzy sensor networks, and designs an optimization scheme of data query method based on fuzzy theory. In this study, according to the characteristics of sensor network data, the data storage mechanism and query distribution mechanism are studied. On this basis, the features of the data to be queried are identified by fuzzy theory, and a data management model is established after the features are classified. In the model, a plurality of query areas are divided, and the data characteristics of each area are returned to the base station according to the query route. The base station obtains the final fuzzy theory value through sorting, so as to achieve the data query goal.

W. Fu and L. Yun (Eds.): ADHIP 2022, LNICST 468, pp. 56–67, 2023.
https://doi.org/10.1007/978-3-031-28787-9_5

2 Data Query Method Optimization

2.1 Fuzzy Data Feature Extraction Management

With the development of fuzzy theory and technology, a special intelligent sensor that can accept, generate and process fuzzy signals-fuzzy sensor came into being [3]. Fuzzy sensor, also known as symbol sensor and fuzzy symbol sensor, is an intelligent sensor that outputs the measurement results in the form of natural language symbol description after fuzzy reasoning and integration based on the numerical measurement of classical sensors. The research of fuzzy sensor is a late branch in the application of fuzzy logic technology [4].

With the maturity of fuzzy sensor theory, many fuzzy sensors have been applied, such as distance fuzzy sensor, chroma fuzzy sensor, temperature and humidity fuzzy sensor, comfort fuzzy sensor and so on. In order to obtain the results of fuzzy query, there are often two basic methods: one is to establish a fuzzy database model and modify its database structure by adding fields containing fuzzy attributes. The second method is to establish an accurate relational database based on fuzzy logic and fuzzy calculation of query conditions by constructing membership functions or fuzzy expansion of SQL statements, Transform it into a fuzzy range, and then conduct accurate SL query. The second method is used to define the fuzzy elements in the feature data management database, and through Visual FoxPro 6.0 converts fuzzy query into equivalent SL query statement to realize fuzzy query instance [5].

Addition, deletion and modification are the basic operations of the database. In order to complete these operations, you need to query the corresponding records in the database and locate them quickly. The current basic query of relational database is very easy to implement for precise fields such as "gender". Part of the information cannot be located correctly due to its fuzziness [6]. Based on fuzzy mathematics and fuzzy set, a SQ query method combined with fuzzy theory is proposed according to the method of reasonably self-defining membership function, and a query example is realized on MicrosoftVisualFoxpro6.0 platform to optimize the feature extraction process of fuzzy data. The specific process is shown in Fig. 1.

It is assumed that all data are deterministic. That is, the data item of each sensor is the most real situation of the detected external environment [7]. In the process of query, these exact values are used to return the results. Although the query is satisfied by calculating the true and exact value, this situation will still lead to incorrect query results, as shown in Fig. 2.

It is assumed that the user queries the temperature value of a sensor. It is assumed that the temperature detected by the sensor is very accurate, and the transmission will not cause the imprecision of the data. However, when a user queries the database and finds that the current data are A0 and B0, the ambient temperature may already be A1 and B1, so the query is inaccurate.

Because the point uncertainty model often leads to accurate data and inaccurate results, an interval uncertainty query model is proposed. The data in the database is different from the traditional data. It replaces the original point form in the form of an interval. The uncertain data interval is defined as U. In particular, U_i is a real valued function of a closed interval with respect to t, which is used to limit the value of Ta at

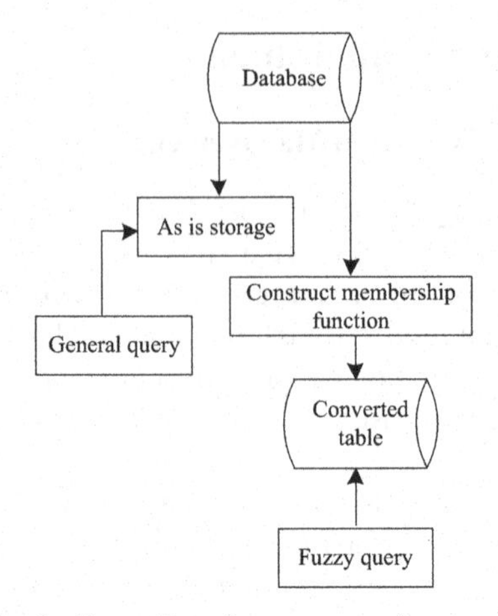

Fig. 1. Fuzzy Data feature extraction process

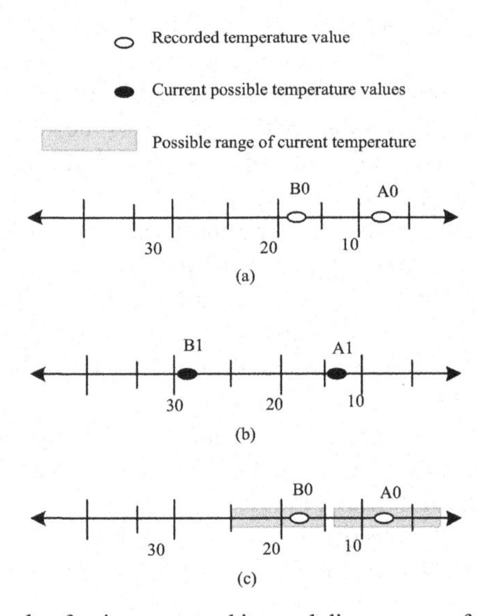

Fig. 2. Example of point query and interval direct query of uncertain data

time t. The model describes the imprecise data in the form of interval, so that U can change linearly according to time until the next update arrives.

2.2 Fuzzy Data Processing Algorithm

Fuzzy data processing is based on the transformation from digital information to symbolic information. In order to realize symbolic measurement, it is necessary to clearly give the corresponding relationship between symbols and numbers to ensure that two different symbols cannot have the same meaning [8]. Based on this, the change principle of fuzzy semantics and fuzzy description is displayed, as shown in Fig. 3.

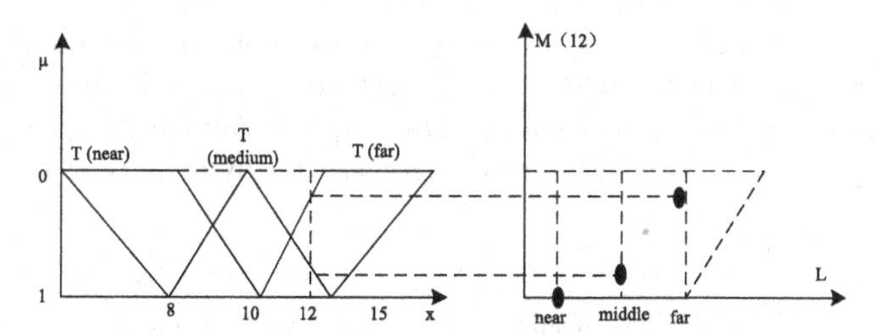

Fig. 3. Fuzzy Semantics and fuzzy description

In this process, there are five ways to query according to various classification methods: according to the query time, it can be divided into query at a certain time, query in a time period and continuous query. According to the value returned by the query, it can be divided into value based query and entity based query [9]. According to aggregation or not, it can be divided into spatial aggregation query, temporal aggregation query and non aggregation query. According to the filtering method, it can be divided into filtering and non filtering. According to the dimension of perceived data, it can be divided into one-dimensional and multi-dimensional queries. Queries based on these classification methods are divided into 28 categories. Although the model in this paper can be extended to all the above 28 queries, in order to facilitate the discussion later in this paper, this paper focuses on two cases, as shown in Table 1.

Table 1. First level query classification method

QUERY CLASS		Fuzzy query
TIME	Point of time	√
	Continuity	
	Time slot	√
ANSWER	Entity	√
	Value	√
AGGREGATION	Spatial aggregation	√
	Time aggregation	
	Non aggregation	√
FILTERING	filter	
	Non filtering	√
DIMENSIONS	Apile	√
	Multi pile	

The answers returned by all these aggregate queries are in the form of fuzzy language and membership degree, which should meet the constraints of fuzzy membership function. Table 2 summarizes the methods of fuzzy query. This table lists the differences between fuzzy query and ordinary sensor query, and finally gives the values given by the query.

Table 2. Classification of fuzzy sensor query methods

Query type	Entity based query	Numeric based query
Gather	FKNN, FENNQ, FEMinP, FEmaxP	FLAvgP, FLSumP, FLMaxP
Non aggregation	FERP	LSingleP
Membership fuzzy query	Sq, [μ1,μu]	Lq,[μ1,μu]
General query	Sq	V

It can be seen that fuzzy query gives more query result information than ordinary query. The query methods in this table are some of the query methods of fuzzy sensor networks. In practical applications, many have been widely used, such as fuzzy theory query. The random walk mechanism is used to search the data query request node in the process of network communication. The approximate nodes and equivalent nodes of the node are searched twice, and the nodes and data with lower load and higher evaluation value are returned. These two are used as the query target nodes and data required by the user. The specific process is as follows: Suppose that the number of service nodes providing user data query in wireless sensor network is d, and all query service nodes form an overlay network c. the set of user data objects on service node n is represented by $b_{|dc_n|}$, and the number of user data objects is $b_{|dc_n||dk_n|}$. For any service node n in wireless sensor overlay network k, further describe the i user data center, The network bandwidth between data centers of different users in B can be expressed as

$$B_i = \begin{pmatrix} b_{11} & b_{12} & \cdots & b_{|dc_n|} \\ \vdots & \vdots & \ddots & \vdots \\ b_{|dc_1|} & b_{|dc_2|} & \cdots & b_{|dc_n||dk_n|} \end{pmatrix} \tag{1}$$

Membership function is a basic concept in fuzzy mathematics. Fuzzy set is completely determined by its membership function. When applying fuzzy mathematics to solve practical problems, the determination of membership function is very important. It is the first step to apply fuzzy mathematics to solve practical problems. Considering the fuzziness of entities, it must be incorrect to use binary characteristic function to express it [10]. In order to express these attributes more accurately, so that their value in [0,1] means that the value can be unlimited, the data characteristic membership function is obtained as follows:

$$\mu_A = \begin{Bmatrix} B_i & x \leq 30 \\ \frac{b_{|dc_n|}}{10}, & 30 < x \leq 40 \\ 1, & 40 < x < 50 \\ \frac{b_{|dc_n||dk_n|}}{5}, & 50 \leq x \leq 55 \\ 0, & x > 55 \end{Bmatrix} \tag{2}$$

Different types of user queries have a great impact on the query methods and efficiency of encrypted data. This paper divides user queries into the following two categories:

(1) Attributevalue query;
(2) Attributeopattribute join query.

Among them, attribute is an encrypted attribute of the trusted database, WLE is the query value for attribute, and OP is the query operator, which can be "=", ">", "<", "", and so on. For a series of attribute = value queries, set the ciphertext data segment of the current encryption attribute A as Q_K, a total of k segments, the ciphertext tuple size of each data segment is $M_1, M_2, \cdots, M_I$, the number of times the query value falls in each ciphertext data segment is L and Z respectively, and the total number of correct results returned by each ciphertext data segment is $H_1, H_2, \cdots, H$, then the total false positive rate of the query is as follows:

$$P = \mu_A - \frac{\sum_{i=1}^{k} B_i}{Q_K \sum_{i=1}^{k} LZ + M_i} \tag{3}$$

Adaptive index partition strategy is adopted to dynamically adjust the segmentation of ciphertext with the change of user query and data Firstly, the influence of ciphertext data segmentation on the false positive rate of query is analyzed. Suppose that the current i ciphertext data segment is re divided into two new data segments, each containing the number of ciphertext tuples M_{i1} and M_{i2}, carrying the query service Q_{i1} and Q_{i2}, and the total number of correctly hit query results is H. on the premise that other ciphertext data segments remain unchanged, the change of false positive rate before and after segmentation is as follows:

$$\lambda = \frac{\sum_{t=1}^{k} P + H}{1 + \sum M_{i1}Q_{i1} + M_{i2}Q_{i2}} \tag{4}$$

There is only one data segment in the initial state of the database. If only AEI method is used to adjust, it will take a long time, which is not conducive to the stability of service quality In order to speed up the optimization and adjustment of ciphertext segments, other methods such as equi width, equidepth or QOB can be used to divide several ciphertext segments during initialization, and then AEI method can be used for dynamic adjustment and optimization with the loading of data and services.

2.3 Implementation of Data Query Under the Guidance of Fuzzy Theory

Fuzzy sensor network is formed by replacing traditional sensors with fuzzy sensors in traditional sensor networks. Its network composition is the same as that of traditional wireless sensor networks. The data retrieval system of fuzzy sensor network is shown in Fig. 4.

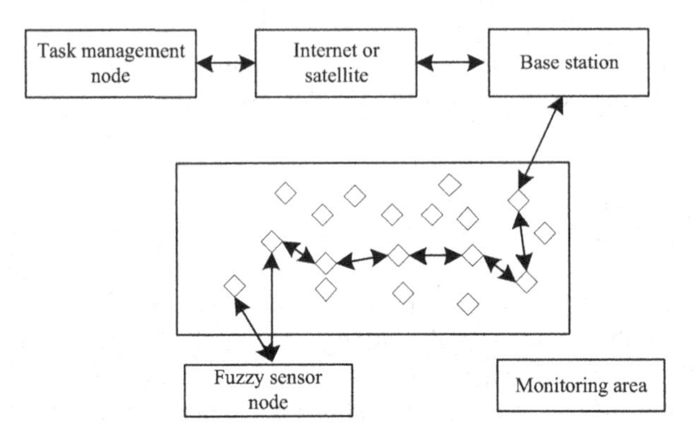

Fig. 4. Fuzzy sensor network data retrieval system

Sensor network is composed of a large number of micro sensor nodes deployed in the monitoring area. The nodes form a multi hop self-organizing network system through wireless communication. The network nodes can cooperatively sense, collect, process and transmit the information of the perceived target in the network coverage area, and send it to the information collector. Fuzzy sensor network can enable people to obtain a large amount of detailed and reliable information at any time, place and under any environmental conditions, so as to greatly expand the functions of the existing network. Fuzzy sensor network usually refers to a wireless network composed of a group of nodes with embedded processors, sensors and wireless transceiver, which collects and processes the target information in the network coverage area through the cooperative work of nodes. The sensor node is deployed in a target area. The information measured by the sensor node, such as temperature, humidity, light, pressure and speed, is transmitted to the convergence point through multi hop, connected through the convergence point, and finally connected to the task management node. The task management node has a man-machine interface, which can intervene, remote control and manage. Convergence point is a system with strong communication and computing power. In some documents and projects, it is also called base station.

Like the general sensor, the fuzzy sensor can sense the measured value determined by the sensing element, but the fundamental difference is that it can output not only the numerical value, but also the linguistic symbol. Therefore, the fuzzy sensor must have a numerical symbol converter. The fuzzy sensor should be able to exchange information with the superior system, so the communication function is the basic function of the fuzzy sensor. The basic function of fuzzy sensor determines its basic structure and implementation method. The basic structure of fuzzy sensor data management is shown in Fig. 5.

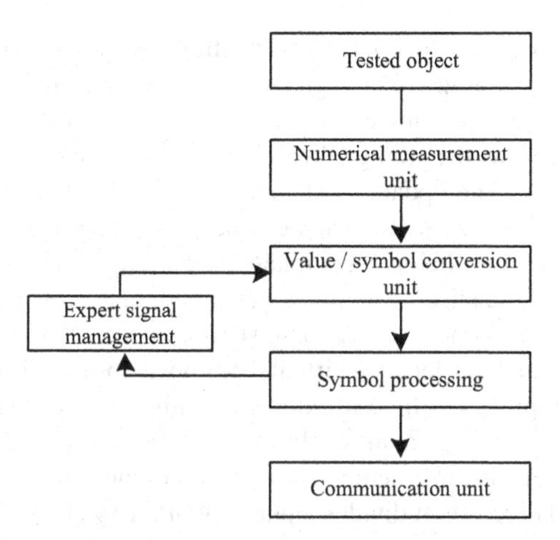

Fig. 5. Basic structure of fuzzy sensor data management

In order to effectively use the dynamic sensing data and accurately express the unique query of sensor networks, it is necessary to study and design a query processing system that can reflect the characteristics of sensor networks, so as to lay a foundation for query optimization and query processing. In the discussion of this paper, the specific uncertainty model is not important. What users need is the possible value of the known data in the query interval. This paper is interested in the query of some dynamic attributes L and the set S of objects in the database. Although the model in this paper can be extended to other domains, this paper still assumes that L is the attribute of a perceived object. For example, very simple integers and coordinates. In this paper, Si is used to represent the ith fuzzy sensor in S (Fig. 6).

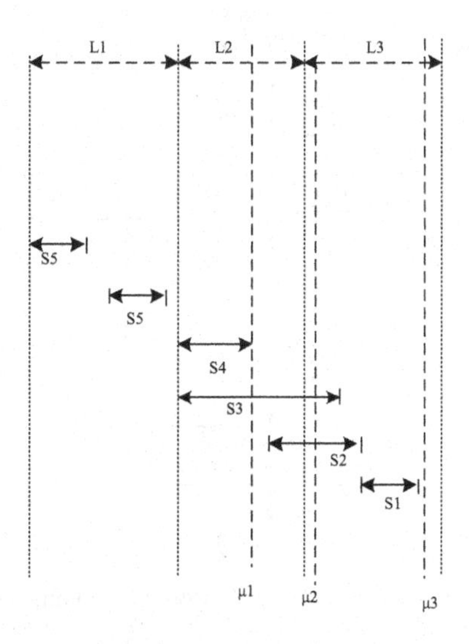

Fig. 6. Example of fuzzy data sensing query

In order to fully describe the query classification method proposed in this paper, a specific query example is given. For a given initial network, the network distribution of the encoded partition center node is obtained. For each partition, a point closest to the diagonal is selected as the storage data node, and the data in the partition will be transmitted to the node. The update mechanism adopts FIU update algorithm, that is, when the data value of fuzzy sensor changes beyond a certain threshold, update the value of database. After the slicing is completed, according to the method in Chapter 4, the virtual topology of the sliced network is generated, that is, K paths with the lowest communication cost are found from the base station to each central node and any pair of chip central nodes, and the network virtual topology is formed. The data between any nodes in the virtual topology will be saved in the routing table of the base station.

To sum up, the design of data query algorithm based on fuzzy theory is completed. The algorithm considers that the membership interval of the data generated by the fuzzy sensor is uncertain. Therefore, it divides a plurality of query areas, and returns the data characteristics of each area to the base station according to the query route. The base station obtains the final fuzzy theory value through sorting, so as to achieve the data query goal. The specific query process is shown in Fig. 7.

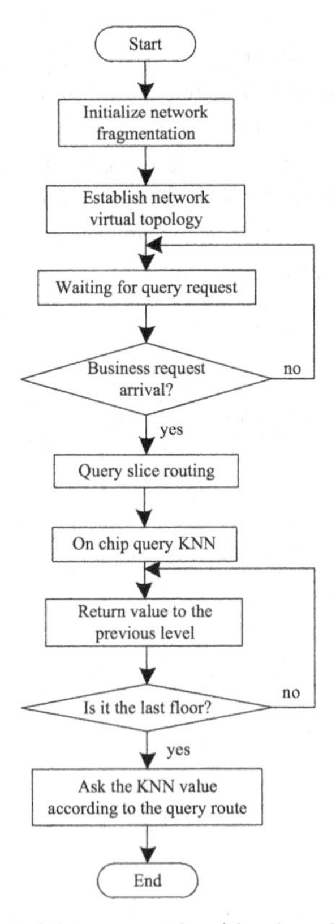

Fig. 7. Flow chart of data query algorithm based on Fuzzy Theory

3 Experiment and Result Analysis

In order to verify the effectiveness of the data query method based on fuzzy theory, the following simulation experiments are designed.

In the experiment, a prototype system is developed using C + technology to realize the development of data query methods. The main development tools are C + + 6.0 and SQL server. 54 sensors are set in the local area network to collect network data. The background database is built using sqlserver2000 technology. The operating environment is: WindowsXP, Pentium (R) d28gcpu, 1g, 160g hard disk. Planetsim is selected as the experimental simulation tool. The node size in the wireless network is set to change within the interval [0, 500], and 5000 query requests are generated randomly for each experiment.

In order to avoid the singleness of the experimental results, this method is compared with the traditional data optimization query method based on ant colony optimization and the data query method based on compression. The test comparison results when the data scale is 150000, 300000, 450000 and 600000 are as follows. In each figure, PQ represents the data optimization query method based on ant colony optimization, RQ represents the data query method based on compression, and FKNN represents the method in this paper.

The query response time results are shown in Fig. 8 and Fig. 9.

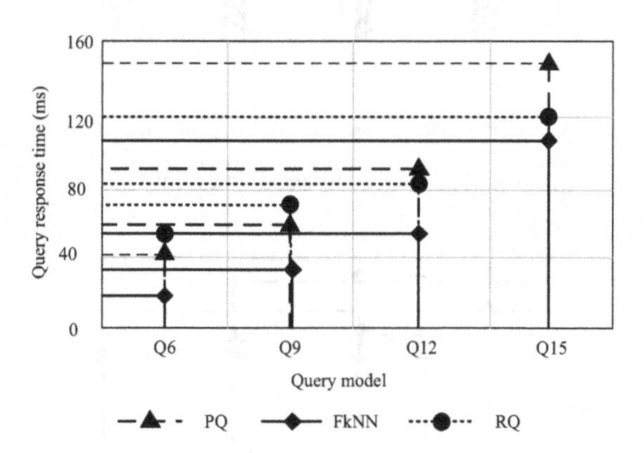

Fig. 8. Data query response time of each region

Because the network of this experiment is a small network composed of 54 sensors, the following methods are used to test the recall and precision of the proposed method: for each query instance, a corresponding data set is generated, including 24 appropriate sensor tuple combinations related to the query, These sensor tuple combinations are formed by removing duplicate tuples and appropriately adding randomly selected sensor tuples from the first 8 relevant query results obtained by the three query methods mentioned above. Then, the user selects 8 sensor tuples most related to the initial query, and tests the recall and precision of different methods on this basis. The results are shown in Fig. 10 and Fig. 11.

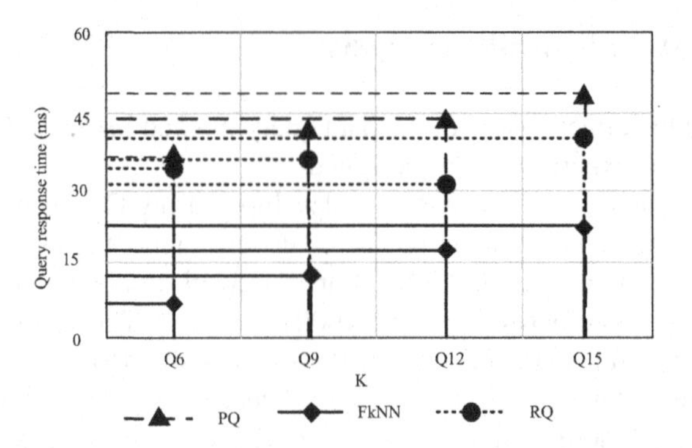

Fig. 9. Query response time when feature value changes

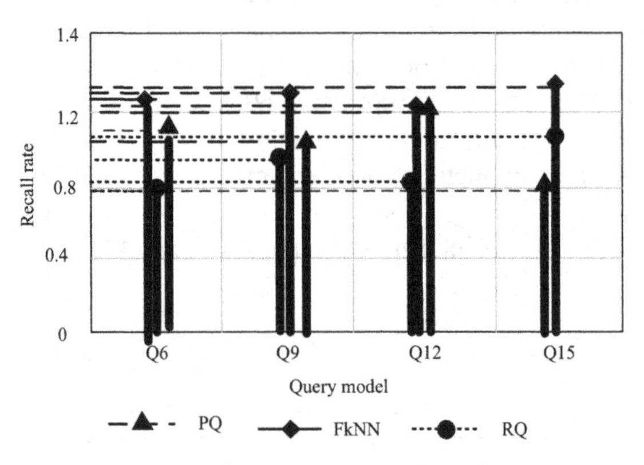

Fig. 10. Data query recall test results

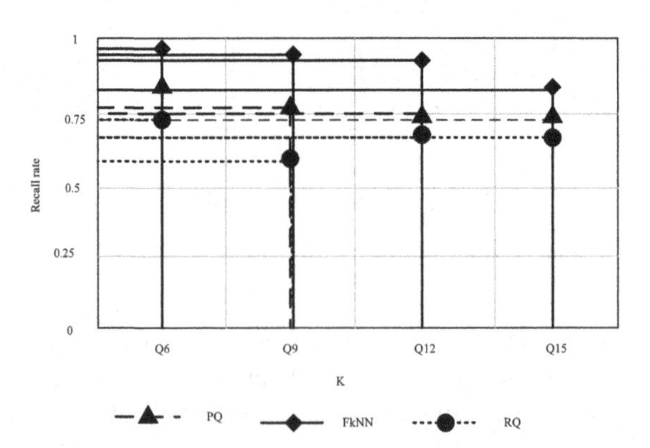

Fig. 11. Recall rate of changing data under interference environment

Based on the above analysis of the experimental results, it can be seen that the fuzzy theory query method proposed in this paper is better than the two traditional query methods in query response time and recall, which also shows that the query method proposed in this paper is feasible.

4 Conclusion

With the development of intelligent database, the processing of fuzzy data and its application in computer will have a better promotion and development space. Therefore, this study proposes a data query optimization method based on fuzzy theory.

In this study, the features of the data to be queried are identified by fuzzy theory. After the features are classified, multiple query areas are divided, and the data features of each area are returned to the base station according to the query route. The base station obtains the final fuzzy theory value through sorting, so as to achieve the data query goal. Through experimental verification, it can be seen that after the application of this method, the data query time is less, and the application effect in recall is also better than the traditional method.

References

1. Xu, B.: Massive incomplete data approximate query system based on improved K-nearest neighbor algorithm. Mod. Electron. Tech. **44**(15), 177–181 (2021)
2. Teng, Z., Liao, Z.: Differentiated service mechanism for data query on named data networking. Comput. Eng. Appl. **55**(9), 17–25+86 (2019)
3. Gao, J., Yang, F.: Semi-structured data query optimization algorithm based on swarm intelligence. Comput. Simul. **38**(8), 381–385 (2021)
4. Lian, J., Fang, S., Zhou, Y.: Model predictive control of the fuel cell cathode system based on state quantity estimation. Comput. Simul. **37**(07), 119–122 (2020)
5. Su, J., Xu, R., Yu, S., Wang, B., Wang, J.: Idle slots skipped mechanism based tag identification algorithm with enhanced collision detection. KSII Trans. Internet Inf. Syst. **14**(5), 2294–2309 (2020)
6. Su, J., Xu, R., Yu, S., et al.: Redundant rule detection for software-defined networking. KSII Trans. Internet Inf. Syst. **14**(6), 2735–2751 (2020)
7. Ma, Z., Yuan, H., Gu, Y., et al.: Research and implementation of document-relational data query execution technology. J. Front. Comput. Sci. Technol. **14**(08), 1315–1326 (2020)
8. Yun, W.: Improved Simulation of Large-Scale Hybrid Network Database Fuzzy Query Algorithm. Computer Simulation **37**(05), 246–249 (2020)
9. Lu, S., Chen, H.: A survey on data query optimization with machine learning. Wirel. Commun. Technol. **29**(04), 5–10 (2020)
10. Zhao, Y., Hu, L.: Design of preference query system based on linked data in open environment. Comput. Technol. Dev. **30**(09), 7–11 (2020)

Study Demand Mining Method of College Students' Physical Health Preservation Based on Cognitive Model

Kun You(✉) and Changyuan Chen

Department of Physical Education, Xi'an Shiyou University, Xi'an 710065, China
youkun2135@163.com

Abstract. In the context of national fitness, contemporary college students physical and mental health of the community focus. In order to solve the problem of low accuracy in the process of learning demand mining, a new learning demand mining method based on cognitive model is designed. Optimizing the structure of PE health preserving curriculum and revealing the laws and rules of scientific PE essence. Health curriculum association model was constructed to classify emotional states and identify the characteristics of individual learning needs of college students. In this paper, we introduce the concept of association rules and use cognitive model to set up the learning needs mining model. The results show that the average accuracy of the proposed method is 74.082%, 64.348% and 65.798%, respectively, which proves that the proposed method is more effective than the other two methods.

Keywords: Cognitive model · College students · Physical education · Learning needs · Mining methods · Physical education

1 Introduction

Based on the guiding ideology of "health first", the reform of physical education in colleges and universities is to re-examine the training objectives of college students' physical education, reform the curriculum system centered on competitive sports in the past, and establish a physical education and health curriculum system with "health first" as the primary objective to promote students' all-round physical and mental development. Through the study of PE health course, it can improve the physical and mental health of college students effectively and make every index tend to normal value. The experimental data prove that the mood index of college students develops well after studying the traditional PE health course for eight weeks. Lifelong physical education is the sports thought of the new century, not only in the society, but also in the school, which is the theoretical basis of school physical education concept renewal and reform. Physical education has more advantages than other sports courses in regulating the indicators of anger, depression, mood, and terror.

W. Fu and L. Yun (Eds.): ADHIP 2022, LNICST 468, pp. 68–80, 2023.
https://doi.org/10.1007/978-3-031-28787-9_6

Nowadays, it is very important to find a way to regulate the health of college students in the age when the health of college students is not optimistic. The profound philosophy and rich training methods of traditional P. E. health preserving suit the physical and mental characteristics of college students. The systematic education of health preserving will promote students to know the origin of health preserving, enhance the awareness of health preserving and explore the training methods suitable for themselves. Sports health preserving is a science with the theme of life, concentrating on the excellent traditional culture of five thousand years. It is a reasonable narration of the way, the aim and the developing course of sports health preserving. It can help college students to develop their skills, expand new fitness space, build strong body, cultivate good psychological quality, and lay a foundation for lifelong sports. The organic combination of traditional Chinese physical education and modern physical education in colleges and universities can enrich the content of physical education in colleges and universities, and actively carry out students' thought of "lifelong physical education", so as to let students understand that physical education is not only physical education, but also to improve students' physical quality and sports ability, and to cultivate students' comprehensive quality and ability. Sports health from its origin, based on the idea of harmony between man and nature, pay attention to the continuation of life, through the scientific way of sports health to prolong life. Sports health from its development, is to maintain the body as the core, so as to achieve physical and mental balance, so that the quality of life can be effectively improved.

2 Learning Demand Mining Method for College Students' Physical Education Health Preservation Based on Cognitive Model

2.1 Optimizing the Structure of Sports Regimen Curriculum

As a brand-new subject in today's society, the definition of sports health preserving can be understood as: sports health preserving is a traditional life science, concentrating on the excellent traditional culture, using the traditional items to exercise the body scientifically, so as to achieve the balance state of physiology and psychology of the body and improve the quality of life. Sports health preservation is a life science, and its aim, means and developing process are all around the center of human life. Looking from the origin, it first take the noble person rebirth as the thought premise, pays great attention lengthens person's life, take the pursue longevity as the goal. From the development point of view, has always been around the maintenance of the body, improve the quality of life, the pursuit of psychological and physiological balance of this line. Taking the traditional culture of Chinese nation as the guiding ideology, the paper defines the sport keeping in good health as a reasonable exercise to achieve the balance of human body. So far, the development of physical education has been the treasure of the Chinese people's practice from generation to generation. It contains the excellent traditional culture of China and has a unique way of exercising. As a new subject, the academic community has not yet reached a consensus on the basic concepts, research objects, research methods and the scope of research, content and other aspects. The concept of Chinese traditional sports health preserving can be understood as follows: Traditional life science with

Chinese characteristics is a special health preserving system based on Chinese traditional philosophy, focusing on maintaining the body, improving the quality of life and pursuing the balance of psychology and physiology in the process of social practice. The main ways to optimize the structure of PE regimen courses are shown in Fig. 1:

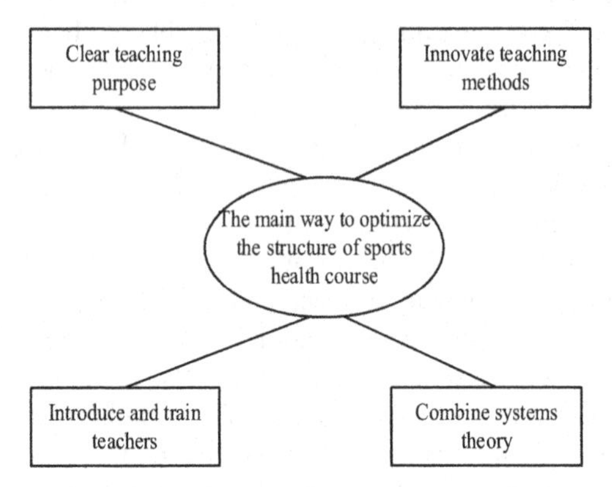

Fig. 1. Diagram of main ways to optimize the structure of PE regimen curriculum

As can be seen from Fig. 1, the main ways to optimize the structure of PE regimen curriculum are: clear teaching objectives, innovative teaching methods, introduction and training of teachers and the principle of system theory. The curriculum of keeping in good health is the core and foundation of the learning demand of keeping in good health. It can show the requirements of teaching, and it is also the main part of teaching plan [1]. Therefore, in order to improve the quality of teaching, so that college students understand our traditional health culture, we must take the curriculum construction as a breakthrough. Traditional physical education relies on the ability of human body to regulate and strengthen the functions of various parts of human body through posture adjustment, breathing exercise and the use of ideas, so as to induce and inspire the internal potential of human body and play the role of preventing diseases, curing diseases, promoting intelligence and prolonging life. It belongs to the field of human science. The traditional Chinese physical education theory holds that the essence, qi and spirit are the three treasures of human beings, and the essence is the material basis of human body and the root of life. Qi is the physiological function of human body and the power of life. On the basis of the essence and qi, the spirit, consciousness, movement and other life activities are dominated by the phenomenon called "God". In the process of PE health preserving teaching, we should not only teach the action of PE health preserving, but also explore a new way on the basis of the original teaching aim. Action does not mean keeping in good health. If it is taught to students mechanically, it is not the completion of the teaching task. Only in this way can students be attracted to learn more about it and not appreciate the inner culture of it.

Spirit, Qi and God are interacted and cause and effect each other, running through all the process of life activities. Full of vigor, the body's function is exuberant and harmonious. Full of vigor, the normal activities of all functions are destroyed. In today's colorful world, students are facing a lot of negative effects. The influence of network and other modern media on students not only makes the young minds of college students become impetuous. In teaching, we should explore new teaching methods constantly according to students' physical and mental characteristics, and fully mobilize students' enthusiasm. We must not adopt the old teaching methods. The final result of this kind of teaching method is that students are more and more bored with sports keeping in good health. Sports keeping in good health is different from other sports in that there is no antagonism or stimulation. Therefore, it is difficult to tap students' learning demand for sports keeping in good health. It is not successful to let students study sports keeping in good health only by "compulsory courses" [2, 3]. Nowadays, many college students are smoking, drinking, playing cards and other phenomena seriously affect their sleep. In physical education, if we can arrange some knowledge of self-cultivation and mental health, we can guide students to establish a correct sense of health, cultivate students' good living habits, and use correct body-building methods to recuperate themselves, so that students can carry forward the traditional sports health culture. Therefore, in the process of physical education teaching, we must continue to innovate in the means.

At the same time, we should fully mobilize the enthusiasm of students and teachers in the classroom, better participation in the classroom, so as to improve the quality of teaching, but also conducive to students' physical goals. The characteristics of Chinese traditional health preserving science are: combining movement with static state, adjusting the whole balance of human body. Its emphasis lies in the whole regulation of the living body and the exploitation of the brain potentiality, and its methods of external movement and internal quiescence and seeking quiescence in movement really reveal the law and law of scientific sports essence, which is a supplement to modern sports methods. The curriculum system of PE health preserving is not a fixed way, nor a single activity of one of the elements, but a dynamic system containing many elements. In this dynamic system, every element of PE health preserving curriculum system is an integral movement of PE health preserving curriculum system. College students are facing increasingly fierce employment pressure and competitive environment. Many College students have different degrees of psychological diseases. Traditional health education can play a certain role in college students' interest in health care knowledge and fitness means. Using the point of view of system theory, the system of sports health care curriculum can be divided into two elements: substantial and non-substantial. These two elements are composed of many concrete elements, which are combined in a unique way to form a unified whole, developing continuously and being relatively independent.

2.2 Establishment of a Health Curriculum Association Model

The health problems we face today are very different from those we faced in the past. In the past, it is generally believed that people's health refers to the physical discomfort, or there is no disease, that is, the so-called "no disease, no injury, no disability is health." This view is based on a purely biological understanding of health issues. In fact, this is one-sided. Because human beings are both biological and social. The regulation of emotion, the regulation of human physiological function and the direct regulation of mental health by sports health preserving is a unified and inseparable organic whole. They influence, promote and restrict each other. Among them, the regulation of emotion by PE health practice is a part of the regulation of mental health, and the regulation of

PE health practice on human physiological function is closely related to the regulation of mental health. Man is not only a biological man, but also a social man. Man, as an extremely complicated and advanced living body, has not only physiological activities, but also psychological activities and adaptation to society, life events, and moral and ethical understanding and compliance. According to the speed, intensity, tension and sustainability of the occurrence, emotions can be divided into mood, passion and stress of the three states. The state of mind is a relatively weak, calm, lasting and emotional state of rendering. Passion is a strong, transient, explosive emotional state: stress is a state of high tension caused by an unexpected emergency. The emotional state of college students has obvious characteristics of richness, instability, impulsiveness, periodicity, externality and violent fluctuation. Scientific practice has shown that physiological activities and psychological activities are not only equally important to health, but also closely related to each other. It should be noted that physical health is the basis of mental health, and mental health is a necessary condition for physical health. Everyone's physical and mental health is the guarantee of social health. The regulation of mood by sports health preserving is mainly realized by the regulation of mood. But it can not be denied that the practice of physical fitness for passion and stress also has a regulatory role. That is, in the face of major events will not occur in the scope of cognitive activities narrowed, rational analysis capacity is limited, weakened self-control, which makes people's behavior out of control, making some reckless behavior or action. Health is defined by the United Nations World Health Organization as not only being free from disease and infirmity, but also being in perfect physical, mental and social condition. That is, health is three-dimensional, including physical health, mental health and social adaptation to the perfect state. In the process of the development of later studies can, but also put forward that human health should include moral health. A large number of studies have proved that physical exercise can help improve people's intelligence. When the level of intelligence has been improved, in the face of emergencies will show a quick-witted, timely action, out of the woods. Instead of being dumbfounded, fussy, or in trouble. In terms of mental health, good mood can not only make college students full of hope and confidence in themselves, but also make them willing to act, quick-minded and eager to learn. Therefore, the sense of health, including physical health, mental health, good social adaptation and moral health of the four aspects. It is impossible to eradicate disease completely, especially in today's society, because the process of life itself constantly changes the environment. Health consists of a relative state of adaptation to the environment. Concentration, open-mindedness, creativity and good interpersonal relationship. The harm of bad mood to college students' mental health is mainly reflected in two aspects: excessive emotional reaction and lasting emotional reaction. Excessive emotional response refers to sudden and strong emotion can inhibit the high-level mental activity of the cerebral cortex, disturb the balance of excitement and inhibition of the cerebral cortex, narrow the range of consciousness, lower judgment, loss of reason and self-control. Lasting emotional reaction means that some students are not good at positive adjustment and elimination of negative emotion after causing such negative emotions as worry, sadness, fear and anger, thus falling into a negative emotional state for a long time, being unable to extricate themselves, feeling pessimistic and painful, and seriously affecting their life and study. Because people are constantly changing the environment, so people continue to be in a bad state of adaptation to the environment. Thus, the search for health is a continuous and adaptive process, not a static state that is always attainable or maintained. In other words, health means adapting to a constantly changing biological and social environment. Relaxation of intense energy through physical activity. Bad emotions can often accumulate into a huge mental and physical energy. This energy if there is no normal channel to dredge, it will be in different forms of externalization, as some physical and psychological symptoms. Sports health is aerobic exercise from the point of view of physical exercise intensity, exercise intensity moderate. Moderate intensity of movement of the limbs with the meridians in the "Qi" movement, so that this energy can be guided, but also be released. Regulate by breathing. This requires that we continue to care for and nurture life. Uses the effective sports keeping fit method to satisfy the people to the healthy yearning. People's health level is affected by the intersection and interpenetration of congenital genetic factors and acquired factors. Heredity is the precondition of physical development and change, and it has great influence on people's health in the future. On this basis, the postnatal factors affecting health status are derived, as shown in Fig. 2:

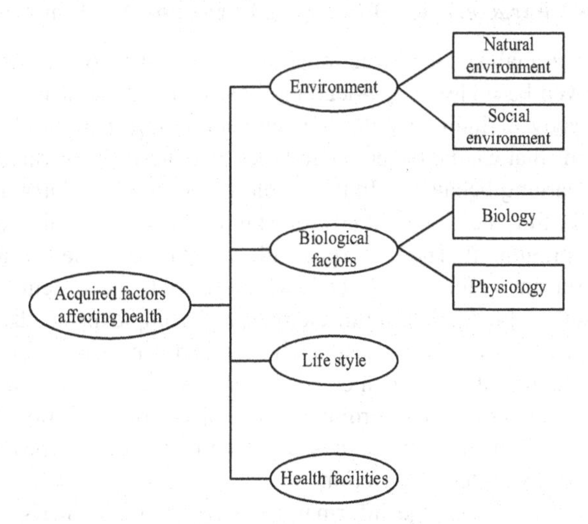

Fig. 2. Acquired factors affecting health

As can be seen from Fig. 2, the acquired factors can be grouped into four categories: the environment (including the natural and social environment), the biological factors (including the biological and physiological factors of the body), the way of life and the health facilities. For individuals, the establishment of a good lifestyle is a positive initiative to improve their health, because the environment, health care facilities and so on to a large extent depends on the level of social development. One of the characteristics of sports health preserving is the regulation of "breath". When people with tension, restlessness, anger and other adverse emotions, to follow the rhythm of Kung Fu practice for regular breathing, can significantly alleviate the adverse effects. Sports regimen is a kind of slow, gentle, both movement and static movement. Most of the exercises of keeping in good health in physical education are embodied in stillness and stillness. In the modern fast-paced life, "quiet and natural" is precisely a unique way to vent bad emotions.

Living standards and quality of life is the way of life in both quantity and quality of reflection. The standard of living focuses on the quantity of life style, which mainly refers to the amount of material wealth possessed by people for consumption. Indicators of living standards, such as the amount of housing area and other means of subsistence owned by each person, can be directly quantified in the form of money or in kind. To "clean and elegant" mind to calm the chest of unhappiness, so as to solve the problem of bad moods in the practice of sports health, often meditates in a beautiful environment, surrounded by some beautiful things. In this process of meditation, one's mind is highly focused, distractions are thrown out, and the negative emotions are ruled out. Quality of life focuses on the quality of life, it refers to a certain social system, people's life to ensure personal health, freedom and all-round development of the degree. The main indicators are: people's health and life expectancy, education, spiritual life needs to meet the degree of free development and creativity to achieve the degree of people.

2.3 Identify the Characteristics of Individual Learning Needs of College Students

Through the further analysis of instructional interaction, it is found that the important factors affecting web-based learning—learners' instructional interaction behavior—can be expressed directly or indirectly through interactive information. Therefore, those concrete operations that can be observed and recorded are of great value for analyzing learners' online learning behaviors. In the whole function of the curriculum system of keeping in good health, the so-called non-substantial factors play an important role in improving and restraining it. The important influence factor of the learning demand of sports health preserving is the entity factor, which restricts the system of sports health preserving curriculum [4, 5]. The multi-element and complexity of the system of PE health preserving course system plays a decisive role in the practicality, comprehensiveness, complexity, controllability and intersection of the system. If we generalize the three forms of information interaction according to the subject-object relationship in teaching activities, the information interaction can be divided into two basic types: human-human interaction and human-content interaction.

The acquisition of knowledge information refers to the learner's online learning behavior aimed at acquiring the target knowledge, which is the interaction of "person" and "content" in the process of information interaction. It is very important for college students to improve their mental health. Physical exercise and mental cultivation are closely combined in the practice of traditional sports, and run through daily life effectively. The acquisition of knowledge information is the most important behavior in e-learning, and the acquisition of knowledge information takes up more than half of the time of all e-learning activities. On the basis of the above, the concept of association rules is introduced to solve this problem by using different minimum support and minimum confidence for different classes. We only need the user to specify a total minimum support, and then distribute it to each class according to the class distribution data as follows:

$$\min D = \frac{|\varepsilon - 1|^2}{D} \times \varepsilon \tag{1}$$

In formula 1, D represents the number of instances in the training data, and ε represents the number of all instances in the training data. Using this formula gives a higher minimum support for frequent (negative) classes and a lower minimum support for rare (positive) classes. This will ensure that enough positive class rules are produced, without producing too many meaningless negative class rules. Information retrieval is one of the important means to acquire knowledge and information. Browsing knowledge information to obtain the most important behavior, whether traditional books or new digital resources, learners need to transfer it to their brain through browsing behavior for knowledge concept interaction and internalization. Therefore, we assume the following overall score function, which is a weighted average that considers the above information (support information in the weights). The following formula indicates the score of the data given one data:

$$L = \frac{\gamma^2}{|H_\alpha + E_{\alpha-1}|^2} \tag{2}$$

In formula (2), γ represents the set of positive class items covering the data instance, H represents the set of negative class items covering the data instance, E represents the confidence of the positive class rule, and α represents the confidence of the conversion of the negative class rule to the positive class rule. Therefore, it is of great significance to adopt a complete learning demand mining model of PE health preserving, and then to improve the teaching objectives, teaching tasks, teaching content and so on. Browsing behavior in the process of online learning includes browsing web pages, browsing digital teaching resources, viewing learning records and viewing communication or email information. Learners download files in various formats for browsing or collecting purposes, and review, archive, and collect resources. These downloaded files include text, videos, and pictures. Through these ways to achieve the mental health of college students to play a beneficial role in physical and mental health in colleges and universities, the implementation of traditional sports can not only enhance the students' interest in sports learning, exercise the body, but also strengthen the development of mental health to make full and rational use of the point of view of system theory, the teaching and learning of sports health preservation as a whole, scientific management of it, and try to maintain the link between sports health preservation teaching and learning as a whole. The expression of knowledge and information refers to learners' expressing their cognitive content in words, language or other ways for the purpose of searching, communicating or recording. The Internet as a central information system, but also to provide people with a platform for self-expression and information exchange.

2.4 Cognitive Model Setting Learning Needs Mining Model

The cognitive model contains not only the information about the learner, but also all the learning behaviors of the learner in the network learning process, including the reference to the data contained in each learning behavior, the sum of browsing time and the sum of visits. Learning needs mining model should be comprehensive, which is the comprehensiveness of students' and society's needs and the comprehensiveness of knowledge, and at the same time teaching students in accordance with their aptitude and roundness [6, 7]. When constructing the curriculum system of keeping in good health of physical education, we must fully consider the factors inside the system, that is, whether the teaching content can meet the needs of students, whether the teaching task can be completed smoothly, whether the teaching stage can be carried out reasonably, whether the teaching objects can meet the requirements of students.

There are many ways of information retrieval, such as directory browsing, keyword searching and so on, but today's information searchers are almost all using the keyword way. Therefore, the information retrieval in this paper refers to the knowledge information retrieval using Google, Baidu, CNKI and other Internet search engines, or the search function built into the network teaching system. The contents of sports health preservation should not only provide gradually complicated technical movements and expand the breadth of skill operation, but also increase the depth of analysis. After the calculation of formulae (1) and (2), the following issues are those for which weights need to be resolved, including negative and positive category weights. After a lot of experiments, it is found that the combination of the two is ideal. The weight formula for the active category is as follows:

$$W_P = \frac{\phi \times u}{G_P} \tag{3}$$

In formula (3), ϕ represents the number of rules in the positive category, u the number of rules in the negative category, G the number of rules in the positive category, and P the number of positive category items. When setting up the pattern of learning needs mining, we should consider the combination of vertical and horizontal and the matching degree between students. Although information retrieval is an important means of knowledge information acquisition, in the process of information retrieval, learners first organize and express the self-cognition and target knowledge as keywords, and then search and browse the knowledge information according to them. Therefore, information retrieval behavior presents the processing and expression of knowledge by learners, which belongs to the part of expression of knowledge information. The information retrieval behavior includes two modes, namely, the query based retrieval behavior and the browse based retrieval behavior. The combination of vertical and horizontal learning needs of PE regimen course means that the teaching contents are arranged in parallel with the PE regimen learning categories and then classified learning. According to the law of students' development, we can arrange the study contents in different periods and stages scientifically and rationally, so as to form a scientific and complete course system. The former refers to the learners have a very clear understanding of the content to be searched on the basis of, to sum up the nature of the problem or keywords, retrieval is to determine access to information. The latter mainly refers to the learners' unclear understanding of the contents to be searched, and the need to grasp the essence of the contents to be searched by browsing the relevant information. On the basis of formula (3), the negative category weights are derived:

$$W_Q = \frac{u}{\phi} \times \sqrt[2]{|G_Q|} \tag{4}$$

In formula (4), Q represents a negative set of class items. In the teaching of keeping in good health, the "choice" teaching mode can best embody the people-oriented teaching concept. It can fully mobilize the students' enthusiasm and make the students and teachers participate in the class actively. In the process of teaching organization, the teacher chooses the existing sports health project, the students have the choice to carry on the sports health study. According to the differences in writing environment and storage format, the behavior of electronic notes is further divided into: through the built-in module of electronic notes in the network teaching system or personal blog, Google Docs, Douban and other Internet sites, record the learning experience online, and write book reviews and papers. Learners write note files locally on their computers and store them on the Internet through the online teaching system and online document storage and parsing services. The electronic note is a kind of record, review, review, induction and arrangement of the learner's knowledge.

In the cognitive model, the acquisition of knowledge information (browsing, downloading and collecting) is the way for learners to acquire knowledge information. Only when the acquired information is internalized by learners, can it be used for learning needs mining. Therefore, before constructing the learner's knowledge model, it is necessary to judge and identify this part of the data of knowledge acquisition behavior. In addition to the traditional text form of expression, electronic notes can also be learners of voice recording, personal video and other forms. Make use of the internet or network teaching system, such as community, forum and chat room to express their own views and opinions to other individuals or the outside world, exchange and transfer their learning experience. Information release behavior, on the one hand, is learners' need for self-expression, on the other hand, is the need to explain the results of personal cognition or a value concept, expecting others to respond and communicate. This teaching mode can fully consider the interests of different students, suit the characteristics of the whole students, help to develop students' cognitive ability and practical skills, and fully guide students to participate in the classroom, thus is conducive to the realization of the demand mining of sports health.

3 Application Analysis

3.1 Test Readiness

Based on cognitive model and J2EE technology, metadata is stored in HashMap, and then the information of teaching resources is transferred to XML file by XML technology. The GUI interface of the SAS Enterprise Miner is data flow driven and easy to understand and use. It allows an analyst to build a model by constructing a visual data flow graph that connects the data nodes with the processing nodes using links. It is used to implement Web Service interface and provide personalized learning requirement mining service for various network teaching systems. In addition, this interface allows processing nodes to be inserted directly into the data stream. SAS Enterprise Miner is running on the client/server. In client/server mode, the server can be configured as a data server, a computing server, or a combination of both. XML data is transmitted to the server back-end for the analysis of learners' learning style, the intelligent mining of personalized learning needs and other subsequent processing. SAS Enterprise Miner performs data access, manipulation, and preprocessing. The data interface runs through the SAS dataset. Data can also access RDBMS and PC-formatted ACCESS through standard SAS data programs. Support for Oracle, Informix, Sybase, and DB2 RDBMS is achieved through ACCESS. The data transmission module adopts Java Socket technology and multi-thread mechanism. Conduct application testing under the conditions of the above test preparation.

3.2 Analysis of Test Results

Choose the learning needs mining method based on neural network, the learning needs mining method based on ant colony algorithm, and the students' learning needs mining method in this paper. Under the condition of different similarity of resources, the accuracy of three kinds of learning needs mining methods for college students' sports health preservation is tested. The experimental results are shown in Tables 1, 2 and 3.

Table 1. Course similarity 0.2 mining method accuracy (%)

Number of experiments	A method of mining the learning needs of college students in health preserving based on neural network	Study demand mining method of college students' sports health preservation based on ant colony algorithm	The method of mining the learning needs of college students in keeping in good health
1	78.465	74.255	86.345
2	74.611	73.109	87.494
3	72.106	73.484	83.408
4	71.334	72.645	86.515
5	70.589	73.008	83.499
6	72.485	75.619	82.616
7	74.694	76.334	83.155
8	74.351	78.948	84.009
9	73.648	79.152	83.466
10	72.157	75.008	85.212

It can be seen from Table 1 that the average accuracy rate of the method in this paper is 84.572%, and the average accuracy rate of the neural network-based method for mining the learning needs of college students' health science is 73.444%. The average accuracy of the mining method of college students' sports health protection needs based on ant colony algorithm is 75.156%.

It can be seen from Table 1 that the average accuracy rate of the method in this paper is 73.674%, and the average accuracy rate of the neural network-based method for mining the learning needs of college students' health science is 64.419%. The average accuracy of the mining method of college students' sports health protection needs based on ant colony algorithm is 66.441%.

Table 2. Course similarity 0.4 mining method accuracy (%)

Number of experiments	A method of mining the learning needs of college students in health preserving based on neural network	Study demand mining method of college students' sports health preservation based on ant colony algorithm	The method of mining the learning needs of college students in keeping in good health
1	65.324	65.348	72.466
2	64.578	66.312	73.221
3	63.156	66.911	74.168
4	65.005	65.337	76.416
5	64.387	66.915	73.499
6	65.125	67.206	72.505
7	66.337	66.339	73.887
8	62.014	65.481	74.515
9	63.287	68.242	73.649
10	64.978	66.317	72.416

Table 3. Course similarity 0.8 mining method accuracy (%)

Number of experiments	A method of mining the learning needs of college students in health preserving based on neural network	Study demand mining method of college students' sports health preservation based on ant colony algorithm	The Method of Mining the Learning Needs of College Students in Keeping in Good Health
1	55.616	53.784	63.495
2	54.468	52.911	64.552
3	53.797	54.109	65.797
4	54.117	56.443	66.319
5	55.006	55.978	65.748
6	56.978	55.922	62.008
7	54.446	56.466	63.745
8	55.464	57.147	62.818
9	56.122	58.009	63.007
10	55.797	57.212	62.499

As can be seen from Table 1, the average accuracy rate of the method in this paper is 63.999%, and the average accuracy rate of the neural network-based method for mining the learning needs of college students' health science is 55.181%. The average accuracy of the mining method of college students' sports health protection needs based on ant colony algorithm is 55.798%. Because the method in this paper optimizes the structure of the physical education course and builds a health course association model. And classify the emotional state of college students to identify the characteristics of individual learning needs of college students. In this paper, the concept of association rules is introduced, and the cognitive model is used to establish a learning demand mining model, thereby improving the accuracy of learning demand mining.

4 Conclusion

By constructing a good cognitive model, we can excavate learners' personalized learning needs of sports health preservation, and lay a foundation for the follow-up personalized recommendation service. For college students, it is helpful to improve their awareness of the importance of physical education to promote their active participation in scientific, healthy and effective sports. At the same time, it can effectively reduce the construction cost and manual participation of personalized network teaching, improve personalized service effect, and expand in new disciplines. Because of the limited capacity, we should carry out a deeper discussion in the application field in the future.

References

1. Yang, Y., Zhong, Y., Woźniak, M.: Improvement of adaptive learning service recommendation algorithm based on big data. Mob. Netw. Appl. **26**(5), 2176–2187 (2021)
2. Wang, L., Soo-Jin, C.: Sustainable development of college and university education by use of data mining methods. Int. J. Emerging Technol. Learn. (Online) **16**(5), 102 (2021)
3. Bissett, J.E., Kroshus, E., Hebard, S.: Determining the role of sport coaches in promoting athlete mental health: a narrative review and Delphi approach. BMJ Open Sport Exerc. Med. **6**(1), e000676 (2020)
4. Denysova, L., Shynkaruk, O., Usychenko, V.: Cloud technologies in distance learning of specialists in physical culture and sports. J. Phys. Educ. Sport **18**, 469–472 (2018)
5. Leyton-Román, M., González-Vélez, J.J.L., Batista, M., et al.: Predictive model for a motivation and discipline in physical education students based on teaching–learning styles. Sustainability **13**(1), 187 (2020)
6. Zhu, Q.: Research on mobile instruction design model based on learning needs theory. J. Hebei Normal Univ. Nationalities **40**(2), 98–101 (2020)
7. Peng, Q..-J..: Emergency decision support system demand data self-service mining simulation. Comput. Simul. **36**(8), 329–332 (2019)

High Precision Extraction of UAV Tracking Image Information Based on Big Data Technology

Xiaobo Jiang(✉) and Wenda Xie

Computer Engineering Technical College (Artificial Intelligence College), Guangdong Polytechnic of Science and Technology, Zhuhai 519090, China
jiangxiaobo2009@126.com

Abstract. The high-precision extraction of UAV tracking image information is of great significance to the development of UAV system, attitude control, real-time flight path planning and correction, accident analysis and payload. In order to improve the precision of UAV tracking image information extraction, a high precision UAV tracking image extraction method based on big data technology is proposed. Collect the tracking image information of UAV, and eliminate the spectral reflectance of ground objects through atmospheric correction; A multi viewpoint dense image matching method is used to generate a feature recognition model of image information; Creatively uses cubic convolution interpolation to establish image matrix; Calculate the world coordinates of the same feature point and the spatial direction of the wing line and symmetry axis to evaluate the image clarity of the UAV; The pyramid image matching strategy is used to automatically match and jointly adjust the points of the same name in each image layer; The size and direction of the key points are determined by fitting the three-dimensional quadratic function, and the gradient and direction of the key points are calculated respectively to achieve high-precision extraction of UAV tracking image information. Finally, the experiment results show that the method improves the precision of UAV tracking image information extraction, has high practicability and meets the requirements of the research.

Keywords: Big data technology · UAV · Image information · Precision improvement

1 Introduction

The 3-D position and attitude of the UAV are the main parameters reflecting its flight status and performance, which are of great significance to the development of UAV system, attitude control, real-time planning and modification of flight path, accident analysis and effective load. Therefore, the precision measurement of UAV image parameters has a strong theoretical research value and broad application prospects [1]. According to the different positions of the measuring sensors, the measuring methods of UAV image

W. Fu and L. Yun (Eds.): ADHIP 2022, LNICST 468, pp. 81–94, 2023.
https://doi.org/10.1007/978-3-031-28787-9_7

parameters can be divided into two kinds: image measurement method and image measurement method. The UAV image measurement method based on inertial navigation system is the most widely used one at present. This method has the characteristics of high reliability and high precision, but the UAV image measurement system has the problems of cumulative error, high cost, and unreliable attitude measurement result when the body is weightless, such as spin. The GPS measurement method has high position measurement precision, but the attitude measurement precision is not high because of the limited length and width of the UAV, and the measurement method depends on the GPS. The high-speed aerial camera measurement UAV image needs to set up a large control field on the ground, and the track range and flight attitude of the UAV are limited [2]. The method of external measurement of image refers to the method of collecting the UAV image, extracting the image features and calculating the image parameters by using the measuring equipment installed on the ground, which mainly includes the method of attitude measurement based on model base, the method of monocular motion track intersection, the method of monocular plus distance information image measurement, the method of multi-station UAV attitude measurement based on line feature and the line intersection measurement based on image point feature. The measurement method of monocular motion trajectory can only get the trajectory of the UAV, and the measurement precision is not high. The measurement method of monocular plus distance information image avoids the problem of multi-view matching, but needs additional equipment to measure distance, and in order to get high measurement precision, the range of target attitude is limited. The method of surface rendezvous measurement of UAV attitude can measure high precision pitch angle and yaw angle, but can not get roll angle and position parameters. The method of line rendezvous measurement based on the feature of image point requires that the stable point is extracted to more than three non-collinear points on the UAV, and the reliability is not high, and the measurement precision is not high because of high flight of UAV and small airframe size.

Different from the above measurement methods, innovatively combined with the image features of UAV, this paper presents a UAV image measurement method based on image feature fusion. The multi spectral remote sensing image is normalized by radiation, and the spectral reflectance of ground objects is eliminated by gas correction; A multi viewpoint dense image matching method is used to generate a feature recognition model of image information; The new image is interpolated innovatively by cubic convolution interpolation, and the image matrix is established; The pseudo change is controlled by radiometric correction, and the world coordinates of the same feature point and the spatial directions of the wing line and the symmetry axis are calculated respectively, which realizes the measurement of the UAV image; The combined adjustment method of multiple images is used to preprocess the tilted image; Combined with the pyramid image matching strategy, the same name points of each layer of images are automatically matched and jointly adjusted; The size and direction of key points are determined by fitting three-dimensional quadratic function; The gradients and directions of the key points are calculated by parallel computing to achieve high-precision extraction of UAV tracking image information.. Experimental results show that this method improves the adaptability and reliability of UAV image parameter measurement and image measurement, and has great significance.

2 High Precision Extraction Method of UAV Tracking Image Information

2.1 UAV Tracking Image Information Feature Recognition

In order to make the radiant signals of different images consistent, radiant normalization of TMETM+, ALI and other multi-spectral remote sensing images is carried out, and the DN value of the original image is converted into reflectivity to correct the radiant signals. Normalized remotely sensed images from different sensors can significantly reduce noise from solar, terrain and atmospheric effects [3–5]. Big Data technology introduced the Big Data COST model for atmospheric correction, correction of atmospheric effects and effects caused by differences in solar-terrestrial distances and solar zenith angles [6]. Firstly, the brightness value of the image is converted into the spectral radiation value L of the sensor.

$$L_\lambda = \mathrm{gain}_\lambda \times Q_\lambda + \mathrm{bias}_\lambda \tag{1}$$

In the formula: λ is the band number; L is the spectral radiant value of the pixel at the sensor, the unit is W; Q is the pixel DN value of the band; gain and bias are the scaled gain and offset values of the sensor, both of which are W.

The radiometric correction parameters used in this study are shown in Table 1.

Table 1. Radiometric correction parameters for TM/ETM and ALI (in W/(m^2, ster, μm)

		TM		ETM + (high gain)		ETM + (low gain)		ALI	
Band name	Band serial number	Gain λ	Bias λ	Gain λ	Bias λ	Gain λ	Bias λ	Gain λ	Bias λ
Blu ray	1P	—	—	—	—	—	—	0.046	−3.5
	1	0.76852	−2.28	0.78598	−6.99	1.182658	−7.39	0.045	−4.5
Green light	2	1.458526	−4.38	0.79586	−7.6	1.265896	−7.65	0.029	−1.8
Red light	3	1.065829	−2.23	0.63586	−5.63	0.956283	−5.95	0.019	−1.4
Near infrared	4	0.86289	−2.38	0.63585	−5.76	0.965871	−6.08	0.012	−0.86
	4P	—	—	—	—	—	—	0.0092	0.66
	5P	—	—	—	—	—	—	0.0084	−1.5
Mid infrared	5	0.126365	−0.48	0.12652	−1.15	0.192582	−1.18	0.0027	−0.7
	7	0.062558	−0.23	0.04523	0.38	0.065586	−0.43	0.00092	−0.22

The ESUM values of the images used in the study at the top of the atmosphere are shown in Table 2:

Table 2. Average solar irradiance at the top of the atmosphere (in W/(m^2.μm))

Band name	Band serial number	ESUN λ		
		TM	ETM +	ALI
Blu ray	1P	—	—	1868
	1	1986	1998	1997
Green light	2	1798	1825	1812
Red light	3	1035	1041	1246
Near infrared	4	—	—	453.6
	4P	—	—	465.8
Mid infrared	5	221.2	131.9	235.2
	7	85.65	85.98	83.65

Figure 1 shows the spectral curves of the typical terrain after FLAASH atmospheric correction for Hyperion images.

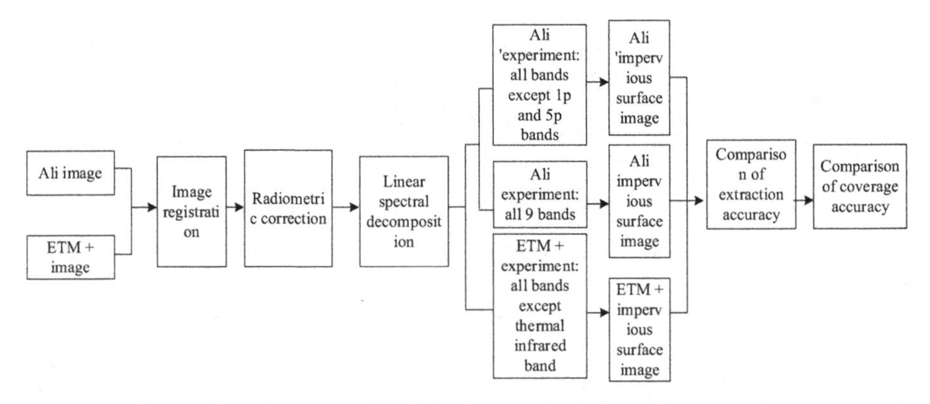

Fig. 1. Technical route of UAV tracking image information comparison

It can be seen from Fig. 1 that, after atmospheric correction, the influence of the atmosphere on the spectral reflectance of various ground objects is obviously eliminated, especially in the range of visible light wavelength, the imaginary high value caused by the elimination of atmospheric Rayleigh scattering and aerosol scattering makes the reflectance in this range obviously decrease, and the spectral curves of various typical ground objects are more close to their real spectral reflectance curves [7]. In order to compare the inversion effect of UAV tracking image information between the ALI and ETM sensors in a more comprehensive way, and to investigate whether the new spectral band added by ALI is helpful to improve the inversion precision of UAV tracking image information, two different schemes are selected to carry out the comparative experiment: one is to participate in the comparative experiment with all 9 wavelengths of ALI, the other is to simulate ALI into ETM+ image for experiment, that is, to remove the new 1p and 5p wavelengths added by ALI, to keep the same wavelength setting as EIM+, and to make the spectral range of ALl and ETM+ wave bands other than thermal infrared one basically consistent. At the same time, in order to compare more objectively, the two aspects of UAV tracking image information extraction and UAV tracking image coverage inversion will be compared respectively.

High precision and resolution digital surface model DSM can be generated by multi-view dense image matching method. The model can express the fluctuation of terrain. This technology has become a new generation of spatial data infrastructure research focus on the object. However, due to the difference of multi-angle tilt images (angle, chromatic aberration, height, etc.), and the serious shadow and occlusion problems in the images, DSM becomes a new difficulty to obtain automatically using tilt images. In order to solve this problem, we can first compute the outer azimuth elements of each image based on automatic aerial triangulation, and then select the appropriate image matching unit to match with the outer azimuth elements, and then introduce the parallel algorithm to improve the computing efficiency.

2.2 Evaluation of UAV Tracking Image Sharpness

After polynomial model is used to transform the coordinate space of the original pixels to be corrected, the repositioned pixels can not be distributed evenly in the original image, but fall among several pixels of the original image. So it is necessary to interpolate each pixel on the new image by resampling, and then a new image matrix is established. Common big data methods include nearest neighbor method, bilinear interpolation and cubic convolution interpolation. The nearest neighbor method assigns the brightness value of the nearest original image pixel near the coordinates of the new pixel to the new pixel. This method can keep the spectral value of the original pixel very well, and the operation is simple and the processing speed is fast. But this method may cause the position deviation of the pixel to be too large, so that the spectral brightness of the new image is discontinuous and the zigzag boundary appears. In bilinear interpolation, the brightness values of the 4 original pixels adjacent to the new pixel are weighted according to the distance between the original pixel and the new pixel, and the weight near the new pixel is higher; then the brightness value of the new pixel is calculated by linear interpolation. The brightness of the image obtained by this method is continuous, which can play the role of smoothing and filtering. But the brightness of the image changes, and the high frequency information in the image is easy to be lost, and the boundary of the ground object becomes fuzzy.

$$x = \sum^{N} \sum^{N-1} a_{ij} X^{i} Y^{j} \tag{2}$$

$$y = \sum^{N} \sum^{N-i} b_{ij} X^i Y^j \tag{3}$$

In the formula: (x, y) is the pixel coordinate of the image to be registered; X^i and Y^j are the values of horizontal and vertical coordinates of each pixel in the standard image respectively; a_{ij} and b_{ij} are polynomial coefficients, which can be obtained by least square regression from the original data; N is the degree of polynomial, the size of which mainly depends on the number of GCP, the degree of image deformation and the size of terrain fluctuation. Camera imaging model often refers to pinhole camera model, that is, the projection center is located in the Euclidean coordinates of the origin in the pinhole camera model. Point X with space homogeneous coordinates (X, Y, Z) is mapped to a point x = (x, y, 1)/2 on the image plane, where the line connecting the projection center X with the image plane intersects as shown in Fig. 2.

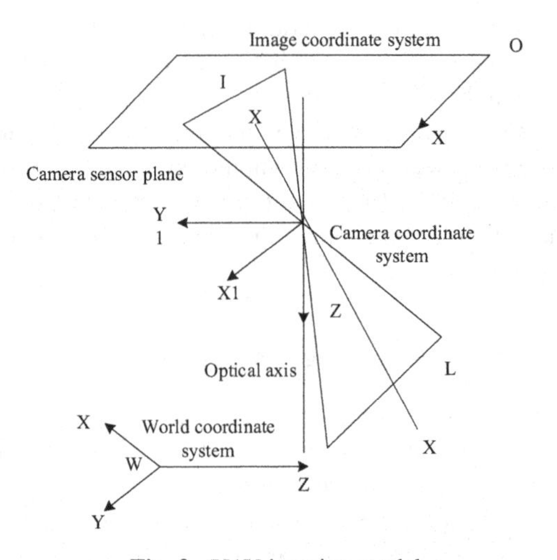

Fig. 2. UAV imaging model

The coordinate system is W-XYZ, the image coordinate system is o-xy, the camera coordinate system is O-XYZ equivalent focal length is f, the rotation matrix from world system to camera system is R, and the center of camera is T. According to the principle of rotation, translation and similar triangle, it is easy to know that the mapping from IR to 2D projective space 2 is IR3 → IR2.

$$k = KR[T|IC - T]L_\lambda - xy \tag{4}$$

In the expression, $K = \begin{bmatrix} f\,0\,f \\ 0\,f\,f \\ 0\,0\,1 \end{bmatrix}$ is the camera calibration matrix, $I = \begin{bmatrix} 1\,0\,0 \\ 0\,1\,0 \\ 0\,0\,1 \end{bmatrix}$ is the 3 × 3 unit matrix, Y and X are homogeneous coordinates, $C = KR[11 - T] = \begin{bmatrix} r_1 & c_1 & c_n & c_4 \\ c_4 & c_4 & c_3 & c_3 \\ c_{i1} & c_{13} & c_{11} & c_{11} \end{bmatrix}$ is the camera matrix, and K is the distance from space point to projection center. According to the imaging model, a line L in 3D space is projected onto a line I on the image plane. Suppose the equation of the line on the image is by c = 0, and the homogeneous representation of the line is l = (a, b, c), then the set of points in the space mapped to line I by the camera matrix C is a plane, and the plane equation is

$$X^T C^T - K = \theta \tag{5}$$

The homogeneous representation of the plane is C, and the expression (2) is the coplanar equation of projection center, image line and space line.

However, it is difficult to detect the detailed feature information of UAV due to the influence of environment, UAV attitude and high speed in the actual measurement process, and the reliability of the detected feature is worth discussing. Corners, wing lines and symmetry axis are the stable points in the UAV image, which mainly refer to the points where the brightness of the UAV image changes dramatically or the maximum curvature on the edge curve is maximized. These points can be located in the neighborhood with rich information such as rotating invariant, scale invariant and illumination invariant. In remote sensing images with different temporal phases, there are always some features whose spectral values do not change with time, such as undisturbed deep water, residential areas, airport runways, etc. The spectral reflectance of an invariant feature does not change significantly over time without interference from other factors. Even if it does, the change may be caused by different atmospheric and illumination conditions, so it is also called "pseudo-change". For historical images, these pseudo-variations can be controlled or reduced by radiometric correction, so that they can be used in conjunction with measured spectral data over the summer period, with sufficient and stable solar light, cloudless skies and small winds. When all the samples are used to measure the spectrum, the GPS position information, time, description and photo record are recorded synchronously. The flow chart of spectrum measurement is shown in Fig. 3.

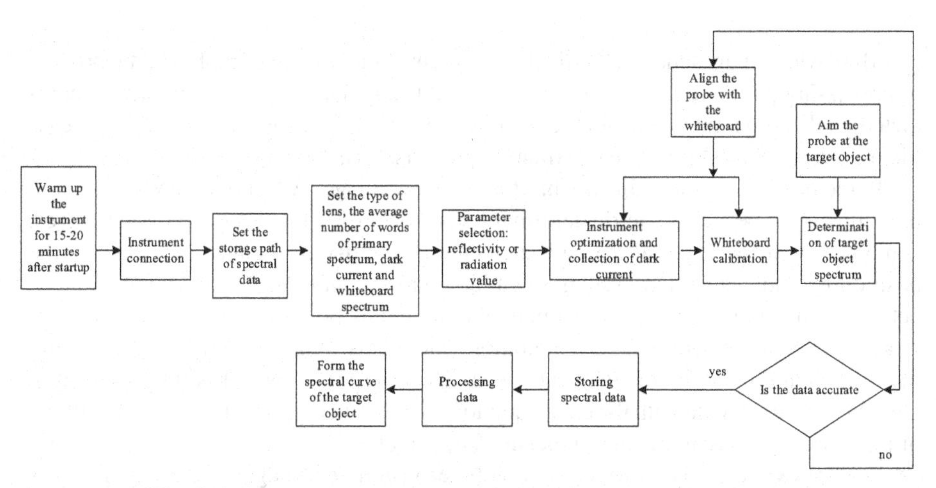

Fig. 3. Flow chart of UAV spectral image information measurement

Wing straight line includes left and right lines. Because of flying altitude and attitude of UAV, the wing line is not easy to be extracted, but the feature of wing line is still valuable. The axis of symmetry is the stable characteristic line of UAV. In most cases, the line of symmetry axis can be extracted from the image of each network camera. The stable extraction of this feature line is also the precondition of high precision image parameters. Based on the above discussion, this paper adopts the method of extracting stable corners, wing lines and symmetry axis to realize the UAV image measurement.

2.3 Realization of High Precision Extraction of UAV Influence Information

Data processing is the core work of UAV low-altitude remote sensing system, which includes image preprocessing, camera high-precision detection, image matching, automatic aerial triangulation, DSMDEM automatic extraction, DOM generation and seamless stitching. The data processing flowchart is shown in Fig. 4.

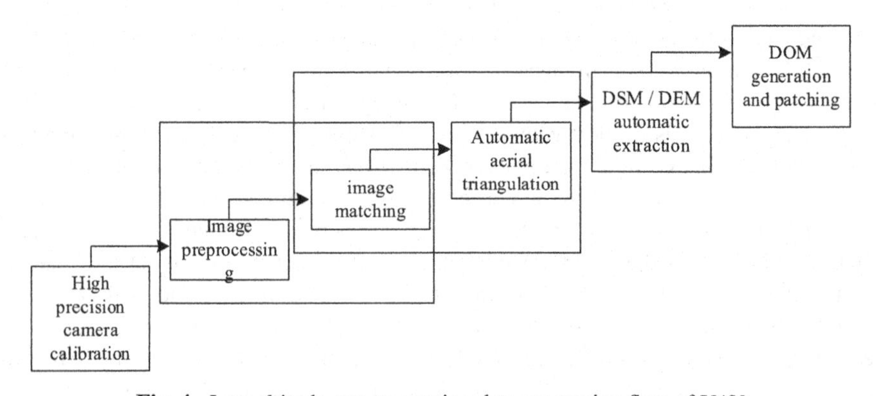

Fig. 4. Low altitude remote sensing data processing flow of UAV

Multi-view images include vertical photography and tilt photography. In the process of processing photographic images, some aerial triangulation systems can not be completed well, so the method of joint adjustment of multiple images is needed to process tilted images. The UAV image information plot steps are shown in Fig. 5.

In the process of joint adjustment of multi-view images, we should pay attention to the following aspects: geometric deformation and occlusion relationship among images, combined with the external azimuth elements of multi-view images provided by the positioning and orientation system, combined with pyramid image matching strategy, automatic matching and joint adjustment of homonymic points on each level of images, so as to get better matching results of homonymic points. When establishing error equation, we should combine the data such as connection point, control point coordinates, GPS/MU auxiliary data with the error equation of self-checking area network adjustment of multi-view images to get high precision adjustment.

Feature extraction is a concept in computer vision and image processing. It refers to the use of a computer to extract the image information of the same name points in the image, which determines the same features in the image. Feature extraction is

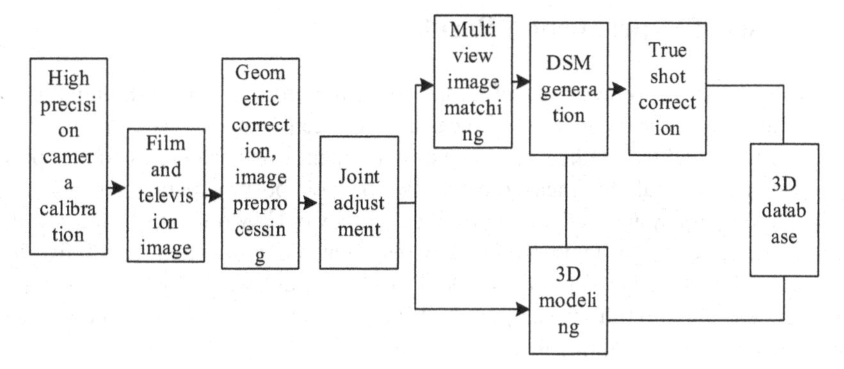

Fig. 5. Step of UAV image information map area

to classify the points on the image into different subsets. These subsets are usually composed of n isolated points, a continuous curve or a continuous region. There are obvious differences and differences between the gray levels of image features and those of surrounding images. Features are usually divided into point features, line features and surface features. The point feature mainly includes the obvious point and line feature in the image, which is the conformation of the edge of the linear object or surface object in the image. The low altitude photogrammetry system of UAV is not easy to operate the UAV under the influence of wind force, and it will deviate from the established flight path. The attitude of a digital camera on an unmanned aerial vehicle cannot remain fixed due to the influence of the wind or the motion of the aircraft, which leads to great changes in the lateral overlaps and heading overlaps between adjacent images. When UAVs fly in urban areas, towering buildings can be distorted in adjacent images, making it impossible to find a matching search area in photographic images. In order to accurately locate the position and scale of the key points, Professor Lowe proposed a method to achieve sub-pixel accuracy, i. e., to fit the 3D quadratic function. This method can remove the key points with low contrast and unstable edge response points, and enhance the matching stability and anti-noise ability.

To determine the size $m(x, y)$ and orientation $\theta(x, y)$ of the key points, firstly the distribution direction of the key points shall be determined according to the local features of the image, and then the gradient $D(X)$ and direction distribution character $L(x, y)$ of the neighboring pixels of the key points shall be used to calculate the gradient and direction of the key points respectively. The specific formulas are as follows:

$$m(x, y) = \sqrt{(L(x+1, y) - L(x-1, y))^2 + D(X)} \tag{6}$$

$$\theta(x, y) = \tan^{-1} D(X)/(L(x+1, y) - L(x-1, y)) \tag{7}$$

However, the SIFT matching algorithm with invariant features can be divided into two types: the matching based on gray level and the matching based on features. In recent years, many scholars have studied the matching based on image features more than the matching based on gray level.

3 Analysis of Experimental Results

In order to verify the correctness and feasibility of the proposed method, digital simulation is carried out for the method of measuring image parameters using point or line features of UAV and the method of image measurement using feature fusion proposed in this paper. The simulation measurement error of UAV position parameters and pixel classification accuracy are used to evaluate the accuracy of UAV tracking image information extraction. The tilt system used in the experiment is a lightweight, handy five-lens tilt camera called the KG2000 Pm. KG2000Pro Tilt Aerial Camera System is a new tilt camera system based on the development of the Dajiang Genie IV, the camera system light pixel height as shown in Table 3.

Table 3. KG2000 pro camera parameters

Overall dimension	Weight	Sensor	Resolving power	Sensor size	Pixel size	Camera lens	Lens distribution
125 × 97 × 81	35	CMOS	4200	1/1.3	1.5 μ	No displacement	Equidistant 29 mm lens group

The SonyQx100 (9009865) used in the experiment was tested on the spot. The contents and results of the test are shown in Table 4.

Table 4. Sorry QX100 (9009S65) results of machine calibration

Serial number	Calibration content	Calibration value	Remarks
A	Main point x0	−0.00S65S mm	Camera interior orientation element
B	Main point Y0	−0.003652 mm	
C	Focal length	10.65862 mm	
D	Radial distortion coefficient K1	2.13E−11	Radial distortion difference coefficient
E	Radial distortion coefficient K2	−S.12E−2S	
F	Eccentric distortion coefficient Q1	3.15E−8	Tangential distortion difference coefficient
G	Eccentric distortion coefficient Q2	−1.65E−8	

In order to solve the problem of UAV image parameters only using corners, the coordinates of the corners in the world system are obtained by the rendezvous of the binocular lines, and then the image parameters of the UAV are obtained by solving the absolute orientation problem by using the linear least squares. In the simulation, the noise standard deviation increases from 0.1 to 1, and the average measurement error Pmn = Pa − Pa/1000 is calculated for each noise level program. The simulation result is shown in Fig. 6.

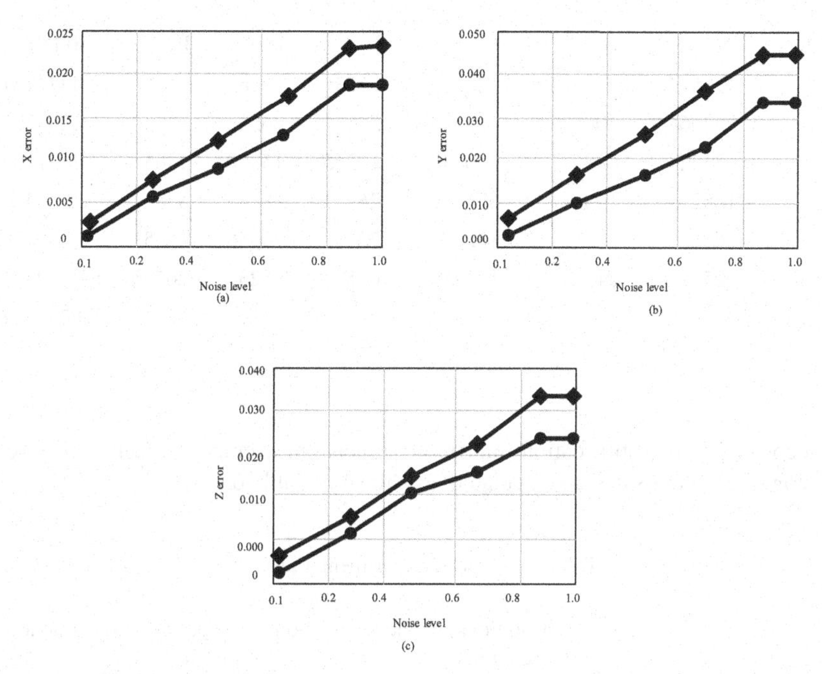

Fig. 6. Simulation measurement error of UAV position parameters

It can be seen from Fig. 6 that the measurement error of each noise level program on the three axes of the simulation results of this method is less than 0.01, which has high anti-interference and improves the accuracy of information extraction of UAV tracking images. This is because in order to improve the image clarity, the method in this paper eliminates the interference of spectral reflectance of ground objects through atmospheric correction, uses multi viewpoint dense image matching method to generate image information feature recognition model, and uses multi image joint adjustment method to preprocess inclined images, so as to obtain low measurement error.

In order to get the relationship between the proposed method and the number of netted cameras, the number of cameras is increased from 1 to 6, and the noise with standard deviation of 1 is added. The same test set samples were used to test the five extracted models respectively. Because the big data technology of this paper is single class extraction model, the results of model extraction need to be merged. In this paper, the classification of each pixel is determined by outputting the probability of each pixel classification, and then the multi-class segmentation is realized. ENVI software is used to extract the conflicting pixels, generate the corresponding position coordinates, compare the corresponding class probabilities, and retain the class values with high probabilities to generate the final confusion matrix, as shown in Table 5. In the confusion matrix, A stands for farmland, B for forest, C for building, D for water, E for road, and F for other land.

Table 5. Confusion matrix of experimental results

Category	A	B	C	D	E	F	Total
A	4125162	213656	12653	16235	5236	782	4373724
B	50256	2068562	126523	596	7652	5688	2259277
C	73652	65856	2008963	11365	15368	2365	2177569
D	1352	1036	625	26521	88	838	30460
E	3652	856	1326	721	72056	1165	89776
F	6253	2156	11355	6253	7523	265861	299401
—	—	—	—	—	—	—	9230207

According to the obfuscation matrix, the precision, average precision and overall precision of each class can be calculated, as shown in Table 6.

Table 6. Table of experimental results

Category	Traditional method extraction accuracy %	Paper method extraction accuracy %
Ploughing pond	76.3	94.1
Forest		91.8
Residential area	87.4	92.5
Waters	77.8	89.5
Road	79.3	91.3
Other land	78.2	89
Overall extraction accuracy	80.6	93.65
Average extraction accuracy	77.8	91.89

As can be seen from Table 6, the overall extraction accuracy is 91.89%. Among them, the main reasons for the higher accuracy of extraction of arable land and forest are the regular shape of arable land and the special texture of forest. But the extraction accuracy of water area and other land is low, mainly because the water area is small, the characteristics are not obvious, other land is complex and diverse, the characteristics of confusion.

Large data method is used to train and extract features from a large number of samples. Finally, the trained model is used to complete information extraction by image classification. Experimental results show that the proposed method can extract the distribution information of objects from the images with high accuracy. Through analysis, the new method combines different objects with model structure reasonably and efficiently, and effectively solves the problem of inaccurate classification of typical objects. This is because the method in this paper normalizes the collected multispectral remote sensing images, eliminates the spectral reflectance of ground objects through gas correction, and provides an accurate data base; The multi viewpoint dense image matching method is creatively used to generate the image information feature recognition model, and the cubic convolution interpolation is used to establish the image matrix, which improves the image information feature recognition accuracy; The accurate measurement of UAV image is realized by controlling the pseudo change through radiation correction; Combined with the pyramid image matching strategy, the same name points of each layer of image are automatically matched, and the size and direction of key points are determined by fitting the three-dimensional quadratic function, which realizes the high-precision extraction of image information.

4 Conclusion

In order to realize the high precision measurement of UAV position and attitude parameters, an image measurement method based on image feature fusion is proposed. In order to solve the problems of the UAV image measurement, the unit-direction vector of the UAV spindle in the world system is calculated by using the stable spindle feature of the UAV, and the image parameters of the UAV are calculated by using the corner feature and wing feature. Compared with the method that solves UAV image parameters by point feature or line feature, the method based on image feature fusion has the advantages of high precision and strong adaptability. High precision extraction of UAV image information based on big data technology is an important means of UAV test and development, attitude control fault analysis and effective payload enhancement. It has important theoretical research significance and broad application prospect. Future research can combine the method in this paper with UAV attitude control, and correct the UAV flight attitude through image information features.

Fund Project. 2021 special project in key fields of colleges and universities in Guangdong Province, research on Constructing "smart cloud" training platform based on new generation hybrid cloud technology (Project No.: 2021zdzx1147).

References

1. Jing, K.L., Si-yu, F., Ya-fu, Z.: Model predictive control of the fuel cell cathode system based on state quantity estimation. Comput. Simul. **37**(07), 119–122 (2020)
2. Zhao, L., Xie, F., Xu, C., et al.: High-resolution simulation analysis of meteorological factors of unmanned aerial vehicle air route in the Beijing-Tianjin-Hebei region. Prog. Geogr. **40**(10), 1691–1703 (2021)
3. Ling, C., Liu, H., Ji, P., et al.: Estimation of vegetation coverage based on VDVI index of UAV visible image —using the shelterbelt research area as an example. Forest Eng. **37**(02), 57–66 (2021)

4. Hu, X., Ni, H., Qi, D.: Tree counts extraction based on UAV imagery. Forest Eng. **37**(01), 6–12 (2021)
5. Zhang, R., Shao, Z., Portnov, A., et al.: Multi-scale dilated convolutional neural network for object detection in UAV images. Geomatics Inf. Sci. Wuhan Univ. **45**(06), 895–903 (2020)
6. Yang, H., Yang, Z., Sun, D.: Research on high resolution satellite image information extraction based on short translation. Geomatics Spat. Inf. Technol. **44**(S1), 168–170+176 (2020)
7. Yan, S., Wen, S., Ren, Z.: Comparative study on fusion methods of quickbird remote sensing images. J. Henan Sci. Technol. **44**(S1), 168–170+176 (2021)

Monitoring Method of Ideological and Political Class Learning Status Based on Mobile Learning Behavior Data

Yonghua Wang(✉)

Sanya Aviation and Tourism College, Sanya 572000, China
wangyonghua00060@163.com

Abstract. In order to improve the quality of ideological and political education, a method for monitoring the learning status of ideological and political courses based on mobile learning behavior data is proposed. Combined with mobile technology to collect ideological and political learning behavior characteristic data. According to the feature recognition results of the data, an accurate stu classification algorithm is designed, and an evaluation system for the learning status of ideological and political courses is constructed. Six characteristic actions in human poses are selected to study learning state classification. Realize the monitoring of the students' learning status in political courses. Finally, it is proved by experiments that the monitoring method of learning state of ideological and political courses based on mobile learning behavior data has high practicability and meets the research requirements.

Keywords: Mobile terminal · Learning behavior · Ideological and political lesson · State monitoring

1 Introduction

With the rapid development of information technology and the advent of knowledge explosion era, the demand for compound talents is increasing, continuing education and lifelong learning become the trend. In order to meet people's ideological and political course learning needs and provide technology and resources support for lifelong learning, we need to expand the scale of distance education. In order to improve the effect of ideological and political education, it is necessary to study the monitoring methods of ideological and political learning state [1]. Based on this, this paper aims to develop a distance learning process monitoring system based on mobile learning technology, and use mobile learning technology to collect part of the learning behavior of learners in real time. Learning process monitoring is a key link in distance education [2]. The distance learning process monitoring based on mobile learning can classify the learning behavior of different learners according to their learning behavior characteristics, and establish a learning behavior classification database; on the other hand, the mobile

W. Fu and L. Yun (Eds.): ADHIP 2022, LNICST 468, pp. 95–106, 2023.
https://doi.org/10.1007/978-3-031-28787-9_8

learning behavior of the dynamically collected learners is entered into the learning terminal through Bluetooth and other wireless transmission methods, and then entered into the distance learning management center through the Internet, and the collected mobile learning behavior is compared with the mobile learning behavior classification database of the learners through the analysis software, so as to obtain the real-time learning status of the learners and provide timely and targeted support services for teachers to intervene and the learning of the learners. Experiments based on mobile learning behavior data show that mobile learning technology can monitor the distance learning process and help managers to master the learning state of learners, and provide personalized learning process and learning resources.

2 Monitoring Learning Status of Ideological and Political Class Based on Mobile Learning Behavior Data

The method of monitoring the learning situation based on mobile learning behavior data is shown in Fig. 1.

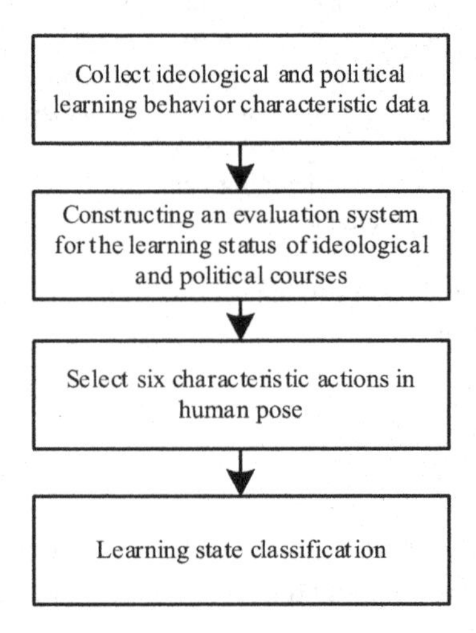

Fig. 1. Identification diagram of student learning behavior

2.1 Feature Recognition Based on Mobile Learning Behavior Data

For learning these two words, people have a lot of views, everyone's understanding is that learning is the process of students to build knowledge system. On the definition of learning state, according to the previous research, several prominent viewpoints have been summarized: learning state is the comprehensive reflection of various functions of the body in the process of learning, specifically including body movements and attention, and learning state is the performance of the body stability of the learners in learning activities [3]. The state of learning refers to the degree to which the learner is grasping the knowledge. The learner's learning state includes three aspects: the state of study preparation before learning activities, the state of study activities and the state reached after learning activities. The specific structure is shown as follows (Fig. 2):

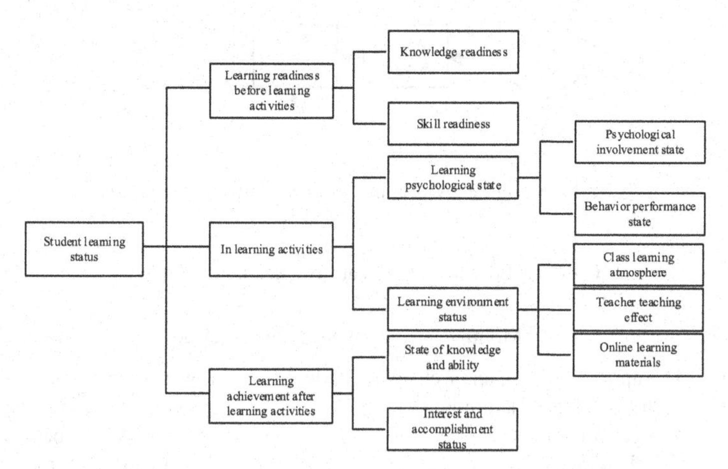

Fig. 2. Identification diagram of student learning behavior

According to the definition of learning state, learning state is the comprehensive performance of physiological and psychological characteristics of learners in learning activities. According to the monitoring requirements of students' classroom performance, learning state is defined as three states of concentration, general and non-concentration, so as to classify the classroom learning state of different students. The recognition of learning state is the recognition of local human motion [4]. In a certain aspect, the local motion state directly reflects the learner's learning state. Therefore, this paper studies the local motion of human body. It is found that local posture can be measured, local motion is relatively limited, and the state of local expression is relatively unlimited. Different people have different local motion characteristics. That is, a series of local posture into local actions, human learning actions constitute the learning state of the evaluation idea, and the use of the corresponding recognition algorithm recognition [5]. The method of multi-level hierarchical learning state pattern recognition is mainly divided into three levels, each level is classified according to the characteristics of the attitude itself. The recognition output data of the former level algorithm is used as the input data of the latter level algorithm. The structure of the designed attitude recognition is shown in Fig. 3:

After several discussions with many educators and experts, the study integrates and analyzes the related materials, and chooses the learning posture features in six classrooms as the basis to identify the learning state (Table 1).

In the technology of motion pattern recognition based on the mobile terminal, the common methods of feature information extraction for the attitude data collected by the mobile terminal are as follows.

From the Table 2, we can see that these three kinds of feature information extraction methods have their own characteristics, different perspectives, the choice of feature extraction methods are different. Based on the experimental results of this paper, it is

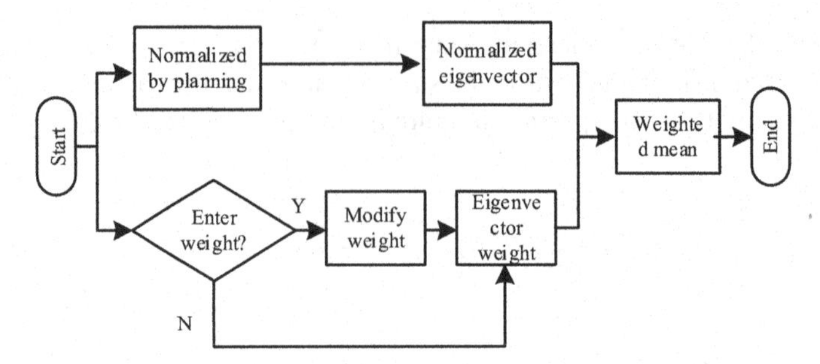

Fig. 3. Learning behavioral posture recognition architecture

Table 1. Learning state feature space model table

State	Eye	Mouth	Visual angle	Head	Body
Absorbed	The upper eyelid is lifted, the eyes do not turn their eyes, the blinking frequency is uniform, and the eye angle becomes larger	Open or close naturally	Static	Towards the teacher, which changes with the position of the teacher	Keep the distance between the front and rear tables
Commonly	Keep your eyes fixed or blink at a uniform frequency, with uniform eye angle and lower blinking frequency	Lips open slightly	Static	Keep the head in the same position for at least 5 s	Keep the distance between the front and rear tables
Inattention	The eyes keep staring at one place, and the angle of the eyes becomes smaller	Yawn and open your mouth	Shake	Nod slowly and often	Close to the front or back table

necessary to recognize the motion patterns of left turn head, right turn head, lower head, raise head, left deviation head, right deviation head and horizontal vision. Considering that the main difference of the motion patterns lies in the direction and angle, the mean of the 32-dimensional attitude angle in the time domain is selected as the feature information of the attitude recognition.

Table 2. Common feature information extraction methods

Feature extraction method	Characteristic parameter
Time domain method	Mean, standard deviation, variance, information entropy and correlation coefficient, etc.
Frequency domain method	Fast Fourier transform coefficients, discrete cosine transform coefficients, frequency domain entropy, energy spectral density, etc.
Time frequency method	Wavelet transform, Fourier transform, etc.

2.2 Evaluation Algorithm for Learning State of Politics Course

In the traditional classroom teaching activities, how to realize the classification of students' learning status in the classroom is an important issue under the background of quality-oriented education. The most direct and efficient place for students to acquire knowledge is the classroom, so the overall learning efficiency of students has a close relationship with their learning state in the classroom. Lack of attention and poor learning condition will directly affect students' mastery of knowledge. Inattentive learning condition will not only result in low learning efficiency, but also cause students' learning enthusiasm to drop [6]. If the students pay attention in the course and pay attention to the lecture in a good study state, they will not only learn the most knowledge in the limited class time, but also make the students more confident in the study. Because of the traditional classroom a teacher facing the situation of dozens of students, teachers simply can not pay attention to every student. Therefore, in order to make each student's learning state known by the teacher accurately, it is very important to design an accurate classification algorithm of student's learning state. The state of learning is closely related to the state of local movement of human body. Therefore, the local motion information of learners is used as the basis to detect and analyze the learning state of learners. In order to realize the detection of local motion information, accurate local position coordinates should be obtained firstly [7]. Because the local motion of human body has the characteristics of large range and low motion frequency, the traditional attitude algorithm can not be completely applied to calculate the local attitude angle. Therefore, based on the traditional attitude description method, the mobile learning behavior and the extended Kalman filtering algorithm are studied respectively. A local attitude solution method is proposed. In the process of attitude description, the angle can be chosen to represent the attitude information, namely Euler angle. Euler Point was first proposed by Euler. The Euler angle consists of roll angle $\dot{\gamma}$, pitch angle $\dot{\theta}$ and heading angle $\dot{\varphi}$. The Euler angle has a range, the roll angle A has a range of (−x, x) radians, and the pitch angle B has a range of (−x/2, x/2) radians turning angle C has a range of (1 − x, 1 x) radians converting the data from the carrier coordinate system into the coordinate system. The Euler angle is used to represent this conversion process. It is assumed that the sum of the head steering angle, pitch angle and roll angle is represented by the moment of inertia alpha. Such as:

$$\omega = A\dot{\varphi} + B\dot{\theta} + C\dot{\gamma} \tag{1}$$

Let the angular velocities in the head coordinate system be expressed as m, n, respectively, and the angular values are differentiated to get the formula.

$$\varpi = \prod \omega(M - N) \tag{2}$$

The differential expressions of the attitude angles y, O, q are shown by the simplification of the expression.

$$\begin{bmatrix} \dot{\gamma} \\ \dot{\theta} \\ \dot{\varphi} \end{bmatrix} = \frac{1}{\cos\theta}\begin{cases} \omega_x - \sin\gamma\cos\theta \\ \omega_y + \sin\gamma\sin\theta \\ \omega_z - \cos\gamma\sin\theta \end{cases} \tag{3}$$

Learning behavior feature point recognition is to automatically mark the key points from the input image data, including eye contour, nose, mouth, eyebrows, face contour and so on. The key point of feature recognition is the accuracy and speed of recognition. A method of feature point recognition using regression tree structure is proposed. In this paper, we use the mobile learning behavior recognition method to design the loss function and modify the error of data marker, the training effect is better, and improve the precision of sparse subset. According to the definition of learning posture, this paper mainly chooses six kinds of feature actions in human posture to study the classification of learning state. They are eye closing feature action, yawn feature action, mouth closing feature action, nodding and dozing feature action, angle of view moving feature action and body leaning feature action. First of all, the above six movements can be relatively easily captured by the camera, there is no undetected problem. Then, according to the angle range of local rotation, we can divide the local learning posture of a certain moment into 17 kinds of learning postures, and define the head-up state as $(-5, 5)$ Because of the selected "front and bottom right" coordinate system, we can divide it into three equal parts, corresponding to the local state of small, medium and large left-turn head. Then the heading angle range of the small left turn head is $(-12, 15)$; the heading angle range of the medium left turn head is $(24, -12)$ the heading angle of the large left turn head is greater than -24.2. Based on this, seven local learning postures are divided in different degrees as shown in Table 3.

The fatigue detection based on weighted average is to fuse the information by summing up the fatigue features according to their weights and calculate the fatigue value f, the calculation formula is as follows:

$$\begin{aligned} & f = \sum_{i=1}^{n} w_i x_i = w_1 x_1 + w_2 x_2 + \ldots + w_n x_n, \\ & i = 1, 2, \ldots n \\ & \sum_{i=1}^{n} w_i = 1 \end{aligned} \tag{4}$$

Variable descriptions are shown in Table 4.

Because the range of fatigue value is different in turn, blink frequency, yawn frequency and nod frequency, it is necessary to quantify these fatigue characteristics as fatigue values from 0 to 1. Xk is the fatigue value of the k component in the fatigue eigenvector, and k is the weight corresponding to the k component xk, and the sum of

Table 3. Attitude classification table

Category		Label	Pitch angle/(°)	Heading angle/(°)	Roll angle/(°)
Turn left	Small range	A	(−5, +5)	(−12.2, −5)	(−5, +5)
	Medium amplitude	B	(−5, +5)	(−25.2, −12.2)	(−5, +5)
	Substantially	C	(−5, +5)	<(−25.2)	(−5, +5)
Turn right	Small range	a	(−5, +5)	(+5, +15.3)	(−5, +5)
	Medium amplitude	b	(−5, +5)	(+15.3, +28.2)	(−5, +5)
	Substantially	c	(−5, +5)	>(+28.2)	(−5, +5)
Bow your head	Small range	D	(+5, +23.3)	(−5, +5)	(−5, +5)
	Medium amplitude	E	(+23.3, 46.5)	(−5, +5)	(−5, +5)
	Substantially	F	>(45.5)	(−5, +5)	(−5, +5)
rise	Small range	d	(−20.2, −5)	(−5, +5)	(−5, +5)
	Medium amplitude	e	(−41.2, −20.2)	(−5, +5)	(−5, +5)
	Substantially	f	<(−41.3)	(−5, +5)	(−5, +5)
Left leaning head	Small range	G	(−5, +5)	(−5, +5)	(−25.2, −5)
	Medium amplitude	H	(−5, +5)	(−5, +5)	(−51.2, −25.2)
	Substantially	I	(−5, +5)	(−5, +5)	<(−51.2)
Right slanting head	Small range	g	(−5, +5)	(−5, +5)	(+5, 27.6)
	Medium amplitude	h	(−5, +5)	(−5, +5)	(+27.6, + 55.2)
	Substantially	i	(−5, +5)	(−5, +5)	>(+55.2)

all weights is 1. The tool supports the user to set weights corresponding to each feature component according to the needs, and the threshold value for judging fatigue can be determined by the user himself or by using the results of several experiments in this case as the default threshold value. When the fatigue value is higher than this threshold value, the fatigue state is determined.

2.3 Realization of Monitoring the Learning State of Politics Curriculum

Since the time of closing eyes increases and the frequency of blinking decreases with the occurrence of yawning and nodding, the detection of fatigue based on behavioral features is mainly based on the recognition of these behaviors. Therefore, we choose PERCLOS,

Table 4. Variable descriptions

Variable	Explain	Variable	Explain	Variable	Explain
Q	PERCLOS value	Z1	Fatigue value corresponding to PERCLOS	M1	Weight corresponding to Z1
W	Blink frequency	Z2	Fatigue value corresponding to blink frequency	M2	Weight corresponding to Z2
E	Yawn frequency	Z3	Fatigue value corresponding to yawn frequency	M3	Weight corresponding to Z3
R	Nodding frequency	Z4	Fatigue value corresponding to nodding frequency	M4	Weight corresponding to Z4

blinking frequency, yawning frequency and nodding frequency as characteristic vectors to measure the fatigue state. Through the mobile end and the SDK provided by Microsoft, we can capture the movement unit of human face, including the degree of eyes closing, mouth opening amplitude, local deflection angle and so on. When the degree of ocular closure is greater than 0.8, it is judged to be ocular closure. The ratio of ocular closure frames to total frames is calculated as PERCLOS parameter. Eyes from open to close to open again, as a blink, statistics within a certain period of time blink times, that is, blink frequency. When the mouth is closed, then opened and then closed, it is regarded as a yawn, and the number of yawns in a given period of time is calculated as the frequency of yawning. Nodding behavior depends on the angle of local downward deflection. When the deflection angle is greater than a certain threshold value, and then returns to the normal value, the nodding behavior is judged to occur once, and the nodding times in a certain time are counted, that is, the nodding frequencies are determined by using the weighted average method and the support vector machine model. Users can choose from the program interface to calculate the fatigue level of the method - weighted average method or support vector machine. If the weighted average method is selected, the user can input the corresponding weights of each fatigue parameter according to the needs, or choose the default combination of weights without filling in. If a support vector machine is selected, the provided trained model is used. In this way, it is very convenient for users to customize their fatigue rules and choose suitable fatigue detection methods for students. The specific flow of the weighted average method for calculating fatigue degree is shown in Fig. 4.

First of all, because the fatigue parameters are different in order of magnitude, it is necessary to map the fatigue parameters to 0–1 uniformly according to certain rules. Here, we do not plan the fatigue parameters according to their values, because the fatigue values do not necessarily increase linearly with the increase of fatigue parameters. For example, the normal range of blinking frequency is 10–15, and too large or too small

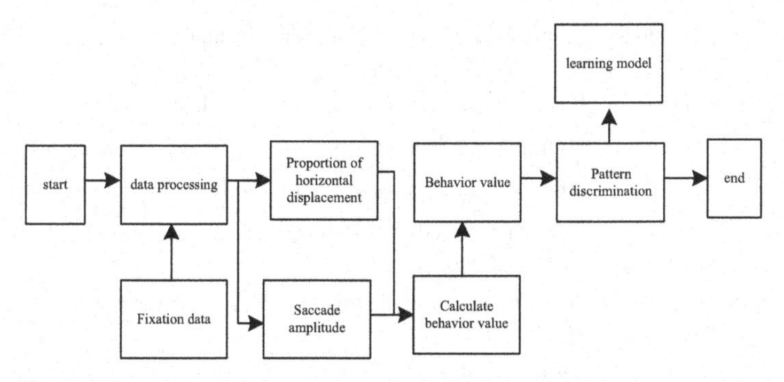

Fig. 4. Flow chart of fatigue value calculation by weighted average method

may be a sign of fatigue. If the user selects to set the weights of each fatigue parameter, modify the corresponding weights, otherwise, the default weights combination is adopted. Finally, the fatigue value can be obtained by summing the normalized fatigue parameters multiplied by the corresponding weights. Users can set their own threshold to judge fatigue can also adopt the default threshold, when the fatigue value is greater than the threshold, it is judged as fatigue. Because reading text is generally slow horizontal movement of the line of sight, until the line, then the next line of text to read. Therefore, in slow reading, the line of sight has a direction, the general level of movement in the majority. Therefore, when watching video, the saccade of the eyes is larger, and the line of sight is not directional, so the typing behavior is similar to slow reading. Therefore, the amplitude of typing mode is smaller than reading mode, and the line of sight is also directional. According to these objective laws and experimental observations, it is easy to distinguish two eye movement parameters of three learning modes, i. e. saccade amplitude and horizontal displacement ratio. Firstly, the ratio of horizontal displacement time in a period is calculated, the average saccade amplitude is calculated, the behavior value is calculated according to the formula, and the behavior mode is divided according to the range of behavior value.

3 Analysis of Experimental Results

The recognition of learning state mainly includes two aspects: fatigue state and inattention state. According to experience and a lot of experiments, a monitoring period of state identification is set to 100 s. In fatigue state, the threshold of nodding times in one cycle is 4, while in inattention state, the threshold of turning times in one cycle is 4. By recording and counting the results of the above 43.2 learning action recognition experiments in 100 s, i. e. the number of times that each action appeared, 5 experimenters were selected to do the experiment, and each experimenter did more than 4 nods in one monitoring cycle. First of all, six kinds of features captured were tested. The experiment selects 20 students to carry on the experiment according to the characteristic movement. The data obtained are shown in tables, and the experimental results are shown in Tables 5 and 6.

In this study, 20 students were selected to collect the video clips of their focused, general and non-focused learning states. A total of 87 groups of video clips were collected, each of which lasted 1 min and was decomposed into 10 frames per second. Then

Table 5. Sample of learning behavior features and actions

Characteristic motion	Positive sample	Negative sample
Eye closing feature action	50 frame eye opening video	50 frame closed eye video
Close your mouth	50 frames mouth closed video	50 frames mouth opening video
Lean forward	50 frames of body not leaning forward video	50 frame forward leaning video
Perspective movement	50 frame view not moving video	Moving video frame 50
Yawn	Video containing 100 yawns	
Nod and doze off	Contains a video of 100 nodding and dozing	

Table 6. Experimental results of characteristic actions

Characteristic action	Recognition accuracy
Eye closing feature action	80%
Close your mouth	85%
Lean forward	98%
Perspective movement	89%
Yawn	92%
Nod and doze off	96%

we use all kinds of feature methods to get the feature actions and record and randomly arrange them. 80% data are divided into training sets and 20% data are divided into testing sets. Used to train training support. The final result of behavioral recognition is shown below (Table 7).

Table 7. Learning behavior recognition results

Learning state	Number of experiments	Correct times	Number of errors	Recognition rate
Fatigue state	510	455	55	90.8000%
Inattention	510	462	48	92.0000%
Population	1200	917	103	91.5000%

From the results of behavior recognition in the table, the recognition rate of fatigue state based on threshold method is obviously higher, which shows that the recognition method designed in this paper has higher recognition effect. After excluding the invalid data of 7 subjects and summarizing the distribution characteristics of blinking signals of 38 valid subjects, it is found that the action characteristics of the subjects mainly focus on the following three conditions. The distribution of the number of actions of 20 subjects (52.6%) is shown in Fig. 5.

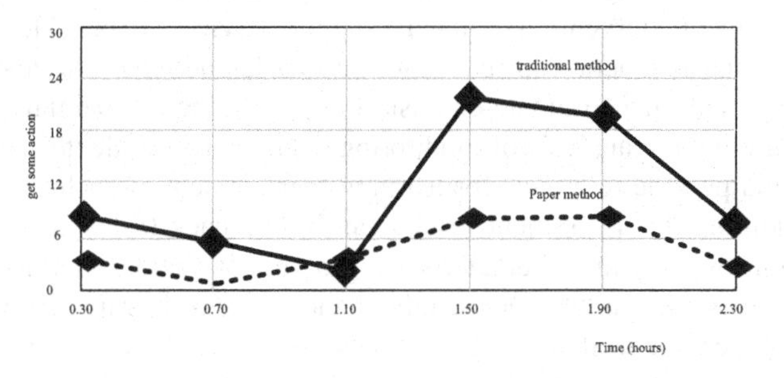

Fig. 5. Distribution of abnormal actions of 20 subjects

In order to compare the performance of the two algorithms, we calculate the mean square deviation of the EKF algorithm and the algorithm designed in this paper in static and dynamic conditions. The results are shown in Table 8.

Table 8. Comparison of mean square deviation of two detection methods

Test content	Static state			Dynamic		
	Roll angle/(°)	Pitch angle/(°)	Heading angle/(°)	Roll angle/(°)	Pitch angle/(°)	Heading angle/(°)
Extended Kalman filter algorithm	0.0521	0.0568	0.0525	0.6752	0.4256	0.7625
Algorithm in this paper	0.0215	0.0265	0.0325	0.3256	0.2236	0.1658

Compared with Table 8, the mean square deviation of the proposed method is smaller than that of Extended Kalman Filter, which shows that the proposed method can track and respond to local attitude changes more quickly and accurately. In the local motion attitude measurement experiment, the moving end can get the accurate local motion attitude. Because this paper combines mobile technology to collect ideological and political learning behavior characteristic data. According to the feature recognition results of the data, an evaluation system for the learning status of ideological and political courses is constructed. Six feature actions in human poses are selected to study learning state classification. In this way, a high-precision student's political course learning status is achieved.

4 Conclusion

To sum up, it is of great significance to monitor the learning process of distance learners by using the mobile learning technology. On the one hand, distance learning managers and teachers can dynamically grasp the learners' learning situation, gradually understand the learners' learning characteristics, and provide personalized learning process support services. According to the abnormal changes of the learner's learning state, corresponding intelligent intervention or manual support services are provided for effective communication and interaction in real time. So that learners feel the teacher's attention and respect, and then stimulate enthusiasm for learning, to maintain a good state of learning. Through the analysis of monitoring data, we can evaluate students' learning behavior and provide reference for learning management to evaluate the learners' process. In addition, the assessment results can be fed back to learners to help them understand their own learning characteristics, progress and effects and to monitor and regulate themselves accordingly. On the other hand, distance learning management can design and push personalized learning resources for learners based on these data.

Fund Project. Research project on education and teaching reform of colleges and universities in Hainan Province: Research on practical teaching dilemma and Breakthrough Strategy of Ideological and political theory course in Higher Vocational Colleges (Project No: Hnjgzc2022-110).

References

1. Xu, N., Fan, W.: Research on interactive augmented reality teaching system for numerical optimization teaching. Comput. Simul. **37**(11), 203–206+298 (2020)
2. Sun, Z., Anbarasan, M., Praveen Kumar, D.: Design of online intelligent English teaching platform based on artificial intelligence techniques. Comput. Intell. **37**(3), 1166–1180 (2021)
3. Martin, F., Ritzhaupt, A., Kumar, S., et al.: Award-winning faculty online teaching practices: course design, assessment and evaluation, and facilitation. Internet High. Educ. **42**, 34–43 (2019)
4. Luo, Y., Han, X.: Research on blended courses classification based on characteristics of students online learning behavior. China Educ. Technol. (06), 23–30+48 (2021)
5. Li, H., Wei, Y.: Research on the construction and application of learning behavior evaluation system in hybrid teaching. China Educ. Technol. **10**, 58–66 (2020)
6. Jing, Y., Li, X.: Characteristics analysis of teachers' online learning behavior supported by learning analytics. China Educ. Technol. **23**(02), 75–82 (2020)
7. Yuan, L., Yuan, Y., Du, X.-F., et al.: The influence of subjective norms on mathematical learning behavior: the mediating effect of mathematical interest. J. Math. Educ. **29**(05), 14–19 (2020)

Real Time Monitoring Method of Exercise Load in Dance Training Course Based on Intelligent Device

XiaoSha Sun(✉) and Xiang Gao

Sanmenxia Polytechnic, Sanmenxia 472000, China
sunxiaosha_1984@126.com

Abstract. Because of the heavy exercise load in dance training class, students' bearing capacity is different. In order to better scientifically guide the movement of dance training classes, a real-time monitoring method of exercise load in dance training classes based on smart devices is proposed. Combined with smart devices, it collects and manages a large number of physical fitness indicators of dance training, and uses heart rate changes as load evaluation indicators. Real-time acquisition of the movement status of dance training courses and real-time monitoring. Finally, it is confirmed by experiments that the real-time monitoring method of exercise load in dance training courses based on intelligent equipment has high practicability in practical application, so as to meet the research requirements.

Keywords: Intelligent equipment · Dance training · Exercise load · Exercise monitoring

1 Introduction

Dance teaching is an indispensable part of physical education reform and basic education. It is not only an important way to enhance students' physique, but also an important part of current quality education. Since the reform and opening up, China's dance teaching has developed continuously and experienced an all-round evolution process. The effect of dance teaching greatly affects the physical development of students [1]. The exercise load of dance training is monitored and analyzed in real time. Exercise load is an important factor affecting the process of body metabolism and energy consumption and recovery. Making effective use of training time and reasonably arranging exercise load is an important aspect to measure teaching quality and improve teaching effect [2]. In club teaching and daily teaching, we should comprehensively analyze the physiological characteristics of objective groups participating in sports, and study the effective physiological load value, threshold and suitable exercise intensity, which is not only conducive to the use of specific quantitative standards to guide dancers in training. Moreover, it is conducive to students' practice, provide reference for reasonable curriculum arrangement, and guide dance lovers to reasonably select dance content with appropriate load for training [3].

W. Fu and L. Yun (Eds.): ADHIP 2022, LNICST 468, pp. 107–122, 2023.
https://doi.org/10.1007/978-3-031-28787-9_9

Literature [4] obtains the health status of community residents and analyzes the influencing factors based on the intelligent health monitoring system. According to the analysis results, the health status of community residents is good, and psychological problems such as anxiety and depression are widespread, which reminds community residents to pay attention to mental health. For the elderly, irregular meals, salty diet, regular drinking and no vigorous exercise, the intelligent health monitoring system detects that they are prone to chronic diseases, and suggests that community residents formulate health improvement measures by referring to the influencing factors of chronic diseases. By analyzing the physiological and psychological changes of the elderly, literature [5] uses 3D software to build a digital model of the health detection integrated machine, analyzes the human-computer interaction behavior under different functions, improves the design of key parts in combination with the force analysis results, and verifies the comfort of the elderly using the improved detection integrated machine through simulation tests.

In order to better achieve the purpose of exercising and strengthening physique, this paper studies the real-time monitoring method of exercise load in dance training classes, collects the load data of dance training through intelligent devices, takes the change of heart rate as the load evaluation index, obtains the exercise state of dance training classes, and realizes real-time monitoring. The experiment verifies that the intelligent device gives a reasonable training scheme according to the formulation of the best load monitoring index. Using intelligent equipment to carry out real-time quantitative monitoring and analysis of key physical indicators has positive significance in practical applications.

2 Real Time Monitoring Method of Exercise Load in Dance Training Class

2.1 Exercise Load Data Management of Dance Training Course

The purpose of dance training is to provide a kind of stimulation, which can effectively improve the competitive performance of dancers, so as to obtain better sports performance. The competitive performance of dancers is the embodiment of their competitive ability level in training and competition, which reflects the harmonious accumulation of the functions of intelligent devices inside the body after training, and shows the sports quality and technical level of dancers [6]. In order to actively meet the requirements of real-time monitoring of exercise load in dance training class, it is necessary to manage the exercise load data of dance training class, and coaches should establish a delicate balance between arranging training load and recovery. Dancers receive intelligent devices and repetitive exercise loads in order to induce adaptability matching the required functions, such as delaying the occurrence of fatigue, increasing power output, improving exercise coordination or reducing the risk of injury. This paper holds that the quality of training results depends on the type and quantity of stimuli, so understanding the causal relationship between training load dose and response is very important for dance training [7]. Taking the heart rate data during dance training as an example, during exercise load monitoring, it is mainly divided into quiet heart rate and exercise heart rate according to different exercise states. The quiet heart rate can be detected with the teacher about

10 min before class and entered into the intelligent device. Since the quiet heart rate is relatively stable, you can also follow the recent normal quiet heart rate. The real-time collection process of exercise load monitoring data is mainly based on the sensors of intelligent devices such as heart rate meters. In order to reuse the equipment, the equipment information needs to be tied with students before class [8]. In order to avoid human error and improve binding efficiency, this paper uses intelligent device RFID tag to simplify the operation. That is, two RFID tags containing student information and heart rate meter information will be distributed before class. Students will stick them on the heart rate meter and bind the equipment through RFID scanning. The heart rate meter is bound with the smart device Bluetooth in advance. The student starts to receive data after wearing the heart rate meter until the student removes and returns the heart rate meter. The data transmission of exercise heart rate meter is completed through the Internet of things. The main transmission process is shown in Fig. 1. The basic data of dance training load data collection and management is that after collecting the state data of dance training, heart rate and other sports data collected by each student, the current age is calculated according to the student's birth date, and the student's age and gender are called from the database to match the parameter values to the core parameters such as the target rate, and then the current load intensity is calculated.

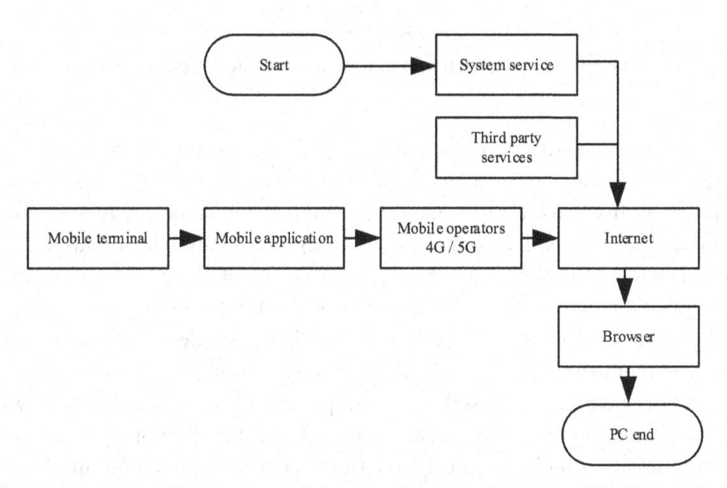

Fig. 1. Data acquisition process of dance training load based on intelligent device

The concept of exercise load based on intelligent devices is based on the mutual adaptation of exercise forms and physical function changes. The fundamental purpose of monitoring and adjusting exercise load is to scientifically and effectively organize and adjust the factors related to the mutual adaptation of physical function and exercise behavior, so as to improve the effectiveness, pertinence and safety of training. The focus of dance training load is the change of body state caused by sports behavior and the stress response of the body to the process of dance training. The stimulation of sports to the body has both physiological and psychological aspects. The stimulation to the body in the process of dance training is higher than that in the state of daily life. The relationship between external stimulation and physiological load is shown in Fig. 2.

External stimulation is divided into physical stimulation and psychological stimulation, exercise load is divided into physiological load and psychological load, and the arrow in Fig. 2 indicates causality.

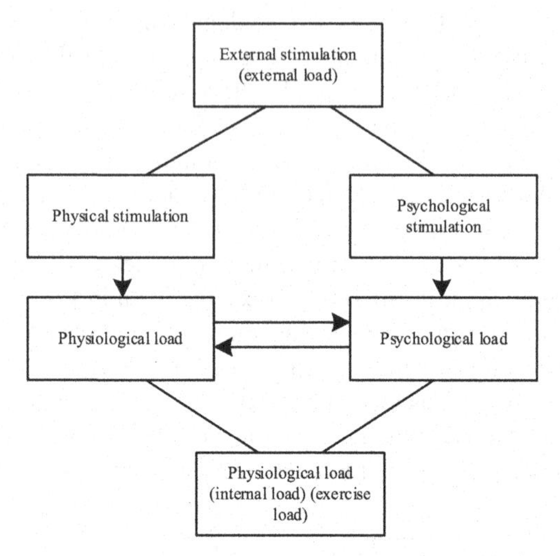

Fig. 2. Relationship between external stimulation and physiological load of dance training

The smart device sensing layer is composed of various sensors and sensor gateway architectures, as shown in Fig. 2. Including carbon dioxide concentration sensor, temperature sensor, humidity sensor, QR code label, smart device label, reader, camera, GPS and other sensing terminals; It also includes vital sign acquisition terminals such as heart rate meter, sphygmomanometer and blood glucose meter. The function of the intelligent device perception layer is equivalent to the nerve endings of human eyes, ears, nose, throat and skin. It is the source of the Internet of things to identify objects and collect information. Its main function is to identify objects. The intelligent device information collection network layer is composed of various private networks, the Internet, wired and wireless communication networks, network management intelligent devices and intelligent device platforms, which is equivalent to human nerve center and brain, Responsible for transmitting and processing the information obtained by the perception layer. The intelligent device application layer is the interface between the Internet of things and users. It combines the needs of the industry to realize the intelligent application of the Internet of things. The application of exercise heart rate monitoring intelligent device belongs to the vital sign monitoring part of intelligent health, which can be divided into nursing (Fig. 3).

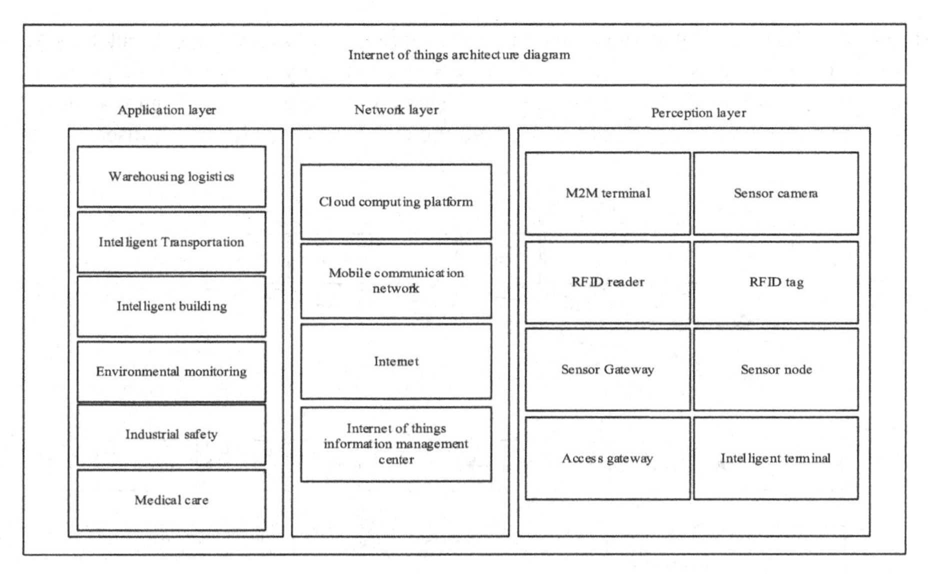

Fig. 3. Intelligent monitoring and management architecture of dance training load

According to the requirements of intelligent devices, flexibly select the appropriate intelligent device architecture. Use intelligent equipment to monitor the exercise load of heart rate. The speed of heart rate can well reflect the intensity of exercise. Next, it shows the rationality of using heart rate index to monitor exercise load. This paper studies the exercise load from the two aspects of "load" and "load intensity", and the load intensity is the most important content of studying the exercise load, and explains the rationality of the phenomenon that the utilization index is widely used to monitor the exercise load according to the research results of predecessors.

2.2 Evaluation Index of Exercise Load in Dance Training Course

As far as China is concerned, professional dancers have no other work and income except participating in competitions and training, but professional dancers are not employed by professional dance clubs, do not participate in professional competitions, and do not obtain additional income through participation. Therefore, the remarkable feature of professional dancers is that they belong to a group, with special intelligent equipment training, special guidance, strong knowledge and professionalism, and a large amount of training. Amateur dancers refer to those who use their spare time outside their own work for the purpose of fitness or leisure hobbies. Their remarkable feature is that they generally do not have the guidance of professional coaches and are purely for personal hobbies. Most of the training funds, equipment and competition fees are obtained by raising or soliciting sponsorship or even donations. They often have their own formal jobs and have to go to work to make a living after the competition, After achieving good results, of course, you can also get some sponsorship or participate in other image endorsement activities. Professional is result oriented, amateur is interest oriented, so professional dancers can earn income through training and competition without considering other work pressure; For amateur dancers, dance is only a way to obtain physical and mental health benefits, not a career or a means of making a living. Figure 4 shows the factors that affect the load of dance training and lead to overtraining, disease and injury.

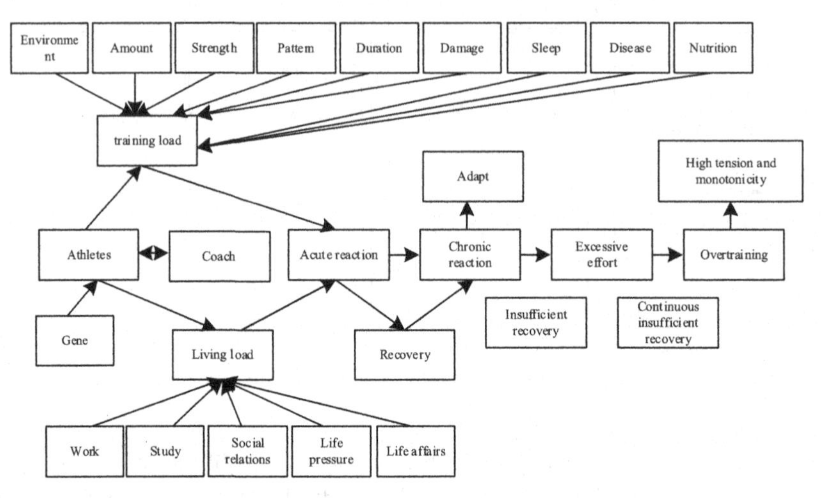

Fig. 4. Factors affecting dance training load and leading to overtraining, disease and injury

Although it is pointed out that load is "the impact of exercise and non exercise stimulation on human biological intelligent devices at different times" until 2020, load monitoring has a history of nearly a century: by the 1980s, indicators such as blood lactic acid, creatine kinase and maximum oxygen uptake can be easily measured by portable devices, which further enriched the means of load monitoring. With the rapid development of science and technology, advanced wearable devices came into being. A series of more convenient physiological and biochemical monitoring technologies have been continuously developed, and the means of load monitoring have been greatly expanded. In training, exercise load refers to the accumulated pressure that individuals bear in competition or training over a period of time. According to the source of monitoring indicators, exercise load can be classified into external load and internal load; According to the attributes of monitoring methods, exercise load is divided into objective evaluation and subjective evaluation; According to the statistical processing method of monitoring, sports load can also be divided into absolute load and relative load. See Table 1.

Taking the heart rate as the load evaluation index for research, in the actual dance training process, the physiological condition YR and psychological condition AR of the human body will affect the exercise load intensity AGE. The exercise load monitoring model needs to establish a self-learning mechanism. The self-learning target rate measurement method of maximum heart rate index MR is:

$$FQ = \frac{YR - AR}{MR - AGE} \tag{1}$$

Table 1. Classification of dance training load and common indicators

Classification basis	Load category	Definition	Common indicators
Indicator source	External load	Individual work	Duration, times, speed, acceleration, distance, times of throwing or changing direction, jump height, weight, etc.
	Internal load	The body's response to certain stimuli	Heart rate, blood lactic acid, testosterone, cortisol, RPE, self recovery scale, emotional state scale, etc.
Method properties	Objective evaluation	Evaluation not based on individual subjective will	Heart rate, blood lactic acid, testosterone, cortisol, speed, acceleration, distance, etc.
	Subjective assessment	The evaluation is mainly based on individual psychological activities and influenced by psychology	RPE, self recovery scale, emotional state scale, etc.
Statistical analysis	Absolute load	Load accumulation over a period of time	Daily load, weekly load, monthly load, season load, etc.
	Relative load	Relative change of load over a period of time	Weekly load change rate, weekly load change amount, acute and chronic load ratio, etc.

The maximum heart rate monitoring identification index is converted to:

$$R = \frac{YR - AR}{MFQ + PFQ} + AGE \tag{2}$$

Among them, exercise heart rate MFQ and quiet heart rate PFQ are the collected data, age is the objective data, and the maximum heart rate index is the reference index. By default, "maximum heart rate index - age" is used. That is, the maximum heart rate calculated in general. Using intelligent devices to calculate the maximum heart rate is an international customary method, which is applied to people of all ages. When applied to students, there may be some deviation. The exercise heart rate monitoring intelligent device involves embedded, Android, web application and other related technologies, and it is necessary to organically combine all parts into a coordinated intelligent device to jointly complete the exercise heart rate monitoring service. On the whole, the intelligent device is divided into three parts: heart rate acquisition module, Android application and web service. Next, the overall operation process of the intelligent device is described from the perspective of the overall data flow. Part of the data flow is shown in Fig. 5.

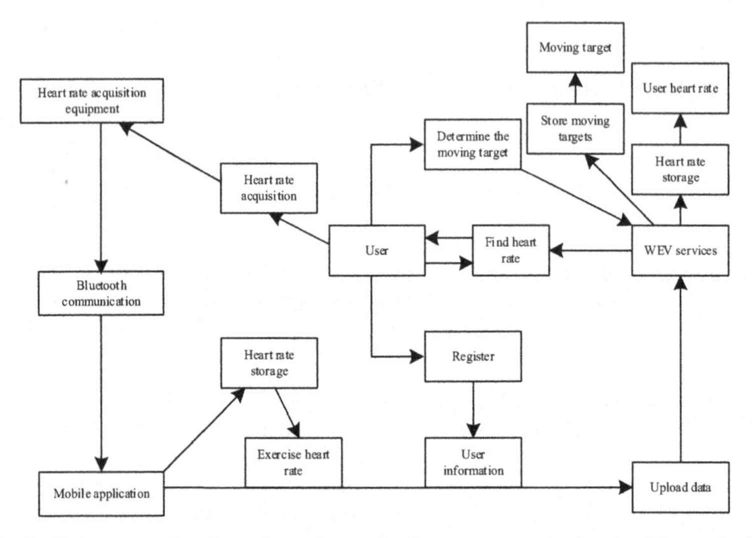

Fig. 5. Data processing flow chart of exercise heart rate monitoring intelligent device

The heart rate acquisition equipment should complete real-time heart rate acquisition. First collect the pulse information, and then obtain the time t for each pulse from the pulse information. The heart rate HR = 60t. The heart rate filtering algorithm is used to filter the collected heart rate information numerically. Finally, the filtered data is sent to the mobile phone heart rate acquisition module through Bluetooth and smartphone communication. The hardware framework design is shown in Fig. 6.

According to the formula converted to calculate the maximum heart rate, it can be seen that the exercise heart rate and quiet heart rate can be directly collected, and the load intensity is obtained through RPE. The maximum heart rate index a_{u1} is calculated through this formula. The data of students of different gender and age are grouped. When the data conforms to the normal distribution a_{zel}, the expected value (mean) is the maximum heart rate index of students of that gender and age, Finally, the maximum heart rate index corresponding to different gender and age is obtained, and a more objective and practical maximum heart rate index is used in practical application. The purpose of this paper is to explore the research progress of sports load monitoring in training from the perspective of external load a_{y-1} and internal load a_{y1}, combined with the differences of method attributes and statistical analysis:

$$HRV = \sqrt{\frac{\left(a_{y1} - a_{y-1}\right)^2 + (a_{u1} - a_{zel})^2 + RPE}{100R}} \tag{3}$$

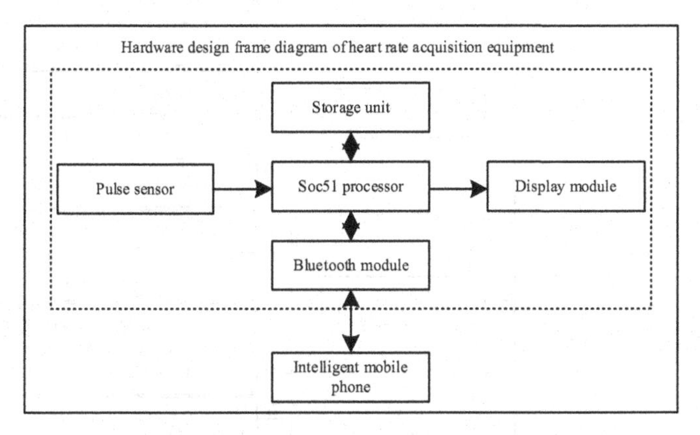

Fig. 6. Heart rate acquisition equipment

When the training load decreases, *HRV* will accelerate; When *HRV* decreases, it is usually related to the increase of training load, especially when the load increases rapidly ($5\% \pm 2\%$), *HRV* decreases, which indicates excessive load. However, some studies have found that when the training load increases, *HRV* also increases. When the subjects are in acute fatigue, *HRV* will increase, while when the subjects are in chronic fatigue, *HRV* will decrease.

2.3 Realization of Load Monitoring of Dance Training Course

The purpose of studying the load monitoring intelligent device of dance training course is to help athletes monitor their exercise status in real time and for a long time, and adjust the exercise load in time according to the exercise heart rate, so as to avoid poor exercise effect due to insufficient exercise intensity or sports injury caused by excessive exercise, so as to achieve good exercise effect. Interpreting the research goal of intelligent device, this paper can get its basic needs: monitoring real-time heart rate, This is the requirement of intelligent device sensor sensitivity and heart rate acquisition algorithm; It can monitor heart rate for a long time, which is the demand for data storage and analysis of intelligent acquisition equipment and dancers' adherence to heart rate monitoring; It can detect the exercise heart rate, which requires the portability or wearability of the exercise heart rate acquisition device; It can view the exercise heart rate in real time, which is the demand for the real-time display of the exercise heart rate monitoring intelligent device. It is the core demand to adjust the exercise load in time according to the exercise heart rate. This demand requires the intelligent device to give a reasonable theoretical model and operable scheme of using the exercise heart rate to monitor the exercise load. The functional module of the intelligent device is designed according to the demand analysis of the intelligent device, as shown in Fig. 7:

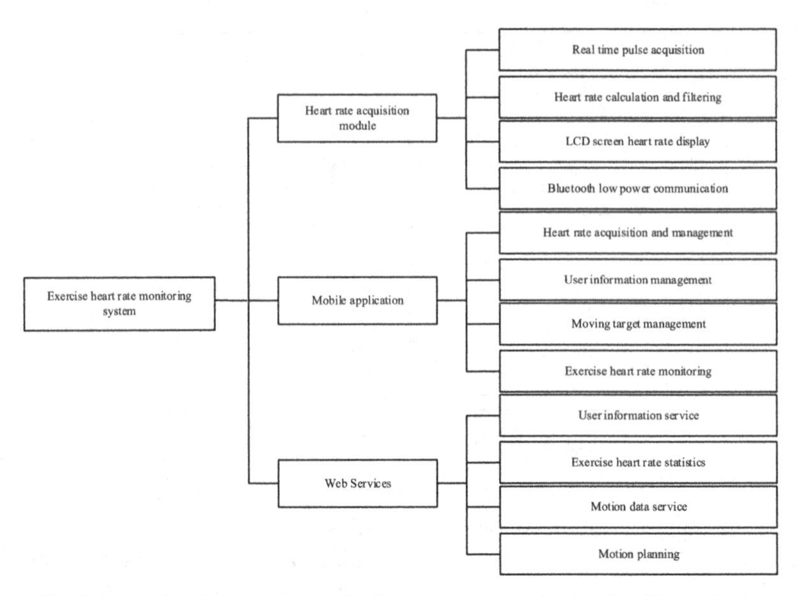

Fig. 7. Functional structure of exercise heart rate monitoring intelligent device

Many experts in the field of dance training in China conduct important research on load. Whether the scientific evaluation of training load, competition load or fitness load should include qualitative and quantitative parts, which are interdependent and inseparable. Quantitative evaluation is meaningful only on the basis of qualitative evaluation. Qualitative evaluation can only have correct evaluation on the basis of quantitative evaluation. The "qualitative" without "quantitative" basis is inaccurate and "qualitative", which may make the training deviate from the correct direction due to the accuracy of evaluation, while the "quantitative" without "qualitative" is also incomplete and lacks the description of key information. Figure 8 shows the selected exercise load evaluation system.

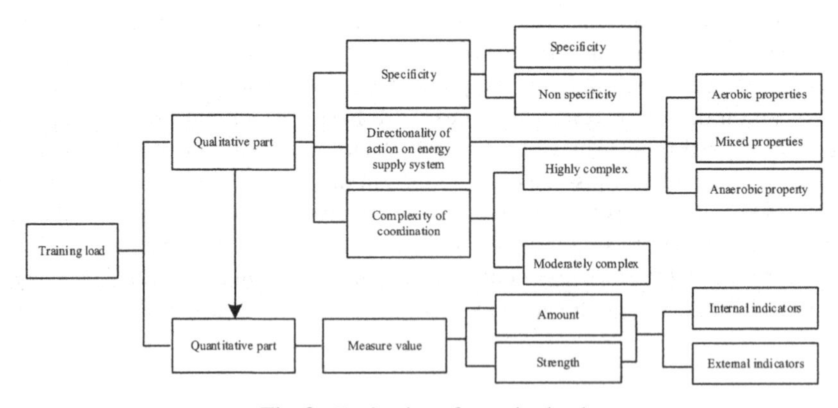

Fig. 8. Evaluation of exercise load

The essence of dance training is to apply appropriate exercise load stimulation to the dancer's organism artificially, purposefully and according to the plan, so as to make it produce the expected adaptive changes. The exercise intensity, exercise duration and exercise frequency are the three factors affecting the load. A full grasp and understanding of them and reasonable arrangement can achieve the best effect of dance training. The duration and frequency of exercise are easy to control, but the intensity of exercise is difficult to grasp. Exercise intensity is not only the external load imposed by the coach in the training class, but also the real reaction of these external loads on the dancer's organism, which is also called internal load. Therefore, the trainer can regard the training load as all the impulse borne by the dancer. According to the historical records of students in all aspects, give early warning of various possible accidents, make relevant suggestions on the intensity of current load, and guide the rationality of curriculum arrangement. Figure 9 shows the flow of sports load monitoring model.

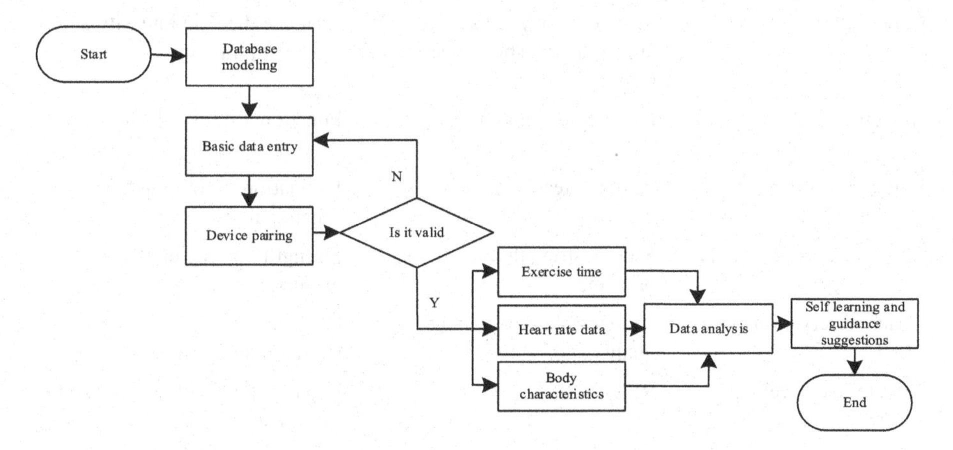

Fig. 9. Flow chart of sports load monitoring model

The model takes the dancer as an intelligent device, which inputs training and outputs exercise ability. The product of exercise duration and intensity is called training impulse (a). Coaches can effectively predict sports performance through this model, so as to design the best training plan according to sports performance. Expected sports performance = body adaptation k_1 - body fatigue k_2, the formula is:

$$a(t) = k_1 w(t) e^{-t/\tau_1} - k_2 w(t) e^{-t/\tau_2} \tag{4}$$

Where $w(t)$ is adaptive impulse and e is fatigue impulse. During training, load stimulation is applied to organisms, and adaptive impulse and fatigue impulse work together to affect the training process. Adaptive impulse plays a positive role in promoting sports performance, and fatigue impulse plays a negative role in promoting sports ability. After training, the body adaptation and fatigue decreased exponentially, but the rate of decline was not consistent. The body adaptive impulse maintained for a long time and the attenuation rate was slow. On the contrary, the attenuation rate was fast. By comparing the expected results with the real results to adjust the attenuation constant and weight factor, so as to produce a more accurate prediction of sports results. It further describes that the changes of dancers' physical functions will be reflected in the intelligent devices of various organs of the body, including cardiovascular intelligent devices, respiratory intelligent devices, immune intelligent devices, endocrine intelligent devices, neuromuscular intelligent devices, etc., as well as the observation of coaches and the self feeling of dancers. Through the comprehensive analysis of multiple indicators, the fatigue degree and recovery of the body are diagnosed, as shown in Table 2.

Table 2. List of commonly used indexes and functions for athletes' physical function evaluation

Organizational system	Index	Effect
Blood system	Blood lactic acid, hemoglobin, creatine kinase, etc.	Evaluate load intensity and functional state
Cardiopulmonary system	Maximum oxygen uptake, anaerobic threshold, heart rate, etc.	Evaluate the load intensity and long-term training effect
Immune system	Leukocytes, immunoglobulins, etc.	Functional state evaluation
Endocrine system	Cortisol, testosterone	Evaluation of training load and functional status
Muscular system	Muscle strength, torque, muscle pain, etc.	Strength training effect, evaluation of muscle load
Coach observation system	Athletes plan to complete quality, etc.	Comprehensive evaluation of sports training status
Athlete's self feeling	RPE et al.	Evaluate the local reaction and overall state of athletes

By integrating the monitoring practice and technology of training load and the literature evaluating the fatigue degree of training and competition in recent years, it is found that 91% of the 55 teams engaged in high-level sports said they had used or had been using some type of monitoring means to monitor them, which proves the value of sports monitoring, as shown in Table 3.

Therefore, the accurate measurement of the training load of organisms can help to more accurately evaluate the adaptability of dancers to the training plan arrangement, and can help coaches adjust the adjustment rate of personalized training plan. Previously, it was considered to be the most appropriate and practical method to quantify the internal load of dance training and other items. Nevertheless, when using training impulse to calculate training load, some technical problems related to weighted heart rate may increase the risk of accuracy error of objective methods. When the number of devices is insufficient, the monitoring of training load becomes the main problem. Therefore, when monitoring training needs to be more simple and effective, the practical application of this method in measuring training load should be paid more attention. However, this suggestion needs to be considered because there are many factors that affect the dancer's

Table 3. Monitoring practice of dance training

Monitoring type	Degree of use	Degree of confirmation	Practical degree
GPS and accelerator	High	Secondary	High
RPE scale	High	High	High
Health Questionnaire	High	High	Low
Biochemical index	Low	Secondary	High
Heart rate measurement	High	High	High
Sports performance test	Secondary	Secondary	Secondary
Motion screening	High	Low	Secondary
Neuromuscular assessment (example: CMJ)	Secondary	Secondary	Secondary

personal perception of physical strength, and these factors need to be considered when using.

3 Analysis of Experimental Results

The intelligent polar S400 telemetry heart rate meter was used to record the immediate heart rate every 5 min, and the heart rate change during exercise was measured in the flow chart of the subject's exercise load monitoring model. The measurement of exercise and exercise center rate starts from the subject standing quietly, and the exercise time is 60 min. In addition, a student majoring in dance monitors the course content and records the course content of each period. Stop the polar S400 telemetry heart rate meter immediately after the exercise. After exercise, the subjects were asked to fill in the rating of subjective fatigue immediately. RPE used relevant analysis software to analyze and statistically process the data. First, use the power load increasing test to select the appropriate subjects, then carry out the power load increasing test on the qualified subjects according to the test plan to obtain HRmax, and carry out the load increasing test of dance training class to obtain another HRmax. Through a large number of experimental comparison, it is found that there are obvious differences between the results measured by the two methods. The results measured by the dance patrol test are significantly higher than the power test. The experimental results are shown in Table 4.

According to the maximum heart rate calculation formula, an initial maximum heart rate is obtained. The athlete starts training according to the exercise load level divided by the maximum heart rate, and adjusts the maximum heart rate according to the subjective physical feeling. Each adjustment can add or subtract 5. Before using this method to feedback and adjust the maximum heart rate, the athlete should first be familiar with the use of the sensory table in this method. The corresponding relationship of each index of exercise intensity is shown in Table 5.

Table 4. Statistics of real-time monitoring results of maximum heart rate

Power car test (people participating in the treadmill)		Fort experiment	
Number of people	Maximum heart rate	Number of people	Maximum heart rate
8	167.35 ± 12.69	8	171.65 ± 11.09
12	164.01 ± 12.36	12	173.11 ± 8.75
21	165.76 ± 13.06	21	172.36 ± 9.76

Table 5. Various indexes of exercise intensity

Correlation strength			
Classification of exercise intensity	Maximum heart rate	Maximal oxygen uptake or maximal heart rate reserve	Supervisor physical feeling scale
Lower strength	<34%	<29%	<9
Low strength	34%–59%	34%–49%	10–11
Medium strength	60%–79%	50%–75%	12–13
High strength	80%–89%	76%–84%	14–16
Super strength	≥90%	≥85%	>16

The exercise heart rate reflects the exercise load intensity of the athlete. The maximum heart rate can reflect the physiological maximum stress level of the athlete. The heart rate reserve obtained by subtracting the resting heart rate from the maximum heart rate reflects the variation range of the athlete's heart rate. The maximum heart rate will not change in a short time, but after long-term training or long-term stop training, the athlete's maximum heart rate will change accordingly. The real-time monitoring method of dance training load of intelligent equipment is proposed, which can well reflect the physical function level of athletes on the basis of ensuring the movement effect and safety of athletes. The content of aerobic dance class belongs to the combination of power routines, which includes warm-up stretching and review. The completion time of the combined action is about 60 min. During the training process, record the immediate heart rate every 5 s, calculate the 60 heart rate values of each member for 60 min in a class in the form of the average heart rate value per minute, and then conduct real-time monitoring, sorting and Analysis on a total of 960 heart rate values of 16 research objects. The variance and standard deviation can represent the dispersion and stability of a group of data and its average value, and the skewness represents the characteristic number of the asymmetry of the probability distribution density curve relative to the average value (as shown in Table 6).

Table 6. Average value of real-time monitoring of intelligent devices

	N	Minimum	Maximum value	Mean value	Standard deviation
Overall data	970	65.00	195.00	142.6529	26.8542
Male heart rate	490	65.00	186.00	139.6628	26.6523
Female heart rate	490	68.00	195.00	143.9858	27.0658

At the same time, frequency analysis shall be carried out for the falling point of heart rate value (as shown in Table 7).

Table 7. Exercise load monitoring frequency of intelligent equipment

Project frequency		Frequency			Percentage		
		Total (970)	Male (490)	Female (490)	Total (970)	Male (490)	Female (490)
Heart rate range	Below 90	55	31	24	5.6	6.5	4.9
	90–110	63	27	36	6.6	5.5	7.6
	110–130	195	107	88	20.2	22.2	18.2
	130–150	208	116	93	21.7	24.2	19.3
	150–170	321	158	161	33.5	33.2	33.6
	170–190	123	45	79	12.8	9.3	16.5
	Above 190	4	0	4	0.4	0	0.7

The results show that the heart rate monitoring value is mainly concentrated in the range of 131–170 times min, which basically meets the actual value, so it can better give the best training scheme according to the dance training state and athlete system.

4 Conclusion

For the best exercise load of dance course, the step experiment method and its exercise center rate data are used as the exercise load monitoring index, combined with the exercise load index as the subjective self feeling evaluation index, this paper analyzes the exercise load index in the process of dance exercise, finally formulates the best load monitoring index, and gives a reasonable training scheme to truly realize the expected goal of strengthening students' physique. At present, intelligent devices can only realize the functions of monitoring and formulating training plans, and cannot give early warning of sports overload. It is hoped that in the follow-up research, the early warning function of overload movement of the research object can be realized.

References

1. Lian, J., Fang, S.Y., Zhou, Y.F.: Model predictive control of the fuel cell cathode system based on state quantity estimation. Comput. Simul. **37**(07), 119–122 (2020)
2. Zhou, X.D., Xiao, D.D., Guo, W.X., et al.: Study on the mechanism of knee joint injury about table tennis players —load characteristics analysis of knee movement by three commonly used pace. China Sport Sci. Technol. **56**(06), 62–67 (2020)
3. Liu, H.Y., Liu, Z.Y., Ha, J.W.: Characteristics, influential factors and monitoring strategies of rugby injuries. J. Wuhan Inst. Phys. Educ. **54**(05), 75–81 (2020)
4. Sun, H.Y., Sun, Y.M., Sun, J.Y., et al.: Analysis of health status and its influencing factors of community residents based on the intelligent health monitoring system. Chin. J. Nurs. **55**(12), 1836–1843 (2020)
5. Lu, N., Zhu, D.X., Li, F.Y.: Ergonomic design and simulation analysis of integrated intelligent health testing machine for the elderly. J. Mach. Des. **37**(10), 128–133 (2020)
6. Li, X.S.: Research on application of flexible strain sensor in human motion monitoring. Adhesion **43**(9), 169–172 (2021)
7. Li, D.B., Fang, G.L., Mi, S., et al.: Study on competition load and sports performance characteristics of Chinese women's 3×3 basketball team. China Sport Sci. Technol. **56**(08), 33–39+81 (2020)
8. Zhang, J.S., Hu, Z.: Automatic tracking method for moving target of intelligent monitoring system based on binocular vision. Manuf. Autom. **42**(05), 142–146 (2020)

An Efficient Compression Coding Method for Multimedia Video Data Based on CNN

Xu Liu[1(✉)] and Yanfeng Wu[2]

[1] School of Media and Communication, Changchun Humanities and Sciences College, Changchun 130117, China
liuxu517@163.com

[2] School of Physical Education, Changchun University of Finance and Economics, Changchun 130122, China

Abstract. Due to the phenomenon of object occlusion and inconsistent motion of different objects in video, high-efficiency compression and coding of video data will generate prediction residuals related to texture structure. To solve this problem, this study proposes an efficient compression coding method for multimedia video data based on CNN. First, the multimedia video coding unit is divided, and the coded frames are arranged in POC order for coding. Then, the coding structure adjustment parameters are calculated, and after coding, the determined reference frame and the bit consumption caused by using the reference frame can be obtained. Finally, an intra-prediction algorithm for video data is established based on CNN. CNN encoder uses a series of down-sampling convolution and ReLU nonlinear mapping to extract and fuse global information, and conducts analysis on areas with low human visual sensitivity on the HEVC transform domain. Frequency coefficient suppression. The experimental results show that the method has good coding performance for test sequences with different contents and different resolutions.

Keywords: CNN · Multimedia video · Video data · Compression coding · Video coding · Efficient compression

1 Introduction

Since entering the Internet era, multimedia technology and communication technology have developed rapidly, and high-definition 1080P and ultra-high-definition 4K, 8K resolution video images are gradually entering our work and life. Compared with text, picture and sound data, video signal has a larger amount of data, especially in the case of limited network bandwidth and storage resources, the research on video coding technology is very important.

HEVC is a new generation of video coding standards. HEVC is based on the traditional hybrid video coding framework, and adopts more technological innovations, including flexible block division, finer intra-frame prediction, newly added Merge mode,

W. Fu and L. Yun (Eds.): ADHIP 2022, LNICST 468, pp. 123–136, 2023.
https://doi.org/10.1007/978-3-031-28787-9_10

tile division, adaptive sample compensation, etc. The core problem of coding technology is to obtain the optimal rate-distortion performance under the condition of limited time, space and transmission bandwidth. Flexible block partitioning, which includes coding units (CU), prediction units (PU), and transform units (TU), improves coding performance the most. These technologies double the encoding performance of HEVC compared to H.264/AVC. Although the existing video coding standards can effectively compress video images, as the resolution of video images becomes larger and larger, the compressed video data is still very large, and video data still faces great challenges in transmission and storage Challenges [1].

At present, the various modules of HEVC are not optimal. For example, in the next generation of VVC, the performance indicators of intra-frame prediction are further improved, which indicates that there is still room for further improvement in the video codec standard of HEVC. At the same time, the existing video coding standards ignore the visual characteristics of the human eye during coding, and do not allocate bit resources according to the importance of the video content. For those areas that do not meet the visual characteristics of the human eye, a large amount of bit resources and computing resources are often consumed. Ensure the image quality of important areas in the video image and affect the subjective experience of the observer.

In recent years, with the vigorous development of the country's most artificial intelligence, people have applied artificial intelligence technology to many real-world scenarios, such as face recognition, machine translation, and speech recognition. The commercial implementation of these applications is an important manifestation of technological innovation in social production. Optimization of the video coding framework has been ongoing since the standard was established. Research workers and industrial designers at home and abroad have made a lot of effort and made a lot of effective work. Under the condition of limited network bandwidth, computing resources and storage resources, it is necessary to study the coding method that can make the video image quality more in line with the visual characteristics of the human eye [2]. In recent years, the development of traditional video codec technology has been hindered, and many performance improvements have to sacrifice a lot of resources and time. Therefore, deep learning technology is introduced into the existing video codec standard (HEVC), and the encoding performance can be further improved. The improvement is a challenging task [3].

Generally speaking, two important factors in video coding: computational complexity and rate-distortion performance, are opposed to each other, so these optimization efforts can be simply divided into two categories. One is to reduce the required computational complexity while ensuring that the coding performance does not drop too much; the other is to appropriately increase the computational complexity to improve the overall coding performance. Introducing deep learning technology into traditional video encoding and decoding standards can solve some problems that traditional algorithms cannot solve, and can improve the encoding and decoding performance of video encoding and decoding.

Based on the above analysis, this paper proposes a new efficient compression and coding method for multimedia video data based on CNN. The design idea is as follows:

(1) Divide the multimedia video coding unit and complete the coding according to the POC sequence.
(2) Calculate the adjustment parameters of the coding structure.
(3) An intra prediction algorithm of video data is established based on CNN, and a series of down sampling convolution and relu nonlinear mapping are used to extract and fuse video coding information.
(4) In the hevc transform domain, the frequency coefficients of the regions with low visual sensitivity are suppressed.

This method accelerates the decision-making process of intra prediction mode from a new perspective, so as to improve the coding quality of multimedia video images.

2 Method Design

2.1 Division of Multimedia Video Coding Units

In the compression and coding process of multimedia video data, after traversing all possible partition modes and optimizing rate distortion, the optimal CU/PU can be obtained, which means that each CU/PU will be encoded many times, which will greatly increase the computational complexity [4]. In this paper, we define the division depth of a 4×4 PU as 4, so the division problem of the entire CTU can be transformed into a combination of division problems at the four levels of depth 0–3. A piece of input multimedia video data has three channels of red, green and blue, and the pixel value of each channel ranges from 0 to 255. Considering that CNN has better performance on data in the range of 0 to 1, in the compressed autoencoder, the encoding end will first normalize each channel of the input video image in the normalization layer as follows: Unification operation:

$$B(x, y) = \frac{A(x, y)}{M} \tag{1}$$

In formula (1), (x, y) represents the pixel position; $A(x, y)$ is the original pixel value at position (x, y); $B(x, y)$ is the normalized pixel value; M represents the value of the pixel value of each channel. The three branches will output feature maps of the same size, and then we combine these three types of features in the depth dimension and pass them through two convolutional layers with small convolution kernels to effectively learn the relationship between these features. Correlation and diverse features can help CNN better understand the content information of the current CU.

The coded frames are coded in POC order, and there is no direct reference between frames. All coded frames are I-frames, which are independently coded using only intra-frame prediction techniques. In this case, the GOP length is 1, that is, each frame is managed as an independent GOP. Correspondingly, in the decoding part of the auto-encoder, the de-normalization layer will de-normalize the input image as follows to restore the original image.

Since there are too many possibilities for CTU division, it is not advisable to predict it directly. Therefore, the scheme we design is to make independent predictions at these four decision-making levels. On the encoder side, the connection layer is a convolutional layer with a convolution kernel of 3×3 and 32 channels, and on the decoder side, the number of channels is set to 128. A 1-unit padding operation is also used in the connection layer to ensure that the size of the input image is not changed during the convolution process. The designed neural network extends the first convolutional layer to three different branches. The first branch is a traditional convolutional layer, using a regular square convolution kernel, while the remaining two branches use asymmetrical A convolution kernel designed to detect texture details in near-horizontal and near-vertical directions. This structure enables our network to extract features more efficiently, the reference chain is a single chain, and the frame POC in the chain is not repeated. B frame in the LDB structure uses a double-chain reference, and the frames in the two reference chains are exactly the same, which requires one more reference process than LDP [5]. Finally, the extracted features are passed through three fully connected layers to obtain the final prediction result.

2.2 Calculate Coding Structure Tuning Parameters

The inter prediction technology of video coding uses reference frames to eliminate the time-domain redundancy of video and reduce the information entropy. The bit consumption of inter prediction is related to the bit consumption of the video source reference frame. The stronger the video motion and the more complex the texture, the more vectors are required for inter prediction.

In traditional video coding frameworks, quantization parameters are used to achieve different tradeoffs between bit rate and video quality PSNR. Under different quantization parameters, the division of CU/PU is different. For videos with non-rigid and complex textures such as water surface ripples, the inter-frame temporal correlation is weak, and the inter-frame prediction residual is relatively significant. To ensure image quality and maintain a low level of distortion, a smaller quantization step size is required. Residual information is retained, which consumes significantly more bits [6].

Generally speaking, the smaller the quantization parameter is, the more bits are used, and the higher the requirement of video quality is. At this time, in order to obtain finer prediction results, the encoder will more likely use smaller blocks. In view of this phenomenon, it is necessary to adapt our acceleration framework to diverse quantization parameters. Each coded frame refers to the previous frame adjacent to its POC and the previous key frame at most 3 GOPs. Then the reference frame set can be expressed as the following form:

$$U = \begin{Bmatrix} \alpha(t-1) & 4\alpha\left\lfloor \frac{t}{4} - 1 \right\rfloor \\ 4\alpha\left\lfloor \frac{t}{4} - 2 \right\rfloor & 4\alpha\left\lfloor \frac{t}{4} - 3 \right\rfloor \end{Bmatrix} \tag{2}$$

In formula (2), U represents the reference frame set; α represents the coded frame; t represents the time. For each coding unit, the determined reference frame and the bit consumption caused by using the reference frame can be obtained after coding. The calculation formula of its coded bits or entropy is as follows:

$$\delta(\beta_t) = \min\{E(\beta_t, \chi)\} \tag{3}$$

In formula (3), δ represents coding bits; β_t represents the coding unit of the t frame of the video; E represents coding; χ represents the reference CU that the frame to be coded finally adopts in the set of reference images that can be used. The spatial neighbors of the current block and the spatial neighbors of the reference block can help to further improve the coding performance.

Except for the two reference blocks, this paper takes the spatial neighbor pixels of the current block and the spatial neighbor pixels of the reference block as the input of the convolutional network. For the current coding CU, the entropy generated by coding the CU consists of three parts, including the entropy of the recorded motion vector, the recorded residual, the entropy of the transform and quantization coefficient matrix, and the entropy related to the recording mode and control information. In order to fully exploit the temporal long-term correlation and short-term correlation of video sequences, the video coding standard adopts a long-term reference mechanism and a short-term reference mechanism. Therefore, there are two types of reference frames for the video coding process - long-term reference frames and short-term reference frames.

In the HEVC coding standard, motion vectors pointing to long-term reference frames and motion vectors pointing to short-term reference frames cannot be cross-predicted [7]. The longer the reference distance, the more times it is indirectly referenced, and the more obvious the attenuation of the indirect reference intensity. The dependency factor is used as the coding structure adjustment parameter to improve the dependency strength of the key frame. The formula for calculating the dependency factor is as follows:

$$\gamma(\alpha) = q^{|\alpha - u|} \tag{4}$$

In formula (4), γ represents the dependency factor; q represents the dependency strength of the reference frame; u represents the distance between the reference frames. The reference block image is expanded to the left and upward to obtain a new image block. In order to utilize the spatially χ adjacent pixels of the current block, the prediction block obtained by the uniform weighting of the two reference blocks is spliced with the spatially adjacent pixels to form a new image block. There are strict reference rules between each level. Frames at higher levels can only refer to frames at lower levels, and frames at the highest level are not referenced. Frames in the lower layers are referenced the most, which enables the encoder to output a stream with frame rate diversity. When the frame rate needs to be reduced, some low-level frame encoding can be selected, and high-level frames are not considered, thereby greatly reducing the bit rate consumption.

2.3 Building an Intra-frame Prediction Algorithm for Video Data Based on CNN

Intra-frame prediction technology plays an irreplaceable role in the current video coding system. It can effectively remove the spatial redundancy of video signals, prevent the spread of coding errors, and improve the random access performance of video streams. Due to the phenomenon of object occlusion and inconsistent motion of different objects in video, high-efficiency compression coding of multimedia video data will generate prediction residuals related to texture structure.

In this paper, an intra-frame prediction algorithm for video data is established based on CNN. The algorithm adopts cross-level direct connection, which can combine deep and global semantic information with shallow and local representation information. Since the image denoising and super-resolution tasks confirm that residual learning can effectively improve the output image quality, in the inter prediction pixel correction technique, we also adopt the residual learning structure to improve the prediction accuracy. CNN-based intra prediction mode for video data which consists of a convolutional autoencoder and an auxiliary trained discriminator. The network structure draws on image inpainting technology for intra-frame prediction. It uses 3 reference blocks to predict the current block to be predicted. Uniformly weighted prediction blocks are used in the skip structure to implement residual learning. Due to the addition of spatially adjacent pixels in the network input, the size of the input image block is larger than the output prediction block, so the network needs to crop the image block and output the target prediction block.

CNN takes the image of the current block and its 3 neighboring blocks as the input image. Among them, CNN uses the intra-frame prediction result of HEVC as the initialization value. The output of the network is the optimized intra prediction result, which is used for prediction coding by HEVC. The convolutional network is trained by extracting the relevant information of all bidirectional prediction blocks according to the code stream information [8]. Uncompressed video sequences are limited, and we increase the number of training samples by downsampling and cropping operations. For intra prediction tasks, we require the convolutional auto-encoder to be able to accurately generate the texture information of the current block to minimize the prediction residual. The loss function of the convolutional autoencoder is as follows:

$$\eta = \frac{\sum_{w_1} \sum_{w_2} (z^* - z)^2}{s w_1 w_2} \tag{5}$$

In formula (5), η represents the loss function of the convolutional autoencoder; w_1 and w_2 represent the width and height of the feature map, respectively; z^* and z represent the MSE of the real map and the predicted map, respectively; s represents the number of prediction branches of the decoder. On sequences of different resolutions, the same block size contains different scales of information. Therefore, it is more universal to obtain training samples generated by sequences of multiple resolutions by downsampling for network training [9]. Following the CNN training mode, a competitive alternating training method is adopted. Then, CNN objective function includes the loss function of the above-mentioned convolutional autoencoder and the adversarial loss function of the generator. The generator's adversarial loss function is calculated as follows:

$$\varphi = \nu\eta + (1 - \nu)\big[-\log K(z)\big] \tag{6}$$

In formula (6), φ represents the adversarial loss function; ν represents the adjustment parameter, which is set to 0.999 in this paper; K represents the discriminator. The loss function of the convolutional autoencoder is not fed back to the encoder, because the decoder is responsible for image synthesis, and the encoder is mainly responsible for extracting and compressing image features. According to the size of the largest CU block, take the center point of the target area as the fixed point, adjust the coordinates of the edge point of the target area, that is, expand the detected target area to the nearest

64 times the pixel boundary. The obtained region is used as the region of interest for subsequent video coding, and the other regions are used as non-interested regions.

The CNN encoder employs a series of down-sampled convolutions and ReLU non-linear maps to extract and fuse global information. We double the number of feature channels when performing each downsampling convolution, preventing a rapid reduction in the number of features. At the top of the CNN encoder, we adopt the Dropout technique with a dropout rate of 0.5 to eliminate and weaken the joint fitness between neurons and enhance the generalization ability of the network.

2.4 Building a High-Efficiency Compression Coding Model for Video Data

After a frame in the video is input to the encoder, the motion vector, division mode, and prediction information such as the residual of each encoded pixel block are obtained through the intra-frame and inter-frame prediction modules, and the residual is subjected to frequency domain transformation and quantization through the transformation/quantization module, and then through entropy coding, the information is synthesized as a binary bit stream.

At the encoding end, after prediction, transformation and quantization, filtering and other modules, a reconstructed image with certain distortion is formed for subsequent images to rely on. In the process of video encoding, it is necessary to control the encoding rate according to the actual situation. In order to make the actual bit rate after video encoding within the set target bit rate range, various encoding parameters of the encoder need to be adjusted during encoding set up. During video encoding, the encoding parameters can be adjusted to ensure that the actual bit rate of the encoding is close to the set target bit rate, so as to achieve the purpose of bit rate control. The principle of rate control is shown in Fig. 1.

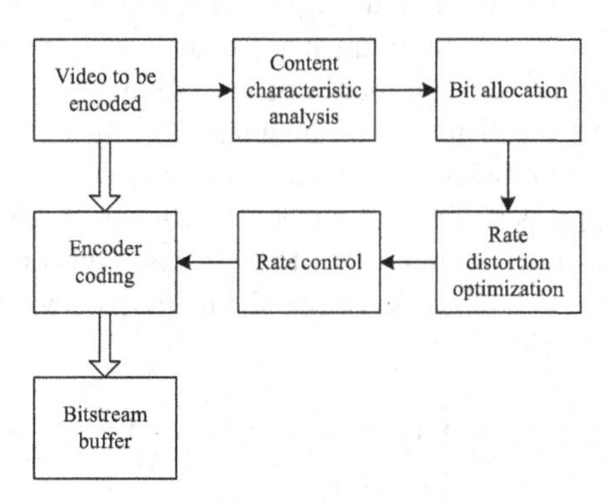

Fig. 1. The principle of rate control

When the code rate changes, the current CTU needs to be observed, and then the quantization parameters used are determined. The CTU will be sent to the encoder in the coding order, that is, from left to right and top to bottom in the current frame. After the prediction residual transformation, most of the energy is concentrated in the low-frequency coefficients in the upper left corner of the matrix, and more detailed information in the image will be scattered in the high-frequency region [10].

Since the luminance component consumes the highest bit rate, we extract the luminance component of the CTU to represent its spatial information. Then, we merge the luminance component and importance map depthwise. Considering that the human eye is not very sensitive to the distortion of high-frequency signals, this section proposes a frequency coefficient suppression method. The areas with high visual sensitivity of the human eye are suppressed by the frequency coefficients of lower intensity.

The bit rate and distortion we use are relative values encoded with the standard quantization parameters, rather than absolute values. This design can better reflect the actual situation through the value of the reward. The frequency coefficient matrix can be expressed as:

$$P = \vartheta \cos\left[\frac{\vartheta\pi(2c+1)}{8}\right]R \tag{7}$$

In formula (7), P represents the transformed frequency coefficient matrix; ϑ represents the compensation coefficient; c represents the encoding transformation size; R represents the residual signal matrix to be transformed. HEVC uses floating-point coefficients to multiply larger values when transforming integer prediction residuals to make the transform results more accurate. Specifically, the floating-point coefficient can be positive or negative. If it takes a negative value, it means that the advantage of reducing the bit rate used is not enough to cover the negative impact of the resulting distortion; if it is a positive value, it may be used the current quantization parameter does not produce additional semantic distortion, or the semantic distortion value is small and can be ignored.

For each CTU to be encoded, according to its texture-aware weight value, calculate the QP parameter down-regulation value DQP. The larger the DQP and the larger the quantization step size, the more low-frequency signals are eliminated in the coefficient matrix, the less accurate the residual matrix of inverse quantization and inverse transformation, and the greater the distortion of the image. For the non-interesting area, each CTU selects a frequency coefficient suppression matrix according to its texture perception weight value to suppress its DCT frequency coefficients to different degrees. With quantization suppression, the residual matrix to be transmitted contains very few significant values and a large number of zeros. The candidate frequency coefficient suppression matrix group is calculated as follows:

$$f = \begin{cases} 1, \left(\omega + \sigma \le \frac{hc}{4} + b\right) \\ 0, \mathit{otherwise} \end{cases} \tag{8}$$

In formula (8), f represents the candidate frequency coefficient suppression matrix; ω and σ are the abscissa and vertical coordinates of the matrix elements respectively; h represents the index of the suppression matrix, which is 1, 2, and 4, and the suppression intensity increases in turn; b is the partial shift. High-intensity pressing for randomly textured areas, medium-intensity pressing for flat areas, and lower-intensity pressing for structured textured areas. High-frequency signals, DC component values are also mapped into smaller intervals, so that the number of bits required to express the coefficient matrix is drastically reduced.

Combining the above processes, the design of an efficient compression and coding method for multimedia video data based on CNN is completed. The design steps are shown in Fig. 2.

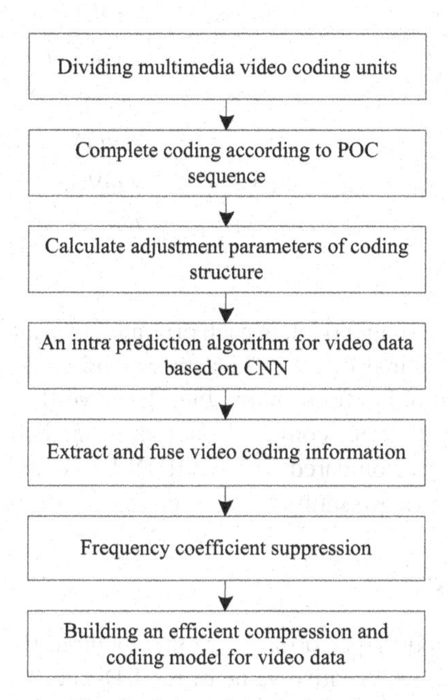

Fig. 2. Steps of method design

3 Experimental Study

3.1 Experimental Environment and Configuration

The following experiments are designed to verify the performance of the CNN-based efficient compression coding method for multimedia video data.

In this experiment, the HEVC encoding software is X265_1.8, the development environment is VisualStudio2012, and the processor of the test platform is Intel Core i5-2520 with a main frequency of 2.5 GHz.

The configuration of the X265 encoder is as follows: frame rate 30fps, IPP mode, I-frame interval is 100, and DCT coefficients are suppressed by odd-numbered frames. The specific configuration of LD is the LDP encoding method, and the quantization parameters are set to 22, 27, 32, 37 and 42 respectively, and the above parameters are used for compression encoding.

The experiments build a dataset containing 90 uncompressed video sequences from the derf and SJTU datasets. For the test set, 18 class A-E sequences with different resolutions proposed by the Joint Video Coding Collaborative Group are used in the experiments. The first 100 frames of each sequence were selected for statistical analysis. The test sequence is shown in Table 1.

Table 1. Test sequence

Serial number	Sequence	Video
1	Class A	2560 × 1600
2	Class B	1080p
3	Class C	WVGA
4	Class D	WQVGA
5	Class E	720p

When training this encoding model, in each original video and its compressed video, we randomly select the original frame and its corresponding decoded target frame and adjacent frames to form training frame pairs. In order to verify the performance advantages of the CNN-based efficient compression coding method for multimedia video data, the analysis results are compared with the RNN-based and LSTM-based efficient compression coding methods for multimedia video data.

3.2 Results and Analysis

In order to measure the coding performance of this method, BD-rate index is used for measurement and analysis. A negative value of the BD-rate index indicates the saving degree of the code rate, that is, the coding compression rate. The experimental results of the A-E test sequences are shown in Tables 2, 3, 4, 5 and 6.

Table 2. BD-rate of Class A (%)

Testing frequency	Compression coding method based on CNN	Compression coding method based on RNN	Compression coding method based on LSTM
1	4.28	0.97	1.38
2	4.26	0.95	1.32
3	4.29	0.98	1.36
4	4.31	0.99	1.35
5	4.30	0.97	1.37
6	4.34	0.96	1.33
7	4.29	0.98	1.31
8	4.32	0.96	1.35
9	4.35	0.97	1.36
10	4.28	0.92	1.41

In the Class A sequence test, the maximum value of BD-rate index can reach 4.35% after applying the method in this paper, while the maximum value of BD-rate index is 0.99% and 1.41% after applying the video coding methods based on RNN and LSTM respectively. It can be seen that the compression effect of this method in Class A sequence is better.

Table 3. BD-rate of Class B (%)

Testing frequency	Compression coding method based on CNN	Compression coding method based on RNN	Compression coding method based on LSTM
1	3.05	1.42	0.92
2	3.08	1.15	0.94
3	3.09	1.48	0.96
4	3.07	1.46	0.94
5	3.05	1.43	0.95
6	3.11	1.36	0.98
7	3.09	1.38	0.97
8	3.10	1.42	0.99
9	3.15	1.37	1.02
10	3.03	1.29	0.98

In the Class B sequence test, the maximum value of BD-rate index can reach 3.15% after applying the method in this paper, while the maximum value of BD-rate index is 1.48% and 1.02% after applying the video coding methods based on RNN and LSTM respectively. It can be seen that the compression effect of this method in Class B sequence is better.

In the Class C sequence test, the maximum value of BD-rate index can reach 2.92% after applying the method in this paper, while the maximum value of BD-rate index is 1.41% and 1.32% after applying the video coding methods based on RNN and LSTM respectively. It can be seen that the compression effect of this method in Class C sequence is better.

In the Class D sequence test, the maximum value of BD-rate index can reach 1.52% after applying the method in this paper, while the maximum value of BD-rate index is 1.23% and 1.17% after applying the video coding methods based on RNN and LSTM respectively. It can be seen that the compression effect of this method in Class D sequence is better.

In the Class E sequence test, the maximum value of BD-rate index can reach 4.47% after applying the method in this paper, while the maximum value of BD-rate index is 1.26% and 1.20% after applying the video coding methods based on RNN and LSTM respectively. It can be seen that the compression effect of this method in Class E sequence is better.

Table 4. BD-rate of Class C (%)

Testing frequency	Compression coding method based on CNN	Compression coding method based on RNN	Compression coding method based on LSTM
1	2.84	1.28	0.96
2	2.86	1.26	0.98
3	2.89	1.23	1.02
4	2.88	1.25	1.04
5	2.85	1.27	0.99
6	2.91	1.31	1.03
7	2.92	1.32	0.99
8	2.87	1.29	1.02
9	2.67	1.41	1.32
10	2.54	1.35	1.32

Table 5. BD-rate of Class D (%)

Testing frequency	Compression coding method based on CNN	Compression coding method based on RNN	Compression coding method based on LSTM
1	2.47	1.19	0.95
2	2.48	1.18	1.04
3	2.49	1.21	0.96
4	2.46	1.22	0.98
5	2.48	1.19	1.03
6	2.51	1.23	1.05
7	2.52	1.22	1.04
8	2.49	1.18	0.99
9	2.46	1.17	1.03
10	2.45	1.21	1.17

The compression coding method designed in this paper significantly improves the coding performance on all test sequences. Compared with the RNN-based and LSTM-based coding methods, the CNN-based efficient compression coding method for multimedia video data has achieved obvious BD-rate gains. These experimental results can confirm that the efficient compression coding method of multimedia video data based on CNN has good performance for test sequences with different contents and different resolutions.

Table 6. BD-rate of Class E (%)

Testing frequency	Compression coding method based on CNN	Compression coding method based on RNN	Compression coding method based on LSTM
1	4.41	1.15	1.13
2	4.43	1.18	1.18
3	4.45	1.22	1.05
4	4.43	1.25	1.07
5	4.46	1.24	1.09
6	4.42	1.19	1.08
7	4.47	1.18	1.12
8	4.45	1.26	1.14
9	4.47	1.15	1.20
10	4.40	1.23	1.18

4 Conclusion

With the continuous development of information multimedia technology, high-definition video and ultra-clear video are gradually popularized in people's lives. In order to relieve the pressure of storage and network transmission, the importance of video coding technology is getting higher and higher.

This paper proposes a high-efficiency compression coding method for multimedia video data based on CNN. Intra-frame prediction is performed by means of a convolutional auto-encoder, which can effectively reduce the prediction residual and improve the coding rate-distortion performance. This paper only designs the intra-frame coding mode, if it can be extended to the inter-frame mode, it will be more valuable. Compared with the intra-frame mode, the inter-frame mode needs to additionally consider the influence of the reference frame, and the lower-quality reference frame will propagate the distortion, so a more reasonable bit allocation scheme needs to be studied.

Fund Project. Higher Education Teaching Reform Research Project of Jilin Province: Construction and Practice of "Online and Offline" Hybrid Teaching Mode for Film and Television Arts Majors in Universities under MOOC Environment (20213F2ENY4001J).

Education Science "Fourteen Five-Year" Project of Jilin Province: Research on the Construction of College of Modern Industry for Film and Television Media Major (ZD21088).

China Association of Private Education 2022 Annual Planning Project (School Development): Research on the Construction of College of Modern Industry for Film and Television Media Major (CANFZG22274).

References

1. Guo, J., Cao, L., Zhu, F.: Progressive compression method of real-time data under load balancing strategy. Comput. Simul. **38**(3), 365–368,429 (2021)

2. Wu, Y., Peng, Y., Lu, A.: Research on image compression coding technology based on deep learning CNN. Comput. Eng. Softw. **41**(12), 18–23 (v)
3. Wang, H., Ma, J., Qu, J.: Design and implementation of full HD video real-time compression coding and storage system. Chin. J. Electron Devices **44**(03), 513–518 (2021)
4. Wang, G., Jin, Y., Peng, H., et al.: Error correction of Lempel-Ziv-Welch compressed data. J. Electron. Inf. Technol. **42**(6), 1436–1443 (2020)
5. Pan, P., Yao, Y., Wang, H.: Detection of double compression for HEVC videos with the same coding parameters. J. Image Graph. **25**(5), 879–889 (2020)
6. Gu, H.: Data compression coding technologies for computer-generated holographic three-dimensional display. Infrared Laser Eng. **47**(06), 42–47 (2018)
7. Wang, T., He, X., Sun, W., et al.: Improved HEVC intra coding compression algorithm combined with convolutional neural network. J. Terahertz Sci. Electron. Inf. Technol. **18**(2), 291–297 (2020)
8. Jiang, W., Fu, Z., Peng, J., et al.: 4 Bit-based gradient compression method for distributed deep learning system. Comput. Sci. **47**(7), 220–226 (2020)
9. Yi, Y., Feng, G.: A video zero-watermarking algorithm against recompression coding for 3D-HEVC. J. Sig. Process. **36**(05), 778–786 (2020)
10. Li, F., Zhan, B., Xin, L., et al.: Target recognition technology based on a new joint sensing matrix for compressed learning. Acta Electron. Sin. **49**(11), 2108–2116 (2021)

Intelligent Push Method of News and Information for Network Users Based on Big Data

Ting Chen[1](✉) and Zihui Jin[2]

[1] Chengdu College of University of Electronic Science and Technology of China, Chengdu 611731, China
chenting08282022@163.com
[2] Tianfu College of SWUFE, Chengdu 621050, China

Abstract. In view of the poor management effect of massive news and information, this paper proposes an intelligent push method of news and information for network users based on big data. Based on big data technology to identify network users news interest preferences, build network user news information classification recommendation algorithm, simplify the network user news information intelligent push steps, the experimental results prove that this paper design method in network user news information push has high practicability, can fully meet the research requirements, has certain application value.

Keywords: Big data · Network users · News and information · Intelligent push

1 Introduction

Online users 'news and information facilitate people's daily life to the greatest extent. Since the intelligent push of news and information has been put forward, it has attracted wide attention. Users want to get the information from their urgent needs in real time, and different users also have different needs for different products, so the needs for information becomes more personalized. However, the current network users' news and information is characterized by complex structure, dynamic change and scattered distribution, which makes information overload and information fan become the key problems that hinder the efficiency of digital Internet [1]. How to quickly and comprehensively get the information needed by users from a large amount of data, improve the ability of news active information service and meet the personalized needs of users has always been a hot topic of information resource experts. The main purpose of information push is to solve the intelligent information push technology proposed by information overload. By reflecting users' interest and preference information, and through personalized recommendation calculation, it can provide different users with different recommendation information. With the deepening of the research and application of personalized information push, this technology has been applied to various industries. For example,

W. Fu and L. Yun (Eds.): ADHIP 2022, LNICST 468, pp. 137–150, 2023.
https://doi.org/10.1007/978-3-031-28787-9_11

Hu Yue and others designed an intelligent push technology for continuing education information based on deep neural network [2]; Huang Weihua and others designed an accurate marketing push system based on intelligent analysis of user characteristic information [3], but the above push method only evaluates user push information by testing user browsing history, which has obvious drawbacks, because when users browse computers, it will cause a large number of browsing history to be biased. If browsing history is used to evaluate push content, it will lead to inaccurate push information, Or the push information data is too complicated to be pushed, resulting in a delay in pushing. In order to solve the above problems, this paper designs an intelligent push method of news information for network users based on big data. A user interest model is constructed to accurately determine the user preference weight and label it. At the same time, depending on the final push mathematical model, the attenuation of user interest can also be observed, so as to better realize personalized push work and give play to the advantages of high push efficiency.

2 Intelligent Push Method of News and Information for Network Users

2.1 Screening of News and Interest Information of Network Users

Traditional media news client relying on the maternal information dissemination, in its specific interest screening process is relatively simplified, "innate genes" and "early development" are good, late interest screening is relatively easy its interest screening process can not well reflect the mobile news client interest screening, so the following research for aggregate news client. And both belong to the commercial website news client, in some interest screening ideas are much the same. In this study, for research convenience, it is divided into internal interest screening and external interest screening; the internal interest screening of mobile news client focuses on the interest screening of the editorial department in the production process of news client, from the detection of news sources, to the later evaluation. According to the mobile news client news interest screening process of time order, the internal interest screening is divided into early interest screening and late interest screening, early interest screening is before the news information push news interest screening, late interest screening refers to the news after the audience interest screening process, late interest screening is obviously different from the traditional media news interest screening process [4]. The flow chart of the internal interest screening and integration of China's mobile news client, as shown in Fig. 1:

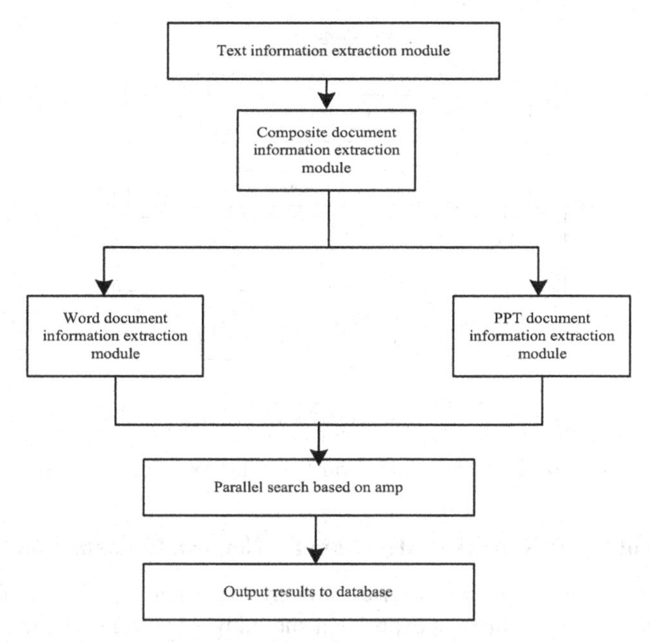

Fig. 1. News dissemination information screening step

In an information system, if the user's evaluation of the information is obtained directly from the user, the feedback will be described as explicit feedback. Although explicit feedback is easy to achieve, in most applications, the user is required to explicitly evaluate the relevance of all documents, which may lead to user disgust, limited learning of the user interest model, and reduced the availability of the whole system [5]. Although it will lead to indirect correlation between feedback and users' usability evaluation of all documents, it has more potential than explicit feedback in supporting user interest model processing, because it is easy to collect and will not affect users' normal lives. The collected temporal behavior preference characteristics are sent to the server and analyzed simply through the data parallel structure. When data is used as algorithm input, the collected data needs to be preprocessed first. The main purpose of preprocessing is to extract algorithm related data from a large number of data and convert it into the required format [6]. The main flow of data preprocessing is shown in Fig. 2.

(1) Extract the relevant site data, and analyze the user behavior data of a certain site;
(2) Filter useless data items, collect the data sent to the server in a specific log format, and be separated by "/ h" characters;
(3) Confirm the recommended range, and filter according to the specific site URL naming rules;
(4) Extract the content page.

After data preprocessing, the parallel characteristics of diversity key data are further analyzed, and big data is used to study the recommendation algorithm.

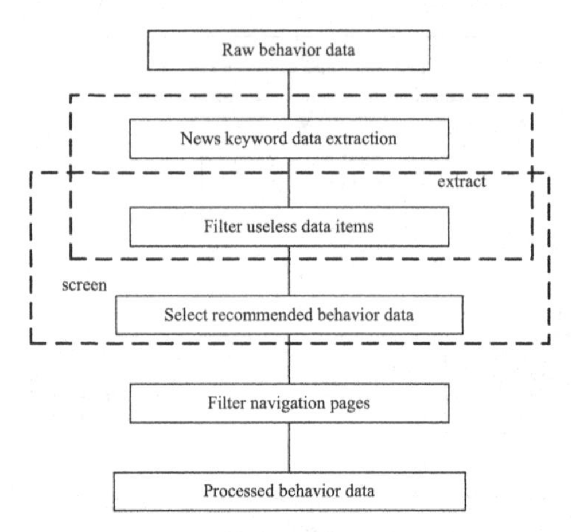

Fig. 2. Main process of news and information data processing of network users

2.2 Intelligent Push Evaluation Algorithm for News and Information

User interest is obtained by using a series of operations such as observing the news of network users, so the user interest expression should be consistent with the expression of network user news, that is, it can express users' interest in certain fields through the vector.When a user has multiple fields of interest, then the user's interest model should be a set of vectors composed of multiple vectors [7]. Set users have N areas of interest, each field of interest with m keywords, then you can use a m dimensional vector V to describe a field of interest, so, the user's interest model can use N vector life to describe t with interest with h network user news with vector, both can use the vector estimation for unified processing. So without the original information about the user's interest, the user's interest can be gradually learned by observing the user's action, so the original set S is empty, when observing the user is interested in a network user news application d, describe vector V by trid and the user's interest in the application R update vector set S. The specific process of user interest model construction is as follows, the original value of vector set S of user interest model is empty for all observed user interested news d for the following processing: application preprocessing, the corresponding analysis of the network user news documents, then with the specific language of document correlation processing, to give them large weights.Estimate the *tid* description vector V of application d. Estimate the user's interest R in application d; Update the *tid* description vector V of application d according to the user's interest R in application d:

$$V_i = Vd - N - htid * R_i \tag{1}$$

A new vector set is formed by the vector in S and the new document vector V, and the similarity between the two random vectors in the new vector set is estimated:

$$\mathrm{sim}\left(V_j, V_k\right) = \frac{V_i}{\left|V_j\right| \times |V_k|} jk \tag{2}$$

Combining the two vectors V with maximum similarity, The time start and completion identification is described by the timestamp, The time interval is mainly used in describing the user's reading time, Then the time ontology weight T (abel) can be described as the ratio of the user's Internet access time and ontology browsing time, Data from the data in the preference analysis were obtained from the data ports in the mobile communication network, In addition to the data in the communication network, User-to-content access logs are also collected in the product operation platform, Taking the access data of the operating platform as the input of two-layer association rule data mining, Therefore, the data mining method, Get the network data of user interest.Let A and B be content item sets, and AT and BT be the types of A and B respectively, then the double-layer association rule set is g:

$$Z = \mathrm{sim}(V_j, V_k)\{A \rightarrow B \rightarrow A \rightarrow BandAT \rightarrow BT\} \tag{3}$$

where $A \rightarrow B$ is a basic content layer association rule, which means that when users access content set A, users will also access content set B with a high probability; $AT \rightarrow BT$ is a content type layer association rule, which means that when users access content type set AT, users will also access content type BT with a high probability. Basic content layer association rules are based on basic content fact table data extraction, and content type layer association rules are based on user access type facts. The behavior collection layer is the most important part of the overall design of the method hardware structure. It is mainly responsible for collecting relevant information between customers and projects and feeding back the recommendation results. For the incentive scoring mechanism designed by the method, it is necessary to actively score the user experience value. The data is relatively sparse, but the corresponding weight is large. Therefore, the following processing should be done for data collection: the corresponding final calculation should be made according to the actual situation, and the evaluation scores should be uniformly processed. Only in this way can the comparison be made in the final calculation, and the recommendation reliability is high. Table 1 shows the main behavior analysis of general users.

Table 1. User behavior analysis

User behavior	Scoring mode	Value size	Remarks
Direct scoring	Active evaluation	0–5	The preference is reflected by the value of the score, and 5 is the favorite
Direct voting	Active evaluation	0–1	No need for excessive user involvement
Direct forwarding	Active evaluation	0–1	Unable to draw a conclusion, the result is inaccurate
Direct comment	Passive evaluation	0–1	Unable to draw a conclusion, the result is inaccurate
Page dwell time	Passive evaluation	–	Not easy to use

The studies on the corresponding implicit scoring of news and information are shown in Table 2.

Table 2. Recommends the information recessive score

Execution name	Score (5-point system)
Search accuracy	3
Simple browsing	1
Browse carefully	3
Favorite page	4

Table 3. Experimental parameter setting

Parameter	Numerical value
node	10
CPU	2
Core frequency	1.9 GHz
Memory	8 GB

After the above behavioral and implicit scoring research, it can be dynamically adjusted through various pages, and jump into the relevant business areas.In the whole dynamic processing stage, large user behavior information will be carried. In order to realize all the above dynamic requests, interceptors should be set up to meet different business needs.The setting of interceptors needs to meet the characteristics of easy to expand. The unified own abstract interceptors mainly have the following three types:

See if the user has their own ID, if not, read the session control ID in the threshold related text file or create the session control ID to the text file and take the user ID as the key to recording all behavioral data.

Since there is a time difference when users query the information, it is necessary to record the dynamic request information and complete the page jump interception.

There are various scoring modes, and the special business logic relationship needs to be intercepted. The timing diagram design for news recommendation is shown in Fig. 3.

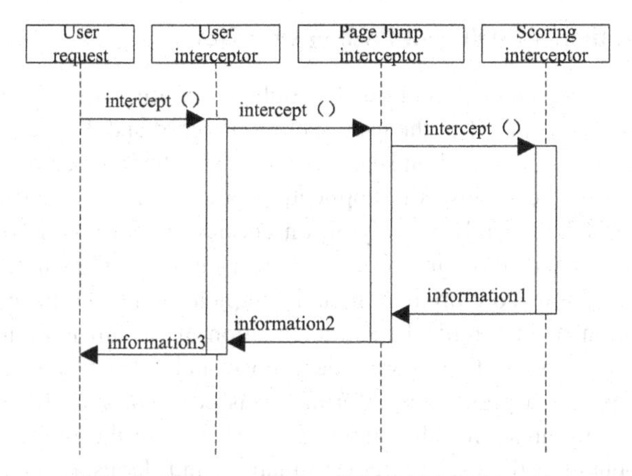

Fig. 3. News recommendation timing diagram

Data parallelism divides the training data into different Windows, and each window has a complete network model, using different data for training.In order to obtain a network model containing all the training information, the different network information needs to be synchronized, as shown in Fig. 4.

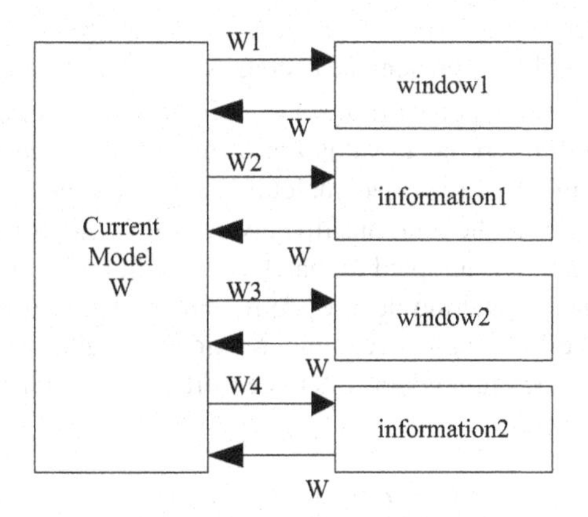

Fig. 4. Data Parallel Structure

After one iteration begins, each window obtains a new network model from the server and is trained. W is transmitted back to the server. The iteration can not stop until all nodes W are updated.If 1/n of size for a single window is used under n windows, then multiple windows are fully equivalent to a single window for training.After iterative processing, if a window data is abnormal, the whole training speed will be slowed down. In order to further improve the data use rate, it needs to be preprocessed.

2.3 Implementation of Intelligent Push of User News

Based on the user interest model, multimedia applications are pushed from two different directions. On the one hand, after the user operation record and the user interest model based on the multimedia application scenario are established, the content that the user may be interested in the multimedia application is searched out, and the content is predicted and pushed vertically by relying on certain branch prediction information score; On the other hand, through the user interest model, similar user interest news applications, application score prediction, and complete horizontal push can be found. The implementation of hierarchical recommendation algorithm is divided into three steps: establishing users, finding nearest neighbors, and calculating recommendation data. The user's personal preference information is counted, and the user preference matrix is obtained by analyzing the similarity according to the multi-level decision-making; After obtaining the user preference matrix, find the user's nearest neighbor; According to the user's preference for the project, set the nearest neighbor set of target user Z as S_{u}, and calculate the user's evaluation result of project B. the specific calculation formula is as follows:

$$P_{\mathrm{z,B}} = P'_z \frac{\sum_{j \in S_{\mathrm{u}}} sim(z, j) \times \left(P'''_{j,B} - P''_j\right)}{\sum_{j \in S_{\mathrm{u}}} sim(z, j)} \tag{4}$$

In the formula: P'_z and P''_j represent the average score of user z and j on item B respectively; $P'''_{j,B}$ is the score of user j on item B; $sim(z, j)$ indicates the similarity between users z and j. Combined with the project evaluation results of users, design the hierarchical recommendation scheme. Before project matching, the project ontology should be established first, so as to describe the relationship between different projects. The multi-level decision analysis algorithm is adopted to match users with the project ontology, and the projects with the highest matching degree are recommended to users.Constructing the project ontology in the hierarchical recommendation of digital news information can realize the correlation recommendation between different digital information.Defines an ontology as a quintuple:

$$Q = P_{\mathrm{z,B}}\left(\mathrm{a}'b', c', d', e'\right) \tag{5}$$

In the formula: a' represents the concept set; b' represents the set of concept instances; c' represents the binary relationship between concepts; d' represents the constraints between examples and concepts; e' represents the inclusion relationship between any two concepts. After building digital news information projects according to the form of five ples, ranking according to similarity, and the projects with high similarity are recommended to the user's intelligent push system into three levels: application layer, processing layer and data layer.The application layer mainly has home page, knowledge scenario, push process interface; the processing layer is the core part of the system, mainly by hybrid push method based on knowledge scenario, is the key formation part of the algorithm (Fig. 5).

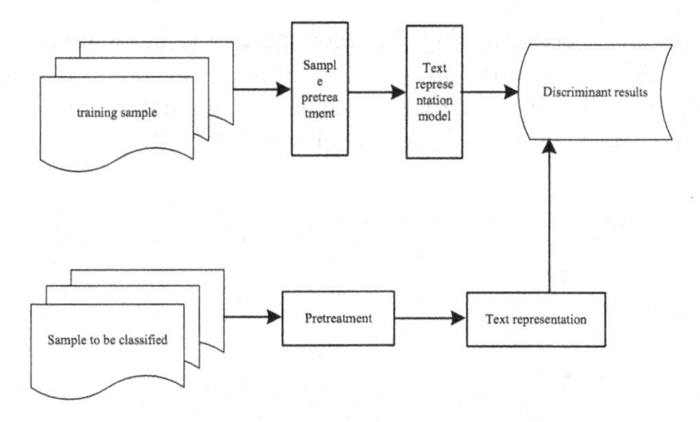

Fig. 5. Mixed push step of news and information

Intelligent recommendation method of network user news and information based on user interest model. From the application level, the intelligent recommendation system here needs to include home page, knowledge situation, push process, resource center, result analysis and other modules, so that users can see the push information when they log in.After the user logs in, the home page shows the most concerned information, push information and user history browsing information. The push information module is the key module of the system. The results are obtained by using the estimation based on the user interest model proposed in this paper.According to the encrypted content, the hierarchical implementation scheme is recommended, as shown in Fig. 6.

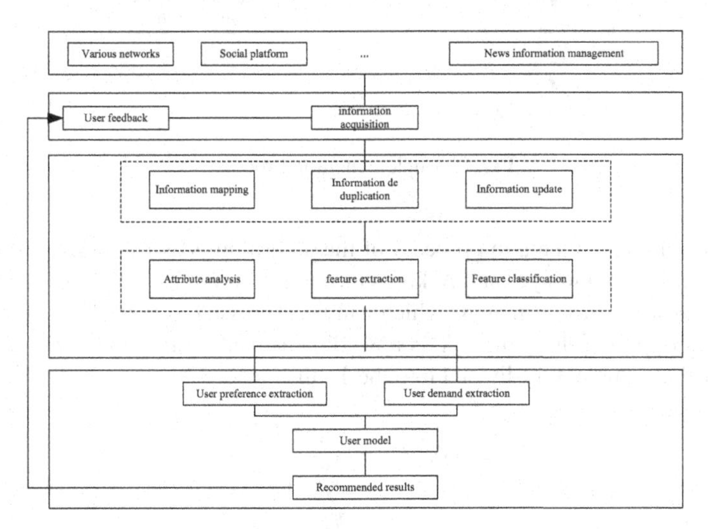

Fig. 6. Recommended implementation scheme of news and information classification

Based on the above preprocessing, the recommendation algorithm is studied under large-scale data distribution conditions, and the recommendation accuracy is low. In order to improve this problem, the recommendation algorithm based on big data is proposed.The implementation process is shown in Fig. 7.

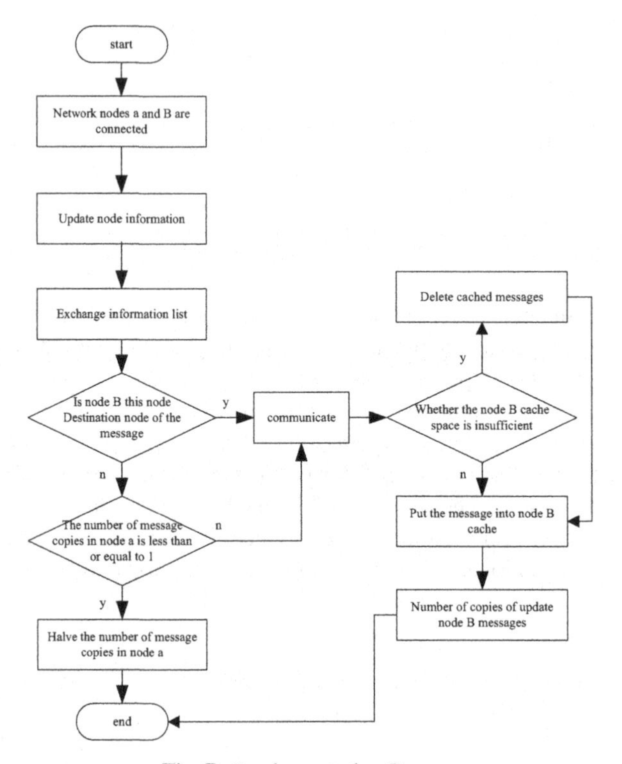

Fig. 7. Implementation Process

The specific implementation process of the algorithm is as follows: assuming that there are K information layer and N kinds of output layers, the weight parameter θ of the output layer is a $A \times B$ matrix, which can be expressed as $\theta \in C^{A \times B}$. The feature obtained after pooling the sample X is a K -dimensional vector, that is $f \in C^{K}$. The probability that sample X is divided into the Y -th category is:

$$P = (\mathrm{Y}|X, C) = \frac{\mathrm{e}^{(c_y \cdot f + q_y)}}{\sum_{h=1}^{N} \mathrm{e}^{(c_y \cdot f + q_h)}} \tag{6}$$

In the formula: q_h represents the h-th offset term of the full connection layer, and the loss function can be obtained by maximizing the likelihood probability:

$$\mathrm{W} = -\sum_{\mathrm{y}}^{R} \log\left(p\left(g_y | x_y, \theta\right)\right) \tag{7}$$

In the formula: R is the training data set, and g_y represents the real data type of the y-th sample. In order to prevent overfitting, the convolutional layer neurons structure should be simplified with a certain probability to ensure that the weights do not work.After feature compression processing, based on the stability of the database storage space, the internal state and behavior control of the data can operate freely.Through the above process, the diversity of key data parallel recommendation schemes based on multi-level decision analysis is realized. The specific implementation process of the digital news information classification and recommendation scheme is: users multi item source into the feature model, store the collected information in the information base through various collection methods, and form a small user database. The core of the user model is established, and the information in the information base is integrated by using the information fusion method to extract the user's needs and preferences. The established project ontology is matched with the demand information, and the recommendation results are obtained, so as to realize the hierarchical recommendation of digital news information.

3 Analysis of the Experimental Results

The experiment mainly evaluated the influence of parameters and dimensions on the analysis of personalized recommendation model, randomly divided the data set with 80% as training set and the remaining 20% as test set, 256 GN memory CPU and 5 NVIDIA Tesla K40 as CPU, with a single precision peak of 4.25 Tflops, display memory of 12 GB and bandwidth of 280 Gbytes/s. Based on the past experimental experience, the value range of the parameter values is selected first, and these values may make the algorithm achieve good results, and then an optimal value is determined for the experiment.The experimental environment settings are shown in the table (Table 3).

Different from calculating the fixed number of nodes, the algorithm of this paper needs to calculate the maximum value of adjacent nodes. All nodes within a certain range are considered as current neighbors. Although the number of neighbors is uncertain, the similarity will not vary greatly, and it is appropriate to handle isolated nodes.The traditional method adopts the display scoring method to make news recommendation for users, while the dual recommendation method adopts the implicit scoring method to make news recommendation for users.The two recommended methods compare the accuracy of the recommended results under the influence of noise interference and human factors, respectively. The specific comparison is as follows.Comparing the traditional recommendation method with the recommended method in this paper, we find a more suitable calculation method for the similarity of news projects in scenic spots, and divide the data set for 20%, 40%, 60%, 80% and 90%, respectively. The data MAE error calculation results using different algorithms are shown in Table 4.

Table 4. Information recommendation error calculation results

Proportion of information	Traditional recommendation method	Recommended methods in this paper
20%	0.12	0.01
40%	0.11	0.02
60%	0.16	0.04
80%	0.16	0.05
90%	0.13	0.06

Based on the algorithm errors collected in the table, the calculation results for different similar algorithms largely depend on the sparsity of the data, and the error results of different methods under 5 experiments are shown in Fig. 8.

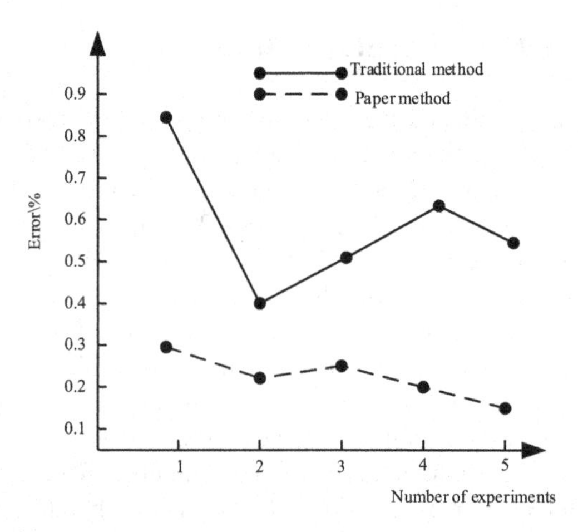

Fig. 8. The two methods compare the results using different algorithms

When traditional news intelligent recommendation algorithms extract news data, they are easy to be affected by noise interference. The final extracted data can not be counted completely, and the error is large, up to 0.9%. It is difficult to recommend effective information. The news intelligent recommendation algorithm designed in this paper is not affected by noise interference when extracting news data, and the final extracted data error is small, up to 0.32%, which proves that the recommended information is more accurate. To further prove the effectiveness of the two methods, the news recommendation accuracy of the two methods under noise interference is compared again, and the results are shown in Fig. 9.

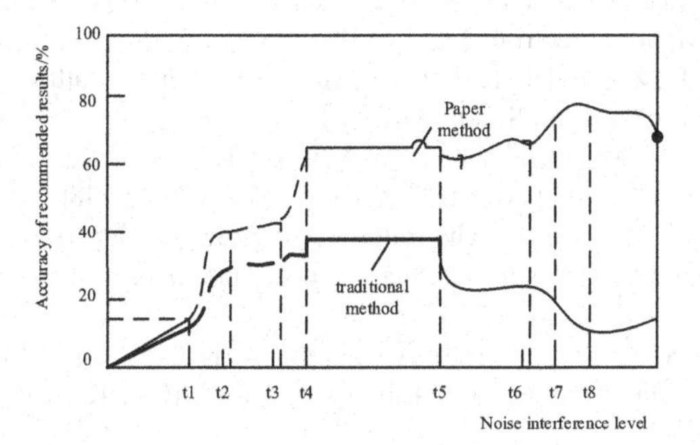

Fig. 9. The two methods recommend accuracy comparison under the influence of noise interference

It can be seen from Fig. 9 that the traditional recommendation method is affected by human factors, and obtains accurate recommendation information, resulting in vague types of news recommendation and low accuracy, with a maximum of only 38%. However, the method designed in this paper can obtain accurate recommendation information through implicit scoring, and uses a dual recommendation mechanism to improve the accuracy of recommendation results, so as to recommend recommendation results under the same environmental background. The recommendation accuracy is as high as 80%, The validity of the design method is proved again.

To sum up, the big data based intelligent push method of news information for network users designed in this paper has high push accuracy and high application value in the field of intelligent push of news information for network users.

4 Conclusion

In order to help users find valuable information more effectively and improve the accuracy of news information recommendation, this paper proposes an intelligent push method of news information for network users based on big data. This method builds user interest model according to different emphases of users, and provides users with the news information they need. Experiments show that the design method is influenced by noise interference, so that all the data are counted, the extraction effect is good, and the recommended results are more accurate. However, the recommendation time is not analyzed this time, which can not guarantee the recommendation efficiency of the design method. In the next study, further analysis is needed.

References

1. Peng-fei, C., Jie, L., Jun, Y.: Research on construction of real-time push service in cloud environment. Comput. Technol. Dev. **30**(03), 204–208 (2020)

2. Yue, H., Xiaonan, L., Bin, W., et al.: Research on intelligent push technology of continuing education information based on deep neural network. Electron. Des. Eng. **29**(14), 42–46 (2021)
3. Weihua, H., Xin, F.: Precise marketing pushing system based on intelligent analysis of user feature information. Modern Electron. Tech. **44**(06), 43–46 (2021)
4. Jing, L., Si-yu, F., Ya-fu, Z.: Model predictive control of the fuel cell cathode system based on state quantity estimation. Comput. Simul. **37**(07), 119–122 (2020)
5. Yingchun, Y.U., Yanjie, W.A.N.G., Xincheng, W.A.N.: Realization of accurate intelligent early-warning push technology based on WeChat. Meteorol. Sci. Technol. **48**(02), 195–199 (2020)
6. Xinnan, H., Bindong, S., Tinglin, Z.: The influence of geographical distance on the dissemination of internet information in the internet society. Acta Geogr. Sinica **75**(04), 722–735 (2020)
7. Wenhong, Z., Yiwen, S., Linxu, D., et al.: A study of the progress and development of web archiving for digital memory preservation. Library Tribune **40**(09), 42–52 (2020)

Sports Simulation Training System Based on Virtual Reality Technology

Xiang Gao(✉) and XiaoSha Sun

Sanmenxia Polytechnic, Sanmenxia 472000, China
gx10298867@163.com

Abstract. Apply virtual reality technology to athlete simulation training, arrangement and innovation of movements, improving the level of sports training technology is conducive to the athletes' rapid mastery of sports technology and can maintain the competitive level. In order to guide sports more scientifically, on this basis, a new motion simulation training system based on virtual reality technology is proposed, improves the function and operation process of system software, constructs the management system of sports simulation training, and the scientific evaluation index of training. Finally, it is confirmed by experiments, The sports simulation training system based on virtual reality technology has high practicability in the process of practical application, and fully meets the research requirements.

Keywords: Virtual reality · Athletic sports · Simulation training · Training evaluation

1 Introduction

Through the interaction of vision, hearing and touch, virtual reality can completely immerse people in the virtual world [1]. Virtual reality technology will be widely used, and it will change people's lives [2]. Using virtual reality technology, it reproduces the coach's teaching experience, the coach's training intention, the coach's organizational plan, and the whole process of athlete training, so as to achieve an experimental technology for the interpretation, analysis, prediction and organizational evaluation of the sports system. In recent years, the system simulation has become a research hotspot [3].

Reference [4] using virtual reality technology to design badminton auxiliary training system, and Kinect The system uses virtual reality technology to carry out virtual modeling of the training ground through 3dsmax tools, and renders the scene of the training ground with cry engine technology Kinect is used to recognize these actions and the user's position changes, and transmit the recognition results to the VR glasses through the Bluetooth module, so that the virtual environment in the VR glasses can make corresponding changes in images, sounds and so on. Reference [5] introduced a virtual training system for upper limb coordination function based on occupational therapy. The system uses a high-precision 5DT data glove and an electromagnetic position

W. Fu and L. Yun (Eds.): ADHIP 2022, LNICST 468, pp. 151–164, 2023.
https://doi.org/10.1007/978-3-031-28787-9_12

tracking system to collect finger bending data and arm motion data respectively. Through socket communication, the system controls the bending and movement of the virtual hand in the game, and realizes the operation training of "hand opening grasping" and "two fingertips pinching" The application of virtual reality technology in occupational therapy has greatly improved the safety of occupational training, increased the enthusiasm of patients to participate in training, and achieved remarkable results in rehabilitation training. Compared with object-oriented simulation, qualitative simulation, distributed interactive simulation, visual simulation and multimedia simulation, virtual reality simulation focuses on multiple perception capabilities such as multiple perception, interaction and immersion, while many sports training only requires the participation of multiple senses of athletes. With the continuous development of virtual reality technology, virtual reality technology will be more and more used in sports training. Based on virtual reality technology, this paper develops a set of motion simulation training system. It optimizes the hardware configuration of the system from three aspects: input system, output system and virtual environment generator. The image processing algorithm based on VR technology obtains sports training information, optimizes the sports simulation training process, and realizes the system software function and operation process.

2 Sports Simulation Training System

The characteristics of VR based sports simulation system are strong immersion, economy and convenience. The VR based sports simulation system consists of input system, output system and virtual environment generator, as shown in Fig. 1.

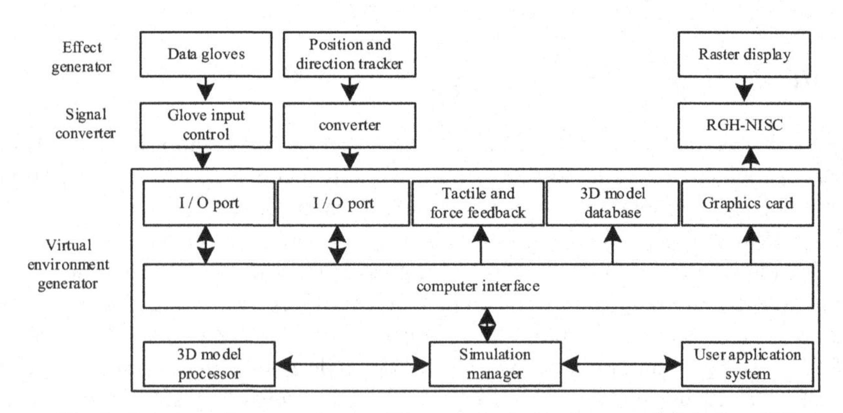

Fig. 1. Structure diagram of competitive sports simulation system based on VR

The input system includes data glove, position tracker, converter, glove input conversion device, etc. The virtual environment generator includes user system, simulation manager, 3D processor, high-performance computer, graphics card, 3D model database, haptic force and other feedback devices, and input/output interface [6]. The output system includes a signal converter and an effect generator (a raster display). The user can experience more realistic stereo and stereo, and can also interact with the surrounding virtual scene naturally, so that the user can be completely in the virtual environment, as if in it.

2.1 Hardware Structure of Sports Simulation Training System

The system uses embedded Linux kernel to process moving images in real time, load them into the process of control information, and transmit them to the root file system of Linux through CAN. In the image processing part, through data processing, image analysis, image capture and other functions, the corresponding script and server configuration are written. In the output part, the virtual reality technology is used to analyze the images and images in the process of sports training, and download them to the development board to complete the analysis and software driving of sports training. Based on the above analysis and description, the overall architecture of the motion image analysis system designed in this paper is shown in Fig. 2.

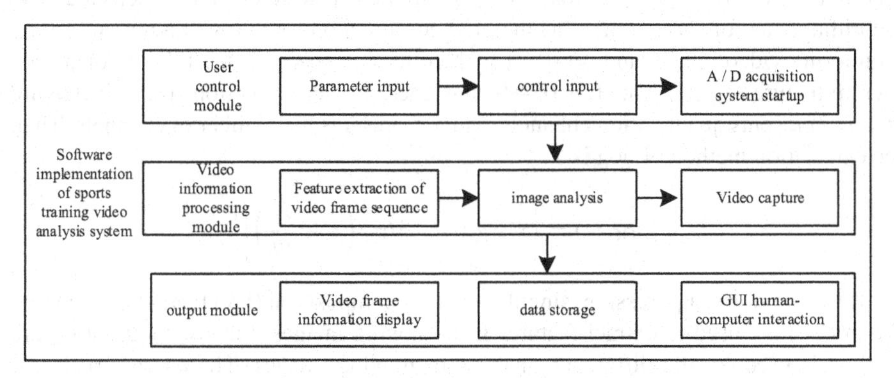

Fig. 2. Overall hardware structure model of the system

The power supply is the main condition for whether the sensor node can work or not, and it is powered by the battery. The data collection module is mainly composed of two sub modules: A/D converter and sensor [7]. The main purpose of the processing unit is to control the operation of sensor nodes, store and process the collected data and other node data. The main goal of wireless communication system is to communicate with other nodes. The structure of a sensor based on virtual reality is shown in Fig. 3.

Virtual reality sensor network software mainly includes application program, operating system, data collection program, bottom driver, etc. the figure shows the software structure of virtual reality sensor node. Among them, network communication protocol belongs to the main part of virtual reality sensor network software structure, which has a direct impact on network performance.

2.2 Software Function of Sports Simulation Training System

This paper introduces an image processing method based on VR technology, which mainly consists of moving image detection and feature extraction [8]. Moving image

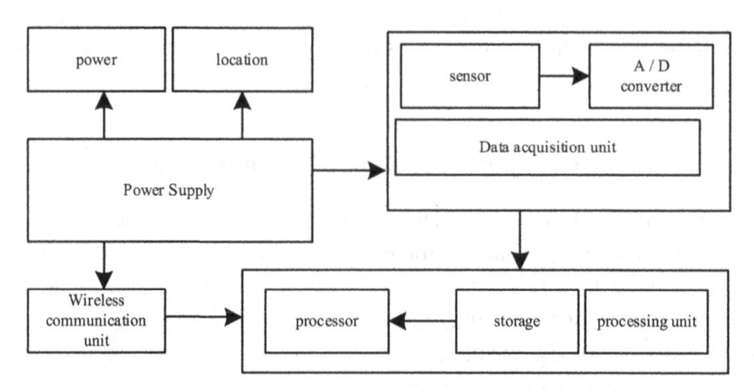

Fig. 3. Hardware structure of virtual reality sensor nod

feature extraction based on gradient histogram [9]. This paper uses solution of EXHOG feature algorithm the problem that the same gradient feature cannot be detected. The algorithm is as follows set M_{HG} as the gradient direction of sports training and sports monitoring video image, and set 0–60 as the direction space; Extract the gradient histogram feature $M_{\mathrm{HG}}(i)$, which is 0°–80° unsigned gradient direction space. Its feature is θ, L represents quantization channels, and S represents the number of channels. Then the calculation method of hog is:

$$M_{\mathrm{HOG}}\,(i) = \left| SM_{\mathrm{HG}}(i) + M_{\mathrm{HG}}\left(i + \frac{L}{2\theta}\right)\right| \tag{1}$$

Based on noise suppression, singular value decomposition (SVD) and phase convolution (PC) are used to extract features from moving images. Finally, the least square method is used for the difference analysis of moving images. The characteristics of machine vision at pixel level are as follows:

$$k_{\mathrm{m}}' = X_{\mathrm{w}} + A_1 - \frac{p_{\mathrm{w}}' - X_n' \sum \rho' + \left(3X_w' - 1\right)}{\left[\frac{X_m'}{\lambda_m}\left(\frac{x_n}{\lambda_n} + \frac{x_a}{\lambda_a} + \frac{x_b}{\lambda_b}\right) - x_m\right]} - M_{\mathrm{HOG}}\,(i) \tag{2}$$

For sit ups, there are left and right bone nodes. The human skeleton nodes of the athlete are left shoulder joint point λ_m, right shoulder joint point λ_n, left elbow joint point λ_a, right elbow joint point λ_b, left hand joint point x_n, right elbow joint point x_a, right knee joint point x_b, left foot joint point x_m, right foot joint point X_m', hip joint point p_{w}' and hip joint point X_w'. The setting α of this system is the angle between EHC and ESL, β is the angle between EHR and ES, γ is the angle between K7L and KFL, and δ is the angle between KTR and KF. The feature matrix is extracted according to the actual motion scene. As shown in the formula, the period of sit ups is represented by T, and the time to complete one sit ups is represented by r.

$$S_T = \left[\alpha_1, \beta_1, \gamma_1, \delta_1\right] \tag{3}$$

The movement is in the key state, so the score ratio of supine state and sitting state is b_1: b_2. Taking a complete sit up exercise cycle as an example, the full score of sit up exercise cycle is k_i, and in practice, k is 100. Suppose the tester completes n sit ups, in which the similarity of supine state is g_i and the similarity of sitting state is m. The final score is obtained according to the formula:

$$S = \frac{S_T m \sum_{i=1}^{n} (b_1 g_i + b_2 k_i)}{4k'_{\mathrm{m}} n^2} \tag{4}$$

Before extracting bone features from athletes, it is necessary to ensure the accuracy of bone data. Extracting abnormal data will cause position deviation of bone data and directly affect the effect of motion recognition [10]. The most likely data abnormality in this system is that there may be interference caused by some accidental factors when collecting data in virtual reality, resulting in errors in bone node data. Therefore, it is necessary to filter the collected data. Because the system needs to identify different movements, the filtering algorithm selected is required to meet the stability of the test data and the high accuracy of the position perception of bone nodes, so the system finally selects the amplitude limiting filtering algorithm to eliminate the jitter of bone node data [11]. In the amplitude limiting filtering algorithm, there are two important attributes, namely, bone jitter radius threshold and bone deviation radius threshold. (1) Bone jitter radius threshold: when the jitter of bone node data is greater than the bone jitter radius threshold, it can be judged that the bone confirmation degree of this point is low, so the bone node must be corrected. (2) Bone deviation radius threshold: when the bone node data movement is greater than the bone deviation radius threshold, but less than the wide value of bone jitter radius, it indicates that it is caused by the new position movement of bone node position data, so as to eliminate the possibility of the influence of bone data node jitter. The system requires the bone node data to meet the bone jitter radius and bone deviation radius at the same time. The limiting filtering flow chart is shown in Fig. 4.

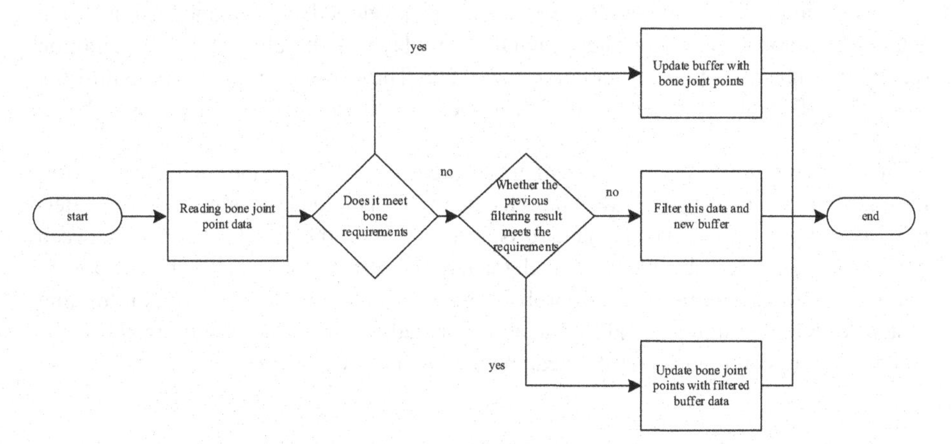

Fig. 4. Flow chart of motion simulation filtering based on virtual reality

Virtual reality algorithm is used in parameter learning. After the parameter initialization in the previous section, the initialization parameters are obtained. On this basis, the initialized parameters are input into the virtual reality model for parameter training. Because the parameters of each action x = {A,B,m) are learned through the process of parameter training, the forward algorithm is carried out on these actions to calculate P(xoy), and finally the motion action with the largest P(O2) can be selected as the recognition result of the test action. The complete process of motion training and recognition is shown in Fig. 5:

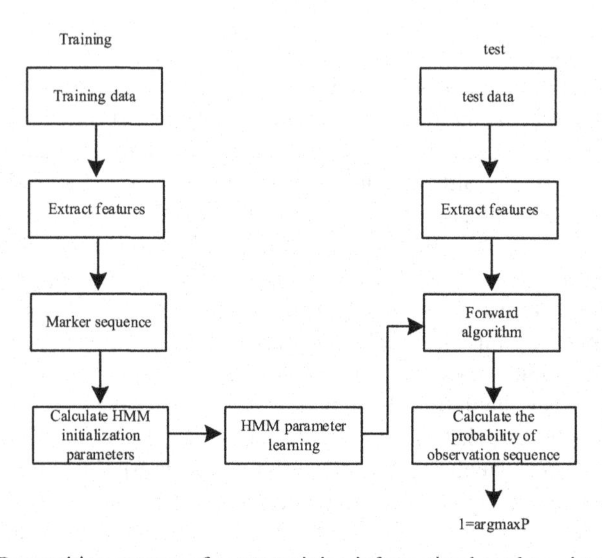

Fig. 5. Recognition process of sports training information based on virtual reality

The system adopts usb 30 interface for data exchange, and uses virtual reality technology to process real-time data of human body color, depth, bone, etc. The application accepts instant data for processing. The system is developed and designed using Visual Studio 2015 software. The system includes data layer, bone algorithm layer, motion recognition layer, motion presentation layer, etc. The client communicates with ECS using TCP/IP communication protocol. The specific software operation process is shown in Fig. 6:

The sports training system uses cameras to measure athletes' bones in real time. After the coordinate transformation, the smooth data of the skeleton is obtained through filtering. Secondly, the space vector method is used to extract the angular features between bone nodes, and the collected motion characteristics are used to train the HMM model. By selecting the maximum output probability, a set of somatosensory motion training system with high reliability and low delay is realized. It can be seen from Fig. 7 that the technical architecture of the system can be divided into four levels:

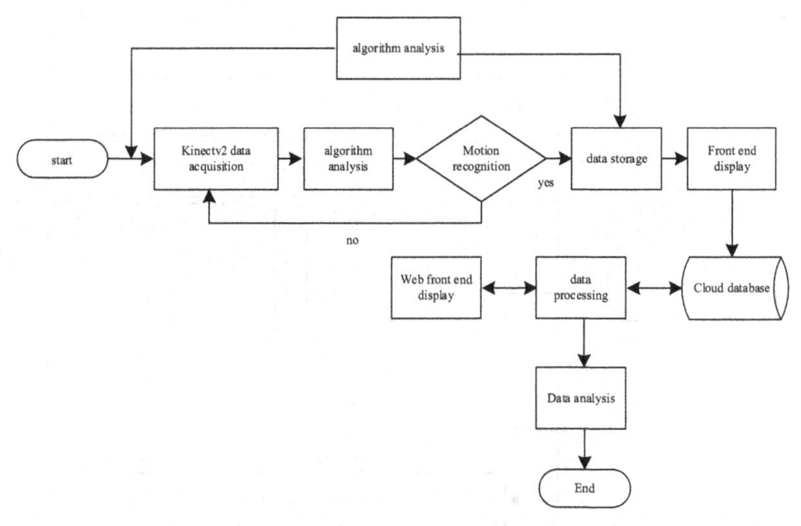

Fig. 6. Operation flow of system software

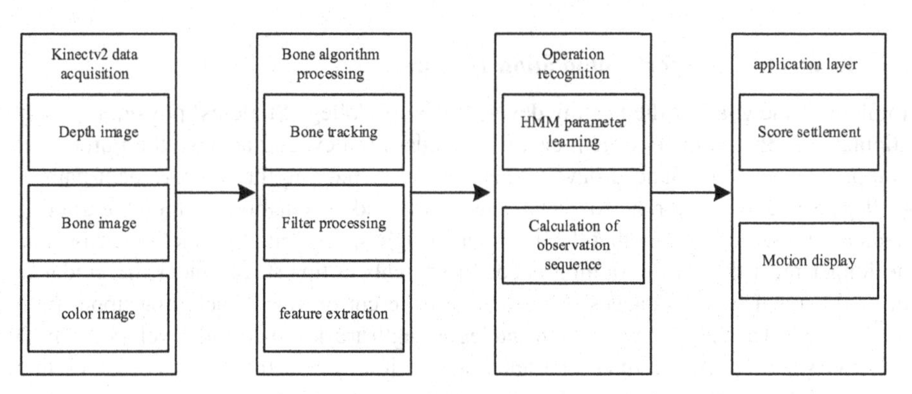

Fig. 7. Framework of physical training system

Through the movement recognition layer, identify the different movements of athletes participating in training, save all data of all actions, including action time, action times, score settlement, etc., and then upload them to the server for scientific analysis and management, and finally form a suitable action training plan [12]. The structure of each functional module of the system is shown in Fig. 8.

As can be seen from Fig. 8, the system consists of four parts. Among them, adding and deleting athletes and adding and deleting training items are two auxiliary function modules; Data analysis of single person multi training program and data analysis of multi person training program are two main functional modules, which will be introduced in detail below.

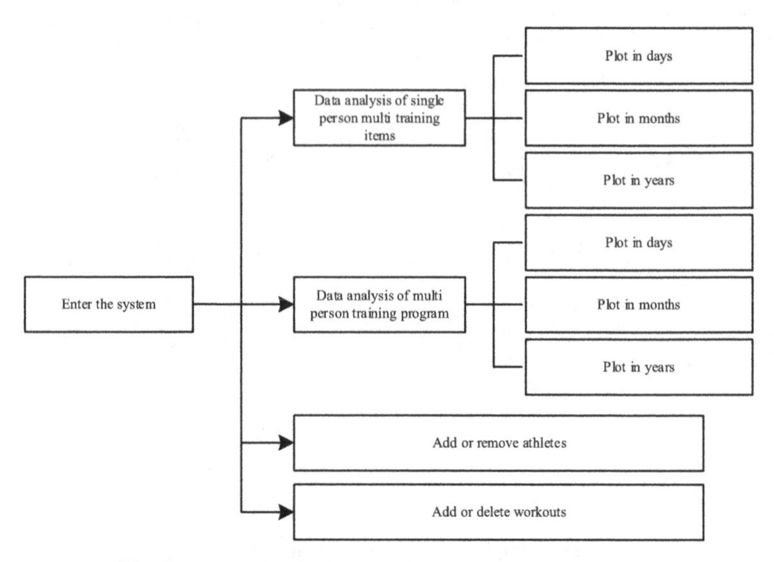

Fig. 8. Structure of each functional module of the system

2.3 Realization of Sports Simulation Training

Implement the goal for the healthy development of College Students' physique clearly stipulated in the national system exercise standard policy, and achieve the purpose of comprehensive and balanced development by fully studying the improvement path of College Students' comprehensive physical quality and reasonably planning the training content. In view of the fact that the traditional single sports training function design can no longer meet the sports training needs of students at this stage, under the guidance of professional fitness coaches, A variety of different exercises including upper limb strength, waist and abdomen strength and leg strength are designed and developed, which integrate sit ups, pull ups, squats and standing long jump respectively. The corresponding sports training purposes and standard movements of various sports functions are shown in Table 1.

The moving area is the projection of the moving object in the video window. The movement of the object causes the gray value change of the pixels in the moving area, which can be reflected in the frame difference diagram of two adjacent video images After binarization, the frame difference image can be used to mark the moving region and the still region with the same size as the frame difference image. In the segmentation, 0 indicates that the pixel at the corresponding position belongs to the still region, and 1 indicates that the pixel at the corresponding position belongs to the moving region. Some early moving region segmentation methods are based on this binarized frame difference image, and the binarized frame difference image can be described as:

$$dk(i,j) = \begin{cases} 1 \ |greyk(i,j) - greyk - L(i,j)| > T \\ 0 \ |greyk(i,j) - greyk - L(i,j)| \leq T \end{cases} \tag{5}$$

Among them, $k(i,j)$ is the binarized frame difference image, $L(i,j)$ is the pixel position, $greyk$ is the gray value of pixel (i,j) in frame k, t is the threshold required for

Table 1. Different sports training functions of sports simulation

Objective name	Training purpose	Standard action
Abdominal curl	Gluteal and abdominal strength	Lean your hands gently against your head, use your abdominal muscles to contract, and swing your arms forward to quickly form a sitting position
Pull up	Upper limb strength, waist strength, gluteal muscles, and many back muscles	Hold the horizontal bar with your palms forward and hands slightly wider than your shoulders. The body will hang naturally. Pull your body up with both arms and the contraction force of back steel muscle. Squat down for one second when you exceed the horizontal bar, and then gradually relax the latissimus dorsi muscle, and the body will fall naturally
Squat exercise	Leg muscles, improve lung capacity and strengthen the heart	Keep the waist and back straight, pay attention, keep the center of gravity stable, and the feet can't move. When squatting, the requirement of hip joint is lower than that of knee joint
Standing long jump	Lower limb explosive strength, hip muscle coordination	During the competition, you can't run up, start from the standing position, and there is no limit to the position of your feet standing. When you start jumping, you are only allowed to jump on the ground once

binarization, which is an empirical value. In this paper, T = 1. When the object moves slowly, the change of the corresponding motion region between adjacent frames is very small, which can be improved by increasing the interval of frame difference image, that is, when detecting the motion region, calculate the frame difference between frame k and k2, L is the number of interval frames. This simple difference algorithm has many problems. A secondary frame difference method is proposed to improve it. Let $d(i,j)$ be the two frame difference images obtained by the above formula, and the secondary frame difference D is defined as:

$$D = \begin{cases} 1\ d(i,j) \leq L \\ 0\ d(i,j) > L \end{cases} \tag{6}$$

The application of computer to the daily training management of athletes will inevitably affect the process of sports training. Based on the traditional training process, this paper proposes a virtual reality assisted training process, as shown in Fig. 9.

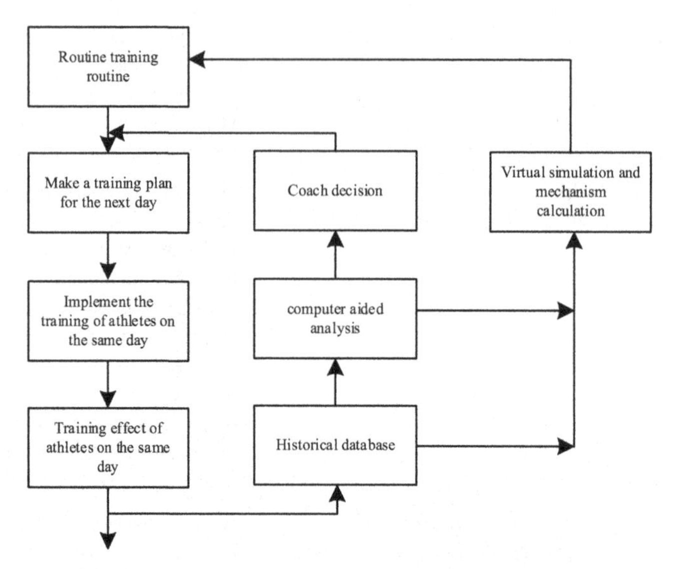

Fig. 9. Flow of Virtual Reality Assisted Training Mode

The process of virtual reality assisted training consists of two closed loops. The inner loop maintains the negative feedback loop of the traditional training mode, and the outer loop is a discontinuous positive feedback loop. The two loops are nested in the inner loop. Compared with the traditional process, only two modules of "historical database" and "Virtual Reality Assisted Analysis" are added to the feedback loop. The "historical database" records the content and effect of all previous training of athletes for analysis; It is an auxiliary method of "physical analysis" and "physical analysis" for the training of athletes, including the application of "physical analysis" and "physical analysis", which is a new training method for coaches, After obtaining various data, the "virtual simulation and mechanism calculation" module simulates the changes of athletes' competitive level after changing the "conventional training routine", so as to change the training elements and jump out of the local minimum. With the help of some software development tools or environments, a simplified model of the actual system can be constructed as soon as possible, as shown in Fig. 10.

Using virtual reality technology, an initial model of the system can be quickly established for developers to communicate with users, so as to accurately obtain the sports simulation needs of users. The gradual refinement method is used to gradually improve the prototype. It is a process of continuous and repeated promotion at a new high level. It can greatly avoid the phenomenon that the prototype of the product can not be seen in the lengthy development process of waterfall model. Compared with waterfall model, prototype model can better reflect people's cognitive and thinking activities. Rapid prototyping can easily improve the communication between the two sides in the early stage

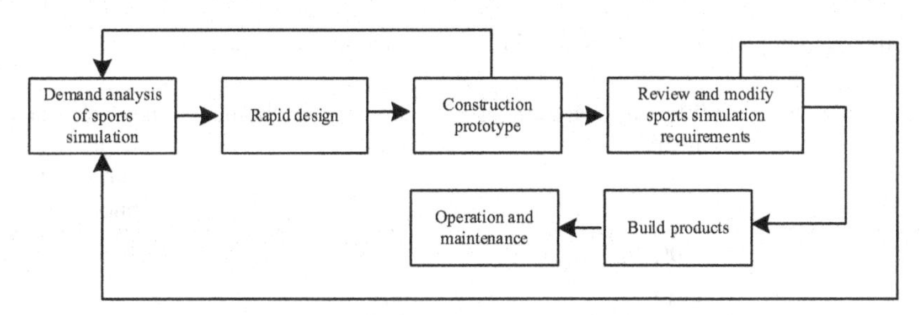

Fig. 10. Simulation training processing flow

of development, and the bow guide stimulates the sports simulation needs of the application side. From the results of practice, it is a more suitable model for the development of sports simulation system.

3 Analysis of Experimental Results

Investigate relevant coaches and athletes in this regard. More than 100 coaches and athletes from sports teams and sports colleges in many places were sampled and investigated by means of telephone consultation and field investigation. Table 2 shows the survey results.

Table 2. Questionnaire on understanding of simulation training

Occupation	Understand the percentage of simulation training	Percentage of simulation training considered important	Percentage of simulation training used	Percentage of people who think simulation training is useful to them
Coach	99%	65%	25%	55%
Old athlete	96%	35%	25%	15%
Sports novice	96%	85%	5%	85%
Leader of training unit	99%	45%	15%	35%

The development of sports simulation system needs the close cooperation of sports technology, training experience, computer simulation technology and other majors. In the process of cooperative development, cooperation faults are easy to appear between disciplines (see Table 3).

Table 3. Sampling questionnaire on skills of both parties in cooperative development

Occupation	With physical training theory	Have physical training experience	Have relevant computer development technology
Coach	100%	100%	<6%
Athletes	55%	100%	<2%
Sports simulation software developer	<55%	<6%	100%

Before the development, coaches do not know what kind of function computer simulation can achieve. Computer personnel are difficult to understand sports technology and experience in place, and there are various objective restrictions in practice, so it is very difficult to cooperate, so the cooperation method is a big problem that needs to be solved urgently. According to the characteristics of physical training, this paper requires to discuss the development mode of each discipline. Before the system is applied, it is necessary to analyze the experimental results of identification and stability of the motion identification algorithm detected by the system. In the experiment, 15 male subjects and 5 female subjects participated in the test for 5 days. During these five days, each participant uses the system for exercise training in the morning and afternoon of each day, and 200 sample data are obtained after the test, The motion recognition rate and accuracy of the motion results of the system are calculated by summarizing the sample data, as shown in Table 4.

Table 4. Motion test results based on virtual reality simulation training

Sports category	Number of movements	Paper system		Traditional system	
		Motion recognition rate	Accuracy	Motion recognition rate	Accuracy
Abdominal curl	200	98.7%	97.3%	67.7%	68.9%
Pull up	197	96.9%	94.7%	76.2%	65.7%
Squat exercise	200	98.1%	96.6%	77.4%	62.3%
Standing long jump	592	95.9%	88.7%	75.3%	66.8%

Through the data analysis of the test results, the results show that the system has a high recognition rate for various functional actions, and the identification function of the system is stable and reliable. In these projects, the accuracy of sit ups, pull ups and squats has met the market demand. Relatively speaking, the recognition accuracy of standing long jump is relatively low. It can be seen from the table that with the increase of training accuracy, the recall rate increases, indicating that there is a positive correlation between the two, and the use accuracy and recall rate of the system in this paper are better than those of traditional methods. It can be predicted that the sports training simulation based on VR has a variety of perceptual abilities, so it can enhance the ability of trainees to interact with the motion simulation system, the effect of simulation training can be effectively improved. The application of VR technology in sports training simulation can effectively promote the scientific and technological training and competitive level of Chinese athletes. And is conducive to the extensive development of national fitness. The application of VR technology can also realize virtual Olympics, Improving the high-tech content in all preparations for the Olympic Games and completing many difficult preparations efficiently. The application of virtual reality technology in sports simulation technology is increasingly extensive.

4 Conclusion

In order to further improve the scientific nature of sports teaching, this paper proposes a new type of sports teaching video analysis system. First, the whole system design is completed, then the specific design is carried out, and the whole system is simulated. The experiment proves that the system can accurately describe the steps of sports training, effectively improve the accuracy of key frame extraction, and achieve good retrieval results. This method is of great significance in guiding physical exercise. The next research will refine the specific sports training content and apply it to the specific sports curriculum training. To further test the effectiveness of the design system.

References

1. Jin, Z., Hao-wen, W., Jun-rong, B.: An optimal design of vocal music teaching platform based on virtual reality system. Comput. Simul. **38**(6), 160–164 (2021)
2. Zhanyi, L.: Analysis on the integration of outreach training in physical education of higher vocational colleges. Vocat. Tech. Educ. **41**(35), 50–53 (2020)
3. Guifang, W.: Formative assessment system of vr teaching in English translation class. Open Access Library J. **09**(02), 1–7 (2022)
4. Xiangyu, D., Keli, H., Hong, D.: A Model and Its Implementation for Distributed Virtual Environment Based on VRML-Java. Comput. Eng. Appl. **01**, 133–134+173 (2002)
5. Jie, L.I.U.: Design of sports assistant training system based on virtual reality technology. Autom. Instrum. **1**, 93–96 (2020)
6. Rui-qing, Z., Ren-ling, Z.: Research of virtual training system for coordination function of upper limbs based on occupational therapy. Softw. Guide **19**(6), 121–124 (2020)
7. ChiYi, T., YuCheng, L.: Design and validation of an augmented reality teaching system for primary logic programming education. Sensors **22**(1), 389 (2022)
8. Junjie, Z., Huichao, L., Chenggen, G., et al.: Modern physical training concepts and approachs into public physical education in colleges and universities. Chin. J. School Health **42**(11), 1605–1608+1612 (2021)
9. Jian, S., Jia-xin, H., Qi, Y.: Research and progress of special physical training in physical education colleges based on "data-driven decision making." J. Guangzhou Sport Univ. **40**(03), 1–3 (2020)

10. Qingkui, C., Xinxin, W., Qin, H., et al.: Reform and practice of experimental teaching model for mechanical specialties based on "VR + cloud platform". Exp. Technol. Manage. **37**(7), 1–4,30 (2020)
11. Zhi, T., Wenming, D.: Problems and solutions of teaching of ideological and political courses in higher vocational education with VR technology. Vocat. Tech. Educ. **41**(5), 64–67 (2020)
12. Xiaoying, Y.: Using VR technology to explore the design method of English vocabulary teaching context. J. East Chin. Inst. Technol. (Soc. Sci.) **39**(6), 588–592 (2020)

A Design Scheme of Data Security for Unmanned Aerial Vehicles

Dongyu Yang[1,2,3](✉), Yue Zhao[1,2,3], Zhongqiang Yi[1,2,3], Dandan Yang[4], and Shanxiang He[5]

[1] Science and Technology on Communication Security Laboratory, Chengdu 610041, China
[2] No.30 Research Institute of China Electronics Technology Group Corporation, Chengdu 610041, China
jobjob2019@163.com
[3] China Electronics Technology Cyber Security Co., Ltd., Chengdu 610041, China
[4] Shanghai Tunnel Engineering & Rail Transit Design and Research Institute, Shanghai, China
[5] Chengdu Luxingtong Information Technology Co., Ltd., Chengdu, China

Abstract. Since the 21st century, informatization, modernization and intellectualization have become an important direction of science and technology development, especially in recent years, with the continuous improvement and perfection of artificial intelligence, 5G, edge computing and autonomous unmanned technology, the UAV industry has made unprecedented development and has been applied to many fields of social life. However, with the development and popularity of UAVs, the development of emerging technologies such as autonomous analysis, unmanned traffic control, UAV swarms, and artificial intelligence continue to increase the complexity of the unmanned systems domain, making cyber attacks against UAVs more and more frequent, and the chances of security threats and potential hazards are increasing. Currently, UAV cyber security has become a very important issue, especially UAV data security, but there are few papers that give systematic solutions for UAV data security. This paper first provides a systematic analysis of UAV security threats from a data security perspective. Next, the existing UAV security protection strategies are analyzed from three aspects: UAV platform security, communication network security, and ground station security. Then, the proposed UAV data security design scheme is introduced in detail. Finally, the full paper is summarized and suggestions for the development of UAV security are given.

Keywords: UAV · Data security · Network security

1 Introduction

Compared with traditional manned aircraft, UAVs (Unmanned Aerial Vehicles) are inexpensive and flexible in use, and in recent years, with the development of UAV swarm technology, they have effectively made up for the shortcomings of smaller individual UAV loads and weaker information sensing and processing capabilities, and have seen

W. Fu and L. Yun (Eds.): ADHIP 2022, LNICST 468, pp. 165–178, 2023.
https://doi.org/10.1007/978-3-031-28787-9_13

explosive development in multiple industry sectors. However, UAVs, as an integrated system with information technology as the traction, face high-intensity information security risks and challenges, especially in complex electromagnetic environments that are more vulnerable to various types of attacks [1], such as information spoofing, false data injection, counterfeit control, signal interference, denial of service, etc.

The traditional UAV security program adopts a divide and conquer approach, mainly focusing on four aspects: sensor security, communication security, software security and network security [2], such as: sensor physical isolation technology, GPS anti-spoofing technology, malware identification technology based on feature codes [3–7], etc. This divide and conquer security scheme can only guarantee the local security of UAVs, while there are many drawbacks, specifically in the following aspects.

1. In terms of application methods, traditional security protection technologies only target individual UAVs and are not applicable to new scenarios such as UAV swarms and manned/unmanned aircraft coordination.
2. In terms of application environments, UAVs and UAV swarm are often required to perform specific tasks in complex electromagnetic environments.
3. In terms of data processing, traditional security solutions focus on security at the communication level, while new scenarios such as swarm and manned/unmanned collaboration are more concerned with the security of multi-source heterogeneous data.
4. In terms of technical means, the traditional "stacked" security protection technology is inefficient and increases the arithmetic and energy overhead of unmanned systems, which is against the principle of lightweighting.

Therefore, this paper proposes a data security service scheme for UAVs, centering on the whole lifecycle process of data collection, data transmission, data storage, data processing, data sharing and data destruction of UAVs, with data security as the core, and proposes a systematic and global security service scheme to realize sensitive information security exchange and sharing, unified authority control, security protocol provisioning and cryptographic algorithm service. The main contributions of this paper are as follows.

1. This paper provides a systematic analysis of the security threats faced by UAVs from the perspective of data security, providing a new approach to the study of UAV cyber security.
2. This paper analyzes the existing UAV security protection strategies in detail and puts forward new suggestions for UAV network security protection.
3. In this paper, a new design scheme for UAV data security is proposed, which can systematically implement UAV data security protection.

This paper is divided into five sections, the second section focuses on systematic analysis of UAV security threats from the perspective of data security. The third section analyzes the current UAV security protection strategies from three aspects: UAV platform security, communication network security, and ground station security. In the fourth section, details the proposed data security service scheme for UAVs. The fifth section summarizes the full text and some suggestions are put forward for the data security of UAVs.

2 UAV Security Threats

UAVs are not designed at the top level with much consideration for security protection, for example, in the design of security protocols, there is only simple link encryption and point-to-point authentication, leading to numerous risks and threats in network information security for UAVs, the main security threats are as follows.

2.1 System Vulnerability

Most UAVs and ground stations are using open source operating systems, and most of the various payloads of UAVs are also standalone operating systems, which may have unknown vulnerabilities that attackers can easily exploit to compromise the UAV or ground station to attack, hijack or steal data [8].

2.2 Malicious Software

Communication protocols in UAVs allow users to control UAVs via wireless remote means (e.g. tablets, laptops and cell phones). However, this approach poses a significant security risk, as an attacker could create a TCP payload of a reverse shell and inject it into the UAV's memory, which could silently install malware on the UAV's ground station system.

2.3 Interference Spoofing

Hijacking of UAV is one of the most important security threats faced by UAV. The most common ways are GPS spoofing, hardware implantation and control signal interference [9–11], among which GPS spoofing includes no-fly zone location spoofing, trajectory spoofing and return point spoofing.

1. GPS Spoofing

   The UAV receives GPS signals with this characteristic: whoever's signal is strong listens to whoever's signal is strong. GPS satellites are so far away that the signal attenuation is very much, so the signal strength will be inferior to the GPS signal faked nearby. No-fly zone location spoofing is by replaying the GPS signal in the no-fly zone near the UAV, so that the UAV will mistakenly think that it has entered the no-fly zone and thus automatically land. The UAV will fly along the selected waypoint, and when the UAV flies towards the next selected location, the trajectory spoofing makes the UAV fly in the direction of the line connecting the spoofed location and the next scheduled location by spoofing the GPS location signal until it reaches the selected point. When the ground station and the UAV lose contact, the UAV will automatically fly toward the return point and eventually return to the return point. Return point spoofing is to control the UAV by tricking it into setting the current position as the return point and continuously changing the return direction.

2. Hardware Implantation

   The UAV determines its own position by using the GPS module. Since some of the GPS modules do not do any encryption processing on the data, the communication between the master control MCU of the UAV and the GPS module can also be effectively changed by hijacking the UAV's position information, thus achieving the effect of deception.

3. Signal Interference

   The control signal commonly used by UAVs is in the 2.4G band, and the graphics transmission is in the 5.8G band. Wi-Fi, ZigBee, Bluetooth, etc. also use the 2.4G band, and the co-channel interference is very serious and unavoidable.

In addition, attackers use jammers to create flight control jamming signals and satellite positioning jamming signals, by blocking the uplink flight control channel and satellite positioning channel of the UAV, so that it loses flight control instructions and satellite positioning information, making it unable to fly normally, and for different types of UAVs will produce the control effect of returning, landing and crashing. An attacker can also hijack a UAV by sending a de-authentication process between the access point and the device controlling the UAV, which can be performed temporarily or permanently, such as jamming the intended UAV frequency and inducing it to connect to the hacker's Wi-Fi, an attack that can be accomplished with a simple Raspberry Pi development configuration.

2.4 Data Security

UAVs are vulnerable to various attacks during data exchange and sharing, for example, data interception, malicious data injection and even installation and insertion of many infected digital files (videos, images) into ground stations.

UAV data security can be divided into three main levels [12] (see Fig. 1), the first level considers the interaction between the UAV and the user, the second and third levels involve wireless data transmission and cloud storage. The main security threats at the first level include: Malicious code/malware virus, Brute force attack, Rootkit booting, Removal of storage media for data theft, etc. At the second level include: Man-in-the-middle, Hacked/no encryption, No ecosystem access control, No network segmentation, etc. And at the third level include: Ransomware, Unverified platform access, Plaintext username & passwords, Insecure admin interface, Insecure cloud backend APIs, etc.

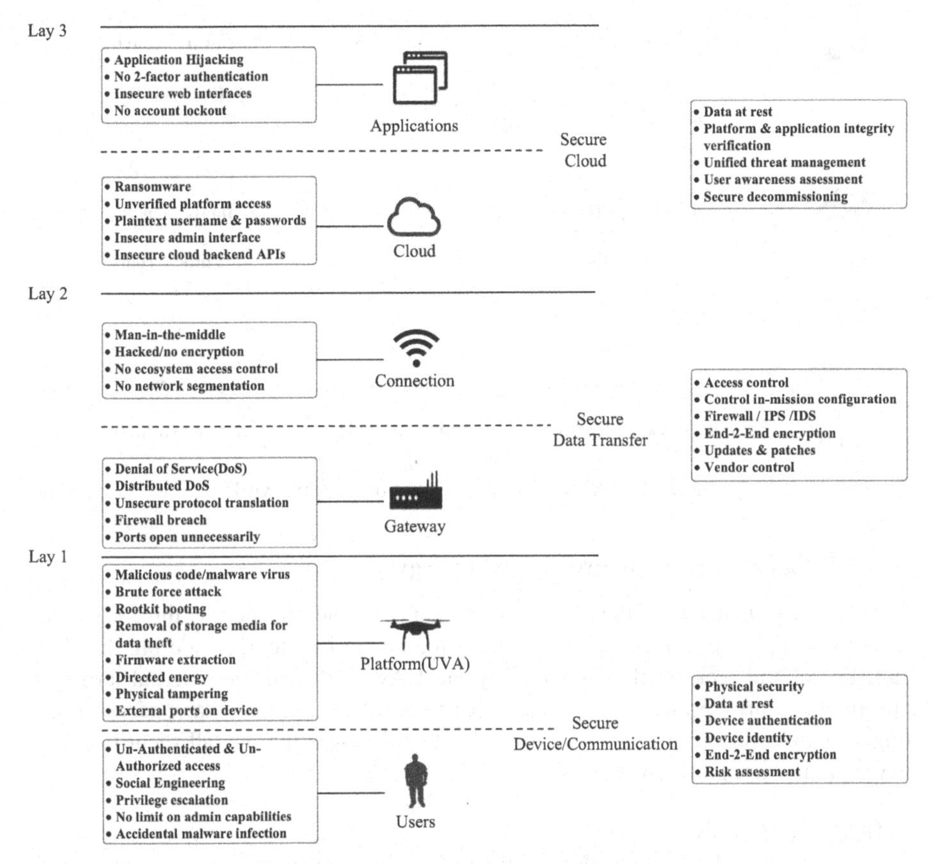

Fig. 1. The data security of UAVs

3 UAV Security Protection Strategy

UAVs are mainly composed of three parts: UAV platforms, ground stations and communication networks [13]. At present, the UAV security protection strategy involves UAV platform security, ground station security and communication network security and other aspects (see Fig. 2).

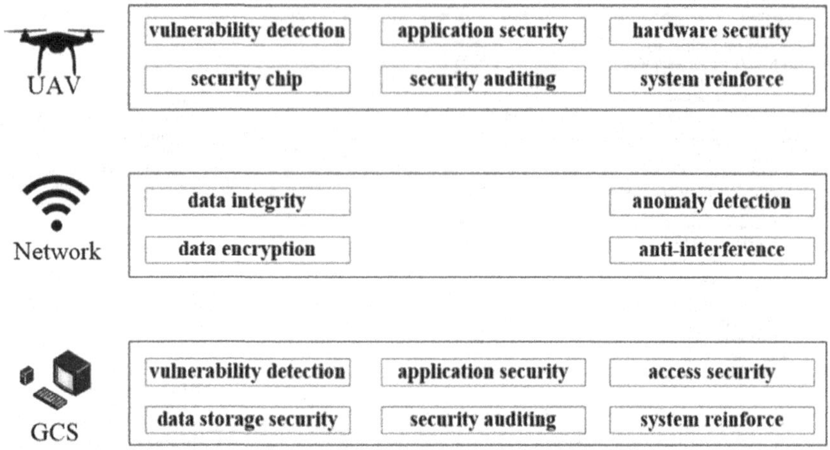

UAV

vulnerability detection	application security	hardware security
security chip	security auditing	system reinforce

Network

data integrity	anomaly detection
data encryption	anti-interference

GCS

vulnerability detection	application security	access security
data storage security	security auditing	system reinforce

Fig. 2. UAVs security protection strategy framework

3.1 UAV Platform Security Protection Strategy

The UAV platform itself and the external terminal face numerous security threats, including viruses, malware, hijacking attacks, phishing networks, etc. Considering the characteristics of the security threats faced by the UAV itself and the external terminal, appropriate security protection measures need to be taken from hardware, access, operating system, vulnerability, application and audit to ensure the overall security of the UAV and the external terminal.

1. Hardware Security

   A security chip can be added to the main UAV control device to achieve secure external access to the main UAV control system and provide a trusted computing environment. The security chip can also provide a globally unique identity ID and an independent high-speed encryption unit without taking up system resources, which can ensure that system programs, terminal parameters, security data and user data within the chip are not tampered with or illegally accessed [14].
2. Vulnerability Detection and Repair

   By conducting security testing activities such as vulnerability scanning and penetration testing on UAVs and external terminals beforehand or on a regular basis, system vulnerabilities and risks in the system are found in a timely manner, repaired and patched, and then system upgrades are performed on UAVs and external terminals to effectively reduce the security risks of system vulnerabilities in UAVs and ensure that UAV systems are safe and controllable [15].
3. Access Security

   Because the UAV will have multiple external terminals, these terminals may become the entry point to hack the UAV to control hijack or damage the UAV. A network relationship whitelist can be established based on the network connection relationship between the UAV and the external terminal, which can be checked based on source address, destination address, source port, destination port and protocol to allow/deny packets in and out.

At the same time, a lightweight access authentication mechanism is developed for external terminals to enhance the ability of the UAV master control system to analyze the abnormal behavior of external terminals, which can prevent illegal terminal access as well as timely discovery of abnormal terminal intrusion behavior [16].

4. Application Security

   To prevent malicious program implantation by attackers on UAVs, UAVs can identify and authenticate the application software to be installed, for example, by using application signatures [17]. The UAV can use a trusted verification mechanism to verify the trusted execution of applications and important configuration files/parameters. The UAV should control the sensitive behavior of installed applications, build a system-level security protection policy, establish an internal data relationship analysis model, unify control over UAV applications, and "end" malicious applications according to the discovered "operation to prevent malicious program infection, illegal access and other attacks.

5. System Reinforcement

   As an embedded terminal, coupled with the fact that many vendors do not have security development capabilities, UAVs can be security hardened by embedding security modules, security SDKs, and other kits in the UAVs [18]. These security kits can make the UAV have the ability to receive and execute security policies, which include network access policies (black and white lists), process operation policies, etc. The kits can also perform security analysis and policy restrictions on some access, data reporting and other behaviors.

6. Security Auditing

UAVs need to conduct security audits of their own system behavior and audit important behaviors and important security events for behavioral analysis or post-event traceability of security events. Important security audits include:

- Inspection and audit of key file directories and files, including file additions, modifications, changes, etc.
- Audit protection of system process behavior to prevent unauthorized interruptions
- Monitoring of various network behaviors to detect abnormal access, system intrusion, abnormal traffic attacks, etc.
- Monitoring of system resources, including monitoring the use of CPU, memory and other resources, and timely alerts for overruns
- Audit and record system behavior, including the date and time of events, users, event types and other audit-related information

3.2 Communication Network Security Protection Strategy

At present, a variety of network communication technologies can be used among UAVs and ground stations, which include radio, Wi-Fi (2.4G or 5.8G), cellular mobile communication networks, satellite communication and other heterogeneous networks, all of which are air-port communication technologies. Also UAVs use technologies such as GPS /GLONAS in order to pinpoint the location. Hackers or attackers usually attack and invade through these open air-ports as attack entrances [19]. So it is very important to protect the communication network security of UAVs.

The communication network security protection can be taken mainly in the following aspects.

1. Network Access Authentication
   For the UAVs communication network, access to the network must be controlled by authority, especially the Wi-Fi communication method. Access to the UAVs communication network access before the need for identity verification and authorization to ensure the legitimacy [20]. Weak passwords cannot be used for access authentication, while strict access control permissions are set for devices accessing the UAVs communication network. For the UAVs communication network, a black and white list of network access can be set, while the ports are open for control and non-essential access ports are closed.
2. Data Integrity Protection
   By establishing a data security channel between the UAV and the ground station and guaranteeing the communication quality, digital signature technology is used to provide a reliability guarantee mechanism for information transmission, guarantee the authenticity and integrity of important data, solve the credibility problem of evidence, and effectively prevent data leakage, communication content from being eavesdropped and tampered.
3. Data Transmission Encryption
   The sensitive data transmitted by the air-port is encrypted, including the use of encryption algorithms and end-to-end encryption between the UAV and the ground station to guarantee the security of data transmission. At the same time, multi-dimensional checks such as device fingerprint, time stamp, identity verification and message integrity can be performed to ensure the security of data transmission to the maximum extent [21].
4. Anti-signal Interference
   At present, the main anti-jamming technologies are divided into three categories: technologies related to improving system reliability and effectiveness, collaborative communication-based technologies, cognitive radio-based anti-jamming technologies, including technologies related to improving system reliability and effectiveness include coding anti-jamming technology, spread spectrum anti-jamming technology, multiple input and output anti-jamming technology, array antenna anti-jamming technology.

3.3 UAV Ground Station Security Protection Strategy

More advanced UAV ground station equipment typically consists of a remote control, computer, video monitor, power system, radio, and other equipment. A simple UAV ground station may only have a remote control or a computer (phone, tablet, laptop) with control software.

The security protection strategy for UAV ground stations is generally similar to that of UAV platform, but since ground stations are control terminals as well as storage terminals, they need to be strengthened in the following aspects.

1. Data Security Protection

   The ground station will store a large amount of UAV flight data, including important data such as aerial photography data. Security protection must be carried out for these data, including identity verification, data encryption, data backup and recovery, etc.
2. System Protection

   UAV ground stations need system security protection in the following aspects, including applications security, system reinforcement, vulnerability detection and repair, security baseline check, anti-DDoS attack, anti-buffer overflow, abnormal behavior detection, security audit and other security construction, which can effectively prevent ground stations from being invaded and causing system damage, data leakage, data tampering and other security problems [22].

4 UAV Data Security Scheme

Due to the numerous types of UAVs and ground stations, large business differences and weak computing power, especially consumer-grade UAVs, traditional security protection means such as firewalls and anti-virus software are not applicable. Aiming at the security threats faced by UAVs and the shortcomings of existing security protection technologies, a data security service solution for UAVs with data security as the core is designed. It realizes sensitive information security exchange and sharing, unified authority control, security protocol provisioning and cryptographic algorithm services in the whole lifecycle process of data collection, data transmission, data storage, data processing, data sharing and data destruction of UAVs, covering all links of UAV data flow, realizing security protection for the whole lifecycle of UAV platform data, and providing guarantee for the safe operation of UAVs.

4.1 UAV Data Security Service System

The UAV data security service solution designs a UAV data security service system (see Fig. 3), which consists of an runtime component, a data security component and a basic security component.

The runtime component includes a runtime library and a resource manager. The runtime library is used to provide the necessary procedures for the security service function of the unmanned platform, and the resource manager automatically configures the system environment required for the security function and manages the runtime resources according to the system properties of the unmanned platform.

Data security components are security functional components used to provide security functions including data source detection, integrity verification, tamper-proof, dense state transmission, distributed storage, data traceability, data access policy, fine-grained level data access authority control and permanent deletion during the whole life cycle of data collection, transmission, storage, exchange and sharing, processing and destruction.

The basic security components include security protocol swarms and cryptographic algorithm library, including efficient dynamic batch authentication protocol, attribute policy control protocol, lightweight group key management protocol, security protocol negotiation protocol and base protocol, which can adapt to different mission scenarios and usage scale of UAV swarms; cryptographic algorithm library is used to provide the required cryptographic algorithms.

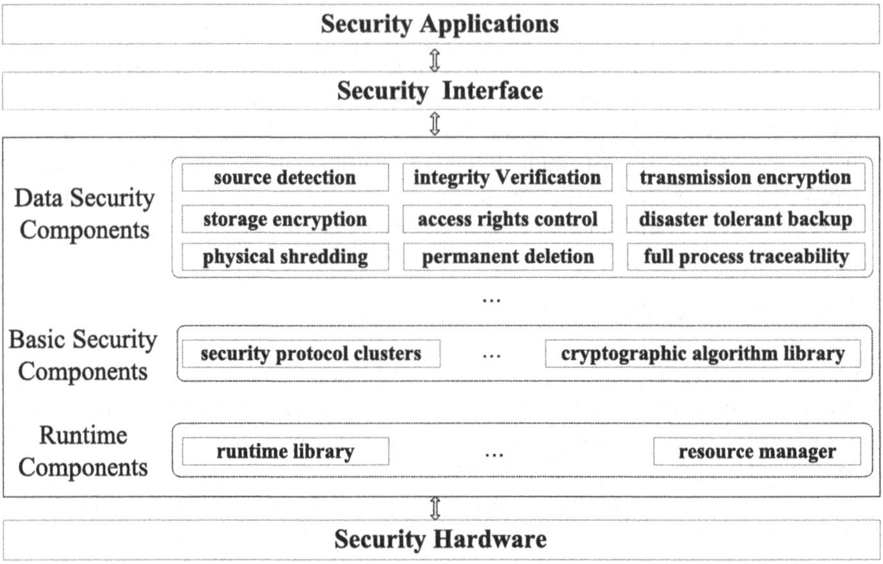

Fig. 3. The framework of UAV data security service system

4.2 System Workflow

The workflow of the UAV data security service system proposed in this paper mainly includes the following steps (see Fig. 4).

Step 1, after the security service system is transplanted to the unmanned platform, a local installation procedure is performed, and if the installation fails, the reason for the error is fed back and the installation is performed again; if the installation is successful, a connection is established with the unmanned platform's own system.
Step 2, run the basic security component to detect the unmanned platform's own system properties and security status, if the security status is abnormal, then feedback abnormal information and alarm; if the security status is normal, then automatically configure the system environment required for the security service function, while the background supervision of the system operating resources.
Step 3, after the system environment is configured, the data security component sets the initial policy for unmanned platform data access.
Step 4, when the unmanned platform task scenario or scale quantity and other circumstances change, the basic security component completes the selection of security protocols such as batch authentication protocol, data transmission protocol, attribute

policy control protocol, and key management protocol, and determines the data encryption algorithm used. At the same time, the data access policy is dynamically adjusted to achieve fine-grained level data access permission control.
Step 5, When the data access policy, security protocol and encryption algorithm are determined, the data security component carries out security protection in the processes of data collection, data transmission, data storage, data exchange and sharing, data processing and data destruction according to the access policy and protocol requirements to ensure the information security of the unmanned platform.

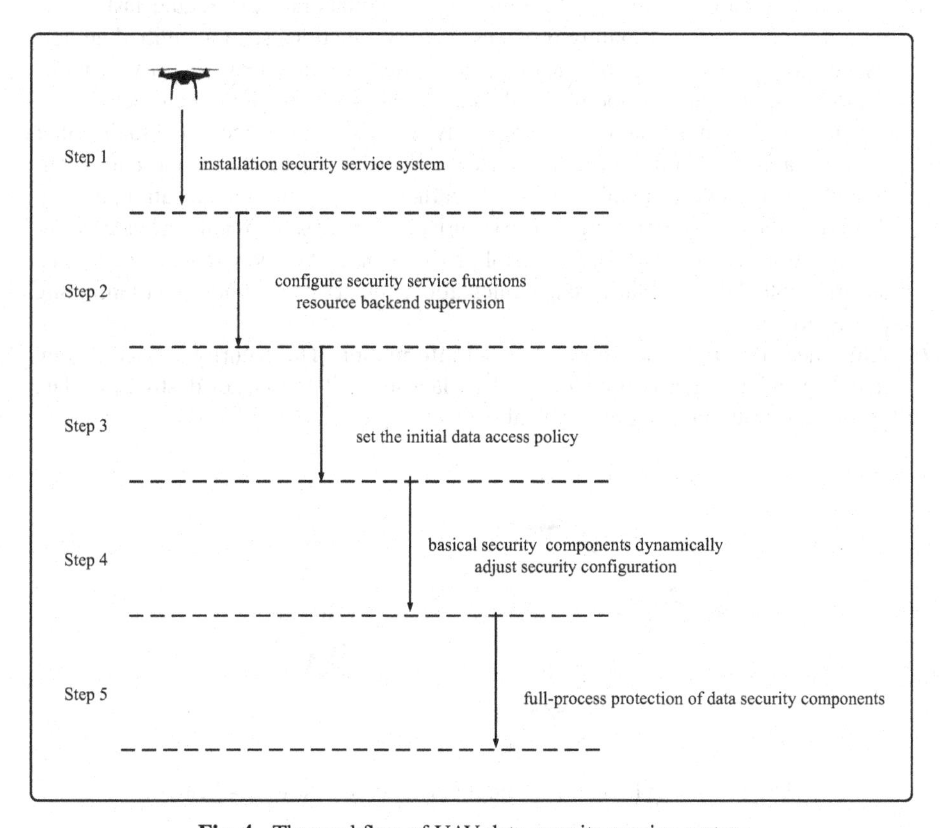

Fig. 4. The workflow of UAV data security service system

4.3 Application Scenarios

One application example of the data security solution between UAV U_1 and U_2 is shown in Fig. 5 and the specific description is as follows.

1. U_1 and U_2 travel to a predetermined target area to collect mission-related information, and the security service system mainly provides security functions such as data source identification, anti-tampering, logging and risk assessment.

2. During the information collection process, U_1 and U_2 respectively store the collected information locally and, at the same time, exchange information with each other for backup. The security service system mainly provides security functions such as data encryption transmission, dense state distributed storage and disaster-tolerant backup.
3. When the information collection is completed, U_1 and U_2 transmit the information back to the ground control center. The ground control center issues task instructions to U_1 and U_2. The security service system mainly provides security functions such as data encryption, highly reliable dense state transmission, and unified authority control.
4. U_1 and U_2 share intelligence information and collaborate to execute tasks. The security service system mainly provides security functions such as unified authority control, data integrity verification, group key management, security protocol deployment, data anti-tampering, and data traceability throughout the process.
5. When U_1 or U_2 finds a target, it immediately establishes a connection with the other party, shares this information, and makes a decision on the most optimal execution plan to complete the mission by comparing and evaluating its location, its own status and the nature of the target. The security service system mainly provides security functions such as security protocol provisioning, group key management, data security calculation, security edge calculation, integrity checking, data tampering prevention, etc.
6. After the mission, U_1 and U_2 destroy all information. The security service system mainly provides security functions such as data desensitization, key destruction, data permanent deletion, and anti-reversal.

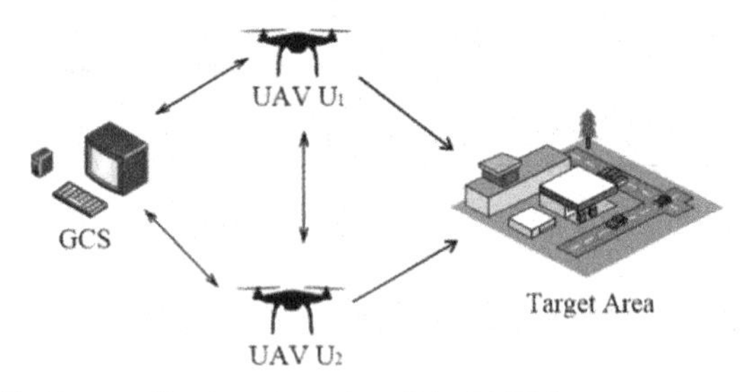

Fig. 5. One application scenarios of the UAV data security scheme

5 Summary

The continuous development of unmanned system technology has also brought new challenges to the security protection technology of unmanned system. On the one hand, the current technology only focuses on point-to-point entity authentication, link level data encryption, anti-electromagnetic interference of communication links, and safe and reliable software operation, without forming a systematic security plan. On the other hand, the safety protection means of the existing system are mostly "stacked", which is inefficient and may affect the system itself. How to prove the safety needs to be considered from many aspects, such as verification and safety confirmation, meeting the specifications and meeting the application requirements.

Therefore, the future unmanned system security technology should focus on the following aspects:

1. Based on the information security theory of "data security as the core", supported by "artificial intelligence", "edge computing" and other emerging technologies, the cryptosystem is constructed around all links of the whole life cycle of data from generation to destruction, rather than simple channel encryption.
2. It is necessary to design a new cryptographic protocol to adapt to the high real-time, high reliability, dynamic changes of network topology and other communication characteristics faced by unmanned systems in various complex task scenarios, rather than a simple modification of the traditional Internet protocol.
3. Realize security protection for the internal data of unmanned system and the data flow between unmanned swarms, and strengthen the endogenous security of unmanned system, rather than the simple superposition of multiple security technologies.
4. With the deepening of the intelligence of unmanned systems, security configuration also needs to have a high degree of autonomy to realize the functions of key independent negotiation and dynamic adjustment of security policies.

References

1. Javaid, Y., Sun, W., Devabhaktuni, K.: Cyber security threat analysis and modeling of an unmanned aerial vehicle system. In: 2012 IEEE Conference on Technologies for Homeland Security (HST), Waltham, MA, pp. 585–590 (2012)
2. He, D., Du, X., Qiao, Y.: A survey on cyber security of unmanned aerial vehicles. Chin. J. Comput. **42**(5), 1076–1094 (2019)
3. Roth, G.: Simulation of the effects of acoustic noise on MEMS gyroscopes (M.S. dissertation), Auburn University, Alabama, USA (2009)
4. Soobramaney P.: Mitigation of the effects of high levels of high-frequency noise on MEMS gyroscopes (Ph. D. dissertation). Auburn University, Alabama, USA (2013)
5. Lee, J.-H., Kwon, K.-C., An, D.-S., Shim, D.-S.: GPS spoofing detection using accelerometers and performance analysis with probability of detection. Int. J. Control Autom. Syst. **13**(4), 951–959 (2015). https://doi.org/10.1007/s12555-014-0347-2
6. Daneshmand, S., Jafarnia-Jahromi, A., Broumandan, A.: A low-complexity GPS anti-spoofing method using a multi-antenna array. In: Proceedings of the 25th International Technical Meeting of the Satellite Division of the Institute of Navigation. Nashville, USA, pp. 1233–1243 (2012)
7. Shabtai A.: Malware detection on mobile devices. In: Proceedings of the 11th International Conference on Mobile Data Management. Kansas City, USA, pp. 289–290 (2010)
8. Shahrear, I.: A Study on UAV Operating System Security and Future Research Challenges. 2021 IEEE 11th Annual Computing and Communication Workshop and Conference (CCWC). IEEE, pp. 759–765 (2021)
9. Gaspar, J., Ferreira, R., Sebastiao, P.: Capture of UAVs through GPS spoofing. In: 2018 Global Wireless Summit (GWS), pp. 21–26 (2018)

10. Riahi, M., Kenney, J.: Detection of GPS spoofing attacks on unmanned aerial systems. In: Proceedings of the 2019 16th IEEE Annual Consumer Communications and Networking Conference (CCNC), pp. 1–6 (2019)
11. Kamkar, S.: Skyjack. https://samy.pl/skyjack/. Accessed 20 Jan 2019
12. Drones and data security: a progressive look into the future. https://droneii.com/drone-data-security. Accessed 23 May 2018
13. UAV Security White Paper 2021. https://www.dbappsecurity.com.cn/. Accessed 24 Apr 2021
14. Gaurang, B., Biplab, S.: S-MAPS: scalable mutual authentication protocol for dynamic UAV swarms. IEEE Trans. Veh. Technol. **70**(11), 12088–12100 (2021)
15. How to analyze the cyber threat from drones. https://www.rand.org/pubs/research_reports/RR2972.html. Accessed 5 May 2020
16. Srinivas, J., Das, A.K., Kumar, N.: TCALAS: temporal credential-based anonymous lightweight authentication scheme for internet of drones environment. IEEE Trans. Veh. Technol. **68**(7), 6903–6916 (2019)
17. Ana, H., Alejandro, Z., Jorge, B.: Security orchestration and enforcement in NFV/SDN-aware UAV deployments. IEEE Access **8**, 131779–131795 (2020)
18. Dominic, P., Thomas, F., Christian, L.: Global and secured UAV authentication system based on hardware-security. In: 2020 8th IEEE International Conference on Mobile Cloud Computing, Services, and Engineering (MobileCloud), pp. 84–89. IEEE (2020)
19. Zhang, J., Cui, J., Zhong, H.: Intelligent drone-assisted anonymous authentication and key agreement for 5G/B5G vehicular ad-hoc networks. IEEE Trans. Netw. Sci. Eng. **8**(4), 2982–2994 (2020)
20. Abhishek, S., Pankhuri, V., Nikhil, P., et al.: Communication and networking technologies for UAVs: a survey. J. Netw. Comput. Appl. **168**, 102739 (2020)
21. Keonwoo, K., Yousung, K.: Drone security module for UAV data encryption. In: 2020 International Conference on Information and Communication Technology Convergence (ICTC), IEEE (2020)
22. Arslan, S., Abid, M., Mourad, E.: Survey of security protocols and vulnerabilities in unmanned aerial vehicles. IEEE Access **9**, 46927–46948 (2021)

Distance Teaching Method of Welding Process Course for Mobile Learning Platform

Juan Song(✉) and Liang Song

Jiangsu Province Xuzhou Technician Institute, Xuzhou 221151, China
sj18952150665@163.com

Abstract. The current distance teaching method is mainly to complete the explanation of basic knowledge by recording course video, students' autonomous learning or with the help of remote communication function. It is difficult for students to fully mobilize their enthusiasm, resulting in poor teaching effect. In order to alleviate the problems of inapplicability and poor teaching effect when the current distance teaching method is applied to welding processing course, this research designs a distance teaching method of welding processing course for mobile learning platform. According to the characteristics of mobile learning platform, the communication path of distance teaching of welding processing course is determined. Then formulate the 5E Teaching Mode and design the teaching process of the course distance teaching platform. Finally, collaborative filtering algorithm and firefly algorithm are used to recommend personalized mobile learning resources for students and generate course learning paths. The experiment shows that after applying this method, the average score of students is improved by about 9.2%, and their learning enthusiasm is significantly enhanced.

Keywords: Mobile learning platform · Welding processing courses · Distance learning · Teaching methods · Collaborative filtering · Firefly algorithm

1 Introduction

With the rapid popularization of mobile Internet and intelligent terminals and the continuous development of educational informatization, students can complete the learning of relevant courses through mobile learning [1]. The development of learning intelligent terminals promotes the dissemination of learning resources, can help students better integrate learning tasks into daily arrangements, and can help teachers effectively respond to students' needs by allowing two-way communication.

With the development of industrial technology, there is an increasing demand for welding processing technicians in relevant fields. Welding processing has become a popular specialty and a key training specialty in many colleges and universities. The course of welding processing not only requires students to master theoretical knowledge, but also requires students to have basic practical ability [2]. However, the practical operation of the welding processing course needs to be carried out under the strict guidance of

W. Fu and L. Yun (Eds.): ADHIP 2022, LNICST 468, pp. 179–190, 2023.
https://doi.org/10.1007/978-3-031-28787-9_14

teachers, so as to prevent students from being in danger due to improper operation during the practical operation. However, in the long-distance teaching of welding processing course, students can not carry out practical training under the guidance of teachers, but can only learn the theoretical content of the course. Therefore, in the theoretical teaching of welding processing course, ensuring that students can master theoretical knowledge and lay a solid foundation for offline welding practice has become the research focus of remote teaching of welding processing course [3].

Relevant scholars have proposed a series of research achievements such as distance teaching methods based on information integration, distance teaching methods based on human-computer interaction, and so on. However, in practical application, it is found that the application effect of the above-mentioned traditional methods still has room for improvement.

Based on the above analysis, this paper introduces information technology to deepen the reform of teaching mode of welding processing course, and uses mobile learning platform to assist students in learning welding processing course, so that these problems can be effectively solved. This paper designs a remote teaching method of welding processing course oriented to mobile learning platform. The specific design ideas are as follows:

(1) According to the characteristics of the mobile learning platform, the communication path of remote teaching of welding processing course is determined.
(2) Design the business process of distance teaching platform and 5E Teaching mode.
(3) Through collaborative filtering, personalized mobile learning resources are recommended for students, and then fireflies are used to calculate and generate curriculum learning paths.
(4) Through experimental verification, it can be seen that after the application of this method, the average test scores of students in welding processing courses have increased by about 9.2%, which indicates that the teaching effect of this method is better and can promote the further development of distance teaching more effectively.

2 Method Research

According to the characteristics of mobile learning platform, this paper defines the communication path of remote teaching of welding processing course. Then, based on the design of the business process of the distance teaching platform, the 5E Teaching mode is formulated. Finally, the collaborative filtering algorithm is used to recommend personalized mobile learning resources for students, and the course learning path is generated based on the firefly algorithm. The specific design steps are shown in Fig. 1.

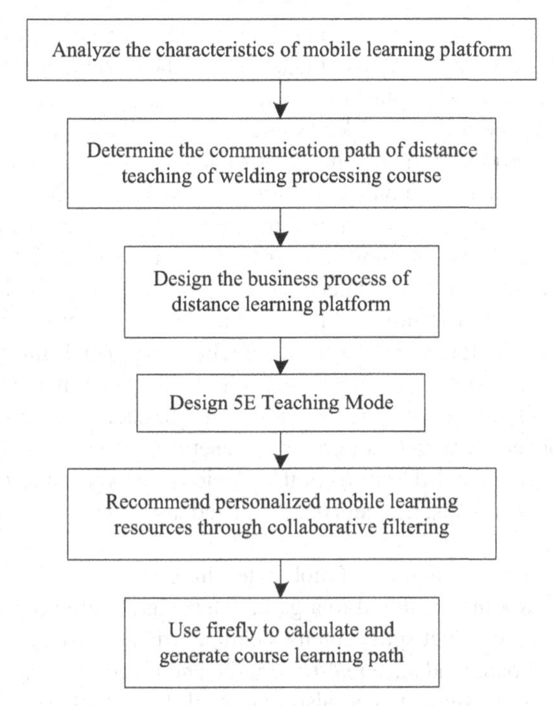

Fig. 1. Design steps of distance teaching method for welding processing course

2.1 Determine the Communication Path of Distance Teaching of Welding Processing Course

Distance teaching includes electronic technology, video, interactive multimedia and other forms, so text, audio, video, PPT, computer software and other modes can be selected for Distance Teaching of welding processing courses. Teachers and students use the mobile learning platform to complete the distance teaching interaction of the course. In the process of distance teaching, analyzing the communication path of the course can effectively grasp the main content of the course teaching and improve the effect of the course teaching. In fact, distance teaching activities are also a kind of information dissemination activities. Distance teaching can be seen as the redesign and combination of the five elements of communication subject, communication media, communication content, communication audience and communication effect.

In the process of offline classroom teaching, knowledge communication is a top-down linear communication mode. Teachers first sort out the knowledge they have mastered, and then spread the knowledge to students. This kind of teaching is carried out face to face. Teachers have a strong control over the overall situation of the classroom. When using mobile learning platform for distance teaching, teachers and students are separated in space. Teachers can't understand their real listening state according to their expressions and expressions. The original teaching design needs to be readjusted. Teachers need to adapt to the mobile teaching platform and the new environment of distance teaching in a short time. Distance teaching breaks through the teaching time and space of the traditional classroom in the past. It not only changes the place and environment of class, but also changes the teaching methods of teachers and the learning methods of students.

In the teaching process of welding processing course using mobile learning platform, teachers are not only the imparter of knowledge, but also the guide of students' thinking, leading them to explore broader and deeper knowledge outside the classroom, and helping them improve their autonomous learning ability. Teachers are facing many challenges in the process of distance teaching, among which the average value of teaching methods, teaching habits and teaching concepts that need to be changed is very high [4]. Distance teaching poses a greater challenge to teachers. In order to achieve good teaching results, teachers need to optimize and restructure the original curriculum content and teaching plan design. Students tend to slack off when they spend time studying on the screen, which requires teachers to change the original teaching methods, highlight the key points in the teaching process, and be good at mobilizing the classroom atmosphere, which requires teachers to spend more time and energy in the process of lesson preparation. Distance teaching is different from the previous classroom teaching. In order to achieve better teaching results, teachers need to spend more time in the process of lesson preparation.

With the rich teaching functions of mobile teaching platform, the teaching environment of offline classroom is restored to a great extent. In the process of live broadcast teaching, teachers can conduct split screen teaching, switch freely between courseware and electronic blackboard, and have real-time audio and video interaction with students. After the live broadcast, students can also play back [5]. Teachers can also organize online examinations. The platform includes single choice, multiple choice, blank filling and other types of test questions. Teachers can freely arrange in class tests and the marking function of the final examination system, which saves teachers a lot of time.

In the final analysis, the teaching objective of welding processing course serves the talent training, and the whole training objective must be implemented into the detailed rules and practice of all courses. Learning is not a process in which learners passively accept ready-made conclusions, but a process in which learners independently construct information structure through real situations. For example, in the teaching of materials and heat treatment related knowledge points in the welding processing course, the main teaching purpose is to let students understand the relevant structure and properties of metal through explanation, form an understanding of the different properties of finance, clarify its internal crystal structure, understand the mechanical properties of metal and the organizational structure of heat treatment, and finally have a certain cognition and understanding of the treated materials in welding, Be able to skillfully select appropriate welding process parameters and effectively control the weld quality according to the nature of processing materials and relevant knowledge of the course in the practice of welding specialty, and be able to objectively and comprehensively analyze the welding defects according to the welding quality inspection results and take corresponding control measures [6].

2.2 Teaching Process Design of Course Distance Teaching Platform

The distance teaching platform provides students with a variety of different businesses, and the basic functions are suitable for Distance Teaching of different courses in all colleges and universities. Figure 2 shows the business process of mobile learning platform during distance teaching of welding processing course.

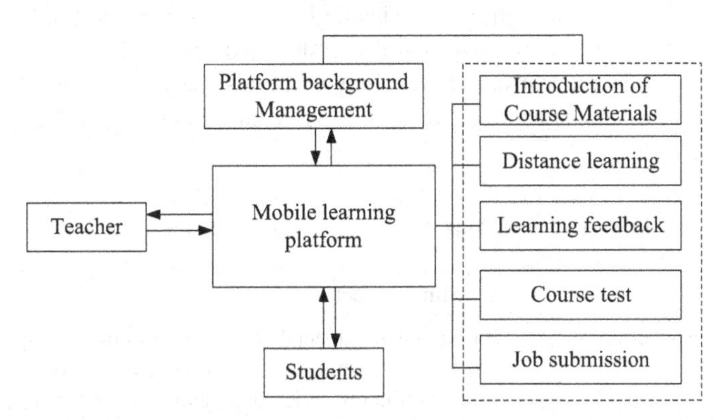

Fig. 2. Business process diagram of mobile learning platform

Homework is the core business of distance teaching management and an effective way to communicate between students and teachers after class. The traditional homework management is mainly through paper, which brings great inconvenience to students and teachers. The long-distance teaching platform greatly simplifies the communication between teachers and students by submitting homework by students through mobile phones and then correcting online by teachers through mobile phones.

Although there are some differences between distance teaching and traditional offline teaching mode, students' learning still needs to follow the principle of step-by-step. Therefore, it is necessary for teachers to make good use of time and communication in the process of classroom teaching, so as to ensure the students' good understanding and mutual acceptance in the process of distance teaching. In addition to the teaching itself, we must also do a good job in the follow-up links, focusing on the quality inspection and evaluation of the whole teaching process. On the one hand, we need to evaluate each student's classroom performance, and also teach students according to their aptitude [7].

Teachers need to combine the characteristics of students' thinking mode and related knowledge, and then choose appropriate methods to show the content of knowledge to students. The first is to do a good job in the relevant theoretical basis, that is, the teacher should fully explain the theoretical content required by the welding specialty at the beginning, fully stimulate the students' interest, summarize and summarize the difficult contents into easy to understand sentences, and try to show the students in a vivid way, so that they can master the welding specialty theory on the basis of full understanding. Through the creation of problem situations and other ways to trigger appropriate information uncertainty, stimulate students' interest in the learning content itself, help them form internal motivation, and pay attention to the expansion of students' original cognitive structure and the cultivation of learning initiative and enthusiasm [8].

Remote online communication is a modern way to realize the traditional after-school Q & A. in the past, students needed to find teachers to answer questions after class. Then this way has the limitations of time and space. The online communication business on the mobile learning platform can help teachers and students have multi person online discussions.

In order to achieve better remote teaching effect of welding processing course, the teaching mode should first highlight the leading role of teachers; Second, we should embody the subjectivity of students; Third, we should diversify the teaching contents and forms. Based on the constructivist teaching view, this study adopts the 5E Teaching mode. Table 1 shows the specific contents of 5E Teaching Mode of welding processing course.

Table 1. 5E teaching mode of welding processing course

Stage	Teaching subject
Knowledge introduction	Based on the students' existing experience, set up welding processing teaching situations and activities to stimulate students' interest in learning, expose wrong concepts and participate in new concepts, so as to lay the foundation for the next stage
Research on knowledge points	Teachers are promoters. Teachers initiate the introduction of knowledge points of teaching content to guide students to operate and think. In addition, teachers can import relevant welding processing materials, welding processing videos, simulation analysis and other tangible materials and specific welding experience
Knowledge point explanation	Guide students to express their understanding of the knowledge points of welding processing course in their own words. Teachers use demonstration videos, simulation teaching and other methods to elaborate concepts to help students have an in-depth understanding of the knowledge points
Refinement of knowledge points	Teachers provide relevant situations, expand students' conceptual understanding, and students apply new knowledge points to the problem-solving process, which is a process of continuous refinement of new knowledge points
Learning effect evaluation	Teachers evaluate students' understanding and application. Teachers can understand the effect of teaching through observation, questioning, paper and pencil test, group discussion and so on

For the welding processing course, students not only need to complete various basic tasks in traditional teaching activities, but also need to be familiar with the processing parameters of different welding technologies, precautions for the use of welding equipment and safety protection in the process of welding processing. These are unable to rely solely on teachers to achieve the best teaching effect by means of explanation or PPT through mobile learning platform. For the above details, this study will use a variety of learning methods to complete the course teaching tasks, such as constructing scenarios, flipping classroom and interactive collaborative learning.

2.3 Personalized Mobile Learning Resource Recommendation and Path Generation

The teaching goal of each chapter is to provide students with time to understand, and each teaching chapter should have its own teaching objectives. Each teaching chapter should have its own teaching objectives. It should not take up too much time for students to explain. Secondly, the differences of students should be targeted. Students' acceptance speed and thinking mode of knowledge are different. Using the mobile learning platform, teachers can set different personalized self-learning schemes for students.

Students are independent individuals, and their will is not transferred by the will of teachers. Their choice hobbies and visual preferences have their personalized characteristics. In order to meet the needs of different students' learning and development, the construction of teaching content should also be diversified. Personalized learning is a learning method in which learners independently select learning resources, formulate appropriate learning strategies, and independently arrange learning time and place according to their own interests and needs. It has the characteristics of active exploration, high autonomy and on-demand learning. This learning method can not only promote learners to learn more effectively and actively, but also give full play to learners' personality and ability in the learning process. In the process of learning using mobile learning platform, learners can select various forms of learning resources according to their own learning basis, which greatly improves learners' autonomy and learning interest. At the same time, it can also better cultivate learners' awareness and self-study ability. Therefore, according to the teaching characteristics of welding processing course, this paper will generate personalized mobile learning resource recommendation and path when students use mobile learning platform for learning.

This paper uses collaborative filtering algorithm to realize personalized learning resource recommendation by analyzing the students' learning data collected on the mobile learning platform. Figure 3 is the flow chart of collaborative filtering algorithm for personalized recommendation of learning resources to students [9].

In the process of distance teaching, the differences of students' different understanding ability and knowledge mastery ability increase the students' demand for personalized learning path. This study will use the firefly algorithm to input teachers into the teaching design of mobile learning platform to generate different learning paths, so as to recommend suitable learning paths for students.

The firefly algorithm uses all feasible solutions in the search space to simulate the firefly individuals in the night sky, simulates the search and optimization process as the mutual attraction and position moving and updating process of firefly individuals, determines the position of firefly individuals by the fitness value of the objective function to solve the problem, and compares the process of survival of the fittest to the iterative process of replacing poor feasible solutions with good feasible solutions in the process of search and optimization.

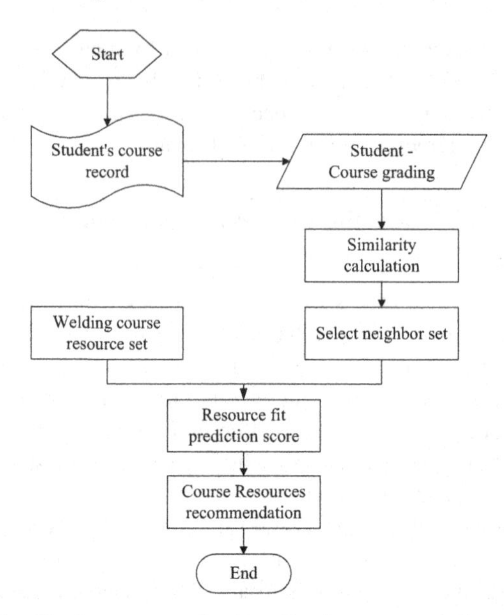

Fig. 3. Collaborative filtering process of personalized recommendation of learning resources

Firefly algorithm involves two key factors: relative fluorescence brightness and relative attraction. The brightness of the firefly's light depends on the target value of its position. The higher the brightness, the better the position, that is, the better the target value. The brighter the fire moves in this direction, the higher the attraction of the firefly. If the luminous brightness is the same, the fireflies move randomly. The brightness and attraction are inversely proportional to the distance between fireflies, and both decrease with the increase of distance. In this paper, the firefly algorithm is improved by using decision attributes to solve the problem of learning path generation in mobile learning environment.

n firefly individuals are randomly distributed in a K dimensional target search space, and each firefly carries luciferin y_i. Firefly individuals emit a certain amount of fluorescence, interact with the surrounding firefly individuals, and have their own decision domain $d_k^i\left(0 < d_k^i < d_s\right)$. In the initial firefly, each firefly carries the same luciferin concentration y_0 and sensing radius d_0. The algorithm implementation includes four important formulas: fluorescein update, firefly movement probability, position update and dynamic decision radius update, as shown in the following [10].

The calculation formula of fluorescence brightness is as follows:

$$I(d) = I_0 \exp(-\xi d) \tag{1}$$

where, I_0 is the maximum fluorescence brightness of firefly, that is, its own fluorescence brightness; Related to the objective function value, the better the objective function value is, the higher its brightness is; ξ is the light intensity absorption coefficient, because the fluorescence will gradually weaken with the increase of distance and the absorption of media, so the light intensity absorption coefficient is set to reflect this characteristic. d is usually the Euclidean distance between firefly i and j.

The fluorescein update formula is as follows:

$$y_i(t+1) = (1-\delta)y_i(t) + \xi F(x_i(t)) \tag{2}$$

where x_i is the component of firefly i in dimensional space, that is, the position of firefly.

The probability calculation formula of selecting individual j to move to the adjacent set domain is as follows:

$$p_{ij} = \frac{y_j(t) - y_i(t)}{\sum y_j(t) - y_i(t)} \tag{3}$$

The individual location update formula of firefly is as follows:

$$x_i(t+1) = x_i(t) + s\left[\frac{x_j - x_i}{\|x_j - x_i\|}\right] \tag{4}$$

The dynamic decision radius is as follows:

$$d_k^i(t+1) = \min\left\{d_s, \max\left\{0, d_k^i(t) + \gamma n_i - \gamma N_i(t)\right\}\right\} \tag{5}$$

where, γ is the attraction of light source to fireflies; n_i is the number of firefly decision-making groups; $N_i(t)$ is the number of fireflies in the neighborhood of the current location.

The essence of the recommendation of learning resources and the generation of learning paths is the process of decision-making, screening and selection of learning resources one by one according to the characteristics and needs of learners. Learning resources include several learning objectives, namely skill objectives. Each skill objective should be carried out according to the set decision attributes.

The four elements that need to be tested before the learner's ability to master the activity are pre test; At the same time, after the completion of mobile learning activities, post test is set for feedback and evaluation to test learners' learning results; By comparing pre test and post test, it can effectively reflect learners' attitude and the rationality of learning path. In order to effectively generate the optimal learning path, according to the ability-based theory, set the following five decision attributes for each skill goal: importance, relevance, pre task, professionalism and learning time, so as to judge the learning value of each goal and its position in the learning sequence.

Aiming at the improvement of firefly coding, a variable sa_i is added to the algorithm to represent the learning ability of learners; Add variable lt_i to represent the lower limit of learners' expected learning time; Adding variable et_i represents the upper limit of learning time expected by learners; Add an array variable $A[]$ to represent the learner's learning objectives. The following personalized learning path decision-making objective function is established from five aspects: the proportion of learning resources covering target knowledge points, the matching of learning resources, students' learning time expectation, the distribution of learning resources and unit learning income:

$$\min F(x) = \sum_{i=1}^{q} w_i F_i \tag{6}$$

where, w_i is the weight of different influencing factors. The improved firefly algorithm combined with decision attributes is used to screen learning resources according to learners' learning objectives. Finally, recommend appropriate learning resources and learning paths to learners for mobile learning. Through the above research, this paper analyzes the current ways and methods of distance teaching of welding processing courses in Colleges and universities for mobile learning platform, which provides theoretical support for the development of online teaching period and subsequent distance learning and further study of relevant personnel.

3 Example Verification

The distance teaching method of welding processing course for mobile learning platform is proposed above. This section will verify the application effect of this teaching method.

3.1 Validation Preparation

Select two classes a and B from 2021 students in a university as the example verification object. In order to ensure the effectiveness and practical application effect of the teaching method, the teaching method proposed in this paper is compared with the traditional distance teaching method, and applied to the teaching of welding processing course of class A and class B students respectively. A total of 50 students in class A as the experimental group, using the teaching methods proposed in this paper; A total of 50 students in class B served as the control group, using traditional teaching methods; The teachers in the experimental group and the control group are the same. The teaching practice of this study lasted 35 weeks, and the learning contents of the two classes were consistent with the teaching plan.

By comparing the test score distribution and learning intention of students in the two classes in the welding processing course, the advantages and disadvantages of teaching methods are evaluated.

3.2 Verification Results

Table 2 shows the score distribution of welding processing course test scores of students in the experimental group and the control group.

By analyzing the data in Table 2, it can be seen that when using the teaching method proposed in this paper to teach the welding processing course, there are 27 students with good grades in the experimental group, which is higher than that in the control group. Moreover, the average test scores of the students in the experimental group in the welding processing course have increased by about 9.2%, which proves that the teaching effect of using this method is better.

Table 3 shows the comparison of students' willingness to learn teaching methods when using different teaching methods.

Table 2. Comparison of score distribution of students' welding processing course test scores

Score section	Experience group		Control group	
	Number of people	Proportion	Number of people	Proportion
90–100	11	0.22	6	0.12
80–89	16	0.32	13	0.26
70–79	18	0.36	20	0.40
60–69	4	0.08	8	0.16
<60	1	0.02	3	0.06
Average	81.54		74.68	

Table 3. Comparison of students' learning enthusiasm for teaching methods

Learning enthusiasm Very positive	Experience group		Control group	
	Number of people	Proportion	Number of people	Proportion
Positive	14	0.28	5	0.1
commonly	20	0.40	11	0.22
Not active	10	0.20	18	0.36
Extremely inactive	6	0.12	12	0.24
Learning enthusiasm	0	0	4	0.08

By analyzing the data in Table 3, we can see that using the teaching method proposed in this paper, students have stronger learning enthusiasm and are more willing to carry out distance teaching, which also reduces the risk of poor learning effect due to students' weak self-control in distance learning.

To sum up, the distance teaching method of welding processing course for mobile learning platform proposed in this paper has good teaching effect, which can significantly improve students' academic performance and mobilize students' learning enthusiasm.

4 Conclusion

Online teaching has promoted the reform of teaching methods and educational concepts, and with the continuous development of network technology, the cost of distance network education has been continuously reduced, which has laid the foundation for distance network education to become more popular.

Facing the mobile learning platform, this paper proposes a distance teaching method suitable for welding processing course. On the basis of determining the communication path of remote teaching of welding processing course, the business process and 5E teaching mode of the remote teaching platform are designed, and the personalized recommendation process of mobile learning resources and the course learning path are specified.

The effectiveness and feasibility of the teaching method are verified by an example. Students use the mobile learning platform for learning. The learning of welding processing course knowledge is not limited to the traditional classroom. Students can carry out learning according to their own situation without the influence of time, place and other factors. At the same time, it can also better increase students' enthusiasm for autonomous learning, improve students' learning efficiency and provide more help for students. In addition, it is of great significance to promote the research of mobile learning in Colleges and universities and promote the in-depth integration of welding processing courses.

However, due to the limitation of research time and other conditions, this method takes a long time to generate the personalized recommendation process of mobile learning resources and the course learning path, which reduces the overall work efficiency. In future research, optimization will be carried out in this aspect.

References

1. Zhang, J., Wang, H., Ban, J.: An optimal design of vocal music teaching platform based on virtual reality system. Comput. Simul. **38**(06), 160–164 (2021)
2. Chen, L., Jing, Y.: Design of mobile learning platform based on deep learning. Mod. Electron. Tech. **43**(14), 177–179 (2020)
3. Shen, X., Zhou, M., Wang, Y., et al.: Exploration and practice of experimental teaching method in distance education. Res. Explor. Lab. **40**(03), 221–224 (2021)
4. Ma, X., Chen, Q., Zhang, Y., et al.: Exploration and practice of innovation and development in welding experiment teaching. Res. Explor. Lab. **41**(01), 245–248 (2022)
5. Liu, Y., Tian, X., Xu, K.: Effect of Web-based teaching on medical education during COVID-19 pandemic in China. Basic Clin. Med. **41**(08), 1242–1246 (2021)
6. Jia, Y., Zhang, Y., Ma, C., et al.: Application of mixed teaching model based on mobile learning platform and seminar teaching method in the teaching of nursing management. Chin. Nurs. Res. **34**(17), 3161–3163 (2020)
7. Fan, J., Peng, Y., Feng, Y., et al.: Teaching experimental platform of high frequency inverter welding power source. Exp. Technol. Manage. **37**(08), 114–118 (2020)
8. Gan, T.: Research on university distance teaching quality evaluation based on decision tree classification algorithm. Mod. Electron. Tech. **44**(09), 171–175 (2021)
9. Wang, X., Wang, J.: Substantial equivalence: realization and transcendence of minimum goal of online teaching during the epidemic period. E-education Res. **42**(03), 42–47 (2021)
10. Song, J., Zheng, J., Ge, Y., et al.: Design and implementation of flipped classroom in the teaching of veterinary operative surgery based on mobile cloud teaching platform—a case study of "Moso Teach". J. Anhui Agric. Sci. **48**(11), 275–276+279 (2020)

Automatic Recognition Method of Table Tennis Motion Trajectory Based on Deep Learning in Table Tennis Training

Liusong Huang[1(✉)], Fei Zhang[2], and Yanling Zhang[3]

[1] Software Engineering, Maanshan Teacher's College, Ma'anshan 243041, China
huangls2016@126.com
[2] Anhui University of Technology, Ma'anshan 243032, China
[3] Department of Computer and Art Design, Henan Vocational College of Light Industry, Zhengzhou 450001, China

Abstract. In table tennis training, in view of the problem of large errors when tracking fast moving targets, this study proposes an automatic identification method of table tennis motion trajectory based on deep learning. The multi-view image of the target object is collected by a multi-eye camera and a stereo image pair is formed. After stereo matching, the three-dimensional coordinate group of the target object is obtained by using the three-dimensional positioning principle of stereo vision. In the three-dimensional coordinate system, the mathematical model of table tennis motion is established. The initial position of the table tennis ball is detected by the Vibe algorithm, and the target area frame is marked, and the frame of the detected target area is used as the first frame of the KCF target tracking to track the table tennis ball. Based on this, a rotating table tennis trajectory recognition network is constructed based on LSTM. The experimental results show that the total trajectory error of this method is 39.00 mm, which can accurately identify the motion trajectory.

Keywords: Table tennis training · Deep learning · Trajectory recognition · Table tennis · Motion trajectory · Automatic recognition

1 Introduction

Table tennis is a sport with strong competitive, ornamental and interesting, it is very popular around the world and has a very strong mass base and enthusiasm in China. In table tennis international competitions, Table tennis players in China often achieve excellent results.

Although table tennis and tennis are both net-confrontation sports, table tennis is small in size and field, so it has higher requirements for the referee. When the tennis ball is rubbed, the referee can easily cause a misjudgment. At the same time, in the table tennis competition, because of the uncertain trajectory of the spinning ball, there

W. Fu and L. Yun (Eds.): ADHIP 2022, LNICST 468, pp. 191–204, 2023.
https://doi.org/10.1007/978-3-031-28787-9_15

are higher requirements for the skills and qualities of the athletes of both sides in the competition.

If the athlete can systematically observe and analyze the running trajectory of the spinning ball in the usual training, it will be of great help to the improvement of the athlete's skills. ITTF and some table tennis event organizers have repeatedly proposed to introduce "Eagle Eye" technology into table tennis competitions. "Eagle Eye" technology was widely used in tennis matches as early as 2003, and then in badminton and volleyball matches. "Eagle Eye" technology overcomes the blind spot and limit of human eyes. Eight or ten high-resolution fast black-and-white cameras are set around the court to quickly capture the movement data of the ball from different angles, and then the flight trajectory of the ball is calculated by the computer to simulate the landing position of the ball, so as to help the referee make a correct judgment.

For "Eagle Eye System" applied to the table tennis game, the main task is to obtain the real-time centroid position information of the table tennis ball and draw the three-dimensional centroid motion trajectory and rotation trajectory of the table tennis ball. In the field of ball games, coaches, ball players and game referees all want to improve their technical level and ability, but all of this is based on fully mastering the law of ball games, and all of this is not clear and accurate to get the law of ball games with human eyes.

In the field of table tennis robot research, it involves many aspects of knowledge learning, such as deep learning, robot kinematics, robot control, etc., which has very high research value and significance. Table tennis has the characteristics of fast moving speed, high-speed rotation, small size and small mass, which brings great research difficulty to the study of table tennis. Table tennis is about skills, with various tactics and increasingly perfect playing methods. Therefore, if a table tennis player or enthusiast wants to improve their skills, the awareness of hitting the ball and the ability to control the ball will be the most important necessary qualities. One is the need for more scientific training.

The traditional research method is mainly based on the color, outline and other characteristics of the table tennis ball, but the surrounding environment in the actual scene will bring great interference to the detection algorithm, resulting in the problem of low detection accuracy, and it needs to be set according to the color of the table tennis ball. The detection threshold is determined, resulting in the limitation of application scenarios. As a substitute for the human eye and human brain, computer stereo vision can obtain the various motion parameters of the ball in the process of movement through the high-speed camera shooting of the ball motion picture and the computer analysis of the shooting picture, so as to master the ball game movement rules. Most of the table tennis trajectory prediction methods are based on the establishment of a physical model, and then the parameters of the model are solved. When predicting the trajectory, the physical model is mainly based on the information of the current moment and the information of the previous moments to make the next position information. Prediction, and then use the predicted information to predict the next moment. When long-term prediction is required, errors are easy to accumulate and the prediction accuracy decreases.

Using scientific and technological means to detect the landing point of table tennis and reproduce the movement trajectory of table tennis to train the awareness of the landing point will effectively improve the ball control ability of table tennis players and contribute to the long-term development of table tennis professional training related fields. In this paper, an automatic identification method of table tennis motion trajectory is proposed based on deep learning, and the method is applied to table tennis training to help athletes better understand the running trajectory of table tennis.

2 Method Design

2.1 Establish the Mapping Relationship from Pixel Coordinates to Three-Dimensional Coordinates

Table tennis has the characteristics of fast speed and small size. In order to ensure that the trajectory of the ball can be captured in real time, a camera with a high frame rate must be used for shooting. Three high-speed industrial cameras are used to extract three-dimensional coordinates of the flight trajectory of table tennis. The "binocular vision system" can also be called "stereoscopic vision", which simulates the principle that the retinas of the two eyes are different when a person observes an object, and the brain fuses them to obtain a sense of depth. Two cameras with a fixed distance are used to obtain two images. Two-dimensional images with different angles, and then calculate the parallax between the corresponding points by the principle of triangulation, and obtain the information of the object in the three-dimensional space.

Let two cameras be in different perspectives, shoot the same target scene at the same time, obtain two two-dimensional plane images through the imaging principle of the camera, and calculate the coordinate deviation between the spatial points mapped to the left and right image pixels, thereby obtaining the spatial information of the three-dimensional scene. Firstly, multi-vision camera calibration is performed to obtain the mapping relationship between image pixel coordinates and world 3D coordinates.

The specific process of multi-vision 3D positioning is as follows: first, the multi-view image of the target object is collected by the multi-eye camera to form a stereo image pair, secondly, the multi-view image stereo matching of the target object is performed, and then the three-dimensional coordinate system of the target object is obtained by using the principle of stereo vision 3D positioning.

The imaging principle of the camera is similar to the ideal model of pinhole imaging. Since the field of view of the camera is very small, it can be regarded as a small hole through which the subject is projected onto the photosensitive element. However, the camera usually needs to be equipped with various convex or concave cameras, which will lead to imaging distortion and become a non-linear model, so the internal and external parameters of the camera need to be corrected. When the left and right cameras are placed completely horizontally, the left and right imaging models of binocular vision will only have deviations in the horizontal direction, but there will be no deviations in the vertical direction. Assuming that the camera coordinate system of the left camera in the stereo vision system coincides with the world coordinate system, and the left camera is set as the main camera, the relationship between the image physical coordinate system and the camera coordinate system is:

$$\gamma \begin{bmatrix} p \\ q \\ 1 \end{bmatrix} = \begin{bmatrix} w_1 & 0 & 0 \\ 0 & w_1 & 0 \\ 0 & 0 & 1 \end{bmatrix} \begin{bmatrix} \alpha \\ \beta \\ \gamma \end{bmatrix} \tag{1}$$

In formula (1), w_1 and w_2 represent the focal length of the camera in stereo vision; p and q represent the coordinates of the spatial target point mapped in the physical coordinate system of the stereo vision image; α, β and γ represent the three-dimensional coordinates. The transformation relationship between the geometric position of a point on the object in the three-dimensional space and the point mapped to the two-dimensional plane image depends on the camera imaging model parameters [1].

In general, camera parameters need to be obtained through experiments and corresponding calculations, and this process is camera calibration. Both the image coordinate system and the pixel coordinate system are coordinate systems on the imaging plane, but the unit of measurement is different from the origin. The unit of the image coordinate system is mm, the origin is the midpoint of the imaging plane, and the unit of the pixel coordinate system is pixel. Upper left corner of the image. The pixel coordinate system can be obtained by discretizing and offsetting the image coordinate system. Assuming that the camera coordinate system of the right camera has a rotation matrix relative to the camera coordinate system of the left camera, the expression of the three-dimensional coordinates can be obtained as:

$$\begin{cases} \alpha = \frac{\gamma w_1}{h_1} \\ \beta = \frac{\gamma w_2}{h_1} \\ \gamma = \frac{w_1(w_1 h_2 - w_2 h_3)}{w_1 h_1 - w_2 h_2} \end{cases} \tag{2}$$

In formula (2), h_1, h_2 and h_3 are the intrinsic parameters of the left and right cameras. For binocular camera calibration, it is necessary to determine the internal and external parameters of the camera itself. The internal parameters are related to the attribute structure of the camera itself, including focal length, the coordinates of the center point of the imaging plane, the physical size of the pixel, and the distortion parameters. The internal parameters are generally fixed. The external parameters describe the positional relationship between the camera and the world coordinate system, including the third-order orthogonal rotation matrix and the three-dimensional translation vector. Three cameras are used for image acquisition, so two-to-two interactions can form up to seven sets of stereo vision pairs, and up to seven spatial three-dimensional coordinate values of target points can be obtained at each moment [2]. The average value of the stereo vision coordinate values of each group is calculated, and finally the three-dimensional coordinate value of the target point can be obtained.

2.2 Mathematical Modeling of Table Tennis

Since the flight motion of the no-spin table tennis ball can be regarded as a free fall in the vertical direction and a uniform linear motion in the horizontal direction, the ideal state of the no-spin table tennis motion is a flat throwing motion, and its trajectory is a parabola. Its kinematic trajectory is shown in Fig. 1.

Under standard conditions, gravity and air buoyancy are generally constant; air resistance is generally proportional to the square of the flight speed of a table tennis ball, and the direction is opposite to the flight speed [3].

In the space three-dimensional coordinate system o-x-y-z, the trajectory of the horizontal throwing motion is only on the x-o-z plane, and the coordinate origin is established at the vertex of the trajectory. At this time, the trajectory equation is related to the

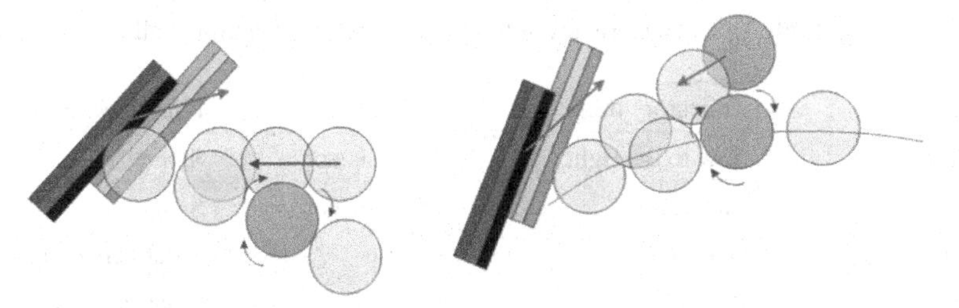

Fig. 1. Kinematic trajectory of table tennis

speed of the horizontal throwing. At present, the commonly used sampling strategies are divided into proportional sampling method and proportional modified symmetric sampling method. Compared with the former, the latter can solve the problem of non-local effects of sampling [4]. Therefore, the sampling strategy for generating the sigma point set in this paper is the proportional correction symmetry method. Among them, the first point is the mean value of the current state, and the other points use symmetrical sampling. Assuming that the coordinate position of the ping-pong ball at the initial moment is (x_0, y_0, z_0), the ping-pong ball moves at the initial velocity u_0, and the angle between u_0 and the xoy surface is ϑ_1, and the angle between the projection of u_0 on the xoy surface and the x-axis is ϑ_2, then the mathematical model of table tennis without rotation at any time is:

$$\begin{cases} x = x_0 + u_0 \cos\vartheta_1 \cos\vartheta_2 \\ y = y_0 + u_0 \cos\vartheta_1 \sin\vartheta_2 \\ z = u_0 \sin\vartheta_1 \end{cases} \tag{3}$$

In formula (3), (x, y, z) represents the position coordinates of the no-spin table tennis ball at any time. In table tennis training, the movement of table tennis is no longer a simple flat throwing movement. Athletes tend to serve spinning balls, and due to the viscous nature of air, spinning table tennis balls are not only affected by gravity, buoyancy, and air resistance, but also by Magnus force. According to hydrodynamics, the pressure on the mountain of fluid with different flow rates is different, so there will be a pressure difference on both sides of the table tennis ball, that is, Magnus force. It can be seen that the Magnus force is not only caused by the rotation speed, but caused by the combined action of the rotation speed and the flight speed.

Compared with the non-spin table tennis ball, the motion model of the spinning table tennis ball is a time-varying nonlinear model [5]. If we ignore the influence of friction on the rotation of the table tennis ball in the air, the attenuation of its rotation speed is very small, so this paper can regard the rotation speed as a constant value.

The trajectory characteristics of the eight kinds of spinning balls compared to the no-spinning balls are shown in Table 1.

When the rotation speed and the flight speed have a large angle and the rotation speed is high, the movement trajectory of the table tennis ball will have a large deviation, that is, the flight trajectory of the rotating table tennis ball is an arc. In the actual movement process of table tennis, it is often not a simple rotation around the coordinate axis, and its

Table 1. Characteristics of the trajectory of the spinning ball

Project	Trajectory trend	Flight arc	Flight distance
Left hand	Yaw to the right	Consistent	Same
Swing ball	Yaw left	Consistent	Same
Topspin	Decline	Little	Relatively close
Backspin	Rise	Big	Relatively far
Left backspin	Turn right, rise	Big	Relatively far
Topspin	Turn right, descend	Little	Relatively close
Right backspin	Turn left, rise	Big	Relatively far
Topspin	Turn left, descend	Little	Relatively close

rotation direction and rotation axis are arbitrary. Therefore, we need to study the rotation around any axis.

The magnitude of air resistance is proportional to the square of the flight speed, and the proportionality factor is determined by the air resistance coefficient, air density and the cross-sectional area of the ping pong ball. The direction of air resistance is opposite to the direction of flight speed. For the rotation of the table tennis ball around any axis, this paper adopts the method of coinciding the rotation axis and the coordinate axis, and then performing the rotation transformation. When the table tennis ball has both flight speed and rotation speed and the directions of flight speed and rotation speed are not parallel, the relative motion between one side of the table tennis ball and the air becomes larger due to the superposition of the rotation speed and the flight speed, and the relative motion between the other side and the air becomes larger. Relative motion is reduced due to the mutual cancellation of rotational speed and flight speed.

Translate the object so that the rotation axis passes through the coordinate origin, rotate the object around the x axis, and make the rotation axis rotate to the xoz plane, the rotation angle is φ_1; rotate the object around the y axis, so that the rotation axis coincides with the z axis, and the rotation angle is φ_2; therefore, the rotation transformation matrix to rotate the angle η around any axis is:

$$B = A_1(\varphi_1)A_2(\varphi_2)A_3(\eta)A_1^{-1}(\varphi_1)A_2^{-1}(\varphi_2)A_3^{-1}(\eta) \tag{4}$$

In formula (4), B represents the rotation transformation matrix that rotates the angle η; A_1, A_2 and A_3 represent the rotation transformation matrix of the point around the x-axis, y-axis and z-axis. The formula for calculating the rotation transformation matrix of any point around the x-axis is given below:

$$A_1(\varphi_1) = \begin{bmatrix} x\ y\ z\ 1 \end{bmatrix} \begin{bmatrix} 1 & 0 & 0 & 0 \\ 0 & \cos\varphi_1 & \sin\varphi_1 & 0 \\ 0 & -\sin\varphi_1 & \cos\varphi_1 & 0 \\ 0 & 0 & 0 & 1 \end{bmatrix} \tag{5}$$

Similarly, the rotation transformation matrix of any point around the Y-axis and the z-axis can be obtained. Given the initial motion state, the continuous motion model can directly calculate the motion state of table tennis at any time without iteration, which not only effectively eliminates the iteration error and cut-off error, but also directly describes the trajectory position time series and the initial motion state. It can effectively use the time series information of the trajectory position of consecutive multiple frames, and can conveniently obtain the gradient of the trajectory position time series relative to the initial motion state [6]. Thus, the mathematical model of the movement trajectory of the table tennis ball is obtained.

2.3 Motion Target Detection in Table Tennis Training

A complete moving target detection mainly includes three parts: preprocessing, detection algorithm and motion area analysis and processing. The specific table tennis detection flow chart is shown in Fig. 2.

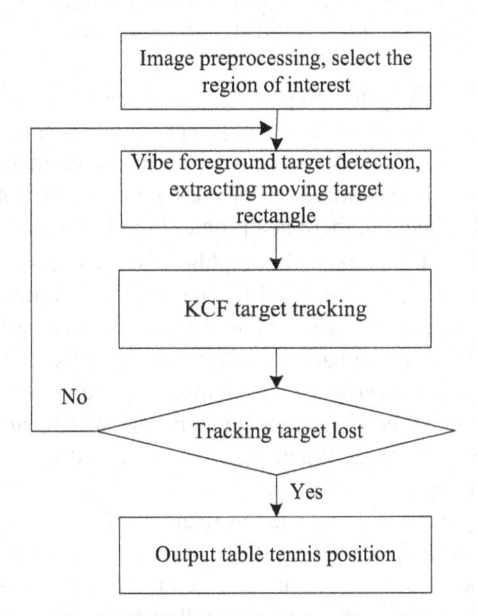

Fig. 2. Table tennis detection flow chart

In this paper, the color information of the image is used to overcome the problem that the result of the difference cannot accurately obtain the moving target area. The image is directly subjected to the difference operation in the color space and the image is binarized by the color threshold, so that the color of the image is preserved information. KCF combines the cyclic matrix in the concept of correlation filtering to solve the problem of a large number of missing negative samples in the target tracking process. The introduction of circulant matrix can increase the number of negative samples, and at the same time can avoid complex matrix inversion operations, effectively reducing the computational load of the classifier.

The Vibe target detection algorithm includes three steps: the construction and initialization of the background model, the detection of foreground moving targets, and the update of the background model set. The image is binarized on the three components of R, G, and B, respectively, and the image is divided by the color threshold, that is, the pixels belonging to different color ranges in the image are distinguished by the set color threshold. For subsequent video frames, the Vibe algorithm uses the two-dimensional Euclidean distance to classify and judge the pixels, and realizes by calculating the Euclidean distance between the pixel value of the current pixel position and the pixel value of the corresponding pixel position in the background model set, namely:

$$H(g_1, g_2) = \sum_{j=1}^{3} (g_1 - g_2)^2 \tag{6}$$

In formula (6), g_1 represents the pixel value of the same pixel point in the background model set; g_2 represents the pixel value of the pixel point of the current input video frame; j represents the three channels of R, G, and B in the color image; H represents the pixel value. Euclidean distance.

Since the area occupied by the ping-pong ball in the image is very small, and the distance between the moving balls in two adjacent frames is small and has continuity and correlation, it will waste a lot of time to process the entire image in each frame [7]. A sub-sampling factor ε is introduced. After the background point in the background model set completes a detection, there is a probability of $\frac{1}{\varepsilon}$ to perform its own update, and there is a probability of $\frac{1}{\varepsilon}$ to update its neighborhood samples [8]. When the number of foreground detections of a certain pixel reaches a critical number, it is immediately set as a background point, and has a probability of $\frac{1}{\varepsilon}$ to update its own background model set. The principle of the dynamic window is to set a fixed-size window. When a moving target is detected in a certain frame of the image sequence, the search range of the next frame can be narrowed to the dynamic window, which can not only reduce the calculation time but also It can eliminate external noise and improve the efficiency of motion trajectory tracking.

Use the spatial propagation between pixels to update the neighborhood to ensure that the update of the background model is gradually diffused outward. When replacing the sample value, a sample value is randomly selected to update to ensure that the sample value has a smooth life cycle. The probability that the sample value remains in the minimal time is:

$$W(\tau) = e^{-\ln\left(\frac{c}{c-1}\right)d\tau} \tag{7}$$

In formula (7), $W(\tau)$ represents the probability that the sample is retained in the minimum time τ; c represents the probability of being updated; e is a natural constant.

The initial position of the table tennis ball is detected by the Vibe algorithm, and the target area frame is marked, and the frame where the detected target area is located is used as the first frame of the KCF target tracking, and the tracking task of the table tennis ball is performed. For the image under the left camera, first determine whether it is the initial frame, if it is the initial frame, first determine the position of the target area, the KCF tracking algorithm obtains from the original image, extracts the features of the moving target, and performs a cyclic shift on it to Training samples to complete the training of the classifier. If the target area is detected in the next frame, continue the tracking process and update the target sample set of KCF. If the target is not detected in the next frame, that is, the target is occluded or lost, the current tracking process is ended. The Vibe algorithm is used again to detect the position of the ping-pong ball, and the frame where the position of the ping-pong ball is detected is taken as the first frame of target tracking. If it is not the initial frame, first determine the position of the candidate region, extract the features of the candidate region and calculate the similarity with the target through the kernel function, and finally select the candidate region with the largest similarity as the final output target, and use it as the tracking target of the next frame [9]. And so on, until the conditions that trigger the system to stop running. In the process of detection and tracking, the coordinates of the center of mass of the tracking frame are always output as the basis for subsequent trajectory recognition.

2.4 Establishment of Automatic Recognition Model of Table Tennis Motion Trajectory Based on Deep Learning

LSTM is a special recurrent neural network, which draws on the long-short-term and forgetting characteristics of human neural memory, and introduces unit state into the neural network, which can solve the long-term dependency that traditional RNN cannot learn and process due to the problem of gradient disappearance. The memory unit of LSTM can consider the connection between distant data, and the flight trajectory of a ping-pong ball is a close connection between long distances. Therefore, this paper attempts to apply LSTM to the trajectory recognition task of table tennis.

This chapter builds an LSTM-based rotating table tennis trajectory recognition network, and trains and optimizes the network. Taking the rotation of the ping-pong ball into consideration, the approximate rotation type of the ping-pong ball is judged by its flight trajectory, which improves the accuracy of the prediction point while ensuring that the detection speed meets the requirements of the vision system. Since the motion state space of table tennis is continuous, the more categories are theoretically classified, the better the adaptability of the extended continuous motion model to table tennis in different motion states [10]. However, the time complexity and space complexity of the algorithm are proportional to the number of categories of the extended continuous motion model. Therefore, considering the accuracy and efficiency of the model, we divide the training data into 8 categories according to the different characteristics of the flight speed decay curve with time. The table tennis trajectory recognition network needs to receive the three-dimensional table tennis position information at the current moment, and output the three-dimensional coordinates of the table tennis ball at the next moment. The basic idea of LSTM is to maintain a state vector internally. During the process of sequence input, the model continuously updates the state vector, and outputs a vector based on the current input and current state. The current moment signal input is combined to output the output value of the LSTM at the current moment.

The overall network structure consists of three parts, namely the input layer, the hidden layer, and the output layer. Input the three-dimensional coordinate information of the ping-pong ball at time t to the network in chronological order. In the LSTM unit, it is necessary to calculate the output of the current state according to the state of the cell and the input, and update the cell information and pass it to the next hidden layer unit, and finally pass the output $t + 1$. The three-dimensional coordinate information of the moment. The output of the LSTM network can be expressed as:

$$Y_t = \theta\left[\varsigma(k_{t-1}, s_t) + \upsilon\right] \tag{8}$$

In formula (8), Y_t represents the output at time t; θ represents the activation function; k_{t-1} represents the output at the previous time; s_t represents the input at the current time; ς represents the update weight; υ represents the bias term. In the research requirements of table tennis trajectory prediction in this paper, it is necessary to predict the long trajectory of table tennis in the future according to the previous piece of table tennis trajectory data. Input at time t + 2, and this cycle can complete the prediction of long-term table tennis trajectory information. Given the same initial motion state, the trajectory prediction based on the continuous motion model only depends on the prediction time and is not related to the motion state of the previous and subsequent frames, that is, the error of the trajectory prediction value of each frame is independent and identically distributed. Then the joint probability density function of the error distribution of the trajectory prediction value of consecutive multiple frames is equal to the product of the probability density function of the error distribution of each frame.

In this paper, the Gaussian mixture model is used to describe the likelihood between the trajectory prediction and the trajectory observation, and the three-dimensional coordinate position of the ping-pong ball with various steps is used as the network input for experiments. In the process of network training, the network evaluates and optimizes the output predicted value according to the known table tennis trajectory information, and calculates the average three-dimensional space distance between the predicted value and the actual value based on the accuracy of the predicted coordinates. The calculation formula is as follows Eq. (9) is shown.

$$L = \sqrt{(F_1 - R_1)^2 + (F_2 - R_2)^2 + (F_3 - R_3)^2} \tag{9}$$

In formula (9), (F_1, F_2, F_3) represents the predicted coordinate value; (R_1, R_2, R_3) represents the actual coordinate value; L represents the three-dimensional space distance between them. The decay curve of flight speed with time of each category has a similar variation law. In the initial stage, the speed in the horizontal direction of the rotating table tennis ball is relatively large, and air resistance plays a major role, so the flight speed is attenuated at this stage. The approximate rotation type and flight direction of the ball are determined by calculating the flight speed in the three coordinate directions by obtaining the first five position information during the flight of the table tennis ball.

As the flight speed decays, the effect of air resistance becomes smaller and smaller until it is close to the effect of gravity. At this time, the size of the flight speed remains basically unchanged (the horizontal component continues to decrease, the vertical component increases, and the total remained largely unchanged). Then gravity plays a major role, and the flight speed starts to increase. Thus, the design of the automatic identification method of table tennis motion trajectory based on deep learning in table tennis training is completed.

3 Experimental Test

3.1 Experiment Preparation

In this experiment, three high-speed black-and-white industrial cameras are used for image acquisition. The camera model is HIKVISON/MV-CA013-21UM, the pixel is 1.3 million pixels, and global exposure is adopted, and the maximum frame number can reach 210 frames. The camera is installed on the straight line where the table tennis net is located, 1 m from the lower edge of the table, and 1.5 m above the plane where the table is located, that is, the upper side of the table. Adjust the shooting angle and focal length of the camera to ensure that the ball can be captured. The complete area of the right table top of the table. Table tennis uses standard game balls with a diameter of 40 mm and a weight of 2.7 g. The table specifications are 2.74 m long, 1.525 m wide and 0.76m high. The experimental simulation platform is Intel®Core™ i7-7700K CPU@4.20 GHz×8 processor, 16 GB memory, TITAN X (Pascal)/PCle/SSE2 graphics card, Ubuntu16.04 operating system.

The 500 ping-pong trajectories collected and the 1,700 ping-pong ball trajectories acquired by the network are used, of which 1,900 are used as training data sets and 300 are used as test data sets. The experiments set the network with stride 10 to train for 1000 epochs. According to the training step size, input the trajectory data of the first 15 frames, and use the errors at 30, 40, 50, and 60 frames to analyze the accumulation of prediction errors with the increase of the number of frames. The prediction speed of 60 frames reaches 85.17 ms, which meets the real-time requirements.

3.2 Results and Analysis

Input the initial 10 trajectory point information of each curve in the test set into the trained LSTM model, and output the three-dimensional coordinate value of the trajectory point in the future accordingly. Calculate the error between the predicted value of the trajectory point and the actual coordinate value.

In order to avoid too single experimental results, the table tennis trajectory automatic recognition method based on deep learning proposed in this paper was used as the experimental group, and the trajectory recognition method based on mean-Shift algorithm and S_Kohonen neural network was selected as the control group. Test the recognition effect of different methods, and compare the average error and total error of different recognition methods on each coordinate axis component. The results are shown in Tables 2, 3, 4 and 5.

In the measurement results of the x-axis component, the average error of the automatic recognition method based on deep learning is 21.05 mm, which is 50.56 mm and 63.14 mm lower than the trajectory recognition method based on the Mean-Shift algorithm and the S_Kohonen neural network.

In the y axis component measurement results, the average error of the automatic recognition method based on deep learning is 21.84 mm, which is 54.98 mm and 62.89 mm lower than the trajectory recognition method based on Mean-Shift algorithm and S_Kohonen neural network.

Table 2. Errors of x axis components (mm)

Testing frequency	Recognition method based on deep learning	Recognition method based on Mean-Shift algorithm	Recognition method based on S_Kohonen neural network
1	21.44	68.42	78.42
2	22.88	72.84	79.84
3	21.65	69.68	82.68
4	21.56	65.20	86.26
5	22.23	78.53	85.03
6	20.32	72.96	85.35
7	20.65	76.39	89.69
8	19.59	74.65	87.22
9	20.92	69.28	84.54
10	19.25	68.14	82.90

Table 3. Errors of y coordinate components (mm)

Testing frequency	Recognition method based on deep learning	Recognition method based on Mean-Shift algorithm	Recognition method based on S_Kohonen neural network
1	18.44	72.03	82.49
2	19.87	73.87	79.88
3	20.68	75.94	84.67
4	25.26	76.68	85.35
5	23.52	79.35	86.23
6	24.95	82.26	88.05
7	19.33	75.53	89.74
8	20.64	78.85	82.51
9	22.50	74.22	85.12
10	23.25	79.48	83.28

In the z-axis component measurement results, the average error of the automatic recognition method based on deep learning is 21.03 mm, which is 55.80 mm and 62.68 mm lower than the trajectory recognition method based on Mean-Shift algorithm and S_Kohonen neural network.

For the table tennis motion trajectory and real trajectory identified by the model, the total trajectory error of the automatic recognition method based on deep learning is 39.00 mm, which is 112.84 mm and 126.59 mm lower than the trajectory recognition

Table 4. Errors of z coordinate components (mm)

Testing frequency	Recognition method based on deep learning	Recognition method based on Mean-Shift algorithm	Recognition method based on S_Kohonen neural network
1	19.64	75.47	82.45
2	18.47	78.81	81.68
3	19.85	75.56	85.32
4	22.68	79.63	86.50
5	24.35	76.35	82.28
6	23.26	74.22	79.64
7	22.53	77.51	85.32
8	21.94	78.04	84.27
9	19.31	77.42	86.41
10	18.27	75.28	83.25

Table 5. Total error of table tennis trajectory (mm)

Testing frequency	Recognition method based on deep learning	Recognition method based on Mean-Shift algorithm	Recognition method based on S_Kohonen neural network
1	35.56	142.08	162.26
2	36.47	145.44	160.44
3	35.84	157.87	158.81
4	42.61	152.55	159.56
5	41.28	161.63	167.37
6	35.53	149.23	168.58
7	38.23	154.52	172.20
8	39.62	138.25	174.54
9	41.54	162.12	175.88
10	43.28	154.74	156.26

method based on the Mean-Shift algorithm and the S_Kohonen neural network. It can be seen from the above research results that the average error and total trajectory error of the table tennis motion trajectory automatic identification method based on deep learning are the smallest compared with the two comparison methods, so this method can be accurately used for training. Table tennis trajectory recognition in

4 Conclusion

The research on the space trajectory of flying objects has great research significance. In this paper, table tennis is taken as the research object, and the research on automatic identification of the motion trajectory is carried out. The method can reduce the average error of the components in each direction of the coordinate axis and the total error of the trajectory, and realize the accurate identification of the movement trajectory of table tennis.

In the training process of table tennis, it is inevitable that there will be some edge balls, tennis touches, etc. This system only considers some common situations. When there are special situations, it is often impossible to accurately identify and judge, thus affecting the applicability of the method. In the follow-up, the special trajectory in table tennis training can be studied to improve the practicability of the automatic trajectory identification method.

Fund Project. Natural Science Research Project of Education Department of Anhui Province "Man-machine comparative verification of table tennis based on multiple feature fusion" (KJ2021A1289).

References

1. Ma, R., Du, Z., Xiao, Y., et al.: Three dimensional trajectory reconstruction of table tennis. Comput. Syst. Appl. **30**(1), 250–255 (2021)
2. Li, J., Li, C., Xie, Y., et al.: Research on the influence of the characteristics of the racquet base plate on the table tennis motion track based on CFD method. Sports Sci. Res. **24**(2), 27–29 (2020)
3. Wang, D., Ruan, C., Yang, Y.: The service trajectory detection method of pneumatic table tennis server based on binocular vision. Autom. Instrum. **3**, 164–167 (2020)
4. Chang, Q., Liu, R.: Intelligent recognition method of bounce trajectory of table tennis under different dynamics. J. Chifeng Univ. (Nat. Sci. Ed.) **36**(04), 54–56 (2020)
5. Ji, Y.: Reverse calculation of rotation using trajectory of table tennis based on genetic algorithm. J. NanJing Sports Inst. **20**(2), 53–60 (2021)
6. Ma, R., Du, Z., Xiao, Y., et al.: Three dimensional trajectory reconstruction of table tennis. Comput. Syst. Appl. **30**(01), 250–255 (2021)
7. Zhang, X., Wang, X.: Continuous correction of moving path error of 3D obstacle space scene target. Comput. Simul. **37**(5), 420–424 (2020)
8. Wang, D., Ruan, C., Yang, Y.: The service trajectory detection method of pneumatic table tennis server based on binocular VI. Autom. Instrum. **2020**(03), 164–167 (2020)
9. Li, B., Jin, P., Wu, Z.: Design of ping-pong recognition based on S_Kohonen neural network. J. Huazhong Univ. Sci. Technol. (Nat. Sci. Ed.) **48**(3), 52–56 (2020)
10. Kong, X., Shi, J., Gao, L.: Detection and trajectory tracking of spherical objects motion. Comput. Digit. Eng. **48**(7), 1606–1610, 1681 (2020)
11. Zhang, X., Wang, W., Tang, Y., et al.: Real-time table tennis motion pattern recognition based on ESP32. Comput. Syst. Appl. **31**(6), 117–124 (2022)

Construction of Social Network Big Data Storage Model Under Cloud Computing

Zihui Jin[1](✉) and Ting Chen[2]

[1] Tianfu College of Swufe, Chengdu 621050, China
jinzihui@tfswufe.edu.cn
[2] Chengdu College of University of Electronic Science and Technology of China, Chengdu 611731, China

Abstract. In order to better store massive network information data effectively, a cloud computing based method for building a big data storage model for social networks is proposed, which combines cloud computing theory to identify user interest data of social network data, constructs a hierarchical user interest conceptual model, and improves the ability of identifying data interest based on Cloud Computing; According to the process of the impact of network information dissemination, the evaluation algorithm of social network data storage is optimized, and the classification management is carried out based on the calculation results, so as to achieve the goal of classification storage of massive data. Finally, the experiment proves that the model built has high practicability in practical applications.

Keywords: Cloud computing · Big data · Social networks · Storage model

1 Introduction

With the development of Internet technology and social media, the network has been deeply integrated into people's daily work and life. People are used to publishing and sharing information using the network platform, and are used to obtaining information and increasing knowledge through the network. The wide use of social network big data and computers facilitates people to obtain and store data, which makes the network data grow exponentially. The amount of data has to be measured in ZB. Many important information and knowledge are hidden behind the surge of data [1]. Social network big data under cloud computing theory is relative to information cognitive ability. Compared with massive data, what is really meaningful to mankind is the knowledge behind it. People hope to conduct a deeper analysis of social network big data in order to make better use of data and absorb knowledge. Therefore, the application of social network big data has become the focus of attention and a scientific problem that people urgently explore.

Reference [2] proposes multi relationship group impact modeling in online social networks. Starting from the types of social relationships among users, it classifies and mines the complex network topology relationships in online social networks, analyzes

W. Fu and L. Yun (Eds.): ADHIP 2022, LNICST 468, pp. 205–216, 2023.
https://doi.org/10.1007/978-3-031-28787-9_16

the online group environment of different dimensions that users may perceive, and proposes the definitions and mining methods of static group environment and dynamic group environment In different online social group environments, the group structure characteristics perceived by users in the environment are quantified from a macro perspective, and the influence mechanism among users is modeled and simulated from a micro perspective. Reference [3] proposes the research on competitive nonlinear dynamic information dissemination model of online social networks, analyzes the internal relationship between the competition mechanism between different types of information on the network, node state transformation and information dissemination evolution law by using Markov chain theory, constructs a probability model of node state transformation from the perspective of probability, and constructs a nonlinear dynamic information dissemination model of network system information diffusion from the perspective of statistics; The equilibrium point of the propagation dynamics differential equation of the proposed model is solved and its stability is analyzed; Through the simulation analysis of the relative change relationship between the parameters in the model, the process of information competition and dissemination is simulated. The era of social network big data has changed the way people collect, disseminate and analyze information. For managers and users, it is more necessary to improve the means and level of traditional data management and use. Clarifying data association, eliminating the deviation caused by data redundancy and data loss, and mining the information, knowledge or wisdom behind the data are the key problems to be solved in the analysis and utilization of cloud computing. For social network big data, the content that people hope to obtain through the network can be divided into two categories: the lowest and most original data set, which can be called "objective reality data" and the most valuable information with the highest matching degree of self cognition, which can be called "objective information". Objective reality data and objective information can be transformed into information, knowledge or wisdom through in-depth mining and utilization, and finally contribute to the improvement of human innovation ability. Objective reality data exist without the influence of any subjective cognition and judgment. The objective information is often the description and judgment of the original data, which has been processed and analyzed by people. To some extent, objective information reflects people's current cognitive ability and knowledge structure, which may be updated by users when it is reused. Based on this, the construction of social network big data storage model under cloud computing is proposed. Combined with cloud computing theory to identify user interest data in social network data, build a hierarchical user interest conceptual model, and improve data interest recognition ability based on cloud computing; The calculation results are classified and managed to achieve the goal of classifying and storing massive data. The research shows that the constructed model has good effect.

2 Social Network Data Cloud Computing Storage Model

2.1 Social Network Data User Interest Collection Model

In order to realize the construction of lower level secondary user interest model in social network big data environment, the following research ideas are proposed, as shown in Fig. 1:

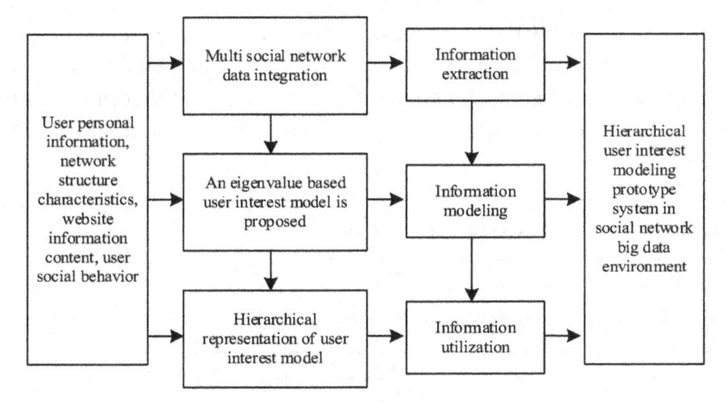

Fig. 1. Hierarchical user interest conceptual model

It can be seen from Fig. 1 that using the hierarchical user interest conceptual model can easily mine user interests from historical records, and conveniently modify and refine the user interest model. The model uses a series of keywords to describe users' interests. In the vector space model, each keyword corresponds to a weight. This method is simple to use and easy to update the user model. At the same time, it requires that these keywords are orthogonal, and does not describe the real relationship between keywords.

Social network big data application is a process of data insight, including "descriptive analysis", "predictive analysis" and "normative analysis" of cloud computing. It aims to analyze valuable "information", "knowledge" or "wisdom" from massive data, finally achieve the goal of improving the ability to support decision-making, and optimize the data interest recognition ability system, as shown in Fig. 2:

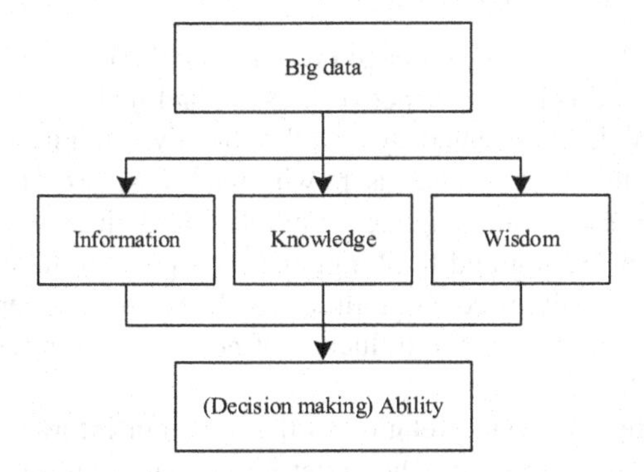

Fig. 2. Improvement of data interest recognition ability based on Cloud Computing

Therefore, for developers and researchers, the goals of social network big data application can be summarized as follows: achieve "insight" based on cloud computing analysis and build a new individual or organizational knowledge structure system; Create or improve the decision-making ability of individuals and organizations [4]. Combined with the goal of social network big data, this paper attempts to decompose the application process of social network big data in stages and construct the application process model, so as to explore the internal evolution logic and external influencing factors of cloud computing applications. We can try to use the corresponding theories of information management and knowledge management to describe its process, as shown in Fig. 3:

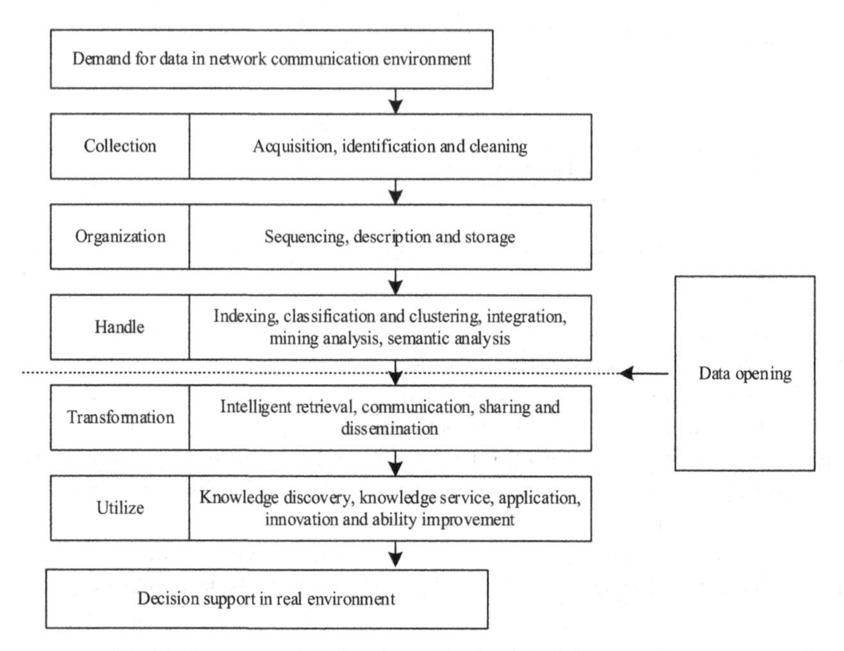

Fig. 3. Process model of network big data acquisition application

In social media, the network language presented in the Pseudo Environment reproduces the cognition and behavior of network users, and realizes the transformation of the subjective reality data presented by users' subjective cognition and behavior into network language symbols [5]. Other users who read these useful data or information will then transform these online language symbols into their own subjective cognition and behavior through reading and thinking, and then produce new continuous offline or online actions, or simply affect cognition, which will have a certain impact on the objective reality. The process of the influence of network communication is shown in Fig. 4.

In the interpretation and interpretation of these data or information, distortion will inevitably occur. For the publisher, when interpreting data or information, the network "non mirror" mimicry propagation leads to the secondary distortion of information. For the receiver, in the process of network data or information interpretation, due to the individual differences of users, there are differences in subjective cognition and behavior, resulting in deviation and secondary distortion. Once propagated again, it will lead to further deviation of distorted information.

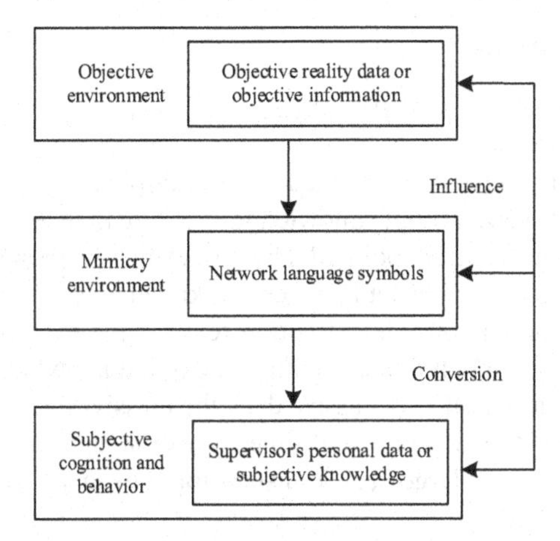

Fig. 4. The process of the influence of network information dissemination

2.2 Social Network Data Storage Evaluation Algorithm

Due to the wide use of the network, the amount of network data is increasing rapidly, but it does not bring the precipitation and accumulation of valuable knowledge, showing a picture of lack of knowledge [6]. Therefore, it is necessary to explore the process and implementation mechanism of knowledge accumulation in the network environment. People carry out information inquiry in cyberspace, which is an information activity consciously engaged in in in order to change the existing knowledge structure. For how to change the existing knowledge structure, this paper puts forward the consistency characteristics of information users' knowledge structure and information needs, and points out that the satisfaction of users' information needs takes self knowledge structure as the starting point. The absorption of information will interact with the recipient's original knowledge structure and produce the recipient's new knowledge structure. The form of Brooks equation is shown in formula (1):

$$K(S) + D_I - \Delta \mathrm{I} \times K(S + D_S) = P \tag{1}$$

In formula (1), $K(S)$ is the user's original knowledge structure; $\Delta \mathrm{I}$ refers to the amount of information absorbed, that is, the information that can be understood and integrated into their own knowledge structure, and $K(S + D_S)$ refers to the knowledge structure formed after absorbing new information. The flow and transformation of knowledge need to be realized among individuals with similar knowledge structure but some differences [7]. Network data or information is transmitted through the Internet platform. When users with data or information needs obtain, clean up, identify, mine and analyze valuable objective data or information with high matching degree from the data set based on the cloud computing processing platform, they can interact with their original knowledge structure, establish a new knowledge structure and form knowledge innovation. Users can also learn and match the old objective information in the data set with the help of artificial intelligence technology, so as to establish a new knowledge structure and form knowledge discovery. Therefore, this study attempts to expand Brooks equation to obtain as shown in formula (2):

$$E = P - K_a(S_a) + K_b(S_b) \tag{2}$$

When users acquire and absorb knowledge for multiple data sets or groups of objective information in data sets, multiple knowledge structure units K_a are transformed into multiple new knowledge structure points K_b by absorbing several matching information or knowledge S_a, so as to establish a new general knowledge structure system S_b. The establishment of this new knowledge structure realizes the explicit transformation of tacit knowledge and the implicit transformation of explicit knowledge [8]. According to the characteristic division results of massive data, the massive data is controlled through protocol transformation to realize the real-time collection, analysis and combination of massive data in the process of large-scale transmission. The model construction is shown in Fig. 5.

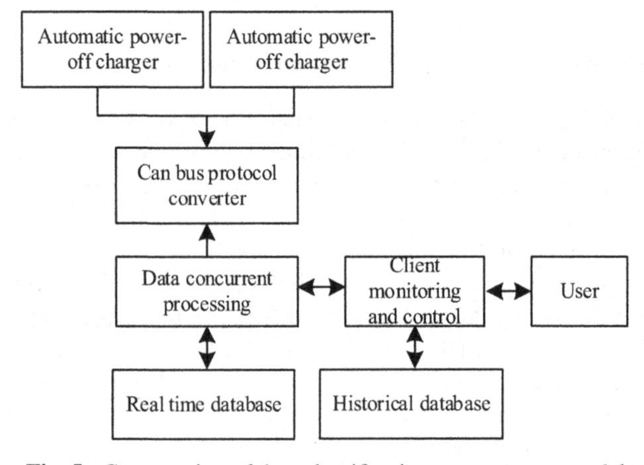

Fig. 5. Construction of data classification management model

Take the server as the core, maintain the connection between the CAN protocol converter and the client, and provide support for data transmission control. Embedded database is to convert all the non current data in the server into the historical database, and read the real-time data of battery and motor through calling and communication interface. There are usually two kinds of data in the historical database: control command data and monitoring command data.

2.3 Implementation of Social Network Data Storage

The social network data storage model is a one-stop platform with spark as the core, including data acquisition, data processing, data mining and data visualization functions. It has good openness, scalability and versatility [9]. It mainly includes four modules: data capture, data preprocessing and storage, data mining and analysis and data visualization. As shown in Table 1, the three types of files collected by wif fence equipment and their corresponding data scales are counted. These three types of files are hotspot real-time information files, terminal real-time information files and virtual identity real-time information files.

Table 1. Statistics of data scale in big data analysis storage model

File name	Growth rate (estimated value)	Estimated total (3 months)
Hotspot real-time information file	168 million pieces/day; 21.3 GB/day	15.5 billion, about 1909 GB
Terminal real-time information file	136 million pieces/day; 17.35 GB/day	12.5 billion, about 1555 GB
Virtual identity real-time information file	139.9 billion pieces/day; 1.63 GB/day	1.4 billion, about 155 GB

In terms of database, it is mainly divided into relational database and non relational database. Among them, relational database includes traditional OSQL database and new Newsql database; Non relational database mainly refers to NOSQ database, which is divided into key value storage database, column storage database, document database and graphic database [10]. Due to the capture limitation of social network API, this paper collects user data by means of distributed capture and analysis of social network web pages when implementing this module. In the capture process, it is not only necessary to simulate the login operation of social network browser, but also need to schedule and manage the tasks of distributed crawler 28. The distributed social network crawler is based on Actor model and implemented by Scaa and Akka. The data storage management process is shown in Fig. 6:

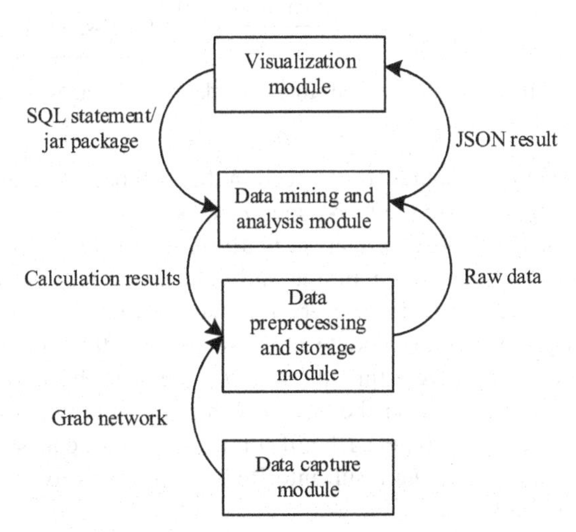

Fig. 6. Data storage management flow chart

The ultimate purpose of establishing user interest model is to realize the accurate push of information service. According to the established multi-level user interest model, the preference description of users on social networking sites can be made, and the user's personal preference information business card can be made, so as to realize the marking of information delivery objects, better meet the personalized needs of social networking users and improve the success rate of push. The hierarchical classification storage model of social networking data is shown in Fig. 7.

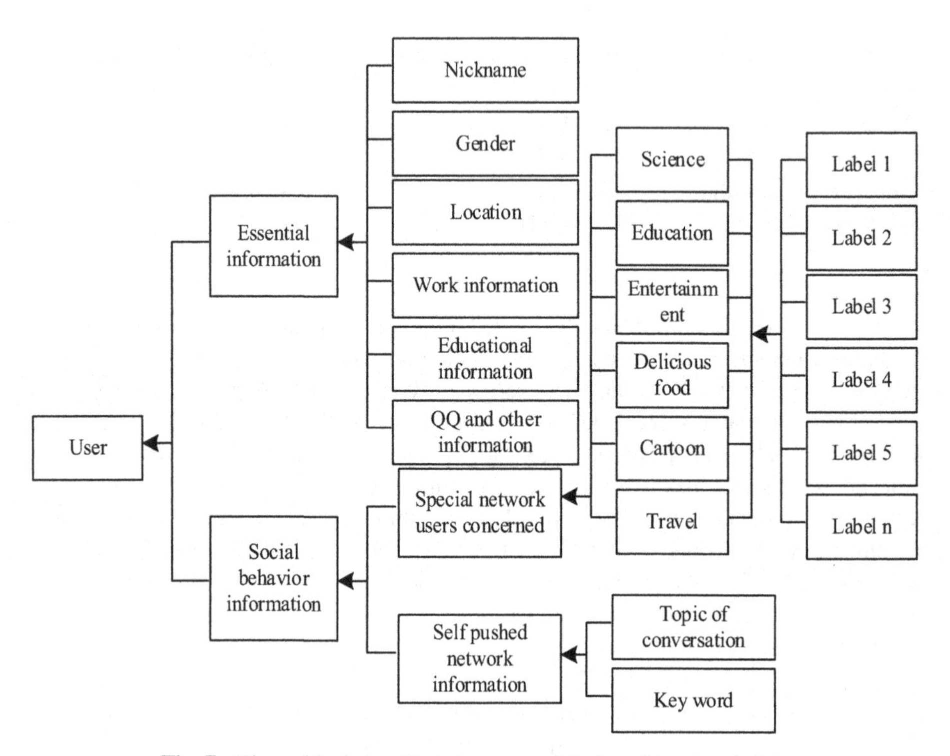

Fig. 7. Hierarchical classified storage model of social network data

In order to speed up the data loading speed in the web page, we also choose to use Reds for data caching, because the web page displays the relevant data and algorithm results of the social network big data analysis platform. These data will not expire within a certain period of time. If there are multiple same data access requests in a short period of time, sending the request to Hive every time is a waste of computing resources. Therefore, the data visualization module in this paper sets the validity period for all visual data when designing. Every time the web page requests background data, it will first go to Reds to find out whether the data is within the validity period. If it is valid, it will be returned directly from the cache: if it fails, route the request to hive, wait for the result to return, and cache the result into Reds again, so as to achieve the goal of classified storage of massive data.

3 Analysis of Experimental Results

Due to the huge amount of data to be processed in the experiment, the social network user forwarding behavior prediction algorithm needs to rely on the fast computing capacity of the social network big data analysis platform when it is implemented. Therefore, the method used in the algorithm to optimize the prediction model under the multi task learning framework has been written and compiled into the algorithm library of the data mining and analysis sub module.

During the experiment, eight sub nodes in the platform provided computing services. Each machine provided 4-core CPU and 10 GB memory. HDFS with a total capacity of 2.5 tb provided file storage services. Taking 20148.31 as the time dividing point, the experimental data set is divided into experimental training set and experimental test set. Randomly select a social network user W and acquire the interest of the social network user according to the above method. Extract the following user list of the social network user W, and obtain the tags of the special social network user w pays attention to (according to the tag tree method), so as to obtain the initial interest tag set of W; Calculate the weight of all interest tags according to the number of sub levels of interest tags. Calculate the value of each interest tag. In addition, the number of forwarded social networks is about twice that of original social networks. Therefore, in order to ensure data balance, we need to sample the original social networks and forwarded social networks, with a sampling ratio of 1:2. Table 2 shows the data sets actually used in the experiment.

Table 2. Data set properties

Number of network users	Number of network forwarding	Number of network originality	Number of concerns
92,068	716,185	9,197,365	1,273,852

Table 3. Characteristics of data sets used

Content	Number of network forwarding	Number of network originality
Experimental training set	589,252	1,169,856
Experimental test set	128,365	250,365

In order to verify the performance improvement of the proposed prediction model, logical regression (LR), support vector machine (SVM) and passive aggressive algorithm (PA) are selected as control algorithms to train and verify on the data set respectively (Table 3).

Table 2 shows the overall experimental results with accuracy and recall as indicators, in which the data storage management behavior prediction algorithm represents the algorithm proposed in this paper (Table 4).

Table 4. Experimental results

Index	PA	Logistic regression	Support vector machine	Forwarding behavior prediction algorithm
Accuracy	0.755	0.865	0.878	0.895
recall	0.389	0.468	0.508	0.522

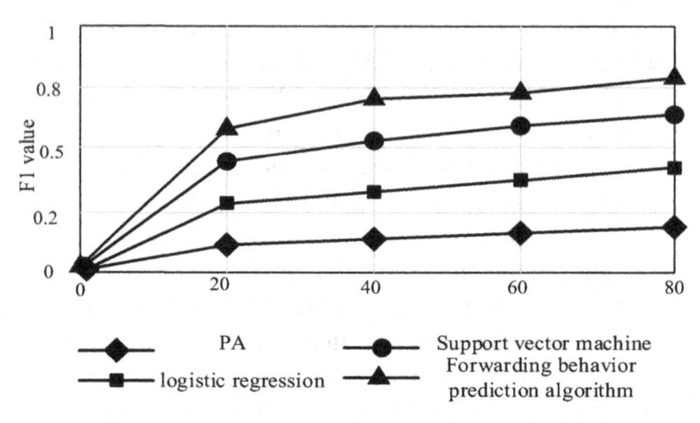

Fig. 8. F1 value of experimental results

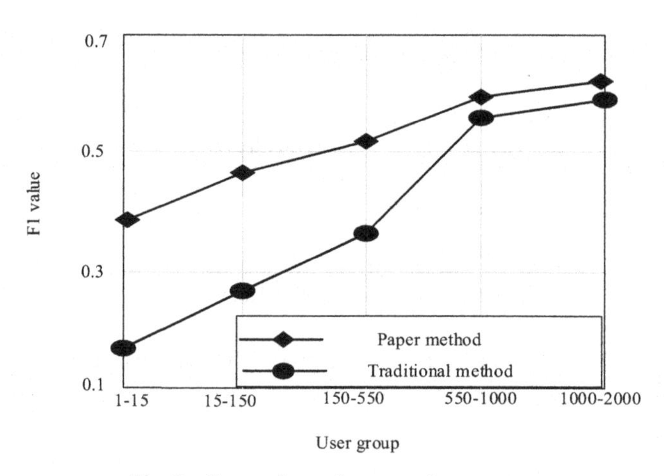

Fig. 9. Comparison of user set data storage

According to the experimental results in Fig. 8 and Fig. 9, on the one hand, because the PA algorithm is too simple, its accuracy and recall are relatively low; On the other hand, compared with the logistic regression algorithm as the benchmark algorithm, the algorithm proposed in this paper can significantly improve the accuracy and recall. In addition, compared with the previous two linear algorithms, although the method in this paper has a better classification effect, the overall classification effect is not as good as the algorithm proposed in this paper because the comparison algorithm is all based on the model trained by global data, and the algorithm proposed in this paper introduces local parameters to improve the classification effect. Through the comparison of F1 values in Fig. 8, it can be seen that the algorithm proposed in this paper can obtain better F1 values. In order to determine the impact of local parameters on the model, the benchmark algorithm logistic regression is used as the control algorithm in the experiment. In order to verify the classification of the model proposed in this paper under different forwarding historical data, users are divided into five groups according to the number of social networks: 10, 10–100, 100–500, 500–1000 and 1000–2000. One user is extracted from each group for 20 times. At the same time, the method in this paper with good results is used as a representative to compare with the algorithm proposed in this paper. The results are shown in Fig. 10:

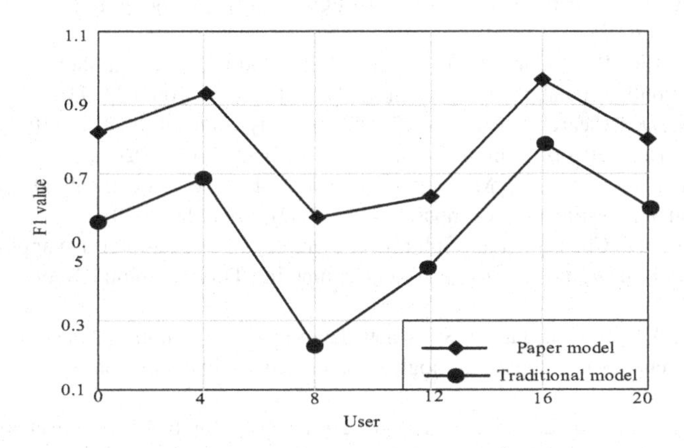

Fig. 10. Comparison of random user set data storage

As can be seen from Fig. 10, the predicted F1 value of the model proposed in this paper is much smaller than that of the logistic regression model, and it can also get better results on several users with poor performance of the logistic regression model, which shows that the model can meet the requirements of individuals.

To sum up, the model in this paper can significantly improve the accuracy and recall of the two performance indicators, and the fluctuation of the predicted F1 value for each user is much smaller than the logistic regression model.

4 Conclusion

The research of this paper is to build a social network crawler as a data source, use cloud computing to store the captured data in a distributed manner, use the fast computing ability of cloud computing to process the stored data, and finally use js visualization tools to display the data and related algorithm results on the web page, so as to finally realize a social network big data analysis platform with cloud computing as the core that can complete this whole process. At the same time, a prediction algorithm for the forwarding behavior of social network users is proposed for the captured social network data. In order to improve the performance of the algorithm, a multi task learning framework is also introduced. Through the obtained prediction model of forwarding behavior of social network users, the new social networks received by users can be predicted, and these social networks can be reordered so that users can easily view the social networks they are interested in. In addition, the algorithm runs on the social network big data analysis platform based on cloud computing, which verifies the effectiveness of the platform and the scalability of the algorithm module.

References

1. Devi, K., Muthusenthil, B.: Intrusion detection framework for securing privacy attack in cloud computing environment using DCCGAN-RFOA. Trans. Emerging Telecommun. Technol. **33**(9), 4561 (2022)
2. Meng, Q., Liu, B., Zhang, H..-Y.., et al.: Multi-relational group influence modeling and analysis in online social networks. Chin. J. Comput. **44**(06), 1064–1079 (2021)
3. Liu, X., He, D.: Research on of competitive nonlinear dynamic information diffusion modeling in online social network. Chin. J. Comput. **43**(10), 1842–1861 (2020)
4. Lian, J., Fang, S., Zhou, Y.: Model predictive control of the fuel cell cathode system based on state quantity estimation. Comput. Simul. **37**(07), 119–122 (2020)
5. Xu, Z., Zhu, D., Chen, J., et al.: Splitting and placement of data-intensive applications with machine learning for power system in cloud computing. Digit. Commun. Netw. **8**(4), 476–484 (2022)
6. Thabit, F., Alhomdy, S., Jagtap, S.: A new data security algorithm for the cloud computing based on genetics techniques and logical-mathematical functions. Int. J. Intell. Netw. **2**(2), 18–33 (2021)
7. Li, Y., Zhou, F., Xu, Z.: Privacy-preserving k-nearest-neighbor search over mobile social network. Chin. J. Comput. **44**(07), 1481–1500 (2021)
8. Ramdani, F., Wirasatriya, A., Jalil, A.R.: Monitoring the sea surface temperature and total suspended matter based on cloud-computing platform of google earth engine and open-source software. IOP Conf. Ser. Earth Environ. Sci. **750**(1), 12–31 (2021)
9. Luo, H., Yan, G., Zhang, M., et al.: Research on node importance fused multi-information for multi-relational social networks. J. Comput. Res. Dev. **57**(05), 954–970 (2020)
10. Xu, M., Zhang, Z., Xu, X.: Research on spreading mechanism of false information in social networks by motif degree. J. Comput. Res. Dev. **58**(07), 1425–1435 (2021)

Anomaly Monitoring System of Enterprise Financial and Economic Information Based on Entropy Clustering

Yu Chen(✉) and Kaili Wang

School of Business, Nantong Institute of Technology, Nantong 226000, China
Cyhsb88@163.com

Abstract. The currently used information anomaly monitoring system has problems of low accuracy and efficiency. In this regard, an abnormal monitoring system of enterprise financial and economic information based on entropy clustering is designed. On the basis of the design of the hardware and control module of the economic information abnormal monitoring system; the crawler tool is used to collect the financial and economic information of the enterprise; the collected data information is cleaned by the mapping operation; the processed data is processed by the abnormal knowledge discovery principle. In feature extraction, abnormal information features are obtained through decision tree; the abnormal information monitoring is realized by using the k-means algorithm of information entropy. The experimental results show that the designed system has an average alarm correct rate of 92.55% and a short response time, which is of practical value.

Keywords: Entropy clustering · Corporate financial economy · Information anomaly · Monitoring system · Data cleaning · k-means

1 Introduction

With the rapid development of modern communication and communication technology and the rapid expansion of the popularization of the Internet, great changes have taken place in the way of information dissemination and reception. Mass information from multiple channels and multiple sources helps people to fully understand events and make correct decisions. However, while enjoying the advantages of multi-channel mass information with multiple dimensions and strong real-time performance, people are also deeply disturbed by it. First, massive amounts of information cannot be obtained and processed by humans alone; secondly, large-scale information from different sources has strong redundancy, making information processing difficult; thirdly, a large amount of noise affects the acquisition of effective information; finally, the efficiency of information center topic judgment is low, the information is difficult to classify correctly. With the continuous deepening of the informatization construction of the financial service industry, the financial industry has become one of the industries with the highest degree of informatization in my country [1]. With the rise of the Internet era, the financial

W. Fu and L. Yun (Eds.): ADHIP 2022, LNICST 468, pp. 217–230, 2023.
https://doi.org/10.1007/978-3-031-28787-9_17

industry has entered the era of information explosion. In addition to paying attention to financial information acquisition channels such as official announcements, professional financial websites, and new financial media platforms, financial institutions and practitioners must also pay attention to financial information sharing platforms such as social networks., it not only consumes a lot of time, but also has low browsing efficiency, and it is easy to miss important information, the news is relatively lagging, it is difficult to respond to risks in time, resulting in financial losses [2]. Perfecting the multi-industry and multi-level financial early warning system, establishing a financial supervision system commensurate with the level of financial development, and enhancing the efficiency of financial supervision are also major issues that must be considered in financial risk prevention. The supervisory department should abandon the traditional post-event supervision method, and use the whole-process monitoring method combining pre-event supervision, in-event supervision and post-event supervision to better prevent financial risks.

Regarding the development of financial business security monitoring information application system, domestic researchers, major banks and system integrators are more active and mature in academic research, system development and system application of financial business security monitoring information application system in the financial industry. However, at present, many domestic financial business security monitoring information application system developers mostly use mobile or China Unicom direct connection. Bank data is landed on mobile or China Unicom, and there are certain security risks in customer financial information [3]. For enterprises, financial and economic information of enterprises is very important for enterprises to guard against financial and economic information. Under the background of the rapid development of modern information technology, the amount of financial and economic information of enterprises is increasing exponentially. By monitoring abnormal financial and economic information of enterprises, it is possible to avoid mistakes in corporate management policies caused by abnormal financial and economic information in the process of operation., or cause corporate financial and economic risks. Therefore, the abnormal monitoring of the financial and economic information of the enterprise is carried out to facilitate the timely adjustment by the business decision-makers of the enterprise, so that the information is more valuable and the information can better serve the business development of the enterprise.

The current methods for abnormal monitoring of information data can be roughly divided into distribution-based methods, graph-based methods and cluster-based methods. The distribution-based anomaly detection method is the earliest detection method [4]. This detection method is based on the assumption that the dataset follows a certain statistical distribution. The statistical knowledge of these statistical distributions is then used to construct mathematical models. Finally, a statistical method is used to detect whether the data conforms to the constructed mathematical model. If there is a significant difference between the detected data and the mathematical model, it is considered that the current monitoring data is an abnormal behavior, and the monitoring of abnormal data is completed. But this method is extremely difficult to estimate or calculate the distribution of the data when the data is in high dimension. The graph-based anomaly detection method [5] mainly maps high-dimensional data to a low-dimensional space through visualization techniques. Outliers appear at special locations in the mapped image. The advantage of using the visualization method is that the possible abnormal points in the data can be found directly and clearly. However, this visualization method requires artificially given five parameters, and human factors will affect the final detection result. The visualization result contains a large number of manual parameters, which requires a lot of manual participation, and the detection results obtained are also different depending on the mapping method of the graph. Therefore, this method has the disadvantages of large workload and uncertainty. Using cluster analysis, the similarity between data can be used to classify unlabeled data objects into different classes without relying on prior knowledge. Cluster-based anomaly detection methods do not require prior knowledge of the data set, and can effectively cluster data and discover anomalies in it. However, this method is prone to the disaster of dimensionality in multi-dimensional situations, and the anomaly detection performance for high-dimensional data is poor. In this regard, this paper designs an abnormal monitoring system for corporate financial and economic information based on entropy clustering. Design the hardware and control module of the abnormal economic information monitoring system, use crawler tools to collect financial and economic information of enterprises, clean the collected data information through mapping operations, and use the principle of abnormal knowledge discovery to characterize the processed data Extraction, obtain abnormal information features through decision tree, and finally use the k-means algorithm of information entropy to realize abnormal information monitoring. Information entropy reflects the stability of the system, and it has been widely used in anomaly detection in recent years. Information entropy is a description of information distribution, which reflects the distribution of uncertainty of random variables in the system. The more stable a system is, the lower the information entropy value is, on the contrary, the more chaotic a system is, the higher the information entropy value is. This paper will design an abnormal financial and economic information monitoring system based on entropy clustering, and reduce the negative impact of abnormal financial and economic information on related businesses by applying the monitoring system. The experimental results show that the average alarm accuracy rate of the system designed in this paper is 92.55%, and the average response time is 10.97s. The accuracy rate is high and the response time is short.

2 Design of Hardware Part of Enterprise Financial and Economic Information Abnormal Monitoring System

This paper optimizes the function module based on the traditional hardware, and improves the system hardware by designing the system control module and communication module. The monitoring system should be extensible for future expansion of network services. The system should accommodate the need to plug in zone controllers and operator terminals anywhere without affecting the normal operation of the existing system. In addition, the monitoring system should be able to ensure long-term stability of various indicators in a long-term stable operation state, thereby reducing system operation and maintenance costs. The selection of the hardware part of the system saves energy as much as possible on the premise of meeting the functional requirements of the design, thereby reducing operating costs. Considering the importance of corporate financial and economic information, the system should be equipped with a permission management function, and users with different permissions can operate different contents, that is, the functions opened to users with different permissions are different. According to the above monitoring system hardware selection requirements, the hardware part of the information anomaly monitoring system designed in this paper adopts a core template of 50 mm * 70 mm, in which the S3C2410A microprocessor is installed, and the MYD-AM335X series development is extended based on the MYC-AM335X core board. The development board has rich peripheral resources including high-speed USB, LCD interface, CAN interface, 485 interface, ADC, SPI, GPMC, LED, 10/100/1000Mbps Ethernet interface, JTAG debugging interface serial port, etc. Figure 1 below is a schematic diagram of the hardware framework of the enterprise financial and economic information anomaly monitoring system.

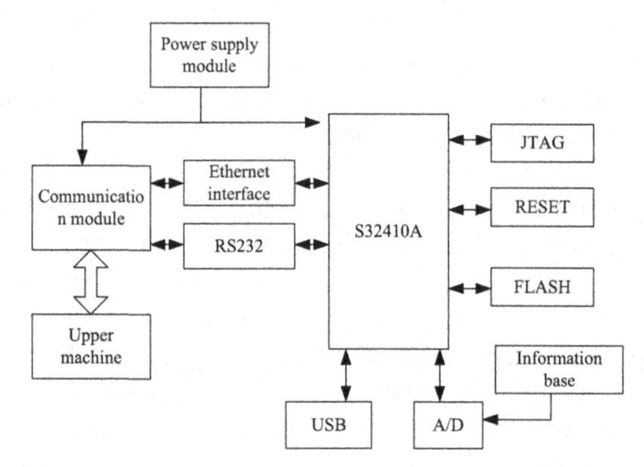

Fig. 1. The hardware frame diagram of the abnormal monitoring system of enterprise financial and economic information

2.1 System Control Module Design

The control module of the hardware part of the system is composed of S3C2410A processor and various peripheral circuits to meet the design requirements. S3C2410A is a low power consumption, high integration, 16/32-bit RISC microprocessor based on ARM920T core, and also adopts AMBA structure. In order to reduce the total system cost, S3C2410A integrates the following resources: 16 KB instruction Cache (cache), 16 KB data Cache, MMU for virtual storage management, LCD controller, NAND Hash controller (supports 4KB code direct boot), SDRAM controller, 3 UART serial ports, 4-channel DMA, 4-channel PWM timer, watchdog counter, 117 multi-function I/O ports, 24 external interrupt sources, RC real-time clock, 8-channel 10-bit ADC and touch screen Interface, USB Host/Device, SD card interface, 2 SPI interfaces, I2S bus interface [6].

Based on the ARM920T core, the S3C2410A processor has built-in Embedded ICE debug module with standard JTAG interface. You can use hardware emulators such as Multi-ICE to perform online real-time simulation debugging through the JTAG interface, or use programming tools such as SJF2410 to program the externally expanded FLASH memory through the JTAG interface. The system hardware microprocessor designed in this paper uses the standard 20-pin JTAG interface to program the test program.

The $3C2410A itself integrates the NAND FLASH controller with the NAND FLASH Boot Loader function, the user can directly expand the large-capacity NAND FLASH memory, and can directly start the user from the NAND FLASH after system reset by configuring the OM [1:0] pins code. The system in this article only uses 64 * 8 bits K9F1209 NAND FLASH memory. After the data has been verified, it can be transferred to the EERAM by the copy scratchpad command to ensure the integrity of the data when changing the memory. The scratchpad is 9 bytes, and the 0th and 1st bytes are the low and high bytes of the temperature code. Bytes 2 and 3 are copies of the temperature encoded low and high bytes, and byte 4 is the configuration register [7].

Among them, the nRTS and nETS signals of UART0 are brought out to support the application of automatic flow control (AFC). In AFC, nRTS is controlled by the receiving condition of the receiver, and nCTS controls the work of the transmitter. Since S3C2410A is a 3.3 V system, the working power supply range is 3.3 V. Call. 5 V MAX3232 for RS. 232 level shifting.

2.2 Communication Module Design

When communicating, first set the communication format, that is, write the D8120 register, and set it to 0C86 under this system condition, that is, the data length is 8 bits, no parity, no start bit and stop bit, baud rate 9600 bps. After modifying the D8120 setting, make sure to power on and off the PLC once. Then use the RS command to transfer data. When processes interact and share data, upstream processes generate new data, and downstream processes consume data. Since each process executes at a different speed, if it is not scheduled, the buffer may fill up quickly, or the same information may be called repeatedly for processing. This problem is generally solved by introducing a circular queue structure. The producer process and the consumer process are located at the head and tail of the queue, respectively, and control the writing of producer data and the reading of consumer data through put operations and get operations, and the rest of the queue is buffer space. The get operation is disabled if the buffer has no data, and the put operation is disabled if the buffer is full. The problem of speed mismatch when processes interact is effectively solved. In this system, the data collection process is responsible for collecting data as a producer, and the data transmission process transmits data as a consumer. The above method is used to complete the process scheduling.

This system uses the RTU mode to transmit the communication protocol, and uses the RS-232C compatible serial interface for communication. The main advantage of this method is that more data can be transmitted than the ASCII method at the same baud rate. When using RTU mode, message transmission begins with a pause interval of at least 3.5 character times. The first field of the transfer is the device address, and the transfer characters that can be used are hexadecimal values. During communication, network devices continuously detect the network bus, including the pause interval, when the first field (address field) is received, each device decodes to determine whether it is sent to its own. After the last transmission character, a pause of at least 3.5 characters is required to mark the end of the message, after which a new message transmission can begin, as shown in Table 1.

Table 1. RTU mode message frame structure

Start Bit	Address	Function code	Data	CRC check	Stop Bit
T1-T4	8 bits	8 bits	N * 8 bits	16 bits	T1-T4
Size is 1 character	The address is 0–247 (decimal)	When a message is sent from the master device to the slave device, the function code field will tell the slave device what actions need to be performed; in response to the objection, the slave device returns a code equivalent to the normal code, but the most important position is logic 1	The data field is composed of two sets of hexadecimal numbers in the range 00…FF	The CRC field is appended at the end of the message, with the low byte followed by the high byte. Therefore, the high-order byte of the CRC is the last byte of the sent message	Least Significant Bit–Most Significant Bit

Using RTU mode, the message includes an error detection field based on the CRC method. The CRC field detects the content of the entire message. The CRC field is two bytes containing a 16-bit binary value. It is calculated by the transmitting device and added to the message. The receiving device recalculates the CRC of the received message and compares it with the value in the received CRC field. If the two values are different, there is an error. The implementation of the CRC algorithm will be described in detail later.

The generation of CRC check bytes is a more critical step, and the process is relatively complicated. The steps are as follows:

(1) Preset a 16-bit CRC register as hexadecimal FFF, that is, all digits are 1.
(2) The lower 8-bit byte of the 16-bit register is XORed with the lower 8-bit of the first byte of the information frame. The result of the operation is placed in a 16-bit register.
(3) Shift the 16 register one bit to the right and fill the high bit with 0.

(4) If the digit shifted to the right (marker bit) is 1, the generator polynomial A001 (1010000000000001) is XORed with this register; if the digit shifted to the right is 0, return (3).
(5) Repeat (3) and (4) until 8 bits are shifted out.
(6) Repeat (3) to (5) until all bytes of the message are XORed with the 16-bit register and shifted 8 times.
(7) The high and low-order bytes of the obtained 16-bit CRC register, that is, 2-byte CRC, are added to the message.

In this paper, the OPB bus is used to connect the hardware-accelerated IP core control logic, because the OPB bus is mainly used for data transmission to external devices. The OPB bus supports multi-bit data, and accepts input from the host when used for peripherals, and performs specified operations, which meets the needs of hardware acceleration. The OPB bus performance is shown in Table 2 below.

Table 2. OPB bus performance table

Serial number	Project	Parameter
1	Data line width	8-bit, 16-bit, 32-bit
2	Address line width	32 bit
3	Architecture	Multiple master/slave devices
4	Data transfer protocol	Single read/write transfer, supports trigger transfer, word, byte, half-word transfer
5	Timing	Synchronize
6	Connect	Multiplex
7	Interconnected	Bus-independent read/write data lines, no tri-state support

The data processing module completes the feature parameter extraction, starts the hardware acceleration module, sets ce to a high level, and the hardware acceleration module starts to work. The three channels of data are processed at the same time, and the respective calculation results prob1, prob2 and prob3 are output. The three output ready signals rdy1, rdy2 and rdy3 are at high level, indicating that the calculation results of the respective components are completed after the operation.

After designing the main core modules of the hardware part of the abnormal information monitoring system for corporate financial and economic information according to the above content, the information entropy clustering algorithm is used to process the collected and acquired corporate financial and economic information data, and the software part of the abnormal information monitoring system is designed to realize Default system functions.

3 Design of Software Part of Enterprise Financial and Economic Information Anomaly Monitoring System Based on Entropy Clustering

3.1 Collection of Corporate Financial and Economic Information

The data acquisition port of the hardware system is connected with the financial and economic information data database interface of the enterprise to obtain the financial and economic information data of the enterprise in a specified time period or a specified business. However, the differences in the scale of enterprises lead to too messy and large amount of financial and economic information data of enterprises. In this paper, crawler tools are used to collect financial and economic information of enterprises.

The distributed crawler tool of this system is based on the Soapy framework, uses an asynchronous non-blocking method to send requests to the target URL, and has a complex structure for the Soapy crawler, does not support distributed crawling, lacks website anti-crawling mechanism response methods, and after downtime The crawler state cannot be recovered and other problems are optimized, the crawler configuration and crawler process are componentized, the crawler configuration parameters are defined by the combination of the preset configuration file and the manual input interface, and the crawler is abstracted into multiple crawler tasks and distributed to multiple machines. Server processing, control of crawlers and extraction of target data through configuration parameters, and solving the problem of website anti-crawling by adding different anti-crawling components, so that new crawlers can be customized simply by modifying configuration files and extracting templates without modifying the crawler logic. Set up a persistent distributed queue of URLs to be crawled to complete the incremental crawling and continuous crawling of the crawler. The crawler task is released through the distributed task queue and assigned to multiple servers for execution to ensure the efficient operation of the crawler and The crawler anomaly is alerted by the crawler monitoring module. The different modules of the system are completely decoupled, which is convenient for further expansion. While ensuring the efficiency and stability of distributed crawling, it is deeply customizable for users. The configuration file is not restricted by the development and running environment, and has a strong generalization suitability. The crawler task scheduling module is responsible for the generation and scheduling of crawler tasks, including crawler task construction unit, task scheduler, task release queue, task execution unit, task execution result queue, task retry queue and task monitoring unit [8].

The crawler task construction unit is used to generate crawler tasks containing various parameters according to the configuration items in the configuration module, which are expressed in the form of key-value and serialized into json strings. Each crawler task has a unique task record. The crawler task construction unit first cannot accept the crawler task parameters by manually inputting the configuration script, finds and parses the static configuration file of the corresponding website through the crawler name, and parses the corresponding running configuration file according to the name of the running configuration file.

The task scheduler obtains the serialized crawler tasks from the crawler task construction unit and the task retry queue, and then detects the task type. For tasks that exceed the maximum number of retries, update their status and send them to the task execution result queue. For common asynchronous tasks and retry tasks within the maximum number of retries, they are directly sent to the associated task release queue. Save periodic tasks and periodically send them to the corresponding task release queue according to the set time interval. After the crawler is used to collect the financial and economic information data of the enterprise, the information data is cleaned and processed.

3.2 Financial and Economic Information Data Cleaning

Since the crawler collects a large amount of financial and economic information and includes many data items, if it is simply written to a text file, the subsequent work such as calculating the information entropy value will require frequent reading and writing operations of the file, which will consume a lot of system resources. The time cost will be huge. The financial and economic information extraction subsystem uses the data flow-data node method to process the collected data in real time, and uses the self-defined analysis plug-in to perform online analysis of the processed data in small batches. Data flow defines the calculation process of data, and data nodes are responsible for executing specific calculation logic, including data cleaning and data standardization nodes. The analysis plug-in is customized according to the business, and provides a plug-in for each analysis business, which is responsible for continuously pulling the unanalyzed data of the business to perform analysis and calculation logic. The plug-ins of different data analysis services are independent of each other. In this paper, the task of extracting financial and economic information is a two-stage task. First, we identify and filter information data that may have abnormal financial and economic information in the information, and then extract information elements from the filtered sentences. If all attributes are used in anomaly detection, it will inevitably increase the amount of computation and consume a lot of system resources, and not all attributes contribute to anomaly detection. Therefore, after the system hardware collects the financial and economic information data of the enterprise, the information is cleaned.

This article will take a mapping operation on the original data and organize it into new clean data. The data selection module queries the required data as the object to be cleaned according to the user's requirements, and fuses the objects to be cleaned by querying set $\{q_1, q_2, \ldots, q_n\}$, and the result obtained is a metadata set. The first step of data cleaning, the form is defined as [9]:

$$CS_{q_1,q_2,\ldots,q_n}() = q_1 \cup q_2 \cup \ldots \cup q_n \quad (1)$$

Because there are many different types of data such as date and time in most business environments, almost all data cleaning implementations must transform these various forms of data into the canonical format of the rule base. For the needs of decision analysis, the data in the data source is generally not stored in a coded form. Before saving into the data source, look up the corresponding text description from the dictionary according to the code, and use the text description to replace the code. Repeatability judgment is mainly completed by three operations: Cartesian product, matching and clustering. In order to improve the efficiency of the data matching operation, we design the Cartesian product operation here, which will only be performed on the data set of the Lai attribute taken out. That is, the data in the data set D output by the data standardization module is subjected to the Cartesian product operation, that is, $D * D$, and the following data pairs can be obtained:

$$D * D = \begin{bmatrix} b & s & b & s \\ b_1 & T & b_2 & T' \\ b_1 & T & b_3 & p \\ b_2 & T' & b_3 & p \end{bmatrix}_v \tag{2}$$

Among them, $b = [b_1, b_2, b_3]$ and $s = \left[T, T', p\right]$ are data sub-items in the dataset. The combination compares the results of matching key attributes, and if and only if the key attributes of an element are all repeated attributes, it can be deduced that the element is a repeated element. After cleaning the repeated financial and economic information data, the information entropy clustering method is used to cluster the financial and economic information to monitor the abnormality in the information data set.

3.3 Clustering of Abnormal Information of Corporate Finance and Economics

The system uses the principle of abnormal knowledge discovery to extract the characteristic data in the financial and economic information of enterprises. Organize the knowledge of corporate financial and economic rules into a rule knowledge tree, which is a directed acyclic tree. The root node in the tree has no parent node, all other nodes have one and only one parent node, and each node can have one or more child nodes, or no child nodes. If a node has no child nodes, it is called a leaf node. Others are called internal nodes. Non-leaf nodes correspond to various possibilities in the value space of each condition attribute in the data set of abnormal financial and economic cases of enterprises. The content of leaf nodes is in addition to the corresponding rules. The conclusion is the attribute value (yes or no) of the decision attribute, including the combination of attribute values reflecting the entire path from the root node to the direct parent node of the leaf node as a condition, and the decision attribute value of the leaf node is the rule of conclusion Whether a valid time-to-live. The mined rule sets are stored and transmitted in the form of decision trees. After receiving the knowledge base subsystem, the initial rule knowledge tree is established by increasing the survival time of the corresponding rules for each leaf node. The rule knowledge tree will be continuously adjusted as the system runs., the number of rule knowledge trees will also increase or decrease. At any time of system operation, the forest composed of all rule knowledge trees constitutes the current rule knowledge base [10]. When a new mining rule comes, adjust and prune the corresponding rule knowledge tree and prune the decision tree corresponding to the new rule set according to whether the new rule set matches the rules in the current rule knowledge base (the premise is the same, the conclusion is the same) or not: The decision leaf node and the rule-related path corresponding to the rule matching the rule in the current rule knowledge base should be pruned, and the TTL value of the corresponding leaf node of the rule knowledge tree where the rule is located should be increased, otherwise it should be retained. For new rules with conflicting (same premise, different conclusions), adjust the decision attribute value of the leaf node of the rule corresponding to the rule tree in the rule knowledge base, and set its TTL value to the initial value max, and then prune the corresponding rule in the decision tree. The leaf nodes and the unique ancestor nodes and paths of the rules, that is, the new rules are discarded for the conflicting rules. The abnormal information attributes in the decision tree are mapped to the feature set, that is, the abnormal information characteristics of the financial economy of the enterprise are obtained.

If the processed enterprise financial and economic information data is $X = \{X_1, X_2, \ldots, X_N\}$, and if the information feature is $f = \{f_1, f_2, \ldots, f_N\}$, it indicates the state of the information data. Assuming that the characteristic attribute f_i of the corporate financial and economic information appears n_i, then the information entropy of the corporate financial and economic information can be defined according to the following formula:

$$H(X) = -\sum_{i=1}^{N} \left(\frac{n_i}{\sum f_i}\right) \log_2\left(\frac{n_i}{\sum f_i}\right) \tag{3}$$

It can be known from mathematical knowledge that the value range of the information entropy value is $(0, \log_2 N)$. When the value of $H(X)$ is 0, it represents the maximum centralized distribution of the measurement data set. When the value of $H(X)$ is $\log_2 N$, it represents the maximum scattered distribution of the measurement data set.. Through the change of the information entropy value, it is possible to judge the orderly degree of a piece of enterprise financial and economic information data, and intuitively understand the changes of the characteristics and attributes of the enterprise's financial and economic information.

The information entropy value of the characteristic attributes of the financial and economic information of different abnormal enterprises is used as the classification and determination standard of entropy clustering, and the k-means algorithm combined with information entropy is used to realize abnormal information monitoring according to the following steps.

(1) Calculate the information entropy value of the characteristic attribute of the detection information, and calibrate the type of the detection information according to the information entropy to obtain a detection data set;
(2) The cluster center finally output in the training phase is used as the initial cluster center in the detection phase, and the Euclidean distance is calculated for the detection samples in the detection data set, and cluster analysis is performed;

The Euclidean distance calculation formula is as follows:

$$dis(t, s) = \sqrt{\sum_{k=1}^{n} (t_k - s_k)^2} \tag{4}$$

Among them, t is the financial and economic information data of enterprises that need to be monitored; s is the financial and economic information data of sample enterprises.

(3) Calculate the distance from the detected data to the cluster center, and compare it with the classification threshold. If it is greater than the classification threshold,

the current sample is considered to be new abnormal information, and the K value of the number of clusters is updated plus 1, and the current sample is regarded as the new cluster. If it is less than the classification threshold, the current sample is classified into the nearest cluster for anomaly detection.

(4) If a certain detected sample is classified as abnormal information, start the abnormal response module. Repeat the above process until all samples in the test sample set are tested;

(5) Output the clustering results, and carry out corresponding monitoring and early warning according to the clustering results.

After combining the software part and the hardware part designed above, the design and research of the abnormal monitoring system of enterprise financial and economic information based on entropy clustering is completed in theory.

4 System Test

This section builds an abnormal monitoring system of corporate financial and economic information based on entropy clustering, tests the feasibility and monitoring effect, and evaluates the system from the aspects of comprehensiveness, real-time and accuracy.

4.1 Test Plan Design

First, the hardware test of the monitoring system is carried out, and the hardware part is divided into the circuit schematic diagram and PCB drawing of each module. Carefully check the connection relationship of each pin before submitting the board, pay special attention to the pins without electrical properties, and carefully check whether the pins are missing in the schematic diagram. When welding components, you need to pay attention to the positive and negative poles of the diode, the pin direction of the chip, etc. When the components are welded and the system is not powered on, use a multimeter to measure whether the power line and the ground wire are short-circuited. After all tests are normal, power on the system to check whether the chip is overheating or smoking. If the circuit board is faulty, quickly disconnect the power supply to check the fault; power on the system after the fault is solved.

After the hardware part is tested, the response time of the system and the monitoring and alarm accuracy of abnormal information are the test indicators. By comparing the system designed in this paper with the monitoring system based on multi-source information fusion, we can intuitively evaluate whether the designed system meets the requirements of the enterprise need.

4.2 Evaluation Index

Accuracy is an indicator used to evaluate data. Here, it refers to the proportion of correct results of abnormal monitoring and alarm of enterprise financial and economic information. The definition is as follows:

$$Accuracy = \frac{TP + TN}{TP + FN + FP + TN} \tag{4}$$

In the formula, *TP* represents the true case, *TN* represents the true negative case, *FP* represents the false positive case, and *FN* represents the false negative case.

4.3 Test Results and Analysis

In order to test the effectiveness of the method in this paper, the system performance of the method in this paper and the traditional clustering method are compared and tested by using the artificial enterprise financial and economic information data set with known abnormal information. The comparison results of the system performance test are shown in Table 3 below.

Table 3. Comparison results of system performance tests

Abnormal information content/%	The method of this paper		Clustering method	
	Response time/s	Monitoring alarm accuracy rate/%	Response time/s	Monitoring alarm accuracy rate/%
0.5	10.73	93.2	15.78	89.6
1.0	10.94	90.6	15.98	85.2
1.5	11.12	92.7	16.57	87.3
2.0	11.06	93.3	23.07	75.6
2.5	10.84	92.4	24.3	79.4
5.0	10.95	94.5	25.76	82.7
8.0	11.07	92.6	27.19	86.1
10.0	11.03	91.1	28.07	90.5

It can be seen from the data in Table 3 that the response time of the method in this paper is significantly shorter than that of the clustering method. The average response time of the method in this paper is 10.97 s, and the average response time of the clustering method is 22.09 s. The method in this paper has an average correct rate of 92.55% for monitoring and alarming of abnormal corporate financial and economic information, and the average correct rate for monitoring and alarming of the clustering method is 84.55%.

To sum up, the monitoring accuracy rate of the abnormal monitoring system of enterprise financial and economic information based on entropy clustering designed in this paper is higher than 90%, and the response efficiency of the system is high, which can effectively improve the operation efficiency of enterprises.

5 Concluding Remarks

In the financial market, liquidity risk is the most common risk. The decision-making basis of economic participants is all kinds of information in the market and enterprises.

However, it is extremely complicated and difficult to obtain relevant information about the industry and the market. Risk is an unavoidable uncertainty in the development and operation of an enterprise, and its occurrence is unavoidable. The abnormal monitoring of corporate financial and economic information can reduce the risk of corporate operation caused by abnormal information. To this end, this paper designs an abnormal monitoring system for corporate financial and economic information based on entropy clustering. Through the system performance test, it is verified that the monitoring system can accurately monitor and warn the abnormal information in financial and economic information. Effective use of financial and economic information reduces the negative impact of abnormal information on business operations.However, in the process of monitoring abnormal financial and economic information of enterprises, due to the complexity of the algorithm, the monitoring time did not achieve the expected effect, resulting in a decrease in monitoring efficiency. In the following research, the algorithm will be improved to shorten the calculation time. Improve the monitoring efficiency of abnormal financial and economic information of enterprises.

Fund Project. Jiangsu Province "14th Five-Year" Business Administration Key Construction Discipline Project (SJYH2022-2/285).

References

1. Zhong, H.: The status quo and countermeasures of high-quality financial development in China—research based on international comparison. Southwest Finan. (02), 74–84 (2021)
2. Deng, C., Peng, B.: Research on the relationship between the financialization behavior of entity enterprises and the information disclosure quality. Theor. Pract. Finan. Econ. **42**(03), 110–117 (2021)
3. Shi, Z..-J.., Zhou, X..-J.., Li, K., et al.: Cyberspace security monitoring technology based on multi-source information fusion. Comput. Eng. Des. **41**(12), 3361–3367 (2020)
4. Shi, X., Qiu, R., Ling, Z., et al.: Spatio-temporal correlation analysis of online monitoring data for anomaly detection and location in distribution networks. IEEE Trans. Smart Grid **11**(2), 995–1006 (2019)
5. Zhao, H.-Y., Liu, K., Wang, T.-M., et al.: Simulation of abnormal information acquisition for network text implication relationship recognition. Comput. Simul. **37**(08), 256–260 (2020)
6. Liu, H., Wang, Y., Chen, W.G.: Anomaly detection for condition monitoring data using auxiliary feature vector and density-based clustering. IET Gener. Transm. Distrib. **14**(1), 108–118 (2020)
7. Li, Y.-T., Guo, J., Qi, L., et al.: Density-sensitive fuzzy kernel maximum entropy clustering algorithm. Control Theory Appl. **39**(01), 67–82 (2022)
8. Kim, H., Park, J., Min, K., et al.: Anomaly monitoring framework in lane detection with a generative adversarial network. IEEE Trans. Intell. Transp. Syst. 22(3), 1–13 (2020)
9. Zheng, W., Yao, J., Wang, J., et al.: Segmentation of MRI brain image based on fuzzy entropy clustering and improved particle swarm. Laser J. **42**(01), 98–103 (2021)
10. Wei, L., Xuesongm, H., Ming, Z., et al.: Abnormal behavior detection method for power monitoring system based on fully connected residual neural network. J. Southeast Univ. (Nat. Sci. Edn.) **50**(06), 1062–1068 (2020)

Classification and Retrieval Method of Library Book Information Based on Data Mining

Xing Zhang(✉)

The Tourism College of Changchun University, Changchun 130000, China
xingxing11852@163.com

Abstract. The current classified retrieval of book information is mainly based on the key words of books. When there are interdisciplinary contents in books, it is easy to ignore the correlation between book information, resulting in a sharp decline in the retrieval accuracy. In order to improve the quality of library information retrieval service, aiming at the problems of the above traditional classification retrieval methods, this paper studies the library book information classification retrieval method based on data mining. After preprocessing the book information, the Apriori algorithm of data mining is used to mine the deep management of information, and the decision tree algorithm is used to determine the classification of information. Using the concepts in book information, this paper constructs the concept map of book information and builds the information retrieval framework. Use Markov network model to realize book information classification and retrieval. In the test, the retrieval accuracy of the proposed retrieval methods is higher than 92%, and the retrieval performance is significantly improved. Through the research and application of this method, the transformation from the traditional library service concept to the personalized information service concept is promoted.

Keywords: Data mining · Library · Book information · Classified search · Apriori algorithm · Decision tree

1 Introduction

As a comprehensive storage center of social information resources, the library not only has the function of collecting and preserving archives, books and materials and continuing social memory, but also should actively serve all aspects of society, such as economy, politics, science and culture, and actively participate in the process of integration and sharing of social information sources. Libraries have also begun to apply various information technologies to improve their hardware and software conditions. Driven by the development of library digital technology, the data scale generated by book borrowing is becoming larger and larger, the data types and items are becoming more and more complex, and the information update speed is fast. The related resources have an explosive growth trend, but the rich and diverse book resources have also brought some problems to people searching books. At the same time, library users and managers have higher and

W. Fu and L. Yun (Eds.): ADHIP 2022, LNICST 468, pp. 231–243, 2023.
https://doi.org/10.1007/978-3-031-28787-9_18

higher requirements for book classification and retrieval information, which is reflected in its stronger pertinence, higher real-time, and more diversified types of needs [1]. The library will update its collection resources every day. Readers use all kinds of resources every day, which makes a large amount of data and information accumulated in the library database. In these information, there are many knowledge worthy of in-depth study by library service workers, such as the association rules between readers' borrowed books. It is found that these rules can realize personalized book recommendation to readers, And optimize the management of bookshelves in the collection. When readers borrow, they need to actively go to the library collection rack to find the books they need or enter the library electronic information system to search for download resources [2]. The traditional management of library book information is to input the year, subject, title and author of the book into the database as the classification basis and search label. Readers need a clear search label to search the book information. For some users, the selection of search words is also a problem to be solved in book retrieval. In book retrieval, the selection of search words requires certain skills. If you want to quickly select the appropriate search words, you need to have certain professional knowledge [3]. For those who have received professional training, although they have good classification and subject indexing ability, they sometimes can't accurately distinguish what they need in the chaotic ocean of information.

Modern libraries use digital technology to provide retrieval services for readers, which has become the primary way for readers to contact the library. How to build a good retrieval system has also become one of the research hotspots of library technology. Using the retrieval system, users can retrieve all the bibliographies in the library and obtain useful information. Information mining technology has attracted more and more attention, and has become a hot spot in the current computer field. Its research focus has gradually shifted from discovery methods to system application, and pay attention to the integration of multiple discovery strategies and technologies and the mutual penetration between multiple disciplines [4]. Nowadays, the application fields of data mining are more and more extensive, from early commercial applications to scientific research, e-commerce, product control, financial industry, education and teaching and other fields. With many successful applications, the emergence of data mining technology and its wide and successful application in the commercial field provide a direction for the research of Library in the development of personalized service [5]. Using data mining technology to mine and analyze the massive borrowing information stored in the library database, we can find the hidden Book association rules and readers' borrowing preferences and habits, and improve the quality of library book management service. In order to meet the higher and higher demand standards of library readers, based on the current construction of Digital Library and the advantages of data mining technology in content information data processing, this paper will study the library book information classification and retrieval method based on data mining, so as to provide library decision-making and information reference in the collection, facilitate readers' knowledge learning and improve borrowing efficiency.

2 Research on Library Book Information Classification and Retrieval Method Based on Data Mining

2.1 Library Book Information Preprocessing

Library book information can be regarded as text information. When processing book information, text preprocessing is first needed. The storage formats of corpus text in are various, including various markers, labels, pictures, numbers and English characters. These noises have no effect on text classification, and some high-frequency useless words will even affect the effect of classification. Different from English, Chinese text has no clear blank between words, and Chinese classification is more complex than English. Therefore, the classification of Chinese information needs a Chinese segmentation system to segment the text. The purpose is to separate the words through spaces and format the text information of library books.

In order to quickly match the Chinese character string, it is necessary to convert the text into the form of Chinese character hash table. The book text information is encoded in double bytes according to GBK code, and the overall coding range is 8140-fefe. The first byte is between 81-fe and the last byte is between 40-fe. Eliminate xx7f one line, and 23940 code points in total. It includes 21886 Chinese characters and graphic symbols, including 21003 Chinese characters (including radicals and components) and 883 graphic symbols. Therefore, the selected hash function value is the first byte of the internal code of Chinese characters. Hash table also known as hash table) is a data structure that can be accessed directly according to key value. That is, it accesses records by mapping key values to a location in the table, so as to speed up the search. This mapping function is called a hash function, and the array storing records is called a hash table. The data structure of hash table can be defined as follows [6]:

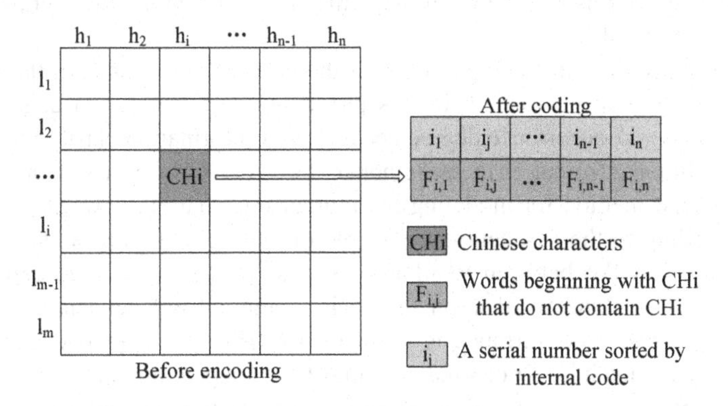

Fig. 1. Schematic diagram of hash table data structure

In Fig. 1 above, h_i is the high byte of the internal code after the book text information is encoded; l_i is the low byte of the internal code after the book text information is encoded. The subscript of the two-dimensional array is used to represent the address of the hash function. Because the inner code of each text information character is different, there is no problem of address conflict. Libraries usually store book information in the library management system database. The data tables related to book information classification and retrieval in the database mainly include: Book Basic information table, book borrowing information table, reader information table, and Book copy table. These data tables store the original information such as book name, book classification, reader information and book borrowing, which are the basis of data mining for readers. Preprocessing the data before data mining can improve the mining efficiency.

Firstly, delete the text marks, such as the author of the book, publishing time, address source and other marks, and present the text as a single word combination to form a consistent text format; Secondly, the text is processed by Chinese word segmentation. Most text mining research usually involves words or sentences. In order to further process these words, they must be separated word by word. Therefore, at this stage, because it cannot represent any category, all words in the sentence will be segmented, which can simplify the calculation process in the next step; Finally, stop words, redundant words and duplicate words are deleted. There are usually many methods to deal with incomplete data such as missing attribute values in records. The main methods to summarize these methods are as follows: the first method is to directly delete the records containing the missing attribute data values when the proportion of missing attribute data values to be processed in the field is relatively small, However, when the proportion of attribute missing data value in the field exceeds a certain degree, the accuracy of mining results will be reduced, resulting in very poor performance; The second method is to fill in the value manually for those missing attributes. However, when there are many missing attribute values in a large data set, if this method is still used for processing, it will take older time and the feasibility is very poor; The third method is to use a predetermined possible value to fill in the missing attributes. This method is relatively simple in operation, but if it is widely used, it may cause large errors in the results obtained by mining; The fourth method is to calculate the average value of the attribute value in other object records with the same characteristics for the missing attribute, and then use the average value to fill in the missing attribute [7].

After the above preliminary processing of the book information text, the book text information features are extracted. In this paper, mutual information degree is used to characterize the co-occurrence degree between text information class and secondary information. Mutual information measurement is derived from information theory, which provides a formal method for modeling the mutual information between features and classes. According to the co-occurrence degree of class c and word w, the pointwise mutual information IW between word w and class c is defined. The expected co-occurrence of class i and word w on the basis of mutual independence is given by $p_iE(w)$, and the real co-occurrence is given by $P(w)E(w)$. In practice, the value of $P(w)E(w)$ may be much larger or smaller than $p_iE(w)$, depending on the degree of correlation between Class c and word w. Mutual information is defined according to the ratio between these two values, and its calculation formula is as follows [8]:

$$I_i(w) = \lg\left(\frac{p_i(w)}{P_i}\right) \tag{1}$$

Among them, $p_i(w)$ represents the normalized probability value of the global distribution of book information in different classes; P is the global probability of book information text. When $I_i(w)$, WORD w is positively correlated with class i, and when $I_i(w) < 0$, WORD w is negatively correlated with class i. Note that $I_i(w)$ is specific to a specific class i. The overall mutual information needs to be calculated as a function of word w mutual information of different classes. These are defined by using the average and maximum values of $I_i(w)$ on different classes.

The use of global probability ensures that the Gini index more accurately reflects the differences between classes in the case of class distribution deviation in the whole text set. In addition, information gain is introduced to further describe the characteristics of book information. Assuming that the book information document contains the word w, let P_i be the global probability of class i and P_i be the probability of class i. $E(w)$ is the score of the text containing w.

$$G(w) = -\sum_{i=1}^{k} P_i * \lg(P_i) + E(w)\sum_{i=1}^{k} P_i * \lg(P_i) \tag{2}$$

The greater the value of information gain $G(w)$, the greater the discrimination of word w. For a text corpus containing n documents, d words and k classes, the complexity of information gain calculation is $O\{n, d, k\}$. After preprocessing the library book information according to the above process, use the data mining technology to process the book information and obtain the deep information relationship.

2.2 Information Data Classification and Mining Processing

Libraries have basically realized automatic management, and the information data of book resources can be found in the automatic management system. The main problem of association rule mining is to find the correlation between items hidden in large databases, so as to help people do well in management and make more accurate decisions. This paper uses the improved association algorithm to mine the library book information. Apriori algorithm decomposes association rule mining into two sub problems [9]:

(1) Find all frequent sets in transaction database d that meet the minimum support minsup;
(2) Using frequent sets to generate all association rules that meet the minimum confidence minconf;

Finding frequent sets is the core problem of association rule discovery. An itemset containing K items is called a itemset. If the itemset meets the minimum support, the itemset is called frequent itemset, which is called frequent itemset for short.

It can be found from the steps of Apriori algorithm that a large number of candidate sets may be generated when generating candidate sets. When there are 10000 frequent sets with length 1, the number of candidate sets with length 2 will exceed 10 million. At the same time, due to the need to calculate the support of itemset, the transaction database will be scanned repeatedly, which will increase the computational overhead. Therefore, this paper improves the Apriori algorithm according to the following process.

1) Read in the compressed transaction set (book information after preprocessing);

2) Scan the transaction set to find each type of frequent single item;
3) Connect all kinds of frequent attribute items and frequent book classification items into 2 a candidate frequency complex item set, $k = 2$.
4) Check the k-candidate frequent itemset and record its support and the support of the predecessor. The connection condition of frequent itemsets is that the first n item is the attribute item of readers, and the content of the attribute item of readers is different, and the last item is the same book classification item.
5) Output the frequent k-frequent itemsets whose confidence and support meet the requirements. The confidence is the support divided by the support of the previous part.
6) $k + 1$-candidate frequent itemsets are obtained by connecting k-frequent itemsets. Pruning can reduce the number of frequently connected itemsets and improve the efficiency of program operation. The following is the rule of pruning connection:
a) If the last item category of frequent itemsets A and B is different, it cannot be connected.
b) If there are different items belonging to the same attribute category, they cannot be connected.
c) Frequent itemsets cannot be connected to themselves.
d) If conf is used to represent the antecedent support, when min (A. conf, B. CONF)/minimum support < minimum confidence, A and B cannot be connected.

Rule $A \rightarrow B$ has credibility. C is the percentage of A-item set and B-item set relative to A-item set, which is the conditional probability $P(B|A)$, i.e.

$$C(A \rightarrow B) = P(B|A) = \frac{|AB|}{|A|} \tag{3}$$

where $|A|$ represents the number of transactions that contain itemset A in the database.

e) Other situations can be connected.
7) $k++$ if the number of candidate frequent itemsets generated is not 0, turn to 4), otherwise it ends.

The implementation of association rule mining is to find all the strong association rules in the transaction database, and the strong association rule $A \rightarrow B$ is the rule that meets the minimum support and minimum confidence at the same time, then the corresponding item set $A \rightarrow B$ also meets the minimum support, that is, the frequent item set, and the confidence of the rule can be calculated from the support of frequent item sets A and $A \rightarrow B$. After using the improved Apriori algorithm to analyze the correlation between library book information, the decision tree algorithm is used to determine the classification category of book information.

This paper uses ID3 decision tree algorithm to determine the classification of book information. ID3 algorithm is a classification technology of decision tree. Its purpose is to select the best attributes as nodes, and take the constructed decision tree as the simplest state or close to the simplest state. The optimal node is determined by the entropy generated by its node, and its calculation method is as follows. If the objects of a certain object set D belong to j different categories, the entropy $S(D)$ of this object set is:

$$S(D) = \sum C_j \ln C_j \tag{4}$$

where, D object set; j is the number of categories, C_j is the total number of objects belonging to category j and the total number of items of D.

Next, select an attribute X_j as the node of the decision tree, establish m child nodes under this node, and assign all objects originally belonging to the node to the appropriate child nodes. The attribute X_j values of objects assigned to the same child node must be equal. According to entropy and information gain, the execution steps of ID3 algorithm are described as follows [10–12]:

1) First, let the root node of the decision tree be D, and all objects belong to the object set of D.
2) If all objects in D belong to the same category, define node D as this category and stop, otherwise continue to step 3;
3) Calculate the entropy of all objects belonging to D;
4) From the root node to the current node, if there is an attribute X_j that has not been a node, the D object set is divided by X_j, and the entropy and information gain of some decision trees are calculated respectively;
5) The candidate attribute with the maximum information gain is selected as the classification attribute of D node;
6) Create sub nodes under node D, which are respectively $d_1, d_2, \cdots, d_m$ (assuming that multiple attribute values are selected as classification attributes), and assign all object sets in D to appropriate sub nodes.

After the classification of library book information is determined by the decision tree algorithm, the book information retrieval framework is designed according to the book concept map.

2.3 Design of Library Information Retrieval Architecture

When searching for books of a certain subject, users do not have a clear understanding of the information of the subject they want to search. Especially for users who are beginning to learn a certain subject, because they do not understand the concept points and knowledge structure of the subject and are still vague about the subject knowledge, it may take a long time to search the subject books by themselves, and the relevant books retrieved may not be comprehensive, Users still need some help from the outside world to learn the subject well and understand the scope of knowledge contained in the subject. At the same time, in the process of book retrieval, users may also face the problem of difficult selection of search words. We need to consider how to select search words. Knowledge nodes, i.e. concept words, can be regarded as the retrieval words of book retrieval, and there is a certain dependency between concepts, which constitutes the knowledge network of books, and the concept map can clearly show the concept information. Concept map is a graphical method that uses nodes to represent concepts and lines to represent relationships between concepts. The theoretical basis of concept map is Ausubel's learning theory. The construction of knowledge begins with the observation and understanding of things through existing concepts. Learning is to build a conceptual network and constantly add new content to the network. In order to make learning meaningful, individual learners must associate new knowledge with the concepts they have learned.

The library information retrieval framework designed in this paper is mainly designed from the perspective of users and librarians. It can be divided into book retrieval module and concept map management module. The book retrieval module also includes concept map module. The book retrieval module is mainly aimed at users. After users input the search words, the information retrieval model provides the concept map of relevant disciplines according to the search words, and retrieves the relevant books according to the index. Users can view these books; The concept map module included in the book retrieval module has the function of uploading concept nodes. Users can supplement concept nodes here. The concept map management module is mainly aimed at librarians. In this module, librarians filter the concept information nodes uploaded by users and update the original concept map.

In this concept map, users can view and understand some discipline branches under the retrieved discipline and the conceptual knowledge points of the discipline, and have a more complete understanding of the retrieved books. In this way, they can also find other required books related to it to help users further study. For some users who need to consider the selection of retrieval words, the presentation of concept map can also help them solve this problem. The conceptual diagram is the "center - surrounding" structure diagram. Figure 2 shows the book information retrieval framework based on concept map.

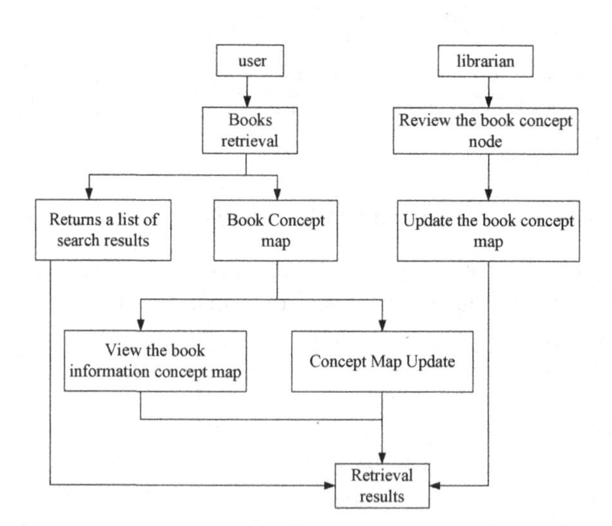

Fig. 2. Book information retrieval framework based on concept map

The central position of the concept map is the discipline name of a discipline. The first layer connected around is the first level concept node, the first level concept node is connected downward as the second level concept node, the second level concept node is connected downward as the third level concept node, and so on. Level 2 concept nodes are included in level 1 concept nodes, level 3 concept nodes are included in Level 2 concept nodes, and concept nodes of the same level are not included in each other. After using the concept map to design the framework of book retrieval, the book information retrieval is realized on the basis of book information classification.

2.4 Realize Classified Retrieval of Book Information

This paper uses Markov network model to retrieve library book information. In the research and practice of information retrieval, there is the uncertainty of retrieval, that is, the exploratory and fuzziness of retrieval. In the behavior of information retrieval, the information needs of users are fuzzy. Usually, the information requirement submitted by the user is a short sentence or a shorter search term. Therefore, most users find that it is difficult to form an information query that can achieve the purpose of effective retrieval. Users may retrieve some information related to their query, but it is impossible to retrieve all relevant information, and the recall rate is always limited. Especially in the case of synonymy or polysemy, the retrieval efficiency is not ideal under the simple matching model between query words and words in book information. However, if multiple book information describes the same topic, there must be some connections between these book information, and some topic words will appear among them to describe a topic or some topics. Through the book information, the initial retrieval results can be corrected to avoid the poor retrieval effect caused by the simple matching of query words.

Given query q, the document set needs to calculate a final retrieval score of each document t_j and query q, record it as $p_f\left(t_j|q\right)$, and sort the retrieval results according to this score. The initial retrieval score of each document without document rearrangement is recorded as $f_o\left(t_j|q\right)$. Make full use of the information contained in the document group and $f_o\left(t_j|q\right)$ to extract the initial score of the document group. A correction score is recorded as $f_n\left(t_j|q\right)$. Finally, integrate the initial retrieval score and correction score, calculate the final retrieval score $p_f\left(t_j|q\right)$ of the document according to the following formula, and rearrange the document in ascending order of $p_f\left(t_j|q\right)$.

$$p_f\left(t_j|q\right) = \xi f_o\left(t_j|q\right) + (1-\xi)f_n\left(t_j|q\right) \tag{5}$$

The retrieval mode used for book information in this paper is full-text retrieval. The specific process is as follows: the index is established by scanning each word in the data set, and the index database is established by the index of each word; The user inputs the search words of book retrieval at the input port, and then preprocesses the semantic of the search words input by the user; After completing the semantic query expansion, match the similarity between the index words in the index database and the semantic expansion words of query expansion; Sort the output results and return the sorted search results to users, so as to complete book information retrieval. So far, the research on library book information classification and retrieval method based on data mining has been completed.

3 Method Test

The above studies the library book information classification and retrieval method based on data mining. This section will test the effectiveness of this method. Through the classified retrieval of book information, we can better understand the advantages and disadvantages of the model, find out what factors lead to the poor performance of the method, and facilitate the follow-up to provide the basis for the improvement and adjustment of the retrieval method. Finally, we can comprehensively, quickly and accurately retrieve the required book information and improve the service quality of the retrieval method.

3.1 Test Preparation

In order to test the performance of the library book information classification and retrieval method based on data mining, experiments are carried out in a comparative way. When users use the book retrieval model for retrieval, the retrieved Book results are generally sorted according to the correlation between the book and the user's query information. In general, the retrieved results are ranked from high to low according to the correlation. The book in the front is the book with the highest correlation with the user's search information, which is also the result that best meets the needs of users. Therefore, for the book information retrieval method, the accuracy of the retrieved Book results needs to be evaluated in the process of use.

The original corpus data used in this paper is the complete book of 3000 humanities and social sciences books downloaded from the database. Because some original corpora are picture documents, and the model can not recognize them, it is also necessary to use OCR technology to convert the books into TXT file format required by the experiment, so as to obtain the initial experimental corpus data. OCR technology is a kind of character recognition technology. It can recognize and extract the characters on the picture and convert them into computer characters.

Under the environment of MATLAB software, it is configured on PC: CPU is 3.6 ghz, memory is 8 GB, and a book information classification and retrieval platform is built. The classification effect of book retrieval information is an important basis for testing the classification and retrieval effect of book information. The information retrieval method based on SOM neural network and the information retrieval method based on Top-k query algorithm are selected as the comparison group to compare with the book information classification retrieval method proposed in this paper. By comparing the accuracy and false detection rate of book information retrieval methods, this paper tests the effectiveness of book information retrieval methods.

3.2 Test Result

Table 1 below shows the comparison of accuracy and false detection rate of classified retrieval of book information of single subject type using three information retrieval methods.

Through the analysis of the data in Table 1, it can be seen that when retrieving the book information of a single subject, the error detection rate of the information retrieval

Table 1. Comparison of retrieval accuracy and false detection rate of information classification retrieval methods/%

Group Accuracy	Information retrieval method based on data mining		Information retrieval method based on SOM neural network		Information retrieval method based on Top-k query algorithm	
	Noise factor	Accuracy	Noise factor	Accuracy	Noise factor	Accuracy
1	94.3	1.39	83.0	2.48	83.2	3.19
2	96.5	2.01	84.1	2.35	82.4	3.45
3	95.7	1.38	82.2	2.63	81.5	3.35
4	97.2	1.85	82.6	2.57	81.8	3.08
5	94.1	1.48	84.3	2.56	80.7	3.24
6	95.4	1.33	85.4	2.34	81.1	3.37
7	96.6	1.87	85.7	2.61	82.3	3.55
8	96.5	1.31	82.5	2.39	83.5	3.49
9	95.8	1.73	84.1	2.65	80.0	3.52
10	96.0	1.86	86.6	2.48	81.6	3.26

method based on data mining is less than 2.1%, and the lowest error detection rate of the two retrieval methods is 2.34% and 3.08%. The information retrieval method based on data mining is far lower than the lowest error detection rate of the other two retrieval methods. The retrieval accuracy of this method is over 94%, and the book retrieval effect is obviously better than the other two retrieval methods. Compared with the two traditional methods, the library book information classification retrieval method based on data mining proposed in this paper has higher retrieval accuracy.

Table 2 below shows the comparison of the accuracy and false detection rate of classified retrieval of book information with interdisciplinary by using three information retrieval methods.

By analyzing the data in Table 2, it can be seen that the book information retrieval is disturbed by interdisciplinary disciplines, resulting in the decline of the retrieval accuracy of retrieval methods and the rise of false detection rate to a certain extent. However, the retrieval accuracy of the information retrieval method based on data mining is still higher than that of the other two information retrieval methods. Summarizing the above test and analysis contents, we can see that the retrieval accuracy of the library book information classification retrieval method based on data mining proposed in this paper is higher than 92%, and the retrieval accuracy is significantly improved. The retrieval accuracy of the two traditional methods is lower than 84% and 80% respectively. Compared with the two traditional methods, the retrieval accuracy of the library book information classification retrieval method based on data mining proposed in this paper is higher.

Table 2. Comparison of retrieval accuracy and false detection rate of information classification retrieval methods/%

Group accuracy	Information retrieval method based on Data Mining		Information retrieval method based on SOM neural network		Information retrieval method based on Top-k query algorithm	
	Noise factor	Accuracy	Noise factor	Accuracy	Noise factor	Accuracy
1	92.0	2.23	80.1	3.34	78.3	3.74
2	93.6	1.91	81.3	3.49	75.1	3.95
3	93.3	2.17	82.9	2.92	77.7	3.76
4	95.4	1.78	82.5	3.38	79.2	3.71
5	93.1	2.05	82.7	2.94	76.5	3.67
6	95.8	2.13	79.6	3.15	79.9	3.32
7	93.9	1.84	79.6	2.93	76.8	4.10
8	95.5	1.85	79.3	3.29	78.1	3.35
9	93.7	1.97	83.9	3.11	77.4	4.08
10	93.2	1.96	82.2	3.36	78.6	4.07

4 Conclusion

With the development of information technology, digital library is more and more valued by the country. The content of digital library is more and more abundant, and the number of books is growing rapidly, which makes the structure of library management system become complex and the content distribution is also very extensive. The information needs and forms of the library for users are becoming more and more diversified. In the field of book retrieval, it is the most fundamental to provide readers with retrieval results that can meet their query needs to the greatest extent, which requires the design of good retrieval methods, and to ensure the quality of retrieval results and the efficiency of retrieval speed. Book intelligent retrieval system is not a simple combination of traditional retrieval system and computer network technology, but needs to comprehensively consider readers' personalized materials and other potential information resources to help readers find the resources they want. How to accurately classify and retrieve book information from library book resources and improve the utilization rate of library resources is a difficult problem to be solved in recent stage. Aiming at this research, this paper puts forward a library book information classification and retrieval method based on data mining. After preprocessing the book information, Apriori algorithm in data mining is used to deeply manage the information, and decision tree algorithm is used to classify the information. The concept map of book information is constructed by using the concepts in book information, and the book information retrieval framework is constructed. Using Markov network model to realize the classification and retrieval of book information. The test results of classified retrieval method show that the accuracy of book information retrieval is significantly improved when the proposed method is applied. Through the research and application of this method, it promotes the transformation from the traditional library service concept to the personalized information service concept. In the future development process, according to the personalized characteristics and personal needs of readers, we will provide more convenient book information retrieval and push services, and constantly improve the speed of book information retrieval.

References

1. Zhao, W.: Design and implementation of book retrieval system based on barcode recognition technology. Mod. Electron. Tech. **42**(17), 124–128 (2019)
2. Hu, X., Lu, C., Qi, B.: Classification model of internet public opinion information based on SOM neural network. J. Ordnance Equip. Eng. **40**(03), 108–111 (2019)
3. Dong, G., Xia, W.: Fast retrieval method for self-integrated information of books based on top-k query algorithm. J. Jilin Univ. (Sci. Edn.) **58**(03), 666–670 (2020)
4. Fan, H..M..: Simulation of accurate retrieval of massive books in library. Comput. Simul. **36**(06), 337–340 (2019)
5. He, M.: Design of library mass information classification and retrieval system based on data mining. China Comput. Commun. **32**(11), 153–154 (2020)
6. Zhang, H.-T., Zhang, X.-H., Wei, P., et al.: Research progress on network user information retrieval behavior. Inf. Sci. **38**(05), 169–176 (2020)
7. Nan, C..-J..: Research on the application of data mining in library user resource management. J. Nanyang Inst. Technol. **12**(05), 125–128 (2020)
8. Chugh, A., Sharma, V.K., Kumar, S., et al.: Spider monkey crow optimization algorithm with deep learning for sentiment classification and information retrieval. IEEE Access **9**, 1 (2021)
9. Maher, A., Supreethi, K.P.: COVID-19 diagnostic system using medical image classification and retrieval: a novel method for image analysis. Comput. J. **8**, 8 (2021)
10. Jasm, D.A., Hamad, M.M., Alrawi, A.: A survey paper on image mining techniques and classification brain tumor. J. Phys. Conf. Ser. **1804**(1), 012110 (8pp) (2021)
11. Kim, D., Kang, H.G., Bae, K., et al.: An artificial intelligence-enabled industry classification and its interpretation. Internet Res. Electron. Network. Appl. Policy 32(2), 406–424 (2022)
12. Karmakar, P., Teng, S.W., Lu, G., et al.: An enhancement to the spatial pyramid matching for image classification and retrieval. IEEE Access **8**, 22463–22472 (2020)

Construction of Cognitive Model of Traditional Sports Health Preservation from the Perspective of Body-Medicine Integration

Fangyan Yang[1(✉)], Chun Zhou[1], and Ying Liu[2]

[1] College of Sports and Health, Changsha Medical University, Changsha 410219, China
xuenxuen152@163.com

[2] School of Information and Control, Shenyang Institute of Technology, Fushun 113122, China

Abstract. The traditional sports health care cognitive model has the problem that it takes too long to deal with, resulting in poor cognitive effect of sports health care. Design a traditional sports health care cognitive model from the perspective of physical and medical integration. According to the health preservation theory of traditional Chinese medicine, identify the biological rhythm characteristics of human body, design the behavior output framework for the integration of sports and medicine, compile the application programming interface for the central core of the model, and construct the cognitive model of traditional sports health preservation. Experimental results: the average processing time of the designed sports health preservation cognitive model and the other two cognitive models is small, which proves that the performance of the traditional sports health preservation cognitive model designed from the perspective of sports medicine integration is more prominent.

Keywords: Integration of sports and medicine · Physical health preservation · Cognitive model · Time rhythm · Health preservation theory of traditional Chinese medicine · Scientific exercise

1 Introduction

Health is the foundation of human existence, the first demand of social development and the eternal theme of building a harmonious society. With the continuous progress of society and the rapid development of China's economy, people's material base is abundant, and health has become a problem that people pay more and more attention to. The concept of integration of sports and medicine originated from the United States. It advocates the combination of sports and medical care, carries out medical judgment and sports intervention on healthy people, sub-health people and patients, urges healthy people to cultivate good physical fitness, enhances disease resistance, and helps sub-health people and patients recover as soon as possible. In recent years, fitness and health preservation have been widely valued by people. The integration of correct and reasonable physical fitness methods and scientific fitness and health preservation has gradually been pursued by more and more people.

W. Fu and L. Yun (Eds.): ADHIP 2022, LNICST 468, pp. 244–256, 2023.
https://doi.org/10.1007/978-3-031-28787-9_19

The integration of sports and medicine is a treatment method that aims at solving health problems or medical problems and combines sports with non-medical means to solve chronic disease problems. However, at present, the vast majority of people do not know how to exercise and keep fit scientifically, nor do they know the practice methods and time of using China's traditional sports to exercise and keep fit. There are even many people who do not realize the need to conform to the biological rhythm of the human body and the time rhythm of traditional sports when using our traditional sports for fitness and health preservation. Instead, they practice at any time according to their subjective ideas, or mistakenly refer to some theories without scientific basis. It is of great significance to organize scientific exercises and promote the improvement of the health literacy of the whole people through the integration of physical construction and physical medicine.

As a new model under the strategic background of healthy China, sports rehabilitation came into being [1–3]. The exercise rehabilitation described in this paper is neither a special means in rehabilitation medicine nor a certain stage in "comprehensive rehabilitation". It is an emerging specialty intersected by clinical medicine, rehabilitation medicine and sports human science. Therefore, the practice time of traditional sports will go against the biological rhythm of the human body, and the practice will often have no effect, or even backfire and affect human health. As an important part of Chinese traditional culture, Chinese traditional fitness culture inherits the essence of Chinese traditional philosophy, traditional Chinese medicine and Confucianism, Buddhism and Taoism. So far, sports rehabilitation is still a relatively unfamiliar term for the general public. In the context of the healthy China strategy, people begin to realize that the integration of sports and medical treatment will greatly improve their health level, and to a large extent, public attitudes can promote or hinder the development of a new technology or emerging industry. On the basis of "harmony between man and nature", Chinese traditional fitness and health preservation sports follow the principle of yin-yang balance in traditional Chinese medicine. Although western sports have entered China, it can not hide the infinite charm of Chinese traditional sports culture, which is still favored by many fitness and health lovers at home and abroad. Therefore, based on this background, this study investigates and analyzes the cognition of traditional sports health preservation. The cognition investigation of sports rehabilitation not only conforms to the social development trend, but also is the theoretical basis for solving a series of current health problems.

2 Construction of Cognitive Model of Traditional Sports Health Preservation from the Perspective of Sports Medicine Integration

2.1 Human Biorhythm Feature Recognition

Chinese traditional health preservation thought advocates the holistic view of "the unity of heaven and man". Heaven generally refers to objective things and their change laws, including the change laws of nature and human society. Man is not only an integral part of natural things, but also one of the members of society. He is the interdependence of nature and society. "Harmony between man and nature" refers to the harmonious unity of man, nature and society. A person's life is a process of gradual change from birth to growth, then to aging and finally to death. In this process, various physiological functions of the human body are constantly changing. Although today's medical treatment, science and technology continue to develop and progress, and various anti-aging methods emerge in endlessly, it still can not stop the natural trend of people getting old gradually. According to the changes of people's psychological and physiological functions over time, modern research divides people's life into different stages with different standards. It is mainly manifested in the unity of man and nature. It is said that man should merge with nature, and man is a kind of natural things. Human life is the essence of natural things, and is constantly gaining the essence of nature. Since mankind comes from heaven and earth, it should follow the law of nature and follow the laws of nature. China's traditional numerology science has also found the existence of human annual rhythm for a long time. Human annual rhythm is affected not only by natural laws and numbers, but also by other factors such as the revolution of the moon. Chinese traditional health experts believe that human life activities should conform to the four seasons according to the natural law. For example, according to the change law of seasonal climate, such as spring temperature, summer heat, long summer, autumn dryness and winter cold, people should adapt to spring, long summer, autumn harvest and winter storage [4, 5]. According to the health preservation theory of traditional Chinese medicine, the process of human life from growth and development to aging is divided into six cycles, as shown in Fig. 1:

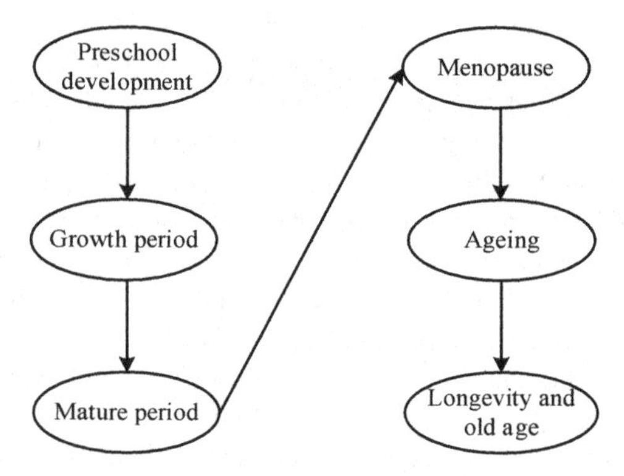

Fig. 1. Human growth and development process

It can be seen from Fig. 1 that the growth and development process of the human body mainly includes: preschool development, growth, maturity, menopause, aging and longevity aging. In addition to the integration of human and nature, the integration of human and society, as one of the members of society, can only improve the quality of life by relying on and integrating into society. There are many kinds of people and things in society. If a person's understanding is not deep enough, his logical thinking is not clear, his guiding ideology and method of adjusting his mind and nourishing his spirit are not correct, and his emotional fluctuations and sudden changes are not handled well, he will often involve the five internal organs and lead to diseases. According to the mutual opposition, dependence, growth and decline, and transformation of yin and Yang, traditional health preservation takes physical exercise as the basic means and method, starts with the Yin and Yang in the internal and external environments that affect human health, and adjusts the balance of yin and Yang of the body through the implementation of various health preservation methods, so as to achieve the goal of sufficient and full body spirit, nourishing yin and Yang, healthy body and prolonging life. For example, according to the laws of the seasons, traditional health preserving exercises in spring and summer are conducted by using some traditional sports functions that nourish Yang and Yin, such as walking, climbing, traveling and other relaxing, relaxing and blood calming health preserving activities. In autumn and winter, some traditional sports health preserving functions that nourish yin and yang are used for training. In the process of human life activities, it is a normal phenomenon that the dynamic balance of life activities can be maintained through biological restraint between various tissues and organs or various physiological functions of the body.

2.2 Design Behavior Output Framework of Sports Medicine Integration

To understand the integration of sports and medicine, it is necessary to define the concepts of "body" and "medicine". "Body" refers to "Sports". Etymologically, it is a physical education activity aimed at health. The integration of sports and medicine literally means the integration of "Sports" and "medical treatment". Its essential connotation is to integrate sports health resources and medical health resources in the context of integrating national fitness into national health, so as to realize the optimal allocation of health promotion resources. In the cognitive model, a complete behavior output process must go through internal thinking activities such as perception, decision-making and learning, as well as the behavior output process of decomposing the behavior scheme into meta behavior.

Next, we study the behavior output framework of how to decompose intention into meta behavior. Specifically, it is a comprehensive application of sports technology, medical technology and other health promotion means to people's scientific fitness, disease prevention, disease treatment and rehabilitation, so as to obtain the whole life cycle process of health promotion. It is a good prescription to promote the construction of healthy China. In a dynamic environment, the agent will constantly respond to environmental changes and make independent decisions, which will produce a large number of unpredictable high-level behaviors. It is impossible to model each complex high-level behavior one by one and store it in the agent's knowledge base. For example, the basic operations of the army are clearly defined in the operation manual, such as when to use radio communication means, how to communicate, precautions, etc.

With the progress of modern society and the acceleration of human urbanization, the development of people's lifestyle has fundamentally changed. Under the living state of being far away from nature, lack of exercise and surplus nutrition, sports participation, exercise load, exercise intensity and scientific degree can not meet the requirements of health. The integration of sports and medicine has the safety of sports, which is reflected in that the safety problem in sports involves many factors. A feasible method is to define a set of meta behaviors, and define a set of behavior operations and a behavior synthesis framework on the meta behaviors. In this way, the high-level behavior can be synthesized by meta behavior through certain operations. This not only greatly improves the flexibility of behavior output, but also reduces the difficulty of system implementation. Based on the above ideas, a behavior output framework is proposed. From the perspective of exercise alone, exercise directly affects the safety of heart, kidney and other organ functions, the normal value of blood glucose, skeletal muscle injuries and other sports safety problems. For example, sudden cardiac death caused by improper exercise intensity often occurs. Rhabdomyolysis caused by excessive exercise intensity is also common. An action plan is a target task set, and the task can be subdivided into the method or process of completing the task, which can form a four-layer behavior output hierarchy, as shown in Fig. 2:

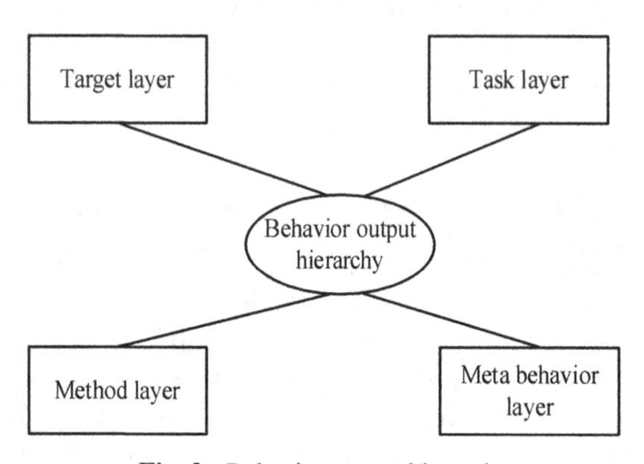

Fig. 2. Behavior output hierarchy

As can be seen from Fig. 2, the behavior output hierarchy includes: target layer, task layer, method layer and meta behavior layer. In the traditional cognitive model of sports health preservation, the effectiveness of sports medicine integration exercise is reflected in that the ultimate effect of exercise is to promote health positively and improve body function through excessive recovery. So as to prevent the occurrence of diseases, or promote the treatment and rehabilitation of diseases. The meta behavior layer can directly interact with the physical model of the subject and the environment model of the system to produce various realistic effects, such as the effects of explosion, movement of the subject and so on. In the cognitive model, the process of computer problem solving is to search the state space and find the knowledge related to the problem in the state space. However, the search space for many interesting problems often grows exponentially and cannot be completed within the time frame that people can tolerate. Heuristic search is the primary tool to deal with this combinatorial explosion problem. The common feature of semiotic models is production rules. The effect of exercise can only be brought into full play through the accumulation of a certain period of time in a gradual process. The continuity of sports medicine integration movement is reflected in that the prevention of chronic diseases or the promotion of chronic disease rehabilitation by sports as a non-medical means all require long-term exercise persistence and life cycle. The lack of continuity of exercise is the biggest obstacle affecting the effect of rehabilitation treatment of various chronic diseases. The rehabilitation effect of exercise is significant and irreplaceable. Scientific exercise can greatly shorten the hospital stay, reduce the drug risk and the possibility of complications [6, 7]. Production rule is the replacement operation of symbol string. In fact, it is an abstract symbol, which is similar to the relationship between mathematical symbols and numbers. Mathematical symbols are the replacement of numbers. Therefore, production system is recognized as a natural structure for modeling problem solving methods, realizing search algorithms and constructing cognitive models. If people want to use cognitive model to solve specific engineering applications, they need to write application programming interfaces for the core of the model. A large part of the efforts made on the interface in different projects are repetitive work. The main reason is that the current cognitive model interface is not encapsulated and is not flexible for our tasks. Adhere to physical exercise, not only for health, but also for physical and mental pleasure. The personalized and emotional design of sports can promote the sustainability of sports, and then produce the sustainable effect of sports on health. The defensive nature of sports medicine fusion disease. The simple health promotion thought of "treating disease without disease" in traditional medicine gives new enlightenment to modern health promotion business. The integration of sports and medicine advocates the use of health resources at all levels of province, city, county, town and village, give full play to the voice of doctors and give full play to the disease prevention effect of sports.

2.3 Build a Cognitive Model of Traditional Sports Health Preservation

Traditional sports health preservation refers to the physical exercise, physical exercise type health physiology and stimulation system that integrates Chinese traditional life science ideas and methods, adjusts human posture, breathing and mind, strengthens body coordination, prolongs body life, enhances body function and excavates body potential. If the external conditions change, the reaction system will also change.

Based on the meaning of sports and medical integration and cognitive governance, sports and medical integration cognitive governance is the practical application of cognitive governance theory in the field of sports and medical integration. It is a cross class, cross department and cross organization horizontal governance pattern. It is a way for government departments, health service departments, the public and basic autonomous organizations to form a governance body to recognize and govern sports and medical integration. The main purpose of most traditional fitness and health preserving sports in China is fitness and health preserving. Under the influence of the theory of heaven and man, most of our traditional fitness and health preserving sports conform to the biological rhythm of the human body. Based on the common value pursuit, all subjects share the governance rights on an equal basis, which is both independent and symbiotic. Therefore, the concept of cognitive governance of sports and medical integration is defined as: it refers to a governance system based on health promotion, in which the government's leading departments, sports departments, health departments, health service departments, social autonomous organizations, social citizens and other multiple subjects strengthen cooperation and interaction between subjects through legal norms, dialogue and consultation, interest checks and balances, responsibility transmission and other governance means, so as to achieve the integration of sports and medical services, so as to maximize national health and promote public interests.

At the same time, some of China's traditional fitness and health preserving sports are also used as carriers of China's national traditional culture and often as carrying tools in festivals. The main function of China's traditional fitness and health preserving sports in this environment is not fitness and health preserving, but carrying China's national traditional culture through festivals. Therefore, some of the time rhythms of traditional fitness and health preserving sports in China are closely related to the biological rhythm of the human body, while others are not closely related to the biological rhythm of the human body, and are more consistent with the festival rhythm. Through combing, it is found that the time rhythm of Chinese traditional fitness and health preservation sports is related to the biological rhythm of the human body. Order parameter and self-organization are the core concepts of cognitive theory. These two concepts and their extensions form a profound and rich theoretical composition of cognitive theory. Combined with corresponding medical resources, the concept of order parameter in Landau's phase transition theory is used as a theoretical guide to solve the self-organization problem. In the traditional cognitive model of sports health preservation, the choice of achievements is determined by the utility value of a set of achievements. The utility of results is defined as:

$$H_p = R_p - \eta + \sum \frac{\|\eta - 1\|}{R_p} \tag{1}$$

In formula (1), R represents a subset of candidate operators, η represents candidate operators, and p represents the result matching probability. According to formula (1), the expression formula of the matching relationship between the possibility and utility of the model results is obtained:

$$R_p = \sum_{n=1}^{m} \frac{\mu + d^2}{\eta_{mn}} \tag{2}$$

In formula (2), μ represents the specific relationship object, d represents the noise controlling the utility, and m, n represents the action objectives of the subject and the object respectively. Order parameter is a macro parameter in the collective operation of many subsystems. This parameter can express the overall behavior of the system. Based on the above reasons, order parameter is introduced into the system as macro parameter, and in the process of system evolution, order parameter can contribute to the formation of new structure of the system. It can be seen that the influence of traditional fitness and health preservation sports on people's sleep has been scientifically confirmed. At the same time, some researchers have found that circadian clock genes are closely related to sleep. To a certain extent, sleep can affect the body's circadian clock. For example, with the accelerated pace of life, staying up late, working overtime and working shifts has become a common phenomenon, which not only affects people's sleep, but also disrupts the body's biological clock and causes physical discomfort. In other words, order parameters are generated in the process of competition and cognition among a large number of subsystems within the body. At the same time, the formation of order parameters plays a role of servitude or domination of subsystems to the whole system, which plays an important role in the evolution of cognition. Since belief is defined as the accumulation of existing evidence, the current belief value of each attribute in the belief database is jointly determined by the belief value of the previous time and the perceived value of the current time (or the result obtained by interacting with other subjects), which is calculated according to the following formula:

$$S = \frac{\|1 - \gamma\|}{\sum m + n} \tag{3}$$

In formula (3), γ represents the attention weight of current belief related elements. Therefore, the order parameter measures and reflects the effect of subsystem cooperation of the whole system. It represents the participation degree of subsystem in cognitive operation and expresses the ordered type and structure of the system. There is also cross time zone travel, which will also affect the body's biological clock due to the confusion of sleep time. In addition, researchers from a biological rhythm research laboratory in the United States pointed out that for most people, sleeping in on weekends is not a good thing, which will delay your sleep time and lead to the disorder of your biological clock. In addition, the occurrence of insomnia, depression, Alzheimer's disease and many other diseases are related to the disorder of the biological clock. In short, the human biological clock is easy to be disturbed by the influence of environmental factors, and then affect human health. Studies have found that many circadian clock disorders are related to sleep, that is, by regulating sleep, we can regulate biological rhythm to a certain extent. In the operation of the human body as a whole, many order parameters promote the orderly evolution of the system in the relationship of competition and cooperation. Self organization is another core concept of cognitive theory. From the perspective of systematics and physics, the whole can be divided into "organized" and "basically unorganized". The whole composed of many interconnected subsystems is an organized whole. This whole shows two forms. One form of whole is that its subsystems can independently complete the interaction and connection, and the other form of whole is that each subsystem interacts under the influence of some external mechanism. From the above, it is known that correct, reasonable and rhythmic exercise, China's traditional fitness and health preservation sports have a certain regulatory effect on sleep. Therefore, we can reasonably use China's traditional fitness sports to regulate sleep, and then indirectly regulate the biologic rhythm of the human body.

3 Application Test

3.1 Test Preparation

This experiment is carried out under the operating system of Ubuntu. The development environment is My Eclipse + Flex Builder plug-in. FlexBuilder develops the presentation layer. My Eclipse uses the framework Spring + Hibernate to build the service development platform. In the computer hardware configuration, the CPU uses Intel Core i7 9600k, and the graphics card uses NVIDIA GeForce RTX 20700 with 8 G video memory. The application server is a famous open source project JDK + Tomcat, the background database is MySQL, and the system modeling and drawing tool is My Eclipse UML.

3.2 Test Results

In order to test the application effect of the traditional sports health preservation cognitive model constructed this time, the experimental test is carried out. The traditional sports health preservation cognitive model based on neural network and the traditional sports health preservation cognitive model based on ant colony algorithm are selected for comparative test with the traditional sports health preservation cognitive model in this paper. Test the processing time of the three cognitive models under different data scales. The test results are shown in Table 1-Table 4:

Table 1. Processing time of 50MB data scale model (MS)

Number of experiments	Cognitive model of traditional sports health preservation based on Neural Network	Cognitive model of traditional sports health preservation based on ant colony algorithm	The cognitive model of traditional sports health preservation in this paper
1	4.152	3.468	2.205
2	3.978	4.511	2.014
3	4.162	5.036	2.336
4	3.645	4.949	2.485
5	4.005	5.032	2.162
6	3.466	4.788	2.754
7	4.877	5.213	2.152
8	3.859	6.948	3.061
9	4.311	4.161	2.345
10	2.668	3.548	2.660
11	3.051	3.154	2.548
12	4.369	3.410	3.212
13	3.461	4.216	3.149
14	5.022	3.502	2.154
15	4.616	3.669	2.221

It can be seen from Table 1 that the average processing time of the traditional sports health preservation cognitive model and the other two cognitive models are 2.497 ms, 3.976 ms and 4.374 ms respectively.

It can be seen from Table 2 that the average processing time of the traditional sports health preservation cognitive model and the other two cognitive models are 4.114 ms, 7.872 ms and 8.152 ms respectively.

Table 2. Processing time of 100MB data scale model (MS)

Number of experiments	Cognitive model of traditional sports health preservation based on Neural Network	Cognitive model of traditional sports health preservation based on ant colony algorithm	The cognitive model of traditional sports health preservation in this paper
1	6.156	7.411	4.315
2	8.163	8.021	3.468
3	7.494	7.336	4.613
4	8.205	8.469	4.006
5	7.691	7.451	3.915
6	8.316	8.116	4.878
7	7.648	7.645	3.942
8	8.123	8.119	4.813
9	7.541	8.346	3.845
10	8.055	8.665	4.613
11	7.613	9.132	4.021
12	8.965	9.065	3.884
13	8.332	8.152	3.672
14	7.649	8.513	4.081
15	8.122	7.846	3.649

Table 3. Processing time of 150MB data scale model (MS)

Number of experiments	Cognitive model of traditional sports health preservation based on Neural Network	Cognitive model of traditional sports health preservation based on ant colony algorithm	The cognitive model of traditional sports health preservation in this paper
1	16.355	15.215	9.156
2	14.166	14.603	10.212
3	13.121	13.991	9.487
4	14.169	14.174	10.342
5	13.558	13.906	9.846
6	14.701	15.229	10.564
7	15.134	14.867	9.711
8	16.315	14.206	10.356

(*continued*)

Table 3. (*continued*)

Number of experiments	Cognitive model of traditional sports health preservation based on Neural Network	Cognitive model of traditional sports health preservation based on ant colony algorithm	The cognitive model of traditional sports health preservation in this paper
9	15.821	13.559	9.548
10	14.906	14.767	10.345
11	13.754	13.207	11.216
12	14.919	14.504	10.714
13	13.202	13.649	12.008
14	14.867	15.203	13.757
15	15.027	14.257	12.366

It can be seen from Table 3 that the average processing time of the traditional sports health preservation cognitive model and the other two cognitive models are 10.642 ms, 14.668 ms and 14.356 ms respectively.

Table 4. Processing time of 200MB data scale model (MS)

Number of experiments	Cognitive model of traditional sports health preservation based on Neural Network	Cognitive model of traditional sports health preservation based on ant colony algorithm	The cognitive model of traditional sports health preservation in this paper
1	19.166	21.021	13.002
2	18.633	20.647	13.417
3	17.525	19.658	13.654
4	18.324	18.603	12.502
5	17.912	17.815	13.6947
6	16.403	18.466	12.505
7	17.348	16.902	13.714
8	18.399	18.466	14.911
9	17.213	19.518	13.806
10	18.566	18.221	14.347
11	17.115	19.636	15.212
12	18.912	18.371	14.611
13	19.613	19.005	13.945
14	18.520	18.346	12.316
15	19.475	19.508	14.209

It can be seen from Table 4 that the average processing time of the traditional sports health preservation cognitive model in this paper and the other two cognitive models are 13.723 ms, 18.208 ms and 18.946 ms respectively. The traditional sports health preservation cognitive model in the expository text takes less processing time and has better application effect.

In order to further verify the effect and feasibility of the traditional sports health care cognitive model from the perspective of physical and medical integration, the cognitive effect of this model is verified by using the coverage rate. The coverage rate is an important indicator of the completion of the test model and a measure of the usability of the test. It is based on the user's test and determines the coverage rate according to the number of users covered in the cognitive process and the comparison with the remaining users, Calculate whether the algorithm coverage cognitive model is effective. If there are no errors or unexpected test results, it indicates that the coverage is good. The calculation formula is as follows:

$$F_{GL} = \frac{R_1 \times R_2}{R} \times 100\% \tag{4}$$

In formula (4), R_1 represents the recommendation list, R_2 represents the user set, and R represents the item set. The comparative analysis results are shown in Fig. 3:

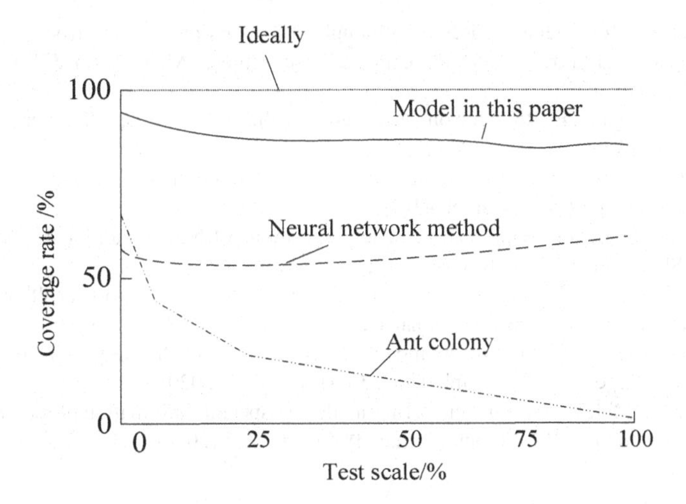

Fig. 3. Comparison results of coverage under different models

It can be seen from Fig. 3 that, in the absence of accidents, the coverage of the model in this paper is high. After comparison, the coverage of the model in this paper is closest to the ideal state, which indicates that the coverage is smooth and complete.

4 Concluding Remarks

The traditional sports health care cognitive model designed in this paper, from the perspective of physical and medical integration, combined with the laws of physical movement, can face the needs of real health, and can be flexibly used in different fields of the life cycle such as health care, body strengthening, heart strengthening and medical treatment. The promotion of sports health preserving thoughts on mental health is realized through the assimilation of individual psychology by sports health preserving culture. At the same time, the theory and method of defense and treatment advocated by the thought of physical health preservation will also play a positive role in avoiding and getting rid of psychological diseases for individuals. Through the above research, the following conclusions are obtained:

(1) From the perspective of physical and medical integration, the construction and processing of the traditional sports health care cognitive model takes less time, and the application effect is better.
(2) The coverage of this model is high. After comparison, the coverage of this model is closest to the ideal state, which indicates that the coverage is smooth and complete.

Fund Project. Supported by Changsha Medical University special funds of Hunan Provincial Key Young Teacher Training Program, Xiangjiaotong [2018] No.574–26.

References

1. Shengwei, D., Tong-gang, F.: Research thoughts on sports prescription from perspective of traditional sports health maintenance. China J. Tradit. Chinese Med. Pharm. **35**(7), 3513–3517 (2020)
2. Yapei, S., Tonggang, F.: Research on the design of traditional sports health prescription library. J. Hebei Sport Univ. **34**(4), 84–90 (2020)
3. Jun, L.: On the extension of traditional sports health culture under the condition of Chinese modernization. Sport Sci. Technol. **41**(4), 67–68 (2020)
4. Guotian, W.: Study on the international communication of Chinese traditional sports health in China-ASEAN. Phys. Educ. Rev. **39**(7), 42–43 (2020)
5. Xiaohua, H.: Research on the cultural characteristics of traditional sports health preservation under the background of healthy China. Wushu Yanjiu **5**(10), 98–101 (2020)
6. Jingyu, C.: The construction of the model of traditional sports health preservation in higher vocational colleges. J. Yuxi Normal Univ. **36**(4), 129–132 (2020)
7. Peng, L., Peng-fei, N.: Optimal clustering model for specific information of massive medical resources based on VSM. Comput. Simul. **38**(6), 383–386 (2021)

Design of Remote Video Surveillance System Based on Cloud Computing

Wei Zou(✉) and Zhitao Yu

Jiangxi Software Vocational and Technical University, Nanchang 330041, China
zouwei6639@163.com

Abstract. Intelligent mobile network remote video monitoring system is an essential technology in the current society. In order to ensure the operation effect of Intelligent mobile network remote video monitoring system, the design method of intelligent mobile network remote video monitoring system based on cloud computing is proposed, and the system hardware structure and software function are optimized. Finally, the experiment proves that the design of an intelligent mobile network remote video surveillance system based on cloud computing has high practicability in practical application, and fully meets the research requirements.

Keywords: Cloud computing · Mobile network · Video surveillance

1 Introduction

With the rapid development of modern network communication technology, more and more enterprises and groups have shown the development of cross-region, and in this context, the use of network to achieve remote monitoring is beneficial for reducing the production costs of enterprises, improving labor productivity, and then improve production safety; on the other hand, with the expansion of production scale, equipment distribution is becoming more and more discrete, and video monitoring is quickly accepted by the majority of users with its real-time intuitive advantages, it is very easy to achieve web-based remote video monitoring. However, in the past, the video monitoring network usually used the common twisted pair or coaxial cable to realize the remote transmission and monitoring, and the video load of the video image with big data usually can easily cause the congestion of the network. Moreover, the network video monitoring system based on this mode has more complicated maintenance in the later period and is difficult to update the system. Therefore, we must try our best to realize the application of new technology in remote network video surveillance system [1].

Reference [2] designed a video surveillance system based on infrared human body detection. In the hardware part of the system, hc-sr501 infrared sensor is used to monitor the infrared signal of human body, and camera and cortex-a9 chip are used to collect and process video information The software environment is opencv and x265 function library, which are programmed with C ++ language. It can sense, record and compress video and remote monitoring of intruders according to infrared signals under unattended

W. Fu and L. Yun (Eds.): ADHIP 2022, LNICST 468, pp. 257–270, 2023.
https://doi.org/10.1007/978-3-031-28787-9_20

conditions. The system has strong flexibility in capturing video, but the data packet loss is large. Reference [3] designed a mobile remote video surveillance system. The system is composed of four modules. The intelligent vehicle based on Arduino system is equipped with a camera, which receives user instructions and is used for mobile video collection. The embedded Linux system realizes the real-time collection of video data through the v4l2 interface. On the one hand, it sends the data to the forwarding server through the network, and on the other hand, it forwards the control instructions from the user to the intelligent vehicle. The server is used to forward video to the client and user control instructions to the Linux system. The system can realize no dead angle monitoring, but the error is large.

The hardware part of the system is composed of the core system platform, the visual display part, the customer control center and the IP network. The software system can be divided into two parts, system layer software and application layer software. Cloud computing is used to analyze the dispersion of distribution characteristics and reduce the monitoring error of the system. Improve monitoring accuracy through cloud computing intelligent device drivers.

2 Intelligent Mobile Network Remote Video Monitoring System

2.1 Hardware Equipment for Intelligent Mobile Network Remote Video Monitoring System

The system can be divided into four parts: the system platform for core management, the visual display part for sensing and control, the customer control center, and the IP network which connects these parts before collecting all kinds of audio-visual data and signal monitoring. The structure of the remote video surveillance system is shown in Fig. 1.

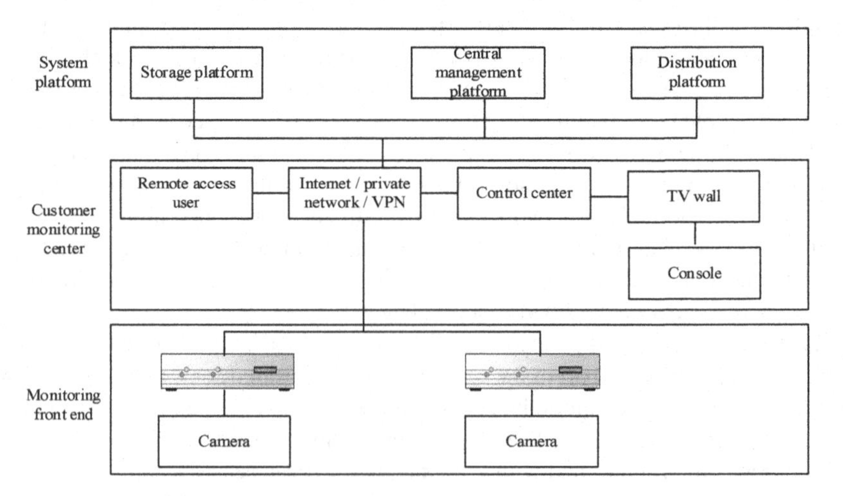

Fig. 1. Remote video surveillance system architecture

For large video-controlled network meters, a special space is usually set up for large mobile display devices for video and control [4]. Call it a customer control center: Generally speaking, for convenience, system platform devices are also placed with display devices. Therefore, there are two functions in the central control center: the system platform and the video sequence control display device. For the convenience of illustration, the "control center" in this paper only refers to the visual centralized control and display center, and the system platform will be described respectively. The core processor is ARM9 (ARM926EJ) processor architecture, adopts DDR2 memory, supports many kinds of Nandflash, and has rich internal resources and interfaces, with the advantages of low power consumption, low cost and high performance. The main hardware resources of the FL2416 development board are shown in Table 1.

Table 1. Main hardware resources for system development

Hardware	Describe
CPU processor	Pentium IV 8
SDRAM memory	32 GB
Flash storage	256M Byte SLC NandFlash
Interface resources	1 network port, 4 USB ports, 2 serial ports, etc

CPU processor is the core accessory of the computer. Its function is mainly to interpret computer instructions and process data in computer software. SDRAM memory is a dynamic random access memory with a synchronous interface. Flash storage is an electronic form of programmable read-only memory, which can be erased and written many times during operation. Interface resources is the channel to obtain resources from the website.

Multi-function network real-time monitoring system is generally divided into two stages, one is the real-time collection of the network, the other is the abnormal data diagnosis. On the basis of these two stages, the optimization of hardware structure and software function of the system is designed [5]. According to the features of multifunctional network, the hardware optimization structure diagram of the monitoring system is designed, as shown in Fig. 2.

As can be seen from Fig. 2, monitoring points 1, 2, 3, and 4 are distributed at various edge positions of the network, and the active monitoring method is adopted to monitor the monitored objects and topological information of the network, so as to facilitate the access of users; the central server refers to the server of the center where the data monitored by different monitoring points are gathered, so as to be able to manage various software configuration information in a timely manner; the data server refers to the data analyzed and monitored, so as to form a data sheet in a uniform format for the convenience of users to inquire; the data receiver refers to the data received by the central server, so as to facilitate the system distribution; the server refers to the important interface that requires users to provide the system with visual graphics, so

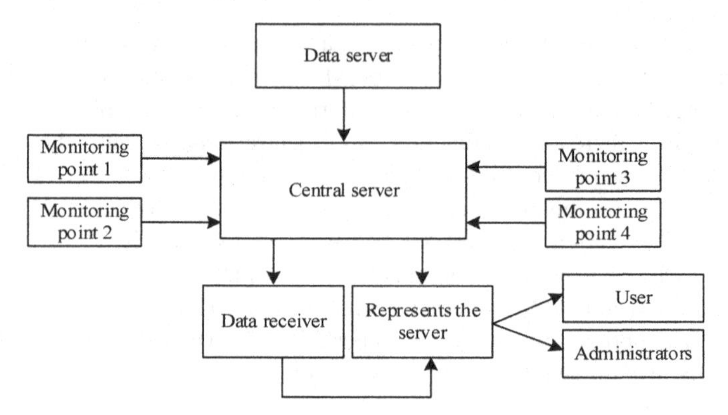

Fig. 2. Hardware optimization block diagram of monitoring system

as to analyze abnormal monitoring results [6]. Front-end equipment refers to the front-end video capture equipment placed in the monitoring site, sometimes including audio equipment and some alarm signal equipment. For a digital video monitoring system, there are generally three types of equipment: the camera, the audio equipment, the alarm equipment, the PTZ, the auxiliary lighting equipment or the infrared equipment and other auxiliary equipment, and the third type is the video server or the coding equipment that is responsible for the analog signals collected for such functions as analog-digital conversion, coding compression, encryption, processing and uploading. The connection structure of the specific front-end equipment of the system is shown in Fig. 3.

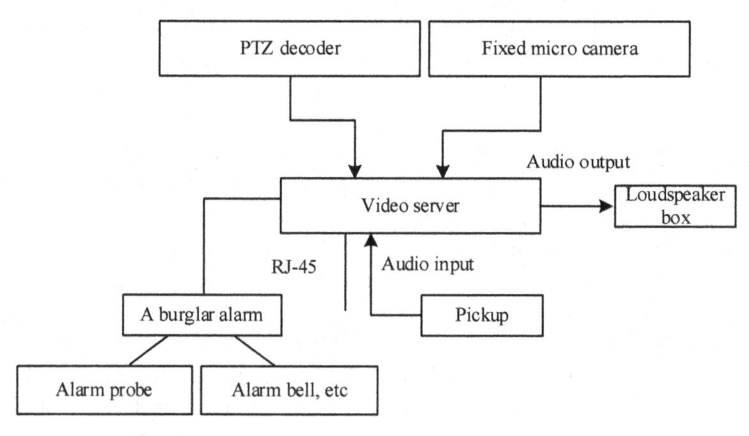

Fig. 3. System front-end device connection diagram

In the map configuration function of the system, it is required to provide the time and video of each video capture device of a certain region for the region attributes, video capture device attributes, alarm device attributes, video capture device group attributes and other functions; provide the video source time, video frequency source device name, video capture device name superposition method; set the backup directory, set the local video directory alarm; in addition, set the sound alarm monitor when the alarm occurs; set the video capture device that needs to be connected when the alarm device alarms; save the local settings in the local configuration file, and obtain [7] from the configuration file when necessary.

2.2 Software Functions of Intelligent Mobile Network Remote Video Monitoring System

The software occupies the core position in the entire intelligent mobile network remote video surveillance system and plays a vital role. The design of software should follow the method of software engineering, analyze the requirements of the whole software system, and then divide the software into different levels according to the functions, so as to make the system structure clear and realize modular design [8]. Based on the idea of software modularization and hierarchical design, the whole software system can be divided into two parts according to its functions: system layer software and application layer software, system layer software is used to separate application program from hardware, and application program operates hardware and controls equipment through the mechanism provided by the operating system. System layer software includes hardware initialization module, embedded operating system module and device driver module [1]. Application-tier software is used to accomplish certain specific tasks for users. The application software can be divided into two parts: video front-end server software and monitoring center client software. The client software includes video image storage module, image display module and video image processing module, in which the video image processing module is divided into video image preprocessing module and moving object detection and display module, as shown in Fig. 4.

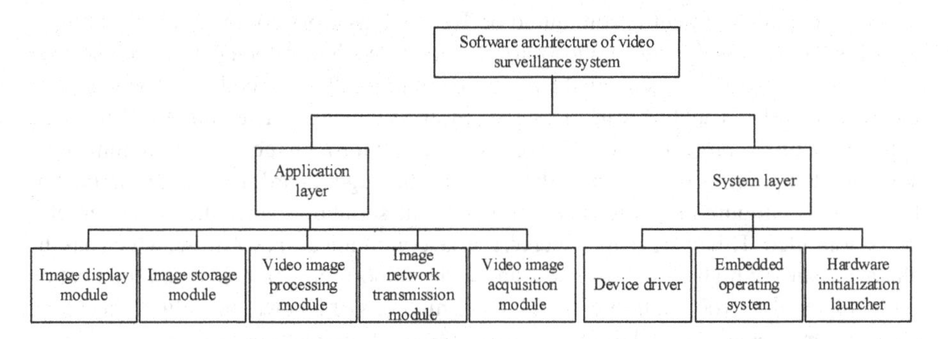

Fig. 4. System software functional architecture diagram

The system has complex implementation functions, and its overall structure shall include several necessary modules, of which the most basic modules are system configuration module, monitoring function and communication module, and the upgrading function module includes equipment management function module, playback function and electronic map module, etc. For example, in the process of MPEG-4 coding, I code and P code are needed to improve the speed and effect of video compression. After the embedded hardware platform is built, the embedded software system can be developed.

The development flow of the intelligent mobile network remote video monitoring system is shown in Fig. 5.

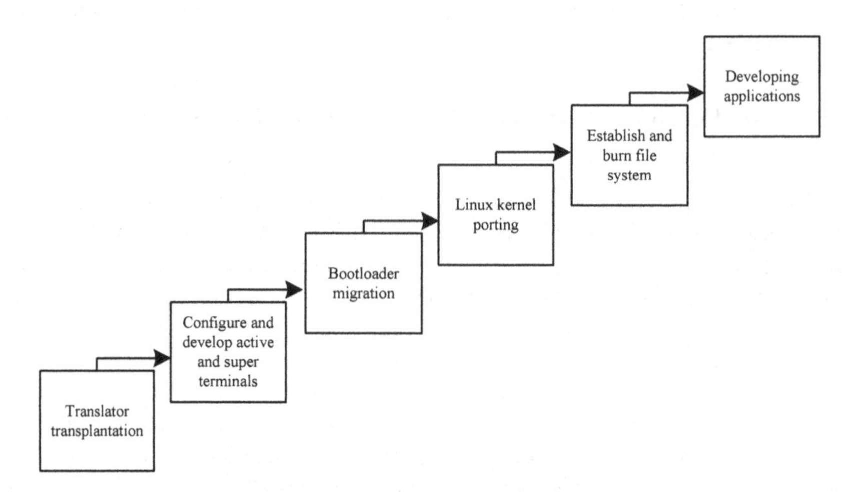

Fig. 5. Intelligent mobile network remote video surveillance system development process

Host is a general purpose computer with powerful software and hardware development resources. Developing embedded application with host environment can greatly improve efficiency. The target board is the actual platform of application program execution, and the software and hardware resources are limited. The host computer and the target computer communicate through a serial port or an Ethernet port. The development flow of embedded server software of video monitoring system is as follows: Firstly, the development environment is configured on PC, such as cross compiler, cross compiler and cross connector, and the application software is developed, then the executable code on the target platform is generated by cross compiling, then the code is downloaded to the target board and analyzed and debugged by cross debugger. The running status of the application on the target board can be seen on the host through the serial communication software, the premise is to connect the host and the target board through the serial port line, and to configure the designated terminal as the serial port when the Linux kernel is cross-compiled. This system uses Secrecrt serial communication software, which has the characteristics of high speed, openness and wide support for all kinds of communication protocols. In this process, frame I has some limitations of its own, in addition, there are other differences between the two, for example, the frame is used forward time, and P frame is two-way time. Background subtraction is a technique that uses the background image as a reference object, stores the background image first, and then subtracts the foreground image and background image to identify moving objects. Generally, because the gray value of the moving object is very different from the gray value of the background, the different image will change greatly only in the region with the moving object. A proper threshold is chosen, when the gray value of the difference image is larger than the threshold value, it is regarded as the moving target point. Otherwise, the background subtraction method is regarded as the background subtraction method. The realization flow chart of moving object detection is shown in Fig. 6.

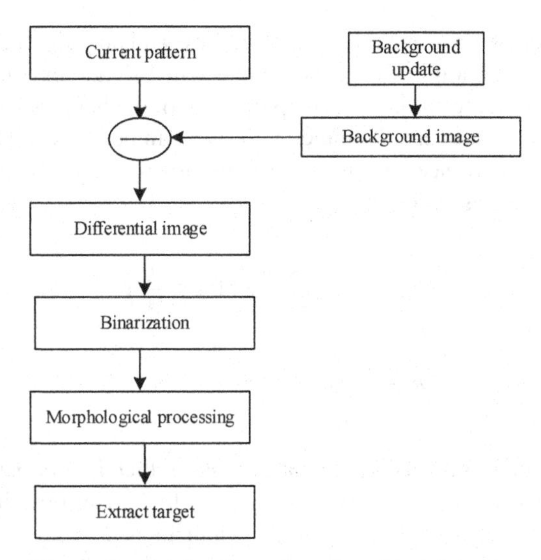

Fig. 6. Realization flow of moving object detection

When the camera is still, the simplest background selection method is to use a fixed frame image as the background, but this method is greatly affected by the external scene, and is not suitable for complex and changing background environment. The acquisition of initial background image plays a vital role in the accuracy of target detection and the update of background image. A good background model requires that the background can be accurately extracted if the target is moved in or out of the background during the initialization process. In this paper, an effective background initialization algorithm is proposed based on the traditional time averaging method. The basic idea is that: in general, the time of moving target in a region is limited, and those points with great difference are caused by the moving target. The moving object region is obtained by the method of difference between adjacent frames, the moving object pixels are discarded, and the non-moving object pixels are averaged by multi-frame cumulation. Firstly, the region of the moving object and the region of the non-moving object are separated by the difference image, and the region of the moving object is invariant. Update the previous background image B_i by binarizing the image B_{i-1}, namely:

$$B_i(x,y) = \begin{cases} B_{i-1}(x,y) \\ \alpha I_t + (1-\alpha)B_{i-1}(x,y) \end{cases} \tag{1}$$

where, (x, y) is the coordinate of the target point, I_t is the monitoring duration, and α is the number of monitoring points.

Once the network is abnormal, then the IP address and port distribution will change. If the network configuration is wrong, both the original IP address and the destination IP address will increase, resulting in a sharp increase of the host packets. Assuming the feature is A, the total number of samples is B, the number of samples selected is C, and the number of occurrences of i for a particular feature is n_i. Therefore, the feature samples can be defined as follows:

$$F(x) = -\sum_{i=1}^{C}\left(\frac{n_i}{B}\right)\log_2\left(\frac{n_i}{A}\right)$$
$$B = \sum_{i=1}^{C} n_i \tag{2}$$

If the values of all the selected samples are the same, then $F(x) = 0$; if the values of all the selected samples are more scattered, then $F(x) = \log_2^C$. This describes the abnormal behavior of the different characteristics, as shown in Table 2.

Table 2. Network abnormal behavior characteristics table

Anomaly type	Exception definition	Abnormal characteristics
Wrong configuration	Device failure caused by incorrect configuration of routing port	Large abnormal characteristic value The normal eigenvalue is large
Service attack	Service attack	The abnormal characteristic value is small Normal eigenvalue is small
Burst access	Multiple hosts send to a single host	Large abnormal characteristic value Normal eigenvalue is small
Worm scanning	A small number of ports on the destination host are detected	The abnormal characteristic value is small The normal eigenvalue is large

There are many ways to control the frame rate, for example, by reducing some images and changing the frame rate of the image. To this end, the original frame rate can be reduced to a certain extent, so that the code rate is reduced, and finally can increase the clarity of image monitoring, but also there will be image cassette or jitter. However, in the real operation of monitoring, it is necessary to analyze the actual situation and explore the appropriate solution. The method to adjust the bit rate is called the slow rise and fast fall method. The main strategy is to improve and multiply. This has something in common with TCP, if the stability of the full load rate can be guaranteed, and once the congestion situation is present, the code rate will be reduced, no congestion will continue to increase.

2.3 Realization of Network Video Remote Intelligent Monitoring

Video surveillance system with its real-time, intuitive features, has been favored by the majority of users, in the traffic, environmental protection, electricity, forest fire protection applications. Video surveillance system will be along the "high-definition, mobile, intelligent" direction of development. The video surveillance system transmits the data through wireless, the wireless communication way has WLAN, the mobile communication network, the satellite communication system and so on. Mobile communication network has the advantages of moderate cost and long transmission distance, and is suitable for long-distance data transmission network. For the specific use of network transmission, it depends on the requirements of the monitoring system for the transmission rate, which is related to the video resolution. The main resolutions of the system are CIF, D, 720P, 1080P. The high resolution images are of good quality, and the required bandwidth is large, which is shown in Table 3.

Table 3. Table of requirements for stable transmission bandwidth at different resolutions

Resolution type	CIF(363 × 289)	D1(705 × 577)	720P(1280 × 720)	1080P(1920 × 1080)
Stable transmission bandwidth	251 kbps	At least 522 kbps	At least 1 Mbps	At least 2 Mbps

The database of the video monitoring system of the power substation mainly records the relevant data during the operation of the system, and the relevant constant data of the system itself, for example: the path of the video file is stored in the database, and when it is called, it first queries the database to find out the corresponding file path, and then browses the monitored video through streaming media technology; the database design idea The database design is the process of building a database suitable for the current application on the basis of the database selection, which includes the analysis of the user's needs, and generally, the database design process mainly includes conceptual design, logical design and physical design, as shown in Fig. 7.

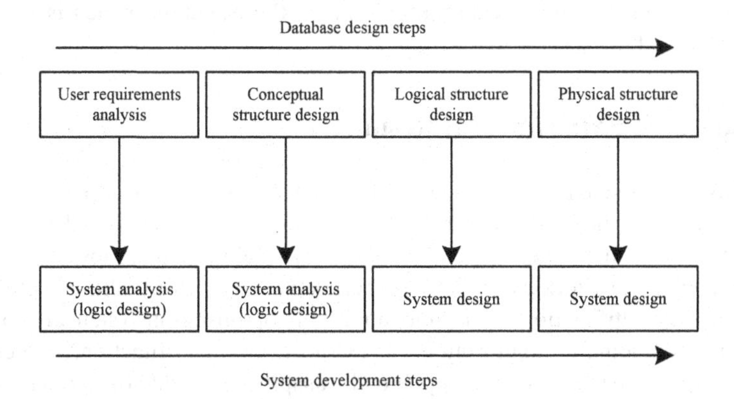

Fig. 7. Relationship between database design and system development

In cloud computing, device drivers are considered to be the interface between kernel and hardware. Cloud computing intelligent device drivers abstract the implementation of specific hardware devices. For users, operating hardware devices and operating files are the same, that is, using standard APIs (system call interfaces) to complete specific operations such as reading, writing, and controlling hardware devices, while device drivers exist to implement these system call functions. The cloud computing smart device driver architecture is shown in Fig. 8.

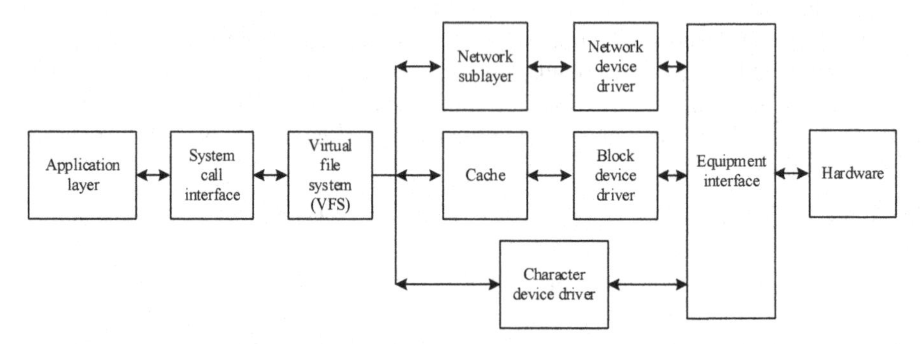

Fig. 8. Cloud computing smart device driver architecture

Writing and implementing cloud computing applications is an important step in the overall system. After the system development environment is built and customized, the development environment and running environment of cloud computing application are established. In this system, a camera driver is added to the core, and the user application program can complete the collection of video image data through the video device programming function of the system. This system video frequency monitoring terminal mainly realizes the video frequency module the collection and the transmission function. According to the software design and server-side program design of the monitoring system, the video image collection and transmission design is implemented by the embedded streaming media server. The JPEG image is first collected and compressed by the streaming media server, and then forwarded to the HTTP server. The HTTP protocol is used to transmit the image data through the TCP connection, which is parsed and displayed by the browser.

3 Analysis of Experimental Results

In order to verify the feasibility of multi-function network real-time monitoring system optimization, under the same experimental conditions, the multi-functional network real-time monitoring optimization system and the traditional system are tested. Taking the network anomaly monitoring error as the experimental object, the traditional monitoring system and the multifunctional real-time monitoring optimization system are compared to the network anomaly monitoring error, PC host: the main function is to establish a cross-compiling environment, compile and transplant embedded Linukou operating system and other applications, as well as development and debugging work. The software environment is as follows: A. Operating system: Windows 7; B. Virtual machine: Vmware 10; C. Linux operating system: Ubuntu 12.04/Linux3.2.0–23 kernel; D. serial debugging tool: remote management software Recrect; E. Java operating environment: JRE 1.6, server: FL2416 development board and its peripheral devices. All application programs on the server side are running on the FL2416 development board. The operating environment configuration is shown in Table 4.

Table 4. System run environment configuration

To configure	Parameter table
CPU	Inter(R) Core(TM) i5-2430M 2.4 GHz
SDRAM	16G
FLASH	NandFlash 256 MB
Chip card	DM9000
USB camera	Hamedal
USB wireless network card	Wing joint EDUP
Peripheral memory card	SDHC card 16G
Embedded operating system	Linux3. 1.8 kernel

The compatibility test of browsers is mainly to test whether the interface function is realized normally when the system uses different browsers. The compatibility of different browsers is tested using different mobile terminals with different operating systems (Windows, Android, iOS). The test results of several browsers in common use are shown in Table 5.

Table 5. Browser test results

Browser type	Interface display	Function operation	Test result
Internet Explorer	Normal	Normal	Normal
360 browser	Normal	Normal	Normal
QQ browser	Normal	Normal	Normal
Firefox browser	Normal	Normal	Normal
Sogou browser	Normal	Normal	Normal

The test results show that the system works well in most browsers. The results can be used as the basis for evaluating the feasibility of system optimization. If there is noise interference in the network monitoring, the whole system will be affected and the accuracy of monitoring will be greatly reduced. In the process of system optimization, the monitor is optimized and the abnormal output signal is amplified by the amplifier with larger resistance, which can effectively reduce the impact of noise interference and thus reduce the monitoring error. In view of the abnormal monitoring of the network, the output signal trend is analyzed, as shown in Fig. 9.

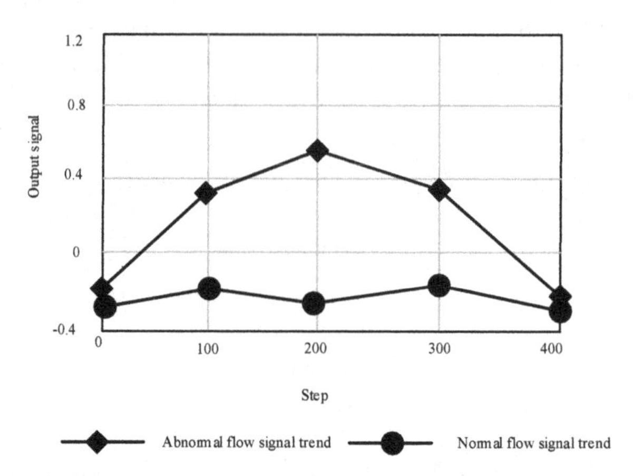

Fig. 9. Abnormal or not monitoring output signal trend

The monitoring error value is an important indicator to measure whether the system can accurately capture data. The smaller the error value, the higher the accuracy of capturing data. Under the influence of noise, the error of network anomaly monitoring is compared between traditional monitoring system and multi-function network real-time monitoring optimization system, and the result is shown in Fig. 10.

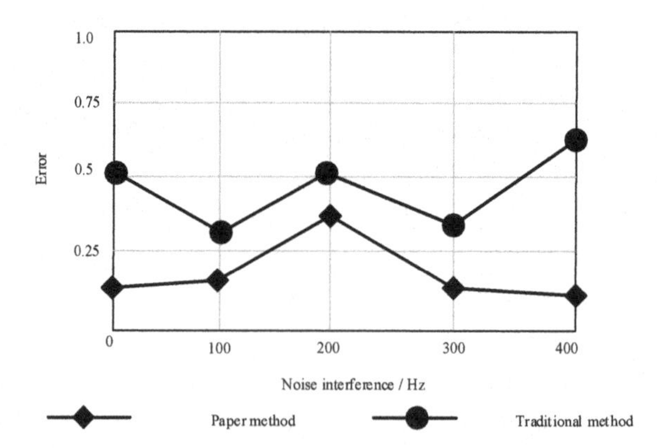

Fig. 10. Comparison of monitoring error between two systems under noise

As can be seen from Fig. 10, the maximum monitoring error of the multi-function system is 35%, and the maximum monitoring error of the traditional monitoring system is 50%. In the case of noise interference, the monitoring error of multi-function system is small. The traditional network monitoring system can't accurately capture the data, resulting in a large increase in the loss of data packets. Therefore, when optimizing the system, the packet capture function is designed. Packet capture is to intercept, retransmit, edit, transfer and other operations of data packets sent and received by network transmission, and is also used to detect network security. Packet capture is also often used to intercept data.

In order to verify the effectiveness of this function, the error of network monitoring between the traditional monitoring system and the multifunctional real-time monitoring optimization system is compared, and the result is shown in Fig. 11.

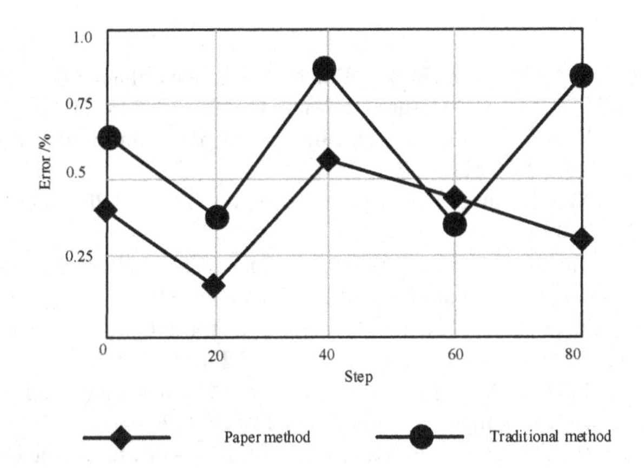

Fig. 11. Comparison of error of network anomaly capture in two systems

As can be seen from Fig. 11, the maximum error rate of the multi-functional network real-time monitoring optimization system is 55%, and the maximum error rate of the traditional monitoring system is 90%, which proves that the optimal design of the grabbing function has less error in real-time monitoring the abnormal network problems. Based on the above experimental contents, the experimental conclusion can be drawn: under the condition of noise interference, the error of multi-function system is small. After the optimization design of monitoring function, the error of multi-function system is small. Therefore, the method of this paper is feasible.

4 Conclusion

Embedded network video monitoring system based on cloud computing technology has more advantages than traditional remote network video monitoring system, such as suitable for long distance transmission, simplifying system structure and development cost. Therefore, in recent years, the mobile intelligent network video surveillance system based on cloud computing technology has been widely used, so the network video surveillance system based on cloud computing technology has become the inevitable trend of the development of network remote surveillance system. This paper discusses the development and realization of network remote video monitoring system in detail with cloud computing technology. It has a good guiding significance for the research of network remote video monitoring. Of course, the embedded network video surveillance system designed in this paper is only designed from the embedded point of view. With the improvement of mobile network transmission speed, the performance of intelligent mobile network remote video monitoring system should be constantly improved. Which need to be the joint efforts of the majority of technical personnel, can finally realize the rapid development and application of network video surveillance technology based on cloud computing.

References

1. Wang, B., Liu, Y., Sun, Q., et al.: Design of mobile video surveillance system based on GB/T 28181 and WebRTC. Electron. Measure. Technol. **43**(18), 112–116 (2020)
2. Chang, F., Ji, S.: Design of video monitoring system based on human infrared detection. Electron. Design Eng. **30**(8), 62–65,70 (2022)
3. Li, S., Zhang, K.: Mobile surveillance system based on remote APP control. Comput. Syst. Appl. **30**(6), 82–87 2021
4. Xia, Z., Xiang, M., Huang, C.: Hierarchical management mechanism of P2P video surveillance network based on CHBL. Comput. Sci. **48**(09), 278–285 (2021)
5. Zeng, T., Huang, D.: A survey of detection algorithms for abnormal behaviors in intelligent video surveillance system. Comput. Measure. Control **29**(07), 1–6+20 2021
6. Lian, J., Fang, S., Zhou, Y.: Model predictive control of the fuel cell cathode system based on state quantity estimation. Comput. Simul. **37**(07), 119–122 (2020)
7. Ou, Y., Wang, H., Ren, T., et al.: Storage design of tracing-logs for application performance management system. J. Softw. **32**(05), 1302–1321 (2021)
8. Jiang, H.T, Chen, X.H, Shi, Y., et al.: Surveillance video analysis system based on SpiralTape summarization. J. Graph. **41**(02):187–195 (2020)

Physical Movement Monitoring Method of College Physical Education Students Based on Genetic Algorithm

Daoyong Pan[1](✉) and Wei He[2]

[1] Sports Institute, Kashi University, Kashgar 844006, China
ppandeng789@163.com
[2] Hunan Vocational Institute of Technology, Xiangtan 411100, China

Abstract. In order to better improve the physical fitness and comprehensive quality of college physical education students, this paper puts forward the physical fitness mobile monitoring method of college physical education students based on genetic algorithm, constructs the physical fitness evaluation index of college physical education students combined with genetic algorithm, optimizes the physical fitness monitoring management algorithm of College physical education students, optimizes the physical fitness mobile monitoring equipment of college physical education students, and simplifies the monitoring process, improve the physical fitness monitoring accuracy of college physical education students. Finally, the experiment proves that the physical fitness movement monitoring method of college physical education students based on genetic algorithm has high practicability and fully meets the research requirements.

Keywords: Genetic algorithm · College physical education · Physical movement monitoring

1 Introduction

As an important type of higher education in our country, colleges not only have the attribute of higher education, but also have the attribute of vocational education. How to train high quality and high skill applied talents according to social needs, vocational characteristics and students' characteristics is the starting point and goal of colleges. The quality of talents training is the lifeline of colleges and the main body of connotation development. With the reform and popularization of vocational physical education in colleges all over the country, vocational teaching quality evaluation system has gradually formed.

College Students' physical exercise monitoring is an important link in sports training, which is helpful to analyze students' body shape and sports ability and guide teachers' specific teaching arrangements. At present, literature [1] proposes a human motion monitoring method based on flexible pressure sensors. A flexible high-sensitivity resistance pressure sensor is fabricated, and the designed strain coefficient is 3.5. According to

W. Fu and L. Yun (Eds.): ADHIP 2022, LNICST 468, pp. 271–285, 2023.
https://doi.org/10.1007/978-3-031-28787-9_21

the sensor array data such as finger bending, arm bending, throat vibration and pulse vibration, the human motion posture is judged. With the help of the acquisition circuit, the change of internal resistance caused by the force of the sensor is displayed on the computer in real time to monitor the human motion state. Document [2] proposes a method for dynamic target detection of human motion with multiple degrees of freedom. The Gaussian background model is constructed to separate the background area and the target area of the image, and the target area is transformed into HSI space; Combining the small area removal method and mathematical morphology processing, the shadow in the target area is determined according to the characteristics of high saturation and low brightness of the shadow area; The shadow part in the target area is removed by matching compensation to realize multi degree of freedom human motion monitoring.

The monitoring management algorithm of the above method is insufficient, and the monitoring results have low accuracy. Therefore, the monitoring method of physical movement based on genetic algorithm is proposed. Based on the genetic algorithm, the physical fitness evaluation index is constructed, and it improves the monitoring algorithm and physical monitoring accuracy.

2 Monitoring Method of Physical Fitness Movement of College Physical Education Students

2.1 Physical Fitness Monitoring Indicators of College Physical Education Students

The construction of students' physical fitness test and evaluation system can effectively improve students' physical quality and develop their physical ability in an all-round way: they can reasonably form good habits and form healthy lifestyle [3]. Have a healthy body. Therefore, the construction of student teaching quality evaluation index is not only the need of teaching technology, but also the goal of physical education and health courses in colleges. Based on the characteristics of sports events, the author holds that "students' special physical ability refers to their ability to bear the load and adapt to the changes of environment in sports, and is the comprehensive embodiment of specialization in the aspects of students' body form, body function and sports quality", which constitutes the body form, body function and sports quality of physical energy structure, and each has its own independent function, mutual restriction and mutual influence [4]. From the functional point of view, physical form and function are the material basis of physical energy. Athletic quality is the external expression of physical energy and the core of physical energy. Body shape and body function are the guarantee conditions to improve sports quality. Analyzed from the interaction, the body shape and the body function affect the sports quality, conversely, the sports quality directly affects the body shape and the body function; the body function affects the body shape, but the body shape will have a minor influence on the body function. It can be seen that the students' special physical ability can be regarded as a multi-dimensional and multi-level organic system composed of body form, body function and sports quality. Any one of the elements of the problem will affect the functioning of the entire system. The architecture for assessing physical fitness is shown in Fig. 1.

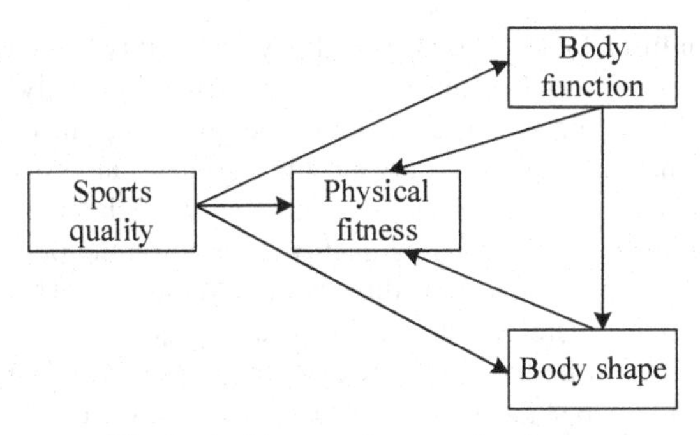

Fig. 1. Architecture for assessing fitness

As can be seen from Fig. 1, the ultimate goal of education is to cultivate useful talents for society. Students in colleges should have the skills for their posts, the appropriate knowledge and the quality for social development [5]. Physical education curriculum has been listed as the school's public basic courses, if viewed from a single health objective, this is understandable. Sports, as a skills-oriented project, is the player's skill ability to achieve excellent results in training and competition, so the final evaluation of students' physical education quality must serve the improvement of students' technical level. Nowadays, the application and development of fast rhythm, strong rotation and mixed multiple technical movements undoubtedly put forward higher requirements for students' physical function and sports quality. And these special requirements, only by special sports technology is impossible to achieve, only through a systematic evaluation of the quality of special sports teaching to meet [6]. The relationship between sports skills and the evaluation of sports teaching quality is shown in Table 1.

Table 1. Relationship between sports skills and physical education quality evaluation

Technical factors	Physical factors
Control technology of body posture	Muscle strength produced by forced static contraction of neck, shoulder, arm, trunk, waist and crotch and other muscle groups; The neck, chest and waist muscles exert force, and the flexibility of women's neck, chest and waist in standard dance
Fast and bouncing technology of body center of gravity	The calf muscle group's degenerative contraction and centripetal contraction force - muscle endurance; Degenerative contraction and centripetal contraction of human leg muscles (force) - muscle endurance; Ankle - I - flexibility
The technique of strong twisting and swinging of the thigh and waist and abdomen	Coordinated exertion of hip, waist and abdominal muscles – muscle endurance, coordination and flexibility of medullary joints
Rhythmic technique of chest, back, shoulder and arm	Trapezius muscle strength, deep back muscle strength - muscle endurance and coordination; The lower arm muscles exert the flexibility of a joint and the rapid strength of muscles
Continuous movement of the body	Reaction speed (resilience on the dance floor) - sensitivity

As can be seen from Table 1, good special physical ability is the basis for students to complete all kinds of special technical movements. Therefore, only by systematically evaluating the quality of special sports teaching and promoting the comprehensive and coordinated development of students' organizations, organs and systems, can students master more complicated, advanced and reasonable sports techniques.

The evaluation index of PE teaching quality is based on the professional post [7]. Most of the students take up their posts directly after graduation. The employment rate of their majors is high. Therefore, the goal of teaching quality evaluation in colleges should rest on urging and improving the realization of the goal of physical education. The aim is to promote the students to do physical exercises actively by analyzing the physical condition of the students in different ages and combining the characteristics of their employment in the future; Promote the students to do physical exercises actively and actively by monitoring the changes in the physical condition of the students in the process of physical exercises; Train the students to learn the ability and habit of monitoring the physical condition of themselves or others step by step, and train the students to do physical exercises purposefully according to their majors so as to meet the needs of future work.

2.2 Algorithm for Student Teaching Quality Evaluation

In the teaching of physical education in colleges and universities, the design of the evaluation algorithm of teaching quality plays an important role. Therefore, the evaluation content must be comprehensive and systematic, the evaluation methods must be diversified, and the learning attitude in the learning process and its learning effects can be systematically evaluated, and be combined with the teaching objectives in each stage, so as to fully embody the role of the evaluation. The evaluation of physical education quality, as an important content of physical education, should naturally be embodied in the teaching evaluation. As a teacher in colleges and universities, the evaluation must be stimulated as a means to stimulate students' learning enthusiasm, and the individual differences of students must be fully taken into account, so as to guarantee the objectivity and effectiveness of the evaluation. The content involved in the physical education evaluation may include the following points: For example, the improvement of students' learning attitude, the physical quality and fitness level of students, the combination of the vertical and horizontal evaluation methods, and the positive role of the evaluation of physical education quality in the teaching of physical education [8, 9]. The establishment of evaluation system is a highly technical and wide-ranging work. In order to ensure the orderly process of the preparation of evaluation system and make the content of indicators and standards reach a relatively ideal level, certain scientific procedures and technologies must be adopted for operation. To establish an evaluation system, the following four steps are usually adopted: determine the evaluation object and target, determine the evaluation index system, determine the branch weight of the evaluation index, and design the evaluation criteria. The specific content is shown in Fig. 2.

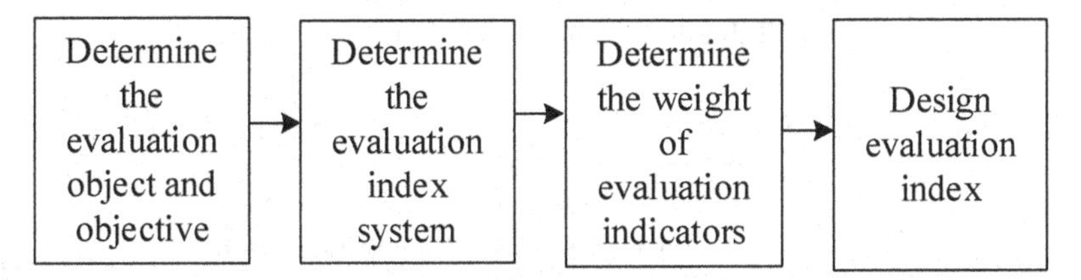

Fig. 2. Evaluation System

As shown in Fig. 2, the evaluation usually adopts the following four steps: determining the evaluation object and goal, determining the evaluation index system, determining the weight of evaluation index and designing evaluation standard.

Questionnaires were distributed to 12 relevant experts, who were asked to assign the importance of indicators of each level by adopting the five-grade scoring method in combination with their characteristics, and the relative assignment of indicators i by experts J was calculated as follows:

$$w_{ij} = \frac{p_{ij}}{\sum_{k=1}^{m} p_{kj}} - m \tag{1}$$

In the above expression, p_{ij} represents the assignment of expert j to the first index i, p_{kj} represents the assignment of expert j to the first index k, and m represents the number of the first index. Find the weight coefficient of index i

$$w_i = \frac{1}{12} \sum_{j=1}^{m} w_{ij} \tag{2}$$

After confirming the body shape, body function and sports quality, we can determine the first and second indexes of the model, but there are many other indexes under the model. If we select them all, the calculation will be increased, which is not conducive to the impact assessment. Therefore, it is necessary to select the indexes. This section adopts the expert interview method to select the third index. The selected indexes are shown in Table 2.

Table 2. Model indicators selected

Primary index	Secondary index	Tertiary indicators
Physical fitness changes of athletes	Body shape	Height, dimension and weight
	Physical function	Biochemical indexes, cardiopulmonary function
	Sports quality	Sports ability

Since each indicator has a different impact on physical fitness, a judgement matrix needs to be defined by quoting 1–9 and its reciprocal (as shown in Table 3).

Table 3. 1–9 and its reciprocal scale

Scale	Meaning
1	Both indicators are equally important
2	The former is slightly more important than the latter
3	The former is obviously more important than the latter
4	The former is more important than the latter
5	The latter is more important than the former
6/7/8/9	Two adjacent intermediate values
Reciprocal	$1/x_{ij}$

Note: x_{ij} is the relative importance of the two indicators.

The judgment matrix built from Table 4 is as follows:

$$(x_{ij})_{n\times n} = \begin{bmatrix} x_{11}, & x_{12}, & x_{13}, & \ldots, & x_{1n} \\ x_{21}, & x_{22}, & x_{23}, & \ldots, & x_{2n} \\ \ldots, & \ldots, & \ldots, & \ldots, & \ldots \\ x_{n1}, & x_{n2}, & x_{n3}, & \ldots, & x_{nn} \end{bmatrix} \tag{3}$$

In the formula, x_{ij} is the ratio of importance and n is the index. According to the contribution rate of each principal component (or principal factor) in the previous principal component analysis, this paper obtains the weights of the secondary indicators (each principal component) and the tertiary indicators (each representative indicator) through calculation. As a result, a system of physical fitness indicators and a weight table for male and female students in physical education institutions has been established, as shown in Tables 4 and 5:

Table 4. Physical fitness index system and weight table of male students in physical education institutions

Primary index	Weight	Secondary index	Tertiary indicators	Weight
Body shape	0.33	Body composition	Ketole index	0.38
		Body length	Height	0.32
		Proportion of body longitudinal structure	Lower limb length	0.18
		Proportion of body transverse structure	Finger spacing - height	0.15
Physical function	0.31	Motor function	Heart rate immediately after exercise	0.58
		Balance factor	Heli test	0.42
Sports quality	0.38	Special factor	Lift	0.29
		Flexibility	Comprehensive flexibility	0.23
		Trunk factor	Abdominal curl	0.21
		Speed force	Vertical jump	0.16
		Sensitive coordination factor	Lateral jump	0.13

Table 5. Index system and weight table of physical fitness of female students in physical education institutions

Primary index	Weight	Secondary index	Tertiary indicators	Weight
Body shape	0.33	Body composition	Ketole index	0.38
		Body length	Height	0.27
		Proportion of body longitudinal structure	Lower limb length	0.23
		Proportion of body transverse structure	Finger spacing - height	0.16
Physical function	0.31	Motor function	Heart rate immediately after exercise	0.56
		Balance factor	Heli test	0.46
Sports quality	0.38	Special factor	Lift	0.30
		Flexibility	Comprehensive flexibility	0.22
		Trunk factor	Abdominal curl	0.20
		Speed force	Vertical jump	0.16
		Sensitive coordination factor	Lateral jump	0.13

According to Tables 4 and 5, the thresholds (or interval values) on which the rating is based are referred to as standards in surveying. Designing evaluation criteria is an important task in establishing evaluation system. At present, the main evaluation criteria are scoring criteria and rating criteria (or rating criteria) [10]. For example, when we evaluate a person's physical stamina, we will say that the person's physical stamina is excellent, good, general, unqualified, poor and so on. Therefore, the criteria for evaluating the quality of physical education are set as follows.

$$M = \{m_1, m_2, m_3, m_4, m_5\} \tag{4}$$

In this set, M is the set of evaluation criteria; m_1 is excellent; m_2 is good; m_3 is general; m_4 is unqualified; m_5 is poor. But evaluation standard concentration's outstanding, good, general, unqualified, poor and so on. Here, the weights of each index calculated above can be used to determine the evaluation criteria determined as shown in Table 6.

Table 6. Criteria for Evaluation of Indicators

Index	Excellent	Good	Commonly	Unqualified	Difference
1	≥95	(86, 91)	(76, 85)	(61, 75)	≤60
2	≥85	(86, 91)	(72, 85)	(66, 75)	≤60
3	≥90	(80, 80)	(76, 80)	(71, 75)	≤60
...	...	...	...	...	...
n	≥90	(86, 91)	(76, 85)	(66, 85)	≤60

After the establishment of the impact assessment model, we can know the changes in physical stamina before and after the special endurance training.

The scoring standards refer to the index values or interval values corresponding to different grades of each index (such as 0, 5, 10, etc.); the rating standards refer to the index values or interval values corresponding to different grades of each index (such as upper, middle, middle, lower and lower). Scoring standards can be divided into individual scoring standards and comprehensive scoring standards, the same reason, rating standards can be divided into individual rating standards and comprehensive rating standards.

2.3 Realization of Physical Education Teaching Quality Evaluation Method

In order to achieve the goal of teaching quality evaluation, the equipment and structure of students' teaching quality evaluation are optimized. The system consists of a video capture terminal, a video server and a display monitoring terminal. With the development of miniaturization of sensor technology, it is possible for people to use micro cameras anytime, anywhere. The structure diagram of the collection equipment of the mobile teaching terminal of the mobile system for teaching quality evaluation is shown in Fig. 3.

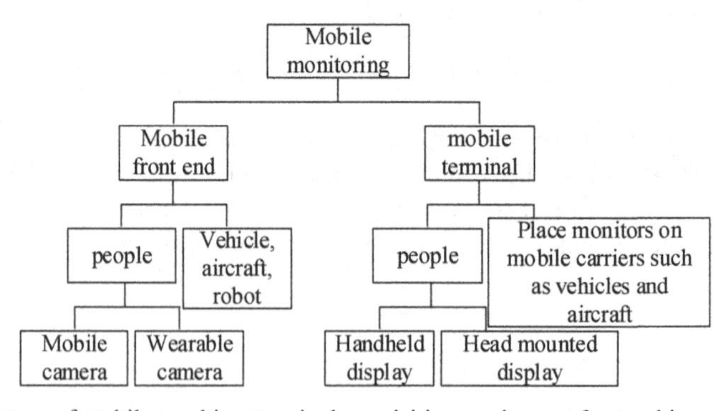

Fig. 3. Structure of mobile teaching terminal acquisition equipment for teaching quality evaluation of mobile system

As can be seen from Fig. 3, the portable intelligent mobile monitoring system with video transmission, storage, analysis and other computing functions can follow the patrol at any time and any place to "observe" the surroundings or specific targets, and assist the human to perform some visual functions when the human visual intelligence drops, so as to improve the reliability of the biological system in completing visual tasks. Intelligent mobile surveillance will become the development direction of manned mobile surveillance. In this paper, the wearable visual system is used to do this task. When a person is found to be in a passive state of attention, a comprehensive analysis is carried out, and when a suspicious/interesting object appears, an independent intelligent surveillance is carried out.

The results of capturing the physical status and monitoring information are shown in Table 7.

Table 7. Physical status and monitoring information capture results

Human mental state	Human attention state	Performance behavior	Main posture of head	Field of view characteristics	Capture/analysis strategy
Sober	Active attention	To glance	Move freely	Frequent scene transfer	No capture, no analysis
Sober	Active attention	Follow	Short pause	Similar scenes	Capture analysis
Dozing/distraction	Passive attention	Dull	Long pause	Scene too long fixed	Do not capture full analysis

From Table 7, we can see that the computation of scene similarity is similar to that of content-based image retrieval, matching the current frame with the key frame. If the similarity is high, the scene is considered as similar. Otherwise, the scene dissimilarity is considered as an abstract representation of an image, and an appropriate image feature can appropriately express the content of an image. At the same time, image feature is also the basis of image matching, which plays a key role in image retrieval. Image features can be divided into three levels.

The first layer is the visual features of the image, including color, texture and shape. These features can be obtained by extracting the pixel information of the image.

The second layer is the abstraction of the objects in the image. Human logic is added to the extraction of this feature.

The third level is semantic features of images, including scene semantics, emotion semantics and behavior semantics. Semantic features can express the rich content of an image.

Low-level features are easy to extract, but the description of image content is not comprehensive. The hierarchical features can fully express the image content, but it is difficult to extract. Because of the high requirement of real-time, the system calculates the similarity of the first level image, which is simple, robust and effective, and calculates the similarity of the collected image and the key frame. Based on the pause effect of human attention behavior, there is less interframe motion caused by head movement. Head pause makes the background "stable" to a certain extent. When attention behavior occurs, moving objects in the scene will be detected. Therefore, it is feasible to confirm the state of concern by estimating the motion region between frames, which can also be summed up as a moving target detection problem. When a moving target can be detected, it shows that a person is in the attention behavior at this time, so as to achieve the goal of accurately monitoring students' physical ability.

Thus, the final score of PE teaching quality evaluation is obtained.

3 Analysis of Experimental Results

Our country specialized sports team sports teaching quality appraisal way is many and varied, each province youth training method is not identical, therefore, the training way effect is also not same. Many coaches only evaluate the progress of students' physical stamina through the grades of students. This evaluation method lacks certain scientific basis. Setting up the monitoring system of special teaching process for adolescent students is helpful to evaluate training effect and students' status scientifically and provide reasonable basis for coaches. Ten adolescent students were selected in this study, none of them had a history of serious injury or illness. The basic information can be shown in Table 8.

Table 8. Basic Information of Students

Team member	Age	Height (m)	Weight (kg)	Training years (years)	Sport level
OCN	16	182.0	75.6	2	Second level
CINE	15	180.3	67.5	2	Second level
XIA	16	193.9	86.3	2	Second level
CYB	15	173.2	68.2	1	Second level
COAN	15	186.2	62.3	1	Second level
XON	16	179.5	73.5	2	Second level
XAG	14	170.3	59.9	2	Second level
QON	16	172.6	57.1	1	Second level
CAN	16	175.3	62.2	2	Second level
CGU	15	169.6	66.5	1	Second level

Human body in the movement of the speed, strength, sensitivity, flexibility and other basic state of the body and functional capacity, known as sports quality. Good sports quality is the basis of mastering sports skills and improving sports achievements, and the lack of good sports quality will affect the formation of perfect and difficult sports skills. According to the development and evaluation method of sports quality in Chinese Youth Training Syllabus, combined with some training forms and experts' opinions, the items and indexes in this study are selected as shown in Table 9.

Table 9. Results of principal component analysis of sports quality index

Principal component	Contribution rate	Sports quality	Representative indicators	Other indicators
A	35.526%	Speed endurance	15 m 13 turn back run (4 groups)	Maximum vertical jump, standing long jump, triangle run, pull-up, breakthrough layup, etc
B	16.558%	Shoot	1.6 min self shooting	
C	10.356%	Sensitive	1 min double swing rope skipping, hexagonal jump	
D	8.658%	Speed	3/4 full court sprint	
E	6.985%	Power	Squats, push ups, sit ups	
F	1.608%	Pliable	Sitting body flexion	
Cumulative	79.691%	Kmo value = 0.553	–	

Traditional teaching quality evaluation standard of collegecollege students uses unified absolute evaluation standard and quantitative standard to measure different individuals with uneven physical and mental development level. The vocational teaching quality evaluation standard of colleges should combine the absolute standard with the relative standard. Let the students see the position of their own physical condition in their peers by absolute standard, make clear their own advantages and disadvantages, and find out the direction of exercise. Through comparative understanding, through exercise, their physical fitness is progress or decline, how much range, and then according to the evaluation of specific information to adjust the exercise program. However, it is very difficult to evaluate the attitude, emotion and the determination and modification of evaluation in the process of physical exercise. Only using quantitative evaluation standard can't reflect the situation of physical exercise, so that the evaluation of teaching quality deviates from the aim of education. Therefore, it is necessary to make a clear longitudinal comparison between the students themselves in the process of strengthening their professional physical strength by adopting the quantitative evaluation standard. The index scores (not weighted) shall be calculated according to the weighted scores of each single index of the special students. For example, if the score of the morphological index of a male student is $A = A11 + A21 + A31$, the weighted scores of all the single indicators (secondary indicators) included in the first indicator are summed up, as shown in Tables 10 and 11.

According to the different weights of the first class index, the weighted scores of the first class index of special students are calculated. The weighted scores of all the first-grade indexes are summed up to get the comprehensive physical fitness scores of special students. Calculation formula: Overall fitness score = A B + C.According to

Table 10. Teaching Process Monitoring Scores of Male Students

Full name	A11	A21	A31	A41	B11	B21	C11	C21	C31	C41	C51
	Score	Score	Score	Score	Score	Score	Score	Score	Score	Score	Score
1	7.05	1.56	2.75	1.06	8.35	4.2	3.1	2.65	3	0	0.66
2	3.15	3.52	1.68	1.46	1.78	0	0.8	3.16	1.7	1.8	1.18
3	0.79	4.66	2.89	0.79	0	1.65	3.9	1.92	2.5	0.76	1.68
4	0	2.18	1.52	2.22	8.86	3.65	2.5	2.36	2.7	0.54	2.7
5	1.58	0.95	1.31	1.96	5.32	5.85	3.5	1.76	0	2.5	0.27
6	5.86	2.85	0.97	2.7	8.86	2.06	1.9	2.28	3.3	2.2	1.18
7	4.69	1.89	0.65	0.38	1.78	6.15	1.9	2.23	3.5	5	0.79
8	5.86	0	0.49	2.32	10.32	6.65	0	2.95	5	1.5	1.83
9	3.52	0.95	2.58	1.68	3.65	4.36	0.7	2.85	5	1.06	2.65
10	2.65	2.79	2.11	1.05	2.95	5.85	3.8	3.65	0.6	1.3	1.52

Table 11. Monitoring Scores of Female Students in Teaching Process

Full name	A11	A21	A31	A41	B11	B21	C11	C21	C31	C41	C51
	Score	Score	Score	Score	Score	Score	Score	Score	Score	Score	Score
1	6.67	4.21	3.53	1.36	8.26	2.26	2.05	3.13	1.15	1.51	1.83
2	4.82	2.87	2.21	0.91	2.21	3.61	1.75	2.81	0.77	0	1.68
3	0.75	1.05	3.75	1.96	3.86	4.15	0.88	2.41	1.91	1.06	1.61
4	5.93	3.65	3.97	2.86	10.46	3.61	0.28	3.38	0.38	1.21	0.92
5	5.21	2.09	0.89	0.76	11.01	4.96	3.49	3.77	3.11	2.10	1.57
6	1.12	4.52	1.32	2.11	8.81	2.71	2.91	0.73	2.51	0.76	2.22
7	1.52	1.57	2.87	2.11	3.31	3.61	3.49	3.25	1.72	1.81	0.27
8	2.23	4.95	3.97	1.66	5.51	2.71	0.59	2.17	0.96	0.91	2.23
9	2.97	1.31	1.99	0.16	1.66	1.36	0	0	0.39	0.61	1.43
10	4.08	5.21	4.41	0.31	0	5.50	1.17	1.29	1.35	0.46	0.92

this formula, the comprehensive scores of special physical ability of the students can be calculated, and the development levels of special physical ability of male and female students can be ranked as shown in Tables 12 and 13.

Table 12. Body Shape Index Test Results

Test items	Height (CM)	Percentage of body fat (%)
First test	176.7 ± 6.9	0.23 ± 0.05
Second test	177.1 ± 7.3	0.18 ± 0.05**
Third test	177.1 ± 6.8	0.19 ± 0.04**
Fourth test	177.5 ± 6.8	0.18 ± 0.03**
Fifth test	177.8 ± 6.8	0.18 ± 0.02**
Sixth test	177. ± 6.2	0.18 ± 0.01**
F	0.078	5.493
P	0.997	0**

Table 13. Test Results of Physical Function Indexes

Test items	Creatine kinase (U/L)	Testosterone	Hemoglobin (g/L)
First test	131.78 ± 51.32	20.28 ± 13.65	132.25 ± 5.65
Second test	249.85 ± 142.659**	23.85 ± 13.65	128.65 ± 14.65
Third test	126.65 ± 57.85ΔΔ	22.62 ± 15.65	129.52 ± 5.85
Fourth test	132.85 ± 48.65ΔΔ	28.65 ± 12.82	125.65 ± 5.65
Fifth test	108.32 ± 48.65ΔΔ	42.52 ± 16.65**Δ^{oo}	128.65 ± 6.85
Sixth test	107.65 ± 36.01	44.15 ± 13.65**ΔΔ^{oo}	128.95 ± 6.32
F	7.88	9.12	1.05
P	0	0	0.50

There was no significant difference in body fat percentage ($p < 0.01$). The results of the second, third, fourth, fifth and sixth tests of body fat percentage were significantly lower than those of the first test ($p < 01$). However, the average percentage of body fat of the six tests did not decrease linearly, but showed a fluctuation curve. The results of comprehensive physical ability grade evaluation for male and female students are shown in Tables 14 and 15.

Through the above evaluation results, we can find different students in the sub-index and comprehensive physical differences, can clearly define each student in the whole group in the position and level. It is helpful for teachers to know and master the differences of students' physical ability, so as to provide a basis for scientific selection and selection of special students and for rational planning of teaching and training so as to realize the overall improvement of students' competitive ability. This is because this method combines the absolute standard and the relative standard, and constructs the quantitative evaluation index of College Physical Education Students' physique based on genetic algorithm, which lays the foundation for the clarity of evaluation; Optimize the management algorithm of physical fitness monitoring of college physical education

Table 14. Results of comprehensive physical fitness rating of male students

Evaluation grade	Inferior	Middle and lower class	Secondary	Medium and superior	Superior
Corresponding percentile	10%	10%–26%	26%–76%	76%–91%	91%
Scoring criteria	7.86	7.87–8.95	8.96–11.38	11.39–12.59	12.60
Special student	2, 3	5, 15	1, 5, 7, 8, 9, 11, 15, 18, 19	8, 12, 16	18, 21

Table 15. Results of Comprehensive Physical Fitness Rating of Female Students

Evaluation grade	Inferior	Middle and lower class	Secondary	Medium and superior	Superior
Corresponding percentile	10%	10%–26%	26%–76%	76%–91%	91%
Scoring criteria	7.62	7.63–8.09	8.10–11.87	11.88–12.59	12.60
Special student	9, 23, 29	2, 3, 11, 28	1, 4, 7, 8, 9, 11, 15, 17, 18, 1, 21, 23, 27, 90	13, 16, 26	6, 13, 27

students, and ultimately improve the accuracy of physical fitness monitoring of college physical education students by simplifying the monitoring process.

4 Conclusion

The physical education in colleges has the dual attributes of higher physical education and vocational physical education. It is not only necessary to fulfill the task of improving students' physique, but also to fulfill the task of enriching and perfecting the knowledge of physical education, the reserve of vocational physical ability and the reserve of sports skills that are beneficial to vocational activities, and to strengthen and develop the physical ability and related abilities that are important to vocational activities. The evaluation system of vocational teaching quality should not only evaluate the general physical ability related to health, but also the occupational physical ability closely related to occupation. In the actual PE teaching and teaching quality evaluation, we should highlight the students' professional characteristics, and guide the students to exercise their physical strength through the evaluation results. In this paper, a mobile monitoring method based on genetic algorithm is proposed for college physical education students. Innovatively based on genetic algorithm, the evaluation index of physical fitness of college physical education students is constructed; Optimize the management algorithm of physical fitness monitoring of college physical education students, and design mobile physical fitness monitoring equipment; By simplifying the monitoring process, the monitoring accuracy of physical fitness of college students is improved.

References

1. Han, C., Pan, P., Wang, Q., et al.: Flexible pressure sensor based on carbon black sponges in human motions monitoring. J. Tianjin Univ. Technol. **38**(2), 38–44 (2022)
2. Tang, K., Wang, S.: Simulation of object detection method in multi degree of freedom human motion dynamic image. Comput. Simulat. **9**(199–202), 437 (2021)
3. Lian, J., Fang, S., Zhou, Y.: Model predictive control of the fuel cell cathode system based on state quantity estimation. Comput. Simulat. **37**(07), 119–122 (2020)
4. Su, J., Xu, R., Yu, S., Wang, B., Wang, J.: Idle slots skipped mechanism based tag identification algorithm with enhanced collision detection. KSII Trans. Internet Inf. Syst. **14**(5), 2294–2309 (2020)
5. Su, J., Xu, R., Yu, S., Wang, B., Wang, J.: Redundant rule detection for software-defined networking. KSII Trans. Internet Inf. Syst. **14**(6), 2735–2751 (2020)
6. Jiang, L.U.: Research on the feasibility of constructing big data public service platform for physical health testing based on college resources. Adhesion **42**(06), 68–72 (2020)
7. Jian, S.U.N., Jia-xin, H.E., Qi, Y.A.N.: Research and progress of special physical training in physical education colleges based on "data-driven decision making." J. Guangzhou Sport Univ. **40**(03), 1–3 (2020)
8. Chen, H., Yang, H., Liang, B.: Governance practice and enlightenment of British family sports in health promotion. J. Beijing Sport Univ. **43**(04), 82–89 (2020)
9. Shi, Y., Huo, X.: Quality of research based on grounded theory in the field of sports in China: Systematic review and control. China Sport Sci. **41**(07), 67–78 (2021)
10. Zhu, M., Chen, J., Yang, X., et al.: A fast algorithm for respiratory rate detection based on motion feature estimation. J. Hefei Univ. Technol. Nat. Sci. **45**(5), 610–619 (2022)

Research on Large Data Mining for Online Education of Mobile Terminal Based on Block Chain Technology

Zhitao Yu(✉) and Wei Zou

Jiangxi Software Vocational and Technical University, Nanchang 330041, China
yuzhitao1210@163.com

Abstract. In order to manage the massive online instructional resources effectively and realize the goal of quickly mining massive instructional resources, this paper proposes a research method of online instructional large data mining based on block chain technology. Based on the block chain technology, this paper constructs the recognition model of online education of mobile terminal, optimizes the management system of big data of online education of mobile terminal. Experimental results show that the block chain-based mobile terminal online education large data mining method has high practicability in the practical application, and fully meets the research requirements.

Keywords: Blockchain technology · Mobile terminal · Online education · Data mining

1 Introduction

With the development of the Internet, people begin to spread and learn content quickly and receive online education through information technology and Internet technology. Currently, building a high level of interaction is an important way to promote online education [1]. At present, the interactive analysis system of online education does not involve the analysis of unstructured data, which is the key factor that restricts the development of online education. Take this as a guide to improve online education services and achieve a two-way balance between online education services and learners' needs [2].

Reference [3] proposes the application and practice of Apriori based data mining algorithm. First, it analyzes the problems existing in mathematics teaching activities, and then analyzes the construction of information-based teaching model with Apriori algorithm in big data as the main idea, data mining is completed. This method realizes the fast mining of educational resources, but the mining accuracy is low. Reference [4] proposes the research of terminal online education data mining technology based on model driven. Using association analysis data conversion method to convert data, and then using model driven crowd behavior modeling method, the task flow of terminal

W. Fu and L. Yun (Eds.): ADHIP 2022, LNICST 468, pp. 286–299, 2023.
https://doi.org/10.1007/978-3-031-28787-9_22

online education data mining is designed After completing the above work, the key technology of model driven data mining is optimized through screening, selecting data subsets, coding, setting thresholds, and evolution steps to achieve efficient data mining for terminal online education, but the mining recall rate is low.

To this end, the research on mobile terminal online education big data mining based on blockchain technology is proposed. According to the characteristics of educational analytics and educational data mining, a scoring model of curriculum review is constructed to objectively evaluate students' learning ability. Process teaching data through blockchain technology and extract important educational interaction information. The knowledge-based information recommendation algorithm completes data mining and improves the accuracy of data mining.

2 Mobile Terminal Online Education Big Data Mining

2.1 Identification of Online Education Characteristics of Mobile Terminals

According to the different research perspectives and emphases, the research of educational data analysis for online learning platform mainly includes two research fields: educational data mining and educational analysis. These two areas have a lot in common, but also have their own key concerns. Educational data mining focuses more on analytical methods for online educational data, while educational analysis focuses more on analysis of learning patterns based on online data to help the development of pedagogical research. Through modeling and analyzing various data variables and their relationships, educational data mining hopes to explore the factors that affect learning goals, learning interests and learning effects. The main goal of educational data mining is to improve learning software, online learning sites and learning models, while educational analysis focuses more on instructing teachers and students to improve teaching and learning methods [5]. Both of them make use of pedagogy theory and data mining analysis technology synthetically. There are many common points in technical means, but the emphasis is different from the final goal. They are both the new cross research fields of pedagogy and computer science. The characteristics of educational analytics and educational data mining are shown in Table 1.

The user's comments on the course can reflect the user's preference for the course to some extent. Through collecting the comments information of users and courses, using a text classification model to score course comments, the preference matrix of users is constructed, and on this basis, the recommendation algorithm is used to recommend personalized courses to users. This paper first describes the comment scoring model, then introduces the recommendation algorithm, and finally explains the effect of the algorithm on the dataset. The value range of course review features is shown below.

$$y_i = \begin{cases} 1, \text{ if } P_i > \Theta \\ 0, \text{ other} \end{cases} \tag{1}$$

In formula (1), P_i is the possible probability of positive and negative evaluation, Θ represents the threshold value of the classification model, which is 0.45 during the experiment. Based on the definition of the problem, a specific neural network model is

Table 1. Characteristics of educational analysis and educational data mining

	Educational analysis	Educational data mining
Research means	Manual analysis is the main method and automatic mining method is the auxiliary tool	Automatic mining method is the main method, and manual analysis is the auxiliary tool
Research method	Pay more attention to the law analysis as a whole	Decompose the system and study the influence of various factors and the relationship between factors
Origin	Educational output prediction and system analysis in "intelligent course" on Semantic Web	Student modeling in learning software, such as course output prediction and other problems
Educational improvement	Focus on guiding educators and learners to improve	Focus on the automatic improvement of learning software and learning model
Main technology	Social network analysis, emotion analysis, influence analysis, discourse analysis, learner effect prediction, etc.	Classification, clustering, Bayesian model, relationship mining, pattern mining, visualization, etc.

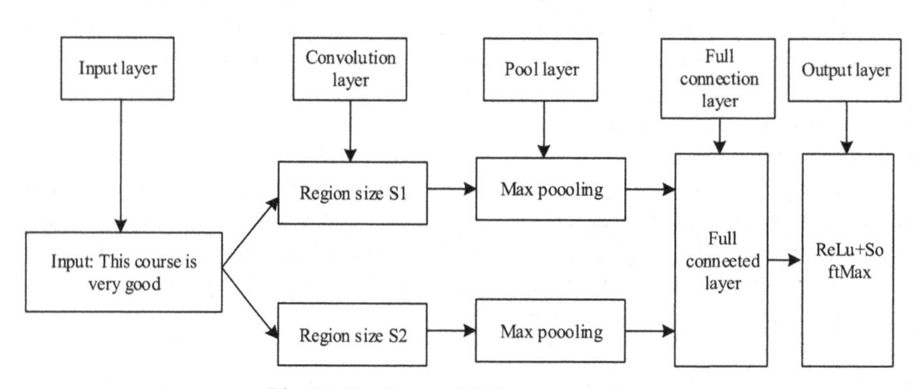

Fig. 1. Grading model for course reviews

designed to grade the curriculum review. The deep learning model for grading course reviews is shown in Fig. 1.

The field of educational data mining focuses on the research of online educational data mining and analysis methods, including prediction, structure mining, relationship mining, model discovery and so on. Prediction, refers to the mining of online education data to get a model about a variable, so as to predict the future trend of this variable, such as data trend prediction. At present, the common forecasting methods include classification, regression and potential knowledge assessment. In particular, the potential knowledge assessment, as an evaluation method of students' knowledge mastery, can evaluate students' knowledge mastery and ability more objectively. It has been widely used in online education and even traditional education.

2.2 Mobile Terminal Online Education Management Model

The forms of data online education are very complex and diverse, which can be divided into structured data and unstructured data according to the different forms of storage, explicit data and implicit data according to the different ways of obtaining, and video information, user information, user behavior information, course information, teaching information and forum information according to the different sources. The various types of data are intricate, so the management requirements are extremely high. Centered on the needs of learners, learn from the learning support service model, integrate the human and material resources needed by students, information, advice and advisory support, emotional support and other services. And try to integrate the individualized learning function into the learning process support service module provided by the network teaching platform, on the basis of the traditional service function, combine the decentralized and single individualized learning service with the learning process support service, design a continuous and multi-angle individualized learning support service system for distance education, provide individualized course selection guidance, course learning process counseling and consulting service, multi-dimensional evaluation and other services for students, create personalized network learning environment for students, and solve all kinds of difficult problems encountered in the learning process. The system mainly includes four modules: management service, information and consultation service, resource service and learning process support service. The system of distance education personalized learning support service is shown in Fig. 2.

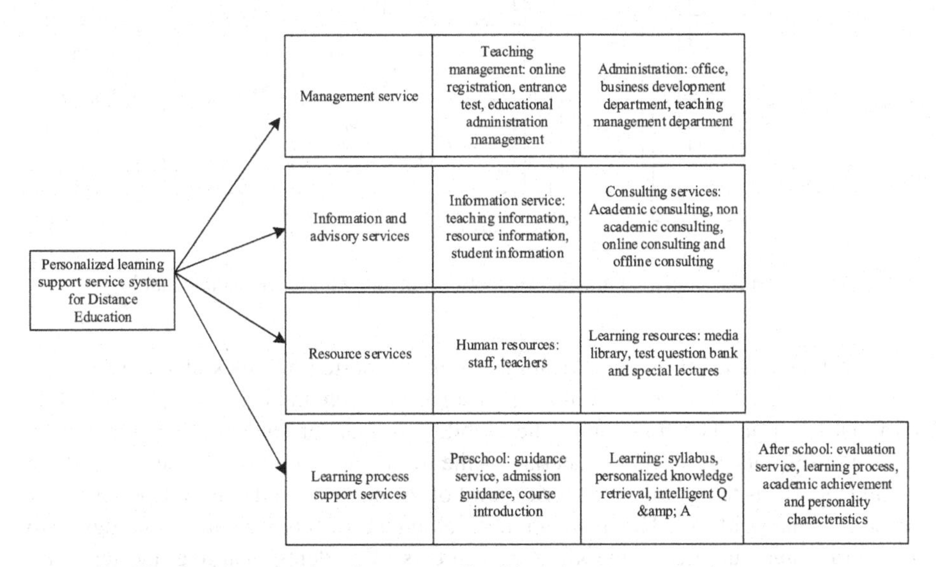

Fig. 2. Distance education personalized learning support service system

Large data mining generally includes data clean-up, transformation, integration, selection, mining, assessment mode and knowledge, etc. In order to ensure the effective integration and utilization of data and facilitate the development of online education platform, the relevant personnel need to mine or extract effective knowledge from a large amount of original data. However, due to the complexity and diversity of the sources of large data generated by school education platforms and the different storage forms, the connection, clean-up and integration of data are very complicated and cumbersome. In the process of data mining, the relevant personnel shall not only comprehensively consider the relationship between multiple characteristics of data, but also carry out independent analysis of data. In addition, due to the differences in analysis methods and tools for different data, it is necessary to have a sound and good data model for processing large data [6]. With the increasing homogeneity of online education, the interactive mobile terminal platform of online education based on big data technology can provide differentiated education services, and online education will be transformed into personalized services. The users of the interactive platform include teachers and learners, and the services provided for them mainly include online teaching content, teaching management, exchange and interaction and learning management. The interactive platform model of online education based on block chain technology is shown in Fig. 3.

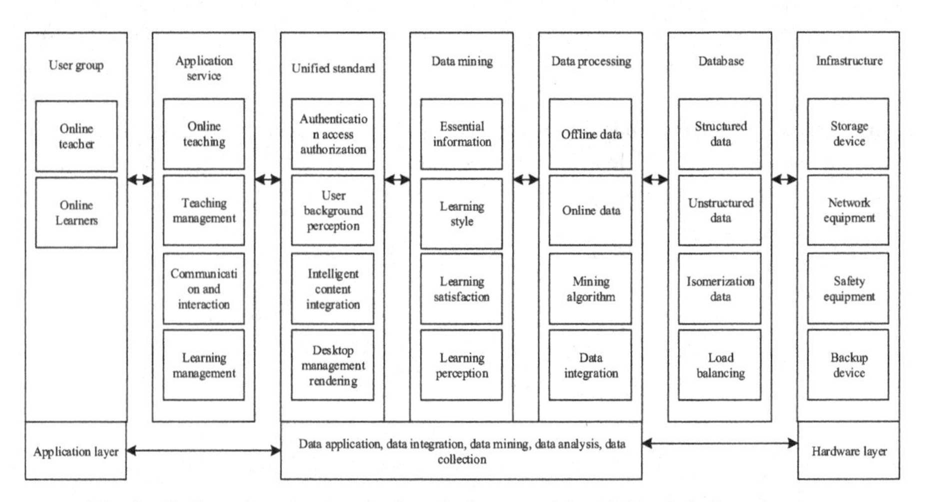

Fig. 3. Online education interaction platform model with blockchain technology

The applications of online education system supported by block chain technology mainly include student oriented mobile terminal function and teacher oriented mobile terminal function. The function of the mobile terminal for students from knowledge map, courses, videos, teaching materials to the forum, all to improve the learning effect of students, so as to improve the service level of the platform and consolidate the development of the platform. The application development of block chain technology is to summarize and summarize the knowledge points of students' learning and teachers' teaching from the angle of students' learning and teachers' teaching, so as to construct the knowledge framework and the knowledge modules of each course for the convenience of learning and teaching [7]. In addition, they still strengthen the connection between each knowledge dot, facilitate the student to study systematically. The application of personalized course service in online education platform is developed to provide better service for users, which is set up according to the characteristics of human beings. On online education platform, students can selectively watch videos according to their interests and learn what they want to learn, which is different from the overall learning offline and fully respects the personalized development of students. Blockchain technology can make course recommendations based on the test results and learning choices of students during the learning process, in addition to providing similar courses. The caption location of knowledge points is based on the main learning form of the online education platform. The application developed from video teaching is mainly based on caption skipping function for clips of interest to students in the video, and attached with professional vocabulary links and knowledge atlases, which can help students learn in an all-round way [8]. Mobile terminal teaching data training is mainly used to consolidate students' learning with knowledge atlas and teaching data base. Firstly, we set up a teaching data base according to different knowledge points, and pick out relevant knowledge points according to the students' answers, then pick up teaching data randomly from teaching data base to facilitate students' consolidation practice. The service application layer is the window for the external interaction of resources and the bridge for users to use resources, which directly affects the user experience. Therefore, the application service layer requests the reconstruction of information resources according to the user's needs to provide users with personalized service resources. Users do not need to know the integration process of background data resources, and the data resources processing layer of the platform completes completely. In view of the teacher, the platform feedback learner's analysis, especially the learning style and preference, tracking the whole learning process, and the students' behavior on the platform. Using SPSS Clementine 12.0 as a mining tool, C5.0 decision tree algorithm, through data collection, data preprocessing, data classification and rule generation steps, to achieve personalized path recommendation rule building and application. The process of teaching data processing based on blockchain technology is shown in Fig. 4.

According to the four-in-one learning mode of learners, constituting learning, answering questions, testing and evaluation and interaction, enjoy autonomous learning, personalized real-time notes, review and evaluation of targeted courses, and online interactive learning services in multiple ways; and formulate reasonable learning progress and personalized learning programs according to the results of backstage data mining. The preprocessed data are related, classified, clustered and biased, from which valuable interactive information of online education is found. By using the network analysis method of block chain technology, the integrated online educational interactive data are formed by integrating the data nodes and links, and the links reflecting the mutual relations are integrated to form a data network. The commonness of the data and the overall characteristics of the network are shown. Through visual analysis, ranking, classification, clustering and link prediction, we can find a large number of common patterns of educational interaction and extract the value of data relations. Analysis mining involves four steps: Map, sort, partition, and Reduce. The function of each step is independent,

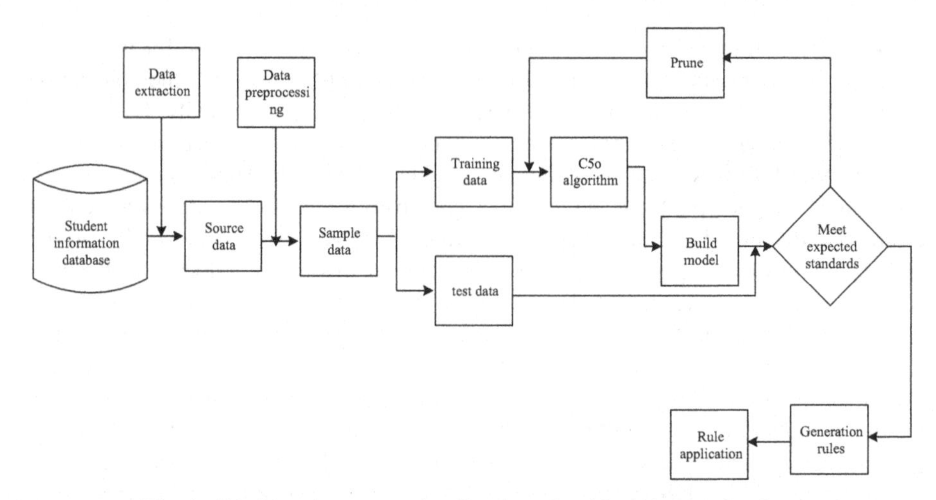

Fig. 4. Teaching data processing flow based on blockchain technology

can not be called each other, between the steps through the "key-value pair connection, the output of the previous step is the next step input." The result of data mining of online educational interaction is unknown and potential, and it can't be used as decision-making basis directly. Only by combining with education quality evaluation system, can it be transformed into the content needed by the platform. Through data filtering, analysis and integration, multi-resource classification results are established, and decisions are made according to users' different needs to facilitate users to access and use services. The purpose of the integrated data is to prepare for the integration of users, to analyze the similarity of users' information resources, to classify the similar users, to provide the allocation of similar information resources according to the basic information, learning style, learning satisfaction and learning perception of online learners, and to realize the customization, personalization and precision services of users. Finally, the analysis results of online education interactive data are presented to the corresponding users simply and clearly, from abstract data to visual structure mapping, allowing users of online education platform to directly observe the relationship between education factors, deducing and reasoning in a very short time, making accurate decisions, and ensuring that the data results provided must be comprehensive, timely and sustainable. At the data processing level, the experience assignment is mainly based on the teacher's experience, but as the user data accumulates, "the algorithm model based on Big Data and Ebbinghaus memory curve will play a role, so that each user's education plan is different."

2.3 Realization of Big Data Online Mining in Education

The process of educational data mining is actually the process of mining knowledge from a large number of data. At the same time, some people call knowledge discovery as data mining, but they are not the same as data mining 3. The main process of data mining includes data collection, data preprocessing, feature extraction, feature selection, data mining and model evaluation. The data mining process is shown in Fig. 5.

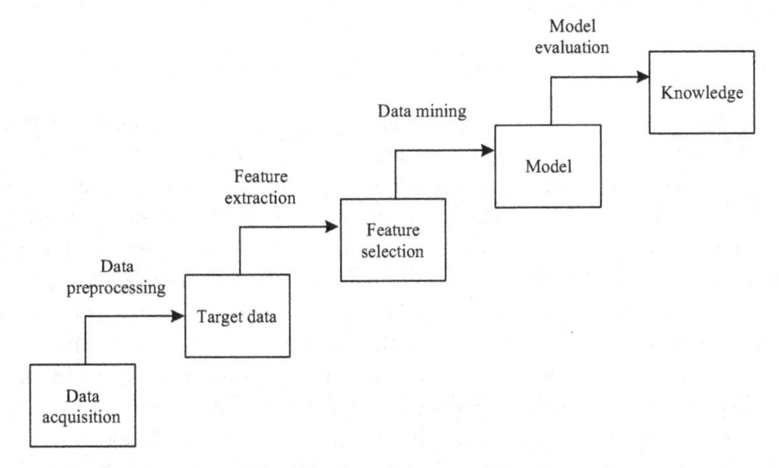

Fig. 5. Data mining procedure

Whether the process of data mining is successful or not depends on the quality of data to a great extent. Therefore, data preprocessing technology plays an important role in model prediction and generalization. In fact, the original data often appear missing value, noise data, incomplete data and abnormal data. These "dirty data" will affect the efficiency of data mining process is not conducive to model training, and may even lead to the deviation of experimental results. Therefore, preprocessing the data before mining, including data cleaning, data integration, data reduction and data transformation, makes the process of searching knowledge and discovering information value more meaningful. Data cleaning. In the information age, every minute and every second of data is generated quickly through the Internet. Dirty data may exist in the data obtained from different ways. In order to achieve high quality data, it is necessary to clean these data. Data cleaning is to solve the abnormal situation that may be encountered in the original data, such as data inconsistency, missing, outliers and so on. In addition, data cleansing is a necessary step in data mining analysis and the most important work in data preprocessing. Different data cleansing tasks aim at different types of errors. Data integration. In order to make the process of data mining more effective and utilize data from multiple sources as much as possible, data with different attributes, dimensions and structures can be integrated together to store and manage them in a unified way. This is why data integration plays a key role. Machine learning is driving the automation of data integration, reducing the cost of data integration in general and improving experimental accuracy. Sometimes the dimensions of different features in the preprocessed data may be inconsistent, and the differences between values may be very large, and the results of data analysis may be affected if not processed. In the various datasets managed, there are also significant differences between the eigenvalues, such as a maximum of 10000 and a minimum of 0.0001. The most common way to generalize attribute values is as follows:

$$\text{value}_1 = \frac{x - x_{\min}}{x_{\max} - x_{\min}} \tag{2}$$

$$\text{value}_2 = \frac{x - x_{\text{mean}}}{x_{\text{stand}}} \tag{3}$$

In formulas (2) and (3), the formula shall be the smallest and largest normalized function and the z-score normalized or naturalized function in the interval. x represents the pre-processed eigenvalues, value_1 and value_2 represents the reduced eigenvalues, x_{min} and x_{max} represent the pre-processed minimum and maximum eigenvalues, respectively. In addition, x_{mean} represents the mean of the eigenvector and x_{stand} represents the standard deviation. You can see that as the feature dimension is set to increase from 5 to 100, the performance of the model improves up to 75% in accuracy and over 60% in recall. It is also found that with the increase of the number of hidden units, the curves of the two indexes show a gentle trend. The reason for the low performance in the lower feature dimension may be that the hidden layer with less training times is not accurate enough to represent the students' behavior effectively. The more hidden layer feature dimensions you set, the more likely you are to get a fuller picture of your students. However, the accuracy of the model is more stable than that of the 50-dimensional model, and some repetitive features may appear in the training process. Therefore, it is helpful to set up a suitable feature dimension for students' behavior model, and it also helps educators to understand students' behavior and explore the information contained in data. The principle of knowledge-based information recommendation algorithm is shown in Fig. 6.

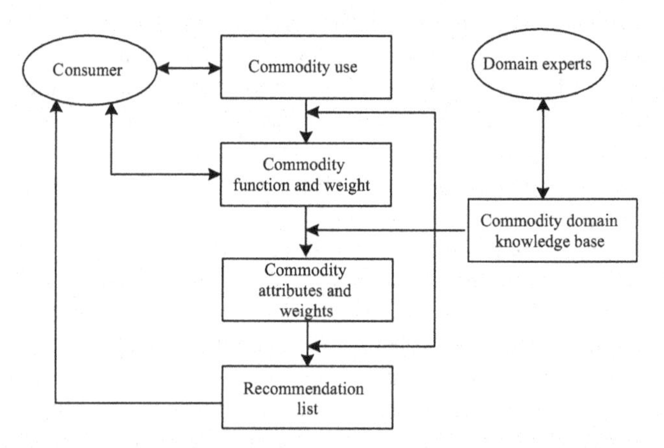

Fig. 6. Schematic diagram of knowledge based information recommendation algorithm

Based on the students' historical information search records, generate the students' historical score matrix (recorded as the A matrix) and then generate the teaching data-knowledge point marker matrix (recorded as the B matrix) based on the knowledge point information involved in the teaching materials marked by experts. These two matrices serve as the original input information to the system. Using the information of Matrix A and Matrix B, combined with the related models of cognitive diagnosis field, the comprehensive cognitive level information of each student was calculated. Similarly, the information of A and B matrices is used to compute the information of each student's knowledge level. Using the information of the students' comprehensive cognition level and the information of the knowledge points, and combining the traditional collaborative filtering method, we can predict the possible correct rate of each teaching material to be recommended by students. If the final recommended list of teaching materials is Q, then we immediately intersect with List R for each knowledge point less than μ. If the intersection is not empty, we add the intersection to List Q. If the intersection is empty, we iterate through all the predictive accuracy of the relevant teaching materials for this knowledge point, and take the teaching materials closest to [beta 1, beta 2] to add to the list Q. In this way, not only ensure the integrity of the knowledge training, but also ensure the accuracy of the teaching materials within a reasonable range, not because of too simple or too difficult to lead to poor experience, teaching materials recommendation list generation process is shown in Fig. 7.

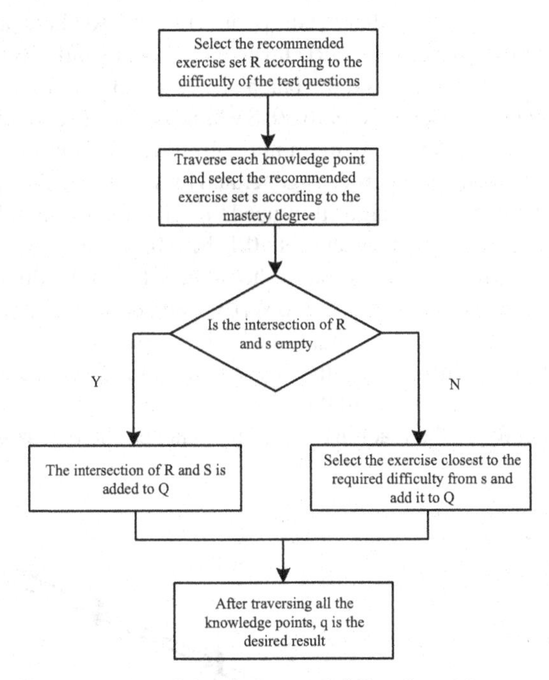

Fig. 7. Generation of the recommended list of teaching materials

There are some defects in the application of data mining technology in online education platform. Although the online education platform has changed the previous teaching model and provided new learning space for students, it is still lack of systematicness and systematicness. In order to improve the application effect of large data mining technology in online education platform, the author thinks, first of all, we should analyze and process the basic operation of online education fundamentally, classify and analyze the data produced by online education, establish corresponding big data model for each kind of data, which is convenient for analysis and processing, reduces the time of data mining and enhances the value of data. Secondly, the application of large data mining technology should be further built, such as teaching database should be updated with the times, so as to better platform development. Finally, we should strengthen the cooperation with government departments in order to obtain the corresponding information technology support and financial support. The information technology involved in large data mining technology is developing continuously.

3 Analysis of Experimental Results

Use the Python and the no libraries along with Nvidia Tesla K8o GPU. Here are some implementation details: The proposed two-stage classifier model has a 50D embedding layer for campus card devices, and the dimensions of each GRU are set to 50 hidden units. Optimized with Adam 6, the initial learning rate is 0001, and the minimum batch size is fixed at 128. To prevent RNN from over-fitting, two dropout layers are applied in HRNN based on attention: one is that the dropout between GRU layer and GRU layer is set to 25%, the other is 50% between GRU layer and bilinear similarity layer. The BPTT is also truncated to 19 time steps and the batch size is set to 512. In the most advanced method, 6, the number of epoches is set to 30. SVM is used to set c to 0.8, gamma to 0.1, formula 21 is the minimum and maximum normalized function in the interval, and the list of recommended teaching materials is generated by setting the upper and lower limit of recommended difficulty. The algorithm needs to give the mean value of difficulty, beta, and then the algorithm will select [beta 0.1, beta 0.1] difficulty range of teaching data reasonable algorithm, the real problem should be set with the difficulty of the line. Here, we use the SR index in Formula 5.2 to verify whether the recommended difficulty of teaching materials meets the set situation. In this experiment, cognitive modulator was set at 0.35. The experimental results were based on the Exercise dataset, after 100 experiments, take the average value of the results for comparative experiments. And the recommended difficulty of the teaching materials matched the experimental results for example, Fig. 8.

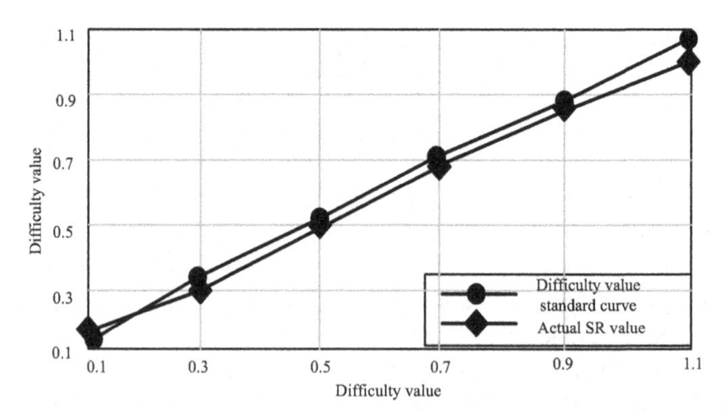

Fig. 8. Comparison chart of recommended teaching data with experimental results

The difficulty value of experiment setting is between 0.1 and 1.1, the straight line in Fig. 8 shows that the expected difficulty value matches the actual difficulty value exactly. Another curve shows the actual results of the experiment. From the coincidence of the two lines, we can see that the actual difficulty of the teaching materials is basically consistent with the difficulty we set in the algorithm. In the case of high or low difficulty, the error will be greater. This is because at this point in our algorithm according to the knowledge coverage and do some of this treatment has a greater impact, and this affects the real reflection of the difficulty of teaching materials. This also proves that the algorithm in this paper can truly according to the intention of topic selection, select the teaching materials suitable for students' difficulty. Recall rate and accuracy rate are commonly used to evaluate the classification performance. The recall rate and the accuracy rate are the evaluation indexes in the classification task reference information retrieval task. In information retrieval, precision rate and recall rate are usually used to measure the quality of retrieved information. Relevant documents are generally called positive and unrelated documents are called negative examples. In the whole process of information retrieval, there are generally four kinds of results: TP, TN, FP and FN. TP refers to that the search engine retrieves the relevant documents correctly; TN refers to that the irrelevant documents are not retrieved correctly and the irrelevant documents are filtered correctly; FP refers to that the irrelevant documents are retrieved incorrectly and the irrelevant documents are deemed as the relevant documents; FN refers to that the relevant documents are not retrieved incorrectly and the irrelevant documents are not retrieved. The four results in the information retrieval process are shown in Table 2.

Table 2. Four results from the information retrieval process

	Correlation (positive class)	Irrelevant (negative class)
Retrieved	True Positive (TP)	False Positive (FP)
Not retrieved	False Negative (FN)	True Negative (TN)

In order to verify the performance of this method in students' educational data mining task, the following benchmark methods are compared. The experimental results for different data length mining accuracy and recall are shown in Fig. 9.

According to Fig. 9, the highest accuracy and recall of the proposed method can reach 77% and 84%, respectively; The highest accuracy and recall of the comparison method are 63% and 64%, respectively, which are lower than the proposed method. The curves of the two indices show a gentle trend as the number of hidden units increases. When the performance of feature dimension is low, the reason may be that the hidden layer with less training times is not accurate enough to represent the students' behavior effectively. The more hidden layer feature dimensions you set, the more comprehensive you may be about your students. Therefore, it is helpful to establish the model of students' behavior, and it is helpful for educators to understand students' behavior and explore the information contained in data.

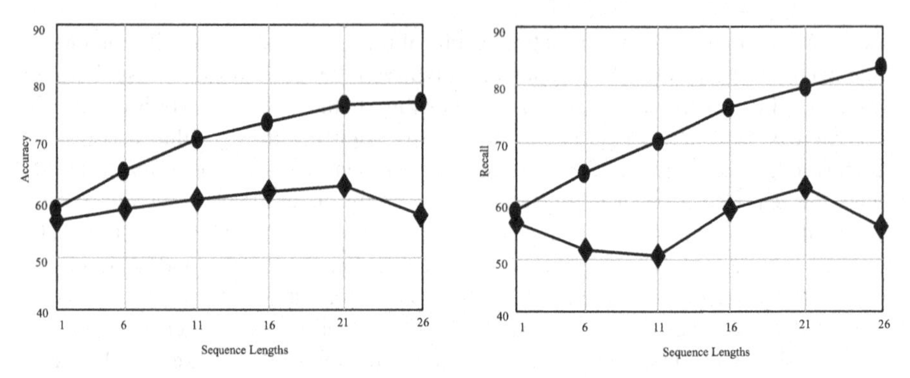

Fig. 9. Experimental results of accuracy and recall of different data length mining

4 Conclusion

In order to accurately mine the required teaching resources from massive data, the research on mobile terminal online education big data mining based on blockchain technology is proposed. Through the characteristics of education data, combined with blockchain technology, education data resources are extracted, and effective management of education resources is realized. The emergence of blockchain has changed our understanding of data. People can find the world hidden behind a large number of data through big data. It also brings new ideas for education and brings opportunities for personalized learning. At the same time, how to guarantee the quality and effect of personalized learning also faces challenges. On the one hand, because of the characteristics of massive data, data processing is different from the traditional data processing, how to effectively process these data is facing challenges. On the other hand, although the education big data theory and technology have a wide application prospect and great development potential in the education field, the application of big data needs the technical achievements of many interdisciplinary fields, and some big data education applications are still in the research and exploration stage. With the progress of network technology and the increase of data volume, in the future work, data mining technology needs to be constantly updated to meet higher-level needs.

References

1. Jin, Y., Chen, H.: Value orientations, realistic limitations and effective strategies of big data mining for education. Theory Pract. Educ. **41**(19), 3–8 (2021)
2. Zhang, T., Zhang, S.: Research on design and computing of learner model of educational big data mining. E-educ. Res. **41**(09), 61–67 (2020)
3. Xu, J., Zhang, G.: Application and practice of data mining algorithms based on apriori. Comput. Technol. Develop. **30**(4), 206–210 (2020)
4. Liu, W., Li, X.: Research on terminal online education data mining technology based on model driving. Modern Electronics Technique **43**(16), 112–114+118 (2020)
5. Lian, J., Fang, S., Zhou, Y.: Model predictive control of the fuel cell cathode system based on state quantity estimation. Comput. Simulat. **37**(07), 119–122 (2020)

6. Wang, D., Liu, H., Qiu, M.: Analysis method and application verification on teacher behavior data in smart classroom. China Educ. Technol. **05**, 120–127 (2020)
7. Wang, C., Cheng, J., Sang, X., et al.: Data privacy-preserving for blockchain: State of the art and trends. J. Comput. Res. Develop. **58**(10), 2099–2119 (2021)
8. Liu, F., Yang, J., Li, Z., et al.: A secure multi-party computation protocol for universal data privacy protection based on blockchain. J. Comput. Res. Develop. **58**(02), 281–290 (2021)

Separation Algorithm of Fixed Wing UAV Positioning Signal Based on AI

Zhihui Zou(✉) and Zihe Wei

College of Humanity and Information, Changchun University of Technology, Changchun 130122, China
yongg30110@163.com

Abstract. Unmanned aerial vehicle (UAV) is an unmanned aircraft remotely controlled by radio, which is widely used in reconnaissance. However, during the operation of UAV, the positioning signal is easily disturbed by noise, which leads to low separation accuracy and poor positioning effect of fixed wing UAV. To this end, a fixed wing UAV positioning signal separation algorithm based on artificial intelligence is proposed. The fixed-wing UAV positioning signal denoising algorithm is constructed by collecting the feature information of fixed-wing UAV, and the denoising of fixed-wing UAV positioning signal is completed. In order to reduce the signal separation error and realize the fixed wing UAV positioning signal separation, signal separation is processed according to the positioning signal algorithm. Experimental results show that the proposed algorithm can effectively separate the UAV location signal from the noise, and has high accuracy and good location effect under serious multipath interference.

Keywords: Artificial intelligence · Fixed-wing UAV · Positioning signal · Signal separation

1 Introduction

UAV has been widely used in many fields such as national defense, agriculture and military because of its flexibility, maneuverability and small size. Signal recognition and separation of UAVs play an important role in their stability and safety. When the UAV is located passively, the multipath noise caused by obstacles on the ground is uncertain and irrelevant in different positions, so it is impossible to locate the UAV in real time. Therefore, it is of great significance to study the separation technology of UAV positioning signals.

Reference [1] proposes the research of UAV positioning signal separation algorithm based on support vector machine, which obtains information entropy by calculating the Euclidean distance between adjacent data sets of UAVs, and provides model data for SVM to map high-dimensional space. On this basis, the threshold soft boundary of the mapping function is added to make the model have the ability of parameter adaptive adjustment to adapt to the data difference caused by the flexible movement of

W. Fu and L. Yun (Eds.): ADHIP 2022, LNICST 468, pp. 300–312, 2023.
https://doi.org/10.1007/978-3-031-28787-9_23

UAV. Finally, the observer operating characteristic curve is constructed to obtain the separation results of UAV positioning signals. This algorithm can effectively separate the UAV positioning signal and noise, but when the signal-to-noise ratio is small, the positioning performance of the algorithm is poor.

According to the characteristics of fixed wing UAV, the signal is collected. Through the features of the fixed wing UAV, the positioning feature signals are collected. The momentum term is combined with the adaptive algorithm to improve the signal denoising effect. After the UAV positioning points are arranged, the positioning data is obtained. According to the similarity between the execution trajectories, the fixed wing UAV positioning signal is separated, which improves the accuracy of UAV positioning.

2 Positioning Signal Separation Algorithm for Fixed-Wing UAV

2.1 Characteristics of Fixed Wing UAV

Fixed-wing UAV is widely used in remote sensing and aerial survey because of its convenience and flexibility. Fixed-wing UAV can be used in remote sensing and aerial survey. Fixed-wing UAV remote sensing is composed of three parts, namely, control, fixed-wing UAV remote sensing platform and image processing. Its main functions include: using control to complete the route planning and flight route control of the fixed-wing UAV, the route planning can set flight routes and specify flight tasks, and the flight route control can be used for real-time flight control and interactive operation; using the fixed-wing UAV remote sensing platform to carry the transmission of data by sensors, which is mainly composed of four-wing UAV, camera, pan tilt and GPS positioning, can complete direct photography to the ground, and its flight position data can be transmitted in real time; image processing mainly carries out image processing, including image rectification, fusion and mosaicing, and on this basis, can be expanded, for example, direct query and browsing of images.

2.2 Fixed Wing UAV Positioning Signal Acquisition

The process for collecting the positioning signal of the fixed-wing UAV is shown in Fig. 1.

As can be seen from Fig. 1, before each operation, the fixed-wing UAV needs to locate the probing area, determine the route, and then inject the route into the remote sensing flight platform; with the assistance of GPS positioning, shoot according to the planned route, and obtain the image sequence; if the aerial photography is completed, the acquired image information shall be transmitted to the processing center, and a series of image processing such as rectification, fusion and mosaic shall be completed, and the image shall be stored so as to further deal with the feature processing of the large impact. The fixed wing UAV needs to operate with zero error in all the above steps. To do this, a fixed wing UAV positioning coordinate system must be established first, and then the target can be located [2]. In the process of UAV positioning, the entropy data of UAV positioning information is extracted as the core function of positioning signal to map dimensions. The more uniform the probability of the random distribution of the

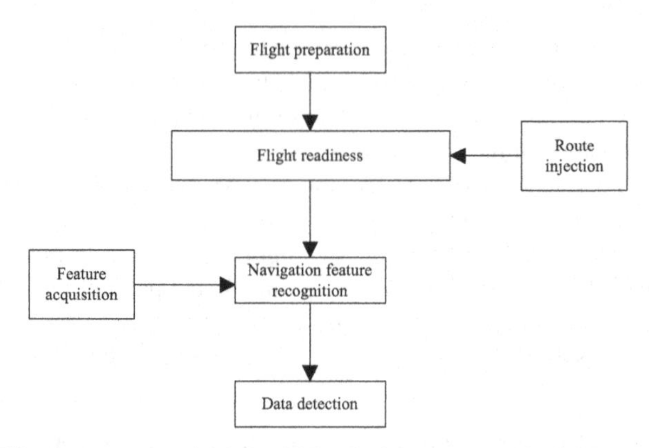

Fig. 1. Feature acquisition of fixed wing UAV positioning signal

UAV positioning signal, the greater the entropy, and vice versa. Therefore, the UAV positioning signal is obviously different from the normalized information entropy data and multipath noise data.

Artificial intelligence refers to the technology of presenting human intelligence through ordinary computer programs, for mapping high dimensional space by calculating the Euclidean distance between adjacent UAV datasets. On this basis, the soft boundary of mapping function is added to make the model have the ability of parameter adaptive adjustment to adapt the data difference caused by UAV's flexible motion. Finally, the observer operation characteristic curve is constructed to obtain the separation result of UAV positioning signal. By mapping the information entropy data in the high dimensional space, the artificial intelligence technology realizes the extraction and separation of UAV positioning signal. In the process of UAV positioning using mobile UAV signals, it is impossible to separate the reflected signals of UAV due to the lack of directionality, low power, low sampling rate and complex environmental noise [3]. In the open environment, there is no barrier such as wall, so the UAV reflected signal received from the base station can be divided into three parts, UAV reflected direct wave signal, multipath interference signal and Gauss noise signal. Therefore, the UAV positioning signal model is established as follows:

$$\hat{y} = Py + m\sum_{m=1}^{M} e_m + N_m \tag{1}$$

In formula (1), e_m is the receiving signal, y is the direct wave signal reflected by the UAV, N is the m multi-path interference signal caused by non-line-of-sight factors, and P is the noise error. After analyzing the signal delay and phase shift in the process of UAV positioning signal propagation, the formula (2) can be refined as follows:

$$\hat{y}(t) = \alpha_0 F\cos(\theta_0) + \hat{y}\sum_{m=1}^{M} \alpha_m F \tag{2}$$

After analyzing the model and the actual test data, it is found that α_0 and N can not separate the positioning signal based on signal power or phase because of the low F

base station power and the small α_0 of UAV. The momentum principle can improve the convergence speed of the adaptive learning algorithm, and provide a new way to improve the performance of the localization signal separation algorithm. But the application of momentum technology in localization signal processing is still at the exploratory stage, and its theoretical and practical effects still need to be further extended and improved. First, the accuracy and rationality of momentum factor selection have an impact on the convergence performance of the learning algorithm. Secondly, in the design of the gradient algorithm, how to make the momentum item and learning steps can be used reasonably and cooperatively, so as to make the algorithm achieve the best performance [4]. These are the problems to be solved to some extent in the future. In order to solve this problem, the information entropy of UAV is taken as the kernel function and mapped in high dimensional space to complete the UAV positioning signal acquisition.

2.3 Denoising of Fixed Wing UAV Positioning Signal

Fixed-wing UAVs carry image sensors and laser imaging sensors, which cannot fly along the prescribed route due to the influence of air flow and propeller in the course of operation. When the UAVs carry out positioning, the deviation of fine pitch angle, yaw angle and roll angle is produced. At the same time, the continuity of the positioning process is limited and the target position cannot be accurately located. Fixed-wing UAV can capture the origin of take-off coordinates in the range of view, establish coordinate system, so that the fixed-wing UAV in the range of view can accurately capture the target after flying for a certain distance, then complete image processing and recognition, and then locate the target position. After coordinate system conversion, the remote sensing positioning model of fixed-wing UAV can be obtained as shown in Fig. 2.

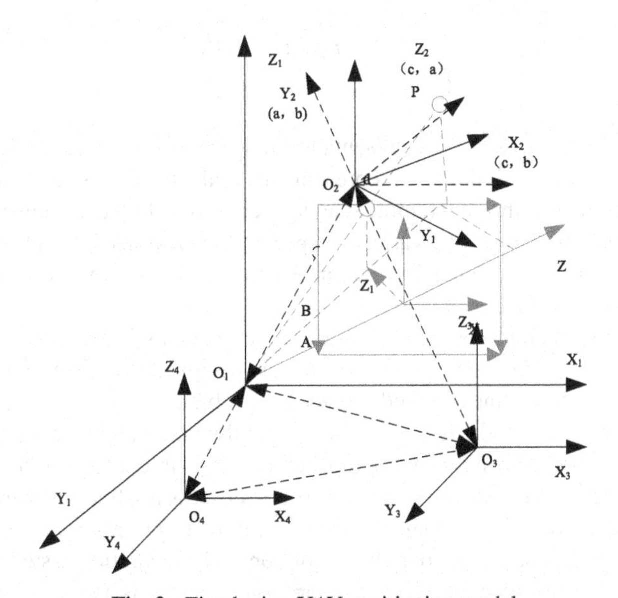

Fig. 2. Fixed wing UAV positioning model

From Fig. 2, the latitude and longitude information is confirmed by the fixed-wing UAV remote sensing images, where a, b and c show the deviation of the fixed wing UAV from the slight pitch, yaw and roll angle caused by the air flow and propeller respectively. For the sensor selection of electromagnetic field sensor, most of them use three-axis orthogonal ECG or three-axis orthogonal ECG or magnetoresistive sensor as sensors, and some of them use Hall sensor or magnetoresistive sensor to show the measuring range of all kinds of sensors. It can be seen from the diagram that the induction coil or inductor can be used for detection within a large range of magnetic field intensity [5]. The disadvantage of using coil as electromagnetic field sensor is that the volume and weight of the sensor chip is larger than that of the magnetoresistive sensor chip. But considering that the coil is easy to make, only enameled wire is needed to wrap around the frame.

According to the artificial intelligence positioning model, the target combined with laser imaging and TV image can be located, the pixel coordinate position can be obtained by coaxial TV image, and the target coordinate position can be detected by fixed-wing UAV. Most of the existing adaptive localization signal separation algorithms belong to gradient algorithms. The main disadvantage of these algorithms is that their convergence rates are not satisfied in some real-time applications. In order to improve the convergence performance of the BSS algorithm, this paper proposes a localization signal separation algorithm based on the minimum mutual information criterion and the momentum learning principle [6]. Firstly, a localization signal separation cost function based on minimum mutual information criterion is introduced. Based on the natural gradient principle, the adaptive updating rules with recursive structure including momentum terms are derived. In addition, many existing BSS algorithms can only separate source signals with the same fourth-order cumulant (gradient) symbol. When a fixed-wing UAV has a posture error, the target position shall be corrected, and the result shall be:

$$\begin{cases} A = A'\hat{y}(t) - (Z - zDc) \\ B = B' + P(yDb + xDa) \end{cases} \tag{3}$$

In formula (3), Δa, Δb, Δc represents the random error caused by deviation from the route respectively, and A' and B' represents the real-time feedback of target position information parameters through remote sensing technology. In order to ensure the normal operation of the deployment of a variety of specific business needs, fault location as the core function to meet the emerging characteristics, based on this further build UAV heading location structure as shown in Fig. 3.

Based on the structure, the UAV positioning information is collected and managed, and the positioning points of the preset position are designed to realize signal separation. In addition to the above linear mixed model, the problem of location signal separation under nonlinear mixed model is also studied. A nonlinear localization signal separation method based on perceptron model is proposed for reversible nonlinear hybrid systems. A three-layer perceptron network is used to construct a nonlinear separation system to separate the source signals. The maximum output information angle is used as a separation criterion to adjust the parameters of nonlinear separation system. In the unsupervised adaptive learning of perceptron separation system, it is necessary to adjust three parameters: the weight matrix of hidden layer and output layer, the threshold vector of

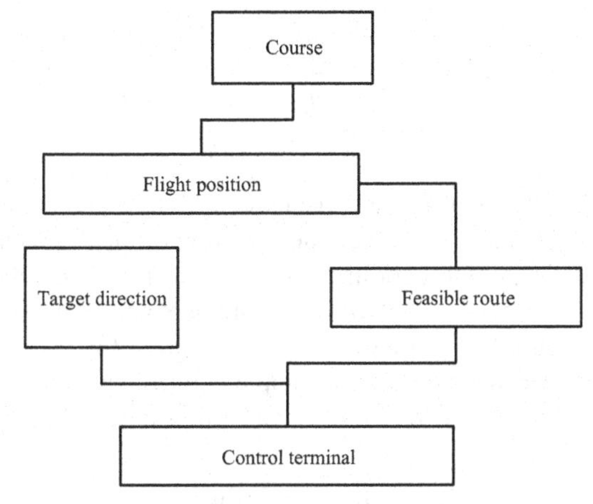

Fig. 3. Schematic diagram of UAV positioning information management structure

neural node. In order to improve the training speed, a conjugate gradient optimization algorithm is introduced to train the weight matrix of the perceptron network. In addition, the Sigmoid function is chosen as the probability distribution function of the separated signal, and an adaptive parameterized probability density estimation method is used to estimate it [7]. By computer simulation, it can be seen that the signal interference ratio of the separation result can reach about 25 dB when the signal mixing process is not instantaneously completed, that is, when the delay of the source signal in the transmission process of the sensor is taken into account, the mixed model of the location signal separation problem is transformed into a convolved mixed model of multi-dimensional signals. A joint approximate diagonalization method for the separation of linear convolution mixed signals is proposed. First, the convolution mixed model is transformed into an instantaneous mixed model to obtain a new instantaneous mixed signal.

2.4 Implementation of UAV Positioning Signal Separation

2.4.1 Positioning Point Layout of UAV TT & C Preset Position

With the development of the research on location signal separation, the research on BSS focuses on two aspects: the choice and exploration of cost function and the design of different optimization algorithms. On the basis of the previous section, this section proposes a well-posed hybrid model localization signal separation algorithm based on conjugate gradient optimization algorithm. The algorithm is still based on mutual information, and the conjugate gradient algorithm is used to search the separation matrix. As the key to the success of the algorithm, the kernel probability density estimation method is used to estimate the probability density and its derivative of the separated signal. Instead of selecting a single nonlinear activation function based on experience, the effectiveness of the proposed algorithm is verified by computer simulation. Set up Q_1, Q_2 and Q_3 to represent respectively the first type of data information, the second type of data information and the third type of data information. Combined with the above physical quantities, the definition of the preset stage of UAV measurement and control can be expressed as follows:

$$E = \frac{\sqrt{(xQ_1^2 + yQ_2^2 + zQ_3^2) - \hat{y}(t)}}{|A + B|\cos(\theta_0)} \tag{4}$$

According to the structural characteristics of different predetermined stages, the orientation of the orientation vectors in the action coordinate system is planned. In the weighted hyper-positioning coordinate system, A_0 represents the initial position information of the preset position for UAV measurement and control, and A_n represents the completion position information of the preset position for UAV measurement and control, In the simultaneous formula, the principle of overall planning of the positioning points of the preset position may be expressed as follows:

$$\overrightarrow{A_0 A_n} = \int_0^n \frac{\beta^2 |A_n \times A_0|^2}{E\dot{u}} - E \tag{5}$$

In formula (5), $\overrightarrow{A_0 A_n}$ represents the UAV marker vector from the initial position to the completion position, n represents the specific actual value of the UAV TT&C preset position from the initial position to the completion position, β represents the given curvature condition of the navigation curve in the weighted hyper-positioning coordinate system, and $\dot{u}$ represents the planned regression coefficient within the TT&C navigation path. On the premise that the overall planning position of the preset position remains unchanged, taking the given time T as the counting condition, from all possible curvature values in the weighted overposition coordinate system, a vector is randomly selected to be defined as β', and R_0 is set to represent the boundary parameters of the lower bound layout, and R_1 represents the boundary parameters of the upper bound layout, and the simultaneous formula, so that the positioning point layout discriminant of the preset position of the UAV can be expressed as follows:

$$W = \frac{\left|1 - \beta' \sum_{R_0}^{R_1} T \cdot \overline{p}^2\right|}{\chi' \left|\overrightarrow{A_0 A_n}\right|} \tag{6}$$

In formula (6), $\overline{p}$ represents the average navigation speed of the UAV measurement and control device in the weighted over-positioning system, and χ' represents the layout permission parameters of the positioning points.

2.4.2 Obtaining UAV Positioning Data

Calculate the Euclidean distance between any two A and B data samples according to the original data set. If B is within the radius of A, connect A and B, set the length of the connecting line to d; if not within the radius, set the length of the connecting line to ∞. This repetition constructs the UAV's undirected course location primitive dataset, as shown in Fig. 4.

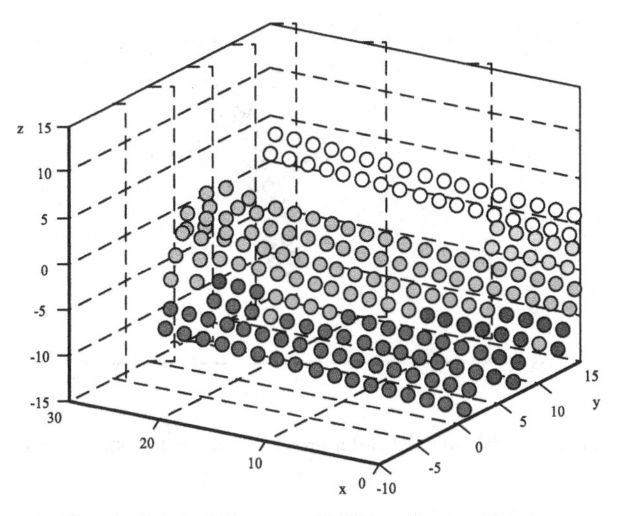

Fig. 4. Original dataset of UAV heading localization

The shortest distance between any two samples is further calculated, and the shortest distance d_1 between any two data in the original data set is calculated by using the Isomap algorithm. The shortest distance matrix obtained therefrom is:

$$d' = \mathrm{W} - \overrightarrow{A_0 A_n}\{d_1(A, B)\} \tag{7}$$

Because the shortest distance obtained by the above formula is affected by noise, the calculation of distance is not accurate. Therefore, the Isomap algorithm is applied to the distance matrix of the formula:

$$D = d'^2 = \left\{d_1^2(A, B)\right\} \tag{8}$$

Since the distance matrix has smooth sample data, we need to use nonlinear dimensionality reduction method to obtain the dimension of the sample data, but this process will be disturbed by noise, so the original data will be distorted in the dimension reduction space. So we use the Isomap equidistant feature mapping method to reduce the dimension, and the reduced dimension and denoised heading data is shown in Fig. 5.

According to the data set after dimensionality reduction and noise reduction, the long-range navigation can be located in real time to ensure the effectiveness of the design.

2.4.3 Calculate the Distance Between Execution Tracks

In the process of execution, there are a lot of redundant test cases, which form a lot of redundant information and reduce the positioning accuracy. Similar execution trajectories of test case execution can be used to accurately locate multiple bugs in a program. The smaller the distance between execution tracks, the more similar the execution tracks, and the multi-valued vectors can be sorted according to the value of vector elements.

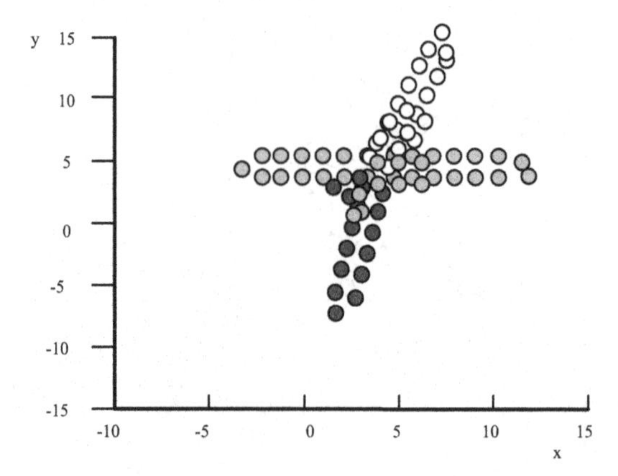

Fig. 5. Course positioning data after dimension reduction and noise reduction

The distance between execution paths is obtained by calculating the number of interconversion steps. Because the selected trajectory can only focus on locating the relationship between new blocks, it is necessary to calculate the distance between the execution trajectories. Set the execution track as follows:

$$\begin{aligned} w_1[j] &= (sus_3, sus_4, sus_1, sus_2, sus_5, sus_6) \\ w_2[j] &= (sus_1, sus_2, sus_5, sus_6, sus_3, sus_4) \end{aligned} \tag{9}$$

Direct use of the number of times parameter calculation is greatly affected by the new block cycle, timely detection of the relative relationship between the number of executions, can fully reflect the similarity between the execution trajectories. According to the coordinates, the new positioning blocks are sorted from near to far, and are numbered sequentially. The new positioning blocks are graded according to the actual defect number, and the higher the score, the higher the positioning accuracy. If the reported score exceeds 90 points, it means that only 10% of the new location blocks need to be audited to quickly locate the defect in the software. The formula is:

$$s = \frac{p - q}{w_1[j] - w_2[j]} - D \tag{10}$$

In formula (10), p represents the total number of new positioning blocks, and q represents the separation result of new positioning blocks. Based on this, the target of fixed wing UAV can be separated and the navigation and positioning accuracy of UAV can be ensured. Artificial intelligence technology has been widely used in the field of location signal separation. In general, this method can be used to solve some instantaneous and convolution mixed BSS problems effectively, but sometimes the separation matrix is singular, so the desired separation signal can not be obtained. In this paper, a joint diagonalization method for location signal separation problem based on artificial intelligence is proposed. The method first transforms the convolution mixture model into an instantaneous mixture model, then divides the transformed signal samples into several groups and obtains the second-order statistics matrix of each group. Finally, a joint diagonalization method is applied to the matrix composed of the second-order statistics matrix to obtain the separation matrix, and a constraint term is introduced to construct the separation cost function of the positioning signal, which avoids the situation that the obtained separation matrix is singular. Simulation results show that the proposed algorithm can effectively separate convolution mixed speech signals. Compared with other convolution mixed-location signal separation algorithms, convolution mixed-location signal separation algorithm can achieve better performance. Nowadays, convolution mixed-location signal separation problem will be met in many applications.

3 Analysis of Experimental Results

In order to test the validity of the fault location algorithm of UAV, the experiment was carried out. Experiment is established in the MATLAB environment, in which the hardware environment for Intel Core 3–550 1 GB of memory, operation for Windows 7. Assuming that a large number of UAV nodes are located in 2000 m × 2000 m uniform array area. The range of various magnetic field positioning information measurement sensors and experimental parameter settings are shown in Tables 1 and 2.

Table 1. Range of Magnetic Field Positioning Information Sensors

Sensor positioning technology	Approximate detection range
Induction coil sensor	$le-8$—$le9$
Fluxgate sensor	$le-6$—$le3$
Optically pumped alkali sensor	$le-8$—$le2$
Atomic motion sensor	$le-8-le2$
SQUID sensor	$le-9$—$le3$
Hall Effect Sensor	$le0$—$le4$
Magnetoresistive sensor	$le-6$—$le2$
Bulk transistor sensor	$le-1$—$le7$

Table 2. Experimental parameter settings

Parameter	Remarks
UAV frequency band	3 kHz–8 kHz
Carrier frequency time width	3 ms
Data initial frequency	0.15 Hz
Number of sampling points	256
Signal to noise ratio variation range	−15 dB–15 dB

According to the experiment environment and the result of parameter setting, the experiment content is analyzed. In order to mine the fault data of UAV, it is necessary to sample the time series of the data, and the waveform of the position information obtained is shown in Fig. 6.

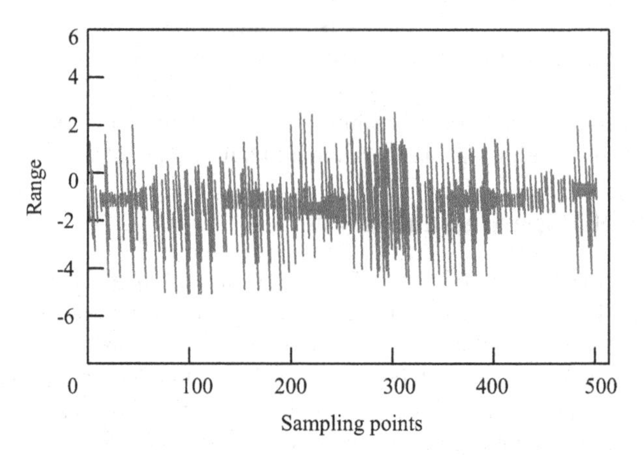

Fig. 6. UAV positioning information waveform

According to the UAV data shown in the graph, the training set is constructed, the fault data mining is tested, and the fault data is analyzed in time domain. Set the default UAV parameters as shown in Table 3.

Table 3. Default Values for UAV Parameters

UAV type	Wavelength	Refractive index	Scattering coefficient scattering coefficient
Single mode	1220 nm	1.521300	−80.0 dB
	1120 nm	1.467200	−83.0 dB
Multimode	1120 nm	1.485000	−72 dB
	450 nm	1.523000	−68 dB

Furthermore, the UAV is taken as the ultimate target, comparing the error elimination results of the technology of reference [1] with the Paper technology, the calculation results of UAV positioning signal separation are shown in Table 4.

It should be noted that the data matrix of positioning signal is always non-negative, so in order to apply the algorithm of UAV positioning signal separation based on artificial intelligence, we need to normalize it to the data of zero mean and unit variance. At the same time, in order to visualize the result of the separation image, the mean and variance of the original signal are recorded. When the separation process is complete, the separation results need to be changed, so that the previously recorded mean and variance data can be restored, while the integer. By further comparing the processing errors with the method of reference [1] and the paper method, and the detection results are shown in Fig. 7.

Table 4. Calculation results of UAV positioning signal separation

Method	Coordinate axis	Maximum	Minimum value	Average value	Error range
Standard coordinates	Axis X	34053	32889	33471	1164
	Axis Y	18113	17432	17772.5	681
	Axis Z	401	−221	90	622
The technology of reference [1]	Axis X	31059	29885	30472	1174
	Axis Y	15442	14381	14911.5	1061
	Axis Z	905	250	551	654
Paper technology	Axis X	34091	33002	33546.5	1089
	Axis Y	17992	16851	17421.5	1141
	Axis Z	398	−198	298	596

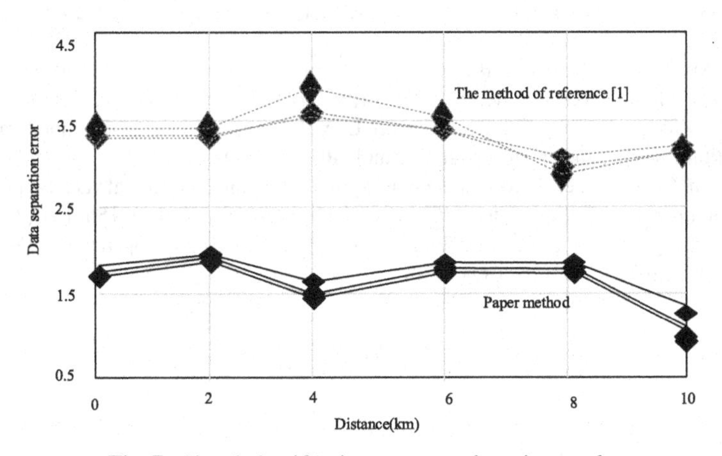

Fig. 7. Signal classification error rate detection results

Compared with the method of reference [1], the proposed algorithm based on artificial intelligence has higher accuracy in the practical application and can fully meet the research requirements.

4 Conclusion

The main problem of passive location of outdoor fixed-wing UAV in multi-path environment is that the UAV location signal can not be separated from noise. A positioning signal separation algorithm for UAV based on artificial intelligence is proposed. By obtaining the information entropy of the original signal, the Euclidean distance is used to balance the data to ensure the validity of the data. When information entropy is used as reference value of positioning information, the positioning signal is processed to reduce space complexity and improve the real-time performance of UAV positioning model.

Combining the information entropy and signal power logarithm, the ROC plane is constructed to realize the separation of the UAV positioning signal by artificial intelligence technology.

Fund Project. National Natural Science Foundation of China Major Project (61890964).

References

1. Li, X., Fang, K., Fan, T., et al.: Research on unmanned aerial vehicle location signal separation algorithm based on support vector machines. J. Electron. Inf. Technol. **43**(09), 2601–2607 (2021)
2. Lian, J., Fang, S., Zhou, Y.: Model predictive control of the fuel cell cathode system based on state quantity estimation. Comput. Simulat. **37**(07), 119–122 (2020)
3. Su, J., Xu, R., Yu, S., Wang, B., Wang, J.: Idle slots skipped mechanism based tag identification algorithm with enhanced collision detection. KSII Trans. Internet Inf. Syst. **14**(5), 2294–2309 (2020)
4. Su, J., Xu, R., Yu, S., Wang, B., Wang, J.: Redundant rule detection for software-defined networking. KSII Trans. Internet Inf. Syst. **14**(6), 2735–2751 (2020)
5. Fan, K., Zhang, X., Liu, H.: Research on UAV localization based on refined radio map reconstitution. Transducer Microsyst. Technol. **40**(11), 47–49+53 (2021)
6. Li, B., Cui, S., Shi, G., et al.: Research on the agricultural unmanned aerial vehicles positioning based on unscented Kalman filter. J. Chin. Agricult. Mechaniz. **41**(09), 156–161 (2020)
7. Ma, J., Huang, D., Wang, J., et al.: Targeting of interest based on cooperative UAV. Comput. Measur. Control **28**(04), 176–180 (2020)

Research on Hybrid Maintenance Cost Prediction of Smart Grid Based on Multi-dimensional Information

Ying Wang(✉), Xuemei Zhu, Ye Ke, Chenhong Zheng, and Shiming Zhang

State Grid Fujian Power Economic Research Institute, Fuzhou 350000, China
Yingwang21212@163.com

Abstract. At present, the change of equipment state is not considered in the prediction of power grid maintenance cost, which leads to inaccurate prediction results. Based on multi-dimensional mixed information, a prediction method of power grid intelligent maintenance cost is proposed. According to the expenses of routine maintenance of various equipment, the intelligent maintenance cost of power grid is divided into routine maintenance and power supply loss cost. The CS algorithm is used to determine the maintenance strategy of power grid equipment, so as to obtain the maximum power grid income under the minimum maintenance cost. The multidimensional mixed information extracted from the daily operation of smart grid determines the maintenance status of equipment in the maintenance strategy. Through the methods of grey prediction and multiple linear regression prediction, the diversified prediction results are output, and then the weighted value of the prediction output results is assigned with the help of the combined prediction model to realize the cost prediction of multi-dimensional indicators. The experimental results show that the intelligent maintenance cost prediction of power grid based on multi-dimensional mixed information can improve the prediction accuracy and contribute to the lean management of power enterprises. Further improve the efficiency and benefit of the multi-dimensional index linkage budget method, promote the digital transformation of power grid enterprises, and provide reference for power supply enterprises.

Keywords: Multidimensional mixed information · Power grid maintenance · Intelligent maintenance · Grid information · Maintenance cost · Cost forecast

1 Introduction

Equipment maintenance is a key link in the planning and operation of power system. On the one hand, it is an important part of equipment asset management. On the other hand, under the planned power capacity and network topology, it largely determines the medium and long-term operation mode of power system. Therefore, equipment maintenance decision-making needs to comprehensively consider the above two aspects and carry out scientifically and carefully. The purpose of cost management of power grid

W. Fu and L. Yun (Eds.): ADHIP 2022, LNICST 468, pp. 313–326, 2023.
https://doi.org/10.1007/978-3-031-28787-9_24

enterprises is generally limited to how to reduce the cost, but not from the perspective of how much benefit the cost can bring to the enterprise. Therefore, the purpose of cost management should bring more benefits to the cost. Power grid enterprises use traditional means to manage costs, while advanced theories such as life cycle theory are not applied and popularized in power grid enterprises in time. These limitations greatly restrict the development and competitiveness of power grid enterprises [1]. The deepening power system reform has put more and more pressure on the cost management of power grid companies, which requires power grid enterprises to face up to the problems in cost control, carry out multi-dimensional lean analysis and inspection on the intelligent maintenance cost of power grid, and specifically analyze the rationality and irrationality of each maintenance cost, which can effectively improve the ability of cost control, meet the requirements of government supervision, improve quality and efficiency. This can help power supply companies solve the problems existing in the reform of multidimensional lean management system. The concept of cost management is relatively lacking in the management of power grid infrastructure projects. At present, the scope of cost management of power grid infrastructure projects is limited to the construction period, while the management of operation cost, maintenance cost and overhaul cost after completion is greatly ignored. The cost of power grid infrastructure projects is characterized by one-time investment (construction cost) accounting for only about 40% of the life cycle cost of power grid infrastructure projects, so the scope of cost management is limited [2, 3]. In today's environment of energy conservation and emission reduction, with the continuous progress of measurement, communication, computer and other technologies and the gradual deepening of the concept of building a smart grid, the equipment condition monitoring and evaluation technology for the purpose of real-time perception of equipment status has gradually developed and matured. At present, with the continuous construction of data acquisition and professional management system in power grid enterprises, the amount of structured and unstructured data related to assets, equipment, personnel, investment, cost, materials, electricity and business activities is increasing exponentially. At the same time, the maturity and application of big data value mining, artificial neural network, semantic recognition and other technologies have provided opportunities and laid a foundation for more multi-dimensional and lean management of maintenance cost prediction of power grid enterprises. Equipment maintenance has rapidly changed from regular maintenance to condition based maintenance, and has been gradually popularized and applied. Power system condition based maintenance decision-making can make full use of the actual equipment condition information from the system decision-making level, and fully tap the potential of equipment condition monitoring and evaluation technology in the decision-making application level. It has important theoretical and practical significance for improving the equipment asset management and operation reliability level of power system. Based on the above background, it is necessary and conditional to use big data technology to connect the industry and finance management link, study and innovate the methods and mechanisms of power grid enterprises in the core budget management links such as target calculation, implementation control, evaluation and incentive, so as to awaken data resources, optimize resource allocation and meet the demands of diversified management, And continuously improve the enterprise's value creation ability and cost lean management level. Reference [4] proposes a template update mechanism to improve the accuracy of visual tracking. First, when the background clutter is detected, the original template is saved. In the background clutter, we use the original template and the current template at the position of optical flow estimation, and select a better template. Then, the original template is reused after the background clutter ends. Finally, the mechanism is applied to KCF and BACF algorithms to verify the effectiveness of the mechanism. Reference [5] proposes a multi-layer template update mechanism to achieve effective monitoring in multimedia environments. In this strategy, the weighted template of the high confidence matching memory is used as the confidence memory, and the unweighted template of the low confidence matching memory is used as the cognitive memory. By alternately using confidence memory, matching memory and cognitive memory, it is ensured that the target will not be lost in the monitoring process. However, the above two literature methods do not consider the change of equipment status in the prediction of power grid maintenance costs, resulting in inaccurate prediction results. Based on multi-dimensional mixed information, this paper proposes an intelligent maintenance cost prediction method of power grid to realize the accurate accounting and control of maintenance information data, so as to promote the cost management level of power companies and provide guarantee for the stability of power system.

2 Intelligent Maintenance Cost Prediction Method of Power Grid Based on Multi-dimensional Mixed Information

2.1 Analysis on Cost Composition of Intelligent Maintenance of Power Grid

The maintenance of power grid infrastructure projects after completion and operation can be divided into two categories: one is routine maintenance, that is, planned periodic equipment maintenance. Such maintenance will not affect the use of power side and cost loss caused by power failure; One is the unconventional maintenance due to the loss of power failure on the power side caused by planned power failure and sudden failure. Routine maintenance cost refers to the cost of routine maintenance of various equipment in the project in order to ensure the stable and safe operation of the completed infrastructure project after the completion and operation of the power grid infrastructure project. The composition of routine maintenance cost is shown in Fig. 1.

The intelligent maintenance cost of power grid has changed from a single accounting subject carrying financial information to a multi-dimensional cross carrying financial information. The dimensions are obtained through the information link carrier, which transmits business and financial information. The accounting is more flexible. It can not only transmit business information in time, but also expand the management dimension according to its own management needs. Due to the large variety of equipment in power grid infrastructure projects and great differences in nature and characteristics, in order to simplify the prediction difficulty, the maintenance fee is regarded as the annual cost, and the ratio of the operation cost of the maintained equipment to its investment is used as the maintenance rate. The calculation formula of single maintenance cost of equipment is as follows:

$$W_1 = \sum_{a=1}^{s} w_a \frac{b(1+b)^{m_a}}{(1+b)^{m_a} - 1} c \tag{1}$$

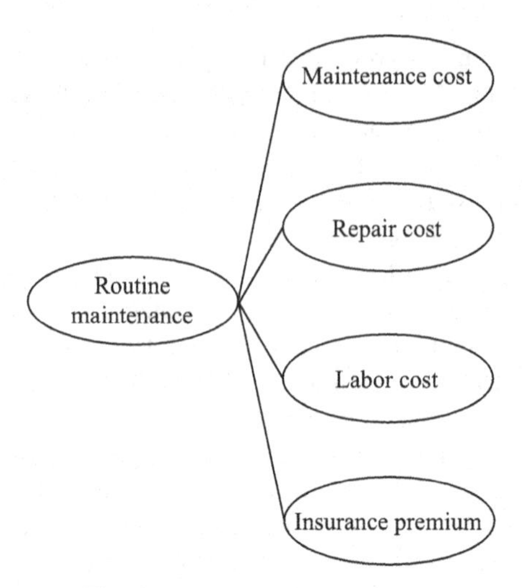

Fig. 1. Routine maintenance cost

In formula (1), W_1 represents the total annual maintenance cost of the project; s represents the number of project equipment; a represents the serial number of the equipment; w_a represents the investment amount of the a equipment; b represents the rate of return on investment; m_a represents the maintenance life of the a equipment; C represents maintenance rate. Through cost collection and allocation, accurate distinction can be realized at the account level, and production costs and management expenses can be calculated. In order to standardize cost accounting, strengthen cost management, meet the needs of service transmission and distribution price and cost supervision, reflect the cost accounting elements in multiple dimensions according to business activities, voltage level, asset type and user category, clearly show the internal value flow of the enterprise, and effectively improve the cost management level of the company. When the power grid infrastructure project encounters an emergency and has a power outage, the power supply loss caused during the power outage should also be regarded as a kind of maintenance cost. For the cost estimation of power supply loss, this paper studies from the aspects of outage time, repair cost and average failure rate. The specific calculation formula is as follows:

$$W_2 = \sum_{a=1}^{s} \chi p_a \tau_a + q_a h_a x_a \tag{2}$$

In formula (2), W_2 represents the cost of power supply loss; χ represents the value coefficient of user's power consumption at the power consumption side; p_a represents the interruption power of the equipment; τ_a represents the fault interruption time; q_a represents the average repair cost; h_a is the annual average failure rate; x_a represents the average repair time. When calculating the single maintenance cost, we can see the maintenance expenditure on each business activity, each voltage level and each asset. According to the corresponding work order, we can analyze the rationality of the maintenance cost and the investment value of each project.

2.2 Formulate Economic Power Grid Maintenance Strategy

Power grid equipment life cycle cost management is based on reliability and life cycle management, which divides the power grid equipment cycle into procurement, installation, operation, maintenance and return. On the basis of ensuring reliability, the goal of this method is to maximize the benefit of power grid on the basis of minimizing the cost of equipment maintenance. In the daily patrol operation of power grid equipment, the operation status of substation equipment is monitored to determine the work content and time of equipment maintenance. This stage is also an important means to reduce equipment failure and improve equipment reliability. With the increase of operation time, the aging degree of transformer becomes more and more serious, and the operation uncertainty also increases. Therefore, during each maintenance, the maintenance cost will increase with the increase of operation time. Therefore, the maintenance cost is:

$$A = (W_1 + W_2)(1 + 0.00625R) \tag{3}$$

In formula (3), A represents the maintenance cost; R indicates the service age of the equipment. The recovery effect of power grid intelligent maintenance always depends on the type of equipment maintenance. According to engineering experience, when the failure rate is small, the equipment is in good condition, and the recovery effect of intelligent maintenance on the failure rate is small. Through a series of latest power technologies, through the collection and analysis of key indicators such as equipment current, voltage and temperature, we can find equipment defects in advance and reduce equipment power failure caused by post-processing methods such as fault power failure. With the aging of equipment insulation, the failure rate becomes larger than before, and its health condition becomes worse. The recovery effect of intelligent maintenance on the failure rate is greater than that in the case of small failure rate. At present, power enterprises have widely used live monitoring and on-line monitoring to replace routine tests to judge the operation state of equipment. The purpose of power grid intelligent maintenance is to ensure that the failure rate of equipment is within an acceptable range and obtain the maximum power grid income under the minimum maintenance cost. The goal of life cycle management in the maintenance stage is to improve the health level of equipment, prolong the equipment cycle and reduce the use cost. The first task of the intelligent maintenance selection model of power grid is to analyze and determine the benefit function and cost function of power grid equipment. The intelligent maintenance process of power grid can be regarded as the model solution of the minimum objective function optimization problem. If the maximum maintenance benefit is obtained, the best maintenance type and the best maintenance execution time can be determined. The traditional maintenance mode is planned maintenance. According to different models of equipment and previous maintenance experience, judge its operating conditions and carry out minor repair or overhaul. There are many types of equipment in the power grid, and the service life of most of them is not uniform. For example, the service life of electrical primary equipment is often long, while the integrated automation system and other secondary equipment are updated quickly and have a short service life due to the characteristics of more electronic components. Under the background of the increasing progress of condition monitoring technology, due to the lack or excessive maintenance of traditional regular maintenance and fault maintenance, the equipment maintenance and repair of power enterprises began to change to condition based maintenance. This paper uses CS algorithm to determine the maintenance strategy of power grid equipment, and its process is shown in Fig. 2.

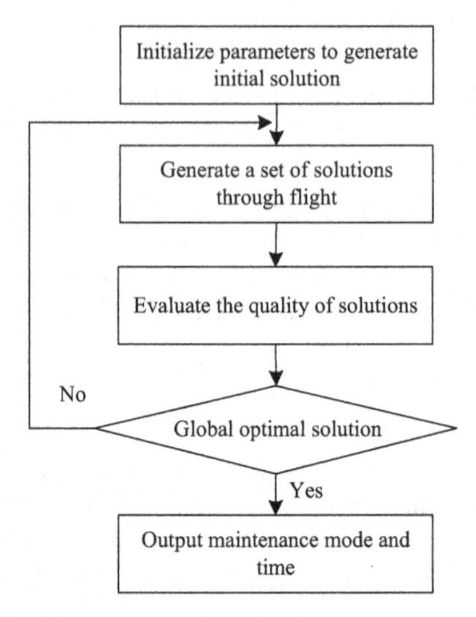

Fig. 2. Decision process of overhaul and maintenance strategy

According to the relationship between the health state, operation life and failure rate of power grid equipment, the influence of different maintenance states on failure rate is determined. According to the economic model, the unknown parameters are determined and the maintenance decision-making model is established. Initialize the parameters of CS algorithm, and set the population size, switching probability and step size. The calculation formula of step length is as follows:

$$D = \frac{\alpha_1}{|\alpha_2|^{\frac{1}{\delta}}} \tag{4}$$

In formula (4), D represents the algorithm step size; α_1 and α_2 represent the parameters of flight search and random walk; δ represents the constant value of controlling random walk, which is taken as 1 in this paper. Arrange the best solutions to find the current optimal solution. Until the number of iterations is reached or the quality accuracy of the solution reaches 10–13, the optimal solution is output [4]. That is, the optimal maintenance mode and time are obtained. Repair sudden equipment faults and equipment defects found in operation, and transform various equipment in the substation at any time according to the latest electrical technology, so as to reduce equipment operation cost and improve safety and reliability [5].

2.3 Determination of Maintenance Status of Power Grid Equipment Based on Multi-dimensional Mixed Information

The maintenance plan in the current condition based maintenance decision model of power system is fixed, and does not consider how to deal with the changes of equipment status in the research cycle. For example, the equipment may fail before the maintenance plan, which will lead to the failure of the maintenance plan. In fact, under the background of on-line monitoring of equipment status, the change of equipment status is visible in the whole process. The data table Association view is represented as a visual data table association model on the front-end page. During the operation of the power grid system, the background of the power grid system will automatically record and precipitate the association information between various data tables and form an association view for automatic modeling [6]. At the same time, it can also provide reference for users when they conduct data analysis or manual modeling. It is not necessary to adjust the maintenance plan after the failure of the equipment. It is absolutely possible to adjust the maintenance plan ahead of time when the deterioration degree of the equipment is exceeded, so as to prevent failure. The multidimensional mixed information extracted in the daily operation of smart grid includes four dimensions: time, branch Item and location. Each dimension table is connected with the fact table to form a star like model. The fields used to describe quantitative analysis indicators in the fact table are called metrics. These fields are usually numerical and are used to perform the calculation of various analysis operators. The dimension table does not contain measures. Its fields are called dimensions. They are usually text, time, classification, geographical location and other types of data, and are the angle of observing data [7]. Therefore, this paper uses multi-dimensional mixed information to determine the maintenance status of power grid equipment. We divide the uncertainties observed in each period into two categories: the first is the combination of equipment states observed at the beginning of the period; The second category is the combination of equipment states observed during the period. In multi-stage decision-making, the gradual realization of uncertainty in multiple periods forms a scene tree, and each node in the scene tree needs to make corresponding decisions. After processing and analyzing the data, the state result of a characteristic quantity of power grid equipment can be obtained, forming a single characteristic quantity evaluation [8]. However, because only one characteristic quantity is used to evaluate the overall state of the transformer, there are some problems, such as the evaluation result is not accurate enough and the conclusion is too one-sided. For the above two types of random data, we need to make two types of maintenance status decisions accordingly. The first type of maintenance status is sequential decision-making, which is determined by the maintenance plan in the decision-making period. This process can be expressed as:

$$z_1 = g(v_t) \tag{5}$$

In formula (5), z_1 represents the maintenance state; g represents the combination of equipment status; v_t indicates the maintenance status of the equipment in period t, 0 indicates no maintenance, and 1 indicates maintenance. The second type is operation scheduling decision, which belongs to non sequential decision. It only appears in a node of the scene tree and will not affect the subsequent decision. This process can be expressed as:

$$z_2 = (\mu, \varphi, \gamma, \eta) \tag{6}$$

In formula (6), z_2 represents the operation scheduling decision; μ represents the output of power grid equipment unit; φ represents the power flow of the branch; γ

represents the voltage phase angle of the equipment node; η represents the load shedding amount at the load point. Taking the power transformer as an example, the health status of the transformer largely depends on the overall health status of its insulation system. In the insulation system of oil immersed transformer composed of paper and mineral oil, deterioration occurs in a complex multi factor process due to the interdependence between aging factors and their additive effect. The cumulative effect of temperature, moisture, acid and oxidation inside the transformer is the main cause of transformer aging. All state transition rates can be estimated by the average transition time between two states, and the state transition rate is constant. That is, the continuous availability on the time axis is discretized into the availability of the period, and its value is equal to the availability of the start time of the period. Assuming that the equipment does not reach the maintenance threshold state before the maintenance threshold time, but fails before the scheduled maintenance plan, in this case, it is necessary to perform post failure maintenance at the scheduled time and cancel the scheduled maintenance plan. Due to the large number of operation states of power grid equipment, the state level can be evaluated by fusing multi-dimensional mixed information, and finally a comprehensive state can be obtained.

2.4 Establish the Prediction Model of Power Grid Intelligent Maintenance Cost

The cost prediction of intelligent maintenance of power grid needs to consider the cost of operation and maintenance period, maintenance period and scrapping period. Some of the costs are determined, some fluctuate with the changes of market and other factors, and most of the factors are uncertain. Human subjective factors are inevitably added in the estimation, which greatly reduces the objectivity of the whole project life cycle cost estimation. The diversity of demands of power grid intelligent maintenance management determines the necessity of multi-dimensional linkage prediction of cost indicators. In the actual operation process, it is necessary to take the historical budget data, final settlement data, price verification data, business activity data, etc. as the basis, and take the business assumption parameters such as electricity, electricity price and investment as the input content. There is a certain correlation between costs and between costs and power generation. Among them, the initial investment cost has a great impact on the cost and power generation in the subsequent stages. Therefore, when calculating the kWh cost, we must consider various parameters of the cost in each stage, and substitute the weight value into the intelligent maintenance cost prediction model of the power grid. This paper uses cloud model to deal with the uncertainty of power grid project cost. A concept within the scope of the universe is a mapping from the universe to the membership [0,1], and it is a one to many mapping, that is, a value in the universe represents the concept, and the membership of accuracy is a one to many relationship, rather than a single membership relationship of other methods [9]. The cloud model can express the concept with randomness, and give the membership degree of each specific implementation to the concept. It reflects the uncertainty of the concept by reaching the shape composed of a certain number of cloud droplets. The total contribution of all elements in the universe to the concept can be expressed as:

$$\vartheta = \frac{\int \sigma(x)dx}{\sqrt{2\pi}\varpi} \tag{7}$$

In formula (7), ϑ represents the contribution degree; ϖ represents the entropy of power information characteristics; σ represents qualitative probability; x represents the information element between any cell. Thus, the randomness of membership can be determined. Through the use of grey prediction, multiple linear regression prediction and other methods, the diversified prediction results are output, and then the weighted value of the prediction output results is assigned with the help of the combined prediction model to calculate the budget scale of multi-dimensional indicators. The statistical data of failure rate has great volatility, so that the data sequence is not a strict exponential form, and because the parameter a in the traditional GM (1,1) model is constant, it will produce a lot of errors. In other words, GM (1,1) model is a linear time invariant system. In this paper, the linear time term is considered to replace the traditional constant parameters, and then the GM (1,1) model with time-varying parameters is constructed. The grey prediction model first assembles the original data into the original sequence, and then uses the accumulation generation method to generate the sequence that can be operated. After revising the residual, the difference differential equation is established for the transformed sequence. Then, based on the analysis of correlation degree and convergence degree, constantly supplement new information, clean up meaningless old information, and solve repeatedly until the purpose of qualitative analysis of system characteristics or quantitative analysis according to correlation factors is achieved. Parameter assumptions in the field of power grid engineering mainly include two items: one is the assumption of project design life, and the other is the assumption of project actual service life; The parameter assumptions in the economic field mainly include discount rate, project residual value, inflation, operation period, etc. Whether the project cost composition analysis is in-depth and whether the kWh cost estimation model is reasonable and accurate will directly determine whether the calculation results are correct, and will also affect the effect of kWh cost management in the intelligent detection cycle of power grid [10–12]. The weight of different combinations of various output results is mainly determined according to the variance covariance method. The prediction combination weight formula can be expressed as:

$$\varepsilon = \frac{\psi_2 - \phi}{\psi_1 + \psi_2 - 2\phi} \tag{8}$$

In formula (8), ε represents the combination weight; ψ represents variance; ϕ represents covariance. Output diversified prediction results, and then use the weighted assignment of the prediction output results with the help of the combined prediction model to calculate the budget scale of multi-dimensional indicators. According to the above process, the design of power grid intelligent maintenance cost prediction method based on multi-dimensional mixed information is completed.

3 Experimental Study

3.1 Experimental Preparation

In order to test the effectiveness of the power grid intelligent maintenance cost prediction method based on multi-dimensional mixed information proposed in this paper, four examples are used to verify it. Select the 2015–2020 power grid maintenance appropriation data as the data source, use matlab simulation to first optimize the initial weight by using CS algorithm, set the population size to 50, the number of iterations to 500, the crossover probability to 0.2, and the mutation probability to 0.01, and then train. The network learning rate is set to 0.05, and the training error is selected as 1×10^{-5}, the maximum number of training is 10^5.The specific settings of each example are shown in Table 1.

Table 1. Example description

Numerical example	System	Constitute	Maintenance period	Safe operation constraints
1	Single bus system	2 units	5	N-0
2	Two node system	3 units and 3 lines	5	N-1
3	Two node system	5 units and 3 lines	10	N-2
4	Rts79 system	30 units, 5 transformers and 50 lines	25	N-3

In these examples, it is assumed that each device contains two elements in series, and each element contains four states: normal, mild deterioration, severe deterioration and failure. The unbalanced power penalty factor is set to 1000 and the convergence gap is set to 10–6. The maintenance cost is calculated by 2% of the purchase cost of transformer every year, and the environmental maintenance cost is calculated by 0.25% of the purchase cost. The service life of the unit is 30 years. When the transformer is scrapped, the residual value is calculated by 10% of the purchase cost and the discount rate is calculated by 8%. According to the selection requirements of the number of transformers, firstly, it can meet the long-term planning load. Secondly, when one main transformer is shut down, the other transformers can meet 70% of the load.

3.2 Results and Analysis

In the above four examples, the accuracy of the power grid intelligent maintenance cost prediction method based on multi-dimensional mixed information is calculated, and the results are compared with the maintenance cost prediction methods based on BP neural network and GA-SVM, so as to verify the superiority of the design method in this paper. The maintenance cost prediction results of each example are shown in Tables 1, 2, 3, and 4 respectively.

In the single bus system of example 1, the accuracy of the power grid intelligent maintenance cost prediction method based on multi-dimensional mixed information is 95.64%, which is 6.11% and 7.56% higher than the maintenance cost prediction

Table 2. Comparison of accuracy rate of example 1 (%)

Number of tests	Intelligent maintenance cost prediction method of power grid based on multi-dimensional mixed information	Prediction method of power grid intelligent maintenance cost based on BP neural network	Prediction method of power grid intelligent maintenance cost based on GA-SVM
1	95.47	88.09	88.24
2	96.58	89.17	87.67
3	94.29	89.48	89.55
4	95.06	90.86	87.86
5	96.32	89.65	86.43
6	96.63	88.23	87.52
7	95.95	90.32	88.20
8	96.28	91.65	87.39
9	94.59	89.54	87.66
10	95.22	88.28	90.23

method based on BP neural network and GA-SVM. Compared with the two traditional maintenance cost prediction methods, the intelligent maintenance cost prediction method based on multi-dimensional mixed information has higher accuracy.

Table 3. Comparison of accuracy rate of example 2 (%)

Number of tests	Intelligent maintenance cost prediction method of power grid based on multi-dimensional mixed information	Prediction method of power grid intelligent maintenance cost based on BP neural network	Prediction method of power grid intelligent maintenance cost based on GA-SVM
1	92.29	86.04	85.49
2	91.56	85.68	84.86
3	92.85	86.21	83.68
4	91.62	87.55	84.25
5	93.23	86.36	83.54
6	92.56	85.22	82.31
7	93.34	86.65	84.25
8	94.65	86.27	82.12
9	92.27	85.51	83.86
10	93.54	86.92	82.43

In the two node system of example 2, the accuracy of the power grid intelligent maintenance cost prediction method based on multi-dimensional mixed information is 92.79%, which is 6.55% and 9.11% higher than the maintenance cost prediction method based on BP neural network and GA-SVM. This is because the method in this paper outputs a variety of prediction results through the methods of grey prediction and multiple linear regression prediction, and then uses the combined prediction model to allocate the weight of the prediction results to achieve the cost prediction of multi-dimensional indicators.

Table 4. Comparison of accuracy rate of example 3 (%)

Number of tests	Intelligent maintenance cost prediction method of power grid based on multi-dimensional mixed information	Prediction method of power grid intelligent maintenance cost based on BP neural network	Prediction method of power grid intelligent maintenance cost based on GA-SVM
1	90.46	82.47	80.94
2	91.87	81.18	82.58
3	92.54	83.09	83.69
4	90.21	82.66	82.26
5	91.62	83.25	81.05
6	92.28	84.54	82.12
7	90.33	82.91	83.53
8	90.56	83.32	82.26
9	90.20	82.20	82.39
10	92.92	84.63	81.22

In the two node system of example 3, the accuracy of the power grid intelligent maintenance cost prediction method based on multi-dimensional mixed information is 91.30%, which is 8.27% and 9.10% higher than the maintenance cost prediction method based on BP neural network and GA-SVM. This is because the power grid intelligent maintenance cost prediction method based on multi-dimensional mixed information uses CS algorithm to determine the maintenance strategy of power grid equipment, so as to obtain the maximum power grid income under the lowest maintenance cost. The multi-dimensional mixed information extracted from the daily operation of smart grid determines the maintenance status of equipment in the maintenance strategy (Table 5).

In the rts79 system of example 4, the accuracy of the power grid intelligent maintenance cost prediction method based on multi-dimensional mixed information is 88.02%, which is 8.62% and 10.32% higher than the maintenance cost prediction method based on BP neural network and GA-SVM. Therefore, the power grid intelligent maintenance cost prediction method proposed in this paper can identify the equipment maintenance status by making full use of the multi-dimensional mixed information of power grid

Table 5. Comparison of accuracy rate of example 4 (%)

Number of tests	Intelligent maintenance cost prediction method of power grid based on multi-dimensional mixed information	Prediction method of power grid intelligent maintenance cost based on BP neural network	Prediction method of power grid intelligent maintenance cost based on GA-SVM
1	87.67	79.74	79.40
2	88.24	80.66	78.16
3	89.58	79.35	77.85
4	88.46	78.22	76.62
5	89.13	78.03	77.33
6	87.82	77.19	76.52
7	88.29	79.55	78.2
8	86.65	80.28	78.91
9	87.36	80.36	76.74
10	87.01	80.62	77.28

operation and maintenance according to the different characteristics of costs in different stages, so as to accurately predict the maintenance cost, which can provide reference for power supply network enterprises.

4 Conclusion

The cost prediction of intelligent detection of power grid is studied, and the optimization of maintenance strategy is effectively supported by multi-dimensional mixed information. The CS algorithm is used to determine the maintenance strategy of power grid equipment, so as to obtain the maximum power grid profit under the lowest maintenance cost. The multi-dimensional mixed information extracted from the daily operation of smart grid determines the maintenance status of equipment in the maintenance strategy. Through the methods of grey prediction and multiple linear regression prediction, a variety of prediction results are output, and then the weight of the prediction results is allocated by using the combined prediction model to achieve the cost prediction of multi-dimensional indicators. In the case of limited human, material and financial resources, priority maintenance projects and resources can be selected according to the urgency of maintenance and the value of line or substation equipment. Using multi-dimensional mixed information to forecast maintenance costs is conducive to the fine management of operation and maintenance costs. In combination with the mature "big cloud intelligent transfer" technology, timely carry out informatization and intelligent application research, further improve the efficiency and efficiency of multi-dimensional indicator linkage budget method, promote the digital transformation of power grid enterprises, and create a world-class enterprise. In the future development, we will introduce advanced information technology and science and technology, integrate various technologies, predict the cost of intelligent detection of power grid, and improve the prediction accuracy.

References

1. Wei, Q., Yang, T., Gao, Y.: Simulation research on multi-energy complementary dispatch based on hybrid power flow algorithm. Comput. Simulat. **38**(11), 77–81 (2021)
2. Meng, F.: Research on overhaul method of relay protection in intelligent substation. Telecom. Power Technol. **37**(12), 109–111 (2020)
3. Wu, S.: Intelligent budget management and forecast optimization analysis of power grid enterprises under the background of big data. Value Eng. **40**(36), 163–165 (2021)
4. Li, H., Yang, Y., Ran, Q.: Progress-cost forecasting of power grid projects based on GA-SVM quality earned value. Mech. Electric. Eng. Technol. **49**(10), 93–95 (2020)
5. Wang, M., Liu, Y., Gao, H., et al.: A two-stage stochastic model predictive control strategy for active distribution network considering operation cost risk. Adv. Power Syst. Hydroelectr. Eng. **36**(11), 8–18 (2020)
6. Pan, J., Chen, Q., Jin, S.: A mathematical method for operation and maintenance cost prediction based on transfer learning under non-stationary power data. Adv. Appl. Math. **10**(1), 98–108 (2021)
7. Xiong, Y., Zhan, Z., Ke, F., et al.: Overhaul operation and maintenance cost prediction of substation based on improved BP neural network. J. Electric Power Sci. Technol. **36**(4), 44–52 (2021)
8. Cao, M., Xu, A., Jiang, Y., et al.: The application of Elman neural network in uninterrupted maintenance of power grid. J. Phys: Conf. Ser. **1673**, 012057 (2020)
9. Xiao, H., Cao, M.: Balancing the demand and supply of a power grid system via reliability modeling and maintenance optimization. Energy **210** (2020)
10. Xu, X., Peng, L., Ji, Z., et al.: Research on substation project cost prediction based on sparrow search algorithm optimized BP neural network. Sustainability **13** (2021)
11. Wu, J., Liu, H., Yang, J., et al.: Tree barrier prediction of power lines based on tree height growth model. IOP Conf. Ser. Earth Environ. Sci. **645**(1), 012008 (2021)
12. Chen, T., Jiang, Y., Jian, W., et al.: Maintenance personnel detection and analysis using mask-RCNN optimization on power grid monitoring video. Neural Process. Lett. **51**(3), 1599–1610 (2020)

Infrared Image Face Recognition Method Based on Signal Interference Technology

Zongren Chen(✉) and Wenda Xie

Computer Engineering Technical College (Artificial Intelligence College), Guangdong Polytechnic of Science and Technology, Zhuhai 519090, China
chenzongren210@sina.com

Abstract. Infrared imaging has the advantages of strong anti-interference, independent of visible light source, anti camouflage and anti fraud. Therefore, an infrared image face recognition method based on signal interference technology is proposed. Based on the in-depth analysis of the characteristics of infrared face image, the characteristics of infrared image face recognition are studied, and a new infrared image statistical face recognition method is proposed. The experimental results show that the recognition method studied in this paper is feasible both theoretically and experimentally, and has good recognition ability. The infrared image face recognition method can solve the difficult problem of face recognition under the change of lighting conditions and face camouflage.

Keywords: Signal interference technology · Infrared image · Face recognition · Anti interference · Visible light source

1 Introduction

Face recognition technology has received more and more attention in the field of computer vision and pattern recognition, and has gradually become a hot topic. At present, most face images are taken under visible light conditions, but in visible light environment, the illumination conditions are changeable and complex, and the performance of face recognition is affected by the environmental illumination changes. Therefore, overcoming the influence of illumination changes has become an important problem in the field of face recognition [1]. Due to the robustness of NIR imaging to illumination changes, NIR imaging technology solves this problem to some extent. In the application of near infrared face recognition, the face images required for registration and detection are taken under the condition of near infrared illumination, while in practical application, a large number of face images are taken under the condition of visible light, such as ID card photos. So it becomes a problem to realize the cross registration and verification of visible light face image and near-infrared face image. Because of the different imaging methods, there are many apparent differences between visible light image and near-infrared image of the same person [2]. However, from the perspective of human cognition, they should be recognized as the same person, which means that there is some

W. Fu and L. Yun (Eds.): ADHIP 2022, LNICST 468, pp. 327–338, 2023.
https://doi.org/10.1007/978-3-031-28787-9_25

correlation between the visible image and the near-infrared image of the same person. This paper will introduce the improvement of the performance of the visible light and near-infrared face recognition algorithm from two aspects. In recent years, the infrared image face recognition is a hotspot of biometrics research, which has a wide range of application value and challenge. Theoretically speaking, infrared image face recognition can solve the difficult problem of face recognition in the case of changing illumination conditions and face camouflage. But at the same time, it should also be seen that the current research on infrared image face recognition is not mature enough, and the mechanism of many factors affecting the performance of infrared image face recognition is not very clear. In addition, infrared image face recognition also has its disadvantages [3]. Therefore, when carrying out the research work of infrared image face recognition, it is necessary to deeply analyze the characteristics of infrared face image and the related factors that affect the performance of infrared image face recognition, so as to explore and study the effective ways and methods to improve the performance of infrared image face recognition.

2 Infrared Image Face Recognition Based on Signal Interference Technology

2.1 Face Feature Extraction from Infrared Image

Face detection is to find out all the regions containing faces in a given image. Visible face detection methods mainly use local feature detection methods, such as detecting eye position. However, the effect of this kind of method in infrared face detection is not very good, because the infrared face image is fuzzy compared with the visible image [4]. Because human skin has high emissivity, and the emissivity is not affected by race and skin color, face detection based on global features is often used in infrared face recognition. The automatic face recognition system includes two main technical links, as shown in Fig. 1: one is face detection and location, that is, find the face from the input image and segment the face from the background; the other is feature extraction and recognition of the planned image.

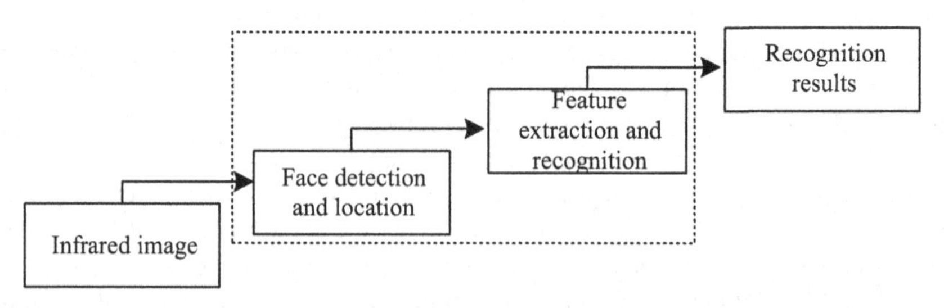

Fig. 1. Feature extraction steps of the planned image

The geometric preprocessing of infrared face image mainly refers to whether the relative positions of the key parts of the face in each infrared face image are the same.

For the original infrared face image without any processing, the position of the face part in the image is offset. When the face recognition method based on overall gray statistics is adopted, it will affect the correct recognition of the face [5]. Therefore, the input infrared face images should be corrected so that the infrared face images in different cases are unified to the same pixel size, and the key parts of the face should be consistent as far as possible. For the extraction of face features, a pyramid recognition method based on signal interference is proposed. Using this method combined with signal interference sub pattern, face can be recognized efficiently [6]. The specific process of extracting signal interference sub mode features is shown in Fig. 2.

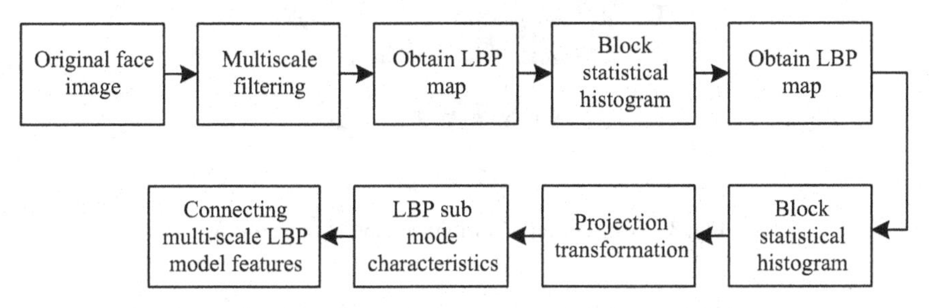

Fig. 2. Flow chart of infrared image feature extraction

The specific extraction steps of sub pattern features can be obtained from Fig. 2: select a filter with strong de drying effect to carefully filter and decompose the face image to remove redundant impurities, and build a low-dimensional image with multi-scale space on this basis; The signal interference operator is used to calculate the dimension of different scale images, so as to obtain the characteristic spectrum data; By dividing different characteristic spectra, the module area with non repetitive attribute can be obtained and the histogram data inside the area can be counted; The operator histograms of different regions are mapped by projection, and the sub module features of different regions can be obtained after transformation; By splicing the sub module features of all regions, a multi-scale sub module sequence can be obtained, which can be used as a feature vector for face recognition. When constructing face recognition image and signal interference map, it is necessary to select the image division level and decomposition area size according to the actual situation of different people [7].

2.2 Infrared Image Face Feature Recognition Algorithm

Different faces have different facial features, but their spatial macro structure is consistent, that is, eyebrows, eyes, nose, mouth and so on have similar distribution. From the local view of human face, the structure of each local feature is also similar. For this rule of face structure, Wiskott 2 A special shape label graph is used to represent the face. The nodes of this label graph are located in the meaningful positions on the face image for recognition. These positions are called the feature points of the face. In the vertical direction of the face, the distance from the hairline to the eyebrow arch, the eyebrow arch to the nose and the nose to the chin is generally the same, while in the horizontal direction of the face, the distance from the outer corners of both eyes to the ears is the same. The distribution map of various organs of the face is shown in Fig. 3.

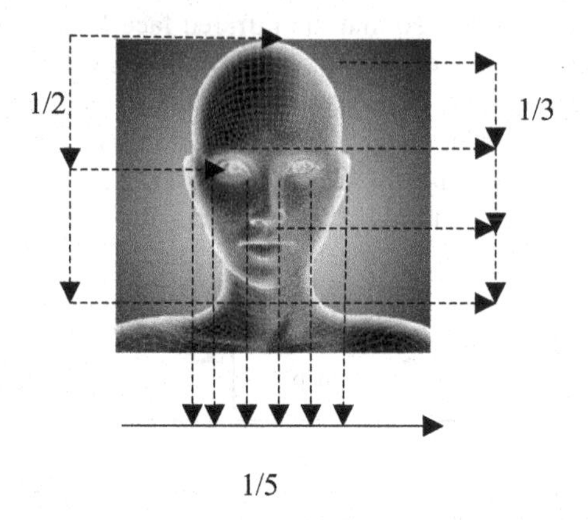

Fig. 3. Distribution of various organs of human face

It can be seen from Fig. 3 that the width of eyebrows is twice the width of both eyes, the nose is located from the center of eyebrows to 1/2 of the bottom of chin, the length from nose root to chin is half of the length of face, the septum of nose is connected with people, the bottom of nose is flush with ears, and the height of ears can be determined by the highest point of eyebrows [8]. The length of the tissue around the mouth is 2/3 of that from the nose to the chin. The distance between the two eyes is consistent with the width of the mouth. The two horizontal lines printed on the upper part of the eyebrow and the bottom of the nose run through the whole ear. According to the response characteristics of two-dimensional Gabor signal interference, it can be seen that the transformation coefficient reflects the change of image gray level, The parts with obvious gray change on the image are easier to locate the feature points than those with gentle gray change. Therefore, the selection of feature points should first consider these parts, such as eyebrows, eyes, nose, mouth and face contour, but ensure that all feature points can cover the face area of the image relatively evenly. By matching with the face moving picture, we can get the label graph representation GM of the training image and the label graph representation g of the image to be recognized, calculate the average value of the similarity between the label graph G of the image to be recognized and the label graph GM of all training images, and use the nearest neighbor rule to maximize the similarity. The training image is the recognized face image. The similarity between label graph G and label graph GM is defined as:

$$S_G\left(G', G^M\right) = \frac{1}{N}\sum_n S_a\left(J_n^l, J_n^M\right) \tag{1}$$

The coarser the position of the feature set is, the more similar the position of the feature set corresponding to the actual sample is to the position of the feature set [9], and the position of the feature set corresponding to the image is formed. The grid structure composed of all edges of the label graph describes the geometric characteristics of the face. For different face images, the shape and size of the grid structure change accordingly. Through signal interference matching, the nodes of the label graph of all face images are unified to the selected corresponding feature points. Therefore, signal interference matching has become a method of locating face feature points. In terms of image processing and analysis theory, any image can be regarded as an element of high-dimensional space. Using statistical model to describe and analyze the target image has become a research hotspot in recent years [10]. By manually locating the eye coordinates, you can judge whether the two eyes in the infrared face image are on the same horizontal line. If they are not on the same horizontal line, rotate and transform the infrared face image to make the two eyes transform to the same horizontal line, so that the infrared face image can be zoomed and cropped next. The rotation transformation formula is as follows:

$$[x, y, 1] = [u, v, 1]\begin{bmatrix} \cos\theta & \sin\theta & 0 \\ -\sin\theta & \cos\theta & 0 \\ 0 & 0 & 1 \end{bmatrix} \tag{2}$$

Formula (2) rotates the point on the u, v plane by degrees counterclockwise relative to the coordinate origin. Among them, angle 8 is the included angle between the connecting line of two eyes and the horizontal line, u, v represents the pixel coordinates in the image before rotation, and x, y represents the pixel coordinate transformation in the image after rotation, which can transform the inclined infrared face image into the face image with two eyes on the same horizontal line, and then carry out the geometric transformation in the following steps. Although the two eyes of the rotated infrared face image r are on the same horizontal line, the distance between the two eyes is different. Therefore, all infrared face images need to be transformed into images with the same size through zoom transformation. The zoom in and zoom out transformation first calculates the zoom in and zoom out ratio in the length and width of the original image and the standard image, and sets it as a, b respectively. When r > 1 (i = x or y), the image in this direction is enlarged; When r < 1, the image is reduced. The scaling transformation formula can be expressed as:

$$[a, b, 1] = [x, y, 1][u, v, 1]\begin{bmatrix} r_x & 0 & 0 \\ 0 & r_y & 0 \\ 0 & 0 & 1 \end{bmatrix} \tag{3}$$

The algorithm uses r for zoom transformation. In the actual infrared face image zoom transformation, the scaling ratio in the horizontal and vertical directions is the same, which is the ratio of the distance between the two eyes of the standard infrared face image and the distance between the two eyes of the original infrared face image. According to the above rotation and zoom transformation, all infrared face images can be transformed into images with two eyes on the same horizontal line and the distance between two eyes is the same. Histogram equalization method is used to preprocess the gray level of infrared face image. The gray histogram of an image is a discrete function about the gray value of the image. It describes the number of pixels with the gray value in the image. Its abscissa represents the gray level of the pixel, and its ordinate is the frequency of the gray level (the number of pixels). Histogram equalization, also known as gray-scale equalization, aims to convert the input image into an output image (that is, the histogram of the output image is flat) network with the same number of pixels on each gray level through point operation. Through histogram equalization, the histogram distribution of a given image is changed into a uniformly distributed histogram, which increases the contrast of the image. For the image with small contrast, the distribution of gray histogram is concentrated in a small range. After histogram equalization, the probability of all gray levels of the image is the same. At this time, the entropy of the image is the largest and the amount of information contained in the image is the largest. Gray histogram is an important statistical feature of image and is considered to be the approximation of image gray probability density function. The probability of occurrence of gray level R in a gray image is:

$$p_r(r_k) = \frac{n_k[x, y, 1]}{[a, b, 1] - n} \tag{4}$$

In formula (4), n_k represents the total number of possible gray levels in the image, and n represents the total number of pixels in the image, that is, the number of pixels with gray level r. $p_r(r_k)$ represents the probability of occurrence of the kth gray level of the original image.

2.3 Realization of Face Recognition in Infrared Image

If the unique face code can be extracted from each face, it can be recognized directly without matching and comparing with each face in the database. On the basis of observing the difference of each person's symmetrical waveform, we can consider using the coding method, which extracts the one-dimensional bar code from the symmetrical waveform. The conversion from symmetrical waveform to bar code seems easy, but it is not. The conversion process requires complex calculation. After 1-dimensional bar code, you can also try 2-dimensional bar code. The method is to establish a database first. In this database, the face is divided into many small units and each small unit is encoded. Calculate the face edge azimuth field T_{global} and the sample edge azimuth field $\{T_{global,n} | n = 1, 2, ..., N\}$ for comparative analysis. The distance between the calculated azimuth fields is $d_{global,n} = D(T_{global}, T_{global,n})$. The similarity between the samples given by set $\{d_{global,n} | n = 1, 2, ..., N\}$ and the overall structure is detected.When recognizing a face image, first segment the face, compare each segmented block with the content in the database, find the most matching block, and then assign it with the code of the block in the database, so as to form a two-dimensional bar code. This method looks promising, but it needs an optimized database containing a large number of face images. The local motion feature points of human body with pseudo feature points removed are processed by windowing method. After appropriate windowing, multiple pieces of information are extracted from one of the window feature points to form a feature vector to characterize human motion behavior. If people are at rest, the acceleration remains unchanged; If people are in motion, the acceleration changes constantly. Therefore, different accelerations are used to distinguish the static and dynamic behavior of the human body. According to the calibrated data, reconstruct the face 3D image. The specific steps are shown in Fig. 4.

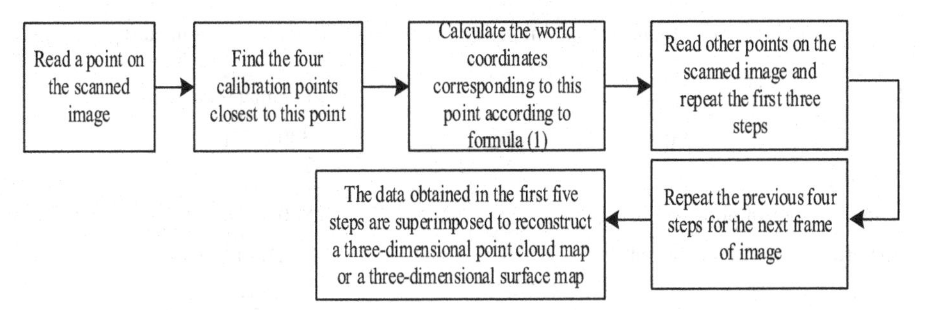

Fig. 4. 3D face image reconstruction process

According to the process shown in Fig. 4, the three-dimensional image of human face can be obtained to provide support for detailed feature location. The face image is an image composed of high-dimensional data, which is unevenly distributed in the high-dimensional space. From the perspective of pattern recognition, the intra class variance of multi-scale LBP sub pattern is large, mainly due to the relatively clear face image, The detailed information contained in it can not distinguish individual characteristics. However, the inter category variance of multi-scale LBP sub model is small, which is helpful to distinguish the different characteristics of the same individual. From the perspective of pattern recognition, the more detailed the details, the greater the intra class variance of multi-scale LBP sub patterns, the greater the probability of sample dispersion, and the fuzziness between categories. At this time, the recognition effect is the worst. In order to obtain face image information, the module needs to be more compact, so as to reduce the intra class variance. At this time, dimension reduction measures need to be taken, Extract more important features with recognition degree. To judge whether the result is valid or not according to the selected features, the category criterion should have different attributes, that is, the distance of the average value vector of different category features is the largest, and the sum of the variance of the average value vector of the same category features is the smallest. Therefore, this method is better for face recognition.

3 Analysis of Experimental Results

In order to verify the rationality of the design of face recognition technology based on signal interference, the data in the standard face database are used to verify the rationality of the technology. Firstly, the traditional PCA + LDA method is compared with the method in this paper. Because the outer face image is fuzzy and its edge features are not obvious, it is difficult to locate the feature points in the signal interference matching; The operation error of manual point selection in the process of constructing face moving map. Face recognition sampling is carried out in the sports wonderful cases section of the measurement website. The specific experimental environment is shown in Table 1:

Table 1. Configuration of simulation experiment environment parameters

Name	Parameter content
Development environment CPU, Inter Pentium 4	3000 MHz
Memory	1.5 G
Visual image sampling sample set	800
Spectrum bandwidth distribution	5 kHz–14 kHz
Interval	0.25 ms
Sampling frequency	24 kHz

According to Table 1, in MATLAB software, the original face images (as shown in Fig. 5) and scanned images in the face database without laser scanning are projected horizontally and vertically respectively, and the face features are located according to the horizontal and vertical projection curves and integral projection gradient curves. The results are shown in Fig. 6.

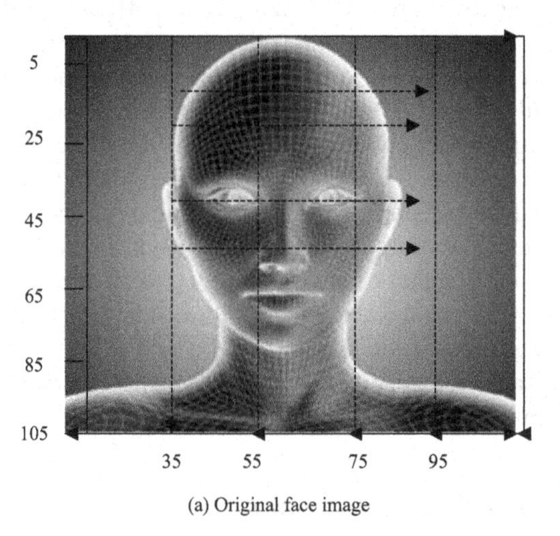

(a) Original face image

Fig. 5. Original face image

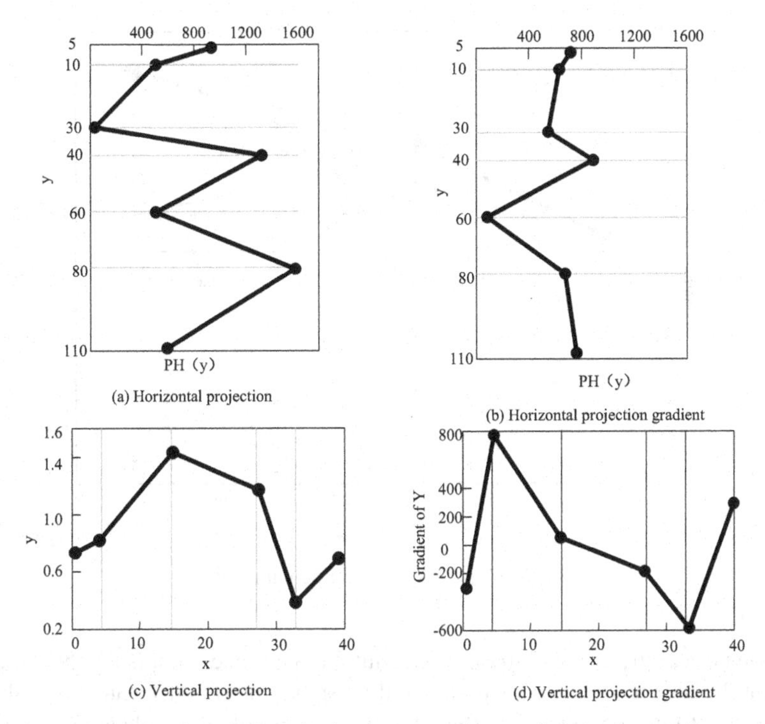

Fig. 6. Horizontal and vertical projection curves and integral projection gradient curves

It can be seen from Fig. 6 that at the boundaries on both sides of the face, the sum of gray values decreases rapidly, forming an obvious peak phenomenon. In the rising process of the vertical integral projection curve, the maximum point is the left boundary of the face, and the falling will follow the right boundary of the minimum point of the gradient value. The maximum ascending gradient of the vertical integral projection curve of the left face is the coordinate position of the left eye, while the maximum ascending gradient of the vertical integral projection curve of the right face is the coordinate position of the right eye. The infrared face image is preprocessed, and then the image in the training database is used to make the adult face beam structure. The registered image and the image to be recognized are matched with the face moving image respectively to obtain the signal interference of the registered image and the image to be recognized. Finally, the similarity between the image to be recognized and the registered image is calculated to realize the classification and recognition of infrared face image. The CMC curve corresponding to the experimental results is shown in Fig. 7:

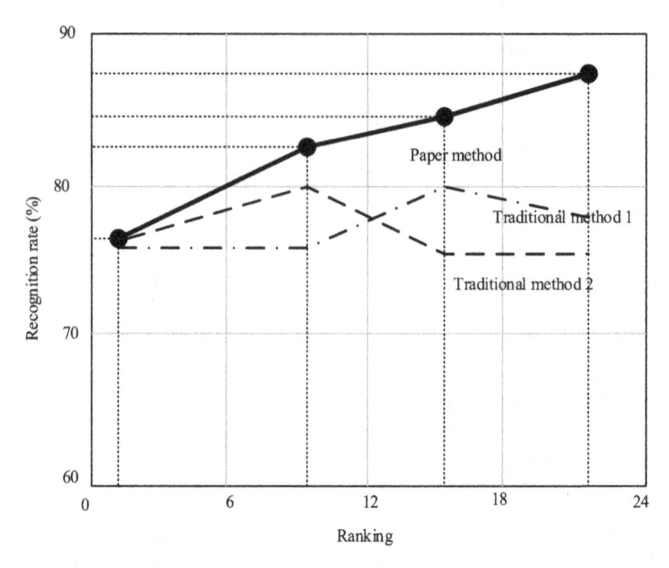

Fig. 7. Analysis of experimental results

It can be seen from Fig. 7 that the recognition rate ranked first is 81.25%. It can be seen that the face recognition method based on signal interference can realize the face recognition of infrared images. Further, the face images of 40 people are selected from the face database, with 10 images per person, including expression changes, posture changes and angle changes. The image size is selected as 90×110. In the experiment, each one, three and five images of the 40 people were used as training samples, and the remaining images were used as experimental test samples. The traditional recognition technology was compared with the recognition technology in this paper. The results are shown in Table 2.

Table 2. Experimental comparison results of different recognition technologies on face database

Training sample	Entry name	Traditional recognition technology	Paper recognition technology
Training sample 1	Recognition rate	0.73	0.83
	Characteristic dimension	397	597
Training samples 3	Recognition rate	0.83	0.95
	Characteristic dimension	498	712
Training samples 5	Recognition rate	0.91	0.98
	Characteristic dimension	1500	2827

Through the experimental comparison results in Table 2, it can be seen that the traditional recognition technology has low recognition efficiency, while the recognition technology based on signal interference has high recognition efficiency for face recognition.

To sum up, under the action of the method in this paper, the sum of the gray values at the boundaries on both sides of the face decreases rapidly, forming an obvious peaking phenomenon. During the rising process of the vertical integral projection curve, the maximum point is at the left boundary of the face, and the falling point will follow the right boundary of the minimum point of the gradient value. The maximum value of the ascending gradient of the vertical integral projection curve of the left face is the coordinate position of the left eye, and the maximum value of the ascending gradient of the vertical integral projection curve of the right face is the coordinate position of the right eye. Infrared image face recognition method based on signal interference technology can realize face recognition in infrared image, and has high recognition efficiency.

4 Conclusion

Face recognition technology has important theoretical value. After years of development, this technology has made great progress. Face recognition technology can obtain acceptable recognition ability in an ideal state, which has been demonstrated in several commercial face recognition systems and has been preliminarily applied. According to the subspace transformation feature structure, the signal interference sub pattern features are extracted, and the features are weighted by face feature matching. On this basis, the classification performance is analyzed. The practice shows that this technology is increasingly mature and has strong robustness and practicability. The experimental results show that this technology has high recognition efficiency.

For future work, we can conduct in-depth research, including:

(1) Infrared face image and visible face image have great complementarity. For example, visible face image is greatly affected by illumination, while infrared face image acquisition is independent of illumination conditions. Therefore, the fusion of infrared image face recognition and visible light image face recognition will be an important direction of face recognition research in the future.
(2) How to further study and analyze the statistical features of infrared face images on the basis of in-depth research on the imaging mechanism of infrared images and the characteristics of infrared face images, so as to obtain the most useful features for face recognition, understand the essential differences between them and the face features of visible light images, and effectively fuse them with the face features of visible light images at the feature layer, How to improve the performance of face recognition system in various practical applications is a topic that needs to be studied in depth.
(3) Because face recognition has a wide range of application value, in order to realize the application of infrared image face recognition in various practical situations, we must achieve the real-time performance of face recognition algorithm and the

automation of face recognition system. Therefore, one of the next research directions of face recognition is how to quickly and effectively express the features of the face itself and carry out face recognition, including the recognition of large-scale face databases, with as few constraints as possible.

Fund Project. 2021 special project in key fields of colleges and universities in Guangdong Province, research on Constructing "smart cloud" training platform based on new generation hybrid cloud technology (Project No.: 2021zdzx1147).

References

1. Ramanathan, S., Basha, R., Banu, A.: A novel face recognition technology to enhance health and safety measures in hospitals using SBC in pandemic prone areas. Mater. Today: Proc. **45**(2), 2584–2588 (2021)
2. Benouareth, A.: An efficient face recognition approach combining likelihood-based sufficient dimension reduction and LDA. Multim. Tools Appl. **80**(1), 1457–1486 (2021)
3. Shen, W.S., Sun, Y.W., Xu, L.J.: Research on will-dimension SIFT algorithms for multi-attitude face recognition. High Technol. Lett. **28**(3), 280–287 (2022)
4. Lian, J.K., Fang, S., Zhou, Y.: Model predictive control of the fuel cell cathode system based on state quantity estimation. Comput. Simulat. **37**(07), 119–122 (2020)
5. Guo, C.: Governance of face recognition technology application in digital human rights era. Mod. Law Sci. **42**(04), 19–36 (2020)
6. Li, X., Zhao, C., Yan, Z.: Research on intelligent classroom based on cloud face recognition technology. Exp. Technol. Manag. **37**(06), 172–175 (2020)
7. Le, Q.D., Vu, T., Vo, T.Q.: Application of 3D face recognition in the access control system. Robotica **40**(7), 2449–2467 (2022)
8. Zhang, X., You, M., Zhu, J., et al.: Face recognition of small-scale dataset based on joint loss functions. Trans. Beijing Inst. Technol. **40**(2), 163–168 (2020)
9. Sarah, B., Kirsten, D., Bennetts, R.J.: Face recognition improvements in adults and children with face recognition difficulties. Brain Commun. **41**(1), 214–219 (2022)
10. Xu, X., Zhang, L., Lang, B., et al.: Research on inception module incorporated Siamese convolutional neural networks to realize face recognition. Acta Electron. Sin. **48**(04), 643–647 (2020)

Remote Tutoring System of Ideological and Political Course Based on Mobile Client

Xiaopan Chen(✉) and Jianjun Tang

Jiangxi University of Software Professional Technology, Nanchang 330041, China
chenxiaopan0154@163.com

Abstract. Long-distance tutoring system can not quickly from the massive learning resources in search of users interested in resources. Therefore, this paper puts forward the design of distance tutoring system based on mobile client. In the hardware part, the FPGA ARM framework is adopted, and the XGMII interface is connected with the harmonic sub-layer of the link layer to realize the continuous transmission of a large number of data streams.In the software part, the interactive structure of the system is designed to form a good docking and interactive relationship between online learning and after-class learning mode. Design module structure based on mobile client, to meet the needs of students' online VOD courseware, curriculum information and teacher information. Based on the hybrid recommendation algorithm, the personalized course recommendation is carried out, and the courses are retrieved from the relevant course recommendation database and returned to users. System test results show that the system designed in this paper can reduce the maximum response time of user query and processing applications, so it is more practical.

Keywords: Mobile client · College education · Ideological and political course · Remote tutoring · Teaching system · System design

1 Introduction

The ideological and political course in colleges and universities is the main channel to carry out the education of Marxist theory and socialism with Chinese characteristics in the new era for college students, and it is the basic task and central link to carry out the moral education in colleges and universities, as well as the important position to cultivate the new generation who shoulder the responsibility of national rejuvenation. The mobile client has brought about a tremendous change in people's lifestyle. Mobile client instead of the original information transmission channels, as long as connected to the mobile Internet, you can access to information, so that people's lives more convenient. With the continuous development of mobile Internet, people can not only rely on the original fixed terminal equipment, but also can not be restricted by the local information access anytime and anywhere. With the continuous development of mobile information technology and social economy, the use of mobile terminals has popularized all aspects of people's lives, the impact is very wide. People use smartphones to accomplish many

W. Fu and L. Yun (Eds.): ADHIP 2022, LNICST 468, pp. 339–352, 2023.
https://doi.org/10.1007/978-3-031-28787-9_26

social activities. Mobile terminal device not only changes people's life style, but also changes people's way of learning. With the development of the times, in the process of education modernization, mobile cloud teaching is popular among college students in the new era, which has a profound impact on the mode of thinking, learning style and behavior of college students [1].

In order to serve more users, it is urgent to find a more efficient and novel learning model. In this environment, mobile learning has become the focus of research. Mobile learning devices rely on PAD, mobile phones and other portable devices, so that users can learn anytime and anywhere. Learners are more and more likely to use mobile terminal devices to participate in learning activities, no longer keen to participate in learning activities through the traditional personal computer, thus creating a mobile learning model. Learners use mobile terminal devices to acquire learning resources more conveniently and quickly through mobile learning platform, and then complete learning activities and achieve learning objectives. The key of ideological and political education is how to make use of the mobile cloud teaching technology to cooperate with the ideological and political class in colleges and universities, accurately convey knowledge goals, guide students to actively master skills and realize that the emotional attitude and values imparted are the key to ideological and political education. At present, colleges and universities should integrate the teaching content based on the modern Internet technology, and deliver it to the students, so as to facilitate students to achieve online testing and autonomous learning, and improve learning efficiency. The remote tutoring system of ideological and political courses in colleges and universities transforms the traditional teaching mode by using the mobile cloud teaching technology. The development of the new era and the popularization of intelligent terminals provide convenient conditions for the effective coordination of mobile cloud teaching and ideological and political courses in colleges [3].

The concept of mobile teaching was first proposed as a form of mobile learning abroad. On the one hand, mobile teaching and mobile learning are only different in terms of subjects. The same system is a learning platform for students and a teaching system for teachers. On the other hand, mobile teaching is a more advanced concept based on the concept of mobile learning. It adds more interactive features as well as personalized study settings. Mobile teaching takes the client as the carrier, allowing students to use the client to learn. In this process, the teaching activities between teachers and students can be reflected more and more. Mobile learning resources present as digital resources, a mobile phone has a very large amount of data, users can freely access mobile devices in the data, compared with the PC and traditional classroom, mobile client is more available. Based on the mobile client, this paper designs a remote tutoring system of ideological and political courses in colleges and universities. It forms a complete, universal and process-oriented teaching mode through the timely learning and simultaneous participation of the mobile end, which can greatly break the constraints of the traditional mode and enhance the effectiveness of ideological and political courses.

2 Hardware Design of Remote Tutoring Teaching System for Ideological and Political Course in Colleges

FPGA ARM framework is adopted in the hardware part of the remote tutoring system. FPGA processes high-speed network data interface, and ARM processor completes configuration management, authentication and encryption.The hardware framework of the system is shown in Fig. 1.

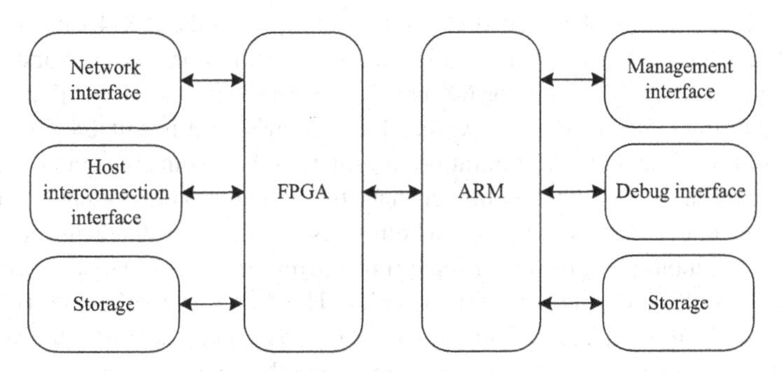

Fig. 1. The hardware framework of the system

The function of the processor module is to receive the data of the system through various communication protocols and different pins. This system uses nRF51822, which is based on ARM Cortex-M0 architecture and promoted by Nordic Company, as the main chip of MCU. The chip has a size of 256K flash and 32K of memory, while the internal integration of 2.4 GHz Bluetooth wireless communications, Bluetooth 4.0 protocol to support low-power standards. According to the requirement of the system, the network data exchange is mainly realized by Ethernet interface, and the management interface is 10/100/1000M adaptive network interface. WIS is added to the WAN physical layer model to achieve seamless connectivity to the SDH. NRF51822 is a multi-protocol SoC with flexible pins, powerful features and low power consumption. There are three sources of low frequency clock signals: external crystal oscillation, internal RC oscillation circuit generation and high frequency crystal oscillation synthesis. The physical layer is connected upwards through the XGMII interface to the harmonic sublayer in the link layer, and is compatible with the previous gigabit Ethernet standards upward from the MAC. So it can support the original upper layer service without changing the Ethernet frame format, which is convenient for unified management and maintenance, avoids the conversion between different protocol standards, and reduces the upgrade cost and improves the application value.

Because the PCB itself will have a certain stray capacitance, this capacitance value is very small and unstable, will cause unstable crystal work, starting difficult, in the crystal two connected with two 12pF load capacitor, crystal work in the resonant frequency state, the circuit is easy to start oscillation, more stable. Encode eight XGMII octets into 66-bit chunks, or reverse decode. Converts encoded data to a 16-bit stream and transmits it to the PMA, or conversely receives a 16-bit stream from the PMA. On the dielectric independent interface, GMII interface is defined in IEEE-802.3 standard, which includes data line, time line and control line. External low-frequency crystal pins XL1 and XL2 share the same pin with GPIO, so a two-pin passive low-frequency chip crystal with a frequency of 32.768 kHz is externally positioned on the P0.26 and P0.27 pins, and the model is FC-135. The XC1 and XC2 pins are outfitted with a 16MHz 2520 passive HF crystal. To further reduce circuit costs, RGMII uses a 4-bit data interface and a 125 MHz reference clock to transmit data along a double-sided edge. Input 2.8 V power supply due to external interference or circuit itself, the superposition of a variety of high-frequency noise, will affect the input signal and lead to load circuit problems.

The AXI bus protocol has also developed two standards, AXI4-Lite and AXI4-Stream. The AXI 4-Lite standard specifies that all transmissions have a burst length of 1 and a data bit width of 32 bits or 64 bits. It is a lightweight AXI bus that simplifies read-write timing control, reduces overhead and latency, and is suitable for passing a small amount of control and configuration information. VDD pin connected with 100nF capacitor, external high voltage is higher than the capacitor capacitor capacitor charge voltage, lower than the capacitor plate on both sides of the plate voltage discharge. In the process of continuous charging and discharging, the high-frequency noise of voltage will be weakened, so that the voltage tends to be stable. The AXI 4-Stream standard eliminates addresses, adopts the concept of data flow, and supports packet transfer patterns, making it ideal for passing large amounts of contiguous data flow between interfaces.

3 Software Design of Distance Tutoring Teaching System for Ideological and Political Courses in Colleges

3.1 Interactive Design of Distance Tutoring System

The characteristics of distance tutoring in ideological and political courses in colleges and universities are as follows: the basic knowledge of the course is explained by online video. Students are required to complete online course assignments, exams, etc. The remote tutoring and teaching system of ideological and political courses, supported by student information database, user portrait system, intelligent evaluation and feedback system, embodies the operation logic and management thought of data collection and verification, algorithm modeling, content supply and learning feedback [4]. The online assessment is mainly carried out by teachers and teaching assistants, but it includes the characteristic link of "peer assessment", that is, learners and learners can also evaluate each other. Students have different learning purposes in the process of using the remote tutoring system of ideological and political courses. This is mainly due to their own learning habits, learning ability and knowledge base, there are differences, so in order to further improve the user experience, the system should be user needs analysis, so that users feel more personalized service [5]. The model diagram for each of these interactions is shown in Fig. 2.

The interactive process between the system and the learners runs through the whole process from the login to the logout. After logging into the system, the learners need to fill out the questionnaire of the system's learning needs. After cleaning, extracting, parsing, converting and verifying the collected heterogeneous unstructured data, a data warehouse is created according to the pre-set criteria and paths. Numerous "data warehouse"

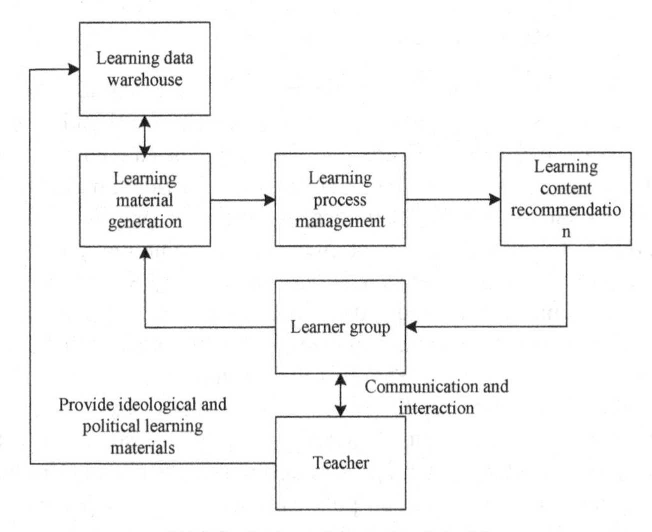

Fig. 2. System Interaction Model

permutations and combinations constitute the big data system of student information, and the stored data constitute the original assets and basic support of the service model of big data precision teaching platform. Students must learn to control time and develop their ability to choose a learning path or learning rate.

The remote tutoring system of ideological and political course can go deep into every student's learning process, track and grasp the whole state of learner's learning, and analyze each student's learning behavior in detail. The remote tutoring system of ideological and political courses quickly locates all the relevant information entries of college students, and constructs the user learning portrait and visual model. In this way, teachers can see the differences and particularities of learning habits, interests, preferences, laws and needs of college students of different ages and majors from user portraits, and scientifically judge the development trend of their thoughts and behaviors [6]. In order to improve the level of miniaturization, the system will process the learning materials and reduce their strength further, which leads to the development of data fragmentation. The system can understand the basic information of learners by testing each fragmented content, and recommend the knowledge and new knowledge that learners have not yet mastered to learners.

Systematic summary of each student's learning characteristics, basic grasp of each student's learning law, and then targeted to guide students, in order to develop personalized teaching program. For those who do not need, or temporarily do not need, or through their own books already understand the video knowledge can be skipped. So that learning with great autonomy, so as not to waste time and energy. The distance tutoring system provides corresponding teaching environment, video, course and supporting PPT, highlights the ideological and moral education value effect of network resources, forms a good docking, interaction and balance relationship between online learning and after-class learning mode, and realizes personalized excavation in the individual development of college students.

3.2 Design Module Structure Based on Mobile Client

The teaching mode of ideological and political course and mobile cloud teaching in colleges and universities is very promising. To some extent, the thinking that should be cultivated and guided by ideological and political course has a common point with the dissemination and openness of mobile interconnection. Curriculum learning is the most important part of students' learning activities, and the main purpose of the curriculum learning function is to meet the needs of students for online on-demand courseware, access to curriculum information and teacher information [7]. Simply put, smartphones are used to learn online lessons and videos. Based on the fast speed of information transmission, wide coverage and far extension of mobile cloud teaching, classroom teaching has greatly expanded and enriched the teaching methods of ideological and political courses by virtue of such distinctive characteristics and the popularization of mobile terminals of college students as a realistic basis, and should give full play to the openness of mobile cloud teaching. Through the analysis of the remote tutoring and teaching system of ideological and political courses in colleges and universities, we can know that the overall structure of the system consists of many modules and the corresponding interfaces are used to connect the various modules. At the same time, the system modules can also be divided into the backstage side of the system server and the mobile client side. The module structure based on the mobile client design is shown in Fig. 3.

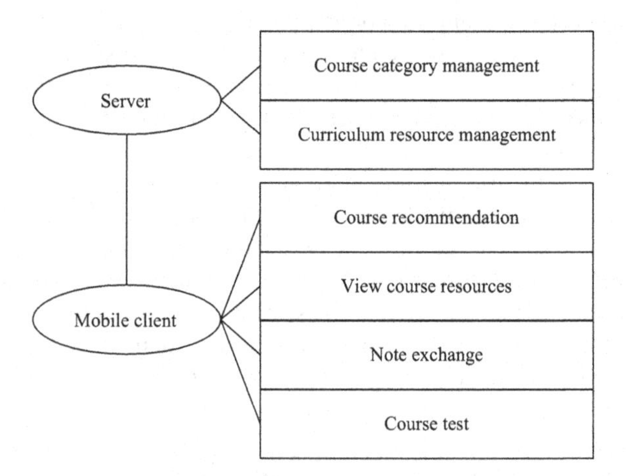

Fig. 3. System module structure

Course information maintenance mainly includes course information addition, course information viewing, course information modification, course information off the shelf and so on. Its main business logic is to add, check, modify and delete the database. According to the teaching plan, the learning time of each chapter can not be less than 95% of the total length of the video. The timing will provide the basis for the assessment of this course, and can save the history of learning, so that the students can continue to learn the video courseware. When adding the course information, first fill in the relevant information of the course on the page, then click the save button to save the course information to the database. Clicking the save button sends an HTTP request to invoke CourseController's addCourseInfo interface. The interface first calls the checkWords method of the CourseService to check for illegal characters in the course information. If so, an error code is returned, prompting for illegal characters in the added course information; otherwise, the addCourseInfo method of CourseService is called to add the course information to the database. Returns a hint of success after a successful addition. Instructional administrators manage all the resources of the course. Teachers use the course resources to design the course, design the course and then generate multimedia courses. Then instructional administrators are required to review the generated multimedia courses. When the course passes the review, instructional administrators are required to post these multimedia courses on the mobile learning platform. Only then can students achieve the goal of online learning of multimedia video courseware.

The course search function is mainly responsible for course retrieval according to the keywords entered by users. Keyword retrieval is an important means for users to search curriculum resources in massive courses. It needs to return retrieval data to users quickly and accurately. Mobile terminal information dissemination itself on the impact of students imperceptibly, for the accumulation of various types of information is the accumulation of more and more, become towers. In order to meet the needs of fast and accurate retrieval, this paper realizes course search function based on distributed search engine ES. First, an IK Chinese word breaker is installed on the ES server for word breakers to process course title and description information for matching when retrieving using keywords; then, an Index and Type are set up on the ES server and the course data in the database is synchronized to the ES using the logstash plug-in; and finally, the course related services are configured.

3.3 Politics Course Recommendation Module Implementation

The main function of the Think Politics Course Recommendation Service is to recommend similar courses to users as they browse through the details of a course. The relevant course recommendation set of each course in the system has been calculated by the course recommendation engine, and the relevant course recommendation service only needs to obtain the relevant course recommendation set corresponding to the course from the relevant course recommendation database upon request, and then return it to the user. Mobile Cloud Teaching collects and collects the dynamic data of students' learning conditions, analyzes the data, and carries out targeted differential teaching, forming a new teaching model. It can provide detailed data for the teachers of ideological and political lesson to teach and prepare for the special topic of teaching and research. It can help teachers of ideological and political lesson to cultivate students' ability to pay attention to politics, understand current events and temper their thinking while teaching knowledge, cultivating ability and shaping value.

Course recommendation module is divided into two sections, the first section is the non-personalized hot course recommendation, hot recommendation is based on the whole system users click on the order of courses, select the whole network to high click on the course recommended to users. The second section is personalized recommendation, through the analysis of user tags set and history browsing records, to recommend users may be interested in the course, add humanized design. The user's implicit feature vectors and the curriculum cluster center feature vectors are used as the input of MapReduce, and the user's affinity for the curriculum cluster center is calculated according to formula (1), then the curriculum cluster center with the highest user's affinity is selected. The calculation formula is as follows:

$$p = \alpha_1\beta_1 + \cdots + \alpha_u\beta_u \tag{1}$$

In Formula (1), p represents the user's preference for each course under that category; α represents the user's implicit feature vector; β represents the course clustering center feature vector; and u represents the number of users. Content-based recommendation algorithm extracts features according to the tags filled in by users when they register and issue courses, so it can effectively solve the cold start problem. Sort the calculated popularity in descending order, select the Top10 courses for each user as their recommended candidate set, and save them in the userlikecoursel.txt file. The cosine similarity between user implicit feature vectors is calculated as follows:

$$\chi(\alpha_1, \alpha_2) = \frac{\alpha_1\alpha_2}{\|\alpha_1\|\|\alpha_2\|} \tag{2}$$

In formula (2), χ represents the cosine similarity between eigenvectors. The cosine similarity is sorted in descending order, and the users of Top-10 cosine similarity are selected as their similar user groups for each user. Project-based collaborative filtering recommendation algorithm solves the long tail problem by recommending courses of interest to users who have operations on the same course. Calculate the user's preference for similar user history scoring courses (the courses that the user has scored are not included in the calculation). The formula is as follows:

$$\gamma = \varpi_1 + \frac{\chi(\alpha, \beta)(\varpi_1 - \varpi_2)}{\chi(\alpha, \beta)} \tag{3}$$

In formula (3), γ represents the likeness of a history graded course; ϖ_1 represents the average grade for all graded courses; and ϖ_2 represents the grade for a single course. Save the recommended candidate set to the userlikecourse2.txt file. Based on the analysis of the algorithms and problems above, this system combines the collaborative filtering recommendation algorithm with the content-based recommendation algorithm as a hybrid recommendation algorithm. The proposed candidate set is sorted according to the degree of preference, and the course of Top-10 is selected as its recommended candidate set for each user. Personalized course recommendation is returned to the selected section of the homepage through the above mixed calculation, and non- Personalized course recommendation is returned to the rotated section of the homepage by counting the number of students. The trigger page for the course recommendation service is the course details page. When the user enters the course details page, the browser triggers the JS script to send an Ajax request to the relevant course recommendation service to obtain the course recommendations. So far, the design of distance tutoring system of ideological and political courses based on mobile client is completed.

4 Experimental Research

4.1 Experimental Preparation

From the point of view of the whole distance tutoring system of ideological and political courses in colleges and universities, the system test is a vital operation method. In order to avoid errors in the design and development of the long-distance tutoring system of ideological and political courses, we should test the system repeatedly in order to reduce the possibility of errors and ensure the system to be in the optimal state.

The deployment environment of system test is based on Windows operating system. The technology used in system development is .NET technology. In order to match with the development technology, IIS server needs to be deployed. The database used in this system is the SQL Server2013 database with many advantages. Using IIS server as the Web server, can deal with a variety of different types of data, but also has a large capacity of data storage. The Web server is configured as follows: CPU: intel core 2.8 GHz; memory: 8 G; hard disk: 2 TB.

Here are some simple test cases: system user login, user registration, testing system page jump function, course recommendation view and teaching case view. Through the above use cases, we test the system in a black box and analyze each function. The test results show that the system basically meets the requirements of the remote tutoring system of ideological and political courses for mobile clients. Functionally, the system can achieve the desired goal.

4.2 Results and Analysis

On the basis of testing the actual functions of the remote tutoring system of ideological and political courses based on mobile client, the performance of the system is further tested. The selected performance test metrics are the maximum response time for user queries and the maximum response time for processing requests. The number of concurrent users for the test settings is 200, 500, and 1000. The performance test results of the system are compared with those of the distance tutoring system based on microservice architecture and deep learning. Under the different number of concurrent users, the comparison results of the maximum response time of each system user query are shown in Table 1 ~ Table 3.

In the test of 200 concurrent users, the maximum response time of user query of the remote tutoring teaching system of ideological and political courses based on mobile client is 1477 ms, which is 413 ms and 548 ms shorter than that of the teaching system based on micro-service architecture and deep learning.

In the test of 500 concurrent users, the maximum user query response time of the remote tutoring system for ideological and political courses based on mobile client is 2253 ms, which is 481 ms and 624 ms shorter than the teaching system based on micro-service architecture and deep learning.

Table 1. Comparison of query maximum response times for the number of concurrent users 200 (ms)

Number of tests	Remote tutoring system of ideological and political course based on mobile client	Distance tutoring system of ideological and political course based on microservice architecture	Distance tutoring system of ideological and political course based on deep learning
1	1549	1875	2037
2	1488	1746	2184
3	1366	1988	1958
4	1533	1804	1946
5	1455	1867	1819
6	1327	1981	1955
7	1514	1752	2021
8	1478	2078	2164
9	1506	1899	2237
10	1553	1906	1925

Table 2. Comparison of query maximum response times for 500 concurrent users (ms)

Number of tests	Remote tutoring system of ideological and political course based on mobile client	Distance tutoring system of ideological and political course based on microservice architecture	Distance tutoring system of ideological and political course based on deep learning
1	2316	2636	2938
2	2107	2528	2806
3	2352	2760	2757
4	2285	2895	2978
5	2390	2859	2882
6	2263	2723	3063
7	2134	2906	2809
8	2028	2877	2754
9	2316	2542	2874
10	2342	2612	2913

Table 3. Comparison of query maximum response times for 1000 concurrent users (ms)

Number of tests	Remote tutoring system of ideological and political course based on mobile client	Distance tutoring system of ideological and political course based on microservice architecture	Distance tutoring system of ideological and political course based on deep learning
1	2623	3409	3865
2	2768	3580	3720
3	2648	3665	3933
4	2592	3727	3854
5	2507	3578	3788
6	2664	3815	3846
7	2586	3582	3710
8	2548	3653	3801
9	2623	3664	3927
10	2465	3732	3865

In the test of 1000 concurrent users, the maximum response time of user query of the remote tutoring teaching system of ideological and political courses based on mobile client is 2602 ms, which shortens 1039 ms and 1229 ms compared with the teaching system based on micro-service architecture and deep learning. For different number of concurrent users, the comparison results of the maximum response times for each system are shown in Tables 4–6.

In the test of 200 concurrent users, the maximum response time of the mobile client-based remote tutoring system for ideological and political courses is 1979 ms, which is 522 ms and 724 ms shorter than that of the system based on micro-service architecture and deep learning.

In the test of 500 concurrent users, the maximum response time of the remote tutoring system for ideological and political courses based on mobile client is 3186 ms, which is 568 ms and 734 ms shorter than the system based on micro-service architecture and deep learning.

In the test of 1000 concurrent users, the maximum response time of the remote tutoring system for ideological and political courses based on mobile client is 3801 ms, which is 496 ms and 814 ms shorter than that based on micro-service architecture and deep learning. The test results show that the system can reduce the maximum response time of user query and processing applications, and is better than the two long-distance tutoring systems in overall efficiency.

Table 4. Comparison of maximum response times for processing requests for 200 concurrent users (ms)

Number of tests	Remote tutoring system of ideological and political course based on mobile client	Distance tutoring system of ideological and political course based on microservice architecture	Distance tutoring system of ideological and political course based on deep learning
1	1842	2305	2643
2	1980	2556	2585
3	1865	2619	2760
4	1931	2463	2634
5	2054	2636	2757
6	2128	2552	2625
7	1815	2480	2816
8	1946	2522	2652
9	2083	2505	2792
10	2150	2367	2763

Table 5. Comparison of maximum response times for processing requests for 500 concurrent users (ms)

Number of tests	Remote tutoring system of ideological and political course based on mobile client	Distance tutoring system of ideological and political course based on microservice architecture	Distance tutoring system of ideological and political course based on deep learning
1	3156	3643	3869
2	3250	3789	3984
3	3185	3896	3957
4	3061	3665	4048
5	3224	3827	3916
6	3258	3851	3923
7	3112	3610	3866
8	3176	3875	3930
9	3283	3758	3725
10	3152	3626	3980

Table 6. Comparison of maximum response times for processing requests for the number of concurrent users 1000 (ms)

Number of tests	Remote Tutoring System of Ideological and Political Course Based on Mobile Client	Distance tutoring system of ideological and political course based on microservice architecture	Distance tutoring system of ideological and political course based on deep learning
1	3627	4142	4520
2	3858	4201	4665
3	3706	4364	4553
4	3563	4357	4647
5	3989	4288	4589
6	3825	4376	4762
7	3764	4453	4463
8	3931	4293	4552
9	3802	4232	4778
10	3943	4265	4616

Analysis of the above experimental results shows that the mobile client-based remote tutoring system designed in this paper has better processing performance when dealing with large concurrent user queries and course responses. Because this paper adopts the FPGA ARM architecture in hardware, and connects the XGMII interface with the harmonic sublayer of the link layer, it can realize the continuous transmission of large amounts of data. And can obtain personalized course resource recommendation results when users query, thus improving students' interest in learning.

5 Conclusions

The coordinated promotion of mobile teaching and ideological and political courses focuses on the use of new information technology to create a new situation in the teaching of ideological and political courses, rather than simply as a tool or means to assist teaching. This paper designs a remote tutoring system based on mobile client. We adopt the FPGA ARM architecture, and the XGMII interface is connected with the harmonic sublayer of the link layer, which can continuously transmit a large number of data streams. The module structure is designed based on the mobile client, which can meet the information needs of students and teachers. At the same time, it also has the function of personalized course recommendation. The system reduces the maximum response time of user queries and processing, and improves overall operational efficiency. However, the recommendation results in the learning system in this paper are obtained by offline calculation, and cannot be recommended according to the real-time behavior of users in the system. The next step is to add a user behavior collection function to the system to recommend course resources based on the user's real-time behavior.

References

1. Xu, M.-W., liu, Y.-Q., Huang, K., et al.: Autonomous learning system towards mobile intelligence. J. Software **31**(10), 3004–3018 (2020)
2. Xu, N., Fan, W.-h.: Research on interactive augmented reality teaching system for numerical optimization teaching. Comput. Simulation **37**(11), 203–206,298 (2020)
3. Yaru, G.: Research on 3D ideological and political teaching system based on multimedia perspective. Mod. Sci. Instrum. **38**(1), 140–144 (2021)
4. Zhu, L.-l.: Application online intelligent teaching system in colleges and universities based on artificial intelligence. J. Jiangxi Vocat. Techn. College Electricity **34**(2), 20–21 (2021)
5. Cui, Y., Wang, F., Chen, K.: Design of distance education system based on VR live broadcast. Experimental Technol. Manag, **37**(6), 132–136,140 (2020)
6. Wu, Juan.: Optimization design of college ideological and political teaching system based on multimedia technology. Techniques Autom. Appli. **39**(9), 180–182 (2020)
7. Zhang, L.: Design of distance teaching system based on artificial intelligence network. Mod. Electronics Technique **44**(2), 131–134 (2021)

Operational Risk Prevention and Control Monitoring of Smart Financial System Based on Deep Learning

Hui Zhu(✉)

State Grid Huitong Jincai (Beijing) Information Technology Co., Ltd., Beijing 100031, China
lihongyu348@163.com

Abstract. With the development of intelligent financial level, intelligent financial system has been gradually applied to most enterprises. Once there is operational risk in the financial system, it will seriously affect the security and reliability of enterprise finance. Therefore, once the operational risk of the smart finance system is prevented and controlled, it is very necessary. In order to ensure the accuracy and effectiveness of risk prevention and control monitoring of smart financial systems during operation, a deep learning-based operational risk prevention and control monitoring method for smart financial systems is designed. First, establish the corresponding relationship between the roles and operations of the smart financial system, establish an operational risk prevention and control model based on deep learning, design a risk assessment structure tree, and complete operational risk quantification. In order to verify the effectiveness of the design method, a performance comparison experiment was designed. The experimental results show that the accuracy of the test samples finally reached 74.6%, of which 21 risk samples were correctly monitored and prevented, indicating that the designed deep learning-based smart financial system operates Risk prevention and monitoring methods have certain effectiveness.

Keywords: Deep learning · Smart financial system · Operational risk prevention and control · Data monitoring

1 Introduction

Intelligent financial management system is the development direction of the financial field. The system uses advanced data entry, data processing and data storage methods to realize real-time synchronous processing of various financial data of enterprises. With the continuous improvement of the national economic system, the economic benefits of all types of enterprises in society have improved, and the industrial results have taken a qualitative leap. This makes the financial data huge and diverse, resulting in the formation of the current big data model in the financial field [1]. In the actual smart finance system, there are many user accounts, so in the process of updating system resources, the current network environment can support multi-level access control rights in communication.

W. Fu and L. Yun (Eds.): ADHIP 2022, LNICST 468, pp. 353–365, 2023.
https://doi.org/10.1007/978-3-031-28787-9_27

Ensuring the security of resources in the intelligent financial system is an important goal of automatic access control. In the process of users accessing the resources of the intelligent financial system, ensuring the security of information can be divided into three aspects: management, operation and network. In the case of changes in the resources of the intelligent financial system, automatic control of updating access rights is a very effective method and means to prevent information leakage and set perfect access rights.

Reference [1 proposes a face recognition based monitoring method for operational risk prevention and control of financial systems. By using dynamic cameras at high points we obtain the face images of financial personnel design a face recognition algorithm based on the combination of Gabor wavelet lbph and PCA dimensionality reduction methods for face recognition draw the corporate portrait of financial personnel and push the risk level. Reference [2] proposes a financial system operational risk early warning method based on support vector machine, and designs the system role, overall architecture and functional modules. On the basis of the above analysis, the collection of basic data is completed through data preprocessing. At the same time, it focuses on the design of the operational risk early warning model of the financial system, and analyzes the risk types with the principle of SVM algorithm. Reference [3] proposes a financial system operational risk early warning method based on AdaBoost strong classifier. As a financial tool for predicting and avoiding risks, the enterprise financial early warning system can effectively avoid the losses brought by potential risks to the enterprise. The company's financial early warning system takes various financial variables of the enterprise as input values and risk results as output values to complete the early warning of operational risks of the financial system. Although the above methods have completed the prevention and control monitoring or early warning of the operational risk of the financial system, they have the problem of low accuracy. In order to effectively solve this problem, a monitoring method for operational risk prevention and control of intelligent financial systems based on deep learning is proposed. The overall research technical route and contributions of the new method are as follows:

(1) Improve the traditional RBAC model, divide the roles of the financial system, build the access management mode architecture, and obtain the corresponding relationship between the roles and operations of the intelligent financial system and the financial early warning hierarchy.
(2) After scaling the financial risk information, the convolutional neural network in deep learning is used to train the information to improve the accuracy of risk prevention and control.
(3) The financial operational risk is quantified by dividing dimensions, and the operational risk prevention and control monitoring of the intelligent financial system is completed.

2 Operational Risk Prevention and Control Monitoring of Smart Financial Systems Based on Deep Learning

2.1 Establish the Correspondence Between the Roles and Operations of the Smart Financial System

The smart financial system is derived by relying on the computer network. With the expansion of the resource information scale of the financial information system, a relatively scalable risk control and monitoring method is required in the operation of access and so on. This paper improves the traditional RBAC model to adapt it to the resource update mode of the financial information system. In the RBAC model, the core idea of control is the role, that is, the different work responsibilities in the financial system, and the role can limit some operations of the user in the financial information system. In the role-based access management model constructed in this paper, the concept of role hierarchy is introduced on the basis of the traditional RBAC model. The architecture of the constructed access management model is shown in Fig. 1:

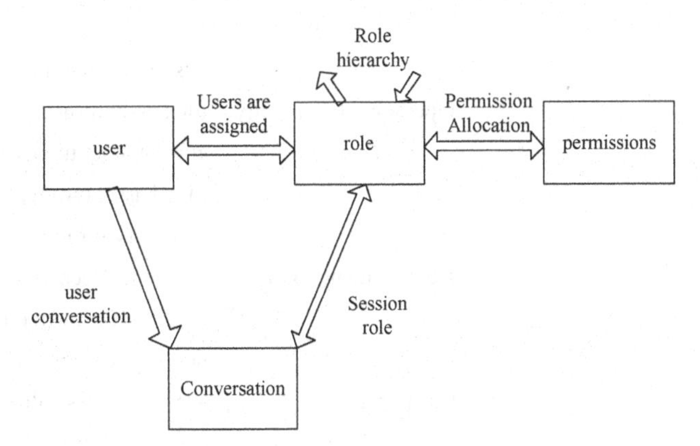

Fig. 1. Schematic diagram of RBAC relationship based on role hierarchy

As can be seen from the structure of the above figure, there is a many-to-many relationship between users, roles and permissions. In the above relations, different relations between different sets are included, and these relations are unified, and the following equations can be obtained:

$$\begin{cases} PA \subseteq P \times R \\ UA \subseteq U \times R \end{cases} \tag{1}$$

In the above formula, P represents the user role permission set of the system after the resource update, A represents the session set, R represents the set in the role hierarchy, U represents all the user sets of the smart financial system, and formula ① represents the assignment between permissions and roles. Relationship, formula ② represents the assignment relationship between users and roles, this relationship can be called role inheritance [4]. There are also several main constraints between role levels, such as mutual exclusion of roles, etc. These constraints can play an important protective role in the resource management of financial information systems. Taking role mutual exclusion as an example, it can support sensitive separation of duties in each relationship of the financial system, and specify the grant status between permissions and roles under relevant conditions. The relationship between roles and operational rights together establishes the permissions of user access rights in the RBAC model. According to the above roles and relationships, the financial early warning hierarchy can be obtained (Table 1):

Table 1. Financial Warning Hierarchy

Target layer	Criterion layer	Factor (indicator) C
Smart financial system operational risk warning	Solvency	current ratio
		Assets and liabilities
		Interest coverage ratio
	Profitability	OPE
		Roe
		cost profit margin
	Operational capability	total asset turnover
		current asset turnover
		Fixed asset turnover
		Inventory turnover
	Development ability	Accounts Receivable Turnover
		total asset growth rate
		net asset growth rate
	Cash flow	Main business value-added rate
		cash flow ratio
		Total Cash Debt Ratio

2.2 Establish an Operational Risk Prevention and Control Model Based on Deep Learning

In the economic stage, the basic demand of the seller's market drives the standardized production of enterprises, and the dominant human value is attributed to the value of goods. In this stage, the financial relationship of enterprises is linear. In the human-oriented economy stage, the individual needs of the buyer's market drive the customized production of enterprises, and the value of the dominant species is attributed to human value. In this stage, the financial relationship of enterprises is nonlinear. This non-linear relationship determines that the shared financial risk of enterprises also presents a non-linear relationship in a buyer's market with asymmetric information. Therefore, the deep learning neural network as the branch of the artificial neural network model is selected to describe the nonlinear relationship of early warning in financial risk operations, and the financial risk early warning variable is used as the input layer variable of the deep learning neural network, and the deep learning neural network is turned into a financial risk early warning model., to establish a shared financial risk early warning model of deep learning [5]. The consistency between deep learning neural network and shared financial risk early warning is mainly reflected in nonlinear mapping features, fault tolerance features and generalization features. In the process of risk prevention and control monitoring, the use of convolutional neural networks to realize multi-modal deep learning can reduce the computational overhead to a certain extent. Before performing convolution, the obtained moving image needs to be scaled, and the scaling formula is:

$$M' = floor(255 * \frac{M - v_{\min}}{v_{\max} - v_{\min}}) \tag{2}$$

In the formula, M' represents the normalized matrix, M represents the image matrix after preprocessing, $v_{\min}$ represents the minimum value of the pixel matrix in the image, $v_{\max}$ represents the maximum value of the pixel matrix in the image, and $floor(.)$ represents the blur function. After the linear transformation of the above formula, the image during the operation of the smart financial system can be captured, and its gray value can be scaled from 0 to 255. Normalize the processed operations into the input format of the convolutional neural network, and input it into the convolutional neural network. After the convolutional operation, the role of the data can be maximized in the process of processing financial system operations [6], in the learning process, the calculation process in the convolutional neural network is shown in Fig. 2:

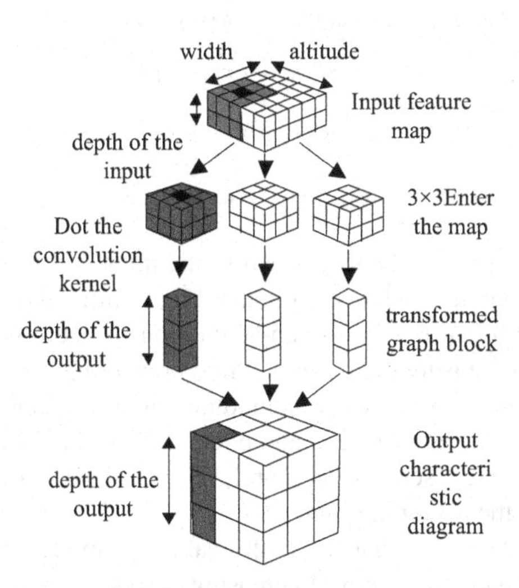

Fig. 2. Convolution process

After the convolution calculation, the feature operation is obtained, and it can be seen that after the convolution calculation process, the output depth of the obtained feature map is increased. After completing the output of the feature map, it is also necessary to perform the maximum pooling operation on the correlation between the local data points [7]. In the risk assessment process of the traditional financial information management system, the amount of data that can be processed is small, and only the method of sampling survey is used to assess the risk of data, ignoring the information correlation between the data. Therefore, in the context of big data, fully consider the influence of system hardware, software, and human factors, integrate all data information, and refer to the description of system functions and devices by previous estimation models. The deep learning of financial information is applied to the human motion data set in this paper, and it needs to be integrated. In the convolutional neural network, the ultimate goal is to realize the monitoring of operational risk prevention and control of the intelligent financial system. Therefore, in the process of deep learning, it is necessary to design multi-modal training samples:

$$\{(x_i, y_i)\}_{i=1}^{N} \tag{3}$$

In the above formula, when the value of y_i is -1, it represents modal class C_1, and when the value of y_i is $+1$, it represents modal class C_2. In the process of classification, it is necessary to use the decision hyperplane equation:

$$g(x) = w^T x + b \tag{4}$$

In the above formula, w represents the decision vector of the hyperplane, and b represents the decision bias of the hyperplane, which can classify the motion patterns of the two classes. Make:

$$w^T x + b = 0 \tag{5}$$

Then the defined function interval can be written as:

$$\gamma_i = y_i(w \cdot x_i + b) \tag{6}$$

The corresponding geometric interval is written as:

$$\chi_i = \frac{1}{\|w\|} \cdot y_i(w \cdot x_i + b) \tag{7}$$

In the actual convolutional neural network, the above parameters are obtained by calculation, the fully connected layer is obtained by training, and the appropriate ratio parameter is set, which can reduce the co-adaptation of neurons in the convolutional neural network, and can ensure the training while preventing overfitting [8].

In the process of risk assessment, each module of the financial information management system completes a common task, and interconnects the data through the communication module to realize a series of comprehensive tasks or a function set of a single sub-task, as shown in the following figure:

It can be seen from Fig. 3 that the built evaluation model takes into account the logical node d of the financial system. When using this node to exchange data or execute instructions with the system, it can abstractly understand the behavior characteristics of system hardware, software and operators, and the communication between nodes. A link is a way of exchanging financial information to illustrate the interactive relationship between data.

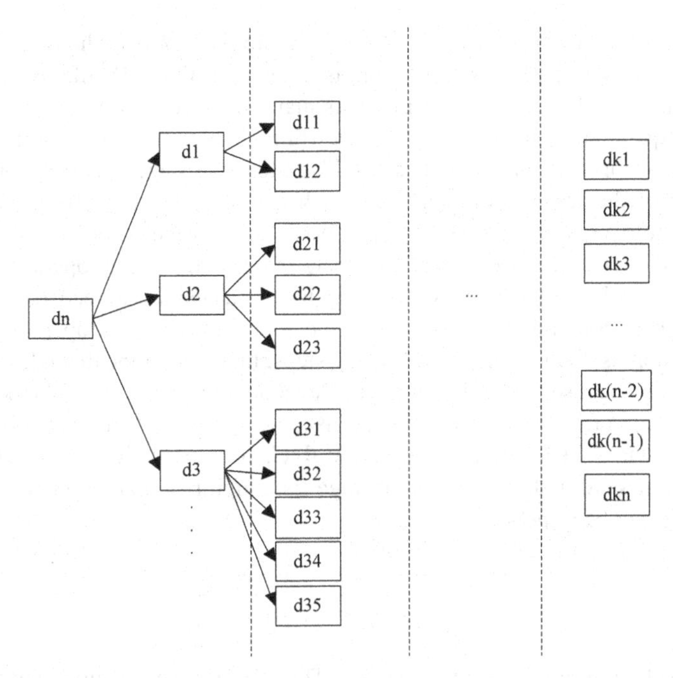

Fig. 3. Risk Assessment Structure Tree

2.3 Operational Risk Quantification

For the intelligent financial system, operational risk can be studied by dividing dimensions in the process of risk prevention and control. For risk quantification, it is necessary to use the questionnaire survey method to collect subjective data from various dimensions, and use key specific objective indicators to measure operational risk. Due to the many factors that constitute the operational risk of the intelligent financial system, and the interaction of various factors forms a complex relationship, this paper starts from four aspects of corporate financial activities, namely operating activities, investment activities, financing activities and allocation activities, from subjective and objective two aspects. This aspect explores the dimensions of the industry environment that are closely related to corporate finance. For the specific investigation of these dimensions, it is difficult to clearly describe them with a single objective indicator due to the rich meaning of each dimension [9]. Therefore, in order to make the results more accurate, this paper will use the subjective and objective evaluation method to design the index system of each dimension of the industry environment. (1) Operational resource risk. The operational resource risk defined in this paper is mainly the richness of permission operations. Its main measurement indicators are as follows: First, quantitative indicators. Second, qualitative indicators. System operation dependencies. (2) Operational competition risk. Operational competition risk mainly describes the degree of competition among enterprises within the operation. Its main measurement indicators are as follows: First, quantitative indicators. Operational concentration, the annual increase rate of enterprises within the operation; second, qualitative indicators. Degree of product differentiation.

(3) Operational life cycle risk. Operational life cycle risk refers to the financial impact of the life cycle stage of an operation on an enterprise. Different life cycle stages of an operation show different growth rates. Its main measurement indicators are as follows: First, quantitative indicators. Sales growth rate, operating fixed asset investment growth rate; second, qualitative indicators. The life cycle stage the operation is in. (4) Operational credit risk. Operational credit risk is essentially a variety of risk factors that affect the overall solvency of an operation. Its main measurement indicators are as follows: First, quantitative indicators. Operating cash flow ratio, operating bad debt ratio; second, qualitative indicators. How easy it is for a business to recover its accounts. Using the fuzzy comprehensive evaluation method to measure the operational risk of financial system is divided into the following steps: first, determine the index domain of operational risk, and secondly determine the level domain of operational risk. Let V be the level domain of industry environmental risk, v_1, v_2, v_3, v_4, v_5 respectively represent the fuzzy subsets of each level domain, and A, B, C, D, E respectively represent five risk level results of small risk, small risk, average risk, high risk and high risk, so the risk level domain can be Expressed as:

$$\begin{aligned} V &= (v_1, v_2, v_3, v_4, v_5) \\ &= (A, B, C, D, E) \end{aligned} \tag{8}$$

Each level corresponds to a fuzzy subset. Then the fuzzy evaluation of each index is carried out, and the fuzzy relationship matrix is established. After constructing the hierarchical fuzzy subset, it is necessary to quantify the indicators of each dimension, which determines the degree of membership of each dimension to be evaluated to each fuzzy subset from the perspective of the indicator u, and then obtains the fuzzy relationship matrix:

$$R_i = \begin{bmatrix} R|u_1 \\ R|u_2 \\ R|u_3 \end{bmatrix} \begin{bmatrix} r_{1,1} & r_{1,2} & r_{1,3} & r_{1,4} & r_{1,5} \\ r_{2,1} & r_{2,2} & r_{2,3} & r_{2,4} & r_{2,5} \\ r_{3,1} & r_{3,2} & r_{3,3} & r_{3,4} & r_{3,5} \end{bmatrix} \tag{9}$$

In the matrix of the above formula, element u_i represents the degree of membership of a certain evaluated dimension to the fuzzy subset of grade v_i from the perspective of index u_i. The performance of an evaluated dimension on a certain index u_i is characterized by a fuzzy vector. The fuzzy vector is represented as:

$$R|u_i = \left(r_{i,1}, r_{i,2}, r_{i,3}, r_{i,4}, r_{i,5}\right) \tag{10}$$

Determine the fuzzy weight vector of the evaluation index. In general, the performance of each indicator has different effects on the whole, so the fuzzy weight vector should be determined. In the fuzzy comprehensive evaluation, let the fuzzy weight vector be expressed as:

$$A = (a_1, a_2, a_3) \tag{11}$$

In the formula, a_i is essentially the membership degree of the index u_i to the fuzzy subset of the importance of the dimension being evaluated, and the weight vector A can be obtained by the AHP method. Subsequent to fuzzy synthesis, there is:

$$\begin{aligned} & B = A^{\circ}R = (B_1, B_2, B_3, B_4, B_5) \\ & B_i = \min\left(1, \sum_{i=1}^{3} a_i r_{ij}\right) \\ & j = 1, 2, 3, 4, 5, \end{aligned} \quad (12)$$

In the formula, B_i represents the degree of membership of the evaluated dimension to the fuzzy subset of grade v_i in terms of indicators. Processing of fuzzy comprehensive evaluation result vector. The results of fuzzy comprehensive evaluation are:

$$B = (B_1, B_2, B_3, B_4, B_5) \quad (13)$$

The above formula is the membership degree of the evaluated dimension to each level of fuzzy subsets. In order to calculate the risk value of each dimension of the industry environment, it is necessary to further process the result vector of the fuzzy comprehensive evaluation. In this paper, the fuzzy weighted average method is used, and the result vector can be single-valued as:

$$C = \frac{\sum_{j=1}^{5} b_j r_j}{\sum_{j=1}^{5} b_j} \quad (14)$$

In this way, the risk index of each dimension in the operational risk process of the smart financial system can be obtained through calculation. So far, the design of the monitoring method for operational risk prevention and control of smart financial systems based on deep learning has been completed.

3 Method Test

3.1 Test Preparation

In order to verify the effectiveness of the deep learning-based intelligent financial system operational risk prevention and control monitoring method designed in this paper, it is necessary to test the method performance in this chapter. The main types of user roles in the smart financial system are accounting, auditing, general ledger, report management, financial system maintenance, bank reconciliation, application management, special query and transaction management. Divide the 162 companies under the total sample into two groups, which are the test samples for judging the accuracy of the model and the training samples for establishing a shared financial risk early warning model for deep learning. There are 136 training samples and 26 testing samples, respectively, among which ST companies and non-ST companies are allocated in equal proportions.

The Matlab R2018b software is used as a tool for programming, and then 19 financial risk early warning indicators data of 136 training samples are input into the program. When a role tries to access out-of-privilege data, the schematic diagram of user identity information interception is as follows (Fig. 4):

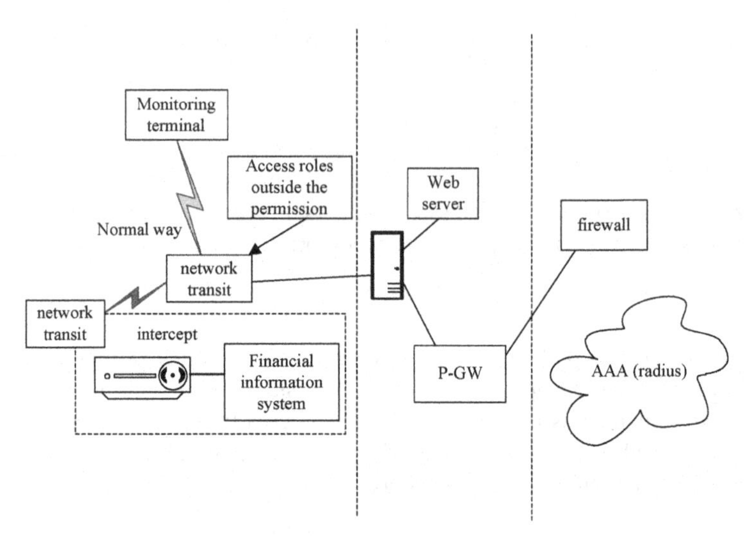

Fig. 4. Schematic diagram of data interception outside of role access rights

In the schematic diagram of the communication interception of the smart financial system shown in the figure above, the interception program can be integrated with the communication network terminal during the authorization verification process. In the design process of the performance test of the authority control method in this paper, it is necessary to use the relevant programming language to realize the authentication of the user identity information. By setting different access rights for different role types in the above, in this chapter, the performance test of the smart financial system operational risk prevention and control monitoring method designed in this paper is carried out. In the process of operational risk prevention and control monitoring, it is mainly divided into two parts, one is the resource update of the smart financial system, and the other is the authentication of user identity information. During the performance test, with the support of the HackRF development board, the network access side tries to access the resources in the system with an access role outside the authority. The processing flow from resource access to system exit when different roles log in to the financial system is shown in the following figure (Fig. 5):

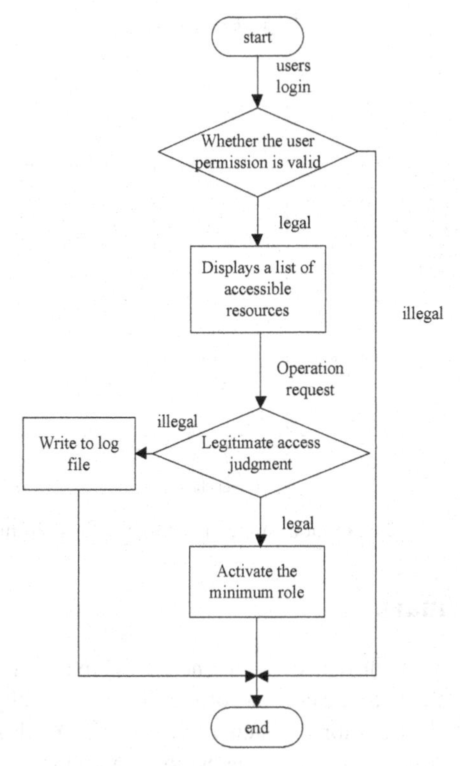

Fig. 5. Flowchart of financial system judgment during testing

During the experiment, the smart financial system runs under the support of various software programs. After the terminal logs in to the accounts of different roles, it will send an authentication request for the identity operation authority to the system.

3.2 Experimental Results and Analysis

Under the above experimental environment and steps, the output accuracy matrix of the test group can be obtained, as shown in the following figure (Fig. 6):

As can be seen from the above figure, the accuracy rate of the test samples finally reached 74.6%, of which 21 risk samples were correctly monitored and controlled, indicating that the deep learning-based intelligent financial system operational risk prevention and control monitoring method designed in this paper has certain advantages. Effectiveness.

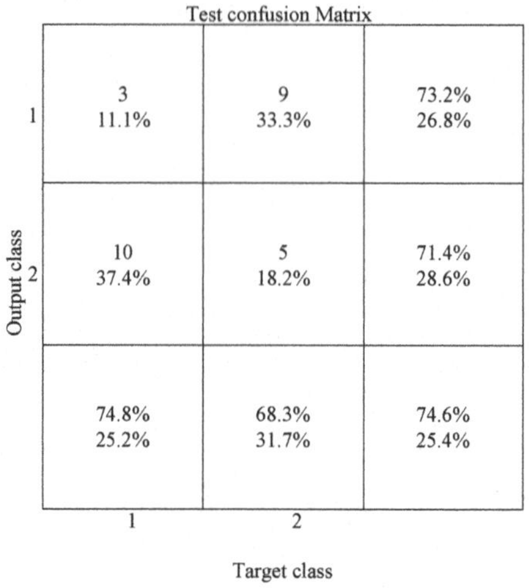

Test confusion Matrix

Output class	1	2	
1	3 11.1%	9 33.3%	73.2% 26.8%
2	10 37.4%	5 18.2%	71.4% 28.6%
	74.8% 25.2%	68.3% 31.7%	74.6% 25.4%

Target class

Fig. 6. The output accuracy matrix of the test group

4 Concluding Remarks

In the context of the Internet, the data of the financial system is more diversified, from the original single and fixed data to the current form of diverse models, multiple structures, and real-time changes. The traditional system risk estimation method can no longer meet the needs of financial system upgrades. The estimation method in this paper takes the hugeness and complexity of financial data as the primary research focus, so that all financial data can be analyzed, evaluated and calculated to ensure the authenticity of financial data. While the system is processing massive data, the proposed method evaluates data for each module, and does not miss every node that processes data, ensuring that all modules and running programs in the system can be estimated. However, there are still some errors in the calculation of the current research. In the future, we will focus on the real-time risk estimation of system modules.

References

1. Tang, S., Lian, X., Zhang, Q., et al.: Research and implementation of tax risk prevention and control system based on face recognition. Comput. Appli. Software **39**(05), 110–114 (2022)
2. Na, C.: Design of enterprise financial risk early warning system based on support vector machine. Microcomput. Appl. **34**(08), 73–77 (2018)
3. Liu, Y., Xu, J.: Financial early warning model based on AdaBoost strong classifier. Mod. Bus. **31**, 187–188 (2020)
4. Lu, X.: Design of financial risk management and control system based on cloud computing. Mod. Electronics Tech. **43**(17), 72–76 (2020)
5. Yao, G.: Financial audit risk analysis and suggestions in the transformation of geological survey scientific and technological achievements. China Mining Magazine **29**(S1), 43–45 (2020)
6. Zhao, H., Cheng, H., Ding, Y., et al.: Research on traffic accident risk prediction algorithm of edge internet of vehicles based on deep learning. J. Electron. Inf. Technol. **42**(01), 50–57 (2020)

7. Wang, J., Pei, L.: Anomaly detection and multi-stage risk pre-warning technology of power grid control system based on deep learning J. Shenyang Univ. Technol. **43**(06), 601–607 (2021)
8. Chen, X., Wu, J., Wu, M.: Individual default risk measurement and prevention of china's credit bonds — based on lstm deep learning model. Fudan J, (Soc. Sci. Ed) **63**(03), 159–173 (2021)
9. Gu, Z., Peng, H.: Grey risk assessment model of industrial control system based on fuzzy sets and entropy. Comput. Eng. Design **41**(02), 339–345 (2020)

Remote Monitoring Method of Athlete Training Intensity Based on Mobile Internet of Things

Qiang Zhang[1](✉) and Hui Xu[2]

[1] Fuyang Normal University, Fuyang 236037, China
zqxh16584@163.com
[2] Anhui Medical College, Hefei 230601, China

Abstract. In order to better design the athlete training scheme, a remote monitoring method of athlete training intensity based on mobile Internet of things is proposed. Combined with a certain logistics network to collect athletes' training body data, build athletes' training body data monitoring and management system, and simplify the steps of athletes' training intensity monitoring and management. Finally, the experiment proves that the remote monitoring method of athletes' training intensity based on mobile Internet of things has high practicability in the process of practical application and fully meets the research requirements.

Keywords: Mobile Internet of things · Athletes · Training monitoring

1 Introduction

The emergence of remote monitoring of athletes' training intensity in the field of competitive training, fitness and rehabilitation mainly stems from the short-term and efficient training effect. This training method has been used in sports monitoring since the 1970s. For example, various versions have appeared in the preparation of football events. At present, the remote monitoring of athletes' training intensity is not only frequently used in the preparation of athletes and sports teams, but also highly sought after in the field of fitness, according to the annual survey of the American Academy of sports medicine. At present, there is a strong interest in the research of remote monitoring of athletes' training intensity in the field of scientific research, including a large number of exploratory research on remote monitoring of athletes' training intensity by scientific researchers in the fields of sports science and sports medicine [1]. The difficulty in the application of remote monitoring of athletes' training intensity lies in how to make a reasonable training plan to achieve the predetermined training objectives. Different variables involved in controlling the training intensity will lead to different metabolism and muscle nerve reactions, resulting in different training effects. There are many variables involved in the remote monitoring of athletes' training intensity, and the combination forms of variables are diverse. Arbitrarily adjusting one of the many variables will lead to the difference of training effects.

W. Fu and L. Yun (Eds.): ADHIP 2022, LNICST 468, pp. 366–380, 2023.
https://doi.org/10.1007/978-3-031-28787-9_28

This paper proposes a remote monitoring method of athletes' training intensity based on mobile Internet of things. On the basis of the athlete's training body data collection, it is processed. Monitor the intensity of sports training based on mobile Internet of things terminals.

2 Remote Monitoring of Athletes' Training Intensity

2.1 Athlete Training Body Data Collection

Athletes' training body data collection is the core function of the whole, and the accuracy and reliability of sports monitoring algorithm also determine the performance of the system. According to the collected data in multiple scenes, after comprehensive comparative analysis, an algorithm for motion monitoring is designed. Firstly, the sensor transmits the generated signal to MCU for preprocessing. The preprocessing link includes signal superposition, digital filtering and other links, of which the digital filtering link is the key. The effect of digital filtering will directly affect the selection of characteristic parameters, and then affect the accuracy of final state recognition. Then, the motion eigenvalues are extracted as input parameters to judge the motion state and step recognition, and the conditional transformation of motion state is realized by using state machine [2]. Motion state recognition essentially realizes the function of classifier, which classifies the motion eigenvalues corresponding to different motion states. At the same time, the motion eigenvalues that have been judged will also be input into the classifier as training data to continuously optimize the parameters of the classifier, improve the performance of the monitoring algorithm and realize the intellectualization of the algorithm [3]. However, in the case of limited existing hardware resources, it is difficult to implement the multi classifier algorithm at the bottom, so the processing method of multi conditional judgment is used for programming. After making the motion state discrimination, enter the state discrimination subroutine to make a more specific and in-depth discrimination of each state. Based on this, optimize the process structure of athletes' training body data acquisition, as shown in Fig. 1:

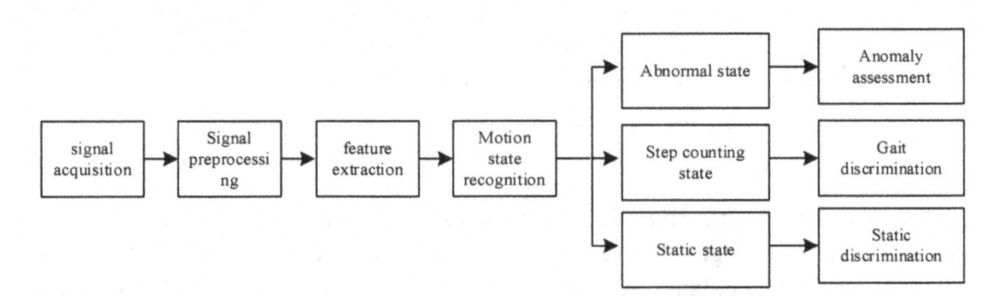

Fig. 1. Structure diagram of athlete training body data acquisition

In the case of movement, the movement of the trunk is mainly forward, backward and vertical. Human walking can be approximately regarded as periodic motion. Here, we can choose to take each step as a cycle or two consecutive steps as a cycle for analysis

[4]. In this paper, we choose one step as a cycle, starting with the heel of the moving foot off the ground and ending with the landing of the moving foot into a supporting foot. During the movement, the foot goes through the process of stepping, landing and supporting. In this process, the vertical acceleration is generated by the combined force of the reaction force of the ground to the foot and gravity, and the forward and backward acceleration is generated by the friction between the ground and the foot. The changes of vertical and forward acceleration are shown in Table 1:

Table 1. Changes of trunk motion acceleration

Trunk movement	Pedal	Stride forward	Landing	Brace
Vertical direction	Decrease → increase	Enlarge	Increase → decrease	Reduce
Fore-and-aft direction	Enlarge	Increase → decrease	Reduce	Decrease → increase

In the process of human swing arm walking, the motion of the upper limb is similar to periodic motion. Assuming that the arm does not exert force in the swing process and does not consider the friction resistance, the model can be simplified as a simple harmonic pendulum, and the acceleration of the sensing equipment can be decomposed into tangential component and normal component [5]. From one deflection angle maximum position to another deflection angle maximum position θ to m or g to e, the tangential acceleration φ decreases first and then increases, and the normal acceleration l increases first and then decreases. The following equation can be obtained from the law of conservation of energy and Newton's law of mechanics:

$$a_t = mge \sin \theta \tag{1}$$

$$\frac{1}{2} a_c^2 = mgl(\cos \theta - \cos \varphi) \tag{2}$$

Thus, the acceleration of any point (e.g. point B) is:

$$\lambda = g\sqrt{a_c^2 \sin^2 \theta - \cos \varphi \cos \theta + a_t^2 \sin^2 \varphi} \tag{3}$$

When you know the amplitude of a person's swing arm when walking φ (i.e. the maximum deflection angle between arm and trunk), the change process of resultant acceleration can be calculated. Assuming point B and point D are equilibrium points, the ideal change of acceleration in a swing arm cycle is shown in Table 2:

Table 2. Variation of upper limb swing acceleration

Arms swing	A → B	B → C	C → D	D → E
Tangential at	Reduce		Enlarge	
Normal ac	Enlarge		Reduce	
Closing acceleration a	Reduce	Enlarge	Reduce	Enlarge

To understand the concept of exercise load, you can't take it for granted. After reading a lot of research on sports load, due to different starting points and professional backgrounds, the description of the basic concept of sports load is also very different. In this paper, there are a lot of fuzziness and contradictions in the understanding of the basic concept of sports load, which shows the complexity of sports load itself. In the design process of heart rate monitoring, the author uses heart rate monitoring to reflect the exercise load, and also uses the heart rate area in the application part of exercise heart rate monitoring, and the root of these designs is exercise load. Therefore, how to make a scientific explanation and definition of exercise load has become an urgent task in the theoretical support of this paper, If sports training ignores the internal functional state of the body, excessive or insufficient exercise, then external training will not achieve the desired effect, and even have a negative impact on the physical health of athletes [6]. In order to better study and control the adaptation process of exercise training intensity and body function, the concept of exercise load is introduced. The concept of sports load is based on the mutual adaptation of sports forms and physical function changes. The fundamental purpose of monitoring and adjusting sports load is to scientifically and effectively organize and adjust the factors related to the mutual adaptation of physical function and sports behavior, so as to improve the effectiveness, pertinence and safety of training. The focus of exercise load is the change of body state caused by exercise behavior and the stress response of the body to the process of exercise training.

2.2 Athlete Training Status Monitoring Data Processing

In order to clarify the relationship between various motion states and state transition conditions, this paper uses the nested state machine model to describe each state and its state transition. Because there are many subdivided States, it is difficult to avoid the disorder of logical relationship by using the same state machine, so the classification method is adopted [7]. In terms of structure, the motion state is divided into three parent states: static state, walking state and abnormal state, and the states with similar logical relationship or similar transition conditions are classified into the parent state. Each parent state can be regarded as a separate sub state machine. The monitored states are reasonably organized in the form of nested state machines to form a clearly descriptive motion monitoring state machine model, as shown in Fig. 2:

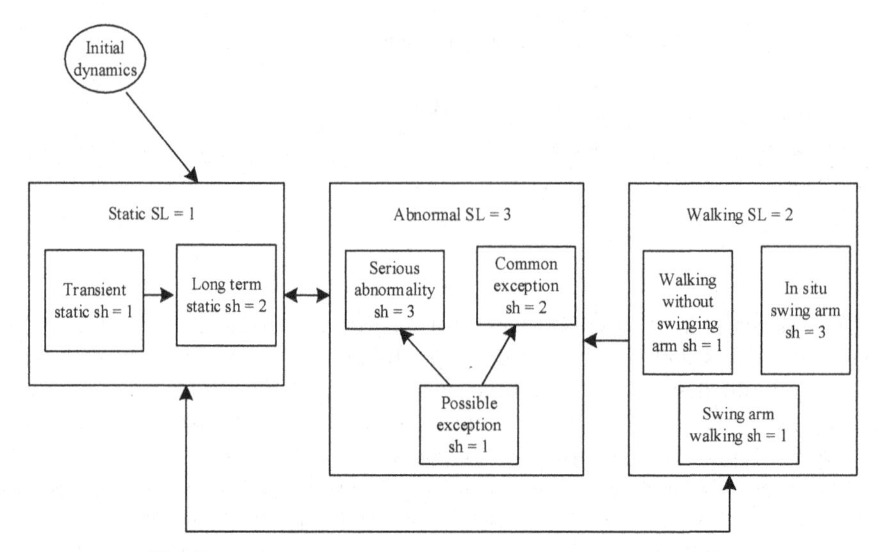

Fig. 2. Athlete training state monitoring data processing model

Among them, the purpose of exercise heart rate monitoring is to help athletes monitor their exercise status in real time and for a long time, and adjust the exercise load in time according to the exercise heart rate, so as to avoid sports injury caused by insufficient exercise intensity and poor exercise effect or excessive exercise, so as to achieve good exercise effect. The basic needs of this paper are: monitoring real-time heart rate, This is the requirement for the sensitivity of heart rate sensor and heart rate acquisition algorithm; It can monitor heart rate for a long time, which is the demand for data storage and analysis of heart rate acquisition equipment and athletes' adherence to heart rate monitoring; It can detect the exercise heart rate, which requires the portability or wearability of the exercise heart rate acquisition device. It can view the exercise heart rate in real time, which is the demand for real-time display of exercise heart rate monitoring; It is the core requirement to adjust the exercise load in time according to the exercise heart rate. This requirement requires us to give a reasonable theoretical model and operable scheme of using exercise heart rate to monitor the exercise load. According to the demand analysis, the functional modules of this are designed as shown in Fig. 3:

When a device is connected, there are only two status roles: master device and slave device. The device that enters the connection state from the initiation state is the master device, and the device that enters the connection state from the broadcast state is the slave device. The master and slave devices communicate with each other and specify the transmission sequence. A slave device can only communicate with one master device, while a master device can communicate with multiple slaves. There may be more than one state machine in the link layer. Of course, there are certain state restrictions when supporting multiple state machines. The multi state combination restrictions of the monitoring link layer are shown in Table 3:

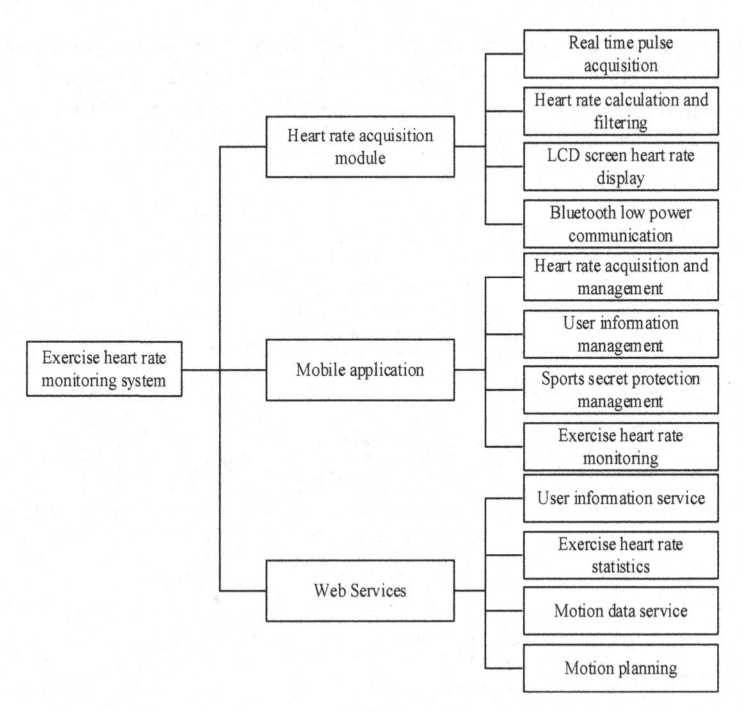

Fig. 3. Functional structure of exercise heart rate monitoring

Table 3. Multi state combination limit of monitoring link layer

Status and roles		Broadcast state	Scanning state	Initiating state	Connected state	
					Main equipment	Slave device
Broadcast state		Prohibit	Allow	Allow	Allow	Allow
Scanning state		Allow	Prohibit	Allow	Allow	Allow
Initiating state		Allow	Allow	Prohibit	Allow	Prohibit
Connected state	Main equipment	Allow	Allow	Allow	Allow	Prohibit
	Slave device	Allow	Allow	Prohibit	Prohibit	Prohibit

Add the motion monitoring program to osal. First, register the task in osal, add the address of the event processing function of the new task to tasks arri, and add the event and event processing function of the new task to the event table and processing function table. Then, add the initialization function of the new task to osallnit tasks, and each task will automatically obtain the assigned task ID. It should be ensured that the sequence of event handlers in the taskbar array corresponds to the order of initialization functions invoked by each task in the salinit Tasks function, so as to ensure that each task's event handler receives the correct task ID. To save the task ID assigned by the osalinittaskso function, to define a global variable for each task to save the event of the motion monitoring service, the corresponding event handler Step Process Evento is invoked. The implementation procedure of the event handler is shown in Fig. 4.

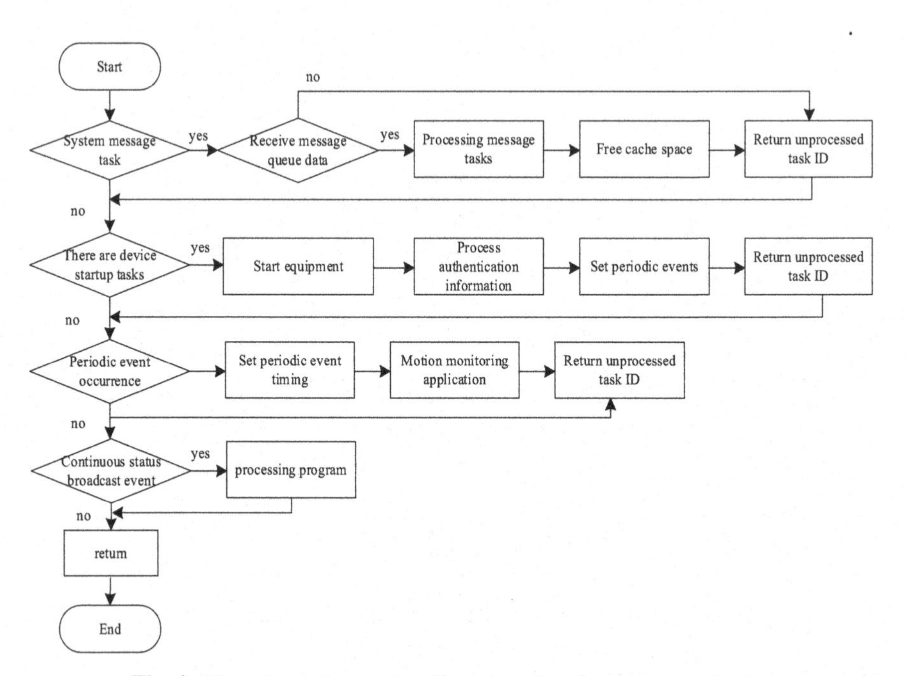

Fig. 4. Flow chart of event handling procedure for sports monitoring

In the process of equipment initialization event processing, the first timing time of periodic event is set, and in the process of periodic event processing, the next timing time is reset. By setting these two parameters, you can adjust the cycle of calling the periodic task processing function, which is the motion monitoring application designed this time. Ideally, regardless of program execution time and other factors, the set values of these two timing times are the cycle of sensor acquisition signal and the cycle of motion monitoring application. In practice, the timing time is less than the sampling period designed by the algorithm.

2.3 Realization of Sports Training Intensity Monitoring

When the human body moves, the values of the three axes of the acceleration sensor represent the sum of the acceleration component and gravity component generated by the human body movement on each axis. The component of gravity direction does not affect the research of this paper. Therefore, the component of vertical direction in this chapter includes gravity component and human acceleration component. Let $a_i(t)$ represent the triaxial acceleration vector of the sensor at time t. in order to calculate the vertical component of the acceleration signal in human motion, first calculate the gravity component when the sensor is stationary, and then use this gravity component and $a_i(n)$ to estimate the actual component in the vertical direction. In order to determine whether the sensor is in a static state, three parameters need to be calculated first, namely, the maximum value and minimum value of the acceleration signal in a sliding window and the average value of each axis:

$$\begin{cases} a_{i\_\max} = \mathrm{Max}(a_i(t))(i = x, y, z, t = 0, 1 \ldots N) \\ a_{i\_\min} = \mathrm{Min}(a_i(n))(i = x, y, z, t = 0, 1 \ldots N) \end{cases} \quad (4)$$

The whole set of motion monitoring sub modules are used to complete the whole set of motion monitoring and real-time control of the whole server. Due to the powerful processing and computing performance of the server, most of the data processing and computing of this system are carried out on the server. The server of motion data real-time monitoring consists of the following six modules: real-time monitoring terminal communication service module, planning task service module, data flow service module, data analysis and processing service module, data base station communication service module and database communication service module. The structure of the server is shown in Fig. 5:

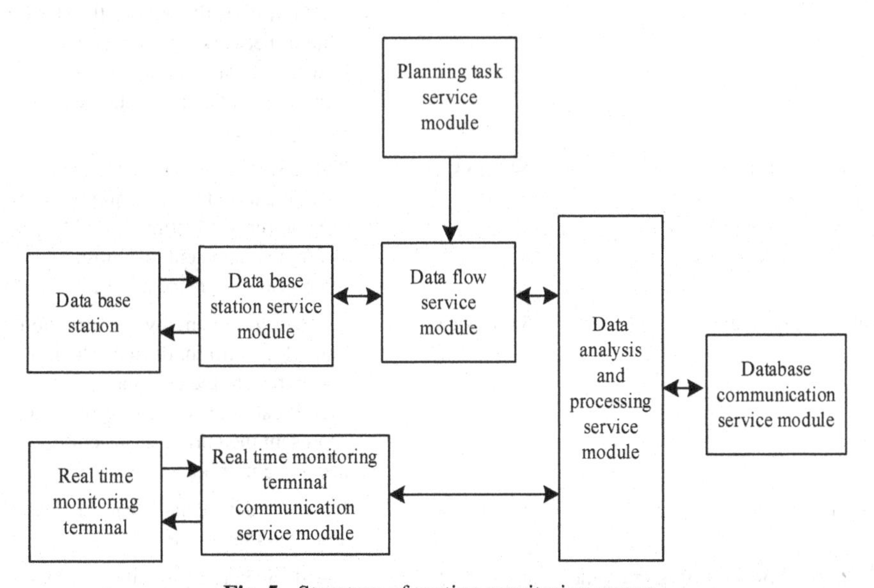

Fig. 5. Structure of motion monitoring server

The real-time monitoring terminal of real-time monitoring of sports data is realized based on Android platform. It is used to present the sports data such as the number of steps and energy consumption generated by the athlete in the process of exercise to the sports observer in real time, so as to grasp the real-time changes of the athlete's physical condition in the process of exercise in time. In order to enable the data acquisition equipment to correctly complete the data communication with the base station, this paper designs a set of data communication protocols to provide guidelines for the data acquisition equipment to send and receive data with the base station. The specific contents of the communication protocols are shown in Table 4:

Table 4. Communication protocol between SCADA equipment and base station

Command name	Command format	Describe
Get the current data of the device	43 44 xx xx	Obtain and set the current data, including device ID, steps, current x, y, z axis acceleration value and remaining power of the device. The first two bytes are the command identification, and the last two bytes represent the current command serial number, which is automatically incremented by PC, the same below
Get movement steps	48 45 xx xx	Obtain the number of operation steps recorded by the equipment, and the equipment will adapt to several data. The return result includes the starting time of network return data, the number of data pieces, and the number of remaining data pieces of the equipment
Set current time	54 55 xx xx	The first two bytes of the current time of calibrating the equipment are the command identification. The last two bytes represent the current instruction sequence number
Set the current exercise index	54 52 xx xx	Set the motion energy consumption data of the current device. The first two bytes are the command identification, and the last two bytes represent the current instruction sequence number
Reset a device	53 54 xx xx	Reset the device and clear the data stored in the device
Address of broadcasting equipment	55 42 xx xx	The device broadcasts its own MAC address to the environment, and adds represents the MAC address of the device

The starting and ending time of sports monitoring is controlled by the sports observer on the real-time monitoring terminal. The monitor can select a specific sports group for monitoring, observe the overall exercise of the class in real time, and also observe the detailed exercise data of specific individuals in the group in real time. Figure 6 describes the workflow of the real-time monitoring terminal:

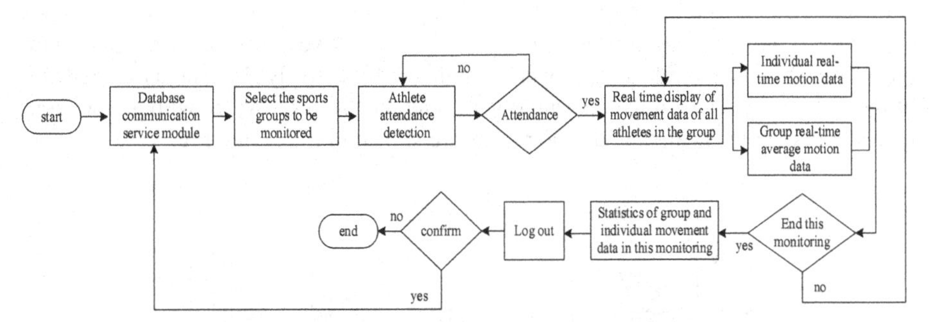

Fig. 6. Work flow chart of real-time monitoring terminal

To sum up, the implementation process of the remote monitoring method of athletes' training intensity based on the mobile Internet of things in this paper is as follows:

Step 1: athlete training body data collection.

Step 2: athlete training status monitoring data processing.

Step 3: perform accurate spatial calculation in common databases.

Step 4: use MBR technology to check the relationship between predicates, eliminate predicate relationships that are inconsistent with the actual situation, and then form a topology data table.

Step 5: calculate the support of the predicate, exclude the items with low support, form the optimal database, and form a new topological relationship data table.

Step 6: complete the training and monitoring according to the topological relationship data table.

3 Analysis of Experimental Results

The proposed method is verified on the intelligent terminal. The intelligent terminal adopts Android operation and is equipped with lis3dh4 acceleration sensor. LIS3DH4 is a three-axis digital acceleration sensor of Italian French semiconductor (st) company. It is packaged in 16 pin plastic, with small and slim overall size and size of only 3mm × 3mm × lmm. It adopts the way of digital output, avoiding the use of other chips for analog-to-digital conversion. The acceleration sensor lis3dh has the acceleration output of x, y and z degrees of freedom, which can sense the motion information of the human body in an all-round way. The measuring range is within ± 2g/ ± 4g/ ± 8g and l6g, and the minimum working current consumption is 2 μA. LIS3DH4 can provide very accurate measurement data output and maintain excellent stability under rated temperature and long-time operation. The sampling rate of the sensor is adjustable between 1-5000Hz. The sampling frequency set in this paper is 50Hz and 50 samples are collected per second. Low sampling rate can not only reduce power consumption, but also reduce noise interference. In order to test the effect of using adaptive hybrid filtering algorithm on improving measurement accuracy. The heart rate synchronously recorded by three channel ECG machine is selected as the reference standard to evaluate the accuracy of the. The measurement results with and without adaptive hybrid filtering algorithm are compared with the reference standard. According to the three groups of measured values of ECG machine and equipment with and without adaptive hybrid filtering algorithm, the average heart rate and standard were calculated respectively. The standard deviation of the difference between the measured value and the reference value with and without filtering algorithm is calculated. The results are shown in Fig. 7 and Fig. 8:

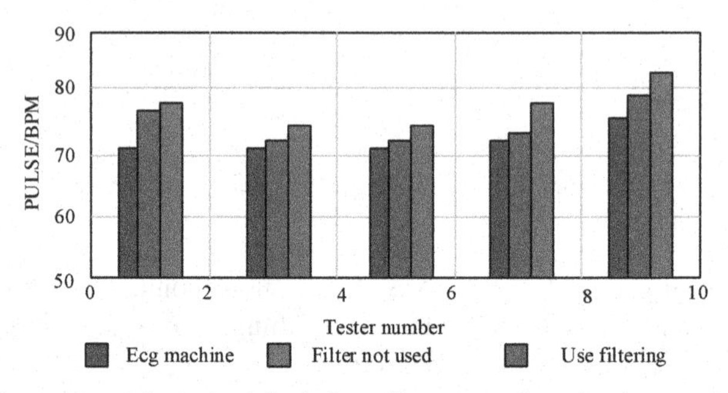

Fig. 7. Comparison of standard deviation of heart rate detection in exercise training

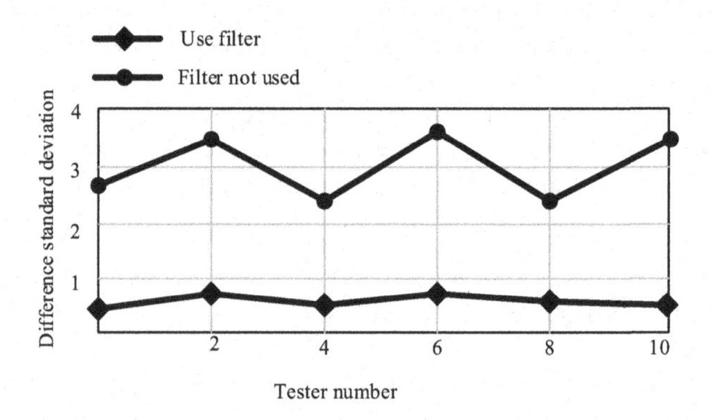

Fig. 8. Standard deviation of difference between detection filter and reference value

The mobile terminal equipped with LIS3DH acceleration sensor is placed in the front chest and coat pocket of the experimenter, and the experimental site is selected in the open outdoor. Each experimenter made three actions of running, taking off and squatting for 10 times respectively, and the interval between actions was more than 20s, so that the experimenter could adjust himself to a normal state. Among them, the take-off and squat are in-situ movements, and the experimenter is tested according to his normal state. This paper has no specific restrictions on the actions made by the experimenter, and is tested completely according to his normal state. In the test process of running, the experimenter is required to run more than 5 steps and test according to his normal running state. The vertical component analysis of acceleration sensor is used to distinguish the two actions of "take-off" and "squat". Figure 9 ~ Fig. 11 respectively represent the variation diagram of the vertical component of acceleration with time in a take-off action and a squat action. The abscissa axis represents time t (unit: s), and the ordinate axis represents the vertical component of acceleration y (unit: m/s2) (Fig. 10).

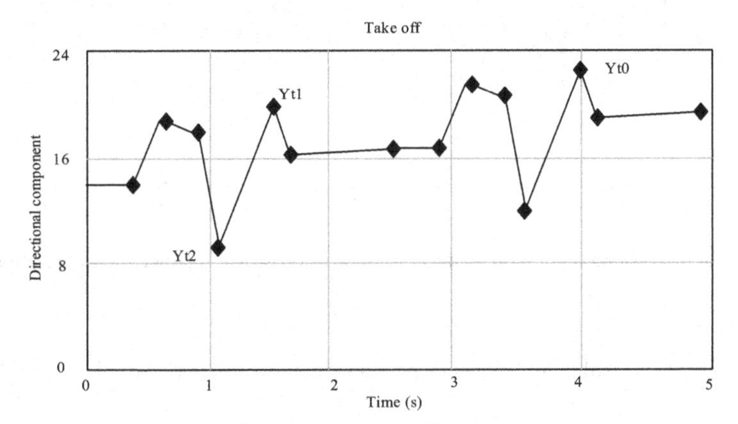

Fig. 9. Schematic diagram of monitoring amplitude of vertical component of acceleration during take-off

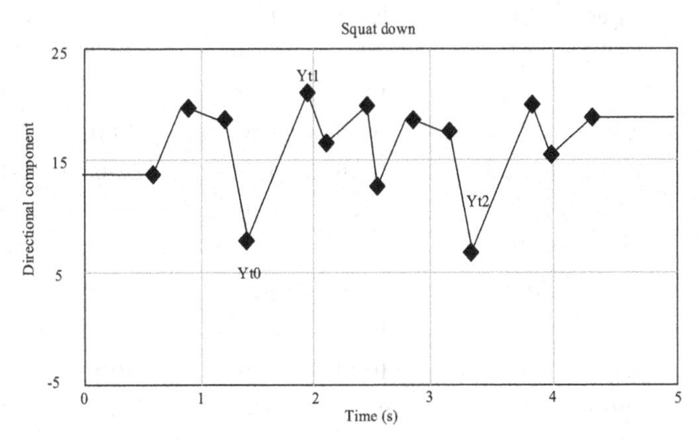

Fig. 10. Schematic diagram of monitoring amplitude of vertical component of acceleration during squatting

According to the experimental scheme designed in this paper, this paper makes statistics on the three types of actions made by a total of 12 male and female experimenters. Each person is required to repeat each action for 10 times. This paper records the correct recognition times of each person's take-off, run and squat. The experimental results are shown in Table 5 and Table 6:

To sum up, the recognition rate of the three movements in this paper has reached more than 90%. Therefore, from the correct recognition rate obtained from the experiment, it can be seen that the method proposed in this paper has a high recognition effect

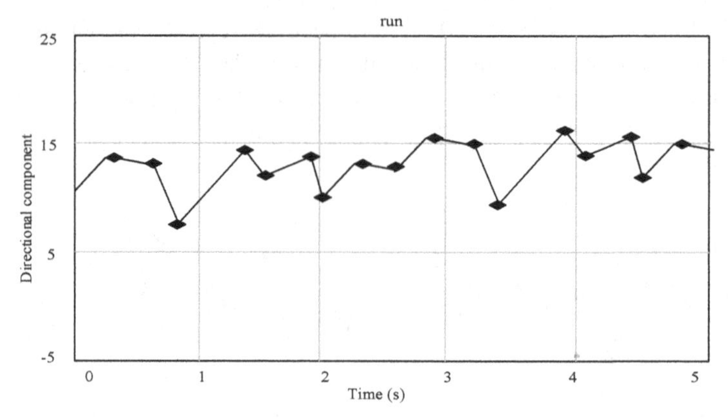

Fig. 11. Schematic diagram of monitoring amplitude of vertical component of acceleration during running

Table 5. Correct recognition times of three actions

Experimenter	Take off (correct recognition times/total times)		Run (correct recognition times/total times)		Squat (correct recognition times/total times)	
	Paper method	Traditional method	Paper method	Traditional method	Paper method	Traditional method
Female 1	9/10	6/10	10/10	6/10	9/10	6/10
Female 2	9/10	7/10	10/10	6/10	10/10	5/10
Female 3	8/10	5/10	10/10	5/10	9/10	5/10
Female 4	10/10	5/10	9/10	6/10	10/10	5/10
Female 5	8/10	6/10	8/10	4/10	8/10	6/10
Male 1	10/10	6/10	10/10	6/10	10/10	6/10
Male 2	10/10	7/10	10/10	4/10	9/10	4/10
Male 3	10/10	5/10	10/10	5/10	8/10	5/10
Male 4	10/10	7/10	9/10	5/10	7/10	6/10
Male 5	9/10	6/10	10/10	4/10	9/10	6/10
Male 6	9/10	6/10	10/10	6/10	9/10	7/10
Male 7	10/10	5/10	9/10	5/10	10/10	5/10

on the three movements of take-off, running and squatting, and has a certain practical application value.

Collect the experimental data of athletes' training intensity risk value monitoring before and after using the method in this paper under the three movements of take-off, running and squatting, and sort them into a table. As shown in Table 7.

Table 6. Statistical table of experimental results of three actions

Action	Total number of experiments	Correct recognition times	Correct rate	Number of missed reports	Underreporting rate	Number of false positives	False positive rate
Take off	120	112	93.33%	8	6.67%	1	0.83%
Run	120	115	95.83%	5	4.17%	0	0.00%
Squat down	120	108	90.00%	2	1.67%	0	0.00%
Total	360	335	93.06%	15	4.17%	1	0.28%

Table 7. Statistical table of experimental results of three actions

Number of experiments	Before use			After use		
	Take off	Run	Squat	Take off	Run	Squat
100	1.51	1.63	2.87	0.61	0.73	1.97
150	1.27	1.39	1.87	0.37	0.49	0.97
200	1.20	1.32	2.47	0.30	0.42	1.57
250	1.28	1.40	3.05	0.38	0.50	2.15
300	1.35	1.47	2.96	0.45	0.57	2.06
350	1.40	1.52	2.67	0.50	0.62	1.77
400	1.53	1.65	3.57	0.63	0.75	2.67
450	1.36	1.48	2.64	0.46	0.58	1.74

As shown in Table 7, the risk value of monitoring athletes' training intensity after the use of this method is significantly lower than that before the use, and the monitoring performance is higher. It can eliminate untrustworthy data from athletes' training data and make the monitoring results more accurate.

4 Conclusion

There is no doubt about the training effect of remote monitoring of athletes' training intensity, but there are still many places to be improved in the theoretical research, practical application and monitoring methods of training monitoring. There are various types of training monitoring, many training variables can be controlled, and the achievable training objectives are very diverse. How to make rational use of different types of training monitoring, how to accurately select and control training variables, and how to monitor training load, load response and other loads to achieve the expected training goals are very difficult problems. In addition, how to incorporate training monitoring into synchronous training and how to integrate training monitoring with special training are also important problems that need to be solved urgently.

Fund Project. School-level general project teaching and research project (brand major self-designed project) Promotion and practical analysis of happy gymnastics in the elective course of performance major (2020JYXM61).

References

1. Lian, J., Fang, S.-y, Zhou, Y.-f.: Model predictive control of the fuel cell cathode system based on state quantity estimation. Comput. Simulation **37**(07), 119–122 (2020)
2. Moreno-Navarro, P., Ibrahimbegovic, A., Ospina, A.: Multi-field variational formulations and mixed finite element approximations for electrostatics and magnetostatics. Comput. Mech. **65**(1), 41–59 (2020)
3. Cunningham, J., Broglio, S.P., O'Grady, M., et al.: History of sport-related concussion and long-term clinical cognitive health outcomes in retired athletes: a systematic review. J. Athl. Train. **55**(2), 132–158 (2020)
4. Bi, X.-c, Zhan, J.-g,: Effect of different incremental load tests on validity of test indicators in rowing training monitoring. J. Beijing Sport Univ. **43**(02), 149–156 (2020)
5. Pan, S., Yuan, M.: System and application of video surrveillance based on edge computing. Telecommun. Sci. **36**(06), 64–69 (2020)
6. Liu, H., Liu, Z., Ha, J.: Characteristics, influential factors and monitoring strategies of rugby injuries. J. Wuhan Inst. Phys. Educ. **54**(05), 75–81 (2020)
7. Fu, Y., Wang, Z., Chen, W., et al.: Infrared human motion target detection based on gauss background model. Autom. Instrument. 34(01), 63–65+69 (2020)

Research on Equipment Management System of Smart Hospital Based on Data Visualization

Yuanling Ma, Xiao Ma(✉), Chengnan Pan, Runlin Li, and Zhi Fang

CCTEG Chongqing Engineering (Group) Co., Ltd., Chongqing 400016, China
mxiao2546@163.com

Abstract. Medical equipment is an important part of the assets of smart hospitals and an important guarantee for clinical departments to complete normal medical treatment. Strengthening the management of medical equipment, giving full play to the maximum benefits of medical equipment, and preventing the loss and idleness of medical equipment have been widely valued by hospitals. However, hospitals are facing high equipment investment, difficult management and low operation and maintenance efficiency. In order to improve the management effect of network equipment, security equipment, guidance equipment and other electromechanical equipment in smart hospitals, an application method of smart hospital equipment management system based on data visualization is proposed. Aiming at the intelligent management of equipment, the visualization technology is used to monitor the operation status of various equipment, and the management strategy of various equipment in the smart hospital is optimized to improve the operation efficiency of the equipment in the smart hospital. Build a smart hospital equipment operation supervision system to achieve efficient management of smart hospital equipment. Finally, it is confirmed by experiments that the smart hospital equipment management function of the system in this paper is perfect, and the operation stability is strong, and it has high practicability and reliability in the actual application process.

Keywords: Data visualization · Smart hospital · Equipment management · Condition monitoring

1 Introduction

With the gradual deepening of hospital informatization construction, the network equipment, security equipment, guidance equipment and other electromechanical equipment in the hospital need to be effectively managed, including energy consumption management, transfer management and other management measures. It is increasingly important to promote the informatization of equipment management. Based on the existing equipment management methods, the equipment management information system uses computer technology and network technology as means to assist equipment management and realize the scientific and standardized management of hospital equipment. Especially in the process of cost accounting, it can accurately and quickly provide detailed

W. Fu and L. Yun (Eds.): ADHIP 2022, LNICST 468, pp. 381–396, 2023.
https://doi.org/10.1007/978-3-031-28787-9_29

data on equipment and consumable materials [1]. Therefore, building a smart hospital equipment management system has important practical significance.

Reference [2] proposes a hospital equipment management system based on artificial intelligence and 5g communication. The system collects the operation information of hospital equipment through sensors, sends these data to the cloud analysis platform in real time through 5g technology, analyzes and processes the data in real time using artificial intelligence algorithms, and transmits the analysis results to the equipment management department in real time. Reference [3] proposes a hospital equipment management system based on the Internet of things mode. According to the distribution of hospital departments, each department is equipped with positioning devices, and the deployment of Internet of things equipment is realized through key points (entrances and exits) and regional coverage (large-scale). The RFID reader is used as a wireless access point device, and the wireless LAN transmits the RFID information and the device location information to complete the management of hospital equipment. Reference [4] proposes a hospital equipment management system based on WeChat applet, develops software with WeChat applet, and conducts related system research and development on the platform. In the above, the WeChat applet was developed based on the WeChat Markup Language (WXML), and the front-end and back-end data transmission followed the JavaScript Object Notation (JSON) format, and completed the research on the medical equipment management system based on the WeChat applet.

Although the above system can detect the information of hospital equipment, it has the problem of insufficient management effectiveness. Based on this, a smart hospital equipment management system based on data visualization is designed. The overall design idea of the system is as follows:

(1) Using Delphic6 as the software development platform, in the Windows server2013 operating system, the overall framework structure of the system is designed.

(2) Design the system hardware structure of the B/S three-tier architecture, and optimize the equipment status monitoring function, equipment positioning and secondment functions.

(3) Analyze the functions of the medical equipment analysis and management module, plan the hospital equipment management process, and complete the hospital equipment management according to the system operation and management hierarchy.

2 Smart Hospital Equipment Management System

2.1 Overall System Architecture

The intelligent hospital equipment management system adopts Delphic6 as the software development platform, and the database adopts the Paradox7 database method in the Borland database engine BDE attached to Delphic. The functions of the system are divided into equipment application management, system management, contract management, acceptance management, equipment planning management, fixed asset management, use management, maintenance management, premium management and other functions. Among them, system management is mainly responsible for backing up the overall data of the system, and assigning relevant roles to better manage the system; application management mainly includes procurement applications, department use application equipment plan management, which is based on the specific situation of the hospital, formulate a unified procurement plan for hospital procurement, etc.; Contract management mainly includes the processing of procurement contracts, procurement information, etc., and provides contract query. The acceptance of equipment includes two parts: equipment acceptance and inquiry; Fixed assets include the storage, delivery, and withdrawal of assets; the use of equipment is mainly based on the use and inquiry of equipment by departments; Equipment maintenance is Including the maintenance of equipment, maintenance and other aspects. Through the design of the above process, combined with the relevant knowledge, the development of the system adopts data visualization technology, and uses Vs2005 as a development tool to develop the system. At the same time, the ADO.NET component is used to realize the connection between the user and the database. The system operating system adopts Windows server2013, and the specific overall system architecture design is shown in Fig. 1:

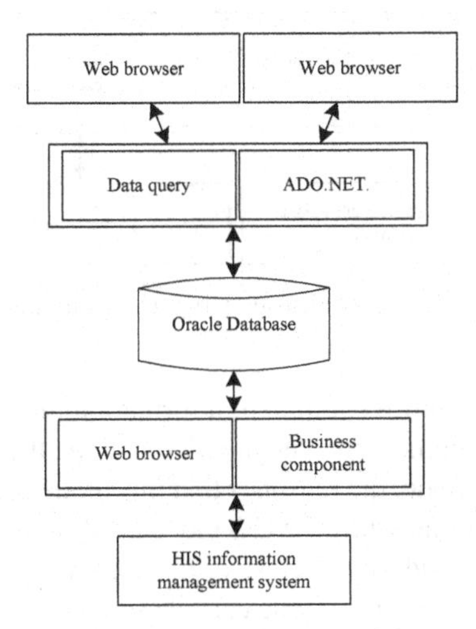

Fig. 1. The overall architecture of the system

The development of medical equipment management system mainly uses computer programming technology, database technology and computer network technology, the purpose is to realize the functions of various information and data input, query, editing, statistical output of medical equipment, and provide medical services for various departments of the hospital. Resource sharing of device information and data. According to the actual work situation of the hospital by the data visualization technology, the user hopes that the equipment management system can not only manage the basic information such as the quantity, attributes, and repair reports of the equipment, but also help the medical staff to grasp the operation dynamics of the equipment in real time, including viewing the equipment generated when the equipment is working. Data, save the alarm-related information generated when the equipment is abnormal or critical.

2.2 System Hardware Structure

Based on this, the system hardware structure configuration is optimized, as shown in Fig. 2:

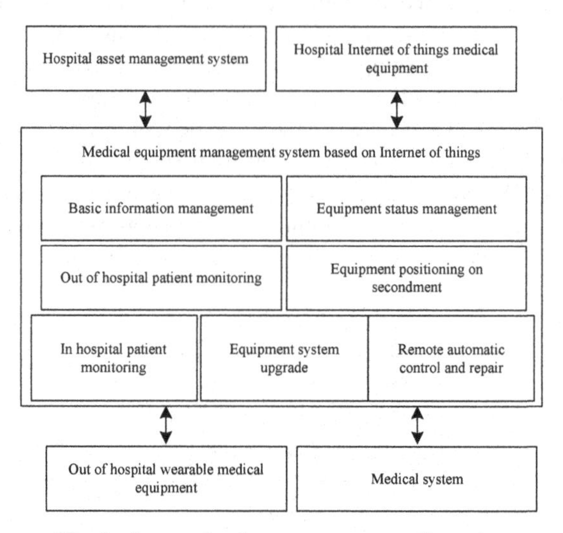

Fig. 2. System hardware structure configuration

The construction of the entire equipment information management system is based on the hospital medical equipment management system. At the same time, the design of the system adopts the B/S mode, the reason is that this mode only needs to maintain and update the server, not the client, while the C/S mode needs to maintain the server and the client at the same time.. Considering the use of the system by the hospital, the decision to adopt the B/S mode is because this mode can be used by all departments only through the Internet. The system takes advantage of the three-tier architecture of B/S mode and divides it into presentation layer, logic layer and data layer [5]. The presentation layer mainly realizes the interactive interface provided to the user and the system. The user realizes the response of the user and the application server by means of the data visualization technology in the web browser. The logic layer submits the relevant requests of the user to the data layer through the logic processing of the application server; the data layer provides the query of the data, and realizes the access to the system through the ADO.NET component.

2.3 Equipment Condition Monitoring Function Optimization

With the improvement of the overall technical level of medical equipment and its wide application and deepening in the medical field, a large number of technologically advanced medical equipment have been introduced into hospitals, which has promoted the progress of hospital medical technology and the improvement of medical level. However, the speed of improving the operation and management level of hospital medical equipment is generally far behind the speed of introduction of medical equipment. The current situation of medical equipment operation supervision is not commensurate with the overall technical level of hospital medical equipment, nor does it meet the needs of modern medical development. Most medical equipment operation management concepts still remain in the stage of medical equipment failure repair and regular maintenance mode, which is also an important reason for this situation.

The basis for the realization of the monitoring function is to collect various sensors of different operating conditions of different medical equipment, using wired or wireless data transmission mode, with the help of the existing local area network in the hospital, the sensors are connected into a network, and the monitoring data collected by each sensor is collected. Send it to the data command server, record, analyze and process the sensor data in the command server, and at the same time publish the operation status of the device data in the form of web, and all the data of the monitored device can be seen through web browsing. In addition, the instruction server finds obvious or potential abnormal operation of medical equipment through comparative analysis of equipment monitoring data, and makes different response strategies for different abnormal conditions. For example, through the GSM/CDMA interface module, an alarm message is sent to one or more mobile data receiving terminals (mobile phones) of medical engineering technicians, so that the technicians can obtain the abnormal situation of medical equipment at the first time, and respond in time to prevent The abnormal operation of the equipment further deteriorates into a fault, or a small fault in a part of the equipment causes the overall fault of the equipment. The structure and function of the medical equipment operating state monitoring system based on the Internet of Things technology are shown in Fig. 3.

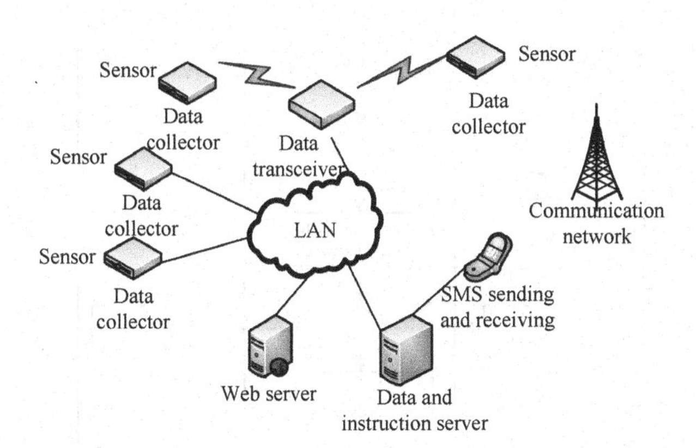

Fig. 3. Functional structure of equipment status monitoring

The data acquisition terminal is mainly composed of a data collector, which mainly collects data including liquid level, pressure, on-off, temperature, humidity, etc., and requires various types of sensors. In order to facilitate the seamless connection between the sensor and the data collector, multiple standard sensor interfaces are reserved on the data collector. As long as the required sensors with suitable accuracy and standard interfaces are used, and the correct collection part of the monitoring equipment is selected, the collection work is carried out normally. At the same time, set the corresponding sensor serial number and acquisition-related parameter settings in the data collector, so that the acquisition parameters correspond one-to-one with the required acquisition equipment, and the whole is clear.

2.4 Equipment Positioning and Secondment Function Optimization

The equipment location and secondment service means that users can check the location of the access equipment through the system and make secondments according to their needs. Because some equipment is small in size, but relatively expensive, the loss will be large when the incident occurs occasionally. Therefore, the equipment management personnel of the hospital hope to manage the location of the equipment in real time through the equipment management system to prevent the loss of the equipment, or when the equipment is lost, In addition to being able to find its location and quickly recover it, sometimes when there is insufficient equipment, it is necessary to borrow data visualization technology from some cooperative hospitals. If you can know the location of a device that needs to be seconded, you can Quickly grasp the usage of the equipment to facilitate secondment. The business process of equipment positioning and secondment is shown in Fig. 4:

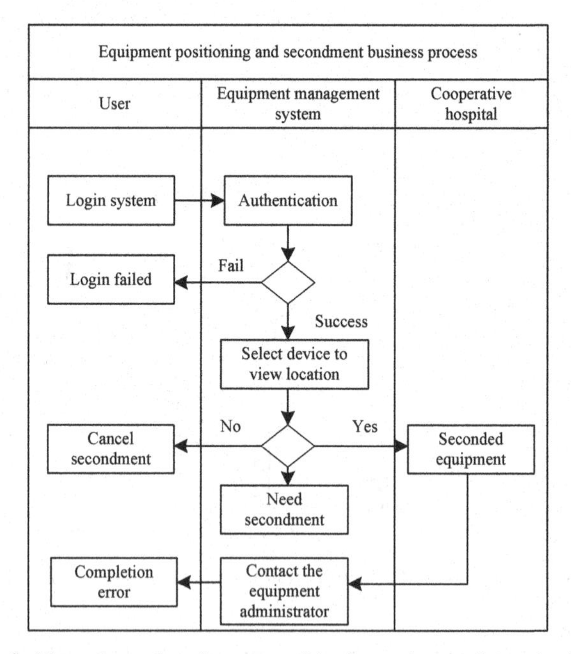

Fig. 4. Flow chart of equipment positioning and secondment business

Taking equipment positioning and secondment as an example, the process of this service is shown in the figure. In this system, the equipment system upgrade business mainly refers to sending the upgrade files to the required equipment [6, 7]. For the medical equipment produced, the system will be upgraded and optimized from time to time, but the upgrade process is more complicated. Professional technicians are required to make the upgrade file into a corresponding upgrade package, and then import the upgrade package into a device through the RS232 serial port to USB port. Among the identifiable hardware middleware, connect the middleware to the device, select the device to enter the upgrade mode, read the files in the middleware, and perform the upgrade. It can be seen from the above description that in this system development, the operation process of upgrading business needs to be simplified and modified to one-key upgrade [8]. That is, when the device needs to be upgraded, the device administrator can log in to the system, click the corresponding device, and select the upgrade option to send the upgrade package. After the device receives the upgrade package, it will detect it, greatly improve the upgrade efficiency, and reduce the difficulty of operation.

2.5 Realization of Smart Hospital Equipment Management

After a detailed understanding and analysis of the process of hospital medical equipment management through data visualization technology, a fully functional medical equipment management system must include five management systems: equipment purchase, equipment management, equipment query, auxiliary data management, and system setting. The functions of the device analysis management module are shown in Fig. 5:

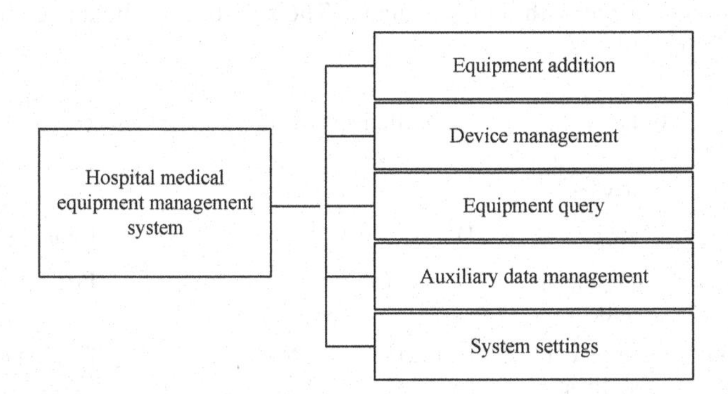

Fig. 5. Functions of medical equipment analysis and management module

The module is mainly aimed at the benefit statistics of large-scale medical examination equipment in hospitals. Since the number of large-scale equipment configured in hospitals is small, but due to its high value, the sum of its costs accounts for a large proportion of the total cost of hospital equipment. Benefit analysis can reflect the actual operation of large-scale medical equipment in a certain period of time, so that problems and shortcomings in the management of medical equipment can be found in time, which is convenient for hospital managers to use the mode and operation direction according to the actual situation. Making accurate and rapid adjustments not only improves the utilization rate of medical equipment, but also provides an effective reference for medical equipment purchase decisions. The equipment supervisor can perform functions such as adding, modifying, deleting, and querying to manage the basic information of charging items. According to the entity relationship design of the conceptual model of the data visualization database, combined with the data relationship and ER diagram in the system, the smart hospital equipment management database table design is shown in Table 1:

Table 1. Smart Hospital Equipment Management Information Sheet

Table name	t_user			
Listing	Name	Data type	Can it be blank	Remarks
ID	User ID	varchar (65)	No	Primary key
Status	state	int (15)	No	0 disable 1 normal

In order to make the equipment better serve the clinic, equipment supervisors can create equipment clinical training or operation training plans through the system, strengthen the equipment use training of front-line medical personnel, and improve the overall medical diagnosis level of the hospital [9]. The equipment information table of the data visualization technology is used to store the basic information of the equipment. The primary key is the equipment number ID, the foreign key is the equipment classification ID, and the supplier name. The device information table has a one-to-many or many-to-many correspondence with multiple tables. The structure of the device information table is shown in Table 2:

Table 2. Device visualization information correspondence

Table name	t_device			
Listing	Name	Data type	Can it be blank	Remarks
id	Equipment number	varcha r(65)	No	Primary key
DeviceType id	Equipment classification	int (22)	No	Foreign key
Stratus	Equipment status	int (22)	No	0 in use; 1 idle; 2 Scrap; 3 external transfer
trder name	Supplier name	varchar (65)	No	Foreign key
Deposit	Storage location	varchar (65)	No	–

After the smart hospital equipment information plan based on data visualization technology is added, you can query the training plan list, supervise the implementation of the training plan, and check whether the training plan has been implemented through the modification function. The flowchart of the training management function is shown in Fig. 6:

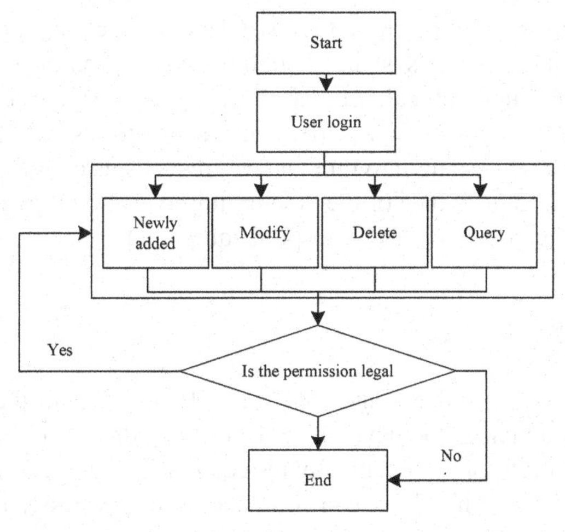

Fig. 6. Flow chart of medical management functions

The system business overview shows that the interaction between users and IoT devices can be divided into four parts, namely, the intelligent medical device layer, the network layer, the system application layer, and the user. The smart device connected to the data visualization technology sends the device data to the management system of the application layer through the network layer, and the user completes the business operation through the system to achieve the purpose of managing the device [10]. The device management system is responsible for the application layer in the entire interaction process, provides the user with an interactive interface, and connects the device layer with the user. The hierarchical structure of the system is shown in Fig. 7.

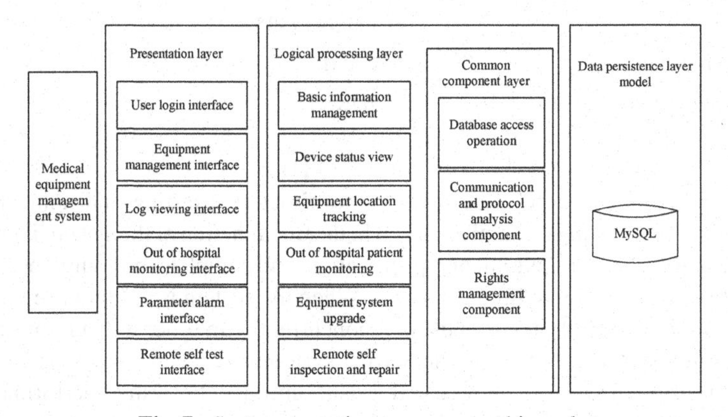

Fig. 7. System operation management hierarchy

Corresponding preparations before testing the data visualization system, including the software and hardware environment required for system testing and the testing methods for functional testing and non-functional testing of the system, since the medical equipment management system is a system based on B/S architecture system, so when users use it, they can access the system through the browser on the computer. It can be seen from the internal hierarchical structure diagram that the system adopts a typical data visualization technology design mode, which separates user interaction, data storage and system logic processing into three layers, reduces its coupling, is conducive to the development and maintenance of the system, improves development efficiency, and realizes the Management objectives of hospital equipment.

3 Analysis of Results

In order to find out the problems and functional deficiencies in the system as early as possible, after completing the development of the medical equipment management system, the system has been systematically tested. There is one server-side device with a high configuration, and the client can meet the system access requirements of the hospital. There are ten devices with a low configuration. The software and hardware environment of the system test server is shown in Table 3:

Table 3. System test server environment table

Server side	
Processor	Inter(R)Xeon(R) CPU E5–1220
Memory	64 G
Hard disk	300 G
operating system	Windows 10
Database software	Oracle Enterprise 11G
The server	Lenovo x3850
Network broadband	200M
Java virtual machine version	JDK1.7.0

System testing can usually be divided into functional testing and non-functional testing. Functional testing is mainly to test whether the function of the system is complete according to the system requirements specification, and complete the comprehensive test of the system function by writing test cases. When the system is put into use, there will be the possibility of multiple users using it concurrently. In order to better evaluate the stability of the system and verify whether the system can achieve the expected effect, the system is optimized through repeated tests to improve the system performance test environment: system Using B/S architecture, the test client and server are carried out in the same local area network, excluding the factors of network speed limitation or unstable network speed, the test environment is as follows:

Table 4. Test environment required for testing

Name	Type	Edition
Server side	Operating system	Windows 10
	Operation platform	Tomcat 9.0
	Operating environment	JDL1.8
	Frame platform	SpringMVC
Database side	Operating system	Windows 10
	Database	Mysql6.5
Client	Operating system	Red Hat Enterprise Linux7.0
	Browser	Baidu browser

Table 5. Terminal login system function test table

Test title	Terminal login system	Test case ID	01
Test purpose	Test whether the system login function is available		
Test object	Equipment maintainer		
Test item	Test content	Testing procedure	Test result
A	Configure server IP address	Click the menu key of the mobile phone to enter the setting interface to set the server IP address	Prompt the set IP address information
B	WiFi on	Click the "turn on WiFi" button	The system prompts "WiFi network card is opening"
C	User login	Staff enter user name and password and click "login"	Enter the correct information and log in to the system successfully

Non-functional testing refers to indicators other than testing functional requirements, mainly referring to the stability and reliability of the system. Through the trial of users, and according to the opinions and feedback of the staff, the system has been adjusted many times to make the system functions more perfect and improve the user experience. The system has carried out functional and non-functional tests respectively. Functional testing is done by developers by writing test cases. The following is an introduction to some test cases of the system:

Table 6. Console query function test table

Console query	Test case ID	02
Test whether the console query function is available		
system administrator		
Test content	Testing procedure	Test result
Query basic information of all devices	Enter information management and click the "query" button	Display basic information of all devices
Query the basic information of a specific device	Enter information management and click the "query" button	Displays the basic information of the number
Query the maintenance information of all equipment	Enter information management and click the "query" button	Display all maintenance information of all equipment
Query the maintenance information of a specific equipment	Enter information management and click the "query" button	Displays all service information for this number

Table 7. Test results of equipment information transfer process

Step name	Test description	Test result
Start process	The supervisor of the transfer out equipment department can start the process instance	In line with expectations
Transfer out hospital equipment supervisor application	Fill in and submit the application report as required; If you fail to fill in as required, you will be prompted that the filling information is incomplete when submitting	In line with expectations
The medical director and financial director of the transferred out hospital shall conduct concurrent review	You can review the application, return or agree to the next step of circulation; If either party executes the return process, it will be returned to the applicant	In line with expectations
The equipment supervisor and financial director of hospitals in districts and cities shall conduct parallel review	You can review the application, return or agree to the next step of circulation; If either party executes the return process, it will be returned to the applicant	In line with expectations

Terminal login system test case:

Table 8. Test results of equipment asset retirement process

Step name	Test description	Test result
Start process	Department personnel can start process instances	In line with expectations
Application submitted by departments of county-level hospitals	Fill in and submit the application report according to the needs; If you fail to fill in as required, you will be prompted that the filling information is incomplete when submitting	In line with expectations
Reviewed by department directors of county-level hospitals	You can review the application, return or agree to the next step of circulation	In line with expectations
The equipment supervisor and the financial director shall review in parallel	You can review the application, return or agree to the next step of circulation; If either party executes the return process, it will be returned to the applicant	In line with expectations

Users within the authority can use the management module functions normally, and can normally initiate the asset transfer approval process. The test results are shown in the table below.

Table 9. System concurrency test data table

Serial number	Response time of 60 simulated users (s)	Response time of 90 simulated users (s)	Response time of 120 simulated users (s)	Response time of 150 simulated users (s)
A	0.9	1.6	2.3	3.5
B	0.7	1.8	1.8	3.4
C	1.2	1.8	1.8	2.9
D	0.7	1.2	2.4	3.3
E	1.3	1.3	2.2	3.6
F	0.6	2.1	1.4	3.1
G	0.9	1.4	2.4	3.2
H	1.0	1.6	1.7	2.7
I	1.0	1.6	2.3	2.9
J	0.7	1.2	1.7	2.6
Average value	0.9	1.56	2	3.12

Users within the authority can normally use the functions of the asset management module, and can normally initiate the approval process for asset scrapping. The test results are shown in the table below.

Table 10. Comparison of debugging before and after the use of the equipment management system

Business content	Before using the system	After using the system	Test result
Approval timeliness	The annual procurement plans of 23 hospitals are generally approved in May	The annual procurement plans of 23 hospitals have been approved in March	Obvious improvement
Asset account management	The format of equipment management account of hospitals divided into districts and cities is inconsistent and the update is slow	Query the equipment asset account of each hospital, master the use status of the equipment in time, and make management measures in time	Obvious improvement
Operation and maintenance management	Maintenance records are not standardized; The maintenance supervision is not in place, and there is a phenomenon of missing inspection	The maintenance record document is standardized, the specific maintenance plan can be formulated, and the maintenance times of the manufacturer can be supervised	Obvious improvement

In order to fully test the concurrency situation, Load Runner was used to assist, and a total of 300 device simulation software were run on ten client devices. Simulates multiple users to perform concurrent tests on system operations such as login, requesting real-time parameters, querying alarms, etc. However, due to space limitations, only the test data of the physical parameters of the requested device is shown here, and this test is used as an example to illustrate the test of concurrency capability. Process. By making a script that requests the physical parameters of the device at runtime, simulate multi-user access to the system, set the initial concurrency to 60 user request operations, and increment by 30 users each time, until the system notices that the response is slow and the operation is not smooth., at this time the number of users is the maximum concurrent amount of the system. 10 tests were performed under each concurrency, and the results were averaged. The test data obtained are shown in the table.

The system operation performance is further compared with the traditional system (reference [2]). The comparison results are shown in Fig. 8:

The medical equipment management system has been put into trial use in 23 hospitals under the jurisdiction of our district, and the equipment management level and approval timeliness have been greatly improved compared with those before the system was used. The specific comparison is shown in the table.

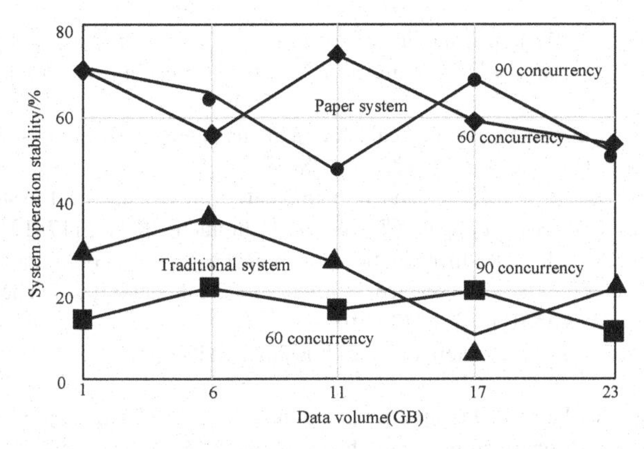

Fig. 8. System performance comparison test results

Using the test method, for the seven functional modules of the system, different test cases were written, using equipment simulation software, sending work data scripts, and comprehensively testing the functions of the system. The results show that the functions implemented by the system can achieve the expected results., In the parameter and alarm analysis, the parameter name/parameter value, alarm name, etc. can also be correctly parsed, business logic can be processed correctly, and the actual operation of the system can be simulated to meet all user needs. With the help of the Load runner test tool software, system performance test, including system compatibility, stability, response time and so on. The test results show that the system has good compatibility and can meet the needs of users.

4 Concluding Remarks

The hospital equipment information management system built on the medical equipment management system can better complement the hospital information management and integrate it into the management of the hospital. In this paper, the more mature data visualization technology is used as the development language to better realize the retrieval and query of system data. At the same time, Oracle enterprise database is used to facilitate the storage of system data. Although the hospital equipment management system studied this time can complete relevant management work, it still faces some problems, such as implementation cost, higher positioning accuracy, more efficient event processing mechanism and data processing ability, etc., which still need to be solved in future research and practice.

References

1. Jiaju, C., Qian, L., Haifeng, Y., et al.: Discussion on the construction of a comprehensive smart reconciliation platform based on hospital financial management. China Med. Devices **36**(09), 118–121 (2021)

2. Zhang, H., Jia, Li.: Design of medical detection equipment management system based on artificial intelligence and 5G communication. Electronic Design Eng. **29**(11),113–116+121 (2021)
3. Zhou, J., Lu, H.: Construction of medical equipment management system based on the Internet of Things model. Chinese Hospitals **25**(06), 88–89 (2021)
4. Shi, Z., Yang, Q., Ni, Y.: Design and implementation of management system of medical equipment based on WeChat applet. China Med. Equipment **18**(11), 117–121 (2021)
5. Wang, J., Qi, J., Sun, J., et al.: Investigation on medical equipment configuration and identification of obstacles in after-sales service of primary medical institutions in zhejiang province. China Med. Devices **35**(01), 13–17 (2020)
6. Li, J.: Energy management system of smart hospital and its utility. Henan Sci. **39**(07), 1137–1142 (2021)
7. Wang, L., Ren, H., Zhang, Y.: Practice of intelligent service construction based on optimization of hospital patients' treatment process. Chinese J. Health Inf. Manag. 17(03), 259–264+284 (2020)
8. Yuan, X., Chen, H., Li, M.: Systemic analysis of introducing supporting medical equipment during construction of proton therapy center. Chinese Hospital Manag. **40**(11), 84–86+96 (2020)
9. Li, L., Chen, Y., Guo, J.: Research on informatization strategy of clinical configuration and scrapping disposition of large medical equipment in hospital. China Med. Equipment **17**(06), 166–169 (2020)
10. Zhang, D., Liu, J., Liu, Y.: Construction of evaluation system of medical equipment management capability and analysis of its reliability and validity. Chin. Nurs. Res. **35**(08), 1499–1501 (2021)

Design of Information Consultation System for the Whole Process of Construction Engineering Based on BIM Technology

Yufei Ye[1(✉)], Xiao Ma[1,2], Zepan Yang[1], Cancan Liao[1], and Leihang Chen[1]

[1] CTEG Chongqing Engnineering (Group) Co., Ltd., Chongqing 400016,, China
yeyufei@cqmsy.com
[2] School of Management Science and Real Estate, Chongqing University, Chongqing 400044,, China

Abstract. The existing building engineering information consulting system has the problems of low reliability and poor performance. This paper designs a new whole process information consulting system of building engineering based on BIM technology. Based on the optimization of hardware configuration structure, BIM technology is used to optimize the system software function and operation process, to realize the collection and classification of the whole process of construction engineering information, so as to facilitate the consultation of users. The system performance test results show that the actual consulting function of the system is strong and can meet the needs of the system.

Keywords: BIM technology · Construction engineering · Information consultation

1 Introduction

With the improvement of economic level and continuous progress of production technology, the construction industry and related industries have also been developing rapidly. The change of architectural design thinking and methods, the improvement of various properties of building materials, the emergence of new construction processes and technologies, and the updating and iteration of building systems and equipment have gradually led to the diversification, complexity and integration of architecture in terms of appearance design, internal space and place spirit [1]. The inevitable result is that the difficulty and dimension of management and maintenance of a building after it is completed and put into use will also be greatly increased compared with the past. Therefore, it is necessary for the unit that manages and maintains the building to get rid of the traditional management mode and introduce new management thinking and methods. The whole life cycle of a building can be divided into four stages, namely, the planning and design stage, the construction and construction stage, the operation and use stage and the scrapping and demolition stage. Therefore, it is necessary to get rid of the traditional

W. Fu and L. Yun (Eds.): ADHIP 2022, LNICST 468, pp. 397–410, 2023.
https://doi.org/10.1007/978-3-031-28787-9_30

management mode for the units of building management and maintenance, and to carry out building information consultation can improve the quality of building maintenance management.

Reference [2] proposes an integrated consulting system for building information based on cloud BIM. Building information is transmitted between different hierarchical structures of the system. After collection, coding and classification, the whole process of building information is stored in the BIM database, component information is extracted through the cloud platform layer, and the information consulting is completed by the adaptive association rule scheduling method. Reference [3] proposes a building information consulting system based on cloud computing platform, which clusters building information using AP clustering algorithm, stores it in a virtual database after virtualization processing, and transmits the consulting results to the service layer for users to view according to users' consulting needs. Reference [4] proposes a building information integrated consulting system based on MVC mode and mysql. Based on MVC mode, the information consulting service system is designed in combination with MySQL, and an online information consulting service platform for public demand information retrieval is developed by applying machine learning algorithm, Django framework and bootstrap responsive layout. The platform can provide users with a fast information retrieval experience, get visual information feedback, and give certain decision support. Although the above system can complete the information consultation, but there are problems of low reliability.

Aiming at improving the performance of building information consulting system, a new consulting system based on BIM technology is designed. The overall design scheme is as follows:

(1) Collect BIM information of buildings and construct the system framework of engineering information management. According to the characteristics of different stages of construction engineering, BIM information is integrated and interactive stored.
(2) Construct a vector space model, transform the text information into a computable model, calculate the similarity of the text information, and form a complete description of the construction life cycle of the engineering information set.
(3) Based on IFC's BIM data integration structure, the attribute set of building information is classified to realize consultation and sharing of building information.

2 Construction Engineering Whole Process Information Consultation System

2.1 Emergency Structure Configuration of Information Consultation System in the Whole Process of Construction Engineering

For the architectural design and construction stage, take a construction project as an example, test the BIM information integration platform, and study the feasibility and effectiveness of BIM-based engineering information integration and management. Effective information management during the life cycle adds value to the construction and use of construction projects. Effective information management refers to the effective creation of information, effective management of information and effective sharing of information. BIM is the key to effective information management. This paper introduces the information management of construction engineering, and analyzes the characteristics, information flow, information characteristics and current information management mode of engineering projects. Analyze the meaning of BIM and its modeling techniques. The system framework and integration mechanism of BIM-based engineering information management are proposed as shown in Fig. 1.

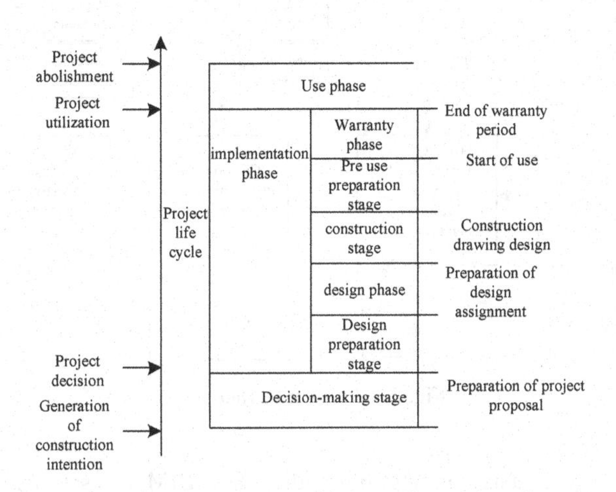

Fig. 1. System framework and integration mechanism of BIM-based engineering information management

DM belongs to the management of the investor and the developer, while FM belongs to the management of the project service life, which may be the owner or the facility management unit entrusted by the owner. Therefore, construction project management is not only the management of the owner, but also involves the management of each participant unit of the construction project, as shown in Fig. 2:

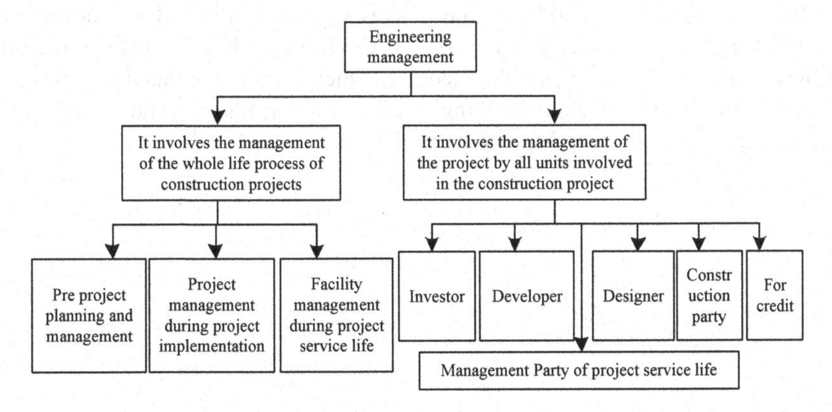

Fig. 2. Construction engineering information management system structure

Considering that many participants will consult information, it is also necessary to consider the information characteristics under different construction periods, as shown in Fig. 3.

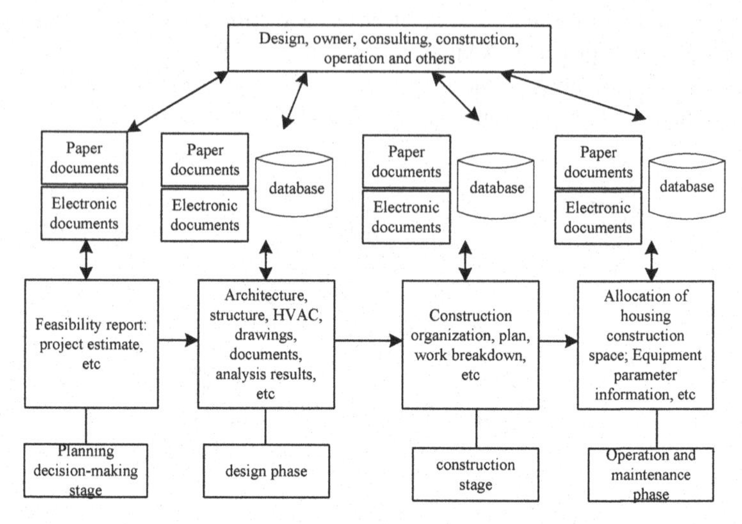

Fig. 3. Hardware structure

In order to process a large number of building data, BIM data sets are constructed to realize operation and maintenance information demand analysis, data function design, data architecture design, database creation and operation and maintenance function module construction [5, 6]. The realization of the BIM data integration platform provides the engineering information and operation and maintenance information required by the building for the integration of the intelligent management system based on BIM.

2.2 System Software Function Optimization

The BIM sub-information model as the core is the phase-oriented and application-oriented BIM information creation method [7]. The BIM database contains information of different stages of construction engineering, which can be integrated and expanded to effectively meet the needs of building information consultation and provide better management services [8]. See Fig. 4 for details:

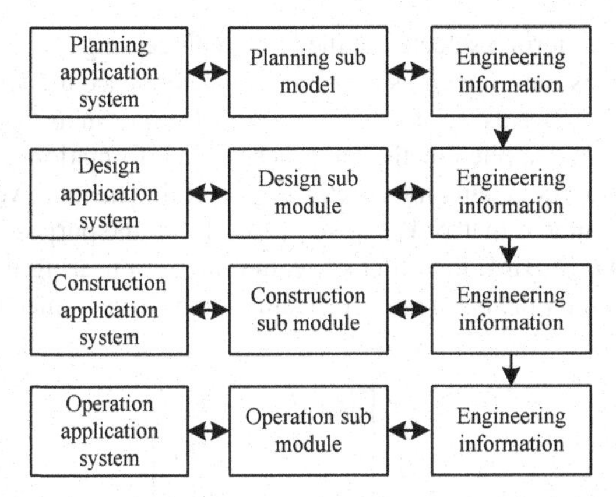

Fig. 4. BIM construction engineering information processing system

After the secondary processing of the BIM as-built model, the optimization of the operation and maintenance model, the creation of the operation and maintenance view, the optimization of the attributes, and the classification of the system, a building information model that is optimized for the operation and maintenance needs is completed. The conversion of the interactive standard format implements a lightweight operation and maintenance model [9]. As shown in the figure, the BIM data lightweight flow Fig. 5:

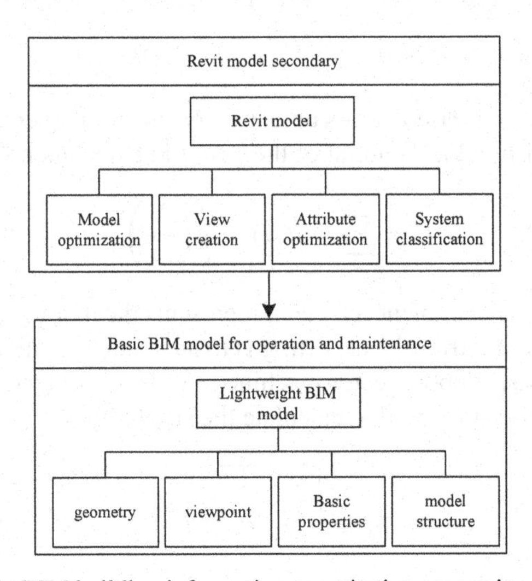

Fig. 5. BIM building information quantitative processing model

The feature items (terms) selected in the text preprocessing stage are collected, and the text information is represented as a structured vector space model (VSM) by using the uniqueness of these terms. Vector space model, also known as "word bag" method, is the most effective and applied method to process text information at present. Its basic idea is to table n text documents into vector d_{nm} in m dimensional vector space, so as to create feature document matrix $V_{m\times n}$, so as to achieve the purpose of modeling and structuring text data. In BIM, the order of feature items appearing in the document is not considered, and only the uniqueness of feature items is guaranteed. The established matrix is as follows:

$$V_{m\times n} = \begin{bmatrix} d_{11} & d_{12} & \cdots & d_{1m} \\ d_{21} & d_{22} & \cdots & d_{2m} \\ \vdots & \vdots & \ddots & \vdots \\ d_{n1} & d_{n2} & \cdots & d_{nm} \end{bmatrix} \tag{1}$$

Through the vector space model, the text information is transformed into a model that can be calculated, and the unstructured information is structured according to the established model, and the inverse document rate statistical calculation of the term frequency is carried out. If the term frequency TF is high in one document and rarely appears in other papers, (that is, the document containing this term has a low frequency of DF in the whole document set), so this term can be considered to have a good distinguishing ability. TF. IDF formula introduces word frequency and inverse document frequency:

$$w(d,t) = V_{m\times n}TF \times IDF \tag{2}$$

$w(d,t)$ is the weight of a term t in document d, TF is the frequency of the term t in a document, and IDF is the local weight of the term t in a document. The formula is:

$$IDF = \sum w(d,t) + \lg\left(\frac{N}{n_1}\right) - t \tag{3}$$

where, N represents the document set, and n represents the number of documents of the term t. . The premise of retrieval and sorting is to analyze the similarity of the text. For text similarity analysis, support vector machine (SVM) or distance formula and cosine formula shown below can be used to calculate the similarity:

$$d = \sqrt{\sum_{i=1}^{N}(p_i - q_i)^2} \tag{4}$$

Among them, p and q are the sub vectors of document vectors p and q in the feature vector dimension [10].

The building classification coding system is shown below.

Table 1. Overall analysis of building information classification system

System category	Name	Classification object	Classification method	Coding mode	Scope of application
Quota system	Valuation form of construction engineering unit	Work item	Line division method	Chapter number	Construction and installation works
	Comprehensive quota of Construction Engineering	Components and design components	Line division method	Chapter number	architectural engineering
Normative system	Quality acceptance standard of Construction Engineering	Mixing of components and work items	Line division method	Nothing	architectural engineering
	Construction technical operation specification	Work item	Line division method	Chapter number	Construction and installation works
Architectural literature	Chinese book classification	Facilities, units, components, etc	Line division method	Subtitle digital hybrid coding	Construction Engineering, civil engineering
Product catalogue	Classification in the valuation table of construction engineering units	Building materials, construction machinery	Line division method	Digital coding	—
	Classification in construction website	Construction products	Line division method	Nothing	—

The application system realizes the integration and sharing of data by extracting and integrating sub models. For example, various document data are mainly generated in the planning stage, which are stored in the form of files. In the design stage, architectural design, structural design, water supply and drainage design, HVAC design are carried out according to the information in the planning stage, resulting in a large amount of geometric data There is a need for collaborative data access between architecture and water supply and drainage and HVAC. These applications generate new information and integrate it into the overall BIM model. The BIM model integrates the engineering information in the planning stage, design stage and construction stage for the operation and maintenance application system to call. For example, the application system based on BIM can easily extract the building component information, building space information, building equipment information, etc. through the sub model. The application of BIM enables the integration and preservation of engineering information at all stages.

2.3 Realization of Construction Engineering Information Resources

The integration and collaboration of BIM data is the most critical basic technology in the BIM system. Data integration and collaboration and sharing mainly refer to the interaction between data: completion BIM is a complete data modeling. Because of the differences in application goals and application scenarios, BIM data cannot be applied to the operation and maintenance process, and it needs to be realized through the unified format of IFC. Data interaction and transformation are completed by using IFC file parsing and data exchange interface. The software logic function is to ensure that users can easily and quickly create, modify, extract and integrate data; In the early stages of BIM, IFC access program was used to extract a large amount of data from BIM database, as shown in Fig. 6 below:

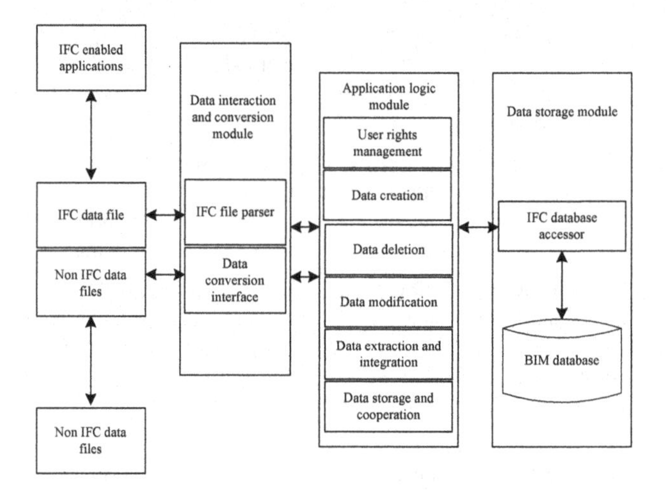

Fig. 6. BIM data integration structure based on IFC

Database is another important carrier of IFC data storage. Different from file based storage, database storage is more suitable for dealing with a complete IFC model with a large amount of data, and provides more powerful information exchange capability for different participants of the project through network-based interface. Databases can be divided into relational databases and object-oriented databases according to their types. Huang Zhongdong and others compared the characteristics of implementation based on relational database and object-oriented database:

Object oriented database has not been widely popularized and applied. Therefore, this section uses relational database to realize the data storage of IFC object model.

Table 2. Comparison of SDAI implementation based on Database

Project	Relational database	Object oriented database
Technology	Mature	Immature
Data mode and interface	Standard	Not standard
Application	Widely	Not extensive
Mode conversion	More complex	More direct
SDAI implementation	Complex	More complex

Attribute set, as the name suggests, is a collection of attributes. The description of things and concepts can be stored in the attribute set through an attribute. Attribute sets provide a flexible way to extend information description. In order to understand the concept of attribute set, this paper classifies the attribute set as follows, as shown in Fig. 7:

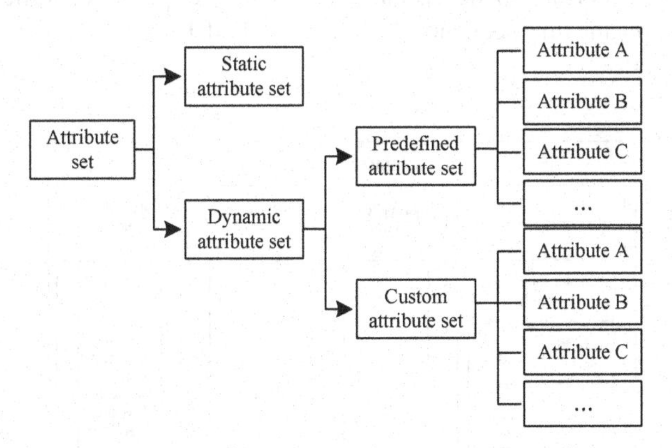

Fig. 7. Classification of information attribute set

The progress error in the S-curve comparison method is linked with the error degree, quantified, and a systematic progress reliability alarm function is established. According to the project progress and the actual situation of the project progress, risks are divided into four levels, and specific risk areas are set with different speed deviation values. In the project implementation stage, if the project progress deviates beyond this range, there will be corresponding alarm information. The prediction scope of project schedule safety risk control is shown in Table 3 below:

Table 3. Construction schedule reliability control warning interval

Progress deviation value t	-3% ~ 3%	3% ~ 4% or - 4% ~ - 3%	4% ~ 5% or - 4% ~ 5%	t > 5% or T < 5%
Deviation level	Low deviation	Low deviation	Moderate deviation	High deviation
Reliable state	High	High	Commonly	Low
Early warning signal	Grey	Brown	Brown	gules

In the reliability management of construction period, because the actual progress of the project does not conform to the plan, it must be predicted according to the actual situation. Therefore, the BIM software must include a prediction function, which can be used to predict according to the current progress and the specific conditions of the construction site, according to the current progress and the specific conditions of the construction site, using the time sequence prediction method. Figure 8 shows the architecture of BIM modeling:

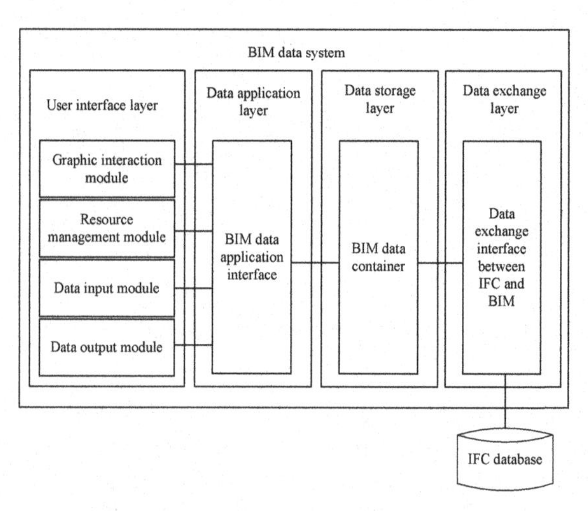

Fig. 8. Structure diagram of BIM model

The first level is the data exchange, which is mainly responsible for receiving and storing data; The second layer is the data storage layer, which is used to classify and store foreign data, and to a certain extent, to achieve the correlation of various data; The third level is the data application level, that is, the application window of BIM model, which can provide users with the ability to query, search and modify relevant project data; The fourth level is the user interface, whose function module is to display the data of BIM model through different charts and pictures according to the needs of customers.

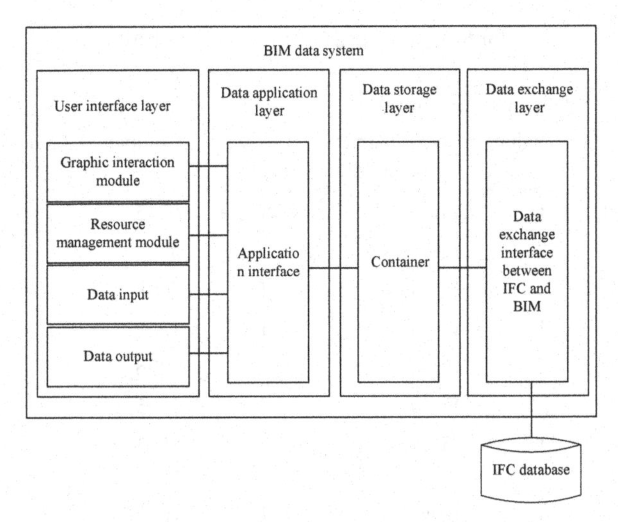

Fig. 9. Structure diagram of BIM model

Based on this, the corresponding project completion time can be calculated when the actual S-shape progress curve of the project is greater than or equal to 100, and the corresponding completion probability, namely progress reliability, can be obtained through computer simulation software simulation. The corresponding project completion time is obtained through computer simulation, so as to obtain the corresponding completion probability, that is, the reliability of the completion time.

3 Analysis of Experimental Results

3.1 Experimental Data

Based on the built BIM database and according to the construction characteristics of construction projects, a project case database of construction projects is formed.

Based on the above 10 construction project cases, the construction information is processed, the original information of the construction project is extracted, and dimensionless processing is carried out to obtain the construction data attributes. The data similarity of construction project cases is as follows:

Test the actual application performance of the consulting system according to the similarity results shown in Table 5.

Table 4. Sample data of project case base

Case	Engineering characteristic attribute					
	Building type	Value	Overall structure of the building	Value	Completed Area	Number of layers
A	Office	1	Frame	1	6235	7
B	Market	3	Frame	3	5968	5
C	Office	4	Frame	1	6698	8
D	Market	1	Frame	1	13652	15
E	Office	2	Frame	2	7236	3
F	Office	3	Frame	3	6895	6
G	Building complex	1	Frame shear	1	8532	2
H	Building complex	1	Brick	1	10252	8
I	Residence	2	Brick	2	7652	6
J	Office	1	Frame	1	7895	10

Table 5. Similarity between the proposed project and each information

sim1	sim2	sim3	sim5	sim6	sim10
0.9652	0.4895	0.9236	0.6895	0.9182	0.8165

3.2 Consulting System Performance Results

The reliability comparison test results of building information consultation are shown in Fig. 10:

Through application and relevant comparative tests, the differences between the intelligent integrated management of Bim and the system of reference [3] are finally obtained. The application advantages of integrated management based on BIM are obtained by comparing the characteristics of the two systems one by one, as shown in the following table. The table lists four main characteristics, namely: three-dimensional visualization, information integration, information processing and automation, which are subdivided under the four categories. Finally, the advantages of BIM intelligent integrated management after comparison are obtained, as shown in the following table.

Table 6. Comparison of advantages of the two systems

Content	Integrated management	Reference [3] system	Advantages of BIM integration
3D visualization	Full module 3D roaming function	Nothing	Convenient and advanced space browsing
	Internal perspective of facilities and equipment	Impossible	Avoid destructive inspection of equipment
	Comparison between model and real scene	Camera monitoring	Quickly confirm the cause of the fault
Information integration	IPad client input	Paper checklist filling	Paperless and complete information record
	Information transmission within the authority is fully transparent	Information transmission is not public	Improve information collaboration
Information processing	Display after system background processing	Manual processing	Mechanization replaces manpower
	Automatic reminder function	Nothing	Reduce the loss caused by human forgetting
Automation	Radio frequency automatic identification	Rely on workers' experience to find	Accurate and efficient identification of fixed assets
	Automated data collection	Workers report to the management center	Intelligent management

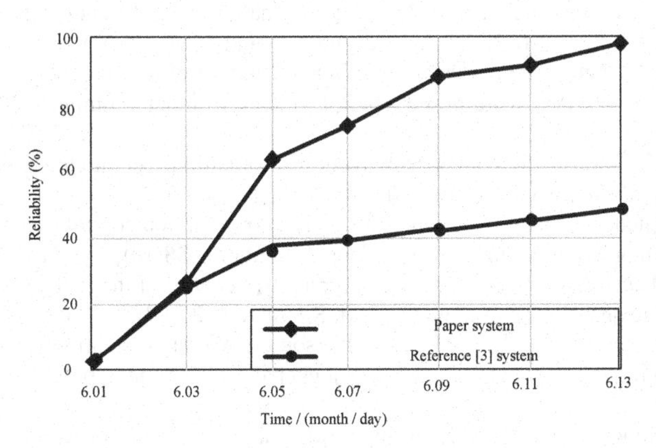

Fig. 10. Comparison of test results

On this basis, through two comparative experiments, it is fully proved that the new BIM information consulting system has obvious advantages. At the same time, we should also see that the traditional operation and maintenance management also has its merits, For example, "query the amount of information within the specified time" "The comparative test of BIM reveals that the breadth and depth of BIM integrated system need to be improved, and also reflects the problems brought by the mechanization and computer programmability in its application process, which can be made up by the traditional flexible management relying on manpower. Therefore, we should make comprehensive use of the advantages of BIM information consulting and the flexibility advantages of manual flexible management in order to achieve a better management trend of building modernization.

4 Concluding Remarks

Using the characteristics of BIM, it is integrated into the cost management of the project. Through the construction of the project cost management system, the daily management of the project cost and various business processes are simplified without losing accuracy, so as to remove unnecessary obstacles for the communication between various departments. At the same time, clarify the business division of each department and provide scientific work guidance to the project department through information management; Provide necessary data support for enterprises during enterprise upgrading and transformation. However, the consulting system in this study does not verify the performance of the system in the case of a large number of concurrent users. In the subsequent research, we will conduct in-depth research on this performance.

References

1. Ma, B.: Design of VR technology based 3D integration simulation system for architectural landscape features. Mod. Electronics Tech. **43**(20), 153–156 (2020)
2. Yang, J.: Informationintegration system of prefabricated building components based on cloud BIM. Syst. Simulation Technol. **17**(02), 123–127 (2021)
3. Luo, X., Fu, C.: Construction project budget information management system based on cloud computing platform. 自Techniques of Automation and Appli. **41**(03), 171–176 (2022)
4. Luo Zixun, X., Peng, H.X.: Design and implementation of information consulting service system based on MVC pattern and MySQL. China Comput. Commun. **34**(09), 184–188 (2022)
5. Xu, J., Ma, L.: Automatic boundary extraction from building point clouds based on virtual grids. Comput. Eng. Appl. **57**(10), 181–186 (2021)
6. Wang, S., Zhou, X., Dong, J.: Simulation energy consumption control system for near-zero energy building based on fuzzy PID. Comput. Simulation **38**(10), 263–267 (2021)
7. K Lian, J., Fang, S.-y., Zhou, Y.-f.: Model predictive control of the fuel cell cathode system based on state quantity estimation. Comput. Simulation, **37**(07), 119–122 (2020)
8. Shen, Y., Wang, J., Hu, D., et al.: Multi-person collaborative creation system of building information modeling drawings based on blockchain. J. Comput. Appli. **41**(08), 2338–2345 (2021)
9. Dong, N., Gong, C., Xiong, F.: Maturity evaluation of building information modeling applications in prefabricated building construction. J. Huaqiao Univ.(Nat. Sci.), **41**(01), 50–59 (2020)
10. Chen, J., Chen, L.: Lean management model for promoting prefabricated building lean construction based on building information model. Sci. Technol. Manag. Res. **40**(10), 196–205 (2020)

The Mobile Teaching Method of Law Course Based on Wireless Communication Technology

Xuejing Du[1](✉), Yi Wang[2], and Meng Huang[3]

[1] Shanghai University of Political Science and Law, Shanghai 201701, China
duxuejing15312@163.com
[2] Tianyi IOT Technology Co., Ltd., Shanghai 200122, China
[3] Jiangsu College of Finance and Accounting, Lianyungang 222000, China

Abstract. In order to better improve the teaching effect of law and improve the teaching efficiency of law courses, this paper proposes a mobile teaching method of law courses based on wireless communication technology. First, optimize the law teaching system, combine wireless communication technology to build a mobile teaching platform for law courses, optimize teaching content and teaching evaluation indicators, and achieve the goal of mobile teaching. The experimental results show that the design method can effectively improve the teaching efficiency of law courses, and has certain application value.

Keywords: Wireless communication · Law courses · Mobile teaching · Learning efficiency

1 Introduction

The informatization of higher education has gradually become a trend, and educational technology has intervened in our education and teaching mode in an unprecedented way. In the field of educational technology, wireless communication technology has become increasingly important as the most cutting-edge resource integration technology. In the prospect of the new higher education teaching model, some scholars believe that "the future classroom must be wireless communication data terminal classroom, including electronic textbooks, electronic desks, e-bags, electronic whiteboard, etc. [1]. In terms of resources, from simulated media to digital media, and then to network media, resources are ultimately on the mobile teaching platform, greatly rich content, so as to meet personalized learning." Then, as far as the transformation of legal education mode is concerned, it is very important to give full play to the role of wireless communication technology to achieve the improvement of education efficiency. In order to improve the efficiency of law course education in the era of wireless communication, this paper will start with the transformation of law teaching mode, point out the shortcomings of traditional law teaching mode, and then explore what wireless communication technology can bring to law education. Combined with the essence of wireless communication technology, it will explore the composition of the mobile teaching system of law. Traditional

W. Fu and L. Yun (Eds.): ADHIP 2022, LNICST 468, pp. 411–425, 2023.
https://doi.org/10.1007/978-3-031-28787-9_31

law teaching forms have their advantages and disadvantages. On the basis of absorbing the latest scientific and technological achievements, After improvement, there can be more technical advantages [2]. For law teaching, the full application of wireless communication technology means the expansion of teaching coverage, the improvement of students' enthusiasm for learning, and the better integration of teaching resources. By reasonably constructing the mobile teaching mode of law courses, we can effectively use the advantages of wireless communication technology in resource sharing, so that law teaching mode can make better use of technology and achieve the purpose of teaching and learning in the tripartite interaction of teachers, students and managers.

2 Mobile Teaching of Law Course in Wireless Communication Technology

2.1 Law Curriculum System Based on Wireless Communication Technology

In the world, there are two kinds of law teaching methods, one is case teaching, and the other is teaching teaching. Case analysis teaching is to take the case as a kind of reasoning and induction material that can clarify the legal theory from it. Teaching teaching focuses on teachers' platform for teaching relatively abstract concepts, principles, basic principles and theories, aiming to impart knowledge to students. Neither the recently proposed clinic teaching or the ancient annotation teaching has escaped both categories. China's current law teaching mode has four characteristics: first, with the teaching of existing knowledge as the main teaching purpose. The teaching activities of teachers and students still focus on the dissemination of knowledge. Second, classroom teaching is the main teaching method. The teaching mode with knowledge as the main goal is still the main teaching method, and self-study, discussion, research and experimental practice are all in an auxiliary position. Third, teachers and teaching materials are the center [3]. It is entirely teachers who decide the teaching content and teaching methods. There are few cases where students actively ask questions in class, not to mention raising different opinions and arguing with teachers. Students are completely passive and subordinate in teaching. Fourth, take the written test as the main evaluation means. The evaluation of merit students and scholarship grades and the selection of talent market are mainly based on the written test results. In college entrance examination, employment and career, written examination is increasingly widely used as a means to ensure fairness, and strengthens the exam-taking characteristics of education at all levels. Although, there is merit for this teaching model, the way knowledge is spread and shared is now changing rapidly [4]. Due to the information network, it will be more convenient and not enough to master the existing knowledge. Competition will mainly depend on the ability to use existing knowledge to creatively solve problems, and the ability to find new knowledge, while the current university law teaching mode is not conducive to the cultivation of students' innovation ability. From the technical point of view, the related technologies of mobile learning are as follows: mobile terminal, mobile Internet mobile learning platform design and development, and the construction of mobile learning resources [5]. The development and progress of these four technologies make the operation interaction, information exchange, resource storage acquisition and processing more convenient and fast, and promote the development and application of mobile learning. As shown in Fig. 1:

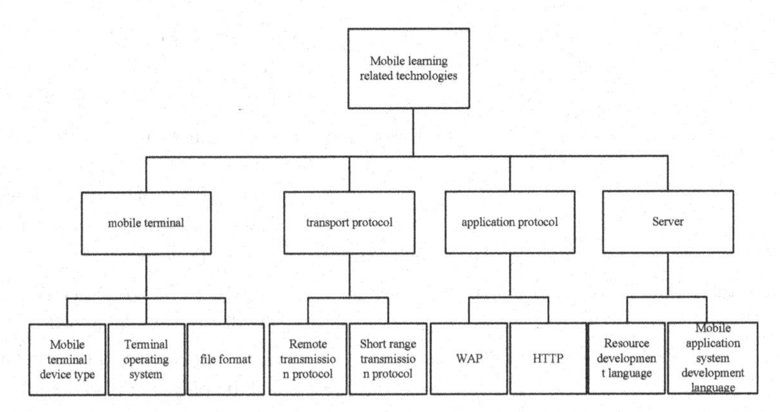

Fig. 1. Structure of wireless communication mobile teaching platform

Mobile teaching platform is a teaching support platform based on wireless communication technology. In the design and development, the resources of the original computer network teaching platform should be used to expand its functions on the basis of the original computer network teaching platform, so as to achieve the purpose of resource sharing and cost saving [6]. In terms of ELearning's digital teaching platform, it should cover all steps of teaching, including online teaching and teaching guidance, online self-study, online teacher-student communication, online homework, online testing and quality assessment and other comprehensive teaching support functions, which can provide real-time and non-real-time teaching interactive support between teachers and students [7]. A network learning platform for teachers and students should generally include teaching design module, learning tool module, collaborative communication module, assignment and review module, online q & A module, learning resource module, intelligent evaluation module and maintenance support module and so on from the current degree of information and the situation of mobile devices, Such as the screen size of mobile devices, mobile device penetration rate, and so on, There is little demand for a complete mobile teaching platform; And based on the auxiliary teaching is only a supplement to school teaching and classroom teaching, Each link of teaching cannot be completely mobile and other reasons It is not necessary for us to implement all of the above modules in the mobile teaching platform, As long as some of the more timely modules are selectively implemented, Such as login, homework assignment reference, online questions and questions video teaching and other modules, Thus, a mobile teaching platform shared by wireless communication network and mobile network is designed for these functions. The development of mobile learning curriculum content resources should reflect the characteristics of diversity. According to the mode of mobile learning, the form of mobile learning course content resources and their production tools can be summarized in Table 1 below:

Table 1. Mobile learning course content resource form and its production tool

Mobile learning mode	Resource form	Resource production method
SMS/MMS mode	Text information	Text editing, tools
	Picture information	Various graphic image making tools
	Short animation information	Flash Lite is a mobile device player, which can enjoy the wonderful content made by flash and make learning content for mobile phones
Web browsing mode	document info	HTML → WML, XHTML and other scripting languages
	View text files online	Text files in txt and PDF formats can be played on mobile learning terminals
	Play audio and video files online	Various audio and video formats change tools
Storage carrying mode	Store read text files	Various audio and video format conversion tools
	Store and play audio and video files	
	Supporting documents	Explanation of necessary audio and video formats for text learning materials

Since the number of users of community wireless communication data is relatively small compared with the public wireless communication data, the cost saving potential of wireless communication data computing is only partially realized. Nowadays, wireless communication technology is mainly a combination of the above methods, which is often called hybrid wireless communication data. Hybrid wireless communication data is a mixture of two or more private, community, or public wireless communication data, but hybrid wireless communication data still requires unique combined entities, thus providing the benefits of multiple deployment modes. This combination expands the deployment options of wireless communication data services, allowing it organizations to use public wireless communication data computing resources to meet temporary needs. Hybrid wireless communication data provides flexibility for internal applications, fault tolerance based on wireless communication data and scalability of services. Teaching methods are often characterized by mixed wireless communication data, as shown in Fig. 2:

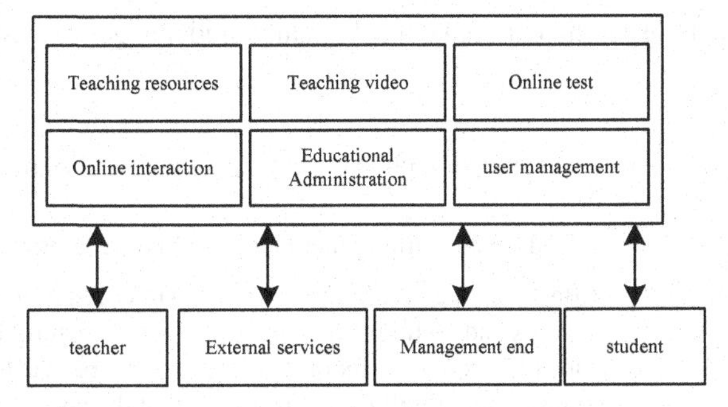

Fig. 2. Mobile education model of law classroom

The model consists of terminal, communication pipeline, and legal education wireless communication data. The terminal is composed of users and their used mobile communication devices. Users are divided into students, teachers, administrators and others according to their functions. Mobile devices can be mobile phones, laptop computers, tablet computers and other communication devices. Communication pipeline refers to the transmission network channel, such as network 3G provided by operators, wireless network WIF covered on campus or in a certain area. Wireless communication data server integrates teaching resources into the wireless communication data of legal education. Students can independently conduct online learning, online testing and so on. Teachers can modify and browse various teaching resources at any time, and communicate with students in real time. The administrator is responsible for managing the wireless communication data server and various users, managing various educational affairs, and integrating teaching resources. This mode mainly focuses on the personalization and portability of students' learning, so it needs to integrate all kinds of users well and reasonably set up mobile document resources.

2.2 Teaching Evaluation Algorithm of Law Course

Mobile learning refers to rely on the more mature wireless mobile network and multimedia technology, network communication technology and other computer means, make teachers and students through mobile terminal equipment, learning at any time, any place, and can make teachers and students for communication, communication, so as to realize the students as the main body of autonomous learning, multimedia has the function of processing, editing images, audio and video. Multimedia teaching resources, can strengthen the students 'attention, with vivid animation and real video, make the students have immersive feeling stored in the teaching server multimedia courseware, can be called at any time, make the platform better service for teaching in the implementation of teaching, as an organic part of the teaching material, teaching media to promote teachers' teaching and students learning plays an important role. Teaching media can make teaching and learning more interesting; It can help students understand knowledge, develop intelligence, improve ability and expand vision; It is conducive to the formation of students' problem consciousness and the cultivation of students' ability to solve practical problems; Cultivate students' cooperative consciousness and spirit. The proper use of teaching media can make teachers and students get twice the result with half the effort.

Table 2. The advantages of mobile learning and traditional learning

	Traditional learning methods	Mobile Learning
Changes in teaching methods	Based on the description of words and pictures, the course location is in the school classroom or online classroom	More descriptions based on sound, pictures and animation, the course can be in any mobile environment
Communication between teachers and students	Delayed e-mail, passive asynchronous communication	Timely communication, strong interaction and initiative
Student to student communication	Face to face, video conference, e-mail, private space, inefficient group discussion	Timely communication, no geographical restrictions, flexible time
Feedback to students	One to one, asynchronous and delayed, popular description, paper feedback	Personalized description, cost saving and more flexible
Operation and testing	Classroom, standard test, feedback effect is not good	Anytime, anywhere, personalized testing, rich and timely feedback
Test evaluation	Based on theory and words, time and place are limited, and individuals and some groups cooperate	Personalized customization and more communication

Analysis the comparison of mobile learning and traditional learning in Table 2, we can find that the characteristics of mobile learning are formally mobile, learners, learning resources and learning environment are mobile; content is interactive, two-way instant interaction; implementation is digital, mobile learning platform for digital learning, mobile learning is a personalized and emotional process, but due to mobile characteristics, mobile learning is also a highly fragmented experience. Distribute the weight allocation W to the evaluation experts, and invite the experts to independently assign the weight N_{ij} of each index in the first round, and fully explain the reasons. To recover the consultation form, it is necessary to make statistics on data n and calculate the average value and deviation of each index.

$$K = \frac{1}{n}\sum A_j - R_{ij} \tag{1}$$

Report the published data and two indicators to experts, ask them to revise their opinions or support their opinions, and then give more weight in the next vote. After recycling, it is classified and counted, and the final conclusion is the expectation of experts and the embodiment of their importance, with the general expectation being at 60% or higher. If not, the index will be cancelled. After testing, the performance indicators meet the needs of system detection and monomer detection. The AHP method can be performed with C.R. To measure whether the weighted results are credible, the results are as follows: C.R. It is calculated as follows:

$$C.R. = K\lambda_{\max} - \frac{1}{n+(R.I.)_n} \tag{2}$$

where: $(R.I.)n$ is the evaluation consistency index of order n matrix; $\lambda_{\max}$ is the average consistency index of the matrix. Mobile education platform is a service platform for integrating educational information integration. It has the characteristics of rich knowledge, classroom virtualization and one-stop. Compared with traditional mobile teaching platform, it can teach and conduct exams online. Mobile education is a service platform integrating educational information and communication between teachers and students. It can teach across regions, which makes students' desire to learn stronger and makes the communication between teachers and students closer. Mobile learning under the mobile education platform makes students' independent learning ability stronger, and students can conduct classified learning of each subject according to their own time. Figure 3 shows that the autonomous learning mode of the mobile education platform is characterized by timeliness and flexibility.

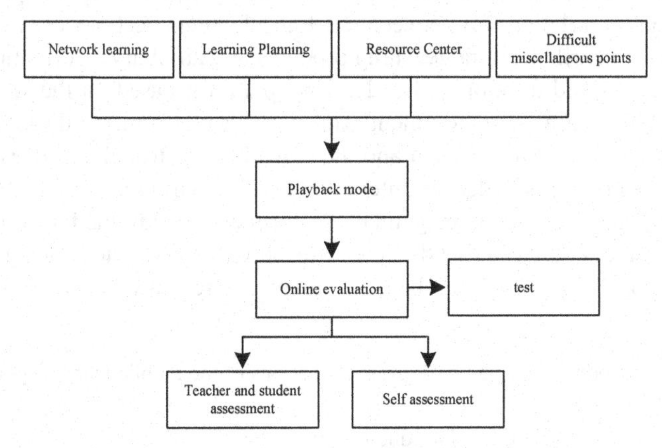

Fig. 3. Independent learning process under the mobile education platform

There are three main conditions for the model to play a role: first, the reasonable setting of courseware. Flexibility and portability is the characteristic of mobile learning, when setting the courseware needs to ensure the short and exquisite courseware, the content is best to 20 min as the node, can not only to avoid the boring courseware, but also conducive to students in the spare time for short learning. At the same time, the courseware should have the function of saving the breakpoints, allowing students to reopen the courseware in other places, without having to play it from scratch. Second, the equal sharing of teaching resources. After the establishment of wireless communication data in legal education, the development of teaching resources is not limited to a certain teaching institution, nor is it restricted by the region. Each law teaching institution can open up a space to enrich teaching resources and provide an equal sharing environment through resource integration. The third is the construction of virtual platform. Real time interaction is also the advantage of wireless communication data in law education. With the powerful computing and storage capacity of wireless communication technology and application services, wireless communication data in law education provides a real-time interaction platform between teachers and students. Students can share their work experience and curriculum related problems. Teachers can organize students to use BBS or other communication platforms to achieve the purpose of teaching.

2.3 Realization of Mobile Teaching of Law Courses

Mobile learning platform is not an independent platform separated from the existing distance education platform. On the contrary, it needs to rely on the rich educational resources in the existing distance education platform to complete its own educational functions. At present, mobile education based on wireless communication technology is still in the initial stage, and there are still many problems waiting to be discussed and solved. At present stage in continue to study the based on wireless communication network learning at the same time, should also be actively to mobile education, try and explore mobile teaching mode based on wireless communication, curriculum design, resource construction and mobile environment students, teachers, teaching content and teaching media the relationship between the four elements such as key issues, developed suitable for mobile learning teaching products and auxiliary information services. This paper designs and develops a law learning platform based on the wireless communication mobile learning environment. Through this platform, we discuss the design and development of teaching content and the realization of teaching process in mobile teaching. The operation is under the Internet network environment, and ASP language and access database dynamic management platform are used in the background. Here, teachers and administrators can set the type of resources, classify and manage resources, add, modify and delete resources. The main functions are shown in Table 3:

Table 3. Introduction of the background functions of the mobile learning platform

Column	Function introduction
Basic settings	Set the types of resources that can be added
resource management	Modify, delete and add resource classification level-1 management system resources
user	Modify the data of existing users or delete users

In mobile learning platforms, the role of media is also very important. Mobile terminal devices have their own uniqueness. Therefore, so the media used based on mobile phones are different. In mobile applications, mobile phones downloading ringtones and watching movies have become a reality, and related technologies have been relatively mature. Integrating this kind of technology into our teaching will play a great role in promoting students' learning and teaching activities. Here, the relevant tools and knowledge of mobile terminal audio and video development are introduced as the development of teaching resources, and used as the teaching resources of the platform. In order to ensure the smooth progress of mobile learning activities, the server side of the mobile learning platform should provide various basic services. According to the role analysis of teachers and platform manager, the server-side functions are divided into two modules: teacher module and platform manager module, as shown in Fig. 4:

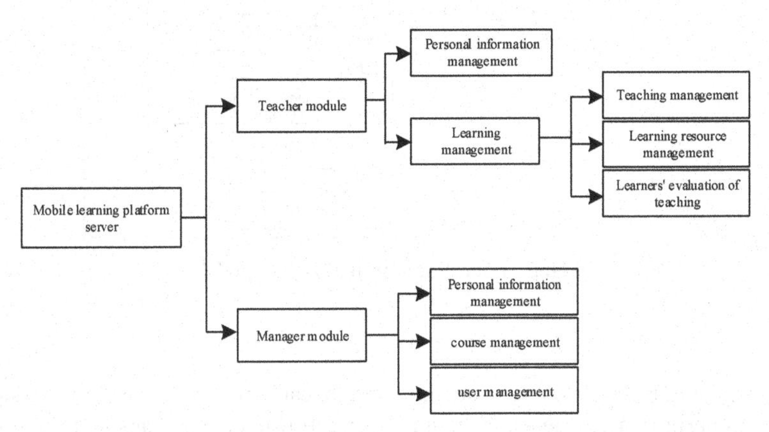

Fig. 4. Functional structure of law teaching management

As shown in Fig. 4, teacher module is to help and support learners by using mobile learning platform. In the platform login interface, the teacher enters his own user name and password and enters the teacher module. To manage the teaching content, upload appropriate learning resources, design teaching activities, evaluate the learning results sent by learners from the mobile learning client, and answer the management module is to support the operation and management of the mobile learning platform. In the platform login interface, the manager enters their own user name and password, and enters the manager module. The manager is responsible for the course arrangement (creation, modification, deletion) and classification, to facilitate the use of learners and teachers, and to conduct information management of learners and teachers, and to timely solve the problems reported by learners and teachers when using the mobile learning platform. Education informatization is a kind of through the use of computer technology, information technology and communication technology to enhance the education of modern education technology, actively develop and make full use of information technology and information resources, to promote the comprehensive modernization reform of education, to adapt to the new requirements for the development of education, to cultivate the process of talents meet social needs, education informatization model as shown in Fig. 5:

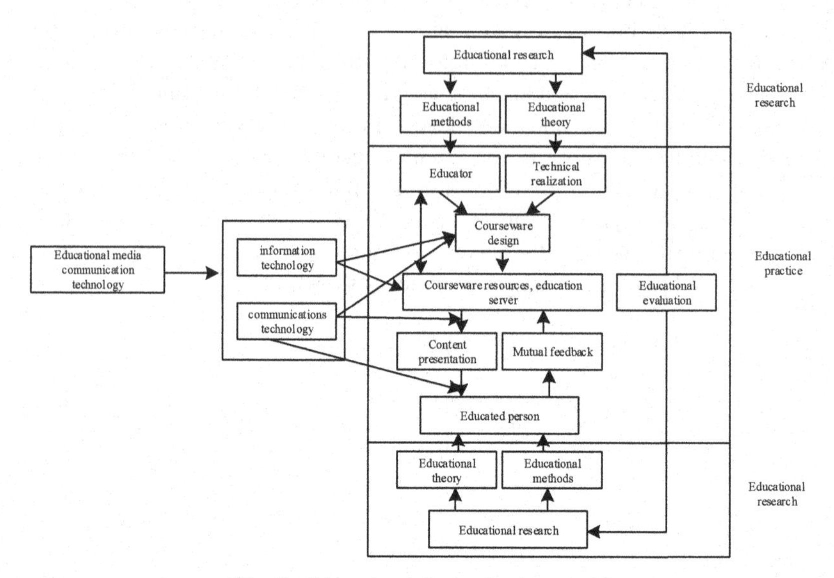

Fig. 5. Education informatization model

The teaching links of mobile teaching include teaching activities such as designing courses, preparing courses, teaching courses, correcting homework, answering questions and learning evaluation. From the perspective of content, the teaching link is very similar to that of traditional education, but its form and focus and teaching design have changed greatly compared with traditional education. First, the separation of teachers and students is the primary feature of network distance education; again, network teaching platform becomes the main means of teaching and learning. In this case, the real-time communication between teachers and students, the implementation conditions of teaching and learning in the network teaching are greatly limited. Therefore, from the perspective of teaching elements, applying mobile learning mode to every link of online teaching can enhance the effect of online learning. A relatively complete mobile learning platform should have the following functional sub modules: resource module, discussion module, question and answer module, test module and other auxiliary modules. The logic structure diagram is shown in Fig. 6:

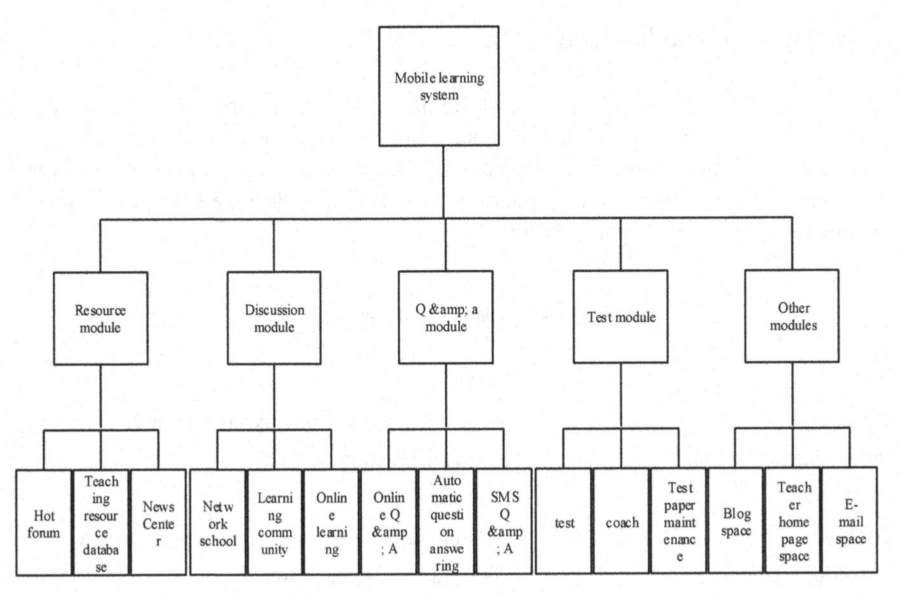

Fig. 6. Function of a mobile teaching platform based on wireless communication

The design and development of learning modules and learning resources in this platform are mainly based on the mobile learning mode of browsing and connection. Due to funds and other problems, the mobile learning mode based on SMS gateway has not been designed and developed, so it is impossible to study the mobile learning of short messages. This part of the function can be performed in further studies on the subsequent platforms. The use object division of the platform design is not detailed yet. At present, there are only two types of users on the platform, but in the real teaching process, the user's needs and functions are different, which are certainly not limited to two types of users. In the follow-up work, different types of users should be divided into different permissions according to specific needs. The improvement of teaching resource database. As it is an exploratory research, the platform does not do much work in the construction of the teaching resource database, but only explores its feasibility. However, as a complete dialogue platform, the construction of the resource library must be perfect. The platform module construction is not perfect enough. As introduced in Sect. 4, mobile learning platform is a complete platform, now part of mobile law platform development module to explore the mobile environment is not enough, such as test module, blog module for the development of the platform and further explore mobile learning is necessary, in the platform follow-up work, can continue to complete the development of these modules and explore the way of teaching in this environment.

3 Analysis of the Experimental Results

According to the research purpose and significance of this topic and the feasibility of the experimental conditions, the respondents were determined to be the teachers and students of a university law school. In order to test the application effect of wireless communication in law teaching, the comparative experiment is designed. The experimental parameters are shown in Table 4 below:

Table 4. Experimental parameters

Project	Data
Power type	High efficiency and energy saving power supply
Processing instruction	Cache instruction
Hard disk	180 GB
Graphics card	ATI mobility radeon standalone graphics card
Detection signal form	TUV radio frequency signal
Cache mode	L3 cache
Data display form	DSP processing signal
Memory capacity	16 GB
Radiation level	Class A
Operation and maintenance voltage	220 V
Power	$\geq$ 200 W

Observation, recording and coding of teacher teaching behavior with Nvivo11 as a research tool. The main interaction types and behaviors of teachers and students in the class are recorded with 30 s as a time unit, and expressed by the corresponding code. Thus, the observation record of the interaction behavior between novice teachers and expert teachers can be obtained as shown in Fig. 7 and Fig. 8:

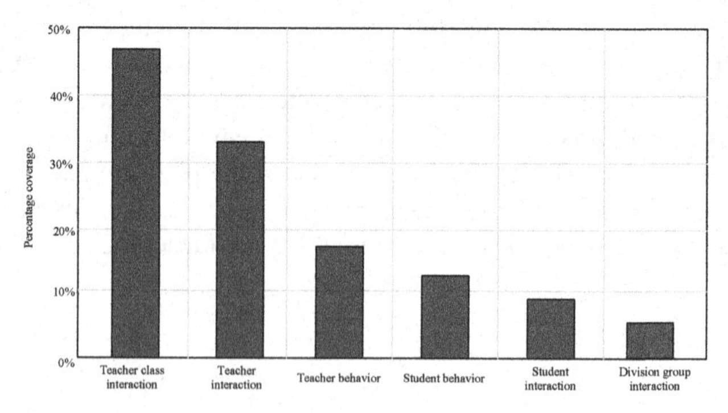

Fig. 7. The interactive behavior coverage of mobile law courses in the classroom offered by novice teachers

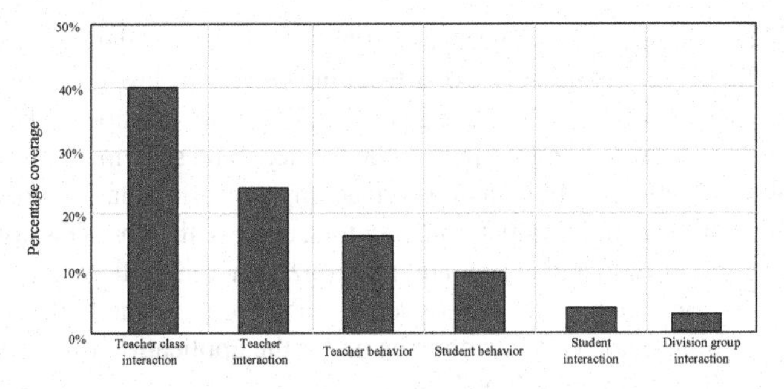

Fig. 8. Coverage of interactive behavior of classroom mobile law courses by expert teachers

As shown in Figs. 7 and 8 above, the coverage rate of the interactive behavior of novice teachers is the largest proportion in the mobile teaching environment. The traditional teaching method and the wireless communication teaching method are selected to compare and analyze the existing deficiencies of the teaching method of wireless communication and improve it. To ensure the consistent experiment time, the student learning efficiency of the traditional teaching method and the wireless communication teaching method is compared and analyzed. The results are shown in Fig. 9.

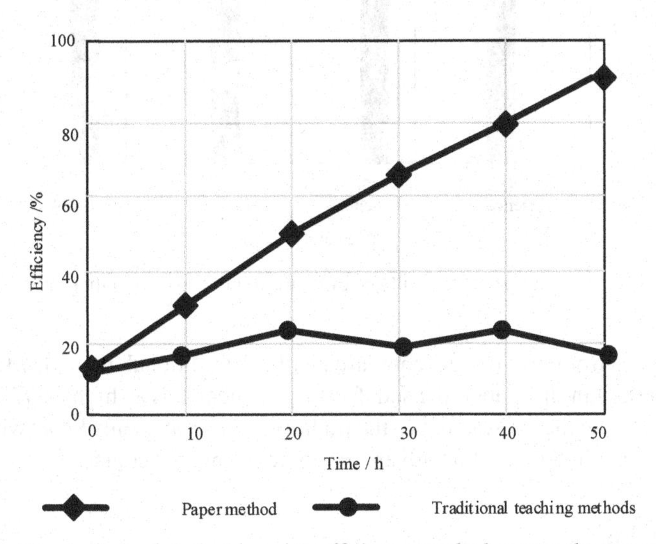

Fig. 9. Comparison of student learning efficiency results between the two methods

According to the above experimental results can get the following experimental conclusion: wireless communication mobile communication teaching method applied to law teaching compared with the traditional method, the method learning efficiency can reach 98%, can improve students 'learning efficiency, strengthen the enthusiasm of learning enthusiasm, strengthen students' practical ability and teamwork spirit, keep the relationship between teachers and students, but also can better application of digital teaching in the classroom. However, under the application of traditional methods, students' learning efficiency of law courses is always around 20%, which proves that the application of traditional methods can not achieve good teaching effects of law courses and can not complete the teaching tasks of law courses in limited teaching hours. The reason for this result is that this method first optimizes the law teaching system, combines wireless communication technology to build a mobile teaching platform of law courses, Optimize teaching content and teaching evaluation indicators, achieve the goal of mobile teaching, break the limitation of time and space, and maximize teaching efficiency.

In order to further analyze the effectiveness of the design method, the evaluation scores of the teaching quality of law courses under the application of the two methods are compared again, as shown in Fig. 10.

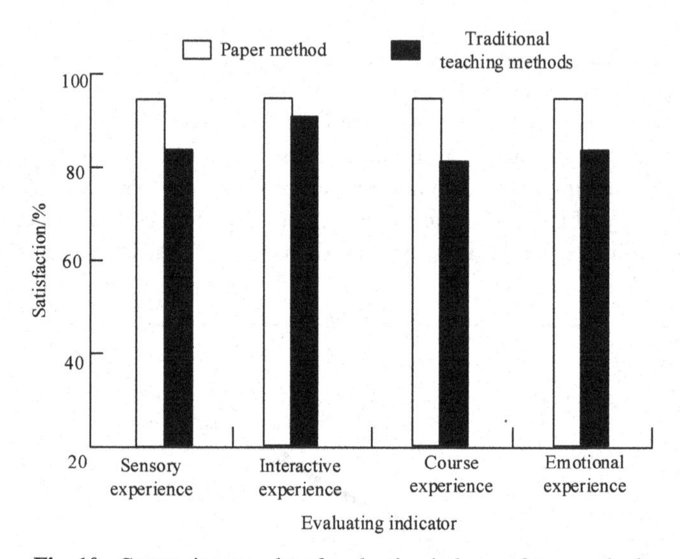

Fig. 10. Comparison results of evaluation indexes of two methods

It can be seen from Fig. 10 that, compared with the traditional method, after applying the design method in this paper, the satisfaction of students is as high as 97%, which is more average, while the satisfaction of the traditional method is only 95%, which proves that the design method has strong advantages in teaching law courses.

4 Conclusion

From the perspective of the development process of multimedia teaching, the integration of high and new technology into education is an inevitable process of educational development and improvement. The application of high and new technology can not change the essence of educational process, but it can change the organizational sequence of educational process, influence the analysis and processing form of education and teaching, thus affecting the teaching effect. With the promotion of new technologies such as wireless communication, mobile learning will certainly be an important form of future development, greatly improving people's learning efficiency and making full use of teaching resources. With the improvement of people's understanding of mobile learning and the deepening of practice, mobile learning will certainly play a greater role.

References

1. Liang, J., Fang, S., Zhau, Y.: Model predictive control of the fuel cell cathode system based on state quantity estimation. Comput. Simul. **37**(07), 119–122 (2020)
2. Su, J., Xu, R., Yu, S., Wang, B., Wang, J.: Idle slots skipped mechanism based tag identification algorithm with enhanced collision detection. KSII Trans. Internet Inf. Syst. **14**(5), 2294–2309 (2020)
3. Su, J., Xu, R., Yu, S., Wang, B., Wang, J.: Redundant rule detection for software-defined networking. KSII Trans. Internet Inf. Syst. **14**(6), 2735–2751 (2020)
4. Yang, Y., Lin, S., Zhou, Y., et al.: The design and practice of the micro-course mobile teaching system based on the smartphone and the WEB platform. Tech. Autom. Appl. **39**(04), 182–185 (2020)
5. Wang, Q., Yan, R., He, Z., et al.: Reform and practice of polymer physics teaching in mobile learning environment. Chinese Polymer Bull. **06**, 59–63 (2020)
6. Wen, H., Yuan, L., An, A., et al.: Exploration and practice of mobile wisdom teaching mode based on rain classroom. Microcomput. Appl. **36**(11), 16–18 (2020)
7. Zhao, J.: Research on comprehensive training teaching model of business subjects in colleges and universities under the mobile learning environment. Microcomput. Appl. **36**(07), 31–33 (2020)

Design of Financial and Economic Monitoring System Based on Big Data Clustering

Kaili Wang(✉) and Yu Chen

School of Business, Nantong Institute of Technology, Nantong 226000, China
xllz13@126.com

Abstract. Aiming at the problem that the selection of financial and economic risk indicators is not comprehensive, resulting in excessive financial and economic operational risks, a financial and economic monitoring system is designed based on big data clustering. Financial and economic risks are risks that are formed and accumulated in the financial system in the process of economic cyclical and financial unbalanced development. According to its formation mechanism, complex network models are used to analyze the dynamic correlation of financial and economic risks. Select financial and economic risk indicators based on big data clustering, and strengthen the supervision of high-risk financial sub-markets such as stocks and foreign exchange markets. Establish a financial and economic monitoring system from three aspects of financial and economic revenue and expenditure, debt and external risks, and jointly determine the development trend of financial and economic risks. The test results show that the financial and economic monitoring system based on big data clustering can reduce the value at risk of operational risk, namely VaR value, and promote the balanced development of the financial market.

Keywords: Big data clustering · Financial economy · Monitoring system · Financial risk · Financial network · Dynamic correlation

1 Introduction

With the increasingly obvious trend of economic globalization and the gradual deepening of financial liberalization, the international capital flows between financial markets are more frequent and closer. The continuous optimization of industrial structure, demand structure, and regional structure has led to the optimization and upgrading of the economic structure, and the coordination and sustainability of economic development have been significantly improved. Economic development has broken through the original single development path, the level of opening to the outside world has been comprehensively improved, the scale of trade and foreign investment has steadily expanded, and the trade structure has been continuously optimized, forming a new economic pattern of economic diversification and globalization. The higher the repercussions caused by the risk fluctuations of one market being transmitted to other markets, the more likely one

W. Fu and L. Yun (Eds.): ADHIP 2022, LNICST 468, pp. 426–439, 2023.
https://doi.org/10.1007/978-3-031-28787-9_32

market is to cause shocks to other markets to varying degrees, and the interdependence between markets increases the uncertainty of the financial system and increases the occurrence of financial probability of crisis. Therefore, in the context of globalization, how to effectively manage risks has gradually attracted the close attention and research interest of financial market regulators and practitioners.

As an important core issue of finance, the systemic risk of the financial system has once again become a hot topic concerning the financial security and core competitiveness of various countries. China's financial and economic risks are multifaceted and wide-ranging, with prominent structural imbalances, many hidden risks of various types, and significantly increased financial system vulnerability. There are four main reasons for the frequent occurrence of financial risks in China: First, China's economy and finance entered a downward cycle after the last round of expansion, posing major challenges to financial development; The phenomenon is serious; third, financial supervision is not suitable for the financial industry, especially the fast-growing technology finance and Internet finance; fourth, it is affected by the spillover effect of loose monetary policies in major developed countries. In order to facilitate the regulatory authorities to track and control the risks of various financial sub-markets in a timely manner, and to better grasp the direction of financial risk management, the most important thing is to choose appropriate measurement methods to identify and evaluate risks, otherwise inaccurate risk measurement will greatly affect the market. Risk management effect. Exploring the correlation and interaction mechanism of risk fluctuations in various financial sub-markets based on accurate risk measurement results will help relevant risk management departments formulate differentiated regulatory strategies in a targeted manner according to the propagation path of market risks, so as to prevent and resolve major financial risks in a timely manner.

At present, relevant scholars have made research on the design of financial and economic monitoring system. For example, the financial and economic monitoring system based on BP neural network [1]. Combined with the operation mode of supply chain finance, it summarizes the risk factors of supply chain finance business, establishes a risk index system with good consistency and stability, introduces the general principles and steps of BP neural network, and uses mat-lab The BP neural network tool builds a risk assessment model, and the validity of the model is proved by establishing a supply chain financial risk assessment model. Financial and economic monitoring system based on Bayesian network [2]. Based on the Bayesian network, the inference analysis and diagnosis analysis of operational risk was carried out, the influence of each key risk inducement on operational risk was evaluated, and prevention strategies were proposed for the key risk inducement, which provided a reference for the financial and economic monitoring system.

This paper designs a financial and economic monitoring system based on big data clustering, selects financial and economic risk indicators based on big data clustering, and strengthens the supervision of high-risk financial sub-markets such as stocks and foreign exchange markets. Establish a financial and economic monitoring system from three aspects of financial and economic revenue and expenditure, debt and external risks, to provide a guarantee for the relationship between economic development and risk prevention, to create a stable and transparent financial market system, and to promote high-quality economic development.

2 Design of Financial and Economic Monitoring System Based on Big Data Clustering

2.1 The Formation Mechanism of Financial and Economic Risks

Financial and economic risks are risks that are formed and accumulated in the financial system during the process of economic cyclical and financial unbalanced development, and are the overall risks of the financial system. In terms of the impact on the financial system caused by changes in political, economic, social and other factors, in terms of internal vulnerability, it is mainly caused by the accumulation, imbalance and unreasonable structure of internal risks in the financial system. Financial supervision is the general term for financial supervision and financial management. On the one hand, it refers to the comprehensive, regular and purposeful inspection and supervision of financial institutions and financial businesses by the financial authorities, so as to promote the sound operation and development of financial institutions in accordance with the law. On the other hand, it refers to a series of activities such as the leadership, organization, coordination and control of financial institutions and their business activities and financial business carried out by financial authorities in accordance with the law. As a new form of finance, Internet finance follows the essential characteristics of traditional financial development. Operational risk events in Internet finance are also derived from operational risk events in traditional finance. The difference is that Internet finance is related to information technology. With the help of the power of the Internet, the impact of some events is weakened, and at the same time, the impact of some events is amplified because of the transmission effect. The subdivision of specific risk events is different from traditional finance. With financial innovation and the development and improvement of the financial system, the objects of financial supervision continue to expand and expand, mainly including participants in financial activities and related financial businesses. Although the external factors of the financial system have a great influence on the risks of the financial system, they do not involve the reasons of the financial system itself. When the risk of the financial system is studied in a closed-system approach, the risk of the financial system specifically refers to the fragility of the financial system. That is to say, in the case of fixed external shocks, the risk of the financial system is the uncertainty of the financial system or the uncertainty of the loss of the financial system caused by the internal instability of the financial system. The premise of effectively supervising and preventing financial operational risk is to understand the formation mechanism of financial operational risk. Basel New Capital Accord once defined the operational risk faced by traditional commercial banks, and believes that operational risk is "the risk of loss due to imperfect or existing problems in internal procedures, personnel, and systems or due to external events". The loss events of operational risk include Internal fraud, external fraud, employment and job security, loss due to customer, product and business operation errors, loss of tangible assets, business interruption and system failure, and errors involving execution, delivery and transaction process management.

Mapping complex network theory in financial system can be used to study the relationship between financial network center and systemic financial risk characteristics. A complex network is generally represented by a node vector and an adjacency matrix. The node vector represents the weight of each node, and the adjacency matrix is used to define whether there is a connection relationship between nodes in a complex network and the weight of the connection. The out-degree of a node in the network is defined as follows:

$$p_a = \frac{1}{z-1}\sum_{b=1}^{z} c_{ab} \tag{1}$$

In formula (1), a and b represent two nodes; p_a represents the proportion of network nodes connected to node a; z represents the total number of nodes. The degree center of a complex network refers to the node with the largest node degree in the complex network, that is, the node with the largest number of connections with other nodes. Corresponding in the financial system, it is the most widely connected financial institution. The central node of the financial network is the "too-connected node to fail", and the failure of this institution will have wider impacts.

2.2 Dynamic Correlation Analysis of Financial and Economic Risks

Effective financial supervision should provide positive externalities for the development of the financial industry while ensuring the financial security of the entire country, making the entire financial market more efficient and dynamic. Using the financial deepening theory to analyze the current financial supervision, it can be found that there are obvious deficiencies. Because from the perspective of financial deepening, it is obvious that traditional static plane supervision cannot adapt to such a new and constantly developing and changing thing as Internet finance. There is a volatility correlation effect among various financial sub-markets such as stock market, bond market and foreign exchange market. Whether there is a Granger causality between variables depends heavily on the choice of lag period, and different lag periods will produce completely different test results [3]. Tracing the source, each financial market does not exist in isolation, and the systemic financial risks in the financial market may be dynamically linked, resulting in multi-market risk contagion. Therefore, looking at potential systemic risks by market is not able to clarify the whole picture of systemic risks. Cross-market or cross-border risk contagion plays an important role in the breeding and contagion of systemic risks. There are differences in the network structure of financial networks, but they generally exhibit the general characteristics of complex networks. In addition, a financial network with high density has a greater impact on systemic risk and a higher risk of contagion. The financial network is a typical scale-free network, and the degree distribution of its nodes satisfies the following equation:

$$w(n) = -\delta \log n \tag{2}$$

In formula (2), $w(n)$ is the probability distribution of the node degree of n in the financial network; δ represents the mutual information coefficient, which is a fixed coefficient for a certain financial network. The calculation formula of δ is as follows:

$$\delta = \frac{\eta(a) + \eta(b) - \eta(a, b)}{\min(\eta(a), \eta(b))} \quad (3)$$

In formula (3), η represents the information entropy. The expression for η is as follows:

$$\eta(a) = -\int q_a(a) \log(q_a(a))da \quad (4)$$

In formula (4), q_a represents the probability density function of a. Using DAG methods to analyze dynamic associations in complex networks of financial economies. DAG is composed of nodes and directed edges, nodes represent variables, and each node is connected by directed edges, indicating that there is a causal relationship in the same period. Starting from an undirected complete graph, assuming that there is a simultaneous causal relationship between any two variables, there are undirected line segments connected between them. Given an undirected graph $R = (U, V)$, (a, b) represents the edge from node a to node b, and $\varphi(a, b)$ represents the weight from node a to node b. If there is a subset where M is V and it is an acyclic graph, so that $\varphi(M)$ is the smallest, then this M is the minimum spanning tree of R, and the calculation formula of $\varphi(M)$ is as follows:

$$\varphi(M) = \sum_{(a,b) \in M} \varphi(a, b) \quad (5)$$

The scale-free characteristics of financial networks can greatly reduce the risk of contagion and systemic risks when the financial system is subjected to external shocks, but external shocks to hub nodes in financial networks will cause greater losses. The stability of the financial system is affected by the hub nodes (financial institutions) in the financial network, which is the main aspect of risk. Based on the residual correlation matrix of variables, the existence of contemporaneous causal relationship between variables is judged. First, test the unconditional correlation coefficient of each two variables. If the unconditional correlation coefficient is significantly 0, remove the connection line indicating causality; then test the first-order partial correlation coefficient of the remaining connected variables. If the partial correlation coefficient is significant If it is 0, remove the connection line indicating causality, and proceed in turn. If there are N variables, the process continues to test N-2 order partial correlation coefficients. From the perspective of the financial network, the contagion between institutions is generally faster and more serious along the direction of the greater degree of association, and the propagation along the edge with the greatest degree of association is the main aspect of risk propagation. Therefore, the study of the maximum spanning tree Transmission is the main aspect of risk contagion. The contemporaneous causal relationship between my country's stock, bond and foreign exchange market risks is adjusted with changes in the socio-economic situation. The global financial crisis in 2008 and the RMB exchange rate reform in 2015 enhanced the contemporaneous causal relationship between risks in different financial submarkets. First calculate the graph composed of the fixed weight minus the weight, then calculate the minimum spanning tree, and then calculate the spanning tree composed of the fixed value minus the weight of the minimum spanning tree, which is the obtained maximum spanning tree. In the event of major financial and economic events and policy changes, regulators should raise awareness of risk prevention and control, strengthen supervision over high-risk financial sub-markets such as stocks and foreign exchange markets, reasonably guide investment expectations in the financial market, and accurately monitor and identify each financial sub-market. The transmission channel between risks, timely curb the spread of financial risks, maintain a reasonable and stable liquidity, and promote the sound development of economic fundamentals.

2.3 Selection of Financial and Economic Risk Indicators Based on Big Data Clustering

When making a comprehensive evaluation of the impact of big data on financial statistics, the key is the selection of evaluation indicators. The selected evaluation indicators must truly and objectively represent a certain aspect of the impact of big data on financial statistics, and reflect the essence of the impact of big data on financial statistics. At the same time, each evaluation index must be independent of each other, cannot overlap each other, and must be measurable, comparable, accessible and operable. The combination of evaluation indicators can reflect the impact of big data on financial statistics from different levels and links. Fully consider the occurrence and transmission of financial and economic risks, and reflect the impact of financial risks on macroeconomics, the financial sector and people's lives, so as to construct impact factors comprehensively, reasonably and scientifically. Then, the big data clustering algorithm is used to classify the indicators that affect financial and economic risks. Big data clustering uses ensemble learning technology to obtain a better and more robust clustering result by learning to fuse the multiple base clustering divisions of the dataset [4]. The big data clustering integration process can be described as: Assuming that data set $Y = \{y_1, y_2, \cdots, y_u\}$ includes u objects, first perform L clustering processes (different algorithms or different parameters of the same algorithm) on data set Y, and obtain L base clusters $F = \{f_1, f_2, \cdots, f_u\}$. f_u means The clustering results obtained in u clustering process [5]. Finally, the L base clustering results are integrated through the consistency function to obtain the final clustering result. The schematic diagram of the big data clustering integration process is shown in Fig. 1.

The first high-accuracy base cluster members are generated based on a criterion that maximizes this information entropy. Serialization iteratively and gradually discovers more base clusters with high clustering quality and strong differences [6]. In the process of generating the new base cluster, the difference between it and the existing base cluster is considered by normalizing the mutual information index. The split hierarchical clustering method first regards the data set to be clustered as a class, and then splits down layer by layer to obtain smaller classes until the split reaches the termination condition of the cluster. The normalized mutual information is used to measure the dissimilarity between base clusters. In order to calculate the expected entropy of numerical data, the kernel-based probability density function estimation strategy is the most commonly used calculation method [7]. The clustering feature CF is a three-dimensional vector used to describe the information of clusters in the clustering process, and it is also a node in the clustering feature tree CF-Tree. Recursively traverse down from the root node, calculate the distance between the data object to be inserted and the current entry, and follow the

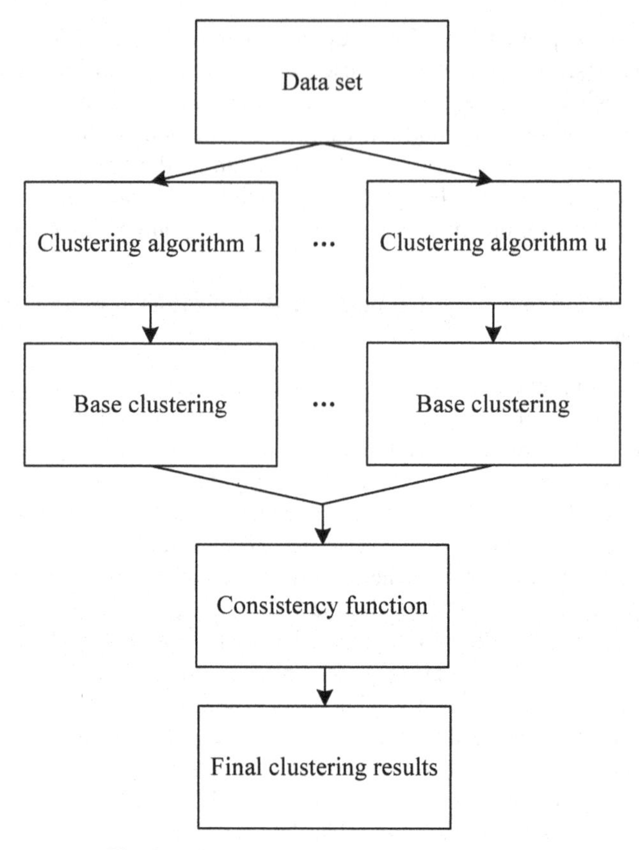

Fig. 1. Big data clustering integration process

path with the closest distance to find the entry in the leaf node that is closest to the data object. In this paper, the MeanNN differential entropy estimation is used to calculate the expected entropy of numerical data. This estimate computes entropy based on pairwise distances between all given data points in a class. The expected entropy of numerical data is described as follows:

$$\theta_1 = \frac{d_1 \log\|x_1 - x_2\|}{s_1(s_1 - 1)} \tag{6}$$

In formula (6), θ_1 represents the expected entropy of the numerical data; d_1 represents the attribute of the numerical data; s_1 represents the number of objects described by the numerical data; x_1 and x_2 are two data points respectively. The time and space complexity required to process high-dimensional datasets will far exceed the processing power of current computers. Feature reduction, also known as data dimensionality reduction, is a method of data preprocessing for high-dimensional datasets. It can identify meaningful features from the dataset. These features can not only reflect the differences between data objects within the same class. The similarity can also reflect the differences between data objects between different classes. For categorical data, uncertainty and ambiguity are measured in the form of complementary entropy. Given categorical data, the complementary entropy on attributes is described as follows:

$$\theta_2 = \frac{d_2}{s_2}\left(1 - \frac{d_2}{s_2}\right) \tag{7}$$

In formula (7), θ_2 represents the complementary entropy of the classified data; d_2 represents the attribute of the classified data; s_2 represents the number of objects described by the classified data. Iterative clustering algorithms randomly divide the given data into different classes. Then, loop through all the data objects and determine whether moving each object from the current class to another class lowers the objective function. The objective function can be expressed as:

$$T = \arg\min[(1 - \varepsilon)\theta_1 + \varepsilon\theta_2] \tag{8}$$

In formula (8), represents the objective function; represents the adjustment parameter. This loop executes iteratively until the objective function no longer decreases or the change is small enough. In the process of generating a base cluster, in order to judge whether the object needs to be reassigned, it is necessary to calculate the update entropy of each class and the NMI value after adding the object to the class. This paper divides the indicators into three levels. The first level is the overall index, and the indicator system framework is constructed from the perspectives of external factors and internal factors that affect financial and economic risks; the second level is structural indicators, including fiscal revenue, expenditure, debt. The third level is the analysis index, which belongs to the sub-index of each structural index of the second level. Evaluation indicators used to reflect the impact of big data on financial statistics. From the perspective of evaluation process and empirical analysis, rating indicators must be measurable. For objective indicators, it is best to obtain them directly or indirectly through the existing data of relevant departments. The relevant statistics can be used for estimation; for subjective indicators, comprehensive statistical survey methods should be used to obtain indicator data, and subjective indicators should be objectified and explained by measurable data.

2.4 Design a Financial and Economic Monitoring System

On the basis of using big data clustering to measure financial and economic risks, establish my country's financial and economic monitoring system. The financial and economic monitoring system established in this paper is shown in Fig. 2.

The reduction of financial and economic income will lead to an increase in the gap between revenue and expenditure, which will have an impact on public financial security. Whether it is central or local fiscal revenue, it depends on local economic development and industrial development. With underdeveloped financial markets, asset prices may not be high enough to trigger macroeconomic volatility. However, now that the economy has entered the era of financial economy, the scale of financial assets in the financial market has far exceeded the scale of the real economy. Fluctuations in asset prices have the potential to trigger macroeconomic changes, especially when large fluctuations in asset prices can trigger economic depressions and financial crises. The increase in financial and economic expenditure will lead to a larger gap between revenue and expenditure, which

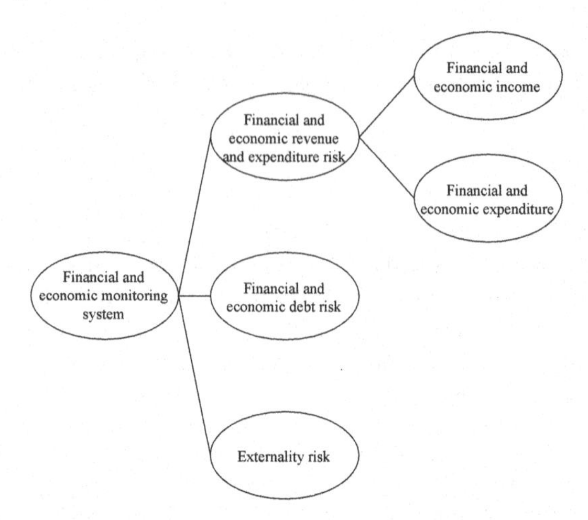

Fig. 2. Financial and Economic Monitoring System

is also not conducive to public financial security. According to the positioning of public financial functions, financial expenditure can be roughly divided into wage expenditure and government operating expenditure, social undertaking development and equalization expenditure, coping with international and domestic Contingency expenditures, etc., the construction of such indicators needs to consider the motivations of the above expenditures. When the cash flow of debt repayment exceeds the production income, the debt will further increase in order to repay the debt. When the economy is prosperous, borrowers take more speculative financing, and both borrowers and lenders have optimistic expectations for the future, underestimate the risk, and financial debt increases. Once asset prices peak or begin to reverse, some economies will be unable to repay principal and interest due to excessive debt, currency liquidity will decline, asset prices will fall, and investment will be sluggish. The volatility of investors' expected consumption and the covariance of expected consumption and asset returns are the main factors that determine asset prices. There is a mutual influence between public debt and economic aggregates, and the influence presents a cyclical state over time. Different stages of the cycle determine whether the government's fiscal risk tends to converge or diverge. Generally speaking, debt risk is a comprehensive manifestation of various risk behaviors, and the monitoring and early warning of debt risk is also a corresponding sequential, level-by-level overall process. The development of the financial market is becoming more and more independent of the real economy, and the changes in asset prices also appear to be separated from the fundamentals of the macro economy, showing a unique law of motion. Macro market movements related to changes in consumption and investment should have pricing power in capital markets. The external influence factors of financial and economic risks mainly consider the transmission effect of macroeconomic and financial risks on the financial field. The impact of macroeconomics on finance is mainly reflected in the decisive role of the economy on finance. The scale of social production and economic benefits determine The scale and growth rate of fiscal revenue, and the structure of social production also determine the structure of fiscal distribution.

Under different macroeconomic conditions, investors tend to change their consumption and investment behavior, and they will trade-off between current consumption and future consumption and savings. Changes in macroeconomic risks can stimulate asset price changes and volatility. Macroeconomic factors that can track the real economic cycle, such as inflation, interest rates, and production technology characteristics, are also important macro factors that can affect the returns of financial assets. The impact of finance on finance is mainly manifested in the liquidity of funds. The financial system provides guidance for the allocation of social resources. Once the financial crisis breaks out, it will greatly restrict the ability of public finance to operate. So far, the design of the financial and economic monitoring system based on big data clustering has been completed.

3 Experimental Studies

3.1 Experiment Preparation

This experiment mainly calculates the value at risk of the financial and economic monitoring system based on big data clustering, and selects the VaR value in the risk loss measurement as the evaluation index. The VaR value reflects the value of the maximum potential loss that may be caused by operational risk in financial activities. Using this value, financial and economic monitoring institutions can predict and control potential operational risk losses, thereby managing operational risk. In order to test the risk prevention effect of the financial and economic monitoring system based on big data clustering, the VaR value obtained by the method in this paper is compared with the VaR value of the financial and economic monitoring system based on BP neural network and Bayesian network. The data in this article is obtained from the Guotai'an database, and the data used is the daily total market value data of financial institutions listed in Shanghai and Shenzhen. Considering the stock market crash in 2015 and the stock market crash in early 2016, this article analyzes the daily data from 2017 to 2021. The experiment included 51 financial institutions in the calculation.

3.2 Results and Analysis

Based on data samples, the VaR values of financial and economic operational risk at 90%, 95%, 99%, and 99.9% confidence levels are obtained. The VaR value of the operational risk of each financial and economic monitoring system is shown in Table 1–Table 4.

At the 90% confidence level, the VaR value of the financial and economic monitoring system based on big data clustering is 10.832 billion yuan, which is 3.118 billion yuan lower than the VaR value of the financial and economic monitoring system based on BP neural network and Bayesian network. And 1.886 billion yuan.

Table 1. Comparison of VaR values at 90% confidence level (100 million yuan)

Testing frequency	Financial and economic monitoring system based on big data clustering	Financial and economic monitoring system based on bp neural network	Financial and economic monitoring system based on bayesian network
1	113.46	132.47	126.65
2	104.85	136.78	124.42
3	106.54	138.69	138.72
4	108.60	142.33	126.83
5	102.22	139.02	122.56
6	113.53	137.25	123.63
7	116.26	145.54	135.24
8	109.18	146.41	120.06
9	105.55	132.82	121.13
10	102.97	143.70	132.52

Table 2. Comparison of VaR values at the 95% confidence level (10,000 yuan)

Testing frequency	Financial and economic monitoring system based on big data clustering	Financial and economic monitoring system based on bp neural network	Financial and economic monitoring system based on bayesian network
1	198.49	265.43	243.54
2	185.56	264.36	240.85
3	192.63	278.65	242.63
4	189.22	266.22	263.30
5	193.30	252.01	256.22
6	199.51	273.54	245.56
7	186.22	262.28	230.18
8	192.53	280.32	242.25
9	201.86	272.63	234.32
10	197.75	265.86	248.23

At the 95% confidence level, the VaR value of the financial and economic monitoring system based on big data clustering is 19.371 billion yuan, which is 7.442 billion yuan lower than the VaR value of the financial and economic monitoring system based on BP neural network and Bayesian network. And 5.1 billion yuan.

Table 3. Comparison of VaR values at the 99% confidence level (10,000 yuan)

Testing frequency	Financial and economic monitoring system based on big data clustering	Financial and economic monitoring system based on bp neural network	Financial and economic monitoring system based on bayesian network
1	326.06	428.65	405.66
2	328.84	436.57	414.47
3	339.66	422.49	408.84
4	346.28	443.63	416.92
5	323.56	422.25	423.28
6	335.72	430.68	412.53
7	324.40	425.23	425.37
8	325.13	428.39	406.68
9	322.22	416.61	403.50
10	336.38	442.48	429.15

At the 99% confidence level, the VaR value of the financial and economic monitoring system based on big data clustering is 33.083 billion yuan, which is 9.887 billion yuan lower than the VaR value of the financial and economic monitoring system based on BP neural network and Bayesian network. And 8.381 billion yuan.

Table 4. Comparison of VaR values at the 99.9% confidence level (10,000 yuan)

Testing frequency	Financial and economic monitoring system based on big data clustering	Financial and economic monitoring system based on bp neural network	Financial and economic monitoring system based on bayesian network
1	528.36	682.41	654.03
2	527.42	688.12	667.48
3	539.18	667.33	649.57
4	546.53	686.57	658.66
5	510.67	693.69	669.34
6	553.59	695.92	636.19
7	552.05	662.10	663.58
8	535.60	685.25	650.63
9	527.44	679.38	642.92
10	536.88	670.46	621.07

At the 99.9% confidence level, the VaR value of the financial and economic monitoring system based on big data clustering is 53.577 billion yuan, which is 14.535 billion yuan lower than the VaR value of the financial and economic monitoring system based on BP neural network and Bayesian network. And 11.561 billion yuan. According to the above results, the operational risk capital that financial institutions need to allocate increases with the increase of confidence level. In general, the VaR value of the financial and economic monitoring system based on big data clustering proposed in this paper is lower than that based on BP neural network and BP neural network. Monitoring system based on Bayesian network.

Financial institutions should still pay attention to and actively guard against operational risks. According to the operational risk value at risk of the financial and economic monitoring system proposed in this paper, financial institutions can allocate operational risk capital as appropriate to effectively prevent operational risks. Optimize the structure of the financial system, increase the proportion of direct financing, and give full play to the financing function of the financial market. Strengthen and improve the construction of the hierarchical system of the financial market, promote the comprehensive and coordinated development of various financial markets such as the capital market, the money market, and the foreign exchange market, upgrade the financial system structure, increase the amount of financing through various channels, and develop a balanced development, reduce the concentration of risks to a certain department, and effectively Defuse my country's systemic financial risks.

4 Conclusion

With the continuous deepening of economic development and the development of the financial system, an effective market supervision mechanism is the cornerstone to ensure the long-term and stable development of the financial market. Give full play to the government's macro-prudential management role in market supervision, set up special agencies to coordinate financial market coordination and supervision, strengthen The comprehensive supervision of the financial system realizes the unification of behavior, function and institutional supervision, and comprehensively prevents the occurrence of systemic financial risks. This paper designs a financial and economic monitoring system based on big data clustering. The construction of this system is conducive to accelerating the informatization and digitization of supervision methods, and provides more complete and accurate information for the monitoring of systemic financial risks.

The method of this paper does not consider more indirect links in the interbank market. For example, the transfer of credit risk in the form of derivatives is an important source of potential contagion. Therefore, in future research, the risks caused by different sources of contagion should be studied in more detail.

Fund Project. Jiangsu Province "14th Five-Year" Business Administration Key Construction Discipline Project (SJYH2022–2/285).

References

1. Sun, X., Lei, Y.: Research on financial early warning of mining listed companies based on BP neural network model. Resour. Policy **73**(2), 102223 (2021)

2. Jya, B., Yue, S.A., Dt, A., et al.: A Digital Twin approach based on nonparametric Bayesian network for complex system health monitoring[J]. J. Manuf. Syst. **58**, 293–304 (2021)
3. Hai-Jun,C., Zhi-xiong, L., Shi-bin, W.: Simulation of mixed attribute data clustering mining based on feature selection. Comput. Simul. , **37**(7), 399–403 (2020)
4. Sun Q., Chen, H., Li, C.: Clustering algorithm of big data based on improved artificial bee colony algorithm and MapReduce. Appl. Res. Comput. **37**(6), 1707–1710, 1764 (2020)
5. Yi, H., Zijiang, Z.: PSO-based big data clustering algorithm in cloud environment[J]. Modern Electron. Tech. **43**(14), 72–75 (2020)
6. Jie-fang, L., Zhi-hui, Z.: Parallel clustering algorithm for big data. Comput. Eng. Design, **42**(8), 2265–2270 (2021)
7. Chunyan, Z., Xinlei, W.: On the measurement of the financial cycle and the business cycle synchronization in China. J. Dalian Univ. Technol. (social sciences) **41**(6), 45–56 (2020)

Recommendation Method of College English Online Mobile Teaching Resources Based on Big Data Mining Algorithm

Yuhong Xia(✉) and Yudong Wei

Chengdu College of University of Electronic Science and Technology of China, Chengdu 611731, China
hi_xiaxiaxia@126.com

Abstract. The current teaching resource recommendation methods mainly recommend resources according to the relationship between users' historical data and resources, do not consider the timeliness of data, and ignore the attenuation of users' interest in learning resources, resulting in low accuracy and poor recommendation performance. To solve the above problems, in order to improve the accuracy of recommendation of College English online mobile teaching resources, this paper studies the recommendation method of College English online mobile teaching resources based on big data mining algorithm. In the big data teaching environment of College English online mobile teaching, data mining technology is used to mine the characteristics of users' preference for resources. Time information is introduced into the collaborative filtering model integrating neural network to realize the recommendation of teaching resources. In the method test, the recommendation accuracy of the resource recommendation method is 97.26%, and the recommendation performance is significantly improved.

Keywords: Big data mining algorithm · English online teaching · Mobile teaching · Resource Recommendation · Neural network · Collaborative filtering

1 Introduction

The rapid development of science and technology has changed people's way of life and learning. People have changed from traditional classroom learning to today's online learning. The development of educational informatization has spawned a large number of e-learning platforms, so that learners can learn anytime and anywhere, and their learning behavior is no longer limited by factors such as site, time and so on. In the process of mobile learning, client users use smart devices such as mobile phones and tablets to install client software and use wireless networks to interact with the system. The cloud environment of mobile teaching mainly provides services such as device support, computing, network, data storage and so on. With the continuous enrichment of the concept of mobile learning, more and more scholars in the society began to study mobile learning. In the context of the wide application of mobile teaching, resource rich education will

W. Fu and L. Yun (Eds.): ADHIP 2022, LNICST 468, pp. 440–452, 2023.
https://doi.org/10.1007/978-3-031-28787-9_33

be more frequently used in auxiliary teaching than ever before, and major mobile online education platforms will constantly update more teaching resources. With the expansion of user scale and information resources, personalized information recommendation technology will also be widely used in the field of Education [1]. Teaching knowledge information is complex and cumbersome, and the user level is uneven. Only by analyzing the user's knowledge level and knowledge system and recommending relevant courses, can we effectively solve the problem of information resource overload, provide good services for users' mobile learning and increase users' loyalty to the mobile teaching platform [2].

As an important part of educational informatization, network teaching resources are becoming more and more important in promoting the knowledge construction of students and teachers, improving practical ability and developing advanced thinking ability. The high-quality resources on the network are increasing rapidly, but new problems have also arisen: the time of searching resources becomes longer, especially for learners who do not have professional search ability, this kind of problem is more prominent. What's more, it is not sure whether the resources searched are really needed by themselves. Therefore, providing learners with the learning resources they need is one of the ways to solve this problem, and it can also realize personalized learning. If we want to select resources suitable for learners from massive data to realize personalized learning, the traditional search engine can not meet the requirements of personalization. The emergence of recommendation methods provides an effective solution for personalization. The recommendation system does not need users to provide clear needs. It mines users' potential interests through users' historical behavior, so as to actively recommend information that can meet their interests and needs. Although the recommendation algorithm has achieved good results in the field of audio-video and commodity recommendation, teaching resource recommendation has some significant characteristics different from traditional audio-video or commodity recommendation, so the traditional audio-video or commodity recommendation methods can not be directly applied to teaching resource recommendation. Its remarkable feature is that each student user has personalized differences in cognitive ability level, learning habits, learning objectives and other aspects. It can not be recommended only through the user's direct interest similarity like audio and video recommendation, but also according to the student user's cognitive ability level, learning objectives, historical response records, current situation and other personality information. At present, there are many methods about learning resource recommendation, such as the multi task feature recommendation algorithm of the fusion knowledge map of curriculum resources proposed in reference [3]. Based on the end-to-end deep learning framework, the knowledge map is embedded in the task; The high-order relationship between potential features and entities is established through cross compression units between tasks, so as to establish a recommendation model. It realizes accurate recommendation of course resources based on learners' goals, interests and knowledge levels. However, in practical application, this method does not fully consider user behavior, only considers the similarity between information contents, which may also lead to over description and low recommendation accuracy; Reference [4] designs an information-based teaching resource sharing system based on multimedia technology.

In the hardware part, S3C6410 processor is selected to build a multimedia embedded processor, and the hardware interface circuit is designed based on the USB interface board. In the software part, e - R diagram is used to contact the entities of teaching resources, delineate the attributes of information-based teaching resources, build databases with different functions according to different processes of sharing teaching resources, and set a data supplement program in the database to finally realize resource recommendation and sharing. However, this method will automatically expand the scale of clustering, make clustering fuzzy, and the recommendation effect is insufficient.

In order to improve the accuracy of recommendation, this paper proposes a recommendation method of College English online mobile teaching resources based on big data mining algorithm. Build a big data teaching environment for online mobile teaching of College English; Using data mining technology to mine users' preferences for English resources; It creatively introduces time information into the collaborative filtering model based on integrated neural network to complete the recommendation of teaching resources. The experimental results prove that the contribution of this method is to solve the shortcomings of existing mobile teaching resource recommendation, improve the accuracy of Learning Resource Recommendation and the effect of mobile teaching.

2 Recommendation Method of College English Online Mobile Teaching Resources Based on Big Data Mining Algorithm

2.1 Build a Big Data Teaching Environment for College English Online Mobile Teaching

The fundamental purpose of College English online mobile learning is to break the time and space constraints of the traditional classroom, and the traditional web page centralizes the management of teaching resources with the help of information technology. Then, through the Internet, users can access these teaching resources for learning with the help of smart phones, tablets, laptops and other intelligent terminal devices connected to the Internet. Mobile teaching is divided into five teaching theories: informal teaching, context teaching, Situational Cognition, experience teaching and activity teaching. The teaching methods of informal teaching are mainly reflected in unconsciously obtaining information, content and communicating and discussing with people. The characteristic is to teach to any object anywhere. Context teaching is based on learners' knowledge structure, learning interest and learning motivation to provide the basis for mobile teaching. Situational cognitive teaching emphasizes the influence of environment on learners' teaching. This theory believes that meaningful teaching can occur only when teaching is embedded in a specific environment. Experience teaching emphasizes thinking and practice. Learners can improve their learning efficiency only through continuous thinking and practice and applying the experience summarized after practice to the subsequent practice process [5]. Activity teaching refers to teaching in activities, focusing on a certain problem, discussing, teaching, practice, activity design, etc. It mainly advocates active teaching and defines the purpose of teaching. According to the characteristics of mobile teaching and the teaching requirements of College English, when carrying out college English online mobile teaching, the mobile teaching platform needs to build a good learning environment for students with the help of big data technology. When students conduct College English online mobile learning, they not only need to interact with teachers through the mobile teaching platform, but also need the platform to meet the needs of students for resources, learning strategies, learning paths and learning emotion regulation in the teaching process. Figure 1 shows the structure of big data teaching environment for College English online mobile teaching.

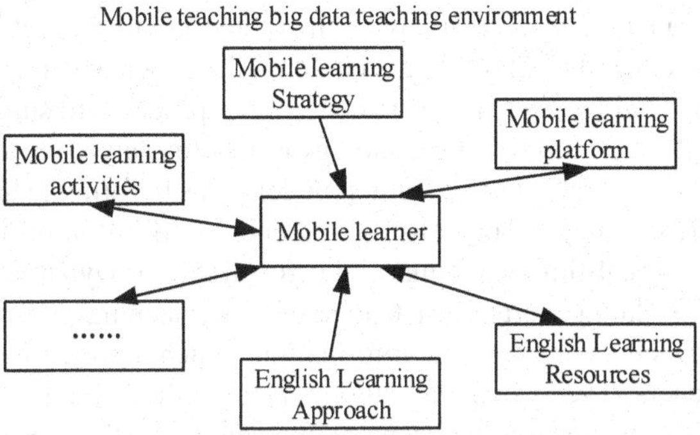

Fig. 1. Online mobile teaching big data teaching environment

Traditional personalized learning will give learners a fixed learning path and provide fixed learning resources, without fully considering learners' autonomy. The learning environment in the context of big data is no longer a closed space. Taking learners as the center makes learners have the right to choose the learning content, gives learners enough space, makes learners have a sense of control and achievement in learning, realizes personalized learning through the choice of learning path and learning content, and enables learners to explore more fields of learning content when they complete their learning easily and happily, Expand knowledge and improve self-efficacy. Personalized learning students supported by big data are no longer isolated learners. They can form learning communities and online virtual learning interest groups in various interpersonal networks. Learners can discuss and exchange with each other and share learning experience. The interaction between learners and others can improve learning efficiency and reduce learning loneliness [6]. The personalized English learning environment built under the background of big data pays more attention to learners' emotional attitude, and the evaluation and diagnosis of learners are more humanized. The emotions for learners are divided into six aspects: boredom, enthusiasm, confusion, frustration, happiness and surprise. Emotional attitude will have varying degrees of impact on the learning process. A pleasant mood is more suitable for completing learning, and boredom is not conducive to completing learning, Through the capture of learners' emotional attitude to assist the current learning diagnosis and resource recommendation, when students show emotional burnout, give positive encouragement mechanism in time to prevent learners from giving up learning halfway. Through the diagnosis of learners' different emotional states, analyze students' interest and preference for College English teaching resources, so as to improve the accuracy of resource recommendation.

2.2 Mining User Preferences for College English Resources

Big data provides technical support for the construction of College English online mobile personalized teaching environment. It can mine the characteristics of teachers, analyze the needs of teachers, predict the behavior of teachers, and make it possible to teach students according to their aptitude. Creating a personalized teaching environment can bridge the boundaries of personal teaching space and teaching space. The construction of personalized teaching environment highlights the characteristics of network teaching in the new era. The personalized teaching environment based on big data can support teachers to carry out independent, individualized and creative teaching, and provide teachers with personalized teaching space based on their own teaching experience. The behavior data generated by the teacher in the teaching process are summarized into the teacher database, and the data mining technology is used to analyze the teacher's behavior data and modify the teacher's static teaching model, so as to form the teacher's dynamic teaching characteristic model. The construction and modification of dynamic teaching model is the core of real-time teaching model construction. Dynamic teaching model studies the behavior data generated by teachers in the teaching process. The behavior characteristics of teachers are invisible information, which does not need the deliberate expression of teachers. The system can automatically collect the data generated in the teaching process, such as the retrieval content of teachers, teaching progress, discussion in the teaching process, etc. In order to make the transition from static teaching model to dynamic teaching model, it is necessary to collect learners' dynamic behavior data and analyze and summarize them through data mining technology and behavior analysis technology.

The cluster analysis technology in data mining is used to locate the teaching level, teaching interest, teaching style, teaching ability and other groups with common teaching aspirations. A large number of dynamic data are more and more accurate for the analysis of teachers, more and more accurate positioning, and more personalized resource recommendation services are provided. If you want to get accurate interest bias, you first need to extract user features, and then judge the polarity of the user's bias towards a feature. According to the characteristics of College English online mobile teaching, this paper puts forward the following seven characteristic dimensions, namely, learning time and learning times, average length of each learning, frequency of viewing course announcements, weekly investment time, frequency of viewing course calendar, fixed degree of daily learning time and fixed degree of learning time interval, and constructs the preference model of mobile learning users [7].

Ontology based model representation has strong expansibility and good adaptability, but the construction of ontology is mainly affected by researchers' knowledge and experience. Especially when the definition domain is large, the effectiveness of ontology construction is difficult to guarantee; Although vector space model is not the best user requirements description method, its good universality makes it the most extensive and mature user model construction method. Therefore, the combination of the two is still used for user preference modeling in this study.

Generally, the representation methods of user interest modeling include keyword list representation and scoring matrix representation between user and project. The former is to extract the keywords of user interest and establish the user interest model based on the keywords. The latter uses the user's score on the project to directly reflect the user's interest in the project. These two methods are easy to understand in use, but they tend to be coarse-grained in interest division. For users' fine-grained interests and the weights of different interest categories, they can not be deeply mined. We need to convert the text data into binary data that can be recognized by the computer. In this paper, a vector space model representation method will be used to solve the above problems. In text classification, it is assumed that simple keywords and sentences can be used to represent the target text. Based on this assumption, vector model can be used to represent the target text, in which vector elements are keywords and sentences of College English teaching resources text.

In the vector space model, the user set is represented by the letter Y as $Y = \{y_1, y_2, \cdots, y_m\}$, and the element set of the user English learning preference model is represented by H as $H = \{h_1, h_2, \cdots, h_n\}$. Where y_i represents the i users, with a total of m users; h_j represents the j element, and there are n elements in total. The user's preference for each element is recorded as the preference weight value ω_{ij}, which represents the weight value of the i user's preference for the j element. The weight value of this study is expressed by Likert five level scale.

In order to provide learners with the required and appropriate personalized learning resources more accurately, we need to et the learning preference of the target learners from the learners themselves, and calculate the learners' learning preference from the three dimensions of learners' knowledge state, online learning participation and resource score. With the decline of students' interest, students' short-term learning interest preference gradually tends to 0. Therefore, the user preference matrix becomes very sparse. User learning preferences are influenced by both short-term learning interests and long-term learning interests. This paper introduces ontology when constructing the learner preference model, and excavates the students' preference characteristics for resources according to the user preference ontology of College English mobile learning shown in Fig. 2 [8].

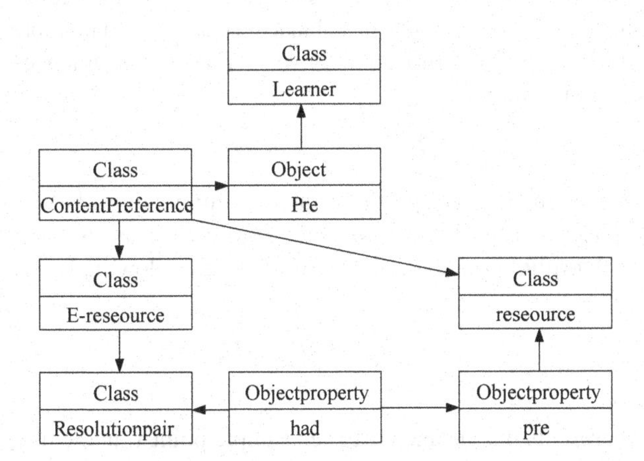

Fig. 2. Schematic diagram of the user's English learning resource preference ontology

According to the above content, the big data clustering mining method is used to mine users' preference characteristics for College English resources. It is assumed that the center of each cluster is c_z and the internal distortion of this class is $\sum \varphi^2(x_i, c_z)$. Here, $\varphi^2(x_i, c_z)$ represents the distortion measure between two data points. Therefore, for all cluster $\{J_c\}_{c=1}^k$, its overall performance can be expressed in the following form:

$$Q = \sum_{c=1}^{k} \sum_{x_i \in J_c} \varphi^2(x_i, c_z) \tag{1}$$

In order to make the weight obtained in the clustering process obey the given feature preference F as much as possible, a penalty term is added to the above objective function to reflect the possible violation of preference. Therefore, for each feature preference tuple $f = \{s, t, \delta\} \in F$, the penalty term is defined as $\max\{\delta - \omega_s + \omega_t, 0\}$. Now, for each preference, m auxiliary variables are introduced to form vector $\gamma = [\gamma_f]$, so as to form the following characteristic preference penalty term:

$$\min \sum \gamma_f + \eta$$
$$s.t. \begin{cases} \omega \in \Delta_d \\ \omega_s - \omega_t \geq \delta - \gamma_f \\ \gamma_f \geq 0 \end{cases} \tag{2}$$

Among them, tuple $\{s, t, \delta\}$ represents the user's preference for College English learning resources. The tuple represents that the importance of feature s is at least δ than feature t, that is, $\omega_s - \omega_t \geq \delta$ and δ are numbers greater than 0, and their values can be automatically adjusted or predetermined in method training. η is the preference penalty parameter; d is the dimension of preference characteristics.

In addition to the given feature preference a priori, we do not want to make unfounded assumptions on other aspects of feature weight to cause over fitting. Therefore, the negative entropy term $-H(\omega)$ is introduced into the objective function and minimized to ensure that the weight is as smooth or balanced as possible. The more balanced the weight, the greater the value of $H(\omega)$, and vice versa. For the convenience of calculation, the entropy used in this study is:

$$H(\omega) = 1 - \omega^T \omega \tag{3}$$

Among them, maximizing entropy is equivalent to minimizing $\omega^T \omega$. In order to establish the local weight clustering algorithm of feature preference, the distance measurement of the clustering algorithm is calculated according to the following formula:

$$\varphi^2(x_i, c_z) = \sum_{j=1}^{d} \omega_{jc} D_{ic}^j \tag{4}$$

Where D_{ic}^j represents the distance between sample point x_i and cluster center c_z on the j feature. The weight of clustering dependence satisfies $\sum_{j=1}^{d} \omega_{jc} = 1$. ω_{jc} is the weight of feature j corresponding to cluster c.

In fact, the distance measures can be defined differently according to the different characteristics of the dataset, with the common Euclidean distance used in this paper. The cluster distance between the behavioral data and the teaching resources is calculated, and the characteristics of the users' preference for college English resources are extracted.

2.3 Realize the Mobile Teaching Resource Recommendation

According to the preference characteristics of users for College English resources mined above, this study adopts the collaborative filtering model of neural network to realize the recommendation of College English online mobile teaching resources. In a neural network, the nodes of each layer will convert the input data into output, and then enter the next node. How to convert the output of the upper layer node into the input of this layer node requires a functional relationship to convert. The so-called activation function refers to this functional relationship, which maps the input of neurons to the output through the activation function. The neural collaborative filtering framework is mainly divided into four layers: input layer, embedding layer, neural collaborative filtering layer and output layer.

The input layer is responsible for the input of users and items, and converts each user and item into n vector. If there are n users, the n users will be converted into the vector of $1 \times n$, which will be converted into a sparse vector. After the input enters the embedding layer, multiply the input vector by the embedding matrix f. If there are n users and the embedding dimension is m dimension, the size of the embedding matrix is $m \times n$, and the row represents the embedding vector of the user preference resource. In order to prevent neuron inactivation, this paper uses prelu activation function in the framework of collaborative filtering model of fused neural network. Prelu activation function adds parameter α on the basis of relu function, which is very small. When $\alpha = 0$, prelu will degenerate into relu function.

$$F(x) = \begin{cases} x_i, x > 0 \\ \alpha x_i, x \leq 0 \end{cases} \tag{5}$$

Loss function is mainly responsible for calculating the difference between the actual value and the predicted value, according to the difference, to evaluate the error between the predicted value and the actual value, by constantly to narrow the difference, to train the model, so as to get a better effect, in general, the smaller the loss function, the performance of the model will be better. In the neural network model, mainly for the current teaching resources recommendation algorithm is not fully mining the effective implicit feature information design, its main purpose is through the neural network in different dimensions of teaching resources depth mining, at the same time, through the output layer of nonlinear transformation seamless into the joint probability matrix decomposition.

To input the feature vectors *yf* and *ef* of users and College English mobile teaching resources into multi-layer neural network to get the final score, it is necessary to connect the user's preference feature vector and the feature vector of mobile teaching resources for input. The output layer is mainly responsible for nonlinear mapping of the output of the previous layer. Therefore, it is necessary to project in the d dimensional space of the joint probability matrix decomposition model to complete the recommendation task.

In real life, user preferences will change over time. Take the time information into account in the recommendation algorithm to better find user preferences. Compared with other recommended objects, the curriculum resource recommendation model constructed in this paper will be greatly affected by time. With the advance of time, the curriculum resources are constantly updated and changed. The content direction of a user to learn at each stage is different, and the user's learning interest will also change with the change of time. The older the courses, the courses that users don't often watch at this stage, the less recommendable they will be to users.

In order to meet the requirements that users' interests change with time, recommend effective teaching resources to users more accurately. In this paper, time factor is added on the basis of neural collaborative filtering algorithm. After unsupervised classification of time information through K-means, the effect of recommendation is improved by adding time information.

Record the mobile teaching information data set, the establishment time of each course, the user viewing time, and the current time, and then calculate the user's relative viewing time for each college English course as time information.

$$T = \frac{T_y - T_e}{t - T_e} \tag{6}$$

Where, T_y is the latest viewing time of mobile teaching users, T_e is the course upload time, and t is the current time.

In terms of time, the closer the user has seen the course, the greater the time information the new course will get. In the process of model training, with the increase of the number of network layers, the convergence speed of training may slow down. Batch standardization can use some standard means to pull the distribution of input values of each layer of neural network back to a certain standard. In this way, the whole training speed can be accelerated. In this paper, batch standardization layer is added to MLP model to speed up the training speed, It can also further alleviate the problem of over fitting. After completing linear learning and nonlinear learning, the obtained potential feature vectors are connected together and output through the activation function. After using the training sample set to train the parameters of the collaborative filtering model fused with neural network, the relevant data of mobile learning users are input into the recommendation model with the determined parameters. After the model processing, the recommendation results of College English online mobile teaching resources are obtained. Based on the above theoretical content, the research on the recommendation method of College English online mobile teaching resources based on big data mining algorithm is completed.

3 Resource Recommendation Test

In the context of the wide application of mobile teaching, the above studies the recommendation method of College English online mobile teaching resources based on big data mining algorithm. This section will test the recommendation performance and effect of the recommendation method of teaching resources.

3.1 Test Preparation and Scheme Design

In this test, the crawler program is used to collect data from a college English mobile teaching platform, and a total of 318860 pieces of data are crawled. Each data information includes the ID of the College English course, the establishment time of the course, the user's ID, the user's viewing date, the user's score of the course, etc., and the data is saved in courses_ and_ users. CSV file to facilitate the subsequent training of the model. In order to get better experimental results, this paper also processed the data as follows. In the crawled data, the user viewing time information and course establishment time information are strings containing "year", "month" and "day". In order to ensure the unity of subsequent data input and facilitate the application of data sets, it is necessary to process the crawled user viewing time information and course establishment time information. Change the form of string into the form of number. At this time, the viewing data is not the date with month, year and day, but a string of numbers. In this paper, the gettimestamp () method is defined through the def function in Python language. Through this method, all the date formats in the data are converted to the time stamp format, which is convenient for the next step. In the process of data analysis, in order to obtain a more accurate experimental result and reduce the impact of inconsistent units of experimental indicators. In this paper, the indicators are standardized and then put into use. These indicators are changed into numbers without units, which can be better compared or weighted, which lays a foundation for subsequent experimental analysis.

In order to prove the recommendation effect of the College English online mobile teaching resources recommendation method based on big data mining algorithm, this paper selects the data crawled and processed according to the above content as the data set. Because some users have too few course records, conduct unified screening before the test, and select the data with more than 50 course learning records for random selection. Finally, 7712 users and all college English online courses in the data set were randomly selected as the data set, and trained under explicit feedback and implicit feedback respectively for experimental evaluation.

The test adopts the form of comparison to ensure the scientific and credible test data. The multi task feature recommendation algorithm (reference [3]) and the resource sharing system based on Multimedia Technology (reference [4]) are taken as comparison method 1 and comparison method 2, respectively, and compared with the teaching resource recommendation method studied in this paper. The experimental indicators are the recommendation accuracy, recall and F-score of teaching resource recommendation methods to evaluate the recommendation effect.

3.2 Test Results

Table 1 shows the comparison of the recommendation accuracy and recall rate of the three resource recommendation methods when recommending teaching resources to different numbers of mobile university English learning users.

By analyzing the data in Table 1, we can see that the accuracy and recall rate of the teaching resource recommendation method based on big data algorithm studied in this paper are higher than the other two recommendation methods. At the same time, with the increase of the number of users, the accuracy of recommendation results is gradually

Table 1. The accuracy and recall comparison of the recommended methods

Number of users	Teaching resource recommendation method based on big data algorithm		Comparison method 1		Comparison method 2	
	Accuracy rate /%	Recall/%	Accuracy rate /%	Recall/%	Accuracy rate /%	Recall/%
10	97.47	90.24	86.78	80.44	85.66	79.63
20	97.36	87.82	89.15	81.45	83.66	77.36
30	96.63	88.45	85.57	83.88	84.31	78.55
40	95.82	90.87	85.13	79.90	85.05	78.41
50	96.98	89.73	84.68	81.96	84.28	75.48
60	96.59	89.59	84.91	83.61	84.17	77.09
70	96.94	89.14	88.03	80.92	83.03	78.84
80	97.06	90.06	86.46	81.43	84.95	78.87
90	97.37	89.27	86.12	80.76	82.04	77.61
100	97.44	89.49	87.51	81.01	81.92	76.46
110	96.96	88.15	89.24	78.86	81.18	77.45
150	96.38	87.83	86.49	80.97	80.82	77.79
200	98.92	90.02	87.43	81.54	80.01	75.62
300	98.33	90.11	86.97	79.95	79.72	76.98
500	98.65	88.86	84.56	82.81	79.63	78.60

stable. From the average recommendation accuracy of the recommendation method, the average recommendation accuracy of this method is 97.26%, and the average recall rate is 89.31%; The average recommendation accuracy rate of comparison method 1 is 86.60%, and the average recall rate is 81.30%; The average recommendation accuracy rate of comparison method 2 is 82.70%, and the average recall rate is 77.65%. The above data shows that the recommended accuracy of this method is at least 9% higher than that of the other two comparison methods.

Figure 3 shows the comparison of F-score values of three resource recommendation methods.

It can be seen from the curve trend in Fig. 3 that the F-score value curve of the method in this paper is always above the F-score value curve of the other two methods, indicating that the recommended performance of the method is more stable.

To sum up, the recommendation methods of College English online mobile teaching resources based on big data mining algorithm in this paper are better than the traditional methods, and the accuracy, recall and F-score are improved. It shows that the method in this paper can predict learners' preferences relatively accurately when predicting

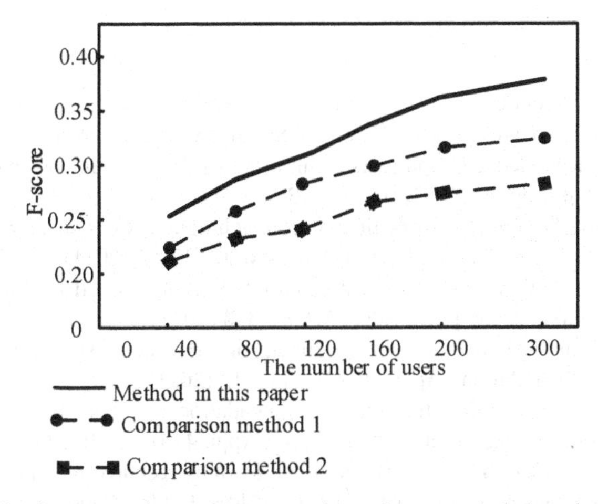

Fig. 3. Comparison of F-score values of recommended methods

learners' preferences for learning resources, and mobile teaching users are more satisfied with the resources recommended by the method.

4 Conclusion

With the continuous progress and development of online education platform, teachers' teaching mode is no longer limited to teaching based and book knowledge, and students' access to knowledge is no longer limited to classrooms, libraries and teachers. Mobile teaching extends the traditional classroom teaching to after class and after class. At present, the existing teaching methods in mobile teaching are relatively single. Each teacher can only obtain teaching resources through active teaching methods such as search, which can not be understood by analogy. However, with the development of educational big data, educational resources show the characteristics of massive resources, information overload and uneven quality, which makes teachers face the problems of information overload and knowledge loss. Therefore, it is necessary to provide more types of teaching resources for teachers, and recommend the contents they may be interested in as far as possible, so as to draw inferences from one instance. This paper proposes a recommendation method of College English online mobile teaching resources based on big data mining algorithm, and tests the recommendation method. The test data of recommendation method show that this method improves the teaching effect and promotes students' teaching on the basis of improving the accuracy of recommendation of teaching resources. The recommendation method of College English online mobile teaching resources based on big data mining algorithm proposed in this study has certain research value in the research and application of College English Education Resource Recommendation. However, due to the limited conditions, the recommendation method of this paper mainly studies to improve the accuracy of resource recommendation, and the improvement of recommendation efficiency is not obvious. Future studies can improve the efficiency of recommendations by ensuring their accuracy.

References

1. Yu, H., Sun, L.: Accurate recommendation algorithm of agricultural massive information resources based on knowledge map. Compute. Simul. **38**(12), 485–489 (2021)
2. Wu, H., Xu, X.: Knowledge Graph-assisted multi-task feature-based course recommendation algorithm. Comput. Eng. Appl. **57**(21), 132–139 (2021)
3. Shi, Y., Zhang, J.: Design of information-based teaching resources sharing system based on multimedia technology. Modern Electron. Tech. **44**(20), 32–36 (2021)
4. Jin, Y., Chen, H.: Value orientations, realistic limitations and effective strategies of big data mining for education. Theor. Pract. Educ. **41**(19), 3–8 (2021)
5. Xie, Y.: Analysis model simulation of multidimensional data de-clustering algorithm based on big data mining. Tech. Autom. Appl. **40**(12), 112–115 (2021)
6. Wang, Z., Liang, J.: Research on the rapid recommendation model of online teaching resources in colleges and universities. Inform. Stud.: Theor. Appl. **44**(05), 180–186 (2021)
7. Zhong, Z., Shi, J., Guan, Y.: Research on online learning resource recommendation by integrating multi-heterogeneous information network. Res. Explor. Laboratory **39**(09), 198–203 (2020)
8. Wang., G., Yuan, H., Huang, X., et al.: Recommendation algorithm for e-learning resources based on improved collaborative filtering. J. Chinese Comput. Syst. **42**(05), 940–945 (2021)

Design of Electronic Communication Power Monitoring System Based on GPRS Technology

Ying Liu[1](✉) and Fangyan Yang[2]

[1] School of Information and Control, Shenyang Institute of Technology, Fushun 113122, China
jjngn123@163.com

[2] College of Sports and Health, Changsha Medical University, Changsha 410219, China

Abstract. If the electronic communication power supply fails, the entire electronic communication system will be paralyzed, resulting in the abnormal operation of the system and increased maintenance costs in the later period. However, due to the slow transmission rate of the electronic communication power supply monitoring system, a GPRS based electronic communication power supply monitoring system is designed. Hardware part: the communication resources of the base station are used for networking, and the AC input is sent to the rectifier module after power distribution; Software part: identify the type of monitoring object, change the electrical signal or non electrical signal into a standard electrical signal, select UDP as the transmission protocol, use GPRS technology to formulate the communication protocol, and optimize the software function of the electronic communication power monitoring system. Experimental results: the average transmission rate of the electronic communication power supply monitoring system designed this time and the other two electronic communication power supply monitoring systems are 63.712 kpbs, 54.586 kpbs and 54.057 kpbs respectively, which shows that the system performance is more superior after the GPRS technology is integrated into the electronic communication power supply monitoring system.

Keywords: GPRS · Electronic communication · Communication power monitoring · Monitoring object · Protocol conversion · Communication resources

1 Introduction

The electronic communication power supply is usually called the heart of the communication system. The electronic communication power supply is extremely important to ensure the smooth flow of the entire communication system. If it does not work properly, it will cause the communication system to fail and even lead to the paralysis of the entire system. With the rapid development of communication business and science and technology, the automation performance of communication equipment has been greatly improved, which provides feasibility for centralized monitoring and management of communication systems. Electronic communication power supply is an important part

W. Fu and L. Yun (Eds.): ADHIP 2022, LNICST 468, pp. 453–466, 2023.
https://doi.org/10.1007/978-3-031-28787-9_34

of communication system. To adapt to the overall development of communication technology, it is necessary to actively adopt centralized monitoring system [1]. In order to ensure the smooth flow of the entire communication system, the electronic communication power monitoring system came into being. The electronic communication power monitoring system is to set up the necessary monitoring points for the distributed electronic communication power equipment and equipment room air conditioners, lighting and other equipment, conduct real-time monitoring, automatically monitor and deal with the faults of various equipment in the system, and achieve 24-h non-stop that cannot be achieved by manual methods. Intermittent automatic inspection [2].

The direction of power supply equipment maintenance reform is centralized monitoring, and gradually realize that there are few and unattended power supply equipment in communication stations. The so-called centralized monitoring of electronic communication power supply is to use a computer control system to reasonably set up necessary monitoring points for the power supply equipment of communication stations distributed in different regions, carry out telemetry, remote signaling, remote control, monitor the operating parameters of the equipment in real time, and find and deal with faults in time. The core and most important function of the monitoring system is fault alarm and real-time monitoring. The key of the monitoring system is mainly to see whether it can monitor and warn the electronic communication power equipment in an uninterrupted, stable, reliable and real-time manner for many years.

The design of electronic communication power monitoring system has important practical significance to ensure the smooth flow of the entire communication system. Reference [3] proposes Design and Research of Communication Power Monitoring System Based on Internet of Things Technology. The system uses ARM chip as the core controller, uses Hall sensors to collect working status information of communication power equipment, and generates DC signals through A/D conversion., Use GPRS DTU module to transmit signals, read signal information in monitoring center, and issue control commands. Complete the optimization of the electronic communication power monitoring system. Reference [4] proposes Design of Communication Power Monitoring System in Power Communication, uses UML modeling technology to analyze system functional requirements, uses C++ to develop power monitoring data acquisition module, uses C# to develop power monitoring system interface, and designs power monitoring system database at the same time. In order to complete the optimization of the electronic communication power monitoring system. However, the above methods all have the problem of low transmission rate of the communication power monitoring system. In this regard, this paper proposes Design of Electronic Communication Power Monitoring System Based on GPRS Technology. On the basis of the hardware and software of the electronic communication power monitoring system, UDP is selected as the Transmission protocol, use GPRS technology to formulate communication protocol, optimize the software function of electronic communication power monitoring system. The experimental results show that the average transmission rate of the electronic communication power monitoring system designed in this paper is 63.712 kpbs. It has a certain technical level and practicality.

2 Hardware Design of Electronic Communication Power Monitoring System

The monitoring system is monitored by computer software through hardware, so the reliability of system hardware is very important. First of all, in the circuit design, it is necessary to ensure that the correct telemetry signal is obtained, and the linearity in the transformation and transmission is guaranteed. Considering the reliability of the AC power supply, the system can be powered by two mains, the two mains work as the main and standby mode, automatically switch over, and have an electrical interlock function. After the AC input is distributed, it is sent to the rectifier module., the DC output of the rectifier module is connected to the DC power distribution unit through the bus bar, and finally the battery pack is charged and power is supplied to each load through the DC power distribution unit. For remote signaling signals, it is necessary to ensure that the logic is correct, and multiple monitoring of important signals can be performed, and the correctness of remote signaling signals can be judged through software. The selected components should have high reliability. The selected control board (1/0 board, AJD board) should choose high-quality and reliable products. The selected computer should be a high-end (referring to reliability) microcomputer. For some electrical signals, limit the amplitude to prevent abnormal signals from damaging circuits and devices. When designing the printed board, increase the wire area for the power line and the ground line to reduce the line resistance. The hardware structure of the electronic communication power monitoring system is shown in Fig. 1:

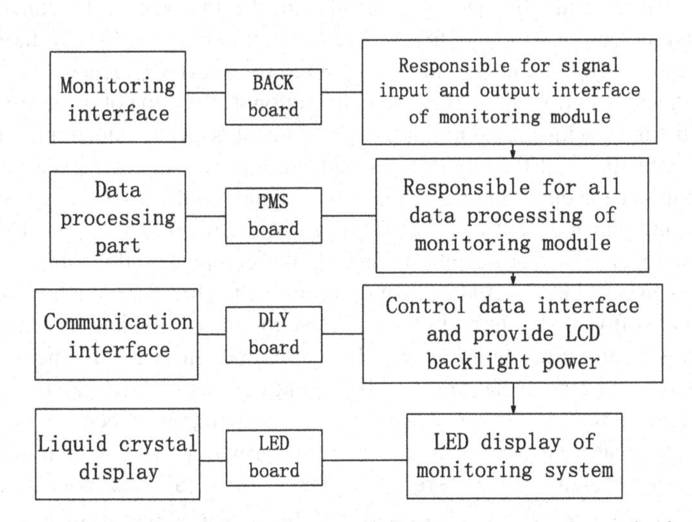

Fig. 1. Hardware function of electronic communication power supply monitoring system

As can be seen from Fig. 1, the hardware can be functionally divided into four parts: monitoring interface part, data processing part, communication interface part, and liquid crystal display part. The whole system structure is divided into four boards: BACK board, PMS board, DLY board and LED board. The connectors used should choose connectors with good contact and a locking device. There are many input and output signal lines in the monitoring system (up to hundreds), how to connect with the system to ensure reliable connection and easy maintenance must be carefully considered. The BACK board is responsible for the signal input and output interfaces of the monitoring module. All analog signals are converted into digital signals on this board and then sent to the PMS board for data processing. The PMS board is responsible for all data processing of the monitoring module and is the core of the monitoring module. The DLY board is responsible for the LCD interface of the system, including the control data interface and the LCD backlight power supply. The LED board is responsible for monitoring the LED display of the system. It can be connected to the system through terminal blocks and connectors. The entire base station centralized monitoring system is divided into three parts, namely pre-data acquisition, long-distance transmission and central local area network. MISU multifunctional integrated monitoring equipment is an embedded microprocessor system. In the input and output circuit design, in addition to considering the I/O resources, factors such as the control form of the switch level, the software processing of the switch level, and the software anti-interference of the switch level signal should also be considered. In order to improve the speed at which the computer samples the switching level signal, the byte processing method is selected in the hardware design of the input interface. It can realize real-time monitoring and alarm processing of various base station power equipment and environmental monitoring signals, and make corresponding control according to application requirements. At the same time, MISU has the networking capability for downlink base stations, which can effectively utilize the communication resources of base stations for networking. The input of switch level should be buffered and isolated and then input to the I/O port of the computer after filtering, and the output will be output by the I/O port after being locked. In the design of the interface circuit, attention should be paid to the anti-interference problem of the digital input, and attention should be paid to the rational allocation of resources. ZXM10 MISU multi-function integrated monitoring equipment is an embedded microprocessor system. It can realize real-time monitoring and alarm processing of various base station power equipment and environmental monitoring signals, and make corresponding relay contact output control according to application requirements. The input of the switch is often accompanied by noise interference. If the computer judges that the state of the switch is uncertain, it will affect the control result. Therefore, the combination of hardware and software should be used to remove the uncertain state (defibrillation) to ensure the reliability of the entire system. In dealing with these disturbances, hardware methods include filtering, isolation, etc. MISU continuously collects monitoring signals of various base station power equipment and environment, and processes the monitoring signals into real-time monitoring and alarm information that can be transmitted by the transmission channel. Under normal circumstances, MISU does not actively report real-time data and alarm information, and only reports these real-time data and alarm information when the transceiver asks the module.

3 Software Design of Electronic Communication Power Monitoring System

3.1 Identify the Type of Monitoring Object

The monitoring objects of the electronic communication power monitoring system include all the power supplies, air-conditioning equipment and environmental quantities of the communication station [5, 6]. The main equipment of the electronic communication power supply system includes high-voltage power distribution equipment, low-voltage power distribution equipment, mains oil generator conversion panel, rectifier power distribution equipment (rectifier equipment, power room AC and DC power distribution equipment and battery pack), converter equipment (UPS, inverter and DC-DC converter) and power station equipment, etc.

The monitoring point is the specific monitoring and control semaphore set by the monitoring object. In terms of data types, these semaphores include analog, digital, state, switch, and so on. From the flow of the signal, there are two types of input and output. The monitoring content, also called monitoring item, refers to the specific monitoring and control semaphore set for the above monitoring object. From the point of view of data type, these semaphores include analog quantity, digital quantity, state quantity, switch quantity, etc. From the perspective of signal flow, they also include input quantity and output quantity. Monitoring objects can be divided into smart devices and non-intelligent devices according to the characteristics of the monitored device itself. Among them, the intelligent device itself can collect and process data, and has an intelligent communication interface, which can be connected to the monitoring system directly or through protocol conversion, and each intelligent device is used as a monitoring module. Telemetry refers to the process of obtaining these data remotely by collecting continuously changing analog quantities of equipment or the environment. The objects of telemetry are all analog quantities, including various electric quantities such as voltage, current, and power, and various non-electric quantities such as temperature, pressure, and liquid level. Therefore, these monitoring items can be divided into four categories, mainly including: telemetry, remote signaling, remote control and remote adjustment [7, 8]. After a series of transformation processing of sensors, transmitters and monitoring modules, it is converted into a digital quantity that is very close to the real value, and is handed over to the computer for further processing. Remote signaling refers to the process of remotely obtaining corresponding state quantities by monitoring the operating state of field devices or environments. Non-intelligent devices cannot collect and process data by themselves, and there is no intelligent communication interface. They need to be connected to the monitoring system through data acquisition and control equipment (data collectors). Each data acquisition and control device is used as a monitoring module. The monitored signals can be divided into power signals and non-power signals. The content of remote signaling generally includes two types of equipment operating status and status alarm information. The values of these states are usually several discrete values, each of which has a fixed meaning and is used to represent an operating state of the device or its part. Some state quantities have only two values, which indicate the presence or absence of certain states, alarm or no alarm. In the monitoring system, the processing of the monitored signal generally needs to go through the process of sensing, transmission and conversion before it can be converted into a digital signal in the computer.

Non-smart devices and environmental quantities cannot be directly connected to the acquisition channel of the data collector for measurement. These power signals or non-power signals need to be converted into standard power signals through sensors/transmitters before they can be connected to the data collector. Remote control refers to the process in which the monitoring system issues specific instructions to the remote equipment to make the equipment perform corresponding actions. The value of the remote control quantity is usually a switch quantity, which is used to express information such as "on", "off" or "running", "stop", etc., and a multi-valued state quantity can also be used to enable the equipment to perform operations between several different states. Toggle action. Monitoring items can be divided into telemetry, remote signaling, and remote control: telemetry refers to data collection of continuously changing analog signals. Remote signaling refers to data acquisition of discrete state switching signals. Remote control refers to discrete control commands issued by a monitoring system. The setting of monitoring points is not the better. When a small number of monitoring points can fully reflect or basically reflect the operating status of the equipment, setting too many monitoring points is not only unnecessary, but will increase the cost and complexity of the system and reduce reliability.

3.2 Optimize the Software Function of Electronic Communication Power Monitoring System

Communication protocol is a set of rules agreed in advance between two computer systems for communication and must be observed jointly. Starting from the problems to be solved by the communication protocol, there should generally be three basic requirements. The communication protocol must be able to achieve accurate communication between intelligent devices and monitoring systems (computers). For the server, it also applies for a fixed socket and starts to wait for the request of the client. Any client can send a connection request and an information request to it. After the connection is successful, both the client and the server can send and receive data to the socket. When the communication ends, close the socket and cancel the connection [9]. Just like the language used by people in conversation, communication protocol is also a set of very strict "language". All the "words and sentences" used must be defined in advance, and ambiguity is absolutely not allowed. The communication protocol must be able to provide reliable communication. Socket skillfully solves the problem of establishing communication connection between processes by using C/S mode. In the monitoring system, UDP is selected as the transmission protocol. Datagram protocol UDP is a connectionless communication protocol based on IP protocol. UDP does not need to establish a communication channel before transmitting data, but directly sends the data to the receiving end. There are three types of communication protocols related to the monitoring system, as shown in Fig. 2:

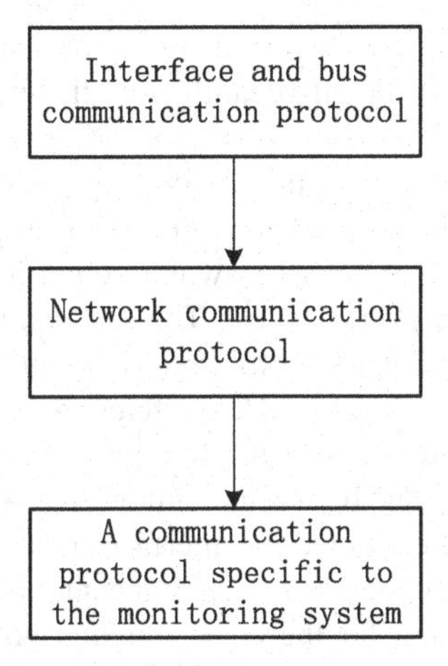

Fig. 2. Communication protocol types

As can be seen from Fig. 2, communication protocol types include: interface and bus communication protocol, network communication protocol, and monitoring system specific (special) communication protocol. The communication protocol should be able to provide efficient communication. This paper uses the designed communication protocol for data transmission.

According to the actual needs, the basic functions of the monitoring system can be divided into monitoring functions, human-computer interaction functions, management functions and auxiliary functions, among which the management functions include data management functions, alarm management functions, configuration management functions, security management functions, Self-management function and file management function, etc. The monitoring module is the lowest monitoring layer in the monitoring system [10]. It is directly connected to the equipment and is used to monitor, collect and process the working status and operating parameters of the monitored equipment to form standardized status, data and alarm information for upward transmission. At the same time, it receives and executes various monitoring and control commands issued by the station center to control the equipment and adjust parameters. For the parameters in the system, the current packet data charging is calculated according to the data above the second layer of the network protocol (that is, IP packet data), and the transmission efficiency calculation is calculated according to the following formula:

$$T = \frac{\phi}{G + \eta} \tag{1}$$

In formula (1), ϕ represents the data length, G represents the user data length, and η represents the TCP header length.

Among them, the monitoring function is the most basic function of the monitoring system, which also includes the monitoring function and the control function. The monitoring system can continuously monitor the real-time operation status of the equipment and the environmental conditions that affect the operation of the equipment, and obtain the original data and various states of the equipment operation for system analysis and processing. Since it is directly dealing with the monitored equipment, to meet the reliability of the monitoring system, the monitoring module must have the highest monitoring priority. The core part of the monitoring module is generally served by a single-chip microcomputer. It sets a certain number of interfaces and channels such as analog input, switch input, digital output, switch output and counting input through a certain interface chip and peripheral circuit. The sensors, transmitters and contacts on the controlled equipment are connected to directly monitor and control the controlled equipment in real time. This process is called telemetry and remote signaling. Human-computer interaction function refers to the function of mutual dialogue between the monitoring system and people and between monitoring systems, including the functions realized by the human-computer interaction interface and the function of interconnection and communication between systems. The station center is a centralized operation and maintenance management point for the environmental power equipment of the communication station, which is used to monitor and manage the monitoring modules of each equipment in the station. Its function mainly focuses on monitoring, which sends monitoring and control commands to the monitoring module, including the setting and adjustment of parameters.

On the basis of formula (1), the calculation formula of network load is obtained as:

$$D = \gamma \times \frac{1}{\sum_{\phi=1} |\eta - 1|^2} \tag{2}$$

In formula (2), γ represents the communication time of the TCP protocol stack, and η represents the number of bytes in a single frame of data. In the monitoring center, the front-end computer is responsible for receiving, analyzing and processing all monitoring point data, including performance data, alarm data and system data, etc. The expression formula of point-to-point data transmission is:

$$\mu = \frac{\left\| t - y^2 \right\|}{2} \times \gamma + \eta \tag{3}$$

In formula (3), t represents the initial time when static IP is written into the GPRS module, and y represents the communication rate.

The system interconnection function refers to the horizontal networking function between the upper and lower monitoring systems with jurisdictional relationship through a certain protocol interface, and the vertical networking function between the monitoring system and other systems (such as the network management system), which enables the system to Network flexibly. Summarize the data collected by each monitoring module for further processing and storage, and display or print if necessary: transmit real-time data, historical data, equipment parameters, status information, alarm information and statistics to the superior at regular intervals or according to the requirements of the regional monitoring center information, and accept the equipment remote control command issued by the superior. Alarm is also a kind of data. From the particularity of its content and meaning, alarm is the most important monitoring data of the monitoring system, and the alarm management function is also the most important function of the monitoring system. For the same reason, dozens of alarm messages may be generated on the monitoring interface at once (for example, when a power failure occurs, corresponding alarms will be issued by AC power distribution, DC power distribution, rectifier, etc.). The regional monitoring center is a centralized operation and maintenance point used to monitor and manage all the bureau stations in a city-level area, and is the basic operation and maintenance unit of the environmental power equipment in the monitoring system. It has the most powerful functions and the most perfect performance in the whole monitoring system. At this time, the system needs to determine the most critical and fundamental alarms according to the pre-set logical relationship, so as to filter out the other related alarms, so as to achieve the real alarm location. Configuration management refers to editing and modifying the configuration parameters, interface and other characteristics of the monitoring system to ensure the normal operation of the system, optimize the performance of the system, and enhance the practicability of the system. Communication and port parameters, such as communication rate, serial data bits, number of ports and modules, addresses, etc. Taker compensation parameters, such as acquisition point slope compensation, phase compensation, function compensation, etc. Some of these parameters often need some necessary adjustments according to the actual situation, and the monitoring system must be able to provide users with convenient and practical parameter configuration functions. The regional monitoring center is usually a local area network composed of computer equipment such as data servers, communication machines, disk arrays, operation desks, and printers. When communicating with the central station, the communication protocol is consistent with the communication protocol of the central station. In some cases, the communication protocol is inconsistent. At this time, the protocol conversion needs to be carried out by software. System parameters are important configuration information to ensure the normal operation of the monitoring system and to reflect the equipment conditions truthfully. These parameters include data processing parameters, such as data sampling period, data storage period, data storage threshold and so on. Alarm setting parameters, such as alarm upper and lower limit, alarm mask event segment, whether to activate sound alarm, etc.The specific monitoring functions are shown in the following figure:

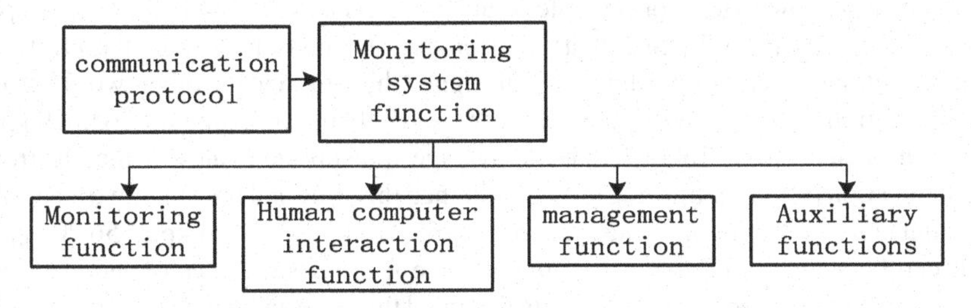

Fig. 3. Schematic diagram of monitoring function

4 Experimental Analysis

4.1 Set up the Experimental Environment

The built test environment: the monitoring center consists of a front-end communication machine, a database server, and a user operation terminal. Because it is difficult to simulate the actual power supply operation data under laboratory conditions, only the parameter data of the computer room environment is simulated and tested in the data acquisition terminal part. To develop application programs on Windows, there are many tools to choose from, such as: VB, Delphi, VC, C++ Builder, etc. The VB man-machine interface is very good, but it is weak in the underlying manipulation ability and the communication speed is slow, while the VC man-machine interface is better, the communication function is strong, and the underlying manipulation ability is strong. The communication module is connected with the central communication machine through the RS232 interface/E1 interface of the digital business channel. The intelligent protocol conversion module performs protocol conversion for intelligent devices (such as switching power supplies, professional air conditioners) through 3 RS485/RS422/RS232 intelligent interfaces and 1 RS232 intelligent interface. In the actual test, two collectors, two smoke sensors, and two access control sensors are used to simulate the operating environment of 82 bureaus and 83 bureaus in the actual monitoring system. Front communication machine: built-in TCP/IP communication module, used to transmit data.

4.2 Experimental Results

In order to test the application effect of the electronic communication power monitoring system designed this time, the experimental test is carried out. The literature [3] and literature [4] are selected for experimental comparison. The transmission rates of the three systems are tested under the conditions of different concurrent users. The experimental results are shown in Table 1–5:

It can be seen from Table 1 that the average transmission rates of the electronic communication power monitoring system designed this time and the other two electronic communication power monitoring systems are: 94.235 kpbs, 84.144 kpbs, 84.161 kpbs.

It can be seen from Table 2 that the average transmission rates of the electronic communication power monitoring system designed this time and the other two electronic communication power monitoring systems are: 76.499 kpbs, 65.936 kpbs, 65.820 kpbs.

It can be seen from Table 3 that the average transmission rates of the electronic communication power monitoring system designed this time and the other two electronic communication power monitoring systems are: 65.212 kpbs, 56.864 kpbs, 56.204 kpbs.

It can be seen from Table 4 that the average transmission rates of the electronic communication power monitoring system designed this time and the other two electronic communication power monitoring systems are: 46.710 kpbs, 37.330 kpbs, 36.574 kpbs respectively.

It can be seen from Table 5 that the average transmission rates of the electronic communication power monitoring system designed this time and the other two electronic communication power monitoring systems are: 35.906 kpbs, 28.657 kpbs, and 27.527 kpbs, respectively. According to the experimental results in Table 1–5, as the number

Table 1. Transmission rate of the system with 50 concurrent users (kpbs)

Number of experiments/time	Literature [3]	Literature [4]	The electronic communication power monitoring system designed this time
1	83.611	82.501	96.748
2	82.004	83.646	95.316
3	83.649	84.745	94.522
4	82.055	82.997	93.748
5	83.747	83.649	92.501
6	84.559	82.554	91.749
7	83.615	83.679	93.545
8	84.517	84.755	93.846
9	83.994	83.902	94.805
10	84.165	84.397	93.648
11	85.912	85.636	94.715
12	86.794	84.122	95.313
13	85.216	85.704	94.705
14	84.773	84.910	93.847
15	83.552	85.223	94.512

Table 2. The transmission rate of the system with 100 concurrent users (kpbs)

Number of experiments/time	Literature [3]	Literature [4]	The electronic communication power monitoring system designed this time
1	65.849	63.845	76.455
2	71.006	65.009	77.311
3	69.546	65.887	76.911
4	66.548	66.9993	75.813
5	65.812	65.213	76.488
6	64.377	64.774	75.412
7	65.696	65.995	76.316
8	65.997	69.822	75.820
9	64.331	67.313	76.117
10	65.819	66.549	75.151
11	66.748	63.215	76.909
12	65.812	66.845	77.463
13	63.774	65.887	78.111
14	62.912	64.233	79.006
15	64.819	65.718	74.203

of concurrent users increases, the data transmission rate of the system decreases, and the data transmission rate of the electronic communication power monitoring system designed this time can meet the needs of system operation.

Table 3. Transmission rate of the system with 150 concurrent users (kpbs)

Number of experiments/time	Literature [3]	Literature [4]	The electronic communication power monitoring system designed this time
1	56.151	56.849	63.021
2	55.846	57.661	64.918
3	59.845	55.774	65.224
4	56.747	56.933	64.833
5	55.822	55.714	65.121
6	54.913	56.880	64.793
7	55.864	55.974	65.812
8	56.893	56.332	66.464
9	55.122	55.818	65.933
10	56.047	54.779	65.714
11	57.448	55.316	63.821
12	58.666	55.822	64.512
13	57.317	56.917	65.315
14	58.916	55.378	66.904
15	57.364	56.919	65.788

Table 4. Transmission rate of the system with 200 concurrent users (kpbs)

Number of experiments/time	Literature [3]	Literature [4]	The electronic communication power monitoring system designed this time
1	36.646	36.385	44.559
2	38.579	37.413	45.877
3	39.845	36.588	46.312
4	36.727	35.811	49.002
5	35.812	36.923	47.616
6	36.914	35.747	48.377
7	37.315	36.499	46.919
8	38.567	35.844	45.213
9	39.454	36.745	46.337
10	36.825	37.614	45.844
11	37.533	36.825	46.918
12	38.166	37.447	47.553
13	37.646	36.914	48.512
14	35.202	35.331	45.211
15	34.714	36.528	46.399

In order to verify the effectiveness of the method in this paper, taking the monitoring accuracy of the electronic communication power monitoring system as the experimental index, the method in this paper, the method in the literature [3], and the method in the literature [4] are used for experimental tests. The test results are shown in Table 6:

Table 5. The transmission rate of the system with 250 concurrent users (kpbs)

Number of experiments/time	Literature [3]	Literature [4]	The electroniccommunication power monitoring system designed this time
1	29.845	28.514	33.616
2	31.546	29.606	34.528
3	28.477	26.774	36.977
4	30.221	25.648	35.844
5	29.466	26.845	36.107
6	28.745	27.451	35.899
7	27.334	26.845	36.071
8	28.549	27.443	37.646
9	29.613	26.887	35.447
10	27.451	27.942	36.919
11	26.448	28.545	35.332
12	27.355	27.994	36.747
13	28.946	26.501	35.218
14	27.313	27.365	36.901
15	28.544	28.541	35.332

Table 6. Comparative experimental results of system monitoring accuracy

Number of experiments/time	Literature [3]	Literature [4]	The electroniccommunication power monitoring system designed this time
10	98%	85%	89%
20	97%	86%	89%
30	98%	85%	88%
40	99%	84%	87%
50	98%	86%	88%
60	97%	87%	89%
70	96%	86%	87%
80	98%	87%	87%
90	97%	86%	86%
100	99%	84%	88%

From the results shown in Table 6, it can be seen that the accuracy rate of the method in this paper is up to 98%, the accuracy rate of the method in the literature [3] is up to 87%, and the accuracy rate of the method in the literature [4] is up to 89%. It can be seen that the accuracy rate of the method in this paper is significantly higher than that of the method in the literature [3] and the method in the literature [4]. It shows that the method in this paper has good monitoring performance of the electronic communication power monitoring system. The technical level and application value of the method proposed in this paper are proved to be high.

5 Concluding Remarks

This paper studies and expounds the relevant theory and technology of data transmission using GPRS, and then analyzes the development status and characteristics of the electronic communication power supply monitoring system. On the basis of the data transmission design scheme, a communication protocol is formulated. In addition, the electronic communication power monitoring system function has been optimized. At the same time, a good client program has been developed and constructed, which makes the monitoring system interface good, easy to use, easy to maintain, and simple to expand, laying a good foundation for further development and improvement in the future.

References

1. Jia, Y.: Research on the application of embedded technology in electronic communication energy-saving. Telecom Power Technol. **37**(15), 104–106 (2020)
2. Xue, Y., Li, Z.: Extraction of logo design elements in the field of electronic communication. Packag. Eng. **43**(2), 216–220 (2022)
3. Li, X., Zhou, X., Wang, S.: Design and research of communication power monitoring system based on internet of things technology. Telecom Power Technol. **39**(1), 47–49 (2022)
4. Ye, D.: Design of communication power monitoring system in power communication. Telecom Power Technol. **37**(1), 249–250 2020
5. Li, F., Liang, F.: Elimination of Redundancy of Data Flow in Multithreaded Electronic Communication Network. Comput. Simul. **38**(11), 158–161,167 (2021)
6. Sharma, V.K., Tripathi J.N., Shrimali, H.: Analysis of power supply noise in AMS circuits including the effects of interconnects using estimation by inspection method. AEU: Archiv fur Elektronik und Ubertragungstechnik: Electron. Commun. **139**, 139–150 (2021)
7. Jia. X., Pei, M., Fu, R., et al. Robust secrecy rate optimizations for healthy monitoring system MISO cChannel with D2D Communications. J. Phys.: Conf. Series **1518**(1), 012079 (6pp) (2020)
8. Wu, L., W., Hou, J.: Mobile oil tank monitoring system based on gprs and gps technology. Process Autom. Instrument. **41**(7), 61–64,68 (2020)
9. Wang, F., Huang, X., Yang, F., et al.: Internet of lamps for future ubiquitous communications: integrated sensing, hybrid interconnection, and intelligent illumination. China Commun. (English) **19**(3), 132–144 (2022)
10. Wang, J., W., Xiang, S., Shan, X., et al.: Design of battery pack cloud monitoring system based on GPRS. Chinese J. Power Sources **44**(6), 905–907 2020

Online Interactive Platform for College English Intensive Reading Teaching Based on Cloud Service

Lin Fan(✉)

Yantai Vocational College, Yantai 264670, China
huhu21220@163.com

Abstract. With the continuous development of science and technology, online teaching has become an important form of teaching in all stages. In the application process of College English intensive reading online teaching interactive platform, there is a problem that the courseware upload time is too expensive. Therefore, a college English intensive reading online teaching interactive platform based on cloud service is designed. Hardware part: select the mode of automatic clock phase control to obtain analog power through digital power supply; Software part: Taking discourse as the basic unit, identify the characteristics of College English intensive reading course, set up online teaching interaction mode, reflect the multidimensional and extensibility of interactive content, and optimize the function of the platform by using cloud service. Experimental results: the cost of uploading courseware between the designed College English intensive reading online teaching interactive platform and the other two college English intensive reading online teaching interactive platforms is 7.812 s, 12.123 s and 12.127 s respectively, and the code rate is low. It shows that after fully integrating with cloud service technology, the use effect of College English intensive reading online teaching interactive platform is more prominent.

Keywords: Cloud service · College English · Online teaching · Interactive platform · Intensive reading courses · Teaching objectives

1 Introduction

Classroom teaching interaction is the most active element in classroom teaching activities, which directly affects the effect and quality of classroom teaching [1–3]. College English intensive reading course is a compulsory course for English majors. College English teaching is an important part of Non-English Majors in higher education. The teaching goal of College English is to cultivate students' comprehensive application ability of English, so that students can communicate effectively in English in their future study, work and social communication, and enhance their autonomous learning ability to meet the needs of China's social development and international communication [4]. The word "interaction" comes from the computer term, which refers to the process in which

W. Fu and L. Yun (Eds.): ADHIP 2022, LNICST 468, pp. 467–480, 2023.
https://doi.org/10.1007/978-3-031-28787-9_35

the system receives the input from the terminal, processes it, and returns the result to the terminal, that is, the so-called human-computer interaction process. With the introduction of computer-aided instruction, interaction has gradually become a teaching term, which can be understood as the activities of mutual communication between teachers and students, students and students in the teaching process.

Reference [5] proposed the research of online course live teaching platform based on real-time interaction mode, combined with the monitoring method of learning process, to show students' learning effects in the whole online course learning process, and finally to build an online course live teaching platform. Reference [6] proposes the application of video interaction technology in online education, researches on enriching the interactive forms of online video to improve the learning experience of learners, and realizes a video player with rich interactive functions and a front-end platform for interactive data visualization, which provides a new development direction for online education video. The traditional classroom teaching interaction is mainly face-to-face interaction between teachers, students and students. Its advantages lie in solving problems in time, improving teachers' and students' emotions, and building a harmonious collective learning atmosphere. However, there are also some shortcomings in the traditional face-to-face classroom teaching interaction. The network provides a more advanced interactive environment than the traditional classroom, but there are also some unfavorable factors. How to make rational use of the interactive characteristics of online teaching, It is a problem that all online course designers and teachers must seriously consider. Therefore, an online interactive platform for College English Intensive Reading Teaching based on cloud service is proposed. In the hardware part, the automatic clock phase control mode is selected, and the analog power is obtained through the digital power supply; In the software part, text is used as the basic unit to identify the characteristics of College English intensive reading course, set up online teaching interaction mode, reflect the multidimensional and extensibility of interaction content, and optimize the platform function by using cloud services. According to the conditions of the University and the English level of students, colleges and universities try to explore and establish the listening and speaking teaching mode under the network environment, and carry out listening and speaking teaching and training directly on the LAN or campus network.

2 Hardware Design of Online Interactive Platform for College English Intensive Reading Teaching

The working mode of 1vixt2001 is configured through SPI serial port to control the circuit. The configuration of several key registers of mxt2001 and the settings related to the configuration of 2GS/s real-time sampling rate are as follows: enable the bilateral edge sampling function. In the first mock exam, the DEN enable register D15 bits are high, enabling MXT2001 to enter the bilateral sampling mode. In this mode, the two sub ADC samples and converts the same analog input signal through time interleaved mode, achieving two times the sampling frequency of the input clock frequency. Choose the automatic clock phase control mode, because if only the CPU core is powered, and the peripheral I/O is not powered, it will not damage the chip, but there is no input and output capacity. On the contrary, if the peripheral I/O is powered on and the CPU core is not powered on, the triode of the DSP buffer driving part works in an unknown state, which may cause the peripheral pins of the DSP to act as the output terminal at the same time. The low-speed ADC is used for time alternating parallel sampling. The phase difference must be only 180° as far as possible to achieve the best equal interval sampling. When the D14 bit of Des enable register is enabled, the clock phase control function can be started. The mxt2001 internal phase detection circuit continuously adjusts the sampling clock edge of channel I and channel Q to make their phase difference 180° to realize high-precision equal interval sampling. At this time, if the output values of both sides are opposite, there may be a large current due to reverse driving, which is very dangerous, which will not only affect the service life of the device, but also damage the device. Similarly, when the power is turned off, if the core power is powered off first, a large current will also be generated.

If two sub ADCs want to realize equal precision voltage mapping for the same measured signal, the full amplitude voltage of the two sub ADCs must be consistent. Otherwise, when two sub ADC data are spliced, the waveform will be sawtooth, resulting in uneven waveform splicing. When designing the power supply system of different DSP chips, we should according to their different power supply characteristics, otherwise the whole power supply system may be damaged. Considering the above characteristics, the special power chip tps70345 provided by TI company is selected to design the digital system power supply.

The full range voltage adjustment is adjusted by the input swing registers of channel I and channel Q (the addresses are 3H and 8h respectively). In this subject, they are set to ox807f, that is, the full range voltage is 700 mVpp. Due to different ADC chips, the input offset of two sub ADCs may still occur after the full range voltage adjustment register is adjusted, so it is still necessary to fine tune in combination with the I-channel offset adjustment and Q-channel offset adjustment registers to obtain the analog power supply through the digital power supply. Among them, while the typical working current of the I/Q pin is 58 mA, which fully meets the requirements even if 50% redundancy is considered respectively. 1VIXT2001 can adjust the offset value of 45 mV at most, so at least 0.176 mV single step adjustment (45 mV/256) can be obtained each time. Through coarse adjustment and multiple fine adjustment of the full range voltage and phase of the two sub cores of ADC, the real-time sampling rate of 2GS/S is finally realized.

3 Interactive Platform for Online Teaching of College English Intensive Reading

3.1 Identify the Characteristics of College English Intensive Reading Courses

As a highly communicative and comprehensive course, English Intensive Reading emphasizes the co construction of group knowledge. Each learner is not only the acquisition of knowledge, but also the creator of group knowledge. In other words, basic language skills are the premise of developing communicative competence.

Using the principle of cloud service, the English intensive reading course is designed into a common learning community. Each learner is not only the acquisition of knowledge, but also the creator of group knowledge. Compared with other subjects, English learning needs to conform to the law of second language acquisition. At the same time, the teaching purpose of the university stage focuses more on the development of students' self-study ability. To complete the classroom task, we need to find information, analyze information and extract effective information, which is undoubtedly beneficial to the development of their self-study ability. The design of course content needs to meet the understandable input principle of knowledge.

The existing intensive reading textbooks for English majors generally take the text as the basic unit and the text as the unit. From the perspective of language learning, in order to further promote the accumulation of learners' language knowledge, we should make the input content meet the basic needs of learners, that is, understandable input conducive to the development of learners' cognitive ability. In order to enable learners to understand the language form from the language meaning and regard the learning of language form as a means to achieve communication rather than set up for form, the best teaching method is to take the text as the unit of teaching. Learning input is not only a multi-dimensional structure, but also has different but highly interrelated aspects. The four levels of learning input are shown in Fig. 1:

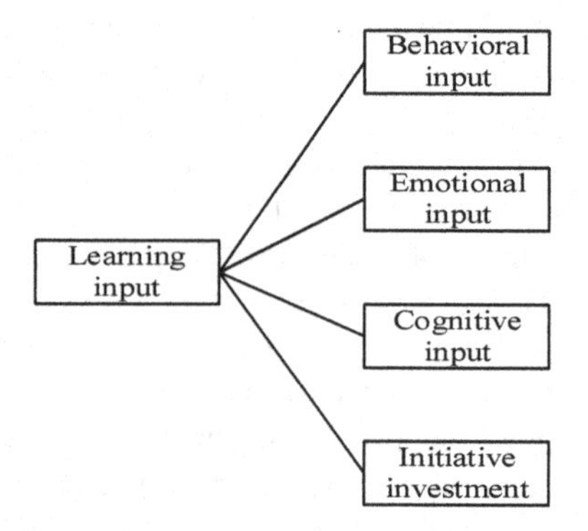

Fig. 1. Schematic diagram of four levels of learning input

As can be seen from Fig. 1, the four levels of learning investment are: behavioral investment, emotional investment, cognitive investment and initiative investment. Each of these four aspects can be connected with other links. Therefore, judging learners' active involvement in learning activities will include assessing attention. When the teaching unit rises to discourse, it pays more attention to the meaning of language communication than the form of language, which has been reflected by the communicative teaching method. As the continuous development of communicative teaching method, interactive teaching focuses on the teaching of language meaning rather than the mechanical training of language form. The discourse teaching system of intensive reading teaching is suitable for interactive teaching. Interactivity is one of the essential characteristics of interactive teaching. Realizing real interactive classroom teaching is an important link to improve the quality of English classroom teaching. Interaction is not only the transmission of information, but also the process of information understanding and processing.

3.2 Setting Interactive Mode of Online Teaching

From the perspective of pedagogy, "interaction" refers to the communication, communication, connection, influence and interaction between teachers and students in the teaching process, taking the teaching content as the media and in order to complete the teaching task under the environment of modern educational technology. It emphasizes that the communication between teachers and students is based on the teaching content. The term "interaction" comes from computer terminology. It refers to the process in which the system receives the input from the terminal, processes it, and then returns the result to the terminal, that is, the so-called interaction between man and machine. At the same time, it also focuses on the interaction between teachers and students and the learning environment, including the interaction between teachers and students, learning resources and students, as well as the interaction between teachers and students.

With the introduction of CAI, interaction has gradually become a term of online teaching, which can be understood as an activity of mutual communication, mutual communication, mutual dialogue, mutual understanding and mutual learning between teachers and students, students and students in the process of online teaching. Online teaching interaction not only emphasizes the interaction between online teaching participants, online teaching interactive behavior and online teaching resources and environment, but also cannot ignore the importance of online teaching objectives and solving online teaching problems. From the perspective of interaction object, the interaction in distance education is defined as the interaction between distance learners and all distance education resources, mainly including the interaction between distance learners and learning materials, and learning support organizations (including tutors, consultants, administrative personnel, institutional facilities, etc.). Online teaching objectives play a guiding role in online teaching activities and online teaching tasks. Online teaching interaction needs to adopt appropriate interactive behavior and content around online teaching objectives. The emphasized online teaching design changes from simple presupposition to dynamic generation, and dynamic generation is the biggest feature of classroom interaction. Therefore, online teaching interaction refers to that in a certain online teaching situation, mainly in classroom teaching, in order to achieve the expected teaching objectives, teachers carry out the communication and discussion between students and teachers and students with the help of technical means based on the teaching content, expand from interpersonal interaction to information interaction, further guide students to think on this basis, and finally lead students to the interaction of new and old concepts within individuals.

The interaction between learners and learning content is a key feature to define education. Because the interaction process between learners and learning content changes learners' understanding, the formation of views or the cognitive structure in learners' mind, there is no education without the interaction between learners and learning content. Interaction and learning promote each other. Students' learning is based on one-dimensional thinking of learning content, while teaching interaction reflects the multi-dimensional and extensibility of interactive content due to the scalability of the number of participants and the diversification of background environment.

The feedback given by teachers is of great value in the interaction between teachers and learners. Educators should organize online teaching plans to maximize the benefits of various types of interaction, and ensure that the types of interaction provided are most appropriate for different online teaching tasks in different disciplines and learners at different stages of development. The online teaching interaction between teachers and students embodies the modern online teaching thought of taking teachers as the leading and students as the main body, and also conforms to the people-oriented educational concept. After students interact with teachers, peers and collectives, it helps to connect individual internal knowledge with the surrounding knowledge network, so as to optimize the knowledge structure. Divided from the participants of online teaching interaction, online teaching interaction can be divided into two types: teacher-student interaction and student student interaction, which can be divided into the following six types, as shown in Fig. 2:

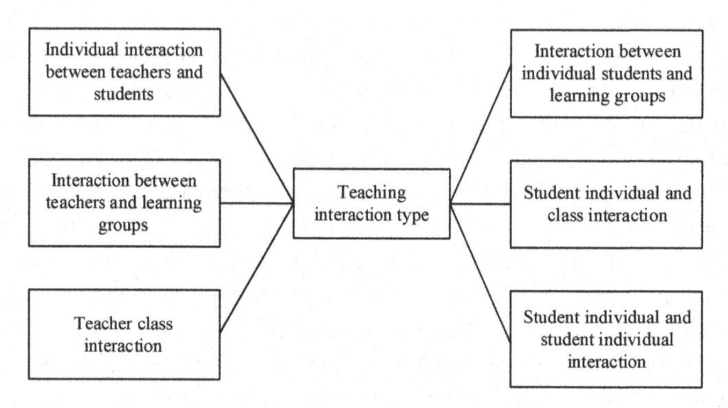

Fig. 2. Types of teaching interaction

As can be seen from Fig. 2, the types of teaching interaction include: interaction between teachers and students, interaction between teachers and learning groups, interaction between teachers and classes, interaction between students and students, interaction between students and learning groups, and interaction between students and classes. The first three kinds of interaction are mainly from the perspective of teachers, and the last three kinds are mainly from the perspective of students, and the interaction between teachers and students and between students are carried out around the learning theme. The teaching interaction under the guidance of teachers can be realized through the interaction between teachers and students, learning group and the whole class. The factors involved in online teaching interaction involve not only teachers' control over content transmission, but also students' control over the presentation of online teaching content and related processes. For social interaction, the interaction between students and teachers and between students and students may sometimes have nothing to do with online teaching, but it can still help create a positive or negative learning atmosphere. The interaction between teachers and individual students can stimulate students' interest in learning and communication and inspire other students to think. However, due to the limitation of classroom time, it is difficult for teachers to communicate and discuss with each student. When the teacher divides the students into groups, the interaction between the teacher and the learning group will help to guide the learning of the group, cultivate the students' cooperative ability and solve the learning problems. At the same time, teachers also have relatively sufficient time to give guidance or suggestions to each group, so as to make the objectives of group activities more clear and improve the efficiency of group interactive learning. These interactions can also give feedback on learners' progress in achieving teaching goals, and some kinds of social interaction can directly promote teaching interaction. For example, the group discussion in class is highly interactive. At the same time, students also actively participate in the comparison of views on the content and objectives of the core curriculum. After completing the teaching interaction or teaching task, teachers can summarize and evaluate through online voting, questionnaire survey and homework feedback, and further interact with the whole class to have an in-depth understanding of the learning situation of all students.

3.3 Cloud Service Optimization Platform Functions

With the rapid development of cloud computing, more and more individuals or enterprises choose to use cloud services, which can not only save costs, but also easily outsource data to cloud services for storage and management, which can greatly save local storage costs and improve data flexibility. The cloud service security agent is located between cloud service users and cloud service providers, and can encrypt and protect the above cloud data. Then in the cloud service environment, the encryption coefficient calculation formula of data transmission is:

$$G = \sum_{q=1}^{p} \frac{1}{D_{pq}} \tag{1}$$

In formula (1), p, q represents the fading coefficient of data and broadcast signal respectively, and D represents the transmission coefficient of source end. On this basis, it is concluded that the data received by the cloud service destination terminal is:

$$l = \frac{\varepsilon^2}{2} \times \sum |D - 1| \tag{2}$$

In formula (2), ε represents the noise power. Similarly, according to Eq. (2), the received power is:

$$R = \frac{\sum\limits_{\varphi=1} Q - \varphi^2}{H} \tag{3}$$

In formula (3), Q represents the upper limit of transmission power of each data set, φ represents the maximum signal-to-noise ratio at the receiving end, and H represents the identity matrix.

In practice, this technology requires a lot of manual maintenance work and the data encryption function is unstable. CASB is deployed between users and cloud services to intercept network data, perform protocol analysis on traffic data, and parse and encrypt on-cloud data. The online teaching interactive platform has simple operation and comprehensive teaching functions, which is convenient for students to participate in teaching interaction easily. The online teaching interactive platform combines its own devices and uses wireless network technology, touch screen technology and other operation technologies familiar to students in daily life to achieve convenient operation and help students easily participate in online classroom teaching interaction. From the homework function, forum function, class appointment function and evaluation function in detail:

(1) Homework function: In this function, students can see the homework published by teachers and upload their own homework in the form of attachments.
(2) Forum function: This function is an independent forum system, where teachers and students can communicate in study, life and other aspects. Online teaching interaction platform, on the other hand, students sign in, add documents, add discussion, add questions, questionnaire survey, in-class test, submit homework, ranking and other functions into an organic whole, comprehensive teaching auxiliary function to help the teachers and students to carry out a variety of teaching activities, diversified teaching interactive content, ensure the interactive teaching between teachers and students.
(3) Class appointment function: Students can apply for class appointment face-to-face to the teacher and wait for the teacher to provide information about the time and place of group tutoring. If the number of students is full, the application for class appointment of individual students may be rejected. The online teaching interaction platform supports multiple synchronous and asynchronous interactions between teachers and students [7]. Online teaching interactive platform is an interactive learning platform that supports real-time synchronous interaction of all staff. It has professional online interactive functions and can realize multi-screen real-time interaction of mobile phones, computers and other devices.
(4) Evaluation function: Students can evaluate teachers' teaching. Students can directly choose the score on the teaching evaluation index, which is helpful to improve the quality of network teaching. The teacher enters the correct user name and password to log in to the system and enter the interface of the teacher operation platform. During class discussion, all students can participate in real-time online synchronous interaction, including text, speech and pictures, etc., and use the visualization technology of online teaching interaction platform to project the interactive process and results in the form of text cloud, charts and other forms of synchronous feedback on the electronic screen for teachers and students to share in real time.

In addition, the online teaching interactive platform can also support asynchronous interaction and help students complete personalized interactive learning independently. The general menu lists the four sub functions provided by the system for teachers, namely "system home page", "teaching management", "teaching resources" and "personal information". The platform administrator can log in to the system and enter the system administrator operation platform interface by entering the correct user name and password. After class, students can find resources about learning topics on the Internet at any time through their own equipment, independently express their learning views and questions through the online teaching interaction platform, and have asynchronous interaction with teachers' partners. The system administrator is mainly responsible for managing and maintaining information in the system. The general menu lists the four sub functions provided by the system for the system administrator, namely "system user management", "academic year teaching management", "teaching auxiliary application" and "system setting management". The system administrator is divided into "super administrator" and "ordinary administrator". The "super administrator" has the maximum authority and can register an account for the "ordinary administrator". The online teaching interactive platform automatically records and saves the interactive content, and reserves in-depth thinking time for students who interact after serious thinking, so as to facilitate effective in-depth interaction at any time.

Teachers and students can conduct asynchronous interaction with their peers by praising or replying to others' views, so as to further promote the learning interaction. The online teaching interactive platform supports interactive data visualization and helps teachers analyze students' learning and interaction. There is a grading test system in the network teaching platform of College English intensive reading, which can help students make diagnostic evaluation at the beginning of course learning and choose suitable learning content. The unit test function in the network test system can make students know the test results and correct answers in time, provide timely feedback for learning, and help teachers collect their usual results [8–10]. The online teaching content and interactive data of teachers and students will be timely presented through the online visual teaching content and interactive information platform. Online question answering system enables teachers to answer students' questions online, and understand students' individual and overall needs in time through summarizing and analyzing these questions, so as to adjust teaching methods and improve teaching effect in time. Secondly, teachers can use the key technologies of learning analysis technology, such as discourse analysis and content analysis, to understand the contents and ideas of students' communication and interaction, pay attention to the progress of online learning, and analyze students' learning and interaction, so as to facilitate targeted teaching and guidance. The learning progress management system has detailed records of students' learning situation, including autonomous learning time and learning progress, which can make teachers master students' situation more comprehensively.

4 Platform Test

4.1 Test Preparation

According to the experimental test needs, Ajax technology is adopted in the client browser, based on CSS HTML language, using JavaScript, interacts with the business logic layer, and the business logic layer adopts PHP scripting language. The database layer encapsulates the access details of the underlying database, and ADODB is responsible for accessing MySQL database. The virtual machine running in the cloud browser is mainly composed of a virtual machine running in the cloud browser. The development environment of the platform is free and rich. The server is based on Linux operating system and adopts MySQL database. The web server is built by apach, and the foreground development language is PHP. The virtual machine running the browser is configured with Intel i7 2.20 GHz CPU 4-core and 16 GB memory, and the virtual machine running c1oudcrypt is configured with Intel I72 20 GHz CPU dual core, 4 GB memory. Database access adopts ADODB technology, which is the abbreviation of active data objects data base. It is an abstract class library for PHP to access database. The core function of cloudcrypt in cloud services is encryption and decryption of sensitive data. The encryption and decryption operation phase mainly occurs at the JavaScript wrapper and security gateway. The JavaScript wrapper encrypts the data before the client code reads it, and the security gateway encrypts the data before sending it to the ECS. PHP supports a variety of database systems, mysql, Sybase, Informix, SQL server, Oracle, DB2, etc. ADODB defines standardized database access interfaces to hide the differences between databases, and the conversion of accessing different database systems is transparent to users.

4.2 Test Results

The reference [5] platform and reference [6] platform are selected for experimental comparison with the online teaching interaction platform for College English Intensive Reading designed this time to test the time cost of the three platforms when uploading courseware of different sizes. The experimental results are shown in Tables 1, 2, 3, 4 and 5:

Table 1. Time cost of uploading courseware 50 MB (s)

Number of experiments	Reference [5] platform	Reference [6] platform	An interactive online teaching platform for College English Intensive Reading
1	3.515	3.974	2.164
2	4.984	4.152	1.978
3	4.112	3.948	2.140
4	3.697	4.366	2.006
5	3.648	4.815	1.996
6	4.031	3.787	2.316
7	2.948	4.315	2.505
8	3.667	4.669	1.848
9	4.251	3.525	2.312
10	4.306	4.077	3.161

It can be seen from Table 1 that the cost of uploading courseware between the designed College English intensive reading online teaching interactive platform and the other two college English intensive reading online teaching interactive platforms is 2.243 s, 3.916 s and 4.163 s respectively.

Table 2. 100 MB time cost of uploading courseware (s)

Number of experiments	Reference [5] platform	Reference [6] platform	An interactive online teaching platform for College English Intensive Reading
1	6.948	7.2021	4.645
2	7.164	6.309	3.312
3	6.009	7.255	3.994
4	6.487	6.784	4.158
5	6.158	7.648	2.370
6	7.664	6.594	4.102
7	6.521	6.123	3.225
8	7.306	7.159	2.109
9	6.412	7.984	4.307
10	7.597	6.316	2.549

It can be seen from Table 2 that the cost of uploading courseware between the designed College English intensive reading online teaching interactive platform and the other two college English intensive reading online teaching interactive platforms is 3.477 s, 6.827 s and 6.937 s respectively.

Table 3. 150 MB time cost of uploading courseware (s)

Number of experiments	Reference [5] platform	Reference [6] platform	An interactive online teaching platform for College English Intensive Reading
1	12.484	13.112	8.02
2	11.506	12.104	9.154
3	13.219	11.516	9.008
4	13.3337	12.309	8.121
5	12.544	12.874	9.304
6	13.206	13.548	9.715
7	11.108	11.669	8.646
8	13.337	12.547	9.825
9	12.455	11.633	9.497
10	12.306	13.825	8.316

It can be seen from Table 3 that the cost of uploading courseware between the designed College English intensive reading online teaching interactive platform and the other two college English intensive reading online teaching interactive platforms is 8.961 s, 12.550 s and 12.514 s respectively.

Table 4. Time cost of uploading courseware 200 MB (s)

Number of experiments	Reference [5] platform	Reference [6] platform	An interactive online teaching platform for College English Intensive Reading
1	15.541	14.154	9.662
2	16.362	16.260	12.515
3	15.554	15.306	12.202
4	17.202	14.509	11.649
5	14.615	15.487	10.548
6	15.299	16.354	10.411
7	17.784	16.556	9.347
8	14.655	15.327	11.519
9	14.021	17.152	12.202
10	15.034	15.741	10.199

It can be seen from Table 4 that the cost of uploading courseware between the designed College English intensive reading online teaching interactive platform and the other two college English intensive reading online teaching interactive platforms is 11.025 s, 15.607 s and 15.685 s respectively.

Table 5. Time cost of uploading courseware 250 MB (s)

Number of experiments	Reference [5] platform	Reference [6] platform	An interactive online teaching platform for College English Intensive reading
1	21.645	21.203	14.466
2	20.845	22.062	13.090
3	21.162	20.051	12.154
4	22.314	21.497	13.263
5	21.468	20.646	14.155

(continued)

Table 5. (*continued*)

Number of experiments	Reference [5] platform	Reference [6] platform	An interactive online teaching platform for College English Intensive reading
6	22.311	22.548	13.195
7	21.562	21.421	12.456
8	20.164	20.039	14.315
9	22.784	22.874	12.241
10	22.915	21.046	14.221

It can be seen from Table 5 that the cost of uploading courseware between the designed College English intensive reading online teaching interactive platform and the other two college English intensive reading online teaching interactive platforms is 13.356 s, 21.717 s and 21.339 s respectively.

The platform of this article, reference [5] and reference [6] are used to compare the bit error rate (BER) caused by online teaching interaction of College English intensive reading, and the comparison results are shown in Fig. 3.

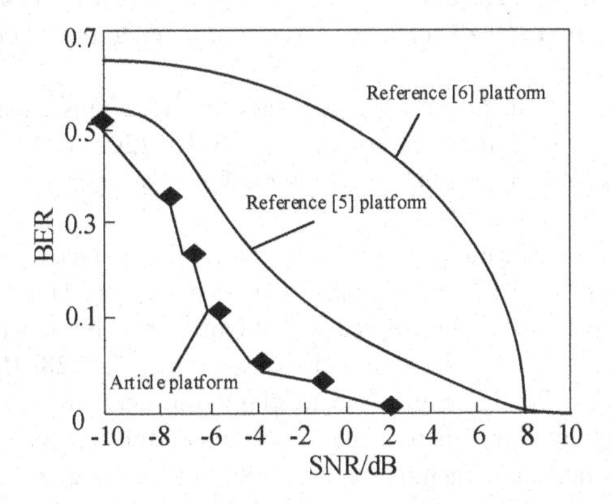

Fig. 3. Comparison results of bit error rate under different platforms

It can be seen from Fig. 3 that compared with reference [5] platform and reference [6] platform, the platform selected in this paper has a low bit error rate in online teaching interaction of College English intensive reading. The main reason is that this method sets up an online teaching interaction mode. In order to achieve the expected teaching objectives, teachers communicate and discuss among students, teachers and students based on the teaching content and with the help of technical means, and expand interpersonal interaction into information interaction. On this basis, they further guide students to think, and finally lead students to the interaction of new and old concepts within individuals, which is conducive to reducing the bit error rate to a certain extent.

5 Conclusion

The designed platform enables the teaching of College English Intensive Reading to mobilize learners' active participation to the greatest extent, which is conducive to the formation and improvement of learners' internal motivation. It can cultivate students' ability to find and solve problems, and enable learners to learn evaluation and self-evaluation in the process of reflection. At the same time, the interaction design in online online online courses is mainly carried out from three aspects: the interaction between learners and teachers, the interaction between learners and learning content, and the interaction between learners and learners. Finally, combined with the function principle of cloud service, the function of online interactive platform for College English intensive reading teaching is optimized. Experience the significance of success by completing tasks and promote the exertion of learners' own potential. Due to the limitation of time and specialty, the research of this paper still has many limitations, and there is no need to invest more energy in the accuracy of the platform in the future.

References

1. Wu, L., Ma, Y.: Study on the interaction mode and quality of online teaching:taking ningxia medical university as an example. jiaoyu jiaoxue luntan **28** 32–35 (2021)
2. Li, H.: Research on interactive teaching methods based on online teaching platform:a case study of Industrial Design History. wuxian hulian keji **18**(10), 139–140 (2021)
3. Zhu, Y., Yao, J., Guan, H.: Blockchain as a service: next generation of cloud services. J. Softw. **31**(1), 1–19 (2020)
4. Zhang, J., Wang, H., Ban, J.: An optimal design of vocal music teaching platform based onvirtual reality system. Comput. Simul. **38**(6), 160–164 (2021)
5. Shang, J.: Research on online course teaching platform based on real-time interactive mode. Microcomput. Appl. **36**(10), 18–20 (2020)
6. Wu, Z., Ke, J.: The application and implementation of interactive video technologies for online learning. Comput. Knowl. Technol. **17**(2), 184–185 (2021)
7. Mai, H., Zhou, X..: Research on the realistic dilemma and influencing fac-tors of college students'discussions in online teaching. Sci. Educ. Article Cult. **28**, 45–48 (2021)
8. Zhang, X., Chen, L.: College english smart classroom teaching model based on artificial intelligence technology in mobile information systems. Mob. Inf. Syst. **2021**(2), 1–12 (2021)
9. Yin, H.: The recommendation method for distance learning resources of college English under the MOOC education mode. Int. J. Continuing Eng. Educ. Life-long learn. **32**(2), 265–278 (2022)
10. Zhang, Y., Yang, Y.: The evaluation method for distance learning engagement of college English under the mixed teaching mode. Int. J. Continuing Eng. Educ. Life-long Learn. **32**(2), 159–175 (2022)

Intelligent Platform for College English Blended Teaching Based on Mobile Learning

Lin Fan(✉)

Yantai Vocational College, Yantai 264670, China
huhu21220@163.com

Abstract. There is a large amount of heterogeneous data in the hybrid teaching platform, and the amount of data is large and various, resulting in low efficiency of database retrieval, which in turn affects the response time of platform query. Design an intelligent platform for college English blended teaching based on mobile learning. The hardware part adopts the combination of FPGA and CPU to realize data collection and processing, and the read-write control module sends data to the process control module to complete the read-write. In the software part, an intelligent platform software architecture is established according to teaching needs, a platform database is designed, and data is stored and accessed through relational databases and SQL. Based on mobile learning, the functional modules of the teaching software are implemented. The client directly faces the user. It receives and calculates and processes the user data, and then sends a request to the server. The test results show that the average response time of the English blended teaching intelligent platform designed in this paper is lower than that of the traditional teaching platform, and the overall cost acceleration is lower than that of other platforms.

Keywords: Mobile learning · College English · Blended teaching · Intelligent platform · Teaching platform

1 Introduction

With the rapid development of information technology, people's learning environment and learning methods are constantly enriched and improved, and various educational concepts and educational viewpoints based on the information environment are constantly emerging and innovating. College English teaching requirements include three levels, namely general requirements, higher requirements and higher requirements, and the teaching mode of English courses should also be improved. The current learning methods can no longer meet the students' learning requirements. Therefore, English needs a new learning method to meet the students' individual learning needs. Under such a wave, the combination of traditional instructional design and information technology is imperative [1].

The design and research of the intelligent platform for college English blended teaching is a powerful supplement to the current domestic research on mobile learning and

W. Fu and L. Yun (Eds.): ADHIP 2022, LNICST 468, pp. 481–493, 2023.
https://doi.org/10.1007/978-3-031-28787-9_36

blended learning, and it also has certain practical value for blended teaching research with specific learners and subject characteristics. In the teaching of college English in the information technology environment, various information technology equipment and mobile tools play a pivotal role. At present, the information technology tools used in college English teaching include PPT manuscripts, Blackboard platform, Moodle platform, WeChat, teaching APP, etc. The use of these mobile tools is an indispensable part of realizing hybrid college English teaching. The designed and developed college English blended teaching intelligent platform can help learners use their spare time to study anytime and anywhere, which will undoubtedly bring great convenience to learners. Mobile learning platforms can provide students with rich digital learning resources, as well as good interactive functions and smooth user experience [2]. Use the mobile learning platform to combine online teaching resources with offline courses, effectively use new media technology, and make up for the lack of deep integration with the education field.

This paper designs an intelligent platform for college English blended teaching based on mobile learning. The hardware and software parts of the platform are designed respectively. The combination of FPGA and CPU is used to realize data acquisition and processing, and a read-write control module is designed to realize signal reading and writing. Design the platform database to store and access data through relational database and SQL. The realization of teaching software function modules based on mobile learning. The above design provides support for the actual teaching activity practice.

2 Hardware Design of Intelligent Platform for College English Blended Teaching

The key to the hardware part of the intelligent platform for college English blended teaching designed in this paper is to store the collected data and perform a series of subsequent processing on the stored large amount of data, so as to meet the large-capacity storage requirements of blended teaching data. This design uses a combination of FPGA and CPU to achieve data acquisition and processing, which is composed of multiple ADCs and multi-level FPGAs. The overall framework of the hardware part is shown in Fig. 1.

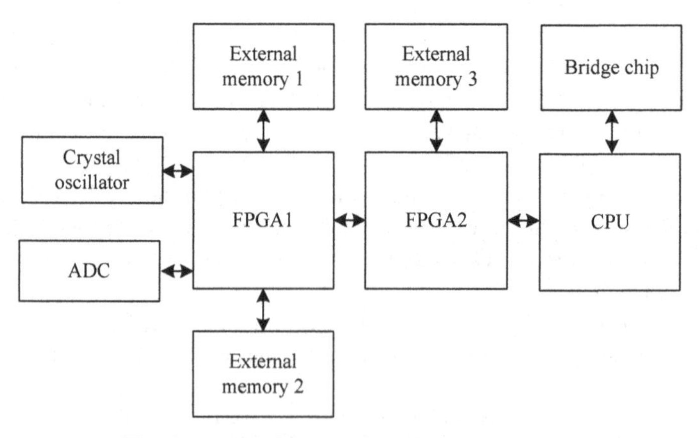

Fig. 1. Overall framework of platform hardware

As shown in Fig. 1, the externally input analog signal enters the signal conditioning channel. After proper gain and attenuation, the signal is sent to the ADC. The ADC converts the analog signal into a digital signal, and the phase-locked loop generates a sampling clock and sends it to the ADC. This design realizes the design of lOGsps digital storage oscilloscope through four ADCs with 2.5 GSPS sampling rate and 12 bit vertical resolution. Each ADC corresponds to one acquisition channel, and the total acquisition data flow is 120 Gbps.

This platform design requires signal cross-clock domain processing. The read-write control module sends a read-write completion signal to the process control module, the process control module sends a request signal to the serial port sent by the serial port sending module, and the serial port sending module sends a serial port sending completion signal to the process control module. Since the above-mentioned signals are all pulse signals, a pulse synchronization circuit is used to complete the cross-clock domain transmission of these signals. The input of SSTL is a differential comparison circuit, as shown in Fig. 2, one end is the input and the other end is the reference voltage VHF.

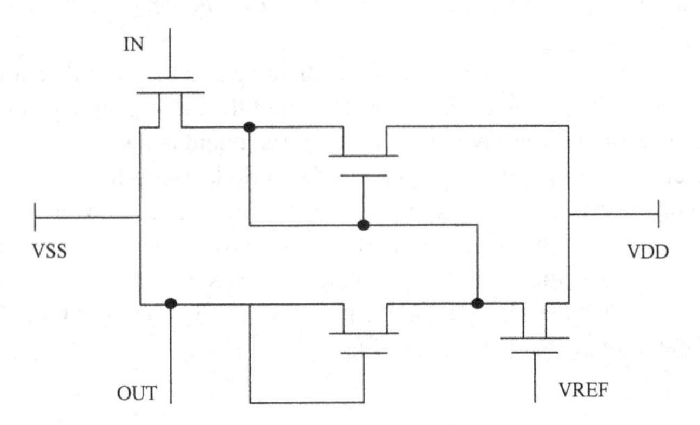

Fig. 2. SSTL interface circuit

Therefore, the input stage provides a better voltage gain and a more stable threshold voltage, which makes it more reliable for a small swing input voltage. SSTL interface can be single-ended IO, and SIO can also be differential IO. SIO uses one signal line for data transmission, and DIO uses two signal lines for data transmission. DIO can be simply understood as the splicing of two SIOs from the structure. For the high-speed mass data stream transmitted by the ADC, it is very critical to choose which logic device to receive, buffer and process. The commonly used logic devices are ASIC and FPGA. For the high-speed data stream collected by ADC, both devices have the ability to process a large number of high-speed data streams.

3 Software Design of Intelligent Platform for College English Blended Teaching

3.1 Software Architecture Design of Intelligent Platform for College English Blended Teaching

Learning is a process in which learners actively participate, and knowledge is actively constructed by learners. Combined with the characteristics of English blended teaching, the software design should take students as the main body, and give full play to students' subjective initiative. Therefore, it is required to consider enhancing students' learning flexibility when designing. Requirement analysis plays a key role in software engineering. It is the basis for design and development, and determines the functions to be implemented by the software. System requirements analysis is to answer the question of "what the system does" to ensure that the developed application can attract users and meet the needs of users. In addition, based on the characteristics of mobile learning theory, English blended teaching can connect teachers' teaching end and students' learning end through the network, teachers can upload resources at any time, and students can receive learning resources and content regardless of whether they are in the learning state [3].

According to the full consideration of the characteristics of mobile learning and the grasp of user needs, the design of the software part of the intelligent platform for college English blended teaching mainly includes two parts: client and server. In the process of UI and interaction design and development of mobile learning teaching software, it is necessary to consider the user's preference and habits during actual operation [4]. Simple and convenient interaction and user interface can play a role in promoting learning. Because the C/S architecture technology is mature and its main features are fast response and strong interactivity, the platform in this paper adopts the C/S structure. The platform software architecture design is shown in Fig. 3.

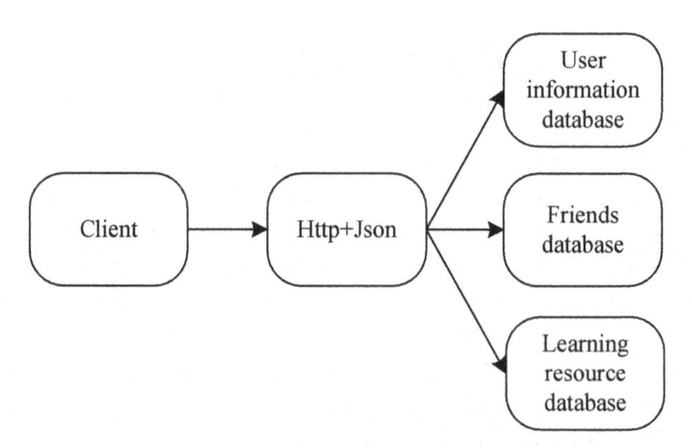

Fig. 3. Platform software architecture

The clients of the English blended teaching intelligent platform mainly include Android smart phones and other mobile devices based on the Android operating system. Users can learn English reading and check their own learning through these mobile terminal devices. The client directly faces the user, it receives and calculates the user data, and then sends a request to the server. Knowing Cloud Server is selected this time, and Knowing Cloud provides developers with various versions of cloud server packages. Due to the low number of users in the early stage, the free developer package provided by Knowing Cloud was selected first. After the number of users grows, it can be adjusted and optimized as needed. After receiving the request sent by the client, the server can quickly obtain the response request, and the client then calculates and processes the received response and data, and finally presents the processed data to the user. The developer completes the service initialization in app.js. After completion, the cloud service can verify whether the current teaching intelligence platform can be accessed. After passing the verification, the teaching intelligence platform can use all the functions provided by the server. Servers include resource servers and database servers. The resource server is mainly used to store learning resources and process the interactive information of teachers and students, that is, the release of learning tasks and the question and answer of teachers and students. The module interface analyzes the use request of the function module after the user platform operation, and requests to call the data information or software function of the database in the server [5]. The server has two major functions: first, to provide data and content for the software; second, to provide teachers with resource management and data monitoring. The function of the database server is to store the personal information of the students and to check the accuracy of the information submitted by the feedback users. When using a mobile device to click to send out a module use request, the mobile teaching platform client sends it to the server, and after the service is processed, it is returned and displayed to the user. The communication protocol of the English hybrid teaching intelligent platform is the HTTP protocol. The client uses HttpClient to connect with the server, and the server receives data through HttpPost, and then uses HttpResponse to transmit data.

3.2 Platform Database Design

The college English blended teaching service needs to be supported by a powerful database, and has high requirements in terms of scalability. The goal of database design is to enable the software system to store relevant data according to the user's needs to achieve the user's use purpose. The problem that developers must consider is to properly design the database so that data resources can be used efficiently.

Intelligent English teaching usually involves a lot of data. In the process of platform calculation and analysis, a large amount of structured and unstructured data information will be generated. It is necessary to have enough database as support to ensure the stable operation of the platform, and the database must also carry and store this information. Part of data. In addition, the realization of functions such as storage and retrieval needs to rely on the database. The resource materials provided and pushed by teaching resources and databases can not only assist students in learning the course content, but also extend and supplement the learning content appropriately along the context of courses and themes [6]. According to the demand analysis, the database entities used in the intelligent platform for college English blended teaching are planned: student entity, administrator entity, user information entity, learning resource information entity, and task information entity. The data table of the relational database is in the form of two-dimensional table.

For the relational database, the design of the data table affects the operation efficiency of the database to a certain extent. Due to space limitations, only the smart platform user information data sheet is shown in this article. The design of the user information data table is shown in Table 1:

Table 1. User information data sheet

Serial number	Code	Name	Types of	Length
1	Username	Username	Varchar	50
2	Registration	Registration number	Int	50
3	Password	Password	Varchar	30
4	Age	Age	Int	10
5	Gender	Gender	Varchar	10
6	Mailbox	Mail	Varchar	50

The teaching resource database is composed of relevant texts, subject-related cultural knowledge, text-derived grammar, English special skills, basic English exercises, extended cultural resources, and course information notices. The interaction between teachers and students, and between students and students can be strengthened through the medium of teaching resources, allowing more information to grow and circulate in such a resource structure [7]. For unstructured data, because the field length is variable and repeatable, and the content and form are flexible and changeable, it is impossible to effectively express the data information by establishing a two-dimensional table, such as text, image, audio and other factors. Difficult to express. There is no small difference between it and structured data, both in terms of structure and semantics. In the field of application software systems, in order to improve the efficiency of distributed database query, in addition to optimizing the database system architecture according to the distributed database cluster structure, it is also possible to design a more efficient and appropriate distributed database query strategy and algorithm. The computers on each network site implement data operations through the data bus, and the computers connected to the bus can communicate with each other. The communication cost estimation formula is:

$$\chi = \chi_0 + \beta\delta \tag{1}$$

In formula (1), χ represents the communication cost incurred in the process of transmitting δ units of data from one network site to another network site; χ_0 is the time it takes to initialize communication between communication network sites once; β represents data transmission quantity. The genetic algorithm is used to optimize the distributed database query process, and the fitness function of chromosomes can be expressed as:

$$F(a) = \sum_a (\chi_0 + \beta\delta) \tag{2}$$

In formula (2), $F(a)$ represents the fitness function of a chromosome. The probability of each chromosome being selected is obtained by comparing the fitness function value of each chromosome with the sum value, as shown in formula (3):

$$\varphi(a) = \frac{F(a)}{\sum_{a=1}^{m} F(a)} \tag{3}$$

In formula (3), $\varphi(a)$ represents the probability of each chromosome being selected; m represents the total number of chromosomes. The database query algorithm has been improved to shorten the data query time. The platform adopts a distributed storage structure, and uses relational databases and SQL to store and access data. At the same time, reasonable arrangements are made on each computing node, which can form an effective connection with ordinary users, and users can easily access data. Upload or download, and also support the storage and access of large-scale data information.

3.3 Realization of Functional Modules of Teaching Software Based on Mobile Learning

The registration and login modules are mainly aimed at users who initially use the intelligent platform to learn, so that the background management system can obtain the user's information according to the account information, and some of the user's learning track or historical records can be well preserved. The teaching resource module is mainly used to show learners rich teaching resources related to information technology, such as documents, pictures, audio and video. Learners can search for the learning content they need in the teaching resource module (categorized by display form). For developers, no account system means that the user's information cannot be obtained. It is difficult to make further decisions based on some basic information of the user, and some historical records of the user cannot be well preserved. The teaching module is equipped with a search function, and learners can also fill in the text related to the desired content through the search bar to search, and there are extended resources, such as audio, video and other resources, to improve learners' interest in learning about information technology. Through background management, managers can update teaching resources in a timely manner and provide a convenient way for learners to obtain rich teaching resources. After the English mixed teaching intelligent platform is launched, it will directly enter the registration and login interface. If the user has already registered the intelligent platform, he can directly enter his user name and password on the login interface to log in.

The after-class answering module is used to answer the practice questions provided by the after-class practice module after learning through the mobile learning platform, corresponding to the knowledge learned. Short answer questions etc. While learning knowledge to expand the amount of reading and pageviews, it is also necessary to pass a certain practice test, so that you can intuitively understand the solidity of the learner's content, so as to fill in the gaps, guide the learner's learning in time, and help the learning. Develop good practice and study habits. The learning module is divided into three parts: course recommendation, latest recommendation and vocabulary learning.

In the course recommendation part, the latest online courses related to English learning are mainly recommended to learners; in the latest recommendation part, the detailed explanation and answering skills of English test-related questions are mainly recommended to learners; in the vocabulary learning part, learners are mainly provided with vocabulary classification learning., and provides phonetic transcription, pronunciation and example sentences of related words. After the learner submits the test paper, the system will automatically give the score and the number of correct answers, and also provide the correct answer and answer analysis for each question to help learners better understand their mastery of knowledge. Managers can analyze the overall answer data of learners, and after communicating with teachers, they can delete and update existing practice questions to provide learners with better and more suitable practice questions. The tool module mainly provides learners with vocabulary query function, unfamiliar vocabulary collection function, self-test function and contact us. The vocabulary query function can facilitate learners to query related English vocabulary and obtain relevant Chinese definitions, phonetic symbols, pronunciations and related example sentences; the vocabulary collection function allows learners to add unfamiliar vocabulary to the new vocabulary book during vocabulary learning.

Compared with traditional teaching forms, mobile learning platforms cannot enable teachers and learners to communicate face-to-face. The interactive platform module is provided as a small forum, which is mainly used for learners to ask questions about the difficulties encountered in the learning process. They can also express their own thoughts on the content published in the forum. After the publication, other learners can comment on it. The district receives information, and also expresses its own opinions and exchanges in this way. The self-test function can test the learners' vocabulary learning effect accordingly; if the learners encounter related problems when using this research platform, they can use the contact us function to send the problems to the corresponding mailboxes, and the system staff will help the users as soon as possible. Solve the problem. Then the client will send the user's login request to the server, which will verify the login information. If the server returns a pass, the user can enter the homepage of the smart platform. In addition, after the learners publish their own remarks, if the questions raised are resolved or they feel that the remarks are inappropriate, they can delete them by themselves or contact the administrator to delete them, but they have no right to the remarks made by other learners. Deleted, but can be viewed, liked and commented on. So far, the design of the intelligent platform for college English blended teaching based on mobile learning has been completed. The software flow of the intelligent platform for college English blended teaching based on mobile learning is shown in Fig. 4:

4 Experimental Studies

4.1 Experiment Preparation

In this test experiment, a total of 6 devices including client, data service module server, application server and database server are used to form a distributed database cluster. The experimental hardware test environment is as follows: CPU: Intel(R) Core(TM) i7-7700HQ @ 2.80 GHz 4 cores and 8 threads; Memory: 16G DDR4. In this experiment, the client is the machine used by the user when the user requests data through the

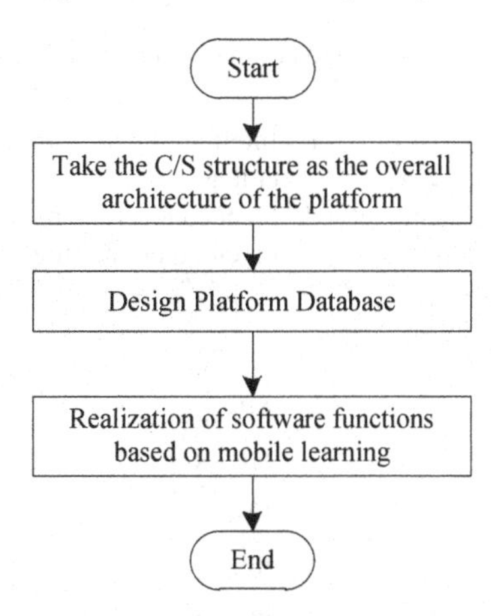

Fig. 4. The software process of the intelligent platform for college English blended teaching

browser; the data report module application program is deployed on the application server, which is responsible for receiving user requests, requesting data from the data service module, and then processing the data further, and then returned to the user. This study chose JavaScript as the development language. The Javascript code of the logic layer is running In X5 JSCore, the view layer is rendered by X5 based on the Mobile Chrome 57 kernel. The data service module receives the SQL statement of the application module, optimizes the SQL query plan, then queries the underlying database, integrates the data and returns it to the application.

4.2 Results and Analysis

According to the actual test environment built, the intelligent platform for college English blended teaching based on mobile learning is tested. Set the index data volume of the platform to 10,000, 100,000, 1,000,000 and 10,000,000 respectively, and count the average response time of students to the teaching data query page of the intelligent platform as a test indicator. The performance test results of the mobile learning college English blended teaching intelligent platform are compared with the college English blended teaching intelligent platform based on cloud computing and data mining. Tables 2, 3, 4 and 5 shows the comparative test results of the response time of each teaching platform under different data amounts.

In the index query of 10,000 data, the average response time of the mobile learning-based college English blended teaching intelligent platform is 200.52 ms, which is 175.71 ms and 192.11 ms lower than the cloud computing and data mining-based college English blended teaching intelligent platforms.

In the index query of 100,000 data, the average response time of the mobile learning-based college English blended teaching intelligent platform is 294.76 ms, which is 190.08 ms and 211.38 ms lower than the cloud computing and data mining-based college English blended teaching intelligent platform.

In the index query of 1 million data, the average response time of the mobile learning-based college English blended teaching intelligent platform is 423.51 ms, which is

Table 2. Response time comparison of 10,000 data (ms)

Testing frequency	Intelligent platform for college English blended teaching based on mobile learning	Intelligent platform for college English blended teaching based on cloud computing	Intelligent platform for college English blended teaching based on data mining
1	191.66	358.62	363.02
2	203.88	374.82	385.48
3	186.55	385.55	398.84
4	188.21	370.48	372.67
5	196.10	402.10	366.26
6	195.42	389.26	382.55
7	214.85	367.53	414.92
8	203.53	348.92	408.39
9	208.26	369.41	421.63
10	216.69	395.63	412.56

Table 3. Response time comparison of 100,000 data (ms)

Testing frequency	Intelligent platform for college English blended teaching based on mobile learning	Intelligent platform for college English blended teaching based on cloud computing	Intelligent platform for college English blended teaching based on data mining
1	289.43	469.67	496.45
2	285.65	482.47	508.29
3	276.27	495.76	482.63
4	294.01	473.18	493.17
5	298.50	468.05	505.43
6	301.86	497.81	517.42
7	304.35	486.39	504.51
8	298.48	493.24	512.06
9	296.92	488.32	523.27
10	302.12	493.53	518.19

282.38 ms and 306.74 ms lower than the cloud computing and data mining-based college English blended teaching intelligent platform.

In the index query of 10 million data, the average response time of the mobile learning-based college English blended teaching intelligent platform is 656.49 ms, which is 426.77 ms and 474.46 ms lower than the cloud computing and data mining-based

Table 4. Response time comparison of 1 million data (ms)

Testing frequency	Intelligent platform for college English blended teaching based on mobile learning	Intelligent platform for college English blended teaching based on cloud computing	Intelligent platform for college English blended teaching based on data mining
1	419.73	683.97	692.47
2	408.48	696.68	724.84
3	425.84	678.06	715.68
4	416.67	682.53	721.26
5	435.21	695.86	736.55
6	427.55	727.44	769.91
7	414.86	734.19	755.64
8	428.12	721.65	746.38
9	432.08	710.38	734.56
10	426.60	728.09	705.23

Table 5. Response time comparison of 10 million data (ms)

Testing frequency	Intelligent platform for college English blended teaching based on mobile learning	Intelligent platform for college English blended teaching based on cloud computing	Intelligent platform for college English blended teaching based on data mining
1	658.86	1012.46	1039.46
2	663.47	1008.18	1065.65
3	674.18	1026.80	1186.87
4	687.53	1036.55	1057.54
5	651.67	1128.28	1224.21
6	642.58	1214.64	1148.15
7	650.86	1207.37	1072.42
8	635.29	1056.85	1255.06
9	648.38	1049.28	1226.28
10	652.04	1092.19	1033.82

college English blended teaching intelligent platform. It can be seen from the above results that as the amount of data increases, the transmission overhead increases, and the response time of each teaching platform increases. The average response time of the English blended teaching intelligent platform designed in this paper is lower than that of the traditional teaching platform, and the overall cost speed is lower than that of other platforms. Therefore, the platform designed in this paper improves the query performance of data under multi-threading.

On this basis, compared with the intelligent platform for college English teaching based on cloud computing and data mining, the query accuracy of the three methods for college English teaching resources was tested. The experimental results are shown in Fig. 5:

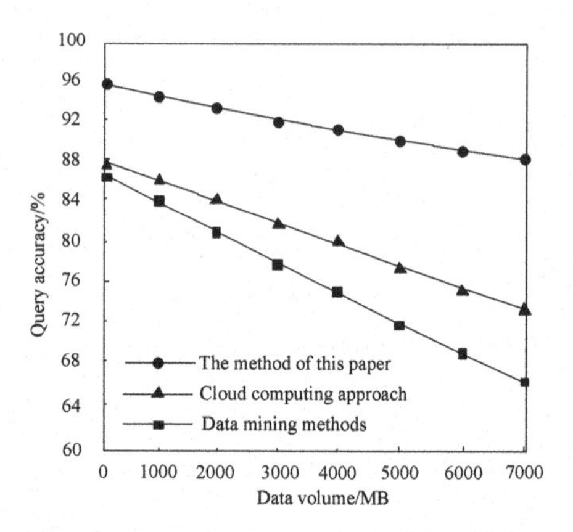

Fig. 5. The query accuracy of college english teaching resources

Analysis of Fig. 5 shows that the method in this paper has a high query accuracy for college English teaching resources, up to 96%, which is much higher than the other two comparison methods. It can be seen that this paper designs a college English blended teaching intelligence based on mobile learning. Platform performance is excellent.

5 Conclusion

With the rapid development of information technology, college English teaching is no longer traditional classroom teaching, because it is impossible for students to acquire English knowledge only through classroom reading teaching in schools. Students need to learn English reading anytime and anywhere according to their own conditions. This paper designs an intelligent platform for college English blended teaching based on mobile learning. The experimental results show that the platform can reduce the average response time of data query and has good performance.

References

1. Chen, H., Zhang, F.: Construction and application of mobile teaching platform based on "Super Star Learning" from the perspective of big data. Wuxian Hulian Keji **17**(1), 38–39 (2020)
2. Du, J., Wu, F.: An analysis of interactive teaching effect based on mobile teaching platform——taking xuexitong as an example. Heihe Xueyuan Xuebao **12**(8), 95–98 (2021)

3. Xu, N., Fan, W.: Research on interactive augmented reality teaching system for numerical optimization teaching. Comput. Simul. **37**(11), 203–206, 298 (2020)
4. Zhang, L.: Design of distance teaching system based on artificial intelligence network. Modern Electron. Tech. **44**(2), 131–134 (2021)
5. Yang, Y., Lin, S., Zhou, Y., et al.: The design and practice of the micro-course mobile teaching system based on the smartphone and the WEB platform. Techn. Autom. Appl. **39**(4), 182–185 (2020)
6. Zhang, L., Xie, Y., Fu, Y.: Research on the construction of multimedia platform for sports nutrition based on mobile learning. Sport Sci. Technol. **41**(4), 125–126, 128 (2020)
7. Wang, W., Zhang, Y., Yao, H., et al.: Research on spacecraft attitude dynamics simulation experiment teaching based on simulink. Comput. Simul. **38**(12), 176–181 (2021)

Cloud Service-Based Online Self-learning Platform for College English Multimedia Courses

Guiling Yang(✉)

Yantai Vocational College, Yantai 264670, China
kkjjn2200@163.com

Abstract. Because the traditional college English multimedia course network independent learning platform has the problems of slow response time and low student satisfaction, a cloud service based College English multimedia course network independent learning platform is designed. Hardware part: simulate the maximum frequency of input signal and design a complete power on reset (POR) and power off reset (PDR) circuit; Software part: increase the investment in multimedia network teaching, improve the management structure of College English multimedia courses, take the Internet as the main carrier, build a network independent learning model, and optimize the software functions of the platform by using cloud services. Experimental results: the average response time of the College English multimedia course network autonomous learning platform in this paper and the other two autonomous learning platforms are 8.464 s, 13.276 s and 13.697 s respectively, which shows that the application effect of the College English multimedia course network autonomous learning platform is better and the satisfaction of students is improved after making full use of the cloud service technology.

Keywords: Cloud service · College English · Multimedia course · Online self-learning · Classroom teaching · Teaching quality

1 Introduction

Multimedia network autonomous learning has caused profound changes in traditional education methods. In the past traditional teaching activities, teachers as the main body of teaching activities, the main task is to impart knowledge to students. Today, as a new helper, the acquisition of knowledge is constructed by students according to their own cognitive structure and existing knowledge structure. In educational practice at all levels, people pay more and more attention to the main role of learners, and more and more attention is paid to the cultivation of students' independent thinking ability and innovative spirit. Allowing learners to conduct autonomous learning through the network has become an important means of cultivating students' comprehensive quality. The focus of the online self-learning platform for college English multimedia courses is the close

W. Fu and L. Yun (Eds.): ADHIP 2022, LNICST 468, pp. 494–509, 2023.
https://doi.org/10.1007/978-3-031-28787-9_37

integration of education and technology [1, 2]. While the application of technology has brought about changes in educational facilities and teaching methods, the corresponding teaching methods and teaching models will also be innovated, and these reforms will inevitably lead to changes in educational thinking and educational concepts. With the continuous expansion of the scale of higher education in my country, the level of students has also undergone major changes. Students have more individuality. Traditional elite education can no longer meet the needs of students. In college English teaching, traditional classroom teaching still dominates status, teachers dominate the classroom, and students learn passively. The essence of technology application is to serve human beings, and its role is to assist human information organs to complete the acquisition, storage, processing, publishing and expression of human information. Computer technology, multimedia technology, virtual reality technology and the diversity of information carriers enable learners to overcome time and space barriers and independently arrange their own learning time and speed. Therefore, foreign language autonomous learning has become a hot research topic in foreign language education circles at home and abroad in recent years. From the learner's point of view, with the changes in learning objectives, learning content and learning form, learners can no longer learn completely in accordance with the traditional teaching mode, and can arrange their own learning flexibly and autonomously. Radical change. Multimedia course network self-learning will open up a global knowledge dissemination channel, realize mutual dialogue and exchange between learners and teachers in different regions, not only can improve the efficiency of education, but also provide learners with a relaxed and rich content. Learning environment. From the perspective of educators, teachers are no longer the leaders of learning, but truly become participants and instructors of learning. Information technology, represented by network technology, will have a profound impact on traditional teaching modes, teaching content and teaching methods. Influence. The application of network multimedia teaching has changed the teaching status of teachers in the past, advocated the teaching ideas of students' autonomous learning and individualized learning, and made the teaching process more colorful. Therefore, in addition to the innovation of educational ideas, educational methods and teaching methods, the multimedia course network autonomous learning is more important to bring about changes in educational and teaching models, and it will certainly be a revolutionary change.

Most of the traditional methods use mobile terminals and cluster analysis to design the College English multimedia course network independent learning platform, but the self-learning platform based on mobile terminals has low student satisfaction, and the operation of the self-learning platform based on cluster analysis is complex, resulting in a long response time of the platform. Therefore, this paper designs a web-based autonomous learning platform for College English Multimedia Courses Based on cloud services, and verifies the effectiveness of the platform designed in this paper through simulation experiments, which solves the problems existing in the traditional platform.

2 The Hardware Design of the Online Self-learning Platform for College English Multimedia Courses

According to the configuration requirements of the self-learning platform, the hardware of the network self-learning platform for college English multimedia courses is designed.

The power supply scheme of the hardware: Vdd, the voltage range is 2.0–6 V: the external power supply is provided through the Vdd pin for I/0 and internal voltage regulator. Vssa and Vdda, voltage range is 2.0 Femto. 6 V: External analog voltage input for ADC, reset module, RC and PLL, within Vdd range (ADC is limited to 2.4 V), Vssa and Vdda must be connected accordingly to Vss and Vdd. In high-speed data acquisition systems, parameters such as phase noise, phase jitter, phase error, frequency error, and signal-to-noise ratio of the sampling clock are critical. Because the jitter of the sampling clock will lead to sampling at unequal intervals, resulting in the deviation between the output of the analog-to-digital conversion circuit and the theoretical sampling value. Vbat, the voltage range is 1.8–3.6 V: when Vdd is invalid, power supply for RTC, external 32 kHz crystal oscillator and backup register (through power switching). The STM32 series of products continue the energy-saving and consumption-reducing features of the ARM Cortex-M3 core and support high-precision power management functions. When the sampling signal is still processed according to the nominal equal interval, noise caused by sampling jitter error is bound to be introduced, resulting in a decrease in the ADC output signal-to-noise ratio, which in turn has a certain impact on the detection performance of the measured signal. The voltage regulator has 3 modes of operation: Main (MR), Low Power (LPR) and Power-down. When the reset circuit works, the minimum power consumption is 2μA in standby mode, and in normal operation mode at 72 MHz, the STM32's The current consumption is as low as 27 mA. A complete power-on reset (POR) and power-down reset (PDR) circuit is designed on the circuit device. The resolution of the analog-to-digital converter 1VIXT2001 selected in this topic is 8, and the voltage peak value of the analog input signal is equal to The full-scale voltage range is 700 mVpp, the maximum frequency of the analog input signal is 200 MHz (−3 dB bandwidth), and the maximum jitter is estimated to be about 3 ps. This circuit is always valid to ensure that when starting from 2 V or falling to 2 V Some necessary operations are performed. Among them, the hardware phase-locked loop structure is shown in Fig. 1:

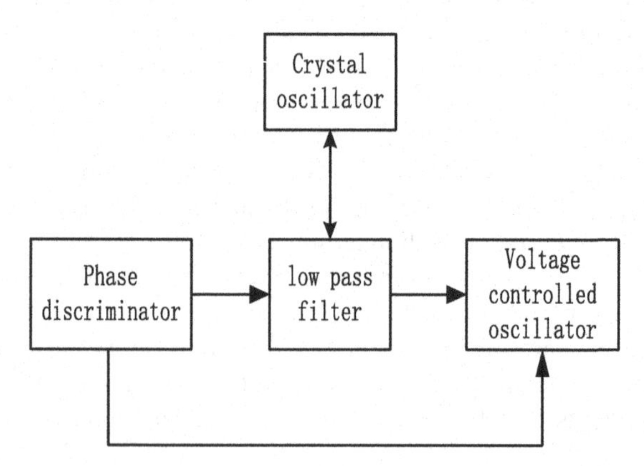

Fig. 1. Hardware phase-locked loop structure

As can be seen from Fig. 1, the hardware phase-locked loop structure includes: a crystal oscillator, a loop low-pass filter, and a voltage-controlled oscillator. When Vdd is below a certain lower limit Vpor/pdr, no external reset circuit is required and the device can remain in reset mode. In addition, STM32 also has an embedded programmable voltage detector (PVD), PVD is used to detect Vdd, and compared with Vpvd limit. An interrupt is generated when Vdd is lower than Vpvd or Vdd is greater than Vpvd. Analog Devices' clock generation chip AD9517–3 has a sub-picosecond output jitter, which meets the design requirements. The AD9517–3 significantly enhances ADC data conversion performance with sub-picosecond low jitter performance and low phase noise. It can provide multiple clock outputs, and integrates an on-chip phase-locked loop PLL and a voltage-controlled oscillator VCO, and its tuning frequency range is 1.75 GHz–2.25 GHz. The interrupt service routine can generate a warning message or put the MCU into a safe state, and the PVD is enabled by software. To sum up, we choose the power supply chip AP1117 provided by Anachip, which is a low-dropout (1.4 V) positive voltage regulator, which can provide a maximum output current of 1 A, built-in overheating and overcurrent protection, and has a fixed output voltage of 1.5 V, 1.8 V, 2.5 V, 3.3 V and 5 V, select the fixed 3.3 V voltage output. At the same time, it also has an external VCO with a frequency of up to 2.4 GHz, which is convenient for the expansion of the external clock of the board.

3 Software Design of Online Self-learning Platform for College English Multimedia Courses

3.1 Improve the Management Structure of College English Multimedia Courses

There are many departments involved in the implementation of the multimedia network teaching model, such as the Office of Academic Affairs, the Office of Planning and Finance, the Office of State-owned Assets Management, the Office of Logistics Management, the Modern Educational Technology and Information Management Center, the Electric Classroom, and the Language Laboratory. The research and development of general English (commonly known as college English textbooks) must be differentiated in terms of language difficulty, so as to adapt to the choice of students with different foundations. The selection of materials should be as close to the actual life of students as possible, combined with their life or learning experience, to make language learning more life-like and make language learning truly meaningful. Teachers should carefully organize and implement multimedia classroom teaching to improve teaching effect [3, 4]. At the same time, efforts are made to cultivate students' self-management ability and improve their independent learning ability using multimedia network. On the basis of the above description, the basic structure of online self-learning of college English multimedia courses is obtained, as shown in Fig. 2:

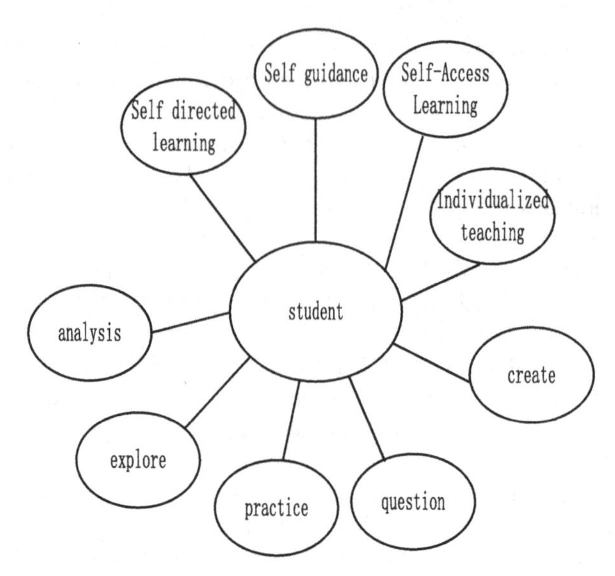

Fig. 2. Basic architecture of autonomous learning

It can be seen from Fig. 2 that autonomous learning is a modern learning mode corresponding to traditional receptive learning. Students are the main body of learning. Students achieve their learning goals through self-guidance, self-directed learning, self-directed learning and personalized teaching through independent analysis, exploration, practice, questioning, creation and other methods. Advocate students to actively participate, be willing to explore and be diligent in doing things, and cultivate students' ability to collect and process information, acquire new knowledge, analyze and solve problems, and communicate and cooperate. According to the characteristics of the language, the learning of English requires continuous intensive training, and the improvement of a student's ability is proportional to the intensity of his training. Compared with online teaching, traditional classroom teaching is also affected by factors such as the number of students, teaching time and teachers' energy, and intensive training in this environment is relatively insufficient. The selection of materials should be as short as possible, concise in content, and strong in language demonstration. This will not only help improve the efficiency of teaching and learning, but also help protect students' enthusiasm for learning. There are many factors that determine the quality of teaching, among which, teaching management is one of the most basic and most important factors. Multimedia network technology has had an unprecedented and profound impact on college English teaching, and also brought new opportunities and challenges to college English teaching management. The emergence of network teaching software can make up for this deficiency, overcome the problem that the traditional classroom teaching cannot increase the training intensity and training time of listening and speaking, and the training of students can also be carried out anytime, anywhere. Increase investment in multimedia online teaching, build independent learning centers and online learning platforms to recruit senior professionals who are proficient in computer network information technology, serve as managers of independent learning centers, and promptly eliminate technical problems in multimedia online teaching. "Guardian". English learning needs to be placed in a certain situation or environment, and its main purpose is to express one's thoughts and describe facts in language, so as to achieve the purpose of cultivating students' abilities and developing students' intelligence. The two most typical situations in the foreign language environment under the traditional classroom model are: shy and introverted students cannot participate in the real language environment, while those students who are lively and generous, and have a strong desire to express their language application ability will appear. More and more skilled. Therefore, while introducing modern teaching methods, in order to ensure the smooth development of multimedia network teaching, it is necessary to strengthen the training of teachers and students in computer operation skills and multimedia network knowledge, so that they can master the basic knowledge of computers and networks, and be proficient in using multimedia teaching. Software, campus network and Internet. The use of the existing network technology to establish a virtual language environment can make up for the deficiencies of the traditional language teaching environment, and this environment will be beneficial to the cultivation of all students' listening, speaking and expression skills.

3.2 Build a Network Autonomous Learning Model

The main feature of English autonomous learning is that learners start from the initial goal setting, progress formulation, strategy selection, process adjustment, control and remediation, and then to the evaluation and reflection of the results. All processes will be independently chosen and decided by the learners. The learners have high enthusiasm in the English learning process, always have a strong interest in English learning, are full of confidence in making progress, and can obtain positive emotional experience from English communication activities.. English for Special Purposes itself includes three different levels: workplace English, academic English and professional English. Although it is difficult to clearly distinguish some content, the purpose and occasion of language use should always be kept in mind during the compilation of textbooks. In this way, vocabulary, stylistic and language difficulty will naturally be distinguished. From a broad perspective, the online learning environment refers to the continuous situation and conditions that online learners rely on in the process of learning activities. It not only refers to the material conditions that support the learning process, but also includes non-material conditions such as learning strategies, interpersonal relationships, learning psychology and learning atmosphere. Whether it is workplace English, academic English, or professional English, it is temporarily unable to fully meet the actual needs of teaching, because there are too many industries, and the number of disciplines and majors is also very surprising. From a narrow point of view, the online learning environment refers to the continuous situation, conditions and psychological factors that online learners rely on in the process of learning activities in modern online education and open education, with the Internet as the main carrier. Self-directed learning is not only manifested in the technical support at the material level, but also in the broader and deepest spiritual impact on learners. The main process of online self-learning of college English multimedia courses is shown in Fig. 3:

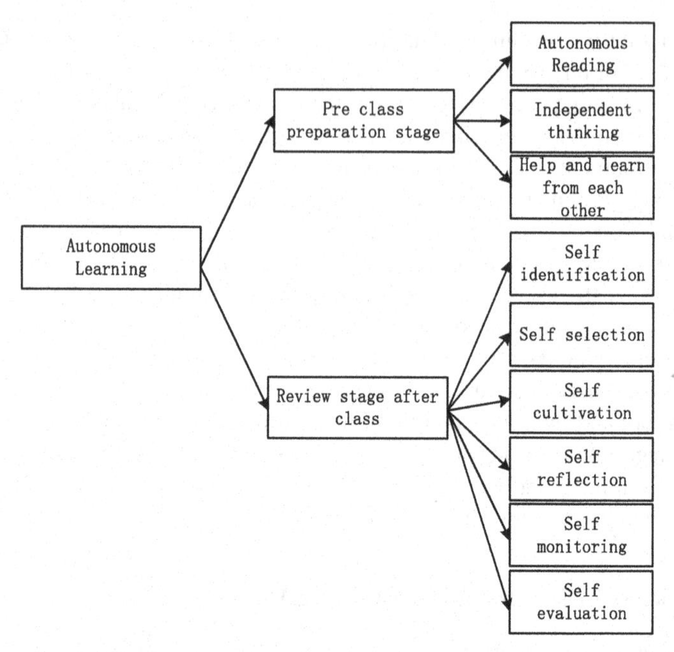

Fig. 3. The main process of autonomous learning

As can be seen from Fig. 3, the basic process of autonomous learning is divided into two main stages. The first major stage is the pre class preparation stage. The basic idea of this stage is "don't teach what you can learn", and the basic principle is "read independently, think independently, and help and learn from each other". I call this stage "pre class". The basic process of "pre class": learning guidance and cooperation. The second main stage is the review stage after class. The main processes are: self identification, self selection, self cultivation, self reflection, self-monitoring and self-evaluation. In addition, academic English can be considered to be compiled with reference to the subject, and it does not have to be subdivided into majors. Because the main purpose of academic English is to train learners' academic oral communication skills and academic written communication skills, the emphasis is not on professional knowledge, but on language skills based on professional knowledge [5]. The compilation of English for Special Purposes should be completed with the help of the full cooperation of language textbook compilation experts, industry experts and subject field experts. Only by integrating the expertise and wisdom of all parties can the quality of textbooks be basically guaranteed. Transmission of information through multimedia pictures, texts, sounds, images and other forms makes teaching more three-dimensional, vivid and vivid, and this effect helps to stimulate learners' interest in learning and improve learning efficiency.

3.3 Cloud Service Optimization Platform Software Functions

Cloud services are developed on the basis of a series of traditional technologies such as grid computing, distributed systems, P2P, and virtualization, and are a new type of platform for sharing information infrastructure [6, 7]. On the basis of cloud computing, the expression formula of the trust computing model of the network platform is obtained:

$$G_{(t)} = \frac{\beta_{t-1} + \frac{1}{\phi}(\beta_{t-1})}{\sum (\beta_t)^{other}} \tag{1}$$

In formula (1), β represents the user's reputation, ϕ represents the number of transactions, and t represents the feedback score for the user's reputation. The optimization of online self-learning in college English multimedia courses is not the optimization without constraints, but the optimization with constraints. In practice, it should be noted that the purpose of network teaching is only to achieve the optimization of teaching effect, not to network for network. The college English course is a course based on knowledge learning and language communication, and cannot use the Internet to completely replace the situational communication in the real environment. The emergence of network platforms has promoted learners' interest in learning. Different from traditional classrooms, in the online environment, learners can choose the right time and study method according to their needs to study "as they want". Therefore, the use of the Internet to carry out independent study of college English should pay attention to the principle of optimization in appropriate aspects. Optimizing the combination and expansion of course knowledge points and network teaching resources cannot be separated from the syllabus of the course to expand the content of network resources without boundaries, which will make learners deviate from the goal of autonomous learning. Whether learning achieves a certain effect depends on the learner's own self-control. Curriculum setting serves teaching goals. In school education, teaching goals are mainly achieved through teaching activities with the help of teaching materials. Therefore, the compilation, selection and use of teaching materials should be carried out closely around teaching goals and learning purposes. Optimize the proportional structure of listening, speaking, reading, writing and other related content. The network is not omnipotent. It is necessary to give full play to the advantages of network technology and multimedia technology, and appropriately increase the parts of listening, speaking and reading. Optimizing the cultivation of autonomous learning ability, the main purpose of using the network to carry out autonomous learning is to cultivate students' autonomous learning ability, and the network content cannot be limited to the simple repetition and reproduction of the basic content of the course, but must be open to a certain extent. In a broad sense, general English, general education English and English for special purposes in the curriculum are set up to achieve different teaching goals and meet the needs of different learners. Secondly, various information tools also play a significant role in cultivating learners' autonomous learning ability and providing learning strategies. To optimize the influence of autonomous learning on students' emotions and emotions in the network environment, it is required that the ease of use, the beauty of the interface and the vividness of the content should be considered when developing the network autonomous learning platform, so as not to affect the students' interest in learning. To optimize the simulation of the situation, the use of the network to learn independently is online and real-time interactive, and it is necessary to design language communication situations that meet the needs of students. Optimize teachers' guidance on learning strategies for students to use the Internet for autonomous learning. A good learning strategy can promote students' learning, but an unsuitable learning strategy will affect students' autonomous learning, and will damage students' confidence in using the Internet to carry out autonomous learning. These information tools include: intelligent tutor system, expert system, virtual classroom, multimedia teaching software, BBS, chat room and network log and a series of countless practical tools. On the basis of formula (1), the local trust value after normalization is obtained:

$$L = \frac{\max(Lmn, 0)}{\sum\limits_{n=1} (Lmn, 0)} \times \frac{1}{\phi} \tag{2}$$

In formula (2), L represents the trust value of a node in the network platform, m, n represents the total trust value and local trust value of the node, and ϕ has the same meaning as formula (1). After multiple iterations, the global trust value of the node is obtained as follows:

$$\eta^{(\gamma-1)} = \frac{(\gamma + L)}{2} + \sigma \tag{3}$$

In formula (3), γ represents the global trust value vector, σ represents a constant less than 1, and L has the same meaning as formula (2). If you want to optimize the space and time of teacher-student communication and student-student communication, it is necessary to design a smooth and functional communication platform on the network platform, and teachers are required to have enough online time to facilitate communication with students and problems between students. Discuss, realize the complementarity and mutual assistance between students and teachers, and between students and students. Under the historical framework of the development of English literature, the selection of representative works will help students understand the influence and shaping of these sources on the British and American literary traditions, and will also help students understand the literary characteristics or the spirit of the times in a certain historical period. In the developed information society, there are still many available tools waiting for learners to use and develop. The main medium of autonomous learning in the network environment is realized through the network platform [8–10]. The construction of the network learning platform is an important part of the smooth progress of autonomous learning. The quality of platform design depends largely on whether the platform's architecture is scientific and reasonable, because the architecture is often the basis of a network education platform, and it has a decisive role in the scope, function and performance of the system.

4 Platform Application Test

4.1 Test Preparation

According to the platform test needs, set the test preparation. Select PHP (PHP Hypertext Preprocessor) as the scripting language running on the server side. The PHP code can be interpreted and converted into standard HTML script on the server side and returned to the browser. In addition, this experiment uses the VKS (Vocabulary Knowledge Scale) test designed by Paribakht and Wesche as a pre-test to measure students' familiarity with words. Database access is performed using the database interface provided by PHP to access the MySQL database. The model itself is developed under the Linux system based on the Apache + MySQL + PHP environment, which is also well supported under the Windows platform, so the interactive teaching plug-in is developed using the above development environment. Except for the VKS test, all data were analyzed by SPSS.

4.2 Test Results

In order to test the application effect of the network self-learning platform for college English multimedia courses designed this time, a comparative test is carried out. Select the online self-learning platform for college English multimedia courses based on mobile terminals, and the network self-learning platform for college English multimedia courses based on cluster analysis, and conduct a comparative test with the online self-learning platform for college English multimedia courses in the text. The response times of the three platforms were tested under the conditions of different concurrent users. The test results are shown in Tables 1, 2, 3 and 4:

Table 1. Response time of the platform with 80 concurrent users (s)

Number of experiments	Mobile terminal-based online self-learning platform for college English multimedia courses	A network autonomous learning platform for college English multimedia courses based on cluster analysis	The online self-learning platform for college English multimedia courses
1	1.202	1.411	0.548
2	1.366	1.106	0.631
3	1.058	2.113	0.489
4	1.694	1.257	0.512
5	1.588	1.163	0.648
6	1.721	1.082	0.571
7	2.019	1.201	0.529
8	1.685	1.665	0.488
9	1.346	2.132	0.643
10	2.067	1.541	0.514
11	2.158	2.203	0.499
12	1.926	1.337	0.562
13	1.407	1.246	0.473
14	1.112	1.046	0.526
15	1.055	1.515	0.847

According to Table 1, the average response time of the College English multimedia course network autonomous learning platform in this paper and the other two autonomous learning platforms are 0.565 s, 1.560 s and 1.468 s respectively.

Table 2. Response time of the platform with 160 concurrent users (s)

Number of experiments	Mobile terminal-based online self-learning platform for college English multimedia courses	A network autonomous learning platform for college English multimedia courses based on cluster analysis	The online self-learning platform for college English multimedia courses
1	4.615	5.229	2.615
2	5.166	5.131	2.014
3	4.915	4.825	2.546
4	5.613	5.009	1.698
5	4.845	4.758	2.162
6	5.221	4.615	2.337
7	4.965	2.337	2.151
8	5.847	5.022	1.649
9	6.032	6.948	2.013
10	5.144	5.166	1.455
11	6.209	4.209	1.331
12	5.777	5.223	2.021
13	4.949	4.674	2.005
14	5.122	5.199	1.142
15	4.969	4.848	2.131

According to Table 2, the average response times of the online self-learning platform for college English multimedia courses and the other two self-learning platforms are: 1.951 s, 5.293 s, and 4.880 s, respectively.

Table 3. Response time of the platform with 240 concurrent users (s)

Number of experiments	Mobile terminal-based online self-learning platform for college English multimedia courses	A network autonomous learning platform for college English multimedia courses based on cluster analysis	The online self-learning platform for college English multimedia courses
1	10.615	13.152	8.145
2	11.289	12.748	7.948
3	10.314	11.646	8.616
4	12.545	13.825	8.553
5	11.722	13.466	7.498
6	11.649	12.899	8.022
7	12.031	13.285	7.645
8	11.088	13.407	8.113
9	12.615	12.516	7.468
10	11.714	12.884	8.319
11	12.166	13.615	7.255
12	11.361	12.711	7.316
13	12.058	13.060	8.005
14	11.999	12.388	8.198
15	15.246	13.051	7.554

According to Table 3, the average response time of the online self-learning platform for college English multimedia courses and the other two self-learning platforms are: 7.910 s, 11.894 s, and 12.978 s, respectively.

Table 4. Response time of the platform with 320 concurrent users (s)

Number of experiments	Mobile terminal-based online self-learning platform for college English multimedia courses	A network autonomous learning platform for college English multimedia courses based on cluster analysis	The online self-learning platform for college English multimedia courses
1	19.646	21.505	15.202
2	20.105	22.114	14.613
3	18.443	23.619	15.812
4	19.219	22.502	14.203

(continued)

Table 4. (*continued*)

Number of experiments	Mobile terminal-based online self-learning platform for college English multimedia courses	A network autonomous learning platform for college English multimedia courses based on cluster analysis	The online self-learning platform for college English multimedia courses
5	18.533	23.121	14.719
6	17.409	22.599	15.922
7	19.914	23.466	16.345
8	18.205	22.108	15.848
9	19.633	23.648	13.747
10	21.011	21.554	12.025
11	22.492	23.606	12.994
12	20.001	22.516	11.633
13	21.758	23.849	12.055
14	22.166	22.644	13.448
15	21.025	21.336	12.504

According to Table 4, the average response time of the online self-learning platform for college English multimedia courses and the other two self-learning platforms are: 14.071 s, 19.971 s, and 22.679 s, respectively.

Table 5. Response time of the platform with 400 concurrent users (s)

Number of experiments	Mobile terminal-based online self-learning platform for college English multimedia courses	A network autonomous learning platform for college English multimedia courses based on cluster analysis	The online self-learning platform for college English multimedia courses
1	26.316	29.484	16.212
2	25.818	28.515	17.301
3	26.144	27.616	15.718
4	27.911	26.335	18.966
5	26.533	26.108	15.844
6	27.482	25.319	17.606

(*continued*)

Table 5. (*continued*)

Number of experiments	Mobile terminal-based online self-learning platform for college English multimedia courses	A network autonomous learning platform for college English multimedia courses based on cluster analysis	The online self-learning platform for college English multimedia courses
7	26.901	26.405	16.877
8	28.516	26.331	17.901
9	29.311	25.818	19.362
10	27.503	25.088	18.011
11	28.519	26.316	18.206
12	29.646	25.822	19.115
13	28.533	26.915	18.612
14	29.417	26.144	19.344
15	26.410	25.013	18.227

According to Table 5, the average response time of the College English multimedia course network autonomous learning platform in this paper and the other two autonomous learning platforms are 17.820 s, 27.664 s and 26.482 s respectively.

To sum up, when the number of concurrent users gradually increases, the response time of the platform also increases. However, no matter how the number of concurrent users increases, the response time of the platform designed in this paper is still faster than that of the other two traditional platforms, which shows that the network independent learning platform of College English multimedia courses designed in this paper runs more efficiently.

In order to further verify the effectiveness of this platform, the College English multimedia course network autonomous learning platform, the mobile terminal based College English multimedia course network autonomous learning platform and the cluster analysis based College English multimedia course network autonomous learning platform designed in this paper are used to carry out a comparative analysis of student satisfaction. The comparison results are shown in Fig. 4.

According to Fig. 4, the student satisfaction of the College English multimedia course network independent learning platform designed in this paper can reach 99%, which is higher than that of the other two platforms.

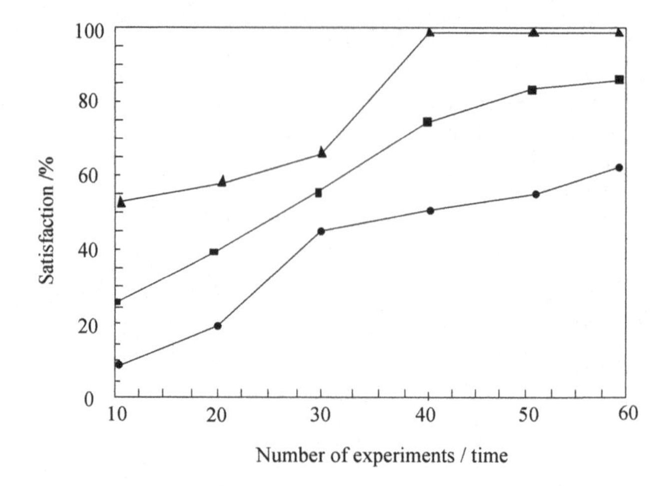

Fig. 4. Comparison results of student satisfaction

5 Concluding Remarks

The autonomous learning platform designed this time uses a variety of relevant theories to analyze several teaching modes widely used in colleges and universities, and proves that the autonomous learning theory and its related theories are not as effective as college students' English learning in the network environment. Popular. Therefore, self-directed learning and self-directed learning supplemented by question-answering mode cannot become the main means of foreign language teaching and learning in colleges and universities, and their use must be effectively monitored. In terms of theory, it deeply analyzes the theoretical basis of autonomous learning under the network environment and the basic concepts and connotations of autonomous learning under the network environment, and analyzes the relevant elements of autonomous learning under the network environment. At the practical level, use a systematic method to look at the process of college English classroom teaching, especially to demonstrate the effect of modern multimedia technology in college English teaching from a holistic perspective, and to analyze the causes of problems and clarify disorders through experimental tests. Phenomenon, and propose an optimization strategy. It has become the driving force for establishing and improving the teaching management system of online self-learning and using a variety of multimedia tools to improve information literacy. The follow-up research will focus on the direction of data diversification and improve the accuracy of the self-learning platform.

References

1. Hu, L.: Application of mixed teaching mode of mobile internet platforms in english courses. J. HeiLongJiang Inst. Teach. Dev. **39**(8), 139–141 (2020)
2. Yang, Y.: Practical research on the "Autonomous and Cooperative" learning mode of college english online courses. Guide Sci. Educ. (21), 49–50 (2020)
3. Ji, S.: On the teaching of college english linguistics with the help of multimedia. Guide Sci. Educ. (29), 129–130 (2020)
4. Meng, X.: A probe into the application of multimedia teaching in college english teaching. J. Jiangxi Vocat. Tech. Coll. Electr. **33**(10), 83–84 (2020)
5. Gao, Z., An, L.: Practice teaching research of mathematical english curriculum in normal colleges and universities. Theory Pract. Innov. Entrepreneurship **3**(21), 3–5 (2020)
6. Jiang, Z., Yi, D., Zhu, G.: Optimal selection of manufacturing cloud services considering elimination of fake cloud service. Comput. Integr. Manuf. Syst. **26**(8), 2020–2029 (2020)
7. Shi, X., Wang, X.: Cloud platform network digital information adaptive recognition simulation. Comput. Simul. 36(12), 387–390, 463 (2019)
8. Choi, H., Kim, M., Lee, G., et al.: Unsupervised learning approach for network intrusion detection system using autoencoders. J. Supercomput. **75**(9), 5597–5621 (2019)
9. Venkatesh, M., Sathyalaksmi, S.: Memetic swarm clustering with deep belief network model for e-learning recommendation system to improve learning performance. Concurr. Comput.: Pract. Exp. (18), e7010.1–e7010.21 (2022)
10. Adegoke, M., Wong, H.T., Leung, C.S.: A fault aware broad learning system for concurrent network failure situations. IEEE Access **9**, 46129–46142 (2021)

Research on Export Trade Information Sharing Method Based on Social Network Data

Guiling Yang(✉)

Yantai Vocational College, Yantai 264670, China
kkjjn2200@163.com

Abstract. Export trade, also known as export trade, refers to the trading activities of selling domestic products or processed products to overseas markets. Due to the large amount of export trade information and the limited storage of resources, the utilization rate of resources is low and the ability of information sharing is poor. To this end, this paper proposes a method of export trade information sharing based on social network data. Through the distributed classification technology of the blockchain platform, we can access trade financing data and information in real time and establish an information service mode. Bayesian estimation is used for data fusion. Establish social network data communication links to transmit information resources. Federal learning algorithm is used to map the original data into the corresponding data sharing model to realize the sharing of export trade information. The test results show that the export trade information sharing method based on social network data can improve the detection rate and shorten the running time, so as to maximize the utilization efficiency of shared information, and achieve better information sharing effect.

Keywords: Social network data · Export trade · Information sharing · Communication resources · Sharing model · Data fusion

1 Introduction

The scale and competitiveness of foreign trade is an important manifestation of a country's economic strength, status and influence in the international economy, and it is also a basic indicator for measuring a trading power. After the global financial crisis, the global economic recovery is still slow, and the international trade situation is still complicated [1]. Social networking is the most popular and trending application on the Internet today. Can fully meet people's social needs [2]. The world economy is accelerating its transformation to an economic activity with the network information technology industry as its important content. Therefore, it is of great theoretical and practical significance to study the impact of social networks on the development of China's foreign trade in the current context.

Reference [3] proposes an interactive sharing method of medical image information based on local weighted fitting algorithm. The generalized feature points of the image are extracted by Moravec operator, and the initial matching relationship is established.

W. Fu and L. Yun (Eds.): ADHIP 2022, LNICST 468, pp. 510–521, 2023.
https://doi.org/10.1007/978-3-031-28787-9_38

The local weighted fitting algorithm is used to add control points, and at the same time, these points are triangulated to ensure that each control point corresponds to a fitting result, and the image distortion points are corrected through the mapping relationship. The shared image is encoded by introducing semantic directivity and feature quantization, and the image is sent to the client. Introduce the above methods into the sharing system, determine the system functions from the main modules such as image source management and encoding format, and deploy the software scheme. This method can improve the throughput of the system, but the system runs for a long time. Reference [4] proposed the research on financial information sharing based on Improved SVM model. Using the correlation test method to test the initial characteristics of financial information, the important characteristics of financial information are determined. The improved SVM algorithm is used to build the financial information early warning sharing model. This method can improve the detection rate, but the algorithm is complex.

In order to solve the above problems, an export trade information sharing method based on social network data is proposed. Establish an export trade information service model and improve the efficiency of export trade information dissemination. Data fusion through Bayesian estimation to promote the conversion rate of trade scientific and technological achievements. Build an export trade information sharing model based on social network data distribution and communication resources, and improve the degree of industrial informatization.

2 Export Trade Information Sharing Method Based on Social Network Data

2.1 Export Trade Information Service Model

Under the influence of the new generation of information technology such as the Internet of things, the boundary between manufacturing and service industry has been broken. Their technical boundary, market boundary and business boundary are integrated into the manufacturing industry [5]. The scientific and technological services composed of information services, knowledge technology services and data services are internalized into the value components of the manufacturing industry. The basis for carrying out supply chain financing services and risk control must ensure the consistency between virtual information and real information, and the proof of transaction authenticity is required to be recorded in the creditor's rights information of the virtual world, and blockchain technology is a very suitable recording method. The Internet brings data agglomeration. As an important part of the new factor theory, big data is constantly being integrated into the production of enterprises. The information it generates can not only be exchanged, but also bring economic value. The high-speed information flow of the Internet increases the degree of openness of information, promotes the flow of factors, and brings together resources in various places, forming agglomeration of enterprises. The clustering of enterprises further promotes the emergence of regional industrial clusters, forms economies of scale, and reduces output costs. The formation of an effective scientific and technological information service mechanism can provide timely and systematic information on product scientific and technological achievements for both sides of the trade, promote China's informatization level, and solve the digital gap of export trade information.

2.2 Information Fusion Process Based on Social Network Data

For export trade information sharing, data fusion is the process of interaction between users and export trade information. Export trade data includes a large amount of user data, not only static data such as user basic information, but also dynamic data such as user behavior, which contains the information needs of users. Information in the network is transmitted from the dissemination end to the audience end at the speed of light, and the reliability is high. The current network technology can ensure that the information is almost error-free during the transmission process. The amount of content stored on the network is huge, and each piece of information has a large amount of repeated storage. The opportunities and times of forwarding and repeated dissemination of information increase exponentially. The process of data fusion is the process of sorting out and utilizing these data. The data fusion features are shown in Fig. 1.

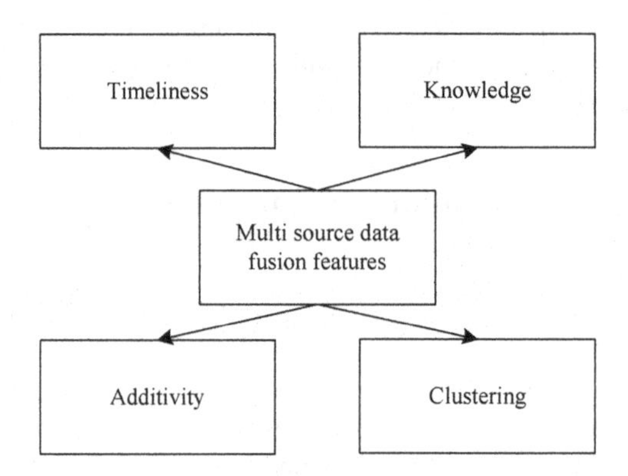

Fig. 1. Data fusion features

Data fusion is an automatic information processing method to assist decision-makers to make decisions and convert the collected information into processable representation values, so as to collect a variety of information sources. The process of data fusion involves many aspects such as data collection, processing, detection, combination and evaluation. After processing the data and fusing the data information, so as to accurately identify the state of events, the security degree of environment and the identity of participants. Data fusion can judge whether to allow the access operation requested by the user according to the constraints of policy and authorization attribute. If the authorization attribute of the user when making the access request does not meet the authorization attribute required in the corresponding policy of the requested access operation, dynamic authorization can assign the user less permission than the access operation requested by the user based on the policy requirements according to the authorization attribute already possessed by the user. In the traditional mode of information transmission, the disseminator and the receiver are fixed two different roles, but the receiver of information transmission in the network can easily forward the received information, change the role from the receiver to the disseminator, and process and express their opinions. In the process of data fusion, massive data are sorted out, identifying corresponding knowledge, and deleting redundant and useless information, so as to obtain target state information. Most social network mining is based on the only relationship between any two nodes, but in fact, there are many kinds of relationships between them, and different relationships have different roles in specific tasks. In multi-level social network communities, the impact factors of different relationships are different under specific needs. The community mining of multi-level relationships can linearly combine different relationships, and the mining results obtained are more accurate and effective. In this paper, Bayesian estimation is used to fuse data. When fusing, the data should be as independent as possible. With independent partitioning, the system can be evaluated for decision using Bayesian estimation. The representation of the Bayesian formula is as follows:

$$w(p|q \cap r) = w(q \cap rp|p)w(p) \tag{1}$$

In formula (1), w represents the probability of making a decision; n represents the total number of decisions; p represents the decision condition; q represents the observation result; r represents the observation result under another source. There is a certain state at a certain moment, these states are recorded, and the whole information sharing system is updated in the form of feedback. Internet users can selectively and targetedly receive information, filter out information that they do not pay attention to and are not interested in, and are no longer passive recipients of information. Social network users may receive the same information transmitted by different friends for many times, forming n-to-1 transmission. After receiving this information for many times, if they are interested in this information, it will increase the credibility of the receiver, and even forward this information to spread the information again [6]. The multi-dimensional digital media resource provider provides the access control device with the multi-dimensional digital media resource, access rights and authorization attributes. The access control device is composed of a judgment module, an authority setting module and a control module. After the access control device obtains the resources, access rights and authorization attributes, the user can realize the self-restraint of the access rights by setting constraints on the basis of the obtained access rights [7]. Ontology-based situational modeling can reveal export trade information and their relationships more accurately. The reasoning rules are set in advance when recommending the rules based on the ontology, so as to discover the relevant relationship between the export trade information resources and recommend the trade information with a high degree of correlation to the users. When recommending, it is necessary to calculate the similarity of information resources. The formula is as follows:

$$\theta(z_1, z_2) = \frac{2\log\alpha(z_1)\alpha(z_2)}{\log\alpha(z_1) + \log\alpha(z_2)} \tag{2}$$

In formula (2), z_1, z_2 represents two export trade information; α represents information entropy, which is the ratio of the number of times the information concept appears in the training set to the total number of the training set; θ represents the similarity. Data fusion expands components such as users, access rights, authorization attributes, etc., and changes the way of information dissemination. The user initiates a resource access request through the access control device. The multi-dimensional digital media provider obtains the resource from the resource server according to the resource name, and obtains the access right from the authority server according to the name and the resource type. Linked data can solve the problem of data heterogeneity on the Internet, so as to save the data in the semantic web.

2.3 Allocating Communication Resources Based on Social Network Data

A social network can be defined as a social structure composed of many nodes, where the nodes are usually considered to be people or organizations, and the people in the social network contain various social relationships. Moreover, with the continuous development of mobile communication technology and the arrival of the era of big data, the society has shown a trend of becoming more networked, and the combination of people and the Internet has become closer. The social network uses the intelligent terminal as the carrier to identify the identity information of the social network user online, and realize various social activities through the data service of the mobile Internet network. In addition to the characteristics of traditional social network information release and group establishment, mobile social networks also have two distinctive features of location attributes and instant messaging. The locatability of mobile terminals enables mobile social networks to have location attributes. In the social network layer, each user can find a corresponding device in the physical network layer. In order to reduce the business burden of the base station and protect the user's personal privacy, users can request the content they need from other trusted users, or send the content directly or with the help of other users to the target user [8]. Node betweenness refers to the ratio of the number of all shortest paths in the network passing through the node, which is used to reflect the role, influence and status of the node in the network. Similar edge betweenness refers to the ratio of the number of all shortest paths in the network passing through the edge, which is used to reflect the role, influence and status of the edge in the network. At the social network layer, each content sent through a link is related to factors such as social relationships, social influence, and similarity between users. In particular, users with friendly relationships, such as users who trust each other, are more likely to establish communication links to transfer data. Only when the probability of the forwarder taking "dishonest forwarding" is less than the forwarding threshold set by the publisher, the forwarding is allowed; otherwise, the system will reject the forwarder's forwarding request. The history record obtaining module obtains the content and decision result requested by the forwarder to forward this time, and records them in the forwarding history record and threshold database. At the physical network layer, the physical communication history between users is collected, which is very helpful for discovering potential communication links. With the help of the base station, every mobile user can detect the neighboring devices in his vicinity at any time. Based on this information, the user is able to know the number of encounters he has with other users. Using the relevant theories of graph theory to study the structural characteristics of social networks, the most common research is to model social networks through graph structures. A graph $H = (G, K)$ is used to represent a social network, H represents a social network, G represents all users on the social network, and node $g \in G$ in the graph is a user and an individual on the social network. $k \in K$ denotes that a link on a social network represents an association relationship. k reflects mutual trust between two users. The probability of two users meeting can be calculated by the following formula:

$$\beta_{xy}(\Delta\tau) = 1 - \frac{\min t_{xy}}{\Delta\tau} \quad (3)$$

In formula (3), β_{xy} represents the encounter probability of user x and user y within $\Delta\tau$ time; $\min t_{xy}$ represents the minimum encounter interval between user x and user y. Edges in a physical network graph can be represented by encounter probabilities. If the base station provides the mobile users with the encounter history information between the devices, each device can calculate the probability, or it can also be calculated at the base station and distributed to the corresponding users. Based on the analysis of the benefits of different game strategies selected by the forwarder and the publisher, combined with the historical data of forwarding operation, the probability of dishonest forwarding by the forwarder is calculated, and the final decision of whether to allow forwarding is given by comparing with the threshold set by the publisher. In order to ensure the successful transmission of data and minimize the transmission delay, the user will prefer to choose the equipment he can meet frequently as his relay. The data rate obtained at the relay node can be calculated by the following formula:

$$v = \frac{d}{2}\min\{\log_2(1+\gamma_1), \log_2(1+\gamma_2+\gamma_3)\} \quad (4)$$

In formula (4), v represents the data rate obtained by the relay node selected by the user; d represents the bandwidth available to the user; γ_1, γ_2 and γ_3 represent the signal-to-noise ratio of the relay node, the receiving node and the sending node, respectively. The final selected relay nodes not only maximize the achievable data rate, but also satisfy both social network and physical network constraints. The execution part is responsible for receiving the forwarder's forwarding request and executing the final forwarding control decision, and supports the publisher to set the forwarding threshold. The forwarding control part obtains the forwarding control decision by executing the relevant algorithm and records the current forwarding content and the forwarding decision of the forwarder in the forwarding history and threshold database [9]. In this way, an export trade information communication network is established with the goal of maximizing the data rate of each node.

2.4 Establish Export Trade Information Sharing Model

Multi-party data sharing is an effective way to alleviate the problem of limited computing and storage resources. In distributed multi-party application scenarios, the risk of data leakage brought by centralized servers is even more significant. Social network registered users who initiate access control effect evaluation upload resources such as texts, pictures, and videos through social networks and formulate their access control policies. To sum up, there are two main obstacles in the current data sharing: first, the centralized management server needs to process a large amount of data from different participants, including some unknown new data; Second, these participants do not fully trust each other (including the management server), and each participant will worry that the data they share will be leaked to other unreliable participants. The existing access control methods based on relationship and game theory only restrict the access operation of resources in social networks, and can not tell users whether the access control policy is appropriate. Inappropriate access control policies will bring serious consequences to resource publishing users, and even threaten the security of themselves and their families. This paper establishes an export trade information sharing method. The new sharing mechanism does not provide the original data, but uses the federal learning algorithm to map the original data to the corresponding data model. It is assumed that all parties involved in export trade information sharing have registered by uploading data retrieval information to the licensing blockchain. The export trade information sharing model established in this paper is shown in Fig. 2.

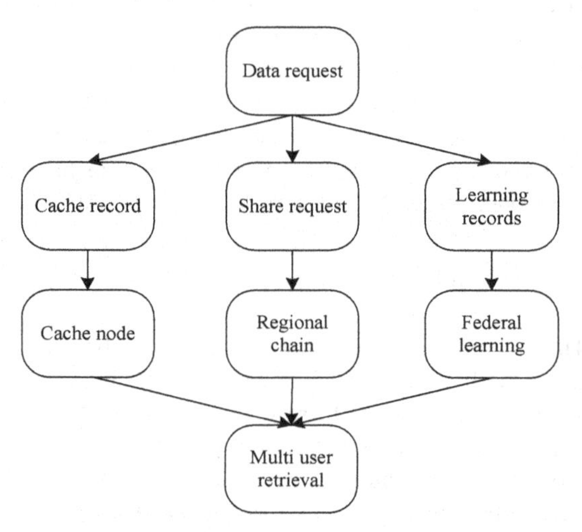

Fig. 2. Export trade information sharing model

When the data requester initiates a sharing request, it sends the request to the super node near it in the licensed blockchain network. The super node first searches the blockchain record to determine whether the request has been processed before. If there is a hit record in the search result, the previously cached calculated data model will be returned directly to the requester. The history includes the resources uploaded by the social network registered user who initiated the access control effect evaluation and the access control policies set for the resources. If there is no cached result, the super node searches for the relevant node of the blockchain for the sharing request through the aforementioned multiple data retrieval processes, that is, the "committee node" of the blockchain. The committee nodes are responsible for implementing the consensus protocol and collaboratively learning the federated data model. The execution part is responsible for receiving the user's evaluation request, calculating the final evaluation result, and allowing the user to modify the access control policy according to the evaluation result. The Lapdace mechanism is applied to the local export trade information data to achieve differential privacy. The calculation formula is as follows:

$$\varpi' = \varpi + L\left(\frac{\mu}{\chi}\right) \tag{5}$$

In formula (5), ϖ and ϖ' represent the local export trade information data before and after the update respectively; L represents the Lapdace mechanism; μ is the sensitivity value in differential privacy; χ represents the privacy budget. μ can be calculated using the following formula:

$$\mu = \max\|\phi(a) - \phi(b)\| \tag{6}$$

In formula (6), a and b represent two adjacent data sets with at most one different data record; ϕ represents the query result. The entire model is trained across distributed export trade information data providers. Sharing behavior shows weak anti-memory, publishing behavior is easy to be followed by a short time interval after a long time interval, and a short time interval is easy to be followed by a relatively long time interval. The membership node calculates the corresponding local data model for the request sent by the requester, and then forwards the data request to other relevant participants according to its local retrieval table. This process is repeated for all relevant parties until all relevant parties have been traversed. The trained data model cycles are returned to the requester as a response to their data sharing request. The receiving module obtains the user's evaluation request, including information about the social network resource requested for evaluation, and provides the information to the evaluation part. The execution module receives and displays the evaluation result provided by the evaluation part to the user, and at the same time allows the user to modify the access control policy according to the evaluation result. Data sharing events between data requesters and data providers are generated in the form of transactions and broadcast in the permissioned blockchain. All records are collected into blocks by blockchain nodes and cryptographically signed accordingly. The consensus mechanism is executed by the relevant nodes selected by the aforementioned multi-party retrieval. The user history acquisition module requests relevant history records from the history database according to the evaluation request and related information provided by the execution part, including resources uploaded to the social network and corresponding access control policies. The keyword generation module generates keywords according to social network resources. The relationship acquisition module acquires the relationship between the user and all the friends in the friend list from the social network relationship graph. The node that wins the competition broadcasts its locally newly generated block to other nodes for verification. After verification, the block is added to the permissioned blockchain. Block records are immutable after being added to a permissioned blockchain. So far, the design of the export trade information sharing method based on social network data is completed.

3 Experimental Studies

3.1 Experiment Preparation

The experiments in this section validate the proposed secure data sharing method on two real datasets, which are widely used to evaluate text-related machine learning algorithms.

The Reuters dataset is a benchmark dataset for classification tasks. Another 20 news groups dataset is a collection of about 20,000 short documents. The data in both datasets are unstructured short text data, which is very different from the structured data in the database. Preprocess each piece of basic data to obtain a four-dimensional vector of eigenvalues and a Boolean-type target value, which together correspond to a piece of training data. The processed 120,000 pieces of training data constitute the training sample set. This experiment uses these two datasets to simulate a large number of unstructured short data fragments in export trade information. This experiment divides the sorted dataset into slice sets and reassembles the slice sets into data subsets to simulate a distributed scenario of multiple export trade transactions. The experimental environment is as follows. CPU: Dual Core i7–3770, 3.4 GHz; RAM is DDR 8 GB; HDD is 500 GB, 7200 r; OS is Windows 10. The programming language is Python 3.5.2 (64-bit); the simulation software is Matlab 7.11.0.

3.2 Results and Analysis

This paper selects the detection rate and running time of information sharing as the test indicators, and trains the export trade information sharing method based on social network data proposed in this paper on the Reuters data set and the 20 newsgroups data set. The detection rate and time consumption results of the method in this paper are compared with the export trade information sharing methods based on random forest and SVM to verify the advantages of the sharing method in this paper. The detection rate results are shown in Tables 1 and 2.

Table 1. Comparison of detection rates on Reuters datasets

Testing frequency	The proposed method	The method of reference [3]	The method of reference [4]
1	0.926	0.886	0.856
2	0.938	0.855	0.878
3	0.945	0.861	0.851
4	0.922	0.854	0.834
5	0.913	0.842	0.827
6	0.936	0.863	0.842
7	0.949	0.858	0.813
8	0.955	0.827	0.823
9	0.922	0.804	0.815
10	0.931	0.821	0.807

In the training of the Reuters dataset, the detection rate of the proposed method is 0.934, which is 0.087 and 0.099 higher than the methods of reference [3] and [4].

Table 2. Comparison of detection rates in the 20 newsgroups dataset

Testing frequency	The proposed method	The method of reference [3]	The method of reference [4]
1	0.941	0.826	0.846
2	0.947	0.849	0.856
3	0.954	0.835	0.867
4	0.968	0.853	0.854
5	0.962	0.866	0.831
6	0.956	0.858	0.818
7	0.943	0.842	0.822
8	0.975	0.871	0.820
9	0.953	0.834	0.833
10	0.952	0.855	0.859

In the training of the 20 news groups dataset, the detection rate of the proposed method is 0.955, which is 0.106 and 0.114 higher than the methods of reference [3] and [4]. Therefore, the export trade information sharing analysis proposed in this paper has strong scalability, and the detection rate is improved compared with other sharing methods. The run time results are shown in Tables 3 and 4.

Table 3. Running time comparison of Reuters dataset (ms)

Testing frequency	The proposed method	The method of reference [3]	The method of reference [4]
1	856	979	1084
2	862	988	1247
3	872	966	975
4	859	925	956
5	860	1054	1062
6	888	1107	1125
7	826	1211	1051
8	853	1042	1198
9	875	1153	1262
10	817	1025	930

In the training on the Reuters dataset, The running time of the proposed method is 857 MS, which is 188 MS and 232 MS higher than that of the methods of reference [3] and [4].

Table 4. Comparison of running time of 20 newsgroups dataset (ms)

Testing frequency	The proposed method	The method of reference [3]	The method of reference [4]
1	816	1096	982
2	808	1056	962
3	824	1182	1095
4	835	965	1158
5	813	937	1126
6	802	1054	1039
7	841	1021	1152
8	827	910	1070
9	835	943	941
10	812	978	1017

In the training on the 20 newsgroups dataset, The running time of the proposed method is 821 MS, which is 193 MS and 233 MS higher than the methods of reference [3] and [4]. The running time of the shared method in this paper is slightly shorter than that of the two contrasting methods. The participation of multiple data providers expands the scale of data used for computation, thereby making the content of data sharing more accurate and improving the quality of applications.

In order to test the memory occupancy of the three methods, 1000 data were randomly selected in the 20 newsgroups dataset for experiments. The test results of the three methods are compared, as shown in Fig. 3.

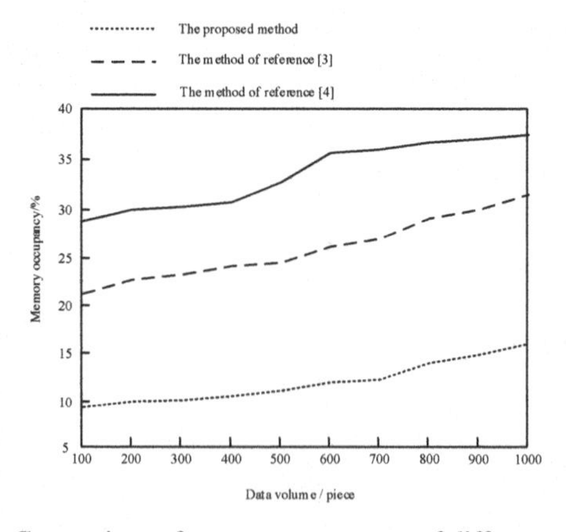

Fig. 3. Comparison of memory occupancy of different methods

As can be seen from Fig. 3, in the 20 newsgroups dataset, the memory occupation rate of the proposed method is 821\,ms. The memory occupancy of the method of reference

[3] and the method of reference [4] are 193 MS and 233 MS respectively, which are higher than the proposed method. It is proved that the information sharing effect of the proposed method is better.

4 Conclusion

Network development not only provides many traditional foreign trade enterprises with opportunities to expand market business and improve service levels, but also provides many small and medium-sized enterprises with an effective way to enter the foreign trade market. Taking advantage of the network opportunities can realize the advantages of China's trade in services, thereby promoting the coordinated development of trade in goods and trade in services. This paper proposes an export trade information sharing method based on social network data. Combined with the comprehensive consideration of detection rate and time consumption, the method utilizes limited communication and computing resources, maximizes the utilization efficiency of resources through reasonable allocation, and improves the sharing performance. Considering the subjectivity of users setting the degree of information privacy, in order to obtain more accurate evaluation results, it is proposed to use machine learning algorithms to classify transaction users, and then use machine learning algorithms to evaluate the effect of shared access control for different categories of users.

References

1. Jiang, S., Deng, X., Zhou, X., et al.: Quantitative predictions of impacts of trade friction between China and the US on wheat trade and its embodied carbon emissions. J. Agro-Environ. Sci. **39**(4), 762–773 (2020)
2. Yang, C.: Research of effect of experienced utility on continuous sharing intention of knowledge in social network. Mod. Inf. **40**(3), 88–102, 110 (2020)
3. Ding, X.: Medical Image Information Interactive Sharing System Based on Local Weighted Fitting Algorithm. Techn. Autom. Appl. **14**(7), 101–104 (2022)
4. Lin, H., Li, R.: Research on financial information sharing based on Improved SVM model. Mod. Sci. Instr. **39**(3), 219–223 (2022)
5. Jia, P., Yin, C.: Research on characteristics and rules of information transmission in blockchain social network. Inf. Sci. **39**(1), 35–40, 47 (2021)
6. Gao, A., Liang, Y., Xie, X., et al.: Social network information diffusion method with support of privacy protection. J. Front. Comput. Sci. Technol. **15**(2), 233–248 (2021)
7. Fang, J., Qian, X.: Information dissemination of social network in improved scir information propagation model. Comput. Eng. Appl. **56**(19), 105–113 (2020)
8. Shi, X.: Simulation of cloud-driven IoT information sharing security mechanism. Comput. Simul. **37**(9), 140–144 (2020)
9. Liu, X.-Y., He, D.-B.: Research on of competitive nonlinear dynamic information diffusion modeling in online social network. Chin. J. Comput. **43**(10), 1842–1861 (2020)

Intelligent Evaluation Algorithm of Undergraduate College English Mobile Learning Efficiency Based on Big Data

Hui Li(✉)

Ordos Institute of Technology, Ordos 017000, China
lq96311@126.com

Abstract. In order to better improve the quality of undergraduate English learning and understand the problems of students in the process of English learning, an intelligent evaluation algorithm of undergraduate college English mobile learning efficiency based on big a data is proposed. Combined with big data technology, a college English mobile learning information management platform is constructed, the hierarchical structure of College English mobile learning efficiency is optimized, and the intelligent evaluation index of College English mobile learning efficiency is constructed, The evaluation algorithm is optimized. Finally, experiments show that the intelligent evaluation algorithm of undergraduate college English mobile learning efficiency based on big data has high practicability in the process of practical application and fully meets the research requirements.

Keywords: Big data · College English · Mobile learning · Learning efficiency

1 Introduction

There are many literatures on the evaluation of undergraduate college English mobile learning education. These literatures emphasize the individual developmental evaluation under the concept of quality education [1]. Through the correlation between the college entrance examination results and the comprehensive results of the University, it is found that after entering the University, the learning efficiency of college students is not only affected by intellectual factors, but also the influence of non intellectual factors is far greater than intellectual factors. However, these evaluation factors are single, which is analyzed through the correlation between the total score and the college entrance examination score. There is no discussion on the impact of the College English curriculum system, which can not reflect the relative effectiveness of its learning. Based on this, using the big data network analysis method, a big data English model is established, and the relative effectiveness of College Students' final academic performance is evaluated on the basis of considering their starting point of admission [2].

The big data network method is to "evaluate" through effective and ineffective, which is especially suitable for the evaluation of "relative advantages and disadvantages" of

W. Fu and L. Yun (Eds.): ADHIP 2022, LNICST 468, pp. 522–536, 2023.
https://doi.org/10.1007/978-3-031-28787-9_39

decision-making among multiple similar sample units. Therefore, the big data network analysis model is used to evaluate the academic performance of different students in the same environment. However, these assessments are based on the overall curriculum as a decision-making unit, and do not distinguish the impact of different curriculum systems on College Students' learning efficiency. Because different curriculum systems have different effects on College Students' learning efficiency, the evaluation results obtained can not fully reflect the problems existing in college students' English learning process. In order to improve the practicability of the evaluation results of College Students' English mobile learning efficiency, this paper first divides the accessible data of the curriculum system, and combines big data technology to optimize the hierarchical structure of College English mobile learning efficiency; The intelligent evaluation index of College English mobile learning efficiency is constructed, and the learning efficiency of college students is studied through optimized evaluation algorithm and generalized big data network analysis. Through different curriculum systems, we can better analyze the impact of different curriculum systems on College Students' learning efficiency, which is very different from the big data network analysis method used to analyze college students' learning efficiency.

2 Intelligent Evaluation of Undergraduate College English Mobile Learning Efficiency

2.1 Undergraduate College English Mobile Learning System

The existing research has hindered the internal dynamic analysis of organizational learning, which is not conducive to people's understanding of the relationship between the overall and local transformation of organizational learning. However, studying the knowledge transformation mechanism between different organizational levels from the perspective of hierarchy has unique advantages to solve this problem [3, 4]. From the perspective of hierarchy, organizational learning includes four levels. Because Inter Organizational learning mainly promotes organizational learning, team learning and individual learning through the external environment, this paper only makes an auxiliary analysis. From individual learning to team learning and then to organizational learning is a knowledge sharing process of continuous inclusion and high-order transition between levels. From organizational learning to team learning and then to individual learning is a knowledge sharing process of continuous inclusion and low-order transformation between levels [5]. Therefore, organizational learning is not limited to a certain level of organizational learning, but includes a dynamic circular learning process in which the three levels of individual, team and organization overlap, promote and transform each other. The occurrence and cycle of learning is like tai chi operation. It can start from any learning level of the organization and induce the operation of other learning levels at the same time. This learning process that starts at one level and leads to or transitions to other levels is essentially the sharing, transition, transmission, attenuation and renewal of knowledge resources and skills and experience among different learning subjects. The cross level transformation of English learning is shown in Fig. 1:

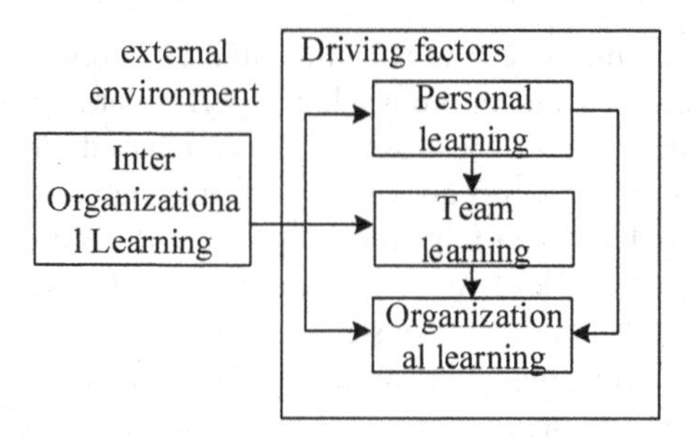

Fig. 1. Cross-level transformation of english learning

For college students English learning efficiency, from intelligence differences, personality differences, teaching methods, learning motivation to stimulate and maintain, teacher characteristics and teachers 'learning efficiency, which summarizes the eight most important factors affecting English learning efficiency: interest, focus, can keep up with the teacher's thinking, logical thinking ability, computing ability, independent homework, confidence, learning attitude "English teaching efficiency theory", specific learning system as shown in Fig. 2:

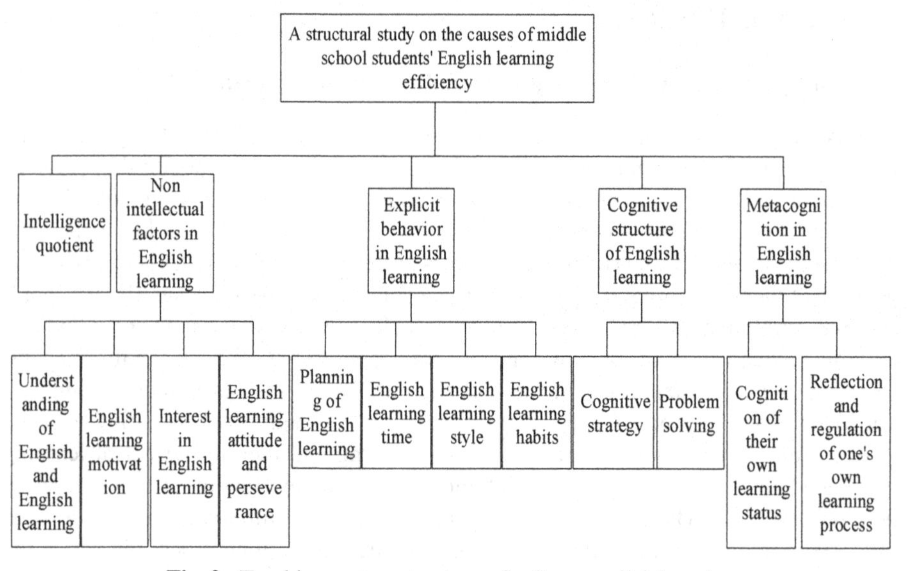

Fig. 2. Teaching system structure of college english learning

According to the structure of Fig. 2, the methods of observation, recording and interview, and the non-intellectual factors of English learning, planning, timing, attitude and habit, English cognitive structure, [6]. The English learning efficiency of middle school students is mainly attributed to the understanding and attitude towards English and English learning; the advantages of English learning methods and habits; whether the English cognitive structure is good: the strength of reflective consciousness and regulation ability in English learning.

2.2 Evaluation Index of College English Learning Efficiency

According to the definition of English learning efficiency, establish the evaluation index of English learning efficiency. The quantity that directly affects and reflects students' English learning efficiency includes three [7]: English learning time, English learning performance and English learning experience. It is also used as a first-level index to measure students' English learning efficiency. On this basis, seven second-level indicators are determined: students 'time use of learning English in class, students' time use of extracurricular learning English, English usual scores, English test scores, feelings before English learning, English learning and English learning. The structure is as shown in Fig. 3:

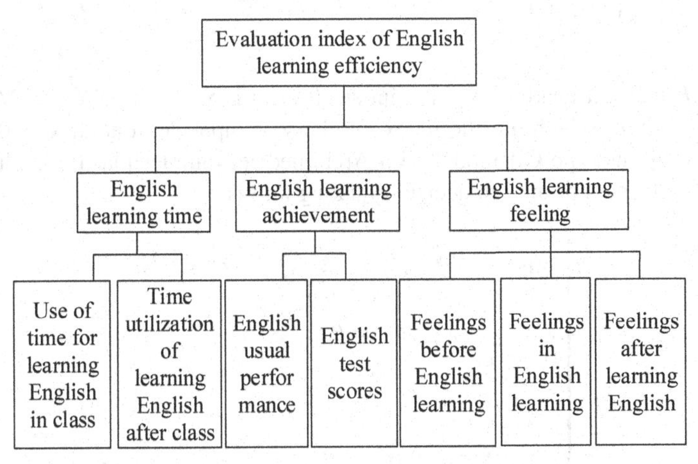

Fig. 3. Evaluation index of english learning efficiency

Considering that the traditional big data method is not feasible in the micro efficiency evaluation, in order to scientifically evaluate the action relationship and strength of different influencing factors and explore the overall evaluation mechanism of organizational cross-level learning transformation efficiency, this paper introduces the DEMATEL method which can effectively analyze factors and identify complex network action associations, which can fully reflect The system quantifies and effectively refines the advantages of subjective decision-making information of different experts, discriminates the influence relationship between organizational cross-level learning transformation efficiency indicators, preprocesses the influencing factor set of organizational cross-level learning transformation efficiency, deletes unimportant or redundant influencing factors, effectively distinguishes and ranks the key influencing factors of organizational cross-level learning transformation efficiency. The specific steps are as follows:

Suppose there are now x_p decision-making unit to be evaluated and y_p sample elements selected. Both the decision-making unit and the sample unit have m input and p output indicators, which are expressed as follows:

$x_p = \left(x_{1p}, x_{2p}, \ldots, x_{mp}\right)^T > 0$ represents the input index value, the value of the p-th decision-making unit;

$y_p = \left(y_{1p}, y_{2p}, \ldots, y_{mp}\right)^T > 0$ represents the output index value, the value of the p-th decision unit.

$\overline{x}_j = \left(\overline{x}_{1j}, \overline{x}_{2j}, \ldots, \overline{x}_{mj}\right)^T > 0$ represents the input index value, the value of the j-th sample unit;

$\overline{y}_j = \left(\overline{y}_{1j}, \overline{y}_{2j}, \ldots, \overline{y}_{mj}\right)^T > 0$ represents the input index value, the generalized CCR model constructed by the m-th sample unit against the decision unit p is as follows:

$$\left(G - C^2R\right) = \begin{cases} \text{Mowize } \mu^{\mathrm{T}} y_p - \overline{x}_j V(d), \\ st : y_p - \overline{y}_j / x_p - \overline{x}_j \qquad \omega\overline{x}_j - \mu^{\mathrm{T}} dy_j \geq 0, j = 1, \ldots, \overline{n} \\ \omega_{x_p} = 1 \qquad \mu, \omega \geq 0, \quad r = 1, \ldots, s; \quad i = 1_1 \ldots, m \end{cases} \tag{1}$$

With d decision-making unit, the input of $V(d)$ is x = (Xij, XY,... XM) and the input is μ^{T}, ω, d, which are the index numbers of input and output x $\geq$ 0 and y $\geq$ 0 respectively, then the CR model with Archimedean infinitesimal for evaluating the overall effectiveness of the j-th decision-making unit is

$$\begin{cases} \min \theta - \varepsilon \sum\limits_{i=1}^{m} s_i^- - \varepsilon \sum\limits_{r=1}^{l} s_r^+ + \left(G - C^2R\right) \\ s.t. \sum\limits_{j=1}^{n} x_{ij}\lambda_j + s_i^- = \theta x_{ij} \\ \sum\limits_{j=1}^{n} y_{rj}\lambda_j - s_r^+ \\ \lambda_j > 0, s_i > 0, s_r^+ > 0 \end{cases} \tag{2}$$

In the comprehensive influence matrix of organizational cross-level learning transformation efficiency calculated in the previous step, element s_r^+ represents the influence degree of influencing factors on r influencing factor θ. The influence value of influencing factor s_i^- on all other influencing factors is the sum of the elements in row j of the comprehensive influence matrix λ_j, and the influence degree y_{rj} is the influence value of all comprehensive influencing factors on other influencing factors.

Similarly, the comprehensive influence value of influencing factor ε affected by all other influencing factors is the centrality of the comprehensive influence matrix. Centrality reflects the role and importance of influencing factor s_r^+ in the evaluation index system, that is, the centrality of influencing factor is directly proportional to its importance. The greater the influence, the greater the role and importance of the influencing factor in the evaluation index system, and vice versa. The cause degree of influencing factor s_i^- is the difference between the influence degree of influencing factor s_i^- and the affected degree, i.e. If $s_i^- > 0$, it is the cause factor, indicating that the effect of other influencing factors on influencing factor s_r^+ is less than that of this factor on other factors;

If $s_i^- > 0$, it is the result factor, indicating that the effect of other influencing factors on influencing factor s_r^+ is greater than that of this factor on other factors [8].

Through the above calculation steps, we can get the centrality, cause, influence and affected degree of the influencing factors affecting the organizational cross-level learning transformation efficiency, so as to screen the key influencing factors of the organizational cross-level learning transformation efficiency in the next step, and take them as the input or output indicators of the organizational cross-level learning transformation efficiency, so as to reflect the learning efficiency or teaching effect of students through these courses, This requires analysis by categories from these courses in order to achieve more detailed analysis results.

2.3 Realization of Intelligent Evaluation of College English Learning Efficiency

The goal of big data analysis is to classify by collecting the similarity of data. The theoretical basis of big data analysis comes from many fields. With the development of science and technology and disciplines, its theoretical basis includes recitation machine science, English, biology, statistics and economics [9]. With the expansion of application fields, big data technology has also been developed. These big data technology methods can be well used to describe data, express the similarity between different data sources, and classify each data source into different clusters. The specific evaluation process of English classroom learning efficiency is shown in Fig. 4 below:

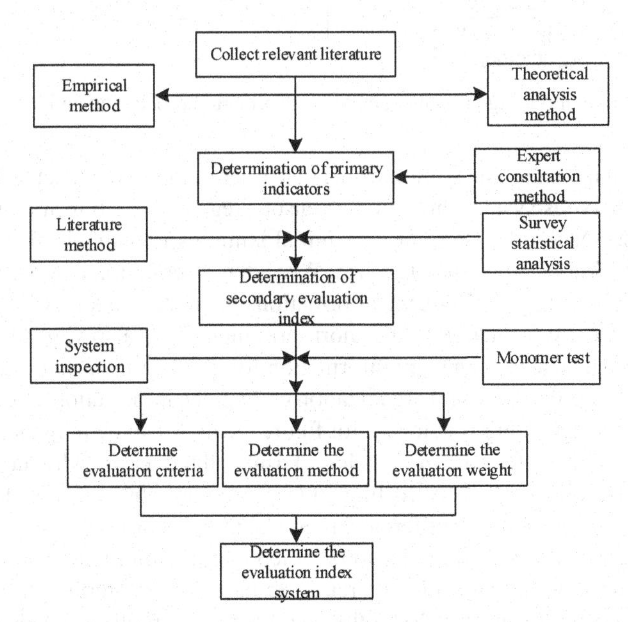

Fig. 4. Evaluation process of english classroom learning efficiency

Big data is a mature and common measurement method in the field of relative efficiency measurement, especially in the efficiency evaluation and improvement closely related to economic input and output. However, the application of big data in the field of education is insufficient, and the research ideas and process specifications lack a unified paradigm [10, 11]. Explain the complex behavior of intelligence with the elements and key stages in the process of information processing, mainly including the perception, selection and acceptance of information, processing, coding, storage, extraction and recovery of information. In particular, it puts forward the concept of "memory" and provides a basic tool for analyzing the thinking process and mechanism. The first mock exam model of learning process is presented based on modern information processing theory (shown in Fig. 5), which shows the information flow in the learning process.

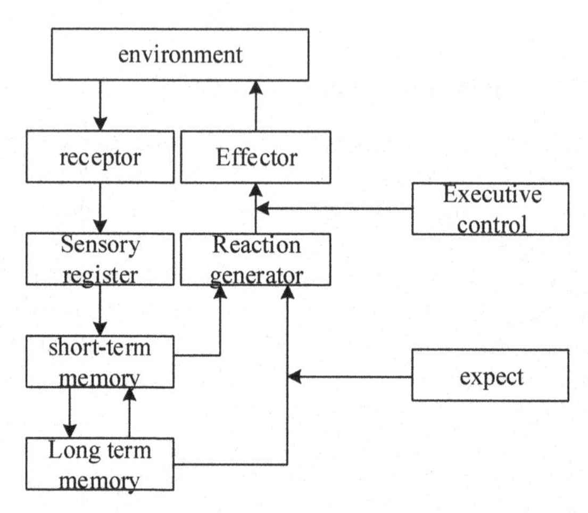

Fig. 5. Basic mode of intelligent management of information processing and learning efficiency

This pattern indicates that stimuli from the learner's environment act on his receptors and enter the nervous system through the sensory register. The information is initially encoded in the sensory register, and the initial stimulus is maintained in the sensory register in the form of image for 0.25–2 s. When the information enters the short-term memory, it is encoded again. Here, the information is stored in the form of semantics. The holding time of the information in the short-term memory is also very short, generally only 2.5–20 s. The learner can copy information for a little longer than a minute. After retelling, surprise processing and organization coding, the information can also be transferred to long-term memory for storage for future recall. Most learning theorists believe that the storage in long-term memory is long-term, and the reason why they can't recall later is due to the difficulty of "extracting" this information. In fact, short-term memory and long-term memory are not different structures, they are just different ways in which the same structure works. It should also be noted that the information from short-term memory to long-term memory may be retrieved back to short-term memory, which is also called "working memory". When the new learning partially depends on the recall of the students' original learning content, the original learning content is retrieved from the long-term memory and re entered into the short-term memory. For English learning, Zheng Junwen and Zhang Enhua believe that the process of English learning is a process in which new learning content interacts with students' original English cognitive structure and forms a new English cognitive structure according to cognitive learning theory (constructivist learning theory). The general process of English learning can be divided into four stages: input stage, interaction stage, operation stage and output stage. First, the so-called input in the input stage is essentially to create a learning situation and provide students with new learning content. In this learning situation, there is a cognitive conflict between the students' original English cognitive structure and the new learning content, which makes the learners have the need to learn new knowledge (i.e. "aspiration"). Second, after the input of the new learning content in the interaction stage, the students' English cognitive structure and the new learning content interact, and English learning is like the interaction stage. This interaction has two basic forms: assimilation and adaptation. Assimilation is the process of incorporating the new learning content into the original English cognitive structure, so as to expand the original cognitive structure. Adaptation is the process that when the original cognitive structure cannot accept new learning content, the original cognitive structure must be transformed to adapt to the new learning content. The result of the interaction stage is the rudiment of a new English cognitive structure. Third, the operation stage is essentially a process of consolidating the newly learned knowledge through exercises and other activities on the basis of the prototype of the new English cognitive structure in the second stage, so as to initially form a new English cognitive structure. Through this stage of learning, students learn certain skills, which makes the new knowledge closely related to the original cognitive institutions. Fourth, the output stage. This stage is based on the third stage. By solving English problems, the newly formed English cognitive structure is improved, and finally a new and good English cognitive structure is formed. Students' ability is developed, so as to achieve the expected goal of English learning. The general cognitive process of English learning is shown in Fig. 6.

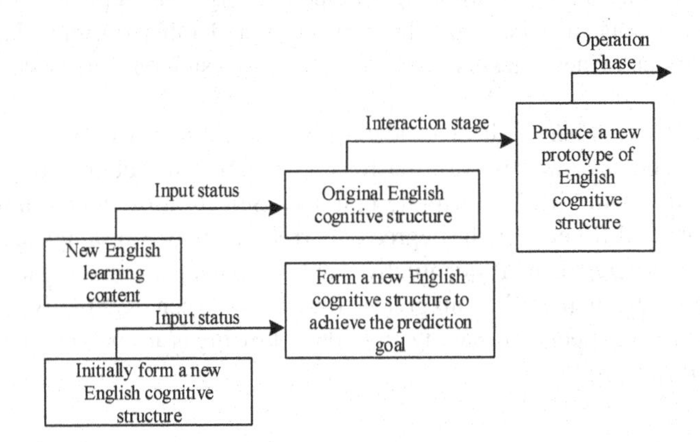

Fig. 6. General cognitive process of english teaching and learning

The above four stages are closely related. Problems in any stage of learning will affect the quality of English learning. Whether new English content is accepted or included depends on the English cognitive structure of students. Therefore, students' existing English cognitive structure is always the basis for learning new English content. On the whole, the principles and methods of big data are applicable to the analysis of learners' learning process. The diversity of big data models meets the analysis requirements of relative learning efficiency at different levels. At the same time, the efficiency analysis among various efficiency models complement each other; Big data realizes the use of learners' learning process data to evaluate relative learning efficiency, avoids the distribution of evaluation index weight in process data, and provides personalized results for the improvement of learning efficiency in terms of input redundancy and insufficient output. However, the big data method also has limitations on the analysis of learners' relative efficiency. On the one hand, the big data model has high requirements for data specification. For example, for the student data with the phenomenon of achievement inheritance, it is impossible to correspond to the relationship between input and output; The number of input and output indicators and the number of students should meet the standard requirements; The guarantee of the correlation between input indicators and output indicators will reduce the integrity of indicators, etc. On the other hand, there is great uncertainty in the interpretation of big data model results. Before efficiency evaluation, introduce big data analysis to determine the sample unit set, and then evaluate the decision-making unit, so as to study the feasibility of this method in teaching evaluation. This method is called the data reaching method of big data analysis, find the sample unit, then calculate the efficiency value of the decision-making unit, and analyze the effectiveness of the decision-making unit from the aspect. The general steps of big data analysis are as follows.

Firstly, the pairwise distance between n sample points is calculated.

The distance matrix D = dam (2) is obtained to construct n classes, each class contains only one sample point, and the platform height of each class is zero at the beginning.

The two nearest classes are merged into a new class, and the distance value of these two classes is taken as the platform height in the new big data graph.

Calculate the distance between the new class and other current classes. If the combined number of new classes is equal to 1, go to step 5); otherwise, go to step 3).

Determine the number of classes and classes through the above process.

Different defined distances are used for classification to obtain different big data results, so as to obtain different sample unit sets. Selection of sample unit set: there should be no great difference in the degree of correlation between samples, so that there will be no great difference in the quantitative analysis of decision-making unit by sample unit set. In this way, there will be no great difference between quantitative analysis and qualitative analysis of decision-making unit. The following is an analysis and calculation through specific examples.

3 Analysis of Experimental Results

3.1 Experimental Preparation

There have 19 freshmen in the same major to analyze and study the learning situation of a basic course.`Students' learning is affected by many factors. After analysis and screening, seven inputs are selected. A is the college entrance examination score of this course, which reflects the enrollment basis of students; B is the pre class preview time. Pre class preview can improve the efficiency of listening and cultivate students' ability of self-study and independent thinking; C is the class time. Classroom teaching is a direct way for students to acquire knowledge and cultivate their ability to analyze and solve problems. It is an important link in learning; The percentage of students' understanding of classroom teaching is D; E is the review time after class, which is an important link to consolidate the knowledge learned in class; F is the homework score, which reflects the students' further digestion and absorption of the knowledge learned in class; G is the time to read the extracurricular guidance books of relevant courses Select an output h as the final grade. 8 items of data of 14 students are shown in Table 1. The college entrance examination score is converted into the score of the hundred mark system.

Table 1. Statistics of Students' English learning

Student number	a	b	c	d	e	f	g	h
1	87	30	100	0.91	10	90	60	62
2	55	60	90	0.91	30	85	60	35
3	56	60	100	0.61	120	70	60	35
4	59	30	100	0.81	120	95	30	63
5	55	0	100	0.81	120	85	30	35
6	61	30	100	0.86	30	80	90	92
7	58	10	100	0.81	30	90	60	75
8	69	30	100	0.76	30	85	60	60
9	89	0	100	0.81	0	95	30	67
10	72	30	50	0.91	60	90	10	76
11	80	30	100	0.71	60	85	0	97
12	69	60	100	0.86	90	90	180	82
13	71	0	100	0.91	60	90	60	82
14	77	30	100	1.00	10	90	60	98

The distribution of students' comprehensive efficiency in the semester course is shown in Table 2.

Table 2. Initial evaluation information of the index of the expert group

Enterprise	I1 English learning time					I2English learning performance				
	C1Use of learning time in class	C2Use of extracurricular learning time	C3Usual performance	C4Examination performance	C5Preview situation	C6Learn attention	C7Logic thinking ability	C8Independent homework	C9Attitude to learning	C10Learning experience
L1	7	3	9	8	8	2	2	8	9	6
L2	6	2	8	5	5	8	1	3	2	8
L3	3	5	5	6	2	6	6	8	1	2
L4	8	7	9	1	3	2	3	7	5	8
L5	4	9	5	2	6	8	8	5	3	6
L6	2	3	9	8	5	2	6	3	8	5
L7	2	4	5	6	8	8	7	6	7	4
L8	5	5	8	5	2	4	2	4	5	7
L9	9	2	2	9	6	8	8	1	3	6
L10	8	4	9	4	8	6	5	2	8	8
L11	3	8	2	8	4	2	6	8	6	5
L12	6	2	8	9	1	8	9	9	8	9

Table 3. Directly affects the matrix

Index matrix		Classroom learning efficiency									
		C1 Learning interest	C2 Use of learning time	C3 Usual performance	C4 Examination performance	C5 Preview situation	C6Learn attention	C7Logic thinking ability	C8 Independent homework	C9 Attitude to learning	C10 Learning experience
Extracurricular learning efficiency	C1 Learning interest	0	4	5	4	5	4	4	5	5	5
	C2 Use of learning time	3	1	3	4	2	0	3	4	2	2
	C3 Usual performance	4	3	0	5	6	2	3	2	5	2
	C4 Examination performance	5	2	2	1	3	2	1	3	2	3
	C5 Preview situation	4	5	4	3	0	5	3	4	4	3
	C6Learn attention	4	3	3	1	2	3	0	2	3	5
	C7Logic thinking ability	5	1	2	3	1	3	2	3	0	1
	C8 Independent homework	1	4	2	2	2	4	2	2	1	2
	C9 Attitude to learning	1	3	3	5	3	1	4	1	2	5
	C10 Learning experience	0	1	5	4	4	5	3	0	5	4

Comprehensive efficiency is the product of pure technical efficiency and scale efficiency. Pure technical efficiency includes students' efficiency in learning methods, skills and time management. Scale efficiency only refers to the degree of energy that students invest in learning. Further, on the basis of combining resources and preliminary research, the teaching effect of 12 foreign language universities in a certain province was evaluated and studied, and the above learning efficiency was evaluated and analyzed. The The directly affects the matrix is shown in Table 3.

The above analysis results show that the overall measurement results obtained by applying the two efficiency evaluation schemes are significantly different. This scheme not only has more reference value for the overall efficiency identification results, but also has more interpretation advantages for the specific measurement information obtained for comprehensive efficiency, pure technical efficiency and scale efficiency. For example, although index u has efficiency in both evaluation schemes, this evaluation scheme measures that it still has deficiencies in scale efficiency, and further combined with its scale efficiency, it is in an increasing state. During the research process, the record of the course learning efficiency of most English students is relatively objective, and the learning effect is basically the same. The scores between courses are basically OK, while there are large differences between the scores of other courses, that is, there is a large gap between the scores. Then, this is different from the way of grey correlation, Grey correlation considers the correlation degree between the course and the grade of graduation design to consider the selected course to reflect the efficiency of students' learning. The big data analysis method classifies students by their course scores.

3.2 Experimental Results

In order to analyze and reflect the evaluation results of students' learning efficiency with big data, the traditional method and this method are compared and analyzed, and the comparison results are shown in Fig. 7.

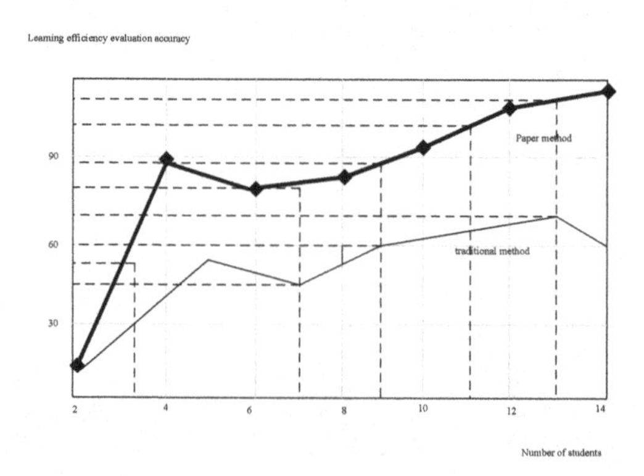

Fig. 7. Comparison and evaluation results of teaching efficiency

The research conclusion of this paper is universal, which is applicable not only to colleges and universities, but also to indicators and scientific research institutes.

3.3 Experimental Analysis

The reason is that organizational cross-level learning is composed of members with different backgrounds, different foundations and different experiences. The starting point of this study is efficiency, which is to realize the horizontal comparability of efficiency at a higher level based on a reasonable perspective, and is the rationality evaluation within a certain range. In this paper, the big data technology is used to optimize the level of college English mobile learning efficiency, construct the intelligent evaluation index of college English mobile learning efficiency, optimize the evaluation algorithm, and obtain the learning efficiency evaluation results are more accurate. The learning efficiency results obtained by using big data can provide a new classification basis for this kind of research. The analysis results of this study can not only enable teachers to understand the overall student learning efficiency and help teachers grasp the overall curriculum, but also enable students to have an intuitive understanding of their learning efficiency and understand the direction they should improve by showing the efficiency data, input redundancy and insufficient output of each student.

4 Conclusion

This paper proposes an intelligent evaluation algorithm for college students' English mobile learning efficiency based on big data. Innovative use of big data technology has built a college English mobile learning information management platform and optimized the hierarchical structure of College English mobile learning efficiency; It constructs an intelligent evaluation index of learning efficiency, optimizes the evaluation algorithm by using big data technology, and realizes intelligent evaluation of learning efficiency. The example shows that the application of generalized DEA model based on big data analysis to evaluate college students' learning efficiency has better theoretical and practical significance. Through the big data analysis method in statistical theory, it is difficult to distinguish the learning efficiency of 28 students, because no matter how many kinds of big data are used, it can be obtained that the vast majority of students are one kind, so it is impossible to distinguish students. However, through the combination of big data and generalized DEA method, we can distinguish the learning efficiency of college students, and obtain the relative efficiency of 23 students of the same class by selecting decision-making unit and reference unit, which can be completely sorted. In this way, we achieve the purpose of investigating the learning efficiency of students.

References

1. Lian, J., Fang, S., Zhou, Y.: Model predictive control of the fuel cell cathode system based on state quantity estimation. Comput. Simul. **37**(07), 119–122 (2020)
2. Su, J., Xu, R., Yu, S., Wang, B., Wang, J.: Idle slots skipped mechanism based tag identification algorithm with enhanced collision detection. KSII Trans. Internet Inf. Syst. **14**(4), 2294–2309 (2020)
3. Su, J., Xu, R., Yu, S., Wang, B., Wang, J.: Redundant rule detection for software-defined networking. KSII Trans. Internet Inf. Syst. **14**(6), 2735–2751 (2020)

4. Zhang, M.: Optimization and turn of college english teaching reform under the background of information technology. J. Hubei Open Vocat. Coll. **33**(10), 143–144 (2020)
5. Shu, D.: College english teaching and the cultivation of international talents. J. Foreign Lang. **43**(05), 8–20 (2020)
6. Cai, Y., Wu, J.: Practice and thinking of "Graded Reading" and "Extraterritorial Cultural Classics" in college english. Libr. Inf. Serv. **64**(08), 64–70 (2020).
7. Huang, Z., Mao, C., Guan, S., et al.: Simulation research on the deformation safety monitoring and evaluation algorithm of coastal soft foundation pit based on big data. Soft Comput. **5**(3), 77-83 (2021)
8. Chen, Y., Wu, C., Qi, J.: Data-driven power flow method based on exact linear regression equations. J. Mod. Power Syst. Clean Energy **10**(3), 800–804 (2022)
9. Ma, X.: Study on college english online teaching model in mixed context based on genetic algorithm and neural network algorithm. Discret. Dyn. Nat. Soc. **2021**(11), 1–10 (2021)
10. Luo, M.: Research on students' mental health based on data mining algorithms. Hindawi Limited **11**(10), 1–14 (2021)
11. Qi, B.: Mobile english learning platform based on collaborative filtering algorithm. Springer, Singapore, **122**(7), 706–715 (2022). https://doi.org/10.1007/978-981-19-3632-6_82

Design of Online Learning Efficiency Evaluation Algorithm for College English Based on Data Mining

Hui Li(✉)

Ordos Institute of Technology, Ordos 017000, China
lq96311@126.com

Abstract. The current learning efficiency evaluation algorithm has low accuracy and speed due to the singleness of the indicators and the neglect of the management of the indicators. To this end, this study designs an evaluation algorithm for college English online learning efficiency based on data mining. After analyzing the factors that affect the online English learning efficiency of college students, the evaluation indicators are abstracted and an evaluation system is established. Then use the analytic hierarchy process to determine the weight of the indicators in the evaluation system, and build a data warehouse according to the indicators. Finally, ES-ANN integrated sampling neural network is used to mine and analyze the data in the data warehouse, and the evaluation results of the students' learning efficiency are obtained. The experimental results show that the evaluation rate of the algorithm is fast and the evaluation accuracy is higher than 93%, which proves that the method greatly improves the evaluation performance.

Keywords: Data mining · College english · Online learning · Efficiency evaluation · Evaluation algorithm

1 Introduction

Online teaching is a teaching mode formed by relying on modern information and communication technology and mobile Internet technology. It can realize a ubiquitous teaching situation with the help of rich teaching resources, so that students can break through the limitations of time and space, and become better, more active, and more effective. Participate more fully in the whole process of online teaching activities.

With the outbreak and development of the new crown pneumonia epidemic in the country and even around the world, online teaching and online learning have moved from behind the scenes to the front desk, and have become the mainstream for a while, and have also become the teaching and learning methods used by teachers and students [1]. However, online teaching has also caused new teaching problems. Compared with traditional offline English teaching, students have higher requirements for self-control and concentration when learning, and online learning cannot provide students with a good

W. Fu and L. Yun (Eds.): ADHIP 2022, LNICST 468, pp. 537–548, 2023.
https://doi.org/10.1007/978-3-031-28787-9_40

learning atmosphere. As a result, students' learning efficiency is low, which seriously affects the quality of English online teaching.

The so-called learning efficiency is the ratio of the time and energy consumed by learning to the quantity and quality of learning obtained. Efficient learning can enable college students to acquire more and better knowledge more easily and happily during online English learning, and achieve a good learning effect of cultivating ability and promoting the all-round development of English ability. Therefore, evaluating the learning efficiency of college students who are studying English online can effectively measure the degree to which teaching objectives are achieved, control the teaching process, and identify the effect of online teaching [2].

The randomness of students' online learning time and learning methods makes management teachers lack strong basis and standards when evaluating students' learning status. At present, the learning efficiency evaluation method adopted by many domestic scholars mainly evaluates the learning efficiency of learners effectively by collecting the data of online learning input and output. For example, the online learning efficiency is analyzed by establishing a data envelopment analysis model (DEA). This method mainly analyzes the learning efficiency of the learners by collecting the input and output data of the learners during the online learning process. However, this analysis method considers a single influencing factor, and the analysis results have a large deviation [3]. However, these evaluation methods are not comprehensive enough to consider the influencing factors of online learning efficiency. It is difficult to directly measure the learning efficiency of college English online learners through the above methods only by collecting the data of learners' online learning input and output evaluate.

With the development of the epidemic, online teaching and online learning have become educational hotspots. More and more scholars pay attention to online teaching and online learning and conduct a lot of research on this theme and opportunity.

The scale and time of college students' English learning continues to expand, and the factors that affect students' learning efficiency continue to increase. The conventional evaluation index selection method can no longer meet the requirements of learning efficiency and accuracy. Data mining processes and analyzes a large amount of business data automatically, through a series of reasoning and classification, which helps to improve the evaluation accuracy [4].

Based on the above analysis, considering the large scale of online teaching and the advantages of data mining technology in evaluation work, this paper will design a data mining-based online learning efficiency evaluation algorithm for college English. It is of great significance to improve the quality of students' English learning, online teaching and other progressive research, and to help teachers of online teaching to adjust their teaching methods in a timely manner.

2 Algorithm Design

2.1 Analysis of Factors Affecting Students' Online English Learning Efficiency

There are many factors that affect students' learning efficiency. This article will analyze the factors that affect college students' online English learning efficiency from multiple dimensions.

Students' interest in English learning, students' attention span and duration of online learning, students' self-control ability in online learning, students' cognitive style, and students' homework completion; teacher factors include teachers' information technology literacy, teachers' class style, teacher's pre-class preparation, teacher's severity, and teacher's inspection and feedback on the content learned; curriculum factors include the duration of online English class, the type of English class, the design style of English class, and the difficulty level of class content; Environmental factors include the degree of family environment interference; equipment factors include class hardware equipment, network quality, teaching platform and web page interference [5].

From the perspective of students, their learning ability and English foundation will affect their English learning efficiency. Students' interest and confidence in English learning, motivation, effort, class participation and family background will also affect their English learning results. In the process of online learning, college students pay more attention to the external needs brought by the fluency of the online teaching process, which motivates them to improve their learning motivation. And the dedication in foreign language learning activities and the positive attitude reflected in the process of this activity. English learning motivation is divided into five categories: effort level, intrinsic interest, extrinsic needs, learning situation and learning value. Online English training makes it difficult for students to concentrate for a long time, and it is difficult to lock in the important information of learning materials, resulting in poor online learning effect. Students' self-confidence in English learning goals, competence in English courses and adaptability to English courses will also affect students' learning efficiency [6].

Due to remote teaching, teachers and students are separated from the screen. Teachers have certain difficulties in the organization and management of online teaching, and the teaching effect is even more difficult to detect. However, the arrangement of teachers' teaching methods and the design of teaching content will affect students' enthusiasm for learning and thus affect students' learning efficiency. The better the comprehensive performance of the learners' emotional participation, cognitive participation and behavioral participation during the learning of knowledge using the online platform, the higher the recognition of the course. The cognitive structure of high-efficiency English learning students has the characteristics of integrity, integrity and connectivity. With the increase of students' age, the influence of learning motivation on learning efficiency is more obvious. From the perspective of the characteristics of students' motivation development, whether it is the level of motivation, the content of motivation, or the intensity of motivation, etc., will change with time and conditions. Fixed place, proper lighting conditions, color, temperature and humidity of the study place also have some influence on study efficiency.

Among the many factors that affect students' online learning efficiency, English learning motivation and self-efficacy, as the two most basic internal influencing factors in students' online learning process, have a significant impact on students' online learning participation.

2.2 Establish an Online Learning Efficiency Evaluation System for College English

According to the above multi-dimensional analysis of the influence of college students' online English learning efficiency, the influencing factors are selected as learning efficiency evaluation indicators, and the corresponding college students' English online learning efficiency evaluation index system is established, and the index weights are determined. Table 1 below is the evaluation system of college English online learning efficiency established in this paper [7].

Table 1. Evaluation system of college english online learning efficiency

Target layer	Criterion layer	Indicator layer	Meaning
Efficiency evaluation of college English online learning	Student's own factors	Interest and Confidence in English Learning	From the perspective of students themselves, evaluate students' learning efficiency from multiple perspectives such as interest, method, purpose and initiative
		Motivation to learn	
		Study method	
		Study effort	
		Learning content awareness	
		Class participation	
	Teaching Impact	Course content suitability	Judging the efficiency of online learning from the perspective of courses according to the contents, progress, teaching methods and other contents
		Teacher teaching method	
		Course content design	
		Reasonable teaching arrangement	
		Attractiveness of teaching content	
	Other factors	Learning environment	Judge online learning efficiency mainly from the perspective of environment
		Learning hardware	
		Interaction with other course schedules	
		Parents concern	

After establishing the college English online learning efficiency evaluation system shown in Table 1, the fuzzy comprehensive evaluation method is used to establish the initial model of the indicators. According to the quantitative evaluation index system and scoring results, mathematical statistics are carried out, and comprehensive evaluation results are given. The method adopted is to quantify the evaluation grades, and obtain the weighted algebraic sum of grade scores with membership as the weight as the comprehensive evaluation value.

After the layering is established, the pairwise comparison matrix should be established from the second layer of the layered model to the last layer by the method of pairwise comparison. In order to compare the accuracy of judgment under different scale conditions, a comparison method with the scale set to 1 to 9 is used.

When calculating the maximum eigenroot and eigenvector of the matrix, if the requirements are not very high and high accuracy is not required, the method of calculating the method root can be used instead. The specific process is as follows [8]:

Step 1: Multiply all elements of each row of the judgment matrix to get the product M_i:

$$M_i = \prod p_{ij} \tag{1}$$

where, p_{ij} represents an element of each row.

Step 2: Calculate the n root of M_i to get $\overline{M}_i$.

The third step: normalize the vector $\left[\overline{M}_1, \overline{M}_2, \ldots, \overline{M}_i\right]$:

$$M_i' = \frac{\overline{M}_i}{\sum \overline{M}_i} \tag{2}$$

After obtaining the normalized comparison matrix, it is necessary to carry out the consistency test, obtain the eigenvectors of the paired matrices through calculation, and then carry out the index consistency test, and then normalize the qualified eigenvectors. Finally, the weight vector of the matrix is obtained. After the weight vector is calculated, the consistency index *CI* can be obtained by the following formula, as shown below:

$$CI = \frac{t_{\max} - n}{n - 1} \tag{3}$$

where, n represents the dimension of the comparison matrix; $t_{\max}$ represents the maximum value of the comparison matrix.

According to the average random consistency index RI table, to calculate the consistency ratio. The weight vector is obtained according to the above content, and then the feature vectors of each layer are aggregated to form a combined weight vector, and then the consistency detection is performed on the combined weight vector. When the calculated result satisfies the consistency index value, it indicates that the combined weight vector is qualified and can be referred to.

Afterwards, the overall ranking of the hierarchy can be used to obtain the priority of each factor relative to the decision-making target, and the ranking result can be used as the basis for decision-making. Therefore, whether the result of the total ranking of the hierarchy is reasonable or not will directly affect the effect of decision-making. To carry out the total ranking of the hierarchy, it is necessary to perform top-to-bottom layer-by-layer operations on the constructed analytic hierarchy, to obtain the importance value of the lowest-level factor relative to the sub-total target level, and to sort each factor according to the obtained results. The weights of the scheme layer relative to the total target layer are obtained through mathematical operations, and these values are sorted, so that the pros and cons of specific schemes can be analyzed, and the scheme can be selected. After determining the above-mentioned evaluation indicators of influencing factors of learning efficiency, in order to facilitate data mining processing, a data mining warehouse of influencing factors of learning efficiency is constructed.

2.3 Building a Data Mining Warehouse for Influencing Factors of Learning Efficiency

Data preparation is the foundation of data mining, and only sufficient data preparation work can ensure the effect of data mining. Determining the goals and objects of data mining is the primary task of data mining. Only when the goals and objects of data mining are established, can data mining work be carried out accurately. Therefore, this paper uses Python language and PyCharm compilation platform as data mining tools to mine and analyze the behavior data of students during English online learning, and establish a data warehouse of influencing factors that affect students' learning efficiency.

As the most popular language now, Python's biggest advantage is that it is easy to use, has an intuitive syntax, and also has many data mining-related class libraries. The Pandas library is a powerful time series processing library that provides the tools needed to efficiently manipulate large datasets; Scikit-learn is an algorithm library for implementing machine learning in Python, providing commonly used machine learning algorithms. Two Python libraries, the Pandas library and the Scikit-learn library, provide the foundation for data mining with Python. After determining the evaluation indicators of students' online English learning efficiency, establish a data warehouse dimension table. The online English learning efficiency evaluation data warehouse dimension table established in this paper is shown in Table 2 below (only part) [9].

According to the teaching practice experience, statistical analysis and design from two different perspectives, vertical and horizontal, can comprehensively grasp the students' online learning situation. Longitudinal refers to taking individual students as the object, and aims to describe the overall situation and details of students' online learning. Horizontal refers to taking the course as the object and aims to describe the learning situation of the students of the course. The horizontal angle is helpful for teachers to understand the overall situation of students' learning in each course, and use this as a reference to adjust the teaching content of the course. According to the objective laws of study, students usually study hard and conscientiously, and correspondingly, they will get better final grades. Therefore, the historical data of students' online learning behavior can be mined to find out the relationship between students' learning behavior and their corresponding final grades, and use this as a reference for evaluating students' online learning efficiency in the future. In the process of college students' online learning of English, the data collected by the online teaching platform is stored in the data warehouse, and the integrated sampling neural network is constructed to conduct in-depth mining according to the established evaluation index system to realize the evaluation of students' online English learning efficiency.

Table 2. Data warehouse dimension table

Field name	Field Type	Field Description
studentno	varchar	student ID
username	varchar	Name
studytype	int	student type
bd_id	bigint_not null	Student Learning Behavior Data Fact ID
courseid	int	course code
svalue	numeric	Starting time
evalue	numeric	departure time
stayt	int	dwell time
lmo	int	motivation to learn
evi	int	learning environment
tme	int	teaching method
tc	int	Teaching content
t date	date	function
tgrade	int	function

2.4 Realize the Evaluation of Online English Learning Efficiency

In this paper, the data mining method is applied, mainly using the analytic hierarchy process and sampling neural network. According to the common steps of data mining, first extract the necessary data sources in the system, then clean the data, and then perform data mining. All the data originally obtained are the data of all students in the school obtained from the school database, and some database tables have different coding formats, which need to be uniformly transcoded for reading, and then filter the required data entries and data items according to the student numbers of the students participating in the experiment., and then check whether there are omissions in each data, some data can be filled, and other missing data can be obtained from other related tables.

Teaching work is a dynamic link, and various factors in each link can affect the quality of teaching, which in turn affects the learning efficiency of students. According to the corresponding data in the dimension table of the data warehouse, extract the most direct relationship between the behavioral characteristics of the students during online English learning and the students' learning efficiency.

This paper introduces ES-ANN ensemble sampling neural network model to mine and analyze the student learning efficiency evaluation index data in the data reference, and obtain the evaluation results of learning efficiency.

After obtaining all the data features required for model training, the specific data is substituted into the model for calculation. For the supervised neural network model, that is, the integrated sampling neural network, the ten-fold cross-validation method is used to divide the data into ten parts. Each time, nine of them are taken as the training set, the other is the test set, and the average is taken as the final result after ten times of training and testing, so that the entire data set can be fully utilized. D represents all datasets, DX train represents the training set, DT test represents the test set, DZ represents the set of positive samples, and DF represents the set of negative samples [10].

The composition of the ten-fold cross-validation dataset is shown in Table 3.

Table 3. The composition of the ten-fold cross-validation dataset

Order	Data set	Make up subsets
1	DT	DT1, DT2, ..., DTk
2	DN	DN1, DN2, ..., DNk
3	D	DT, DX
4	DT	DT1, DT2, ..., DTk-1; DN1, DN2, ..., DNk-1
5	DX	DNk, DTk

Because the problem of unbalanced samples needs to be solved, five weak classifiers are used for ensemble sampling, so the negative samples in the remaining training set are randomly divided into five parts, and then the positive samples are copied into the set of each negative sample to form training set of five weak classifiers.

The neural network is built using the keras library, the tensorflow library as the lower layer support, three hidden layers with 64, 8, and 2 nodes respectively. The hidden layer uses relu as the activation function, and the output uses softmax as the classification. The loss function uses category cross entropy, with o and $\hat{o}$ as the actual output and label results, respectively. The calculation formula is as follows:

$$C(w, t) = -\left[\left(o \ln \hat{o}\right) + (1 - o) \ln\left(1 - \hat{o}\right)\right] \tag{4}$$

where, w represents the network connection weight; t represents the loss parameter.

The model uses the adam optimization algorithm, which is a method that computes an adaptive learning rate for each of its parameters. Its algorithm idea is equivalent to the combination of RMSprop + Momentum algorithm. In addition to saving the exponentially decaying average of the square δ_t of the past gradients like the Adadelta algorithm and the RMSprop algorithm, the Adam algorithm also stores the exponentially decaying average of the past gradients δ_{mt} like the momentum algorithm:

$$\begin{aligned} \delta_t &= \alpha\delta_{t-1} + (1 - \alpha)\varepsilon^2 \\ \delta_{mt} &= \alpha\delta_{mt-1} + (1 - \alpha)\varepsilon \end{aligned} \tag{5}$$

where, α represents the gradient squared bias parameter; ε represents the gradient descent bias parameter.

If mt and vt are initialized to 0 vectors, their biases will be biased towards 0, and bias correction is made to prevent this from happening.

The gradient update rules are as follows:

$$\delta'_{t+1} = \delta_{t+1} - \frac{\alpha}{\sqrt{\varepsilon + \vartheta}}\delta_{mt} \tag{6}$$

where, ϑ represents a hyperparameter. According to the above processing process, the relevant index data of the online English learning of the college students are obtained, and the evaluation result of the online English learning efficiency of the college students can be obtained through the analysis and processing of the algorithm.

All the above is the theoretical research process of the efficiency evaluation algorithm of College English online learning based on data mining. The specific evaluation process is shown in Fig. 1.

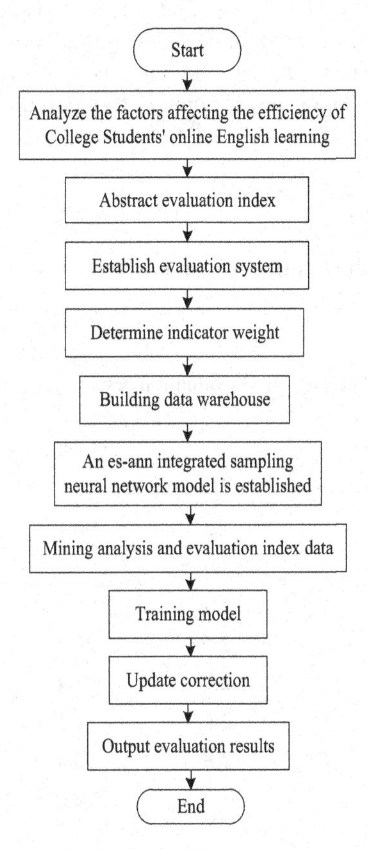

Fig. 1. Evaluation flow chart

3 Experimental Study

Before the evaluation algorithm is applied to the actual college English online teaching platform, it needs to pass the algorithm performance test. Therefore, this section will use a comparative experimental form to test the performance of the learning efficiency evaluation algorithm.

3.1 Experimental Content

Among all the freshman to senior year students who were taught online English in a university, the students were divided into two parts according to English majors and non-English majors, and each part was randomly divided into an experimental group and a control group according to the grades of the students. In the experimental group, the evaluation algorithm designed in this paper was used to evaluate the students' learning efficiency, and in the comparison group, DEA-based evaluation algorithm was used to evaluate the students' learning efficiency. The student's academic performance is used as the evaluation standard of the evaluation algorithm, and the evaluation accuracy of the algorithm on the student's learning efficiency is compared. At the same time, the evaluation time cost of the algorithm is counted to judge the evaluation rate when the algorithm is applied.

3.2 Experimental Results

Table 4 is a comparison of the evaluation results of two evaluation algorithms for English majors.

Table 4. Comparison of evaluation results of english majors

Student number	Test group		Comparison group	
	Evaluation accuracy/%	Time costs/s	Evaluation accuracy/%	Time costs/s
100	98.6	1.3	96.7	2.4
150	98.7	1.5	97.2	2.6
200	98.4	1.9	97.3	34
250	97.7	2.5	96.5	3.8
300	97.2	2.7	93.4	4.1
400	96.9	2.6	94.1	4.6
500	96.5	2.8	92.0	4.9
1000	96.3	3.9	90.6	6.7

Table 5 compares the evaluation results of the two evaluation algorithms for non-English majors.

Comparative analysis of Table 4 and Table 5 shows that the evaluation accuracy of the algorithm in this paper for the learning efficiency of students in different majors in the experimental group has reached more than 93%. There is a big difference in accuracy. Combined with the students' English test scores, the evaluation results of the algorithm in this paper are more reliable. From the time cost of the algorithm, the evaluation rate of the algorithm in this paper is increased by at least 49.6%, and the performance is better.

Summarizing the above experimental analysis content, it can be seen that the evaluation accuracy and evaluation efficiency of the online college English learning efficiency

Table 5. Comparison of evaluation results for non-English majors

Student number	Test group		Comparison group	
	Evaluation accuracy/%	Time costs/s	Evaluation accuracy/%	Time costs/s
100	97.5	1.4	90.6	2.7
150	97.2	1.6	92.4	2.9
200	96.8	1.9	87.5	3.4
250	96.4	2.3	88.6	3.8
300	96.1	2.5	89.3	4.2
400	95.6	2.6	84.1	5.7
500	94.7	3.1	76.5	6.3
1000	93.5	3.7	80.2	8.9

evaluation algorithm designed in this paper based on data mining are relatively higher, which can more effectively help teachers optimize online teaching methods and content and improve English. The quality of online teaching. The reason for this result is that after establishing the evaluation system, this method uses AHP to give weight to the evaluation indicators, and builds a data warehouse according to the indicators, which improves the pertinence of the evaluation process. Finally, the es-ann integrated sampling neural network is used to mine and analyze the data in the data warehouse, and the learning efficiency is accurately analyzed.

4 Conclusion

Traditional classroom teaching has always been dominant in English teaching, and online teaching has always existed as an auxiliary tool as a new teaching method. Under the influence of the new crown pneumonia epidemic, many problems have arisen in students' online learning in the online teaching conducted by various primary and secondary schools and colleges and universities. From the overall situation of online teaching, students participate in online teaching and online learning. The degree is obviously not high, which makes the efficiency of students' online learning very low.

The efficiency evaluation algorithm of college English online learning based on data mining proposed in this paper provides methods and means for objectively, fairly and reasonably evaluating students' online learning, This algorithm utilizes es-ann integrated sampling neural network to mine and analyze relevant evaluation data, with fast evaluation speed and high evaluation accuracy, which can help teachers deeply grasp students' online English learning situation and guide students to reasonably arrange online English learning.

However, due to the limitation of research time and other conditions, the algorithm proposed in this paper still has some shortcomings. In future research, the algorithm will be further optimized from the perspective of improving the diversity of assessment.

References

1. Jiang, M.: Learning assessment: the path choice to improve college students' learning quality. Heilongjiang Res. High. Educ. **37**(8), 45–48 (2019)
2. Yao, Y., Xu, J., Zhu, X.: Online learning evaluation based on processing technologies in 2D and 3D images. Comput. Technol. Dev. **31**(12), 128–134 (2021)
3. Lijuan, D.: Research on university education quality grading evaluation based on data mining. Mod. Electron. Tech. **43**(15), 101–104 (2020)
4. Liu, J., Huang, Y., Y, L.: The mining and analysis of the evaluation data of classroom teaching. J. Educ. Sci. Human Normal Univ. **18**(2), 118–124 (2019)
5. Lu, Z., Liu, Z., Zheng, Q.: Evaluation of the efficiency of online learning learners based on data envelopment analysis. J. Open Learn. **24**(02), 30–38 (2019)
6. Peng, L.: How does evaluation promote learning?——Empirical analysis on the effectiveness of vocational education learning evaluation. Vocat. Tech. Educ. **42**(22), 37–44 (2021)
7. Gou, R., Ye, X., Wang, B., et al.: Design of learning quality of intelligent evaluation system based on deep learning. Microcomput. Appl. **37**(09), 23–26 (2021)
8. Huang, B., Xie, Y., Tang, Y., et al.: Data storage information serialization completeness and efficiency evaluation simulation. Comput. Simul. **37**(4), 159–163 (2020)
9. Huang, T., Zhao, Y., Geng, J., et al.: Evaluation mechanism and method for data-driven precision learning. Mod. Distance Educ. Res. **33**(1), 3–12 (2021)
10. Hui, C.H.E.N.: Application research of multi-objective decision algorithm in evaluation of data mining. J. Guiyang Coll. (Nat. Sci.) **15**(3), 5–9 (2020)

Design of Network Traffic Anomaly Monitoring System Based on Data Mining

Yanling Huang[1](✉) and Liusong Huang[2]

[1] Department of Computer and Art Design, Henan Vocational College of Light Industry, Zhengzhou 450001, China
kjkmgm123@163.com

[2] Software Engineering, Maanshan Teacher's College, Ma'anshan 243041, China

Abstract. The security hidden dangers in the network will affect the normal operation of the network. Therefore, in order to better ensure the security of the network structure, it is necessary to monitor the abnormal network traffic. However, due to the low monitoring accuracy and long monitoring time of the traditional network traffic anomaly monitoring system, this paper designs a network traffic anomaly monitoring system based on data mining. Through the configuration of data acquisition equipment, analysis equipment, exception handling equipment and system management equipment, the hardware structure of the system is designed. On this basis, through the system software functions of acquisition module, data processing module, data analysis module, data application module and infrastructure management module, the abnormal monitoring of network traffic is realized through data mining. Finally, the experiment proves that the network traffic anomaly monitoring system based on data mining has higher monitoring accuracy and shorter monitoring time, which is practical in practical application and fully meets the research requirements.

Keywords: Data mining · Network flow · Abnormal monitoring

1 Introduction

With the continuous enrichment of network functions, the potential security problems in the network are becoming increasingly prominent, which puts forward higher requirements for network security monitoring [1]. Abnormal traffic is a common security risk faced by the network. Insufficient configuration of monitoring equipment or software and the new compiled virus formed by combining computer virus and hacker technology are the main reasons for the frequent occurrence of abnormal network traffic. In order to ensure the safety of network applications, it is necessary to find problems in time and quickly monitor the fault location through real-time monitoring of network traffic, so as to avoid network failure to the greatest extent. Network traffic data information is growing rapidly in the era of big data. The content dissemination and learning process of network traffic data information need to be realized by combining network

W. Fu and L. Yun (Eds.): ADHIP 2022, LNICST 468, pp. 549–563, 2023.
https://doi.org/10.1007/978-3-031-28787-9_41

technology and information technology [2]. The large-scale network based on complex network structure covers computers and network equipment of different manufacturers. Most of the different functions or applications of the network need to be realized based on different protocols [3]. In the network environment composed of different personalized learning networks, channel congestion and unbalanced distribution are easy to occur, which makes it difficult to evenly distribute trust nodes and accurately locate the problems, resulting in abnormal traffic. Real-time and accurate monitoring of network abnormal traffic has become the main means to improve network security.

The network traffic abnormal monitoring system based on Android platform is designed by traditional methods, and the key technologies of traffic monitoring are analyzed. The designed traffic monitoring system includes data extraction module, traffic monitoring module, data management module and user interface module. Finally, the system function is tested The results show that the module can meet the requirements of mobile phone traffic monitoring, but the accuracy of abnormal network traffic monitoring is low. Based on this, this paper designs a network traffic anomaly monitoring system based on data mining, optimizes the hardware configuration of the system, and combines data mining technology to optimize the software function and smooth operation of the system, improve the identification efficiency of network traffic anomalies, and improve the monitoring function of network traffic anomalies. Finally, the experiment proves that the system designed in this paper has high practicality.

2 Network Traffic Anomaly Monitoring System Based on Data Mining

2.1 Overall System Architecture

The network traffic anomaly monitoring system adopts a layered architecture design, which divides the whole system into six layers: acquisition layer, access layer, computing layer, storage layer, service layer and application layer. Each layer has its own independent functions, only depends on the services provided by the next layer to itself and provides its own services to the upper layer. The upper and lower layers coordinate and cooperate with each other to provide complete functions for the whole system. The overall architecture of the system in this paper is shown in Fig. 1.

The layered architecture meets the principle of single responsibility, allowing each layer to focus on its own functions, and the responsibility boundary is clear. It can not only simplify the system design, but also comply with the object-oriented design principle of high cohesion and low coupling. All the same types of businesses are placed on the same layer, and the upper services only rely on the functions implemented by the lower layer. At the same time, the lower layer shields its own internal implementation from the upper layer, as long as the service interface provided by the lower layer to the upper layer remains unchanged, Developers' modifications to the lower layer will not affect the upper layer services [4].

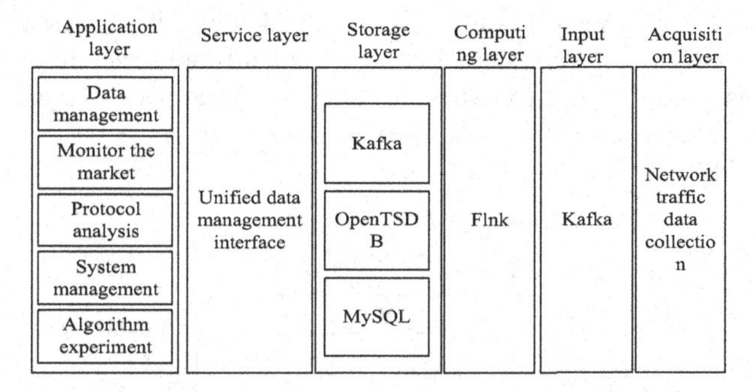

Fig. 1. Overall architecture of the system

2.2 Hardware Structure Design of Network Traffic Anomaly Monitoring System

2.2.1 Hardware Configuration Structure

The hardware structure of network traffic anomaly monitoring system is designed to better visualize the data results in the system, make the data intuitive and clear, and facilitate users to understand and use [5]. It is a complete process to filter and evaluate the results based on data mining technology, present the correct results by using visual methods, mine previously unknown, effective and practical information from large databases, and use this information to make decisions or enrich knowledge. The data mining process is shown in Fig. 2:

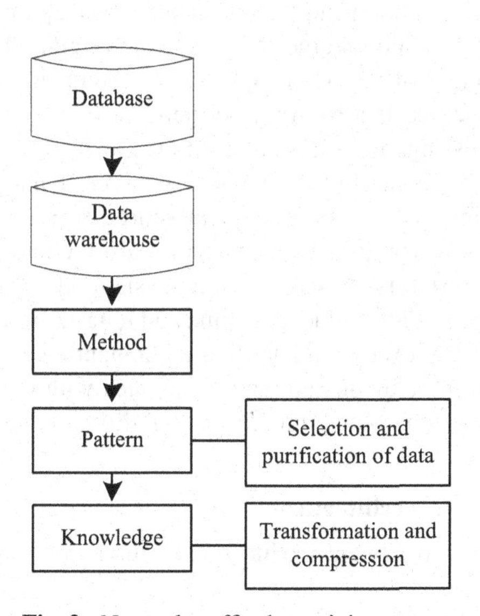

Fig. 2. Network traffic data mining process

In order to ensure the data mining effect of network traffic, it is necessary to optimize the hardware structure of the system, optimize the hardware structure of the monitoring system based on the characteristics of multi-functional network, and improve the convenience of user access. The block diagram is shown in Fig. 3:

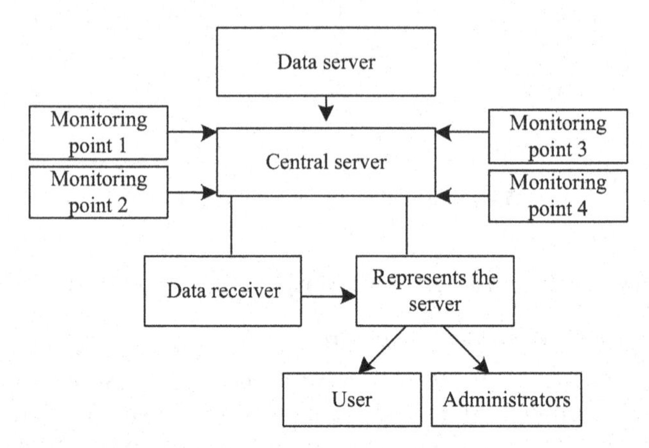

Fig. 3. Hardware configuration block diagram

The system includes four independent and interrelated equipment configurations: data acquisition equipment configuration, analysis equipment configuration, exception handling equipment configuration and system management equipment configuration [6]. The collection equipment is configured to provide the original flow data for the whole system, which is the basis of the whole system. The acquisition device configuration is responsible for collecting data from the set netflow output device, storing it in the database after preprocessing, and submitting it to the analysis device configuration for further analysis and processing. In the database, a source IP address table is generated according to the original data to save frequent source IP addresses. The purpose of analyzing equipment configuration is to conduct real-time and long-term monitoring of the new IP address, establish a reasonable analysis model according to the change law of the new IP address, and conduct real-time and long-term analysis and reasonable prediction of the data. The exception handling equipment configuration is responsible for the alarm of abnormal phenomena, and cooperates with the network firewall and network management system to maintain the availability of the network [7].

2.2.2 System Hardware Architecture

The overall design of the hardware architecture of the monitoring system is shown in Fig. 4.

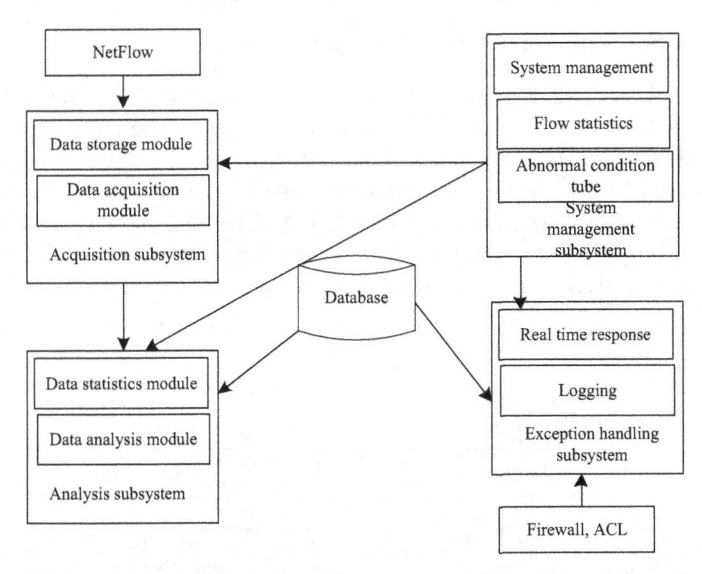

Fig. 4. Hardware architecture of monitoring system

The figure shows in detail the relationship between each equipment configuration and its role in the whole architecture, as well as the main module division of equipment configuration and the relationship between them. D realize distributed computing through RM technology of Java, and run different functional modules on different hosts, so as to make full use of network resources and computing power of multiple hosts and reduce costs [8]. DOS monitor can only be activated when monitoring abnormal network traffic. I filter will be activated when monitoring DoS attack. Therefore, in the case of positive traffic, the DOS monitor is not required. At this time, the system RMI server (data storage server) will learn the new IP address as the database middleware. Just load the JDBC actuator on the server side, and the client can call the JDBC on the server side through the RMI of Java. This three-tier structure realizes the isolation of functional modules, reduces the load of the client and server, and makes the load distribution of the whole system more reasonable. Users download the web page containing applet from the server through the browser, run swing of the base applet on the user's host, and communicate with the server process on the system configuration server through RM, so as to use the graphics community to manage the whole system [9]. The monitoring object mainly adopts the active monitoring method through four monitoring points to complete the effective monitoring process; After the monitored data is processed by the data server, a unified data sheet is obtained for subsequent query; The central server receives and summarizes the monitoring data obtained at each monitoring point, and manages the software configuration information; The traffic anomaly monitoring results are finally managed and presented through the presentation server.

2.3 System Software Function Optimization

According to the user's demand analysis, extract the business problems that the software system can help users solve, define use cases, that is, analyze the user's business problems, and plan the functional modules of the system. This step is the understanding and sublimation of users' business needs, indirectly reflects the system structure of network traffic monitoring and analysis system based on data mining, and lays a solid foundation for subsequent program development. Through the analysis of user requirements, a business use case diagram is established, as shown in Fig. 5:

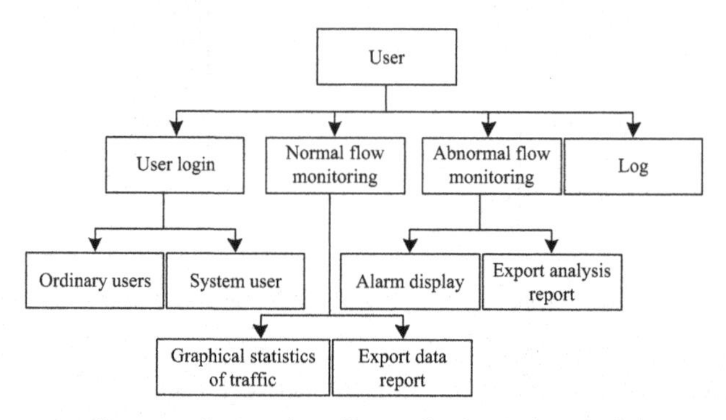

Fig. 5. Business case diagram of network traffic monitoring and analysis based on Data Mining

As shown in the above figure, users can log in, but the attributes of logged in users are divided into two categories: ordinary users and system users. Ordinary users can view the graphical statistics of data flow and obtain the alarm display of abnormal flow monitoring [10]. System users can view the detailed statistical data of individual user traffic and the analysis report of abnormal traffic. The purpose of setting system user permissions is to protect the privacy of personal PC data. The alarm of abnormal traffic and the user's operation record will be stored in the log record in real time for system users to query. Network traffic is represented as a set of time series in large-scale networks. In order to provide accurate characteristic basis for the monitoring system, the traffic characteristics can be extracted by constructing a signal model. In large-scale networks, MAC is accessed through CSMA/CA channel, and the periodic broadcast network model is used to obtain the clustering model of the corresponding data structure for the network traffic information in transmission, which is represented by $n(k)$. It is assumed that the signal received by the cognitive user is represented by $s(k)$, and the traffic information sent by the authorized user is represented by h. The expression of the channel access model is as follows:

$$x(k) = \begin{cases} n(k) & \leq 1 \\ hs(k) + n(k), & \succ 1 \end{cases} \tag{1}$$

It is assumed that for abnormal traffic, β represents its spectrum sensing transmission signal, and $\sigma_{x_1}^2$ represents the channel gain; k represents the number of iterations; n(k) refers to additive Gaussian white noise; p refers to channel equalization index; the variance of R during generation n is represented by i; the expression of selecting adaptive generation step is as follows:

$$\mu_1(k) = (1-\beta)/R_n^i x(k) - p\left[\sigma_{x_1}^2(k)\right] \tag{2}$$

$$\sigma_{x_1}^2(k) = \beta\sigma_{x_1}^2(k-1) + (1-\beta)x_1^2(k) \tag{3}$$

Once the network traffic is abnormal, the IP address and port distribution will change accordingly. If the network configuration is wrong, the original IP address and target IP address will increase, resulting in a sharp increase in host messages. According to this feature, the network traffic matrix method is used to analyze the dispersion of traffic distribution characteristics. Assuming that the flow characteristic is A, the total number of samples is B, and the number of occurrences of a specific flow characteristic i is n_i, therefore, the flow characteristic sample can be defined as:

$$\begin{aligned} G(x) &= -A\sum_{i=1}^{C}\left(\frac{\mu_1(k)}{B}\right) - \left(\frac{n_i}{\sigma_{x_1}^2(k)}\right) \\ B &= \sum_{i=1}^{C} n_i \end{aligned} \tag{4}$$

If the results of all selected samples are consistent, then $G(x) = 1$; If the results of all selected samples are highly dispersed. Thus, the abnormal behavior of different flow characteristics can be described, as shown in Table 1.

Table 1. Identification of abnormal flow behavior characteristics

Anomaly type	Exception definition	Abnormal characteristics
Configuration error	Device failure caused by incorrect configuration of routing port	Large abnormal characteristic value The normal eigenvalue is large
Service attack	Service attack	The abnormal characteristic value is small Normal eigenvalue is small
Burst access	Multiple hosts send traffic to a single host	Large abnormal characteristic value The normal eigenvalue is large
worm scanning	A small number of ports on the destination host are detected	The abnormal characteristic value is small The normal eigenvalue is large

The monitoring and analysis module is used to subdivide the abnormal data traffic in the network into unknown abnormal data and known abnormal data, which is conducive to improving the alarm efficiency and the work efficiency of the network administrator. In its working process, it first receives standardized abnormal data, and then uses data mining algorithm to calculate the similarity of abnormal data. Those with high similarity are classified into one class, and all kinds of direct data association or similarity are relatively low. Judge whether it is necessary to send relevant abnormal data to the alarm module according to the danger value K. if the similarity between the data packet and the same class is greater than the clustering radius K value within a certain period of time, it indicates that it has exceeded the early warning range, Then send the relevant characteristic data to the alarm module. The alarm module displays and alarms the data in detail. Figure 6 shows the operation process of network data monitoring and analysis module:

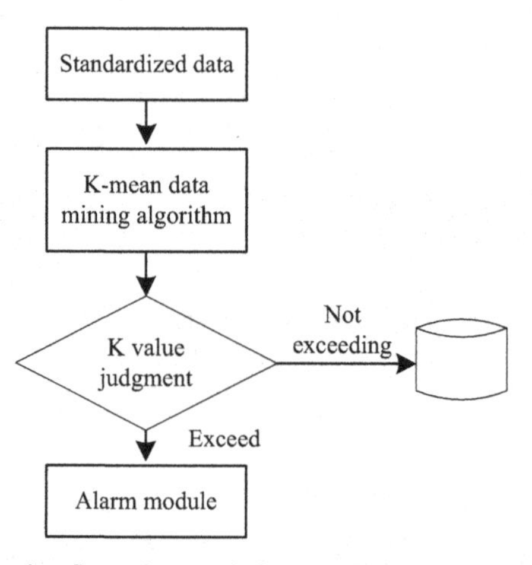

Fig. 6. Operation flow of network data monitoring and analysis module

Using the important characteristic of network traffic - chaos, the model is mainly used to analyze the self similar characteristics of network traffic, and analyze and predict from the perspective of non-linearity of traffic. Because in practice, the flow data is limited and its phase space trajectory is complex and changeable, it is difficult to be applied in reality, and it still stays at the level of theoretical research. Compared with the global method, the local method has stronger practicability and less calculation, but it still has some defects: on the one hand, it needs more storage space in calculation, on the other hand, it needs more time to solve model parameters and construct adjacent state vectors. The characteristics and applicable scenarios of some representative network traffic prediction models mentioned above are summarized in Table 2:

Table 2. Network abnormal traffic prediction model

Model	Complexity	characteristic	Applicable scenario
AINA	low	The prediction of short correlation flow is more accurate	Online prediction and wireless sensor networks
WNE	high	For non-stationary flow, the prediction is more accurate	It has high requirements for prediction accuracy and prediction time, and has sufficient resources to support the network
ABBGE	high	It is more accurate to predict the short-term and long-term related flow	
CO	low	It can better describe the change of sample trend and predict higher	
AII	high	The nonlinear flow prediction is more accurate	
SON	high	For small samples, the flow prediction is more accurate and has strong generalization ability	

The negative impact of abnormal flow on the network is self-evident. Therefore, timely and effective monitoring of abnormal flow wells and taking appropriate methods to control them have become an indispensable part of the network flow monitoring system. According to the different locations of anomalies in the network, combined with the characteristics of data mining technology, the system subdivides the module into abnormal traffic monitoring for the controller and abnormal traffic monitoring for the host.

2.4 Implementation of Abnormal Network Traffic Monitoring

Combined with the system architecture design, the system is divided into five functional modules, namely data acquisition module, data processing module, data analysis module, data application module and infrastructure management module. Each functional module can be divided into smaller functional modules, and each functional module is responsible for a single system function. Describe the design of the five functional modules of the system and their subdivided functional modules, as shown in Fig. 7.

Because the two-stage abnormal traffic monitoring model needs to be applied in the scenario of data imbalance, the abnormal traffic belongs to a few categories, and the data of abnormal traffic is much less than that of normal traffic. Therefore, based on the original two-stage abnormal traffic monitoring model, the data is oversampled by data mining algorithm to avoid the problem of model over fitting caused by unbalanced data set as far as possible. The abnormal flow monitoring model at different stages is shown in Fig. 8:

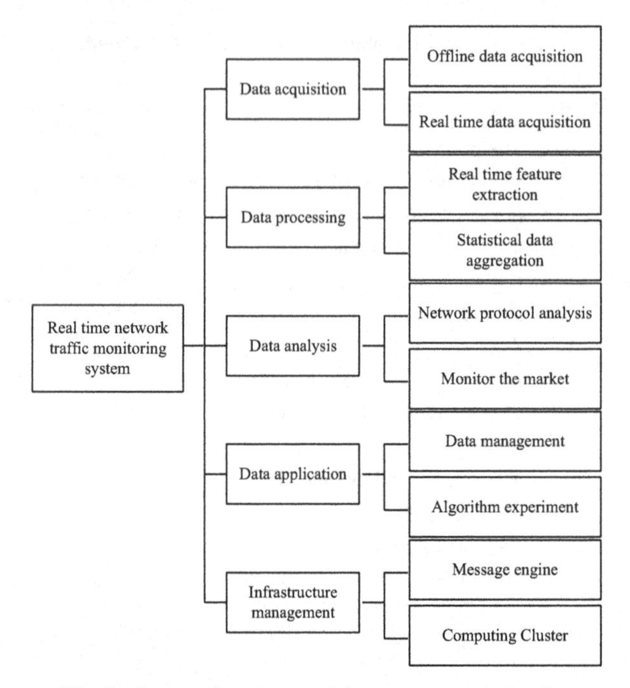

Fig. 7. System function module structure optimization

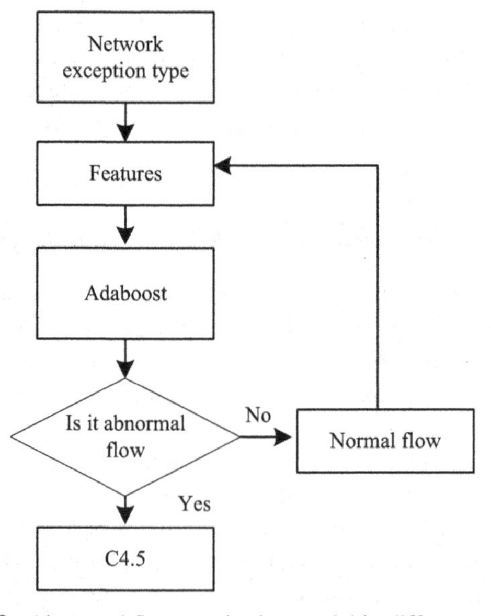

Fig. 8. Abnormal flow monitoring model in different stages

Monitoring is an important function of real-time traffic monitoring system, which is mainly responsible for the visual display of all indicators of monitoring network and key indicators of infrastructure on which the system depends. Users can simply and quickly organize a monitoring market by creating, organizing, managing data sources and configuring forms. By monitoring the market, system users can clearly and intuitively understand the operation of the whole monitoring network and system. If abnormal behavior of network traffic is found, it can be found directly in the monitoring market. For experienced network security operation and maintenance personnel, they can also find potential network security threats from various statistical charts of the monitoring market, and then update the network abnormal traffic monitoring model and network defense response module in time, Increase the monitoring ability of the system to network abnormal traffic, facilitate users to operate and troubleshoot the system, and improve the operation and maintenance efficiency of the system.

3 Analysis of Experimental Results

A series of performance tests are carried out on the system designed in this paper to ensure that the system fully meets the system requirements. The network traffic uses the same configured server in the same computer room. The specific experimental parameters are shown in Table 3.

Table 3. Experimental platform server parameters

Hardware configuration	CPU	Pentium MMX200
	Memory	64G
	Hard disk	300G
Software configuration	Operating system version	Linux 6.5
	Storm version	Storm 0.95
	Kafka version	Kafka 0.8.1
Network environment	Network card	Gigabit Ethernet

The function test mainly focuses on whether the function of the system meets the expectation, and verifies whether the logic and function of each functional module meet the system requirements. The system mainly adopts the methods of unit test and integration test to test the function of the system. The unit test mainly verifies whether the functional logic of the system is correct and whether all exception handling is reasonable. The integration test mainly verifies whether the function of the system meets the system requirements. For example, the test results of the data acquisition module are shown in the table, and the test results meet the expectations, as shown in Table 4:

Table 4. System data acquisition and transmission function test

Test name	Offline and real-time traffic collection function test
Test purpose	Test whether the offline traffic collection script and real-time traffic collection script can collect network traffic and whether the real-time script works normally
Preconditions	Offline traffic packet integrity
testing procedure	Offline traffic collection script starts locally; System registration monitoring equipment; Issue the real-time traffic collection script and start it; Kill real-time traffic collection script
Expected results	Offline traffic collection script and real-time traffic collection script work normally
Test result	The test results are in line with the expected results

In order to test the anomaly monitoring ability of the system to known attack types. In the experiment, we launched eight kinds of network attacks during training from an attack host (202.201.94.190) of network VIAN1 to the host (202.201.95.1–202.201.95.20) of network VAN2. The final monitoring results are shown in Table 5.

Table 5. Traffic monitoring results of known attack types

Data type	Normal data	Tcpport scan	Vulnerability scanning	OS scan	Tcpsyn scan	ACK scan	Land attack	Syn flood	Ddos	Total
Number of records detected	13839	3619	2918	1232	3292	0	38	5575	5749	37491
Actual records	14000	4000	3070	1430	4000	450	90	5760	6000	38800

The system performance test is carried out after the system function test. On the premise that the system function meets the expectation, the performance test is carried out to collect the indicators of the server and application and the use of hardware resources. The performance test will focus on the system throughput, end-to-end delay, system performance index and fault recovery speed of the cluster. The test results are shown in Table 6:

Table 6. System monitoring performance test results

Test object	Performance index	Test result
Kafka	Average throughput of producers	45 MB/s
	average delay	1.3 s
	Average consumer throughput	129 MB/S
	Consumer rebalancing delay	9.3 s
Flink	Average End-to-End Delay	4.5 s
system	Average response time	2.5 s
	QPS	1300
The server	CPU utilization	79%
	Memory	95%
	Disk throughput	420 MB/s
	Mean recovery time	6 min 15 s

In order to verify the effectiveness of the multi-function network monitoring system with the traditional traffic monitoring system, as shown in Fig. 9.

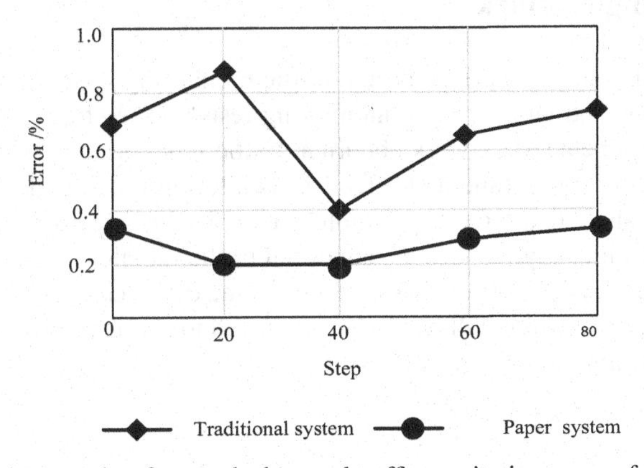

Fig. 9. Comparison results of network abnormal traffic monitoring errors of the two systems

It can be seen from this that the error of the system proposed in this paper in real-time monitoring of abnormal traffic problems in the network is within 0.2–0.4, while the error of the traditional system in monitoring abnormal traffic in the network is within 0.4–0.9. The error of the system in this paper in real-time monitoring is smaller, and the monitoring effect is better, which fully meets the research requirements.

In order to further verify the effectiveness of the method in this paper, the time taken by the system in this paper and the traditional system to monitor abnormal network traffic is compared and analyzed. The comparison results are shown in Table 7.

Table 7. Monitoring time of abnormal network traffic/s

Number of experiments/time	Article system	Traditional system
10	2.3	11.5
20	2.6	13.4
30	3.4	14.7
40	3.9	14.9
50	4.2	15.2
60	4.9	15.9
70	5.6	16.8

According to the data in Table 7, the time taken by the system in this paper to monitor the abnormal network traffic is within 5.6 s, while the time taken by the traditional system to monitor the abnormal network traffic is within 16.8 s. The time taken by the system in this paper to monitor the abnormal network traffic is the shortest and the monitoring efficiency is the highest.

4 Concluding Remarks

Most of the traditional network abnormal traffic monitoring systems are vulnerable to noise interference and large error of monitoring results. Therefore, this paper designs an optimization scheme of network abnormal traffic monitoring system based on data mining to realize the real-time and effective monitoring process of abnormal traffic. The test results show that the optimization scheme in this paper has high monitoring accuracy and significantly reduces the abnormal traffic monitoring error. However, the operating efficiency of the system has not reached the expected effect. Therefore, in the next research, the algorithm will be further improved to shorten the operation time and improve the operating efficiency of the system.

References

1. Ifzarne, S., Tabbaa, H., Hafidi, I., et al.: Anomaly detection using machine learning techniques in wireless sensor networks. J. Phys: Conf. Ser. **1743**(1), 012021–012034 (2021)
2. Yang, J., Hou, X.: Research on data mining algorithm based on positive and negative association rules. Comput. Technol. Dev. **30**(11), 64–68 (2020)
3. Choi, H., Kim, M., Lee, G., et al.: Unsupervised learning approach for network intrusion detection system using autoencoders. J. Supercomput. **75**(9), 5597–5621 (2019)
4. Lian, J., Fang, S., Zhou, Y.: Model predictive control of the fuel cell cathode system based on state quantity estimation. Comput. Simul. **37**(07), 119–122 (2020)
5. Xu, Y., Sun, Z.: Research development of abnormal traffic detection in software defined networking. J. Softw. **31**(01), 183–207 (2020)

6. Xiao, F., Chen, L., Zhu, H., et al.: Anomaly-tolerant network traffic estimation via noise-immune temporal matrix completion model. IEEE J. Sel. Areas Commun. **37**, 1192–1204 (2019)
7. Ma, W., Zhang, Y., Guo, J.: Abnormal traffic detection method based on LSTM and improved residual neural network optimization. J. Commun. **42**(05), 23–40 (2021)
8. Meng, Y., Qin, T., Zhao, L., et al.: Network anomaly detection method based on residual analysis. J. Xi'an Jiaotong Univ. **54**(01), 42–48+84 (2020)
9. Peng, Y., Chen, X., Chen, S., et al.: Cross-domain abnormal traffic detection based on transfer learning. J. Beijing Univ. Posts Telecommun. **44**(02), 33–39 (2021)
10. Liu, Y., Li, J., Zhang, Y., et al.: Network abnormal flow detection method based on feature attribute information entropy. Netinfo Secur. **21**(02), 78–86 (2021)

Research on Operational Risk Monitoring Method of Intelligent Financial System Based on Deep Learning and Improved RPA

Liang Yuan(✉) and Hui Zhu

State Grid Huitong Jincai (Beijing) Information Technology Co., Ltd., Beijing 100031, China
lihongsheng51@163.com

Abstract. The accuracy of traditional financial system operational risk monitoring is low. Therefore, this paper proposes a method of intelligent financial system operational risk monitoring based on deep learning and improved RPA. Set the financial monitoring index and obtain the warning threshold parameters; Using deep neural network method to mine key risk indicators and obtain reconstruction coefficients of data mining errors of financial system; By improving the RPA method to calculate the fit degree of financial risk, it matches the internal business process of the enterprise; The operational risk monitoring algorithm of financial system is designed to realize the operational risk monitoring of financial system. The experimental results show that the risk monitoring accuracy of the design method is 80.3%, and the overall test threshold of the model is 0.5 after the introduction of non-financial indicators, which shows that it can be applied in practice.

Keywords: Deep learning · Improve RPA · Intelligent financial system · Operational risk monitoring · Optimization algorithm

1 Introduction

As an important part of the national financial industry, banking is very important to promote national economic development. With the rapid development of China's economy, the speed of financial system reform is also faster and faster. As an important part of China's banking system, policy banks have played a strong role in promoting China's economic and financial system reform and maintaining economic stability and development. However, according to the research of relevant institutions in the United States, since the 1980s, several major international banks have suffered more than $200 billion in losses due to operational risks, which has brought huge losses to their own economy and banking operations [1, 2]. The huge losses caused by operational risk have attracted the attention of the Basel Committee on banking supervision, and also set off an upsurge of academic research on operational risk. In recent years, some domestic banking institutions, especially small and medium-sized financial institutions, have frequent major and important cases due to imperfect corporate governance, imperfect internal control

W. Fu and L. Yun (Eds.): ADHIP 2022, LNICST 468, pp. 564–577, 2023.
https://doi.org/10.1007/978-3-031-28787-9_42

system, or lack of effective supervision over the implementation of the system [3]. With the deepening of financial globalization, the rapid development of banking business and the continuous progress of information technology, the operational risks faced by the banking industry are more complex.

Due to institutional reasons, there is still a big gap between China's commercial banks and foreign commercial banks in terms of operational risk management. In particular, there are many deficiencies in China's financial institutions in terms of property right system, corporate structure and governance, which leads to the widespread existence of various operational risks in China's banking industry, especially small and medium-sized banks, and presents an increasingly serious trend and signs. Relevant systems and regulations have been issued for the management of operational risk of domestic commercial banks [4]. The prevention of operational risks has been standardized in 13 aspects, such as the construction of rules and regulations to prevent operational risks and the strengthening of audit construction. The establishment of operational risk early warning and monitoring system is the need for commercial banks to meet the needs of domestic supervision. The CBRC clearly requires commercial banks to establish an operational risk management system suitable for the nature, scale and complexity of the bank's business in accordance with the guidelines on operational risk management of commercial banks, so as to effectively manage the bank's operational risk.

Relevant scholars have studied this and made some progress. For example, reference [5] proposes a financial software operational risk prediction method based on nonlinear integration and deep learning. It obtains the types of financial software operational risk through big data mining, classifies operational risk attributes through AHP, trains operational risk data according to the deep learning method, solves financial software operational risk through nonlinear functions, and realizes financial software operational risk monitoring, This method can improve the monitoring time, but the monitoring accuracy is poor.

In view of the above problems, this paper proposes an operational risk monitoring method of smart finance system based on deep learning and improved RPA, which effectively improves the accuracy of operational risk monitoring.

2 Design of Operational Risk Monitoring Method of Intelligent Financial System Based on Deep Learning and Improved RPA

2.1 Calculation of Financial Monitoring Index

Financial risk monitoring index aims to reflect the financial risk status of enterprises. In order to achieve the goal of risk monitoring, this paper follows four basic principles in the setting of indicators: operability principle, comprehensiveness principle, continuity principle and comparability principle. The principle of operability emphasizes that the compilation of the index lies in application, the selected indicators must have reliable data sources and accurate quantitative methods, and the number of indicators should not be too large. The principle of comprehensiveness is that the financial risk of an enterprise is a multi-dimensional and multi-level complex system, which covers many aspects of profitability, development ability, solvency and operation ability. It is necessary to establish a comprehensive and systematic index system for evaluation [6]. The principle of continuity focuses on the essence of the evaluation content. The constructed index system should meet the dynamic monitoring function, so the selected indexes should be continuously available. Comparability principle through the differences of industry, scale, operation mode and other factors, different enterprise indicators are often not comparable in absolute numbers. Therefore, relative number indicators are used to weaken this influence, so as to ensure the same caliber of indicators of different enterprises horizontally and the same calculation method of different years vertically. The financial risk monitoring index in this paper consists of two parts: the individual financial risk monitoring index and the comprehensive financial risk monitoring index (R).

The individual index of financial risk monitoring is a dynamic relative number compared with the "early warning critical value", which measures that the actual financial index of an enterprise exceeds or fails to reach the "early warning critical value" (i.e. the safety margin or danger margin of a financial index). Its calculation formula is:

$$\mu_k = \frac{T_f - T_h}{|\delta_m|} \tag{1}$$

where, μ_k represents the individual index of financial risk monitoring; T_f actual index value of financial monitoring in the current period; T_h represents financial early warning index; δ_m represents the critical value of financial early warning. The calculation formula of the comprehensive index of financial risk monitoring is:

$$\mu_p = \frac{\delta_m - T_f}{T_h} \tag{2}$$

where, μ_p represents the comprehensive index of financial risk monitoring. If the financial monitoring index is the bigger the better index, the financial monitoring index is the smaller the better index or moderate index. Among them, the actual index value of financial monitoring in this period is various financial index data objectively existing in the financial report, and the critical value of financial early warning is the financial data obtained by stripping and calculating the financial data to judge whether there is an alarm. If the calculation result is positive, the result is the financial risk monitoring safety index; If the calculation result is negative, the monitoring and early warning indicators of financial risk can be obtained. Combined with the closed model in the dynamic monitoring factor, two heterogeneity measurement indicators with similarity can be obtained:

$$\mu_k = \frac{w_f \times \mu_f}{w_s \times \mu_s} + \left(w_f \times w_s\right) \tag{3}$$

where, μ_k represents the heterogeneity measurement index obtained after the combination of two similar monitoring data; w_f and μ_f represent the weight distribution and weight index of data respectively; w_s and μ_s represent the monitoring closure weight and closure index. There is an object feature with adjacent relationship in the merged area of the object, and the image structure of each area can be directly operated in the merging process [6]. The region adjacency graph needs to set the merging arc segment in the initial segmentation layer, and after obtaining the initial label, convert the energy at the cost of formula (4).

$$Q_f = \frac{\sum_{r=1}^{n} P_r}{\sum_{i=1}^{n} W_i + P_i} \tag{4}$$

where, Q_f represents the cost conversion energy of regional adjacent data; P_r represents the cost from the initial object to the final object; W_i represents the merging cost function of the divided region in the initial label; P_i represents the merging cost function of the divided region in the termination label. There are a large number of dynamic structures and feedback mechanisms in several existing risk monitoring systems, which determine the operation direction of the system as the final behavior. The system is influenced by internal and external forces and constraints, and develops and evolves according to a certain law. When delimiting the system boundary, we should try to make the boundary include all the quantities closely related to the modeling purpose, so as to ensure the overall integrity of the system, and pay attention to the closure of the system boundary. The operational risk monitoring system of medium-sized commercial banks is a huge complex system composed of multiple internal subsystems and external influence systems. Therefore, it is impossible to cover all the subsystems involved in the operational risk monitoring system of commercial banks and their related influencing factors and feedback paths in this paper. When establishing the system dynamics simulation model, this paper strives to not only complete the system modeling in a limited time, but also fully reflect the essence of the operational risk management of medium-sized commercial banks [7]. Therefore, this paper finally decided to complete two tasks: first, build the operational risk monitoring model of China's medium-sized commercial banks from a macro perspective, and draw the causal feedback diagram of its first-level sub model [8]. Second, select the most typical primary subsystem in the operational risk monitoring system as an example for model construction and simulation test, and get the most common problems of operational risk of medium-sized commercial banks by analyzing the current situation.

2.2 Parameter Based Training Algorithm

To use the deep learning algorithm, it is necessary to first establish a CNN network, in which each level requires more than 32 layers of network functions. If the neuron between the explicit layer and the hidden layer is a feature detection device, it can be represented as the structure diagram shown in Fig. 1:

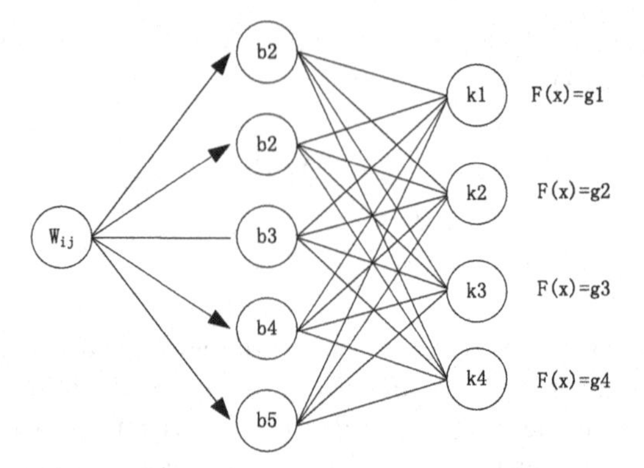

Fig. 1. Deep learning network model

As shown in Fig. 1, the energy function p is introduced, the hidden element state of $k_1 - k_4$ is taken as layer gn, and the explicit element state of $b_1 - b_5$ is taken as layer f(x). Different levels of weight values need to be connected between hidden elements and explicit elements. Then in the probability distribution function, there is the distribution law of energy function:

$$P(x) = \frac{\sum_{j=1}^{n} P(b_j, k_j)}{\sum_{i=1}^{n} g_n} \tag{5}$$

where, $P(x)$ represents the conditional probability distribution value between hidden layer and explicit layer; $P(b_j, k_j)$ represents the joint probability distribution function of the inner explicit layer and the hidden layer of the energy function, where b_j represents the explicit layer and k_j represents the hidden layer; g_n represents the probability distribution function of explicit and implicit elements. By deriving the two-dimensional random unit in the state variable into an activated function, the reconstruction coefficients of cycle times and errors can be obtained, and the training parameters of the financial system operational risk monitoring model can be obtained.

2.3 Financial Risk Fitness Measurement Based on Improved RPA

Guided by the target requirements, it is used to evaluate whether the characteristics of each business process are suitable for the use of RPA financial robot according to the following six aspects:

(1) Business volume. In the actual financial work, there are many businesses with large workload and easy to make mistakes. The application of RPA financial robot in these businesses to replace accountants can not only save human resources, but also improve the accuracy of work.
(2) Repeatability of work content. RPA technology is characterized by clear running scripts with rules. The more repetitive work in financial work means the higher

consistency of rules, the more suitable it is for the application of RPA financial robot.

(3) Degree of digitization. At present, RPA financial robot does not have the human thinking mode, but the embodiment of primary artificial intelligence. Therefore, the subsequent financial work can be carried out only by scanning documents through image technology and extracting useful digital information. Therefore, the process should be digital.

(4) Log in to the external information system. Because in the actual business processing process, it is troublesome to log in to multiple enterprise external information systems respectively, and it is necessary to manually download the relevant data information to be obtained, and then import it into the enterprise internal information system. This series of data transmission and interaction is troublesome, which is not conducive to the development of daily work. At this time, RPA financial robot can automatically log in to heterogeneous systems to complete the integration of underlying data. Therefore, businesses that need to log in to different information systems are more suitable for the application of RPA financial robot.

(5) Risk level and strategy formulation. Many work of financial management has a high degree of risk, and the whole process needs to be monitored or even completed manually. Similarly, some processes need to formulate the long-term development strategy of the enterprise. These two types of business processes that need to rely on the thinking and work experience of accountants are not suitable for the application of RPA financial robot.

(6) System upgrade. Because RPA financial robot belongs to a plug-in deployment and does not change the structure of the original system, it depends on the system operation. Once the system upgrade changes, the operation of RPA financial robot may be wrong and the operation script needs to be modified accordingly. Therefore, the process of preparing the system upgrade does not apply RPAS financial robot for the time being [9]. In different processes, RPA fitness model can be established, as shown in Fig. 2.

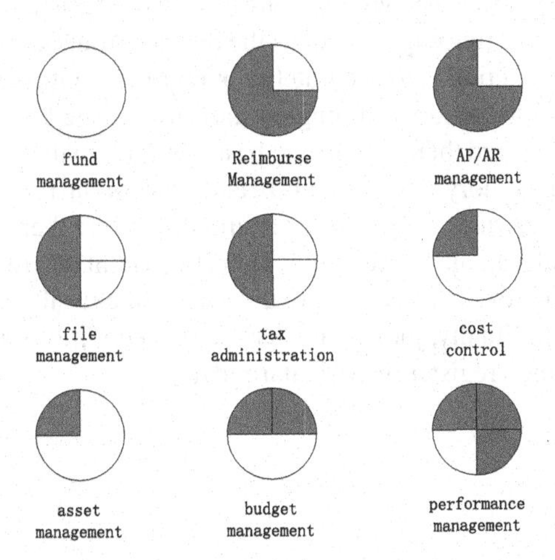

Fig. 2. Model for improving RPA fitness

As shown in Fig. 2, the RPA adaptability of each business process of the enterprise has a certain degree. The RPA adaptability of capital management business process is relatively high, followed by reimbursement management, collection management, payment management, file management, tax management and general ledger management, while the RPA adaptability of cost management, asset management, budget management, performance management and compliance management is relatively low. When determining the RPA suitability of business processes, enterprises should flexibly apply and analyze it in combination with the actual situation of their own business processes.

2.4 Design the Operational Risk Monitoring Algorithm of Financial System

For the financial risk monitoring of enterprises, we should not only estimate the point of the risk prediction value of the financial risk, but also estimate the range of the risk prediction value, and determine that this range includes the credibility of the financial risk, that is, interval estimation of the financial risk prediction of enterprises.

The first step for banks to carry out comprehensive financial risk control and monitoring is to clearly define financial risks. In the work, the definition of financial risk is mainly based on the relevant systems and methods of financial risk control formulated by the head office. However, these systems and methods are formulated based on the level of the head office. The definition of financial risk is very broad. There are many financial risk indicators in the reference library and there is no clear risk threshold. In many financial systems, there is no detailed financial risk, no reference database of financial risk indicators in line with their own conditions, and no differentiated quantitative indicators and thresholds are formulated according to policy business and non policy business, resulting in the lack of a clear grasp of financial risk by the personnel of the centralized Department of risk control. Moreover, most financial departments do not correctly understand their own nature and policy risks, which will directly lead to the difficulty of enterprises to bear the debts caused by poor management [10]. Therefore, the operational risk monitoring algorithm of financial system is designed, as shown in Fig. 3.

As shown in Fig. 3, after receiving the business data, the master program starts the rule engine interface. The rule engine obtains the business transaction code by analyzing the business data. Its function is to accurately obtain the parent rule information of this business from the shared memory. The engine starts to analyze and filter the parent rule, get n sub rule keywords, and then provide the basis for finding the sub rule configuration information of shared memory. Start to analyze and process sub rules one by one, convert the data according to its rules to obtain the result string, and then obtain the operation result through the result calculation function. Then, the parent rule result string composed of N sub rule operation results is returned to the parent rule result operation function for final result calculation. Finally, judge whether the risk conditions are met according to the final results, and then register the risk alarm data.

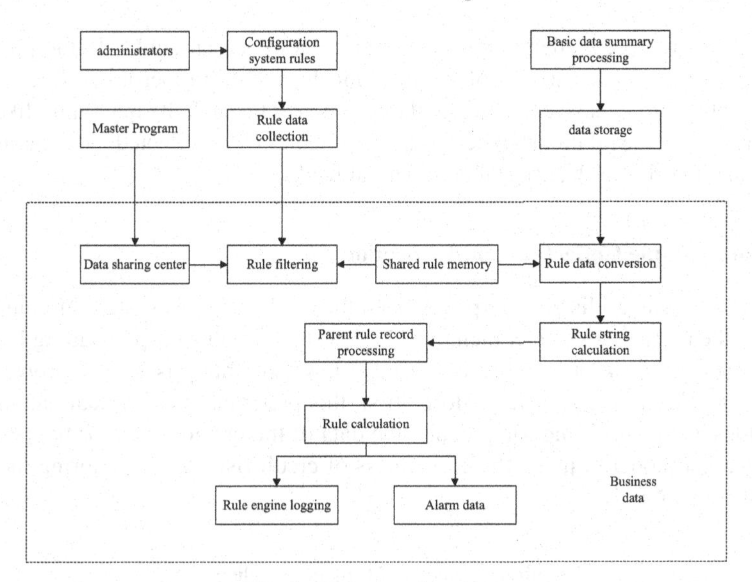

Fig. 3. Rule data flow chart

3 Case Analysis

3.1 Model Adaptability Test

So far, excellent scholars at home and abroad have shown through a large number of empirical analysis that the logstic model has the advantages of high discrimination ability, wide application range, simple results and easy operation in the credit risk monitoring of commercial banks. Although the amount of data processing and calculation required by the model is large and the program is complex, we can still overcome this problem with the help of statistical software and make the analysis results more accurate. And the model has been applied in dealing with the credit risk of China's commercial banks for many years, which is rooted in China's financial industry. As a nonlinear probability regression model, we can set up different classification basis through default samples and non default samples, and analyze the default probability of enterprises on the basis of linear regression equation.

$$P_m = \frac{1}{1 + \exp(-(a_i + b_i x))} \tag{6}$$

where, P_m represents the default probability of the enterprise based on the linear regression equation; a_i and b_i are parameters of independent variable function; x represents the independent variable in the function. Suppose the probability of financial crisis of an enterprise, that is, the probability of credit risk of commercial banks is P(x), then the probability of non occurrence is 1 – P(x). we can get the possibility of enterprise default by calculating the ratio relationship between the two probabilities, and we can estimate the size of credit risk faced by commercial banks.

$$\ln \frac{P_k(x)}{1 - P_k(x)} = a_i + \sum_{i=1}^{n} b_i x \tag{7}$$

where, $P_k(x)$ represents the credit risk faced by commercial banks. In the process of adaptability test, make full use of the relevant theoretical properties of linear regression model, comprehensively estimate the parameters through the maximum likelihood method, explain the relationship between the variables of the selected index parameters, and predict the risk probability of commercial banks.

3.2 Data Sample Collection and Inspection

This paper selects 20 listed enterprises from the credit analysis system of commercial banks in the research example, including 11 enterprises with non-performing loan ratio higher than 8% (risk probability is 1) and 9 low-risk enterprises (risk probability is 0). Combined with the database information, this paper takes the annual data of these companies as the modeling sample, and the data of the previous year as the prediction and detection model to judge the correctness of credit risk. The monitoring indicators are shown in Table 1.

Table 1. Selection of financial indicators

Type	Purpose and principle of selection	Specific indicators
Corporate profitability	Corporate profitability is one of the most important indicators of a company and the best embodiment of the company's development potential. Accumulating profits to increase the value of the company's assets is a way to reflect the healthy financial situation and operating effect of listed companies	Return on net assets
		Profit rate of total assets
		Cost profit margin
		Net profit margin on sales
Solvency of the company	The company's solvency is an indicator reflecting the healthy state of enterprise capital flow. Being able to repay the loan in time and effectively shows that the company is in good operation. At the same time, it will also strengthen the company's credit system to facilitate fund-raising during industrial expansion	Current ratio of net assets
		Cash flow ratio
		Interest cover
		Property right ratio

(*continued*)

Table 1. (*continued*)

Type	Purpose and principle of selection	Specific indicators
Asset management capability of the company	Corporate asset management is one of the most important aspects of corporate development. Reasonable and effective management can make the company have long-term development ability. Reasonable management of assets can effectively improve the speed of asset flow and utilization	Turnover rate of accounts receivable
		Total asset turnover
		Inventory turnover
		Turnover rate of current assets
Operation and development capacity of the company	The operation and development ability of a company represents the development prospect of a company. The change of this index can predict the development direction and trend of the company, which is conducive to a more comprehensive understanding of the state of the company	Growth rate of operating revenue
		Net profit growth rate
		Growth rate of total assets

As shown in Table 1, through financial indicators and non-financial indicators as variables, all testing processes are through spss18 Created by 0. The stock code and risk status information of 20 listed enterprises and the scores of non-financial index scale statistics are shown in Table 2.

Table 2. Company risk and non-financial index scale

Code	Abbreviation	Risk status	Score situation
602563	X1	1	545
604156	X2	0	663
603258	X3	1	425
604652	X4	1	822
603654	X5	1	336
602142	X6	0	234
609634	X7	1	512

(*continued*)

Table 2. (*continued*)

Code	Abbreviation	Risk status	Score situation
604512	X8	0	163
607458	X9	1	242
606325	X10	0	112
602312	X11	1	263
603245	X12	0	321
606328	X13	1	541
606632	X14	1	444
601425	X15	1	263
602362	X16	0	135
601141	X17	0	226
602352	X18	1	332
601252	X19	0	622
603274	X20	0	214

The data test includes two steps, first through the K-S sample data test, and then the mean test. Both processes include significance test. K-S sample data test method was first used in foreign commercial banks to test the probability of data in a certain range, that is, normality test. During this test, if the sample data obeys the distribution within the specified range of the test, the smaller the statistical quantity s is, it belongs to the normal distribution. Otherwise, the larger the s value is, the larger the s value is, it indicates that the data index does not obey the distribution within the specified range and belongs to the non normal distribution. The test results are shown in Table 3.

Table 3. K-S test

Indicator name	S value	Significance P value
X1	1.142	0.007
X2	1.635	0.008
X3	2.415	0.000
X4	1.336	0.006
X5	2.254	0.000
X6	1.156	0.004
X7	2.324	0.000

(*continued*)

Table 3. (*continued*)

Indicator name	S value	Significance P value
X8	1.526	0.006
X9	3.333	0.000
X10	1.748	0.008
X11	1.569	0.005
X12	2.254	0.000
X13	2.145	0.000
X14	2.263	0.000
X15	1.321	0.006
X16	1.123	0.005
X17	2.425	0.000
X18	2.415	0.000
X19	1.635	0.004
X20	1.415	0.006

As shown in Table 3, the larger the statistic s, the lower the significance p value. When the s value is greater than 2.0, the p value is 0.000, indicating that the selected sample data does not conform to the normal distribution. Based on the above research results, it can be found that the indicators of 20 listed enterprises generally do not obey the normal distribution. Through the comparative analysis of literature in recent years, this result is effective. The purpose of mean test on sample data is to distinguish the index distribution of high-risk enterprises and low-risk enterprises. If the sample data follows the normal distribution, the mean test can be carried out through the t-test method commonly used in the parameter test method; If the sample data does not obey the positive distribution, the mean value of the index can be tested by nonparametric test method to obtain higher test accuracy. However, considering that these indicators may have multicollinearity in the test process, the amount of calculation will be increased in the process of model construction, resulting in reduced accuracy and complexity of the model, so on this basis, the indicators are further screened by principal component analysis to ensure that there is no influence of multicollinearity and build the model accurately. At this time, the comprehensive model of credit risk monitoring of commercial banks can be expressed as:

$$P_d = \frac{P_k(x)}{\sum_{i=1}^{n} F_x} \tag{8}$$

where, P_d represents the identification accuracy of risk monitoring; F_x represents the performance proportion of enterprise asset management. Through the above model, the prediction and judgment of data indicators are established, and the corresponding monitoring and judgment results are obtained.

3.3 Risk Monitoring Capability Test

The establishment of the model does not only depend on the application of the sample data, but also the accuracy of the model is the best embodiment of the prediction ability of the model. The accuracy can ensure the correctness of the application in the process of credit risk monitoring. The sample data of 43 listed enterprises in 2016 is taken as the verification sample. The nonlinear integrated deep learning method, artificial intelligence method and the method in this paper are used to improve the accuracy of financial system operational risk monitoring. The results are shown in Table 4.

Table 4. Operational risk monitoring accuracy of financial system

Sample data volume/GB	Operational risk monitoring accuracy of financial system/%		
	Methods in this paper	Artificial intelligence method	Nonlinear integrated deep learning method
100	99.6	82.5	69.2
200	98.5	80.2	72.8
300	99.0	76.9	75.3
400	92.9	78.3	78.1
500	96.3	79.3	82.6
600	95.1	80.0	85.0

From the analysis of Table 4, when the financial sample data is 100\,GB, the accuracy of the financial system operational risk monitoring of the method in this paper is 99.6%, the accuracy of the artificial intelligence method is 82.5%, and the accuracy of the non-linear integrated deep learning method is 69.2%; When the amount of financial sample data is 500\,GB, the accuracy of financial system operational risk monitoring of the method in this paper is 96.3%, the accuracy of financial system operational risk monitoring of the artificial intelligence method is 79.3%, and the accuracy of financial system operational risk monitoring of the nonlinear integrated deep learning method is 82.6%; The risk monitoring accuracy of this method is always high, which indicates that the operational risk monitoring effect of this method is good.

4 Concluding Remarks

Aiming at the risks existing in the operation of smart financial system, taking 20 private enterprises as analysis samples, this paper establishes an operation risk monitoring method of smart financial system based on deep learning and improved RPA. The system adopts a non coupling technology to obtain and analyze the transaction data in the operation of accounting business without affecting the normal operation of various monitored business systems. It is a risk management system for the risk supervisors of grass-roots legal entities of associated press and the personnel of business management department of provincial associated press. It will not affect and embed into any system, nor will it affect the change of computer room, network and other infrastructure supporting the operation of the system and its operation security. Based on the RPA system process monitoring log, user information, financial work scene and other relevant data, the system studies the user portrait construction technology of the financial system to find the system problem users in time. In the integration of financial system operational risk loss event database and RPA system process monitoring log, the system operational risk discovery model based on in-depth learning is studied to realize the accurate identification of loss events. It can also form user portraits based on user static data information, build event maps based on dynamic RPA process operation data, establish a risk prevention and control model based on event transmission chain analysis, and support multiple intelligent risk prevention and control analysis scenarios such as illegal clue analysis and risk transmission analysis.

References

1. Sun, X., Zhang, J., Feng, J.: Hospital financial risk monitoring and early warning method based on monitoring information. Autom. Technol. Appl. **41**(04), 132–135+138 (2022)
2. Yin, L.: Empirical analysis of financial risk prediction of listed companies based on discriminant analysis. Mod. Market. (Xueyuan Edn.) **19**(05), 172–173 (2021)
3. Gao, S.: Research on the application of financial risk early warning of SMEs under the fuzzy analytic hierarchy process. China Bus. Rev. **26**(11), 46–47 (2020)
4. Adewole, A.E., Larry, A.O.: Assessment of financial risk and its impact on an informal finance institutions profitability. Can. Soc. Sci. **18**(1), 133–139 (2022)
5. You, S., Liu, X.: Software module risk prediction based on nonlinear integrated deep learning. Comput. Simul. **38**(11), 305–308 + 318 (2021)
6. Osipov, V.S., et al.: Ecologically responsible entrepreneurship and its contribution to the green economy's sustainable development: financial risk management prospects. Risks **10**(2), 44 (2022)
7. Chen, D.: On the establishment of hospital financial risk monitoring and early warning management index system. Financ. Circles (02), 122–124 (2022)
8. Yang, S., Wu, H.: The global organizational behavior analysis for financial risk management utilizing artificial intelligence. J. Glob. Inf. Manag. (JGIM) **30**(7), 1–24 (2021)
9. Popkova, E.G., Sergi, B.S.: Dataset modelling of the financial risk management of social entrepreneurship in emerging economies. Risks **9**(12), 211 (2021)
10. Qi, Q.: Study on financial risk prediction of enterprises based on logistic regression. J. Comput. Methods Sci. Eng. **21**(5), 1255–1261 (2021)

Mobile Terminal-Oriented Real-Time Monitoring Method for Athletes' Special Training Load

Hui Xu[1] and Qiang Zhang[2](✉)

[1] Anhui Medical College, Hefei 230601, China
xuhui@ahyz.edu.cn
[2] Fuyang Normal University, Fuyang 236037, China
zqxh16584@163.com

Abstract. Aiming at the problem of inaccurate training load monitoring results due to the large individual differences of athletes' special training, a real-time monitoring method for athletes' special training load for mobile terminals is proposed. Use sensors as data collection devices to establish sports scenarios for mobile terminals, and use smart wearable devices to collect real-time data generated by special sports measured by sensors embedded in watches. Select the index reflecting the intensity of training load to find the regularity of athletes' training growth. By recording the time ratios of different heart rate intervals for each exercise, the distribution and variation of the load intensity during the training were analyzed. Capture complete training load data during the movement. Through the real-time monitoring and early warning of special training load, the training status and training load of athletes can be adjusted, which is beneficial to improve the training quality of athletes. The test results show that the proposed method can improve the monitoring accuracy and provide a reference for improving the project training program.

Keywords: Mobile terminal · Athletes · Special training · Load monitoring · Real-time monitoring · Exercise load

1 Introduction

With the continuous progress of science and technology, the special training of athletes in our country has begun to develop and reform in a scientific way. Special training must meet the actual combat requirements, show the distinctive characteristics of sports, and reflect the rich content of special competitive competitions. How to use the existing theoretical support and advanced scientific and technological means to guide training and improve the performance of athletes training, achieve excellent results, and improve the dominant position in the competitive level. "Starting from actual combat" is the fundamental guarantee of scientific training and training, and it is also the basis and premise of the essential characteristics of special sports. With the development of science

W. Fu and L. Yun (Eds.): ADHIP 2022, LNICST 468, pp. 578–591, 2023.
https://doi.org/10.1007/978-3-031-28787-9_43

and technology in the past few years, especially the widespread use of accelerometer, GPS and other technologies, the research on the characteristics of sports activities and events is more scientific and in-depth [1]. Only on this basis can we grasp the intensity and measurement of the training load. Therefore, the training load should be close to the competition load in form and intensity to achieve good results.

A few years ago, as the technology of accelerometer and GPS became more and more mature and accurate, professional clubs in Europe and the United States took the lead in the emergence of wearable real-time monitoring devices, which were mainly used to detect the physical condition of athletes during competition and training, and also detect the physical condition of athletes during competition and training. Some technical indicators of athletes. Although there have been in-depth studies on the field of special training for athletes at home and abroad, on the one hand, the physiological functions of foreign and Chinese athletes are not the same, and their data can only be used as a reference; on the other hand, domestic research is still in qualitative At this stage, there are relatively few detailed quantitative data. Therefore, under the current environment of vigorously developing sports in my country, quantitative data on training load is urgently needed for the reference of coaches. In recent years, a series of professional sports real-time monitoring equipment companies have emerged in China, which have made significant contributions to the popularization of intelligent real-time sports monitoring equipment. And with the improvement and increase of domestic professional events, amateur events, youth events and campus events, it provides a good platform and feasibility for us to use scientific equipment to study the load characteristics of football events.

With the continuous development and deepening of modern sports, special sports have put forward higher requirements for the accuracy of scientific training. Therefore, scientific quantitative data of exercise load is needed to reflect the physical state of athletes during training and improve the physical quality of long-distance mobilization. This paper proposes a real-time monitoring method of athletes' special training load for mobile terminals. Based on the analysis of the collected data, this paper explores the characteristics and laws of the external load of data. According to the competition load of high-level athletes, the scheme of training to improve the level of athletes is designed.

2 Mobile Terminal-Oriented Real-Time Monitoring Method for Athletes' Special Training Load

2.1 Establish a Mobile Terminal-Oriented Sports Scene

Using sensors as data acquisition devices, sports scenes are built on the basis of sensors as perception devices, and intelligent mobile terminal devices as "bridges". Sensors such as temperature, pulse and blood pressure are integrated on the wearable device to monitor the physiological characteristics of the user in real time. A device that provides data is called a GATT server, and a device that accesses the GATT server and obtains data is called a GATT client. A device can act as both a server and a client. Because of the concise definition of the GATT specification, the amount of code required to implement the function is less, which effectively reduces the non-data information generated by the

Bluetooth device communication, which is conducive to reducing power consumption. In athlete-specific training, wear a wearable device on their wrist, finger, or clothing, while keeping a smartphone on their arm or in a clothing pocket. Users can pre-set an exercise goal on the phone. In the process of exercising, users can view the completion status and the state of the body in real time through the mobile phone [2]. The host layer includes general access specification, general attribute specification, attribute protocol, security manager, logical link control and adaptation layer protocol. The general access specification is the bottom layer that the application layer can directly access the Bluetooth protocol stack. The general access specification defines all the basic functions of Bluetooth, such as transmission, protocol and access process. It also includes Bluetooth device broadcasting, discovery, connection, association model, Security authentication and service discovery, etc. Use smart wearables to collect real-time sports-specific data measured by sensors embedded in the watch. In this process, since we need to use the data of each sensor at the same time for the next modules, we need to align the data of each sensor that collects data in time.

During the whole process of exercise, the wearable device collects various physiological characteristic parameters of the user, and sends the data to the mobile phone through Bluetooth transmission technology for processing, analysis and presentation of the results to the user. The general attribute specification is the layer where the real data transmission is located. It includes basic operations such as a data transmission and storage framework, and defines two types of roles: server and client. At the same time, these data can be sent to the remote server through the mobile phone for storage and backup, which can be used for users to view the staged results of the exercise and for further analysis. The interaction process between the wearable device, the smart mobile device terminal and the server is shown in Fig. 1.

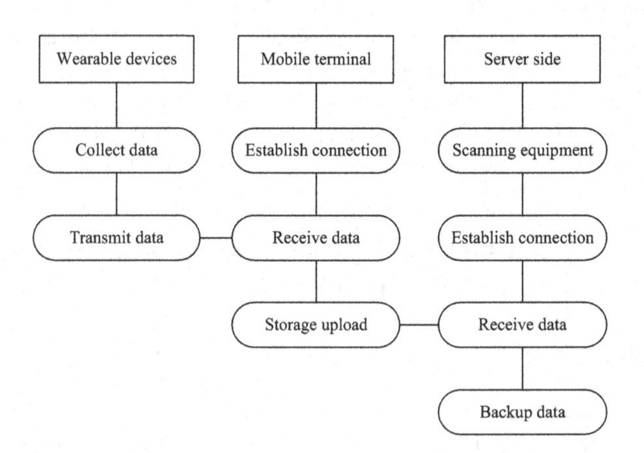

Fig. 1. Interactive process

Use wireless transmission technology as a means of data transmission. Due to the portable and mobile characteristics of wearable devices, the transmission environment is determined to be dynamic, and the use of wired transmission has great limitations. Mobile terminal devices such as smart phones are used to provide users with personal services such as real-time data analysis, display and reminders. When collecting, it is necessary to ignore and align the small time error between the sensor data. Or the sampling rate of each sensor is not specified, but the data of all sensors are collected when the data of a certain sensor changes. The attribute protocol is an end-to-end communication protocol specially used to deal with small data packets, which connects the attribute client and the attribute server. The principle is that the ATT server and the ATT client transmit data through a fixed L2CAP channel. The security manager assigns keys to paired devices. At present, the intelligent mobile terminal device already has a strong processing capability, which can perform preliminary processing and analysis on the data. At the same time, almost all mobile phones are equipped with a Bluetooth module and a wireless network card, which can realize the functions of data reception and upload, as a bridge between the wearable device and the server, and play the role of relay.

2.2 Select the Index Reflecting the Training Load Intensity

Load is the central part of sports training, and it is also the most core factor. The ultimate goal of sports training is to stimulate the body through appropriate load, thereby producing adaptive changes, that is, the process of "adaptation-improvement", so as to continuously develop special athletic ability. Exercise load applies training stimulation by means of physical exercise, so that the body produces functional changes physiologically with a certain load stimulation, which is reflected in the movement rhythm, number of exercises, and degree of exertion in sports training, sports competition, physical education and fitness exercise, range of motion and other aspects of the changes and regulation [3]. Therefore, it can be said that in the process of sports training, the control of exercise load is the most important criterion for evaluating scientific training. The load and intensity together constitute the main part of the exercise load. The two influence each other and are interdependent and inseparable. In the exercise, they exert influence on the difference in the performance of the two. Although the current domestic and foreign reference does not have a completely unified definition standard for the definition of exercise load, the basic point of view is the same: exercise load is a stimulus to the body; the body has a physiological stress response to this stimulus. The load is the basis of the load intensity. Increasing the load intensity within a certain range of the load can effectively stimulate the body level [4]. The selection of athletes' competition load indicators needs to take into account the reliability, relevance, scientificity and rationality of the research, and also needs to refer to the simplicity and operability of the collection, as well as the non-invasive and sustainable principles for athletes. The indicators selected in this paper to reflect the training load intensity are shown in Fig. 2.

The game load data in this study were obtained by real-time monitoring during the game. The equipment worn by the test athletes included a GPS activity acquisition unit and a heart rate acquisition unit to monitor and measure the relevant data of the athlete's game load in real time, using wireless transmission. Modules are sent to researcher and coach terminals. Heart rate is a reliable indicator of the state of human functioning and is easy to measure. In special sports, different intensity of exercise occurs intermittently, and heart rate is more sensitive to changes in intensity, which can more accurately reflect its real-time load characteristics. Moreover, the operation of collecting heart rate with the help of the instrument is relatively simple and convenient. Therefore, it can be said

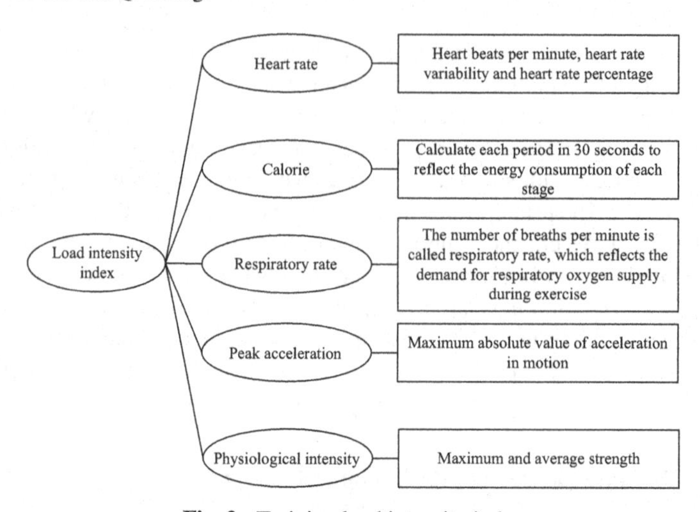

Fig. 2. Training load intensity index

that heart rate is a simple and effective physiological indicator for quantifying special exercise load. At present, there are two main methods for evaluating exercise intensity, and they are both evaluated by methods related to heart rate. The first is to evaluate the exercise time and proportion of athletes in different heart rate ranges. The second is to use the maximum heart rate and average heart rate of different athletes to measure and evaluate sports goals [5]. For this study, we need to reflect the average load intensity and maximum load intensity of athletes through a large number of personal situations, so as to reflect the characteristics of the overall load intensity, so the second method is selected to evaluate with less error. The formula for calculating the average exercise intensity of special training is as follows:

$$A_1 = \frac{B_1}{B_{\max}} \times 100\% \tag{1}$$

In formula (1), A_1 is the average exercise intensity of special training; B_1 is the average heart rate of special training; $B_{\max}$ is the maximum heart rate. The formula for calculating the maximum exercise intensity of special training is as follows:

$$A_2 = \frac{B_2}{B_{\max}} \times 100\% \tag{2}$$

In formula (2), A_2 represents the highest exercise intensity of special training; B_2 represents the highest heart rate of special training. Exercise load includes load intensity and load amount. The load and strength constitute the whole of the load, and the two depend on and influence each other. The strength of any load is based on a certain amount, and any load is based on a certain strength. Changes on the one hand inevitably lead to changes on the other. However, only in the increase of the load, the response of the body is relatively stable; the adaptation process affecting the individual is mainly determined by the size and nature of the load intensity. Different combinations present different load structures and the response to the body is also different [6]. Therefore, the factors that can affect the load intensity must also affect the load amount, and vice versa. The two are interrelated and inseparable unified wholes. In sports training, the so-called arrangement of the load is the arrangement of the amount and intensity of the load. The important guarantee of scientific training is also based on the correct relationship between scientific arrangement of load and load intensity.

2.3 Distribution Characteristics of Special Training Load Intensity

It is self-evident that the method of digital monitoring training plays an important role in improving the effect of training. The use of scientific monitoring methods can reflect the different effects of different training methods; explore suitable training methods and recovery time for athletes; improve the training quality of athletes; clarify the actual level of athletes. The principle of periodic arrangement is one of the basic principles of sports training, which refers to the periodic arrangement of training load and training content in training. As the most important basic unit in the training plan, weekly training also needs to follow the principle of periodicity. Predict the highest level an athlete can reach and avoid undertraining or overtraining during training. Monitoring training is to monitor and control the athletes' physiological indicators, competitive performance, training status, daily diet, etc. within a period of time, so as to adjust the sports training program. Weekly training load and load intensity statistics in this study were monitored by randomly sampling one week of training from the entire training session. Weekly training load monitoring mainly monitors and records the training load value of each training session in a week. By calculating the load of each session, the total training load for a week can be calculated. In the process of sports training, it is mainly to ensure the rationality of training content, training intensity, duration and training density in order to achieve the ideal training effect. The monitoring training can provide a basis for the formulation and adjustment of training programs. In the process of sports training, monitoring training has been paid more and more attention. At the same time, by recording the time ratio of different heart rate intervals in each class to analyze the distribution and variation of the load intensity during training [7]. The content of special training needs to be more comprehensive, and various training contents are scientifically combined.

Since the specific content of monitoring training is relatively complex and the information is generated quickly, it is necessary to grasp it in time and carry out centralized analysis. Through the use of big data, data mining and other technical means, more valuable monitoring data can be obtained. By arranging different training contents, the athletes' physical fitness, technique and coordination ability can be continuously improved, so as to achieve the purpose of training. Based on this feature, in the process of monitoring and training, information technology has been widely used. Monitoring the development of training trends has a profound impact on the habits of coaches when formulating training programs, gradually transitioning from the initial method of personal memory, experience and subjective judgment to the current method of objective analysis based on science and technology.

Monitor training and non-training data collected daily to analyze all factors that determine success or failure. The training load of the week showed a wave-shaped trend, and the exercise load value on Friday afternoon was the highest in the week, and the load on Wednesday morning, Friday morning and Saturday afternoon was significantly lower than other training sessions. According to the one-week training plan provided by the coaches, a total of 1 high-intensity class, 5 medium-intensity classes, 1 medium-intensity class and 3 low-intensity classes are scheduled this week. Saturday morning is a high-intensity class for the actual combat training within the team. Identify the parameters with the most predictive value, integrate performance results and personal statistics along with coaching and expertise to create predictive models. Create interactive tabular content that provides daily training regimens that adapt to changing training and environmental conditions. The main content of the medium-intensity training class is personal technical training and small-scale tactical explanation and application. The low-intensity training class is usually arranged in the morning after the strength training, focusing on personal skills and positional skills. From the perspective of training time, it adopts a mode of one half day and two full days, 2 cycles, and 1 day of rest. The system includes a diagram of the entire solution, a database module, an analysis and classification module, a predictive model module and an end-user solution module. Half-day single training time is more than 100 min. Overall training time is longer. On Monday, because there is only a half-day training schedule, the training time is the shortest. Combined with the training load, there is a characteristic that the longer the training time, the greater the total training load. Using cloud storage, through long-term, multi-cycle and orderly data integration and analysis, a proprietary database is formed, combined with different athletes, different sports, and different training cycles, to find the training growth pattern of athletes, and establish an index growth change model.

2.4 Real-Time Monitoring Model of Special Training Load of Athletes

The unified sorting and analysis of the data and the comprehensive and systematic monitoring of training will help to standardize the evaluation of athletes' training results on the one hand, and help coaches to better arrange training plans on the other hand. Due to time, manpower, material resources and other reasons, the monitoring test is mainly carried out for the athletes' load-recovery system, and the shortcomings are found through practice and experience is summarized. Through the statistics of the training and competition data of the players, it is found that different players in the same position have obvious individual differences in special training and competition. In this paper, we mainly discuss the specific construction and application process of the real-time load monitoring model, give feedback on personal test results, give adjustment suggestions, and find and solve problems existing in training. The real-time monitoring model of athletes' special training load is shown in Fig. 3.

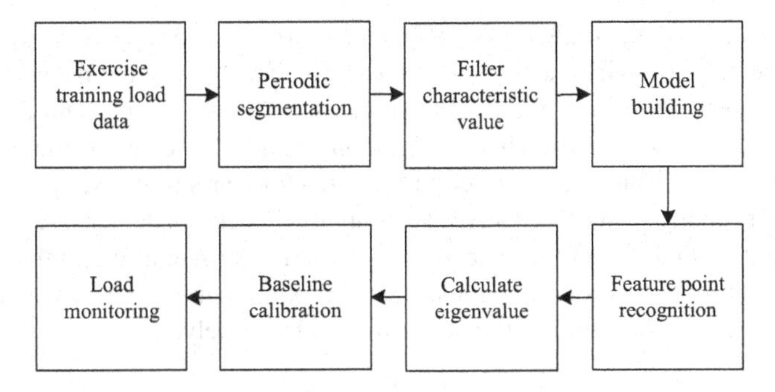

Fig. 3. Athlete special training load real-time monitoring model

The load monitoring steps shown in Fig. 3 are as follows:
Step 1: collect exercise training load data.
Step 2: divide cycle training plan.
Step 3: obtain the characteristic value of the filter.
Step 4: establish a load monitoring model.
Step 5: feature point recognition.
Step 6: calculate genetic value.
Step 7: baseline calibration.
Step 8: load monitoring.

In the real-time monitoring of sports-specific training load, wearable devices are used to collect training load data. In the morning, after the test is completed, part of the test results are fed back to the coach for reference and decide whether to adjust the training plan. Ask the athletes again in the evening to see how the amount of training that day has had on the athletes. A summary is made once a week, and the training status of the athlete for one week is fed back to the coach as a reference to formulate the training plan for the next week. If the regulation is improper or not timely, it may result in a decline in the training status and competitive level of athletes. Due to the limited computing power of the wearable device, it transmits the recorded signal to the smart mobile terminal via Bluetooth for further processing. The mobile terminal minimizes the interference of ambient noise by applying a bandpass filter to the raw data. Perform wavelet time-frequency analysis on the collected signal, and put the received signal $w(m)$ of length m into the wavelet transform with a series of filters for analysis, namely:

$$w(m) = \sum_{i=2}^{n-1} w(i-1)(2m-n)\varphi(n) \tag{3}$$

In formula (3), i represents the data serial number; n represents the total number of data; $\varphi(n)$ represents the output of the low-pass filter and the high-pass filter. Pass the signal through a bandpass filter to remove low and high frequency noise. Athletes need to break the original balance state of the body through different load stimulation, so that the body produces a certain degree of fatigue, and through excessive recovery, the body can transform to a higher level of function, and then continue to increase the load stimulation, so that the body breaks the balance again. A deeper level of fatigue is created, followed by a higher state of balance through over-recovery, over and over again, leading to an ever-increasing level of performance. Hook the infrared pulse sensor module to the PIO 10 port of the Bluetooth module through a wire, and the Bluetooth module can directly detect the change of the high and low levels of the PIO 10 port. After power on, the Bluetooth infrared pulse collector starts to work, open the APP on the mobile terminal, and start to scan the surrounding Bluetooth devices. A finite impulse response filter is a natural choice because its phase response curve is stable and can be designed as a filter that produces a linear phase response curve. The output signal of the FIR filter is the weighted sum of the original signals, namely:

$$w'(m) = \sum_{i=1}^{n} \delta w(m - i) \tag{4}$$

In formula (4), $w'(m)$ represents the filtered output data signal; δ represents the impulse response value of the finite impulse response filter, that is, the coefficient of the filter. Click the discovered device to enter the pairing and service discovery process. When the collector establishes a connection with the mobile phone, the collector will automatically transmit the heart rate measurement value to the mobile phone. The acquisition data is sliced into sliding-window-sized slices, and the average energy of each window slice is calculated. The calculation formula can be expressed as:

$$\kappa(m) = \frac{1}{\vartheta} \sum_{m-\vartheta+1}^{m} w'(m)^2 \tag{5}$$

In formula (5), $\kappa(m)$ represents the average data energy of each window; ϑ represents the window size. The purpose of doing the above is to capture complete training load data during the movement. Through the real-time monitoring and early warning of special training load, the training status and training load of athletes can be adjusted, which is beneficial to improve the training quality of athletes. Through the load-recovery monitoring, the problems existing in the athletes' psychological stress, injury status, and fatigue recovery were found, and a personalized training plan was established. The distribution of TRIMP and heart rate zones can be used to understand the training load and intensity of athletes during special training and competition. Training is the premise of developing competitive state. With the improvement of the level of sports training, athletes can better cope with the competition by improving their adaptability to different levels of load stimulation. According to the test results, the cumulative effect of load and fatigue, the lack of recovery and the severity of fatigue symptoms can all explain the individual performance and readiness of the athlete. As can be seen from the table below, after a unified analysis and evaluation of the overall training load, training performance, and fatigue recovery of athletes, targeted monitoring status, training arrangements and suggestions are put forward for athletes. So far, the design of the mobile terminal-oriented real-time monitoring method for athletes' special training load is completed.

3 Experimental Studies

3.1 Experiment Preparation

This paper takes the athletes who participated in a football team's training camp as the research object, conducts real-time monitoring of the special training load, and analyzes the monitoring results to verify the effectiveness of the method proposed in this paper. The monitoring objects mainly include the main players participating in the special football training and the main substitute players, a total of 16 players. During the preparation of the athletes, the real-time exercise load is monitored on the training and competition of the players, and the experimental data is obtained. The data is mainly used to quantify the training load intensity of athletes and monitor the training status of athletes on the training ground at any time. Three main data of heart rate, running distance and speed can be collected. The Australian-made GPSports wearable device is used to monitor the entire training process of athletes. The GPSports device consists of a GPS module, a heart rate belt, a bra, a gyroscope and other parts. This study mainly uses the heart rate belt to collect heart rate data, and the heart rate collection frequency is 5 times/S, the data is sent to the Team AMS analysis software for comparison and analysis. Through the analysis software, the heart rate variation curve and the heart rate distribution interval of the athletes during the heart rate recording period can be read. At the same time, the refresh rate of the system and computer software is 1 Hz, which means that the data is collected once per second, and the positioning accuracy of GPS is less than 2.5 m, thus ensuring the validity of the data.

3.2 Results and Analysis

The monitoring accuracy of maximum heart rate, average heart rate, average intensity and average speed is selected as the evaluation index to measure the application effect of the real-time monitoring method for athletes' special training load for mobile terminals proposed in this paper. The accuracy results are compared with the real-time monitoring method of athletes' special training load based on big data technology in reference [1] and reference [2] based on virtual reality technology. Through real-time monitoring, it was found that the maximum heart rate appeared within 2–5 min after the start of the warm-up exercise. The reason for the analysis is that the athlete has just arrived at the training ground, and the body suddenly threw itself into the special preparation activity from a static state. The completion of special training movements of athletes requires a certain amount of muscle power, and a personal heart rate peak will appear when reflected in the maximum heart rate. The comparison of the maximum heart rate monitoring accuracy of each real-time monitoring method is shown in Table 1.

According to the test results in Table 1, the maximum heart rate monitoring accuracy obtained by the mobile terminal-oriented real-time monitoring methd for athletes' special training load is 87.52%, which is 11.10% higher than the real-time monitoring method based on big data technology and virtual reality technology. And 12.99%. The monitoring results of the average heart rate can reflect the athlete's cardiopulmonary function and maximal oxygen uptake capacity. After the training, the average heart rate is increasing, which means that the interval has not been adjusted properly, and after high-intensity

Table 1. Comparison of maximum heart rate monitoring accuracy (%)

Testing frequency	Mobile terminal monitoring method	Big data monitoring methods	Virtual reality technology monitoring method
1	86.24	79.67	78.46
2	89.62	75.38	74.83
3	85.80	74.26	71.63
4	87.56	76.05	72.25
5	88.48	78.54	75.58
6	86.75	75.81	73.17
7	85.19	74.98	70.47
8	89.53	76.62	75.51
9	87.81	75.35	76.05
10	88.25	77.54	77.32

training, the heart rate has not been lowered. The comparison of the average heart rate monitoring accuracy of each real-time monitoring method is shown in Table 2.

Table 2. Comparison of average heart rate monitoring accuracy (%)

Testing frequency	Mobile terminal monitoring method	Big data monitoring methods	Virtual reality technology monitoring method
1	89.40	80.49	79.48
2	91.87	81.68	78.87
3	92.64	82.27	76.64
4	88.28	80.54	79.31
5	89.56	83.88	78.05
6	92.12	82.65	81.53
7	93.65	84.36	82.66
8	91.81	82.23	83.28
9	90.54	83.05	82.94
10	89.27	83.12	84.51

According to the test results in Table 2, the average heart rate monitoring accuracy obtained by the real-time monitoring method of athletes' special training load for mobile terminal proposed in this paper is 90.91%, which is 8.48% and 10.18% higher than the real-time monitoring method based on big data technology and virtual reality technology.

The average intensity can reflect the physical quality of athletes. Between repeated intervals and exercise, the average intensity represents good cardiopulmonary function to adjust breathing and heart rate. Coaches can tap athletes' greater potential from other aspects, such as techniques and tactics, or physical strength and quality. The comparison of average intensity monitoring accuracy of each real-time monitoring method is shown in Table 3.

Table 3. Comparison of average strength monitoring accuracy (%)

Testing frequency	Mobile terminal monitoring method	Big data monitoring methods	Virtual reality technology monitoring method
1	85.44	71.49	72.56
2	86.81	73.87	76.42
3	87.68	72.64	75.06
4	85.52	75.31	72.27
5	86.28	74.55	73.84
6	85.56	72.26	76.91
7	88.24	71.03	72.68
8	89.38	73.15	70.32
9	86.91	75.82	71.53
10	85.80	70.88	74.75

According to the test results in Table 3, the average intensity monitoring accuracy of the real-time monitoring method of athletes' special training load for mobile terminal proposed in this paper is 86.76%, which is 13.66% and 13.13% higher than the real-time monitoring method based on big data technology and virtual reality technology. In addition to good technical and tactical ability and spiritual willpower, excellent physical fitness and special speed ability requiring neuromuscular coordination are also the key parts that high-level athletes must have. Therefore, the monitoring results of average speed are directly related to the effect of special training. The comparison of average speed monitoring accuracy of each real-time monitoring method is shown in Table 4.

According to the test results in Table 4, the average speed monitoring accuracy of the real-time monitoring method of athletes' special training load for mobile terminal proposed in this paper is 86.40%, which is 11.85% and 14.16% higher than the real-time monitoring method based on big data technology and virtual reality technology. Therefore, the real-time monitoring method of athletes' special training load proposed in this paper can accurately reflect the training state of athletes. Combined with the training preparation state, it defines the training load pressure of players, and feeds back to the competent coach in time to obtain certain support and recognition. According to the monitoring results of training load, combined with the overall performance and

Table 4. Comparison of average speed monitoring accuracy (%)

Testing frequency	Mobile terminal monitoring method	Big data monitoring methods	Virtual reality technology monitoring method
1	88.46	75.98	70.64
2	86.68	76.86	71.88
3	85.25	74.65	75.96
4	87.54	73.54	74.63
5	88.37	72.41	73.52
6	84.61	71.12	72.73
7	86.84	75.75	71.61
8	86.32	76.58	70.25
9	85.20	75.26	71.78
10	84.74	73.33	70.42

readiness, it is necessary to increase the load moderately in order to continuously improve the overall competitiveness of athletes.

Test the results of real-time data collection rate of athletes' special training load before and after the use of this method under different sampling rate settings, as shown in Fig. 4.

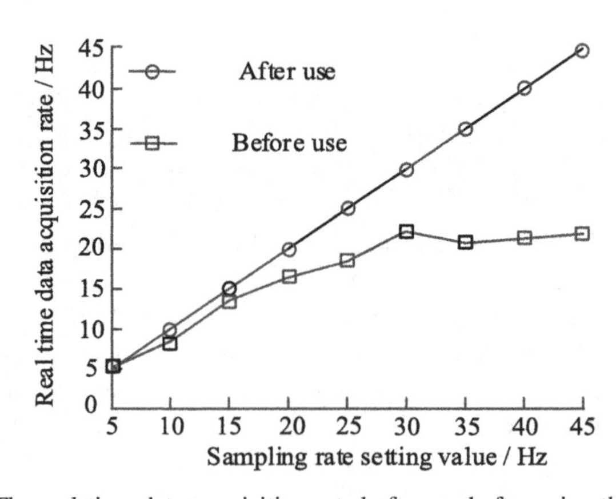

Fig. 4. The real-time data acquisition rate before and after using this method

By analyzing Fig. 4, it can be seen that under the condition that the sampling rate setting value is 5 Hz, the real-time data collection rate of athletes' special training load before and after the use of the method in this paper is the same as the setting value; Under the condition that the setting value continues to rise, the values of the two methods remain the same after the use of the method in this paper, while the deviation between the values of the two methods gradually increases before the use of the method in this paper. Therefore, it can be shown that the method in this paper has excellent data collection effect of athletes' special training load, without data collection delay, and meets the real-time requirements of safety monitoring of athletes' special training load.

4 Conclusion

The daily training load is generally lower than the actual competition load requirements. Athletes are in a low-intensity training state for a long time. Technical and tactical training is very easy to lead to inertia of the body. Once they face the situation of high-intensity competition, the body is very easy to suffer from discomfort and function decline, resulting in the decline of athletes' competitive ability. There are differences in the training load and intensity of different training contents. In the technical and tactical training, coaches should strengthen the targeted training of some players, take the competition load and intensity data as a reference, and formulate an effective training scheme to achieve the training effect. This paper presents a real-time monitoring method for athletes' special training load facing mobile terminal. This method can improve the accuracy of load monitoring and has high application value. Because the server only realizes the data upload function, the function is relatively single. Hadoop processing technology can be connected to the background service layer of the system, which provides a new direction for the application research of real-time monitoring of training load.

Fund Project. 2021 Domestic Visiting and Training Program for Outstanding Young Backbone Talents in Colleges and Universities, Project No.: gxgnfx2021063.

References

1. Duan-ying, L.I., Jie, L.I., Qun, Y.A.N.G., et al.: Research on digital monitoring of physical training of elite athletes in big data era. J. Guangzhou Sport Univ. **41**(5), 104–108 (2021)
2. Wang, Y.: Design of Wushu training action simulation system based on virtual reality technology. Mod. Electron. Tech. **43**(12), 127–129+134 (2020)
3. Ping, G.A.O., Yihai, H.U., Yin, Y.U., et al.: Empirical study on the load arrangement of physical training for elite canoe slalom players. China Sport Sci. Technol. **57**(1), 66–71 (2021)
4. Xu, L.I., Na, Y.U., Jing-wen, L.I., et al.: A moving target tracking method with overlapping horizons and multi-camera coordination. Comput. Simul. **38**(11), 162–167 (2021)
5. Qi, H., Xu, Q., Song, Q., et al.: Analysis of load structure in different time patterns of excellent triathlon athletes. Contemp. Sports Technol. **11**(21), 39–42,46 (2021)
6. Zhao, Z.: Design of training load evaluation model for football players. J. Henan Inst. Educ. (Nat. Sci. Edn.) **30**(4), 74–76 (2021)
7. Ding, Y., Matthew, S.: Analysis on the characteristics of training load of the chinese women's handball players. J. Cap. Univ. Phys. Educ. Sports **32**(2), 186–192 (2020)

Design of Energy Consumption Monitoring System for Group Building Construction Based on Mobile Node

Yan Zheng, E. Yang(✉), Shuangping Cao, and Kangyan Zeng

Department of Architectural Engineering, Chongqing College of Architecture and Technology, Chongqing 401331, China
yang16785@163.com

Abstract. Because the nodes in the data transmission network are generally fixed, once there is a problem of node missing or interference, the reliability of data transmission will be reduced, which will affect the performance of the monitoring system. Therefore, a group building construction energy consumption monitoring system based on mobile nodes is designed. First, the system framework including operation layer, decision-making layer and management layer is designed. Secondly, according to the monitoring content of energy consumption data information of large-scale group buildings, the hardware structure of the system is optimized. Finally, through the wireless transmission technology of mobile nodes, the accurate collection and effective transmission of energy consumption in group building construction are carried out, so as to complete the energy consumption monitoring function of the system. Finally, the experiment proves that the energy consumption monitoring system of group building construction based on mobile nodes has high practicability in the practical application process.

Keywords: Mobile node · Group Architecture · Building construction · Energy consumption monitoring

1 Introduction

With the rapid development of China's economy, the restrictive effect of resources on economic development is becoming increasingly prominent [1]. At present, the proportion of building energy consumption in the total energy consumption of social commodities continues to increase, and the resulting energy shortage will be more prominent. In China, according to the different subjects served by buildings, their energy consumption is divided into rural building energy consumption, urban residential building energy consumption and group building energy consumption. Among them, group building energy consumption accounts for the largest proportion, about 40%. If the energy consumption of various buildings can be monitored and regulated in real time, then the management system and energy-saving measures can be optimized to effectively reduce energy consumption, which will be of great help to create a green and energy-saving environment. At present, there are some problems in the energy consumption of

W. Fu and L. Yun (Eds.): ADHIP 2022, LNICST 468, pp. 592–606, 2023.
https://doi.org/10.1007/978-3-031-28787-9_44

large group buildings in China at the levels of measurement, transmission and energy conservation supervision. Extensive energy consumption management will cause a lot of energy consumption waste to a certain extent. Energy consumption monitoring and energy conservation management of large group buildings have become the focus of energy conservation and consumption reduction in China [2]. Although China has carried out information-based energy consumption supervision at present, there are many problems of concealment in the process of data reporting in many places due to the low degree of automation of data collection and low degree of management informatization, which directly increases the difficulty of energy consumption data statistics [3]. In order to achieve accurate energy consumption data collection and effective dynamic control of energy consumption, this paper will analyze the energy consumption monitoring system of large-scale group buildings in detail.

2 Group Building Construction Energy Consumption Monitoring System

2.1 Hardware Structure of Group Building Construction Energy Consumption Monitoring System

Mobile node technology is an important aspect of information technology. It is widely used in many fields of modern production and life. It is an indispensable basic technical guarantee for the current smart city construction. In terms of building energy consumption control, mobile node technology can realize multi angle and fine-grained system monitoring and form monitoring information for energy-saving treatment, Provide support for taking necessary energy-saving measures [4]. From the current development of building energy-saving technology in China, there are still many aspects that need to be further improved. Only through continuous optimization and upgrading of technology can a more advanced and perfect energy-saving guarantee system be formed and contribute to the healthy development of the city. Building energy consumption monitoring system refers to the installation of classified and itemized energy consumption measurement devices in buildings and the transmission of technology through the network, It is a general term for hardware and software systems that collect and summarize energy consumption data to information processing equipment in time to realize on-line monitoring and dynamic analysis of building energy consumption [5]. It is further evolved from the building automation system, which can comprehensively monitor the overall energy consumption equipment of the building, and automatically control and optimize the management, so as to achieve high efficiency and energy saving; At the same time, the energy-saving technology with intelligent building as the core is adopted to realize the integrated management of power, air conditioning, lighting and other equipment through networking. Taking the energy management system of medical building as an example, it generally adopts a distributed hierarchical structure, mainly including the decision-making level, management level and operation level. See the figure below for details:

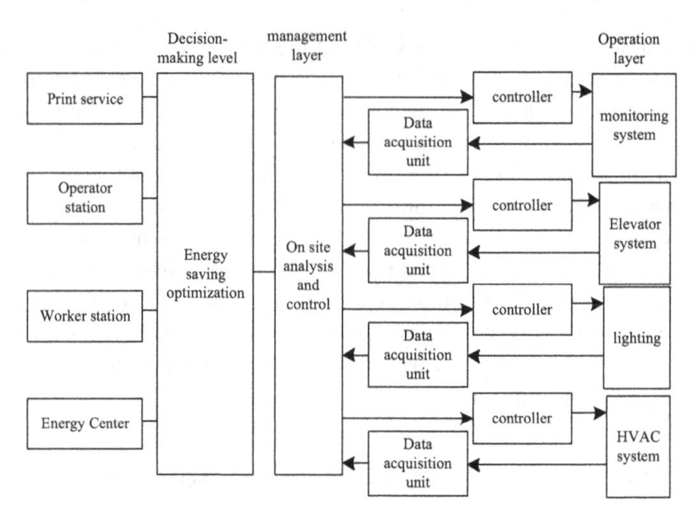

Fig. 1. Energy management system framework

Based on the frame structure of the energy consumption management system shown in Fig. 1, the flow of energy consumption data management is analyzed: first, at the operation layer, the terminal collection instrument automatically collects energy consumption data and reports it to the management layer through the network; At the management level, energy management experts submit the processed data to the decision-making level through data analysis and field analysis. Secondly, at the decision-making level, after the corresponding energy-saving scheme is obtained through relevant optimization methods, the scheme is fed back to the energy management experts; Finally, optimization instructions are generated according to the scheme to control the air conditioning, elevator, lighting, security and other equipment systems on the operation floor through various equipment controllers [6].

The monitoring node is the foundation of the whole building energy consumption monitoring system, with CC2530 chip as the core, supplemented by power consumption detection module, power supply module, alarm module, switch circuit, external memory and other peripheral circuits. The overall structure is shown in Fig. 2.

The internal data system of energy consumption of large group buildings is complex, and there are many energy consuming units. In order to do a good job in the framework of energy consumption monitoring system of large group buildings, it is necessary to control the content of energy consumption monitoring data of large group buildings [7]. According to the requirements of the specific contents of office buildings of state organs and large group buildings issued by China, the monitoring contents of energy consumption data information of large group buildings are shown in Fig. 3.

The energy consumption monitoring system of large-scale group buildings studied is designed in close accordance with the requirements of the software development guide for energy consumption monitoring system of state organ office buildings and large-scale group buildings, which effectively meets the design standards stipulated in China. For the information resources and data layer, it is to achieve the acquisition and transmission

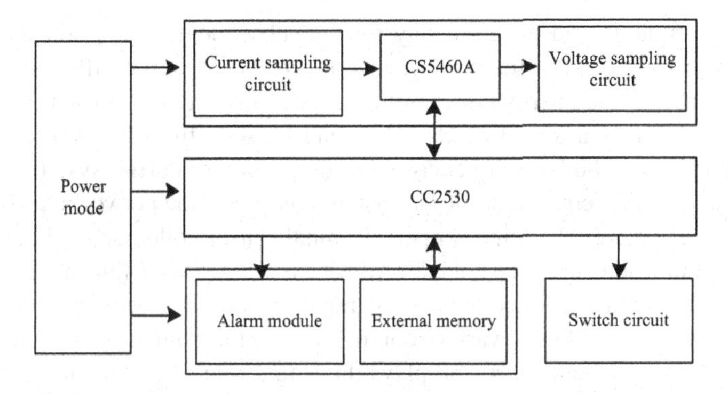

Fig. 2. Overall structure of monitoring node

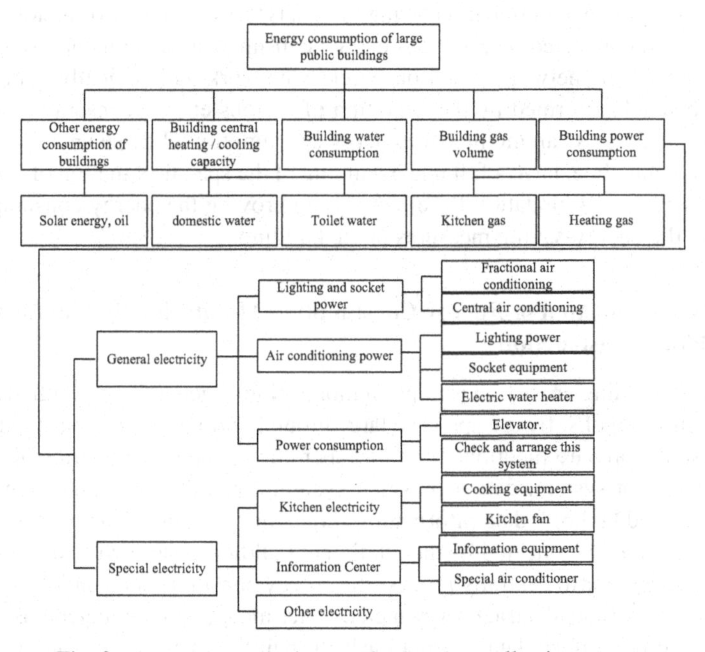

Fig. 3. Energy consumption monitoring data collection content

of first-hand energy consumption data information in large group buildings, and classify the collected energy consumption data information [8]. For the application layer, it covers four levels: data and information management, analysis and display, information service and background management, and many contents are covered in different levels. The function of the application layer in the energy consumption monitoring of large group buildings is used for energy consumption data information processing, display and data information monitoring. In the application layer, each function can be regarded as an independent system module. When designing the application layer module, each processing module should be relatively independent to reduce the mutual interference between each module and pave the way for the subsequent energy consumption data information processing. For the performance layer, it means that different roles in the society can analyze the energy consumption data of large groups of buildings in combination with their own actual needs [9]. It can be seen from the whole networking structure diagram of the building energy consumption monitoring system that the terminal data acquisition equipment of the system completes the networking based on the mobile node. The network includes many terminal sensor collectors and one node collector. The mobile node terminal collection device is defined as an information collection point, and the aggregation collector is the aggregation node. The acquisition node will set the working mode according to various control commands from the sink node. This node is the core of the whole network and plays the actual role of coordinator and gateway. The function realization of the convergence point is to effectively collect the regional network data and transmit it to the data center through the network. The basic structure of the whole energy consumption monitoring system is based on the information collection technology and connected with various energy consumption equipment through nodes, The form of ad hoc network based on wireless network automatically completes the real-time basic data formed by the operation of various energy consuming equipment in the building. Users can monitor the energy consumption of these energy consuming equipment at any time, have a full understanding of the specific situation of energy consumption, and lay a foundation for effectively improving the energy consumption and optimizing the energy-saving measures in the building.

2.2 Software Function of Energy Consumption Monitoring System for Group Building Construction

Large group building energy consumption monitoring is generally controlled and managed with the help of software app. The large group building energy consumption monitoring system software mentioned in this paper covers monitoring terminal, database, data management system, data acquisition system, firewall, communication network, concentrator and building data information acquisition terminal. The energy consumption data information collected by the intelligent building system will be transmitted to the data concentrator to centrally process the energy consumption data information and operation status of the electric energy meter, water meter, cooling meter, gas meter and other energy consumption data information in the building. The concentrator will convert the data information into TCPP protocol data packets. The communication network will promote the operation of the data information processing module under the action of the firewall, Transfer relevant energy consumption data information to the database [10]. The energy consumption monitoring system adopts the way of centralized monitoring of real-time data to collect the equipment energy consumption and display the data to users in the form of graphics, statements and figures, so as to realize the functions of energy consumption data collection, query and analysis. At the same time, the operation authority, operation records and fault conditions of the system shall be managed to provide data support for energy conservation and emission reduction and energy safety. This paper analyzes the functional requirements, performance requirements and development technology of the visualization platform of the energy consumption monitoring system. The graphical user interface of the monitoring terminal adopts C++ language and is designed by using the mobile node development environment. The background uses the embedded database SQLite to realize data management. The overall functional structure is shown in the figure, mainly including authentication, system setting, data display Query and exit five modules. The system setting module is subdivided into serial port setting and monitoring node setting. The query module is also divided into real-time curve query and node data query. The implementation interface and specific description of the main function modules are as follows (Fig. 4).

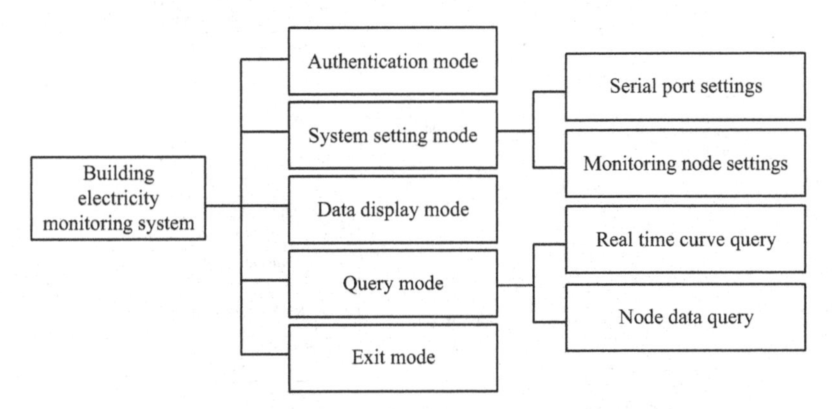

Fig. 4. Overall functional structure of graphical user interface

The data acquisition system manages the building terminal communication protocol in the concentrator and checks the errors in data communication regularly. Carry out relevant data processing for many contents such as data loss and abnormal work, effectively obtain relevant data information in the database, and dynamically monitor, evaluate, analyze, store and display the energy consumption in the building in combination with the energy consumption monitoring indicators set by the system. The monitoring terminal will generally obtain relevant data from the database and obtain the evaluation results for comprehensive analysis and control. The monitoring terminal will analyze the control instructions in combination with the database, data acquisition system, communication network, firewall, concentrator, etc., so as to control the data information status of the acquisition terminal of the building. Due to the different use functions of different buildings, the energy consumption status is also different, even in large buildings, Using different energy consuming equipment, the energy consumption is also different. Therefore, it is necessary to establish a sub item energy consumption data model, so that no matter how different buildings are, the system can use a unified model to realize the comparative analysis of energy consumption of large public buildings. The general itemized energy consumption data model should accurately reflect the energy consumption characteristics of buildings, so that different building energy consumption conditions can be expressed by such a unified model, as shown in Fig. 5.

In the actual application of large-scale group building energy consumption monitoring system, it is necessary to introduce the development technology of this software platform in combination with the overall framework and module layered characteristics of the system. With the help of Java, JavaScript and other programming languages for

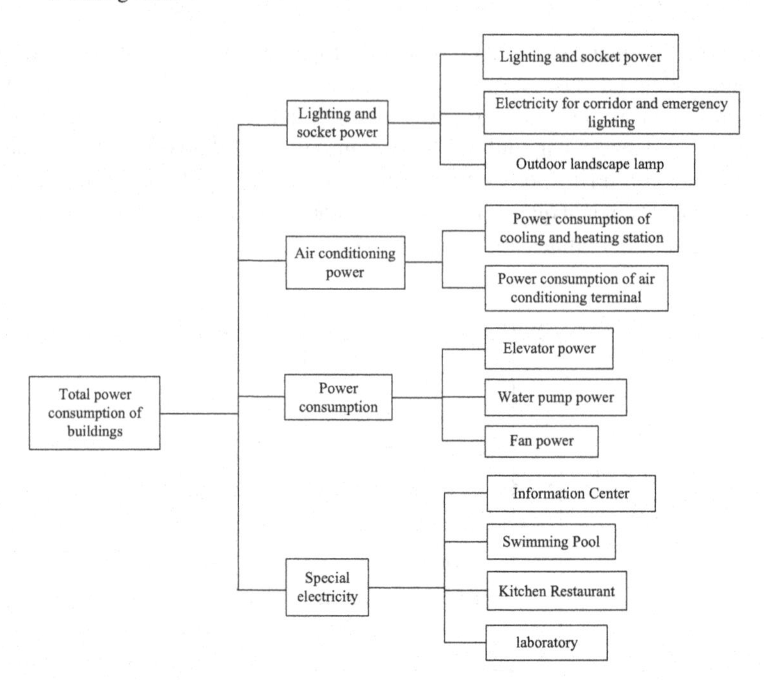

Fig. 5. Reference model of electric energy consumption data

program coding design, the data repository is connected with cloud storage technology to ensure the security and huge capacity of energy consumption data information storage. During data information transmission and communication, we should choose a stable RS485 data communication standard to ensure the high-quality operation of stored data information. The campus energy consumption monitoring and management software is divided into six main functional modules: basic information management, data acquisition, data processing, query statistics, energy-saving control and abnormal alarm. Figure 6 shows the structure of energy consumption monitoring and management software.

Collecting energy consumption data of urban group buildings is the first step of energy conservation and emission reduction. The collection can be carried out manually and automatically. The automatic collection of data covers the energy consumption data of subdivisional and classified buildings. It can be monitored and collected in real time through technologies such as mobile nodes, and then automatically transmit these data to the corresponding data processing system, Conduct in-depth analysis to determine the specific energy consumption and lay a data foundation for the next rectification. The alert management page contains three functions: you can set the threshold of relevant

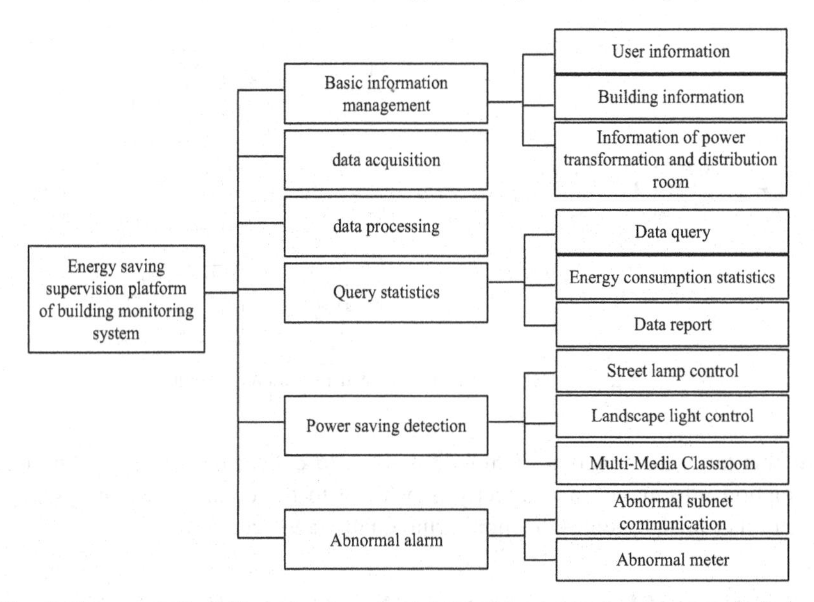

Fig. 6. Structure of building energy consumption monitoring and management software

alert items, including alert settings for a certain type of equipment or energy consumption, as well as alert settings for the overall building energy consumption; Display all real-time early warning information to users, and quickly support users to query each kind of early warning information; Support the query and comparison of historical early warning information to facilitate managers to make relevant reference. Early warning management module is an important window for managers to monitor equipment energy consumption. It can help users realize complex real-time data, historical data and preset data comparison and quick operation. The design of early warning interface includes two categories: current alarm and historical alarm. The user (administrator) can edit and refresh the page according to the specific situation. The page can display the equipment name, equipment number, alarm type, alarm status, alarm time, handling opinions, etc. Early warning information includes equipment early warning and energy early warning. Equipment alarm is aimed at abnormal equipment operation, line fault, machine aging, operation error, etc. energy consumption alarm is aimed at the situation that equipment energy consumption exceeds the set threshold of energy consumption. The early warning management module automatically analyzes the collected data, and then matches the early warning threshold information set in the system. When the collected data approaches or exceeds the alarm threshold, the defined early warning information will be automatically generated. Then, the early warning module sends early warning information in the form of Web pop-up, SMS and e-mail to users through the system to timely notify users to detect and maintain the system and eliminate faults. The program logic flow of the early warning management module is shown in Fig. 7.

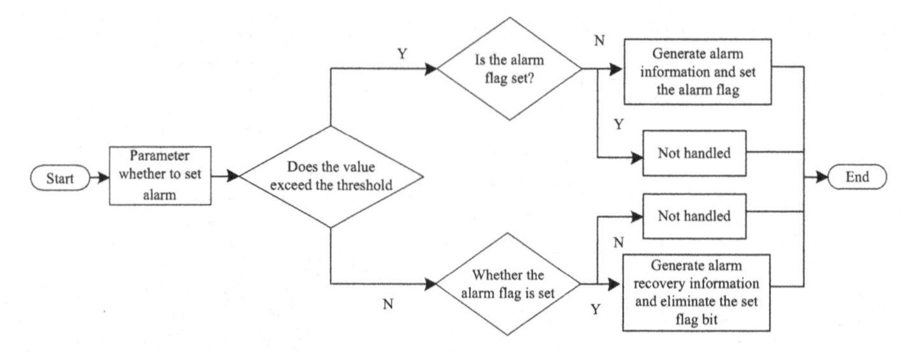

Fig. 7. Design flow of early warning management

At this time, the function of mobile node is to collect the corresponding energy consumption information and directly transmit it to the communication system. The degree of its role determines the monitoring quality and accuracy.

2.3 Realization of Energy Consumption Monitoring in Group Building Construction

The functional characteristics of the energy consumption monitoring system developed based on mobile node technology are based on the relevant information collection and transmission of wireless sensor technology. Its network nodes can complete the collection of temperature and other information parameters, and can also connect with various energy consumption devices to automatically collect relevant data based on ad hoc network, Managers can monitor various energy consumption equipment at any time, and diagnose, evaluate and transform energy consumption through corresponding data processing platform. From the current construction situation, the system has been fully applied to the energy consumption measurement and management of many group construction projects in China. Especially for the professional power distribution system of hotels and other buildings, the energy consumption equipment is scattered, and the wireless sensor system based on mobile node technology has greater application advantages. Basic building information, envelope information, indoor personnel activities, indoor environment, outdoor meteorological factors and the operation status of building electromechanical equipment are important factors affecting building energy consumption. The above information is described in the standard data set of building space energy consumption as shown in Table 1.

In practical application, managers can carry out more intelligent management through computer platform. Including access control, lighting, air conditioning, alarm and other energy consumption systems, which can be mixed for networking and share the bus. This construction scheme can save 75% of the construction cost. According to the four principles of demand analysis and system design, combined with the actual functional modules of the energy consumption monitoring and management system, the overall architecture of the energy consumption monitoring and management system is as follows (Fig. 8):

Table 1. Classification information of building space unit energy consumption quasi data set

Information classification	Subclass information	Quantity
CPN name	-	1
Essential information	-	1
Structural information	-	1
Personnel information	-	1
Indoor environment information	-	1
Electromechanical equipment information	-	1
Electromechanical equipment information	Lamps and lanterns	28
	Socket	23
	Sun visor	5
	Electrically-operated window	13
	Fan coil unit	5
	VAV air conditioner	5
	Radiator	8
	Fresh air fan	8
	Tap water	1
	Domestic hot water	1
	Gas	2

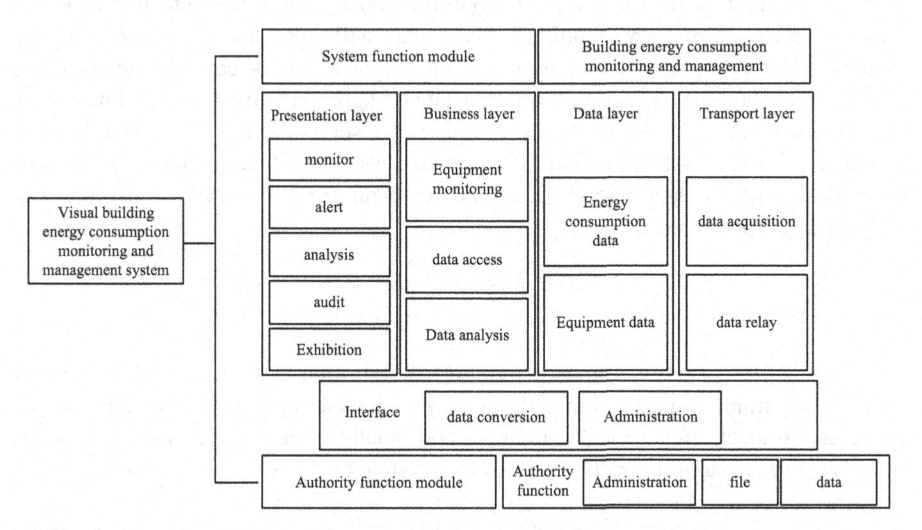

Fig. 8. Overall architecture of energy consumption monitoring and management system

The equipment can complete the linkage function of hardware by using fieldbus, and its operation stability is higher. Because the system platform integrates multiple energy consumption management systems, its operation and application are more convenient and flexible through the management of the software platform. The prediction methods of building energy consumption are mainly divided into two categories: forward model and data-driven model. The former is all kinds of energy consumption simulation software, such as request, Energy Plus, DeST, etc., which is mainly used to optimize the design of buildings and their air conditioning systems; the latter is all kinds of prediction algorithms to simulate and predict their energy consumption with the help of historical data of existing buildings, such as linear regression, artificial neural network, etc. When a node initiates the building energy consumption information summary instruction, such as "building energy consumption summary of a seven storey space unit of a building", the group intelligent building energy conservation monitoring system establishes a undirected graph G(V,E) of a seven storey space single n-node topology of a building based on the automatic topology identification algorithm, in which all CPN nodes in the undirected graph are represented, and E represents the connection connecting all nodes in the undirected graph. The spanning tree algorithm is adopted to establish the minimum tree with this node as the root node. The collection of building energy consumption can be equivalent to the summation problem. Therefore, the information of building energy consumption is shown in formula (1):

$$Q = f(x_1, x_2, x_3, \ldots x_n) \tag{1}$$

where, x represents the data in the standard data set of building energy consumption of corresponding nodes. Accurate building energy consumption data is the basis of building energy consumption data statistics. If abnormal values are mixed into the statistical data, the deviation of corresponding analysis data will increase, and even the chaotic operation of building energy consumption control system will be caused, Therefore, the verification of building energy consumption data is of great significance. Building energy consumption data is based on the law of energy conservation, the law of heat balance and other relevant constraints. There is correlation between several variables of building energy consumption data. Therefore, equality constraints and inequality constraints can be used to check the data:

$$\begin{aligned} h_i(x_i, x_{i1}, x_{i2} \cdots x_{im}) &\leq 0 \\ g_i(x_i, x_{i1}, x_{i2} \cdots x_{im}) &= 0 \end{aligned} \tag{2}$$

In the building energy efficiency monitoring algorithm based on mobile nodes, the equality constraint is that the sum of the corresponding energy consumption data of each space unit or electromechanical equipment node shall be equal to the monitoring value of the energy consumption data of the area or system node, as shown in formula (3):

$$f(x_n) = f(x_{n1}) + f(x_{n2}) + f(x_{n3}) + \ldots + f(x_{nn-1}) \tag{3}$$

The established mobile node combination prediction model belongs to the latter category, and its main modeling steps are as follows for gray processing. Accumulate the original energy consumption data at one time and obtain the accumulation sequence, so as to weaken the randomness of the original data, highlight the overall development trend and further preprocess the data. Normalize the input and output data in the interval of −1 and 1, and the normalization process is as follows:

$$y = 2\frac{x}{x_{max} - x_{min}} - 1 \tag{4}$$

where: x_{max} and x_{min} are the maximum and minimum values respectively; y is the value after pretreatment. In order to evaluate the effectiveness of each model, this paper uses three performance indexes: the absolute value of maximum relative error E_{max}, the average relative error E_{ave} and the root mean square error RMSE to evaluate the prediction accuracy and stability of the three models:

$$E_{max} = \max\left|\frac{x - x_{min}}{Y(i) \text{ - y}}\right| \tag{5}$$

$$E_{ave} = \frac{1}{n}X(i)\sum_{i=1}^{n}\frac{\tilde{Y}(i) - Y(i)}{x - x_{min}} \tag{6}$$

$$RMSE = \sqrt{\frac{1}{n}\sum_{i=1}^{n}y(Y(i) - X(i))^2} \tag{7}$$

where, n is the total number of samples, $X(i)$ is the predicted value of the ith series, and $Y(i)$ is the actual value of the i series. The calculation results of the above evaluation indexes are shown in Table 2.

Table 2. Summary of calculation results of evaluation indexes of each model

Error	GM (1,1) model	Neural network model	Gm-bp combination model
Absolute value of maximum relative error Emax/%	47.32	5.36	0.52
Average relative error Eave/%	7.98	0.98	0.11
Root mean square error RMSE (106)	67.25	2.68	1.65

Building energy consumption prediction is an important functional module of group building energy consumption monitoring system. It is an important source to master the future development trend of energy consumption. Scientific energy consumption prediction is the basis and guarantee for making correct decisions. It has practical guiding significance for formulating building energy-saving operation management measures and carrying out energy-saving transformation. Energy consumption prediction research is one of the methods to optimize the operation mode of each system of the building. It not only helps to explore the energy-saving focus of new buildings, but also provides an important reference for the energy-saving evaluation and transformation of existing buildings. By formulating artificial energy consumption strategies and carrying out energy-saving planning, building energy conservation can be realized. Taking the widely popularized energy quota, excessive price increase and low reward system as an example, its establishment needs to fully consider the energy consumption level of each building, and the research on energy consumption prediction provides an important basis for formulating the benchmark energy consumption level of buildings, Thus, it provides a strong guarantee for the implementation of the energy quota system.

3 Analysis of Experimental Results

Collect the energy consumption data of group buildings as the sample data of this experiment. The group data types include municipal buildings, library buildings, hospital buildings, supermarket buildings and office buildings. The energy consumption of the above group buildings is large, which can meet the requirements of this experiment. The construction energy consumption data of the above group buildings will be collected for 6 months with an interval of 5 min.

Since the core of the operating system and the design interface are completely separated, a graphical window framework should be set before each operation. The graphical window standard in the embedded environment is realized by windows 10. Because the embedded system is developed by the system, the embedded system also has compatibility properties. According to the hardware design, connect all software functions, download the software to the embedded system, and connect the temperature sensor with the fuzzy controller through the serial interface, so as to build the experimental environment and complete the experimental verification. Before the experimental acceptance, the building structure shall be inspected to ensure the safety of building energy consumption monitoring. The traditional detection method shall be compared with the building temperature loss detection results of the system designed in this paper. The results are shown in Table 3.

Table 3. Comparison results of two methods on building energy consumption detection rate

Typhoon series	Traditional system	Paper system
10	32%	85%
15	39%	92%
20	45%	90%
25	35%	89%
30	42%	87%
35	50%	82%
40	55%	90%

Further, take the group building m as an example. The total construction area of the building is about 30000 square meters, with 3 floors underground, mainly parking garage and equipment room, 1–3 floors above the ground are commercial, and there are 6 towers above 3 floors. The building energy consumption monitoring system is equipped with 197 intelligent instruments. During operation, the building energy consumption monitoring system has the following problems: firstly, the system energy consumption monitoring can only realize the itemized measurement of building energy consumption and only display the total itemized energy consumption data. Secondly, the system can only realize the collection and transmission of energy consumption data, and can not realize building energy conservation control and building energy efficiency analysis. Finally, the naming of energy consumption collection points is inaccurate, The user cannot accurately judge the position of the monitoring point. Each data is predicted 20 times, and the average value is taken as the prediction result to predict the energy consumption in the last ten days. The comparison between predicted energy consumption and measured energy consumption is shown in Fig. 9.

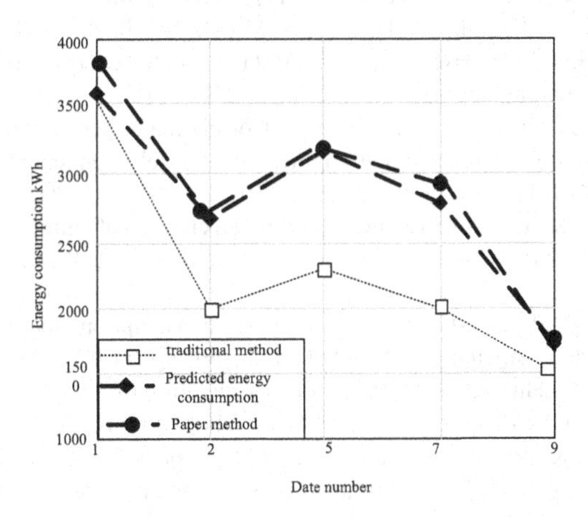

Fig. 9. Comparison between predicted energy consumption and measured energy consumption

Based on the analysis of the above comparison test results, it is not difficult to find that compared with the traditional methods, the operation effect of the monitoring system of this method is obviously better, and is closer to the actual value, which fully meets the research requirements.

4 Concluding Remarks

To sum up, there is a very broad development space for the current energy-saving design of group buildings in China. Various advanced energy-saving and monitoring technologies are used in the energy consumption management of urban group buildings, which can fully ensure the quality and efficiency of relevant management. The successful application of mobile node technology in building energy consumption monitoring system effectively improves the operation efficiency of the whole monitoring system, which is very necessary to optimize and improve. The continuous upgrading of mobile node technology itself also provides more weapons and tools for building energy conservation, and increases the research on relevant aspects. This is the basic work that must be done carefully, and it is a very important attempt to expand the application depth of mobile node technology.

References

1. Fu, Q., Wu, S., Dai, D., et al.: Method of building energy consumption prediction based on transferring deep reinforcement learning. Appl. Res. Comput. **37**(S1), 92–94 (2020)
2. Ji, T., Wang, T.: Building energy consumption prediction based on word embedding and convolutional neural network. J. South China Univ. Technol. (Nat. Sci. Edn.) **49**(06), 40–48 (2021)
3. Xue, B., Gang, L., Lei, H., et al.: Ensemble model of building energy consumption prediction based on feature selection algorithm. Comput. Eng. Des. **41**(10), 2892–2896 (2020)
4. Wang, S., Zhou, X., Dong, J.: Simulation of energy consumption control system for near-zero energy building based on fuzzy PID. Comput. Simul. **38**(10), 263–267 (2021)
5. Moon, J., Park, S., Rho, S., Hwang, E., et al.: Robust building energy consumption forecasting using an online learning approach with R ranger **25**(8), 116–119 (2021)
6. Szustak, L., Wyrzykowski, R., Olas, T., et al.: Correlation of performance optimizations and energy consumption for stncil-based application on intel xeon scalable processors. IEEE Trans. Parallel Distrib. Syst. **33**(9), 19–25 (2020)
7. Liu, X.-J., Hu, S.-K., Li, L.-Y.: Temporal and spatial changes of building energy consumption in China's provinces and analysis of its influencing factors. Math. Pract. Theory **50**(06), 74–85 (2020)
8. Zhang, L., Li, Y.-A., Liu, X.-L.: Research on energy consumption prediction of civil buildings based on grey relational analysis. Archit. Technol. **50**(06), 74–85 (2020)
9. He, L.-H., Cui, X., Hu, Q.-C.: Estimation and simulation of energy consumption of public buildings based on BIM. J. Eng. Manag. **34**(02), 84–89 (2020)
10. Wang, Z.-Q., Guo, H.-J., Wang, S., et al.: Analysis of building energy consumption and discussion on energy-saving reform in an office park. Archit. Technol. **51**(06), 670–672 (2020)

Construction of Mobile Education Platform for Entrepreneurial Courses of Economic Management Specialty Based on Cloud Computing

Huishu Yuan[1] and Xiang Zou[2](✉)

[1] Chengdu College of University of Electronic Science and Technology of China, Chengdu 611731, China
[2] School of Accounting and Finance, Wuxi Vocational Institute of Commerce, Wuxi 214153, China
sln13526@163.com

Abstract. In order to solve the problem of unbalanced course resource scheduling of mobile education platform when the number of users increases, a mobile education platform based on cloud computing is constructed. In the hardware part, the FPGA chip of XC6SLX16 is selected as the platform, and the decoupling network is designed according to different power input to eliminate the noise on power pin. In the software part, according to the requirements of the economic management course, we integrate the scattered teaching to form the rich teaching resource base and realize the unified management of users, roles and organizations. In order to improve the concurrency performance of the platform, the entrepreneurial course resource database of economic management major is scheduled based on cloud computing. Design each function module of the mobile education platform, input the keywords can get the related more detailed development resources, realize the mutual communication discussion. The test results show that the platform has good performance, and can improve the network throughput and meet the design requirements.

Keywords: Cloud computing · Platform design · Economic management major · Entrepreneurial courses · Mobile education · Educational platform

1 Introduction

In the era of "Mass Entrepreneurship and Innovation", how to meet the needs of innovation-oriented national strategy and cultivate innovative and entrepreneurial talents in higher education is one of the key areas in the process of talent training. Vigorously promoting entrepreneurship education in colleges and universities is also of long-term significance to the scientific development of higher education and educational reform. Entrepreneurial activity is mainly concentrated in the retail sector, with less than 2% of

W. Fu and L. Yun (Eds.): ADHIP 2022, LNICST 468, pp. 607–619, 2023.
https://doi.org/10.1007/978-3-031-28787-9_45

entrepreneurs based on high-tech entrepreneurship. As the main group of college students, their entrepreneurship is mainly concentrated in the tertiary industry, mostly in the low-end services, high-end services, emerging technology and high-tech entrepreneurship. Under the current circumstances, it is of great value for institutions of higher learning to carry out innovation and entrepreneurship education: it is the only way to promote the national innovation-driven development strategy and high-quality social and economic development; and it is an effective way to promote the comprehensive reform of higher education and promote higher quality entrepreneurship and employment of college graduates. However, in general, the knowledge superiority and professional ability of college students in the field of engineering science and technology innovation and entrepreneurship are not given full play. With the development of the knowledge economy and innovative society, the mode of knowledge production has been gradually transformed to "marketization" and "capitalization", which have a lot of influence on the goal and link of talent training in colleges and universities, and promotes the continuous enrichment of professional education and the rise of entrepreneurship education. After the concept of "innovation and entrepreneurship education" was put forward, the concept of "facing all students" in entrepreneurship education has become a recognized concept of government, academic fields and institutions of higher learning. With the rapid development of mobile communication technology and the popularization and application of intelligent terminal equipment, mobile teaching platform have become an industry with rapid development and great market potential [1]. Mobile learning content means more and more, more and more rich. Students are also increasingly using mobile devices to learn and communicate. The Internet has changed the traditional model of learning and teaching. From today's information technology development, mobile Internet education will become the mainstream trend. Therefore, the new teaching-learning model will be mainly about mobile learning. The innovation of information technology and the change of information knowledge systems are so fast in the information age that it is necessary to master the knowledge of new fields to solve practical problems. Learning in the mobile environment has become a trend in the information society, and has become more and more popular in colleges and universities. In this way, students can get updated teaching information at any time and anywhere in the school. Not only that, this mobile teaching platform in interactive Q&A, breaking with traditional teaching and learning methods, will not delay the teacher's rest time, will not increase the teacher's burden. Therefore, combined with innovation and entrepreneurship education, it is necessary to develop a complete platform for the mobile learning of entrepreneurship courses. Cloud computing technology can store very large files in the cloud nodes, cloud computing platform can also provide a distributed parallel computing framework. With the cloud computing platform, time-consuming computing tasks can be broken down into smaller tasks, which can then be handled by individual cloud nodes simultaneously, dramatically reducing overall task execution time. Based on cloud computing, this paper constructs a mobile education platform for entrepreneurial courses of economic management specialty to improve the learning flexibility and ease of use of the platform.

2 Hardware Design of Mobile Education Platform for Entrepreneurship Courses of Economic Management Specialty

Considering the expansibility of the mobile education platform, the external circuit connects it to the mobile education platform through the I/O of the extended FPGA. Each part of the circuit has different working voltage requirements, so we need to configure different power circuits for each part of the circuit on the mobile education platform. In addition, due to the use of the MRAM chip, but also for the DDR3 chip to provide a separate memory power management chip. The hardware architecture of the mobile education platform for entrepreneurship courses in economics and management is shown in Fig. 1.

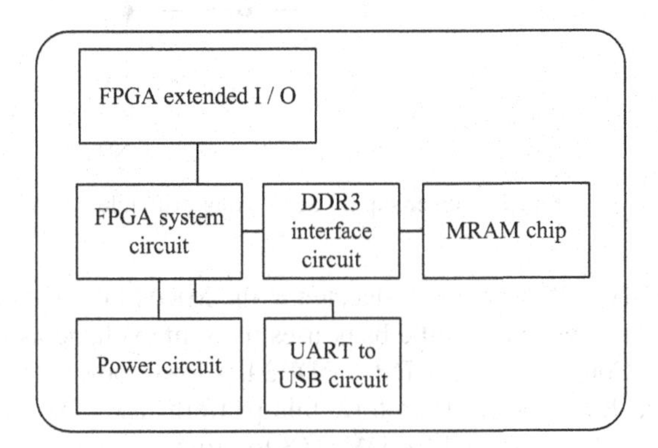

Fig. 1. Hardware architecture diagram

As for the FPGA chip that needs to be selected for the platform, we choose the FPGA chip with the lowest cost when the internal resource of the FPGA chip can satisfy the circuit function design. The FPGA chip of XC6SLX16 is chosen as the platform in this paper. The main function of the peripheral circuit of FPGA chip is to ensure the normal operation of FPGA, in addition, it also includes FPGA configuration chip circuit and FPGA clock circuit. The whole circuit board provides a single 12 V input power supply through a vertical power outlet, which is the source of all the power, and then uses TI's TPS series step-down chips to depressurize 12 V to get the power supply voltages of other chips. In the actual design, the DONE pin is connected to the LED lamp to indicate the configuration of the FPGA; the PROGRAM B pin is connected to an external key so that the FPGA can be reconfigured at any time through the key; the SUSPEND pin is pulled high because it does not need to run in power-saving mode; and the JTAG pin is finally connected to the 14 pin JTAG connector. This platform uses TPS65232 to reduce 12 V voltage to 5 V, 3.3 V, 1.8 V and 1 V, and uses TPS65001 to get 1.8 V AC voltage. The TPS65232 contains a PWM step-down controller and two adjustable synchronous step-down regulators. In order to ensure the stability of the chip, TPS65232 has set the overcurrent protection circuit internally, also may control two step-down controllers through the pin EN _ BCKn when to start. The correct design of the power system can lead to better overall performance, lower clock jitter, and more stable systems. Therefore, when designing the power input circuit of FPGA, it is necessary to design the decoupling network for different power input to eliminate the noise on the power pin, and the decoupling capacitor is used. The principle of power decoupling is shown in Fig. 2.

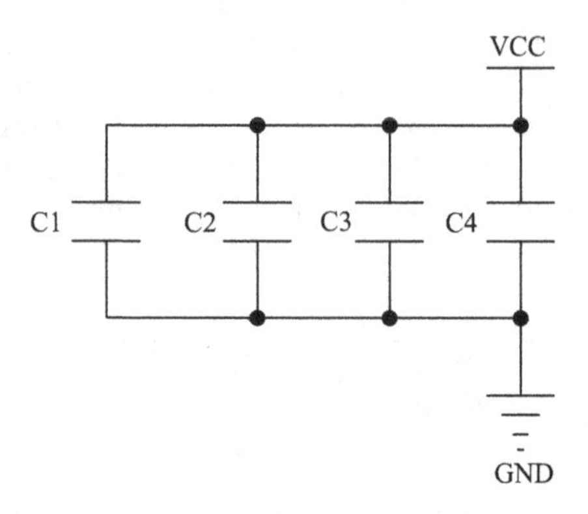

Fig. 2. Power supply decoupling principle

The bank3 of the FPGA chip is connected to the MRAM chip to design the DDR3 interface read-write controller, so the bank uses the same voltage as the MRAM chip for power supply. Compared with DDR2, DDR3 has lower power consumption, faster storage speed and lower price. Therefore, this platform uses the dual 32-bit DDR3 interface EMIF which is reserved for DM8168 to provide more storage space. DDR3 circuit requires the use of 4 pieces of 16-bit DDR3, each channel two, configured as a dual-channel 32-bit, can achieve the highest address 2 GB. In the construction of this platform, DDR3 adopts the cross storage mode, and EMIF0 and EMIF1 can work at the same time when accessing data. Use a clock input frequency of 50 MHz. It should be noted that the power supply terminal of the crystal oscillator needs to be filtered, and the output terminal series resistance can be used to prevent the reflection of high frequency signal from disturbing the signal source and play the role of eliminating the reflection wave from stabilizing the clock signal.

3 Software Design of Mobile Education Platform for Entrepreneurship Courses of Economic Management Specialty

3.1 Analysis of Demand for Mobile Education in Entrepreneurship Courses for Economic Management Majors

The Mobile Education Platform for Business Start-up Courses of Economic Management Specialty aims to provide users with a mobile education platform with richer content and more convenient use by combining the rich educational resources generated by the informatization of domestic colleges and universities. Achieve users to use mobile terminal devices for convenient learning. So that learning is no longer subject to geographical and time constraints, can better use of the fragmented time of users. The mobile education platform based on cloud computing is shown in Fig. 3.

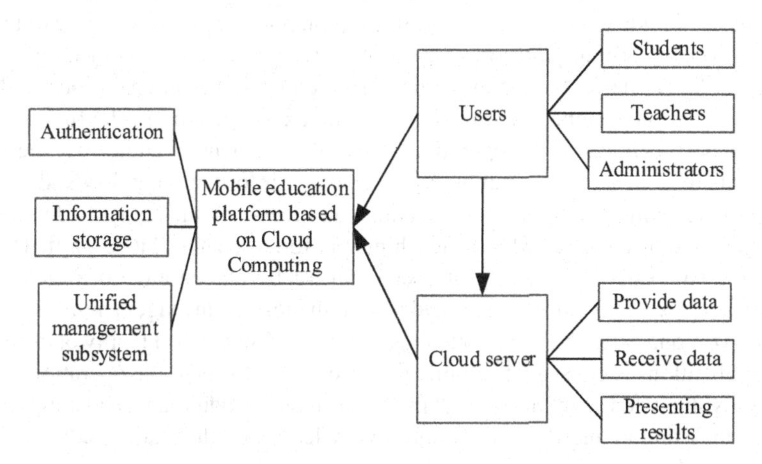

Fig. 3. Mobile education platform based on cloud computing

The construction effect of the mobile education platform is measured by the goal of entrepreneurship curriculum of economic management specialty. The goal of the entrepreneurial curriculum is to express the overall objectives of talent training in different sides and different extension directions. Based on the concept and value orientation of entrepreneurial education, the objectives of entrepreneurial curriculum should be layered and clear, with students as the center, subject knowledge as the basis, and social needs as the orientation. [2] Based on the characteristics of entrepreneurship education, the objectives of entrepreneurship education should include three levels: the cultivation of entrepreneurship awareness, the mastering of entrepreneurship knowledge, internalization and the generation of entrepreneurship experience. The composition of the entrepreneurship education curriculum objectives is shown in Fig. 4.

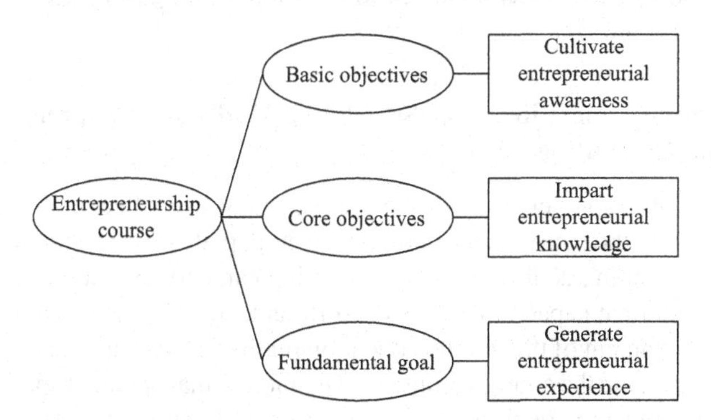

Fig. 4. Entrepreneurship education curriculum objectives

Based on the goal of the entrepreneurial courses for economic management major, the mobile teaching platform integrates the scattered teaching to form a rich teaching resource database. Through personalized curriculum customization, students can customize their own courses according to their own circumstances. The mobile education platform has the functions of education information portal, application store of teaching resources, open learning platform and so on. It will bring users a full range of multimedia services related to teaching, such as teaching information, resource sharing, teaching affairs management and individual course customization. The function menu module mainly belongs to the operation scope of the teacher's authority. The teacher implements the teaching plan, publishes the teaching information, uploads, downloads, updates the teaching materials in this module, and feedback the students' learning and teaching situation on time. Mobile teaching platforms shall achieve unified identity authentication, verify the identity of user login, realize the storage of organizational structure and personnel information, and realize the unified management of users, roles and organizational structure of various subsystems [3]. Ensure the identity of users in different applications is consistent. Mobile phone client belongs to the operation category of students, knowledge module, teaching information module and teaching feedback module, click on the micro video learning, view learning information, upload learning works. After the education portal provides a unified entrance for users, users can enter into the main interface and see the sub-links of each subsystem. Users can enter into each subsystem only with a single click, and do not need to re-enter the user password to submit a login request. Thus, users can directly enter into their respective application systems after a login. At the same time, the backstage can manage the users, and all users are organized by organization management. The cloud server ECS mainly provides data for the PC terminal management system. When the PC terminal sends a request to the WeChat server, the server immediately responds, calls the Database class for database operation, and presents the result on the PC terminal. The mobile teaching platform assigns an organization ID to each organization, under each organization ID can include 1 organization administrator ID and a specified number of ordinary user ID. The organization administrator is the super administrator user of an organization, and can configure and manage all the equipment and ordinary users of the organization. User management shall include user registration management, user inquiry, user addition and deletion, user access rights management, user locking and unlocking, user grouping management, and user access record viewing.

3.2 Establishing Course Resource Scheduling Model of Entrepreneurship Based on Cloud Computing

As the number of users increases, the service needs of users become more diversified and hierarchical. In the data center of mobile teaching platform, the amount of resources available for scheduling is different, the processing capacity of computing nodes is different, and the storage capacity of nodes is different. In order to ensure the stability of the service environment of the mobile education platform, this paper makes use of cloud computing to schedule the course resources of economic management specialty. To some extent, the selection and conciseness of entrepreneurial courses for economic management specialty determine the scientificity and reasonableness of course construction.

The choice of curriculum content should consider the particularity of entrepreneurship education, pay attention to the social value and the dynamic development of the external industry market. In order to improve the resource utilization of cloud nodes and the speedup performance of asynchronous data parallelism, the strategy of starting multiple users in each cloud node is proposed. The scheduling process of cloud computing production environment is divided into four layers: cloud computing user layer, cloud computing task collection layer, virtual resource node collection layer and data center layer. The four levels involve three levels of mapping: one-to-many tasks for users, multi-task corresponding to multi-virtual resource nodes, and the many-to-many relationship between virtual resource nodes and data centers. For example, if a single cloud node starts two users at the same time, ideally, when one user gets a batch, the other user might be performing a GPU calculation, and when that user performs a GPU calculation, the other user might be getting a batch. The two processes of each iteration of the two users complement each other, thus greatly improving the utilization of node resources. To address high concurrency, use clustering to separate applications from static resources, such as static resources and server applications, in static resource servers and application servers respectively. It also uses Nginx as a reverse proxy server and front-end server to process request server integration for forwarding different requests, such as static resource request and application request. Enable load balancing among cluster servers. Firstly, the processing capacity of all resource nodes is collected, and then the 2D time matrix is obtained according to the task length in the task set. The matrix represents the matrix formed by the processing time of the task assigned to the resource node. The minimum execution time of each task at each resource node is extracted from the matrix, and a one-dimensional array of length n is obtained [4]. Based on the array, the task with the shortest expected execution time is selected, and the shortest task is mapped to the corresponding resource node for processing. HDFS is stored in blocks, and files are divided into blocks and stored on HDFS. The default size for Block is 64 MB. Using blocks with HDFS has several benefits. Files can be divided into blocks and stored on different disks so that HDFS can hold files that are larger than a single disk space [5]. Particle swarm optimization is used to schedule cloud computing tasks. Particles are composed of two basic properties: velocity and position. The formula for calculating the velocity attribute is as follows:

$$\beta_{t+1} = \varpi\beta_t + \varphi_1\chi_1(\delta - \alpha_t) + \varphi_2\chi_2(\delta' - \alpha_t) \tag{1}$$

In formula (1), β represents the velocity of particles; t represents the iteration time; φ_1 and φ_2 acceleration constants, respectively, indicate the degree of dependence of particles on their own historical optimal solution and the global optimal solution of the population; random numbers between 0 and 1 for χ_1 and χ_2 increase the randomicity of the particle optimization process and prevent particles from moving in the same direction too early; δ and δ' represent the historical optimal solution and the global optimal solution of the population; α_t represents the search location of particles; and ϖ represents the inertia weight. Assuming that particles are searching in D dimensional space, the formula for calculating the position attribute may be:

$$\alpha_{t+1} = \alpha_t + \beta_{t+1} \tag{2}$$

In formula (2), α_{t+1} represents the position of the iterated particle. In each iteration, the particle updates the direction and velocity of the next flight based on its own and the population's experience. According to the defined utility function, the corresponding value of each particle is calculated, and the global optimal solution is found. Block based management is much simpler than file based management. Because the block size is fixed, the number of blocks needed to store the file can be easily calculated. At the same time, you can use a simple data structure to store the block's metadata. The linear decreasing inertia weight can improve the convergence speed of the algorithm and obtain a better global optimal solution. The formula of inertia weight is as follows:

$$\varpi = (\eta_1 - \eta_2)\frac{\mu_{\max} - \mu}{\mu_{\max}} + \eta_2 \tag{3}$$

In formula (3), η_1 and η_2 are two fixed inertia weights; $\mu_{\max}$ and μ represent the maximum number of iterations and the current number of iterations. By adjusting the size of $\eta_1 - \eta_2$ and η_2, the maximum and minimum inertia weights can be controlled, and the convergence speed, the development range and the local development ability can be improved. Build the cloud platform using Hadoop in the server cluster, and then build Spark on top of that. Cloud computing platforms, with HDFS provided by Hadoop and parallel computing engines provided by Spark, are well suited for parallel computing that requires iterative iterations [6]. In the cloud computing environment, after the user submits the application, that is, the task of cloud computing and the usage request of the virtual machine instance, the task is assigned to the corresponding virtual machine resource node by the task scheduling algorithm, and then the virtual machine is mapped to the data center according to the virtual machine scheduling algorithm.

3.3 Design of Functional Modules of Mobile Education Platform for Entrepreneurial Courses

Mobile teaching platform education learning function, students can click the corresponding chapter for learning, keywords input can get more relevant detailed development resources, exchange, testing, etc. Teachers explain the contents of the corresponding chapters, group messages, exchange of discussion and so on. Login module is divided into the student side and the teacher side. Users can choose according to their own identity, students use their own study number and password to log in, teachers use their own staff number and password to log in. If the wrong choice role, you can return to re-select. If you forget your password, you can retrieve it according to your ID number. Through the unified authentication center to user identity authentication, support the mailbox, user accounts, student numbers and other forms of authentication, and can expand other forms of authentication. Student's personal information module contains the student's picture, name, student number, gender, grade, major, college, school and other information, in addition to the personal information module has "change password", "change mobile phone number" function. The teacher's personal information module is very similar to the student's, including the teacher's picture, name, employee number, title, college, school, mobile phone number, and also "change password", "change mobile phone number" function. Login integration does not need to modify the original subsystem, and is automatically logged in by the Unified Information Portal according to the configured login script. The implementation principle is as follows: The Information Portal configures each subsystem with a JavaScript auto-login script that executes a script to automatically log in when entering the platform from navigation. The idea is to dynamically create a Form on an embedded page, and then automatically submit it. The platform administrator pre-assigns access to different information and customizes the content and layout, depending on the role and level of security of the user [7]. The background management of cloud service application includes curriculum resources arrangement, keyword reply, personalized function menu, test questionnaire, data statistics analysis, platform maintenance and so on. Bulletin module is when the teacher has something to notify, namely the announcement, all students will receive the announcement. Students in the announcement module, click the announcement details, you can see all the announcements list, unread red dot logo, read the red dot to cancel. The teacher end also has the announcement details, the function is the same with the student end. What is different with the student end is the teacher end has the announcement function, sets up the announcement the subject, the date and the content, then releases the announcement. The construction of entrepreneurial courses for economic management needs more practical. The relevant knowledge imparted in the subject curriculum can be applied in practice to support the development of practical activities. This practical theoretical knowledge should be added to the content of the course to educate students with practical theories or practical training courses. The timetable module is designed to make it easier for students to use, similar to the curriculum grid, where students can see information about all of their classes during the week, as well as places and teachers. Set up a number of both theoretical teaching supplemented by practical activities at the integration of theory and practice courses. The essence of entrepreneurship education is to create a platform for practical activities and provide students with more and better practical teaching, which is the essence of entrepreneurship education. The teacher's side of the curriculum is slightly different, in the teacher's side, the teacher can see all the courses they taught, click to see the list of all students in this course click each student can see the personal information of the students. Through the design of hardware and software, the paper completes the construction of mobile education platform of economic management professional entrepreneurship courses based on cloud computing.

4 Experimental Research

4.1 Experimental Preparation

Mobile platform testing is a vital work, we must combine the developed system with other resources, and carry out various tests in the actual running environment to ensure that all parts of the function are normal. Through the test, we can find out the potential problems of the mobile teaching platform and whether the relevant functions meet the requirements in the actual operation, so that developers can adjust and optimize the existing problems in a timely manner, so as to ensure the reliability of the platform and enhance users' experience. Integration testing is the simultaneous testing of software as it is assembled. According to the different ways of assembling modules, there are two kinds of testing methods: top-down combining and bottom-up combining. This system adopts top-down combining method to test. Server Test Environment: Intel Core i5-4210u, dual-core 1.7 GHz CPU, 64-bit Windows 10 Professional OS, 8 GB RAM; Java Language JDK Version 1.7.0, Tomcat Server Version 6.0, MySQL Database Version 5.0.22; Server Performance Test with Apache JMeter 3.0

4.2 Results and Analysis

Taking the mobile teaching platform for example, we set up four different testing scenarios to simulate the number of concurrent requests of 1000, 2000, 5000 and 8000 users, respectively. In the above four tests, the number of successful concurrent requests is counted, and the throughput of the mobile teaching platform is used as the measurement index of the platform stress test. This paper compares the throughput of the mobile education platform based on cloud computing with that based on data mining and personalized recommendation. The test results for the platform are shown in Tables 1–4.

Table 1. 1000 throughput comparison of requests (MB/s)

Number of tests	Mobile education platform for entrepreneurial courses of economic management specialty based on cloud computing	Data mining based mobile education platform for entrepreneurial courses of economic management	Mobile education platform for entrepreneurship courses of economic management specialty based on personalized recommendation
1	946.4	812.4	807.4
2	953.7	826.8	848.8
3	968.8	868.5	859.5
4	959.5	835.6	826.3
5	957.6	822.2	833.2
6	961.3	843.3	862.5
7	942.2	819.6	820.1
8	956.1	855.2	851.4
9	960.5	822.5	824.8
10	943.2	837.1	845.2

As can be seen from Table 1, the throughput of the cloud based management entrepreneurship course mobile education platform for 1000 concurrent requests is 954.9 MB/s, representing an increase of 120.6 MB/s and 117.0 MB/s over the data mining and personalized recommendation based mobile education platform.

As can be seen from Table 2, the throughput of the cloud -based entrepreneurship courses in economic management mobile education platform was 924.4 MB/s for 2000 concurrent requests, representing an increase of 220.3 MB/s and 211.3 MB/s over the data mining based and personalized recommendation based mobile education platforms.

Table 2. 2000 throughput comparison of requests (MB/s)

Number of tests	Mobile education platform for entrepreneurial courses of economic management specialty based on cloud computing	Data mining based mobile education platform for entrepreneurial courses of economic management	Mobile education platform for entrepreneurship courses of economic management specialty based on personalized recommendation
1	923.4	702.4	720.6
2	926.8	697.8	734.7
3	938.5	684.6	721.4
4	922.6	696.2	725.8
5	913.2	703.3	702.5
6	906.3	722.5	716.2
7	932.6	715.2	703.3
8	925.2	709.3	692.6
9	921.5	706.6	685.2
10	934.2	703.2	728.5

Table 3. 5000 throughput comparison of requests (MB/s)

Number of tests	Mobile education platform for entrepreneurial courses of economic management specialty based on cloud computing	Data mining based mobile education platform for entrepreneurial courses of economic management	Mobile education platform for entrepreneurship courses of economic management specialty based on personalized recommendation
1	706.4	647.4	668.3
2	711.8	668.8	626.6
3	712.5	659.7	632.5
4	705.6	646.4	613.4
5	708.2	622.2	605.8
6	723.3	635.8	622.2
7	725.2	613.7	631.6
8	706.5	602.2	644.2
9	711.6	661.6	605.5
10	718.2	624.8	638.3

According to Table 3, under 5000 concurrent requests, the throughput of the cloud based management entrepreneurship course mobile education platform is 712.9 MB/s, representing an increase of 74.6 MB/s and 84.1 MB/s over the data mining and personalized recommendation based mobile education platform.

Table 4. Throughput comparison of 8000 requests (MB/s)

Number of tests	Mobile education platform for entrepreneurial courses of economic management specialty based on cloud computing	Data mining based mobile education platform for entrepreneurial courses of economic management	Mobile education platform for entrepreneurship courses of economic management specialty based on personalized recommendation
1	512.4	458.9	411.2
2	506.8	456.7	419.6
3	515.6	428.4	423.0
4	512.2	426.6	436.5
5	519.5	437.2	433.4
6	508.2	430.5	437.8
7	507.6	459.8	452.2
8	511.3	462.3	449.3
9	506.7	438.2	416.6
10	523.8	441.0	435.2

As can be seen from Table 4, the throughput of the cloud based management entrepreneurship course mobile education platform for 8,000 concurrent requests is 512.4 MB/s, representing an increase of 68.4 MB/s and 80.9 MB/s over the data mining and personalized recommendation based mobile education platforms. The test results show that the mobile teaching platform constructed in this paper has good performance and meets the design requirements.

5 Conclusion

Entrepreneurship education is based on comprehensive quality education to develop and expand the quality of students. It opens up a new path for the cultivation of innovative and entrepreneurial talents. Curriculum is the carrier of education and teaching. The key to improve the quality of entrepreneurship education is curriculum. Mobile teaching platform can provide convenient conditions for the entrepreneurial courses of economic management, so that learning is no longer subject to geographical and time constraints. Based on cloud computing, this paper designs a mobile education platform for entrepreneurial courses of economic management specialty. The test results show that the platform has good performance and can improve the network throughput. Follow-up research can further refine the operation of the platform, sort out more teaching resources and upload them to the platform, which can integrate these accumulated entrepreneurial knowledge with the platform and continuously improve the mobile education platform.

Fund Project. Sichuan Provincial Key Research Base of Humanities and Social Sciences in Higher Education Institutions - Research Project of Newly-built Institutions Reform and Development Research Center: Discussion on Green Innovation and Entrepreneurship Education System of Newly-built Application-oriented Undergraduate Universities under the Background of "Innovation and Entrepreneurship" (Project No.: XJYX2021B07).

References

1. Zhao, D., Xi, Di., Li, Y.Y.: Application application of mobile teaching platform in college biology courses——practical exploration based on speed-learning platform. J. Hubei Open Vocational College **33**(20), 163–165 (2020)
2. Liu, T.W., Sun, H., Fung, W.: An artifact-based simulation method for teaching intellectual property management in an innovation and entrepreneurship course. Asian Case Res. J. **25**(02n03), 193–212 (2021)
3. Xiao, X., Liu, X., Xiao, Z.: Construction and application of computer virtual simulation teaching platform for medical testing. J. Phys. Conf. Ser. **1915**(4), 042074 (7 pp) (2021)
4. Wu, Y.-l., Huang, W.: Real-time task scheduling simulation of cloud computing based on resource delay perception. Comput. Simulation **38**(9), 490–494 (2021)
5. Maghsoudloo, M., Khoshavi, N.: Elastic HDFS: Interconnected distributed architecture for availability–scalability enhancement of large-scale cloud storages. J. Supercomput. **76**(1), 174–203 (2020)
6. Mostafaeipour, A., Rafsanjani, A.J., Ahmadi, M., et al.: Investigating the performance of Hadoop and Spark platforms on machine learning algorithms. J. Supercomput. **77**(2), 1273–1300 (2020)
7. Wu N.Z., Xing, X.S.: The occurrence logic of big data supporting large-scale personalized teaching. Lifelong Educ. Res. **32**(2), 20–28, 39 (2021)

Social Network Real Estate Advertisement Push Method Based on Big Data Analysis

Yun Du[1](✉) and Xuanqun Li[2,3,4]

[1] The University of Manchester, Manchester 999020, UK
duyun332@163.com
[2] Institute of Oceanographic Instrumentation, Qilu University of Technology (Shandong Academy of Sciences), Qingdao 266061, China
[3] Shandong Provincial Key Laboratory of Marine Monitoring Instrument Equipment Technology, Qingdao 266061, China
[4] National Engineering and Technological Research Center of Marine Monitoring Equipment, Qingdao 266061, China

Abstract. In order to solve the problem of large-scale user attribute identification of real estate advertising push, a social network real estate advertising push method based on big data analysis is proposed. First, according to the real estate advertising push strategy of social networks, it is refined and implemented level by level, focusing on specific target customer groups. Given the initial link of the original blog of the advertiser, the text features of the real estate advertising project are extracted by extracting the basic information of all the blogs in a circular manner. Secondly, analyze the social relations of users, mine the characteristics of social network users based on big data analysis, realize the classification and recognition of user attributes, and calculate the similarity between the two using similarity calculation formula. Finally, the calculation results are sorted in reverse order of similarity to generate a real estate advertisement recommendation list for users. The design method is tested on the epinions data set, and the test results show that the design method can improve the accuracy of recommendation and reduce the overall running time.

Keywords: Big data analysis · Social network · Real estate advertising · Advertising push · Text features · User features

1 Introduction

Real estate is a real economy, not a virtual economy, that builds real physical houses to meet social needs. Real estate transactions are accompanied by capital flows. The objects of transactions are physical assets. The cycle for completing transactions is at least one or two years, at most three or five years, or longer. With the development of computer technology and the substantial increase of Internet users, the data volume is growing exponentially. More and more customer data can be mastered and used by enterprises. People are in the era of big data. Following cloud computing and the Internet of things,

W. Fu and L. Yun (Eds.): ADHIP 2022, LNICST 468, pp. 620–633, 2023.
https://doi.org/10.1007/978-3-031-28787-9_46

big data has triggered another technological revolution in the digital economy industry, redefining the decision-making methods and results at different levels of the country, enterprises and individuals. The "high inventory" of the real estate industry nationwide makes real estate marketing more and more difficult. In the context of increasingly squeezed profit space, reducing marketing costs and seeking new profit growth points are the challenges faced by real estate marketers. The development of mathematical science has also helped the development of big data, collecting and mining big data from multiple platforms and applying mathematical algorithms and models for analysis. Big data analysis provides the basis for the prediction and decision-making of enterprise management, can accurately and effectively position the target market, focus on the target customers, and can achieve more accurate results in marketing work. The convenience and efficiency of the Internet are accelerating. This trend has also brought some enlightenment to real estate projects, marketing changes and service innovation. Internet enterprises are pioneers of big data applications, not only optimizing existing businesses through data, but also accelerating data-driven business innovation. Encouraged by the government and driven by the market, a large number of traditional enterprises have opened up the Internet market, and a large number of brand-new Internet enterprises have appeared in the favor of capital for Internet enterprises. The rapid increase of Internet enterprises has brought exponential growth of data volume. In this regard, the development model of the combination of Internet and industry provides the necessary conditions for the smooth transformation of real estate enterprises in the new economic environment. At the same time, consumers are increasingly pursuing diversification, customization and personalization of products, which also puts forward higher requirements for the development of enterprise marketing and the overall strategy of enterprises. How to accurately find customers who need to buy houses and match the real estate products developed by enterprises has always been a difficult problem for traditional marketing.

Reference [1] proposes an Internet advertisement push method based on personalized analysis. Starting from the overview of personalized advertisement recommendation, it analyzes in depth the key technologies of personalized advertisement recommendation in recent years, including data collection and preprocessing, user preference acquisition, personalized advertisement recommendation technology, etc. This paper statistically analyzes a variety of data sets and evaluation indicators used in personalized ad recommendation, and summarizes the current application of personalized ad recommendation in traditional Internet scenarios. Reference [2] proposes an advertising information push method based on user portraits, which uses big data technology to calculate user portraits and build effective research focuses. The information ontology extraction method based on user portrait realizes intelligent user portrait construction and completes the push of advertising information according to the different contents such as object, time and behavior. Reference [3] proposes an art advertisement push method based on user behavior information, which uses a similarity algorithm to calculate the similarity of user browsing media content, and obtains user behavior characteristics, that is, the user's preference for content. The threshold is set, and the content is regarded as the same cluster when the content similarity is above the threshold, and the classification of the content cluster of the user behavior information is completed. The freshness and dispersion of user behavior in the effective cluster are calculated, and the final weight of the effective cluster is obtained by combining the freshness factor. Finally, the final weight of the effective cluster and the user's preference for content are used to calculate the artistic advertisement push score, and the advertisements are sorted to select the advertisements suitable for the user. Although the above method can complete the push of advertisement information, it has the problems of long running time and low accuracy when directly applied to the push of real estate advertisements.

The real estate platform is to build a transaction bridge between buyers, sellers, housing sources and other parties. The real estate transaction market involves the new house, second-hand house and rental market. As one of the research hotspots of contemporary sociology, social networks are widely used in anthropology, epidemiology, biology, communication, economics, geography, information science, social psychology and other fields. Together with other complex networks, social networks promote the development of new fields of network science. With the launch of social networking platforms such as Weibo, wechat, easy to believe, anjuke and SouFun, and the popularity of mobile app applications, customers have increasingly recognized online access to real estate project information and sales services. The focus of applying big data analysis technology to precision marketing is to change the marketing perspective, focus on customer consumption behavior and preferences contained in data information, and comprehensively plan the marketing system. This paper proposes a social network real estate advertising push method based on big data analysis, which can effectively mine the value of data information and seek the sustainable development of real estate enterprises in the market through precision marketing.

2 Social Network Real Estate Advertisement Push Method Based on Big Data Analysis

2.1 Analysis of Social Network Real Estate Advertisement Push Strategy

Big data marketing innovation requires companies to match their ideas at the strategic level. These matching from small to large should cover key strategic elements such as data capabilities, decision-making models, strategic orientation, and organizational culture, and should clarify the matching process and dynamic evolution process between key strategic elements and innovation focus points. Only accurate information can make the dissemination more effective. The traditional push method cannot support the information push because of the inability to accurately push the scope, and instead it becomes a flood of spam. The main reason is that it lacks the support of customer characteristic data and cannot conduct detailed and accurate analysis. Customers are the starting point and end point of marketing activities. Enterprise big data marketing should establish the concept of customer participation in the whole process, especially to explore the customer demand level, which has an important impact on promoting the reform of marketing concepts. The design of the marketing plan needs to be guided by the marketing goals, and be refined and implemented step by step, and finally focus on the specific target customer group. Only by focusing on it can the efficiency of marketing be maximized. Then design targeted marketing campaign ideas (for example, product combination, channel selection and pricing, etc.) according to the target customer group to form the final marketing plan. The advertising push strategy of real estate enterprises is shown in Fig. 1.

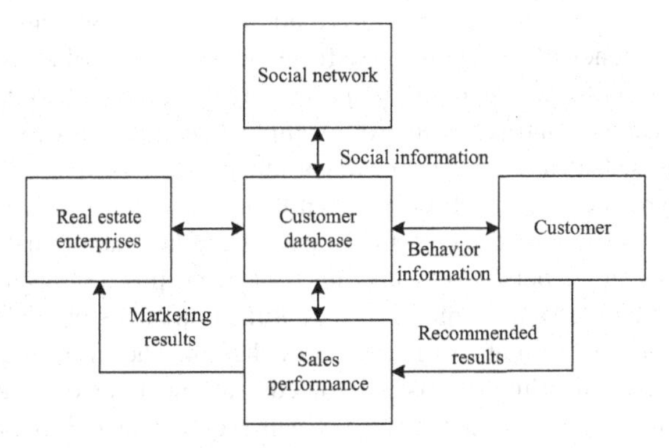

Fig. 1. Advertising push strategy of real estate enterprises

Analyze users' social relationships to distinguish between strengths and weaknesses and build a graph. Taking Weibo as an example, people only follow bloggers that interest them. After following, Weibo will recommend other bloggers in the same field, which is easy to attract users' interest. A social network is a structure composed of a set of binary relationships between social actors and social interactions among social actors. Social network theory helps to study the relationships between individuals, groups, organizations, and even entire social units in the network. A social network can be represented by a graph, social actors are represented by nodes in the network, and the relationship between nodes represents the interaction between social actors. It is important to note that customers are changing, and the design of the marketing plan needs to have a certain width of execution and needs to observe the dynamics of the customer group in real time, so that the developed marketing plan can be effectively implemented. Weibo is a kind of weak-relationship social interaction. The interaction and intersectionality depends on the social behavior of bloggers. The social heterogeneity among users is strong, and the probability of being converted into target customers is low. QQ groups and WeChat Moments belong to strong social relationships, and their social networks are more homogenous, and have a higher probability of being converted into target customers. Social influence is the result of the multiplication of the three variables of individuality, directness and the number of influence sources, and increases with the increase of each variable. If one of these three variables is 0 or significantly lower, the overall social influence of the node The force will be affected, expressed as:

$$Z(a) = f(H, C, S) \tag{1}$$

In formula (1), a represents the social network node; $Z(a)$ represents the overall social influence of node a; f represents the mapping relationship between individual nodes; H, C, and S represent the individuality, directness, and number of influence sources of the node, respectively. The strong relationship network is usually constructed by the user's friends, relatives and other people who have acquainted in reality, and the number is limited. In the weak relationship, there are not only friends of friends, but more strangers who do not know each other, forming a huge social network. Information can be It spreads rapidly in it. The dynamics and complexity of social networks make it impossible to obtain direct and accurate information. Therefore, we can only improve the dissemination efficiency of social influence from the perspective of individuality and the number of influence sources. With the support of big data, the screening of high-quality customers will be quick and accurate. For example, customers' browsing footprints on various shopping websites can determine the products they care about recently, and their life and work dynamics can be observed from their consumption conditions. By using the correlation analysis method, enterprises can obtain key target customer resources. In big data mining, customers' basic demands for privacy protection should be met, and attention should be paid to the exploration and utilization of customers' core values and the cultivation and transformation of customer behavior. The R&D and event planning based on a full understanding of customer needs can meet customer expectations for products, cater to customer needs, and truly achieve customized marketing that suits them.

2.2 Extracting Text Features of Real Estate Advertising Items

Based on the information boundary theory, this paper defines the consumer information boundary as the extent to which consumers allow advertisers to use consumers' personal network data, that is, the precision of online advertising. With the continuous improvement and development of Weibo, more and more users use Weibo to obtain news, and its platform has a lot of valuable user information, which has high research value and commercial value in today's Internet industry. On the one hand, the precision of online advertising will bring convenience to consumers in shopping and have a positive impact on consumer utility; on the other hand, it will cause consumers' privacy leakage concerns and have a negative impact on consumer utility. Utility determines its demand for goods, which in turn affects the profit effect of online advertising. The precision of online advertising requires advertisers to invest more costs, which affects the overall cost of online advertising. Generally speaking, the network service platform provides users with http request links. Users need to fill in the request parameters and the signature provided by the platform, and send the request to obtain service data. The real estate advertising data is generally a JSON object. This paper mainly uses Weibo advertisements to extract the text features of real estate advertisement items. The original text-based blog post information includes the blog post content, publication time, the number of likes and like lists, the number of retweets and retweets, and the number of comments and comment lists. In order to collect the data set of the network service, first collect the data of the third party as the data of the input interface of the service, and then encapsulate the http request provided by the platform based on the Java language, which is convenient to fill in the request parameters, obtain the service data in batches, and store the final running data. A dataset of input interfaces into the service [4]. Given the initial link of the advertiser's original blog post, extract the basic information of all blog posts through a loop. The process of obtaining the text of real estate advertisement items is shown in Fig. 2.

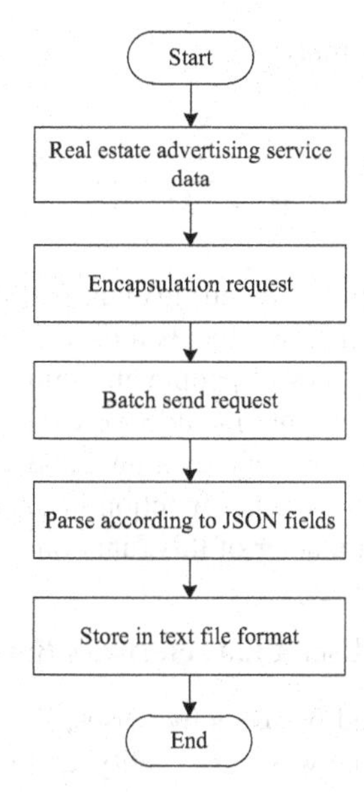

Fig. 2. Extraction process of real estate advertisement item text

Use fastjson to parse the returned JSON object according to the JSON fields, and write the data of the same JSON fields obtained in each request into the same file as the service output interface dataset [5]. Both the input dataset and the output dataset are stored in a text file format. After obtaining all microblog text data, this paper uses Chinese text cleaning technology to remove non-text data in the text. After analysis, the data collected from the Weibo platform contains a large number of HTML tags, text hyperlinks, and some special symbols. This article uses regular expressions to replace these non-text data with null characters. Extract some keywords of the text and convert the keywords into word vectors, then TEXT can be represented by a set of multiple word vectors, each word vector is a 200-dimensional real vector, namely:

$$X = \{\beta_1, \beta_2, \cdots, \beta_M\} \tag{2}$$

In formula (2), X represents the text vector; β represents the keyword vector; M is the number of keywords. The keyword vector can be represented in the following form:

$$\beta_M = [\varphi_1, \varphi_2, \cdots, \varphi_N]^T \tag{3}$$

In formula (3), φ represents a real vector; T represents a transposed matrix; N represents the dimension of the vector. Use Chinese word segmentation technology to achieve word segmentation. Considering the different abilities of articles expressed by different parts of speech, this paper selects "noun", "verb", "adjective" and "adverb" to express the main content of the article. The centroid of the word vector is calculated as the vector of the text [6]. The average value of each dimension value of all keyword

word vectors of the centroid, namely:

$$p(X) = \left[\frac{\sum_N \varphi_1}{M}, \frac{\sum_N \varphi_2}{M}, \cdots, \frac{\sum_N \varphi_M}{M} \right]^T \tag{4}$$

In formula (4), $p(X)$ represents the centroid of the text vector X. In order to make word segmentation more accurate, this article adds a user-defined dictionary, which mainly includes the latest words and names of people and things. A user-defined dictionary is a text file. The text file should not only include words, but also be marked with the part of speech of the word. Each word occupies a line, and all words are arranged in lines [7]. This article calls the ImportUserDict function in the pynlpir package to import the user-defined dictionary. The parameter of this function is the path of a text file.

2.3 Mining Social Network User Characteristics Based on Big Data Analysis

As the number of products used by users increases, if the user's utility in the product increases, it reflects the direct network externality, and if the number and effect of the product's complementary products also increases, it reflects the indirect network externality. According to the similarity of customer characteristics, customers are divided into several subgroups. The internal characteristics of each group are similar, but there are obvious differences between groups. Only by distinguishing different customer groups can we carry out targeted marketing, develop and provide products and services that are aligned with customers. Real estate companies obtain data through big data collection technology, and the sources of data are mainly divided into two parts: intranet and extranet. The extranet mainly captures housing rental and sales information and transaction information from urban housing construction committees and competing websites to provide data support for real estate market research. Before the support of big data, traditional analysis methods, such as demographic factors and geographical factors, can only provide a relatively vague customer profile and cannot achieve customer segmentation in the true sense, so precision marketing cannot be carried out smoothly [8]. The intranet mainly obtains user preference information from the websites of real estate companies, including price, area, apartment type, time, etc., on the one hand, to provide data for market research, and on the other hand to provide data support for marketing products. The main attributes of customer behavior pattern analysis are shown in Fig. 3.

The user interest mining method based on big data analysis is mainly divided into three steps: The first step is to use Python web crawler technology to obtain the target user's watch list, and add the Weibo IDs of the five users with the largest number of followers in the watch list. User collection. These user portrait tags are calculated based on the user's browsing preferences, which facilitates the communication between sales and users and improves communication efficiency. If the customer contacts multiple consultants, the recommended properties that meet the user's expectations will be more trusted by the user. Similarly, obtain the target user's fan list, and add the Weibo ID of the five users with the largest number of fans in the fan list to the user set. It should be noted that the variable elements of customer analysis are not static and need to be updated and optimized at any time according to market changes. The user portrait contains the user's

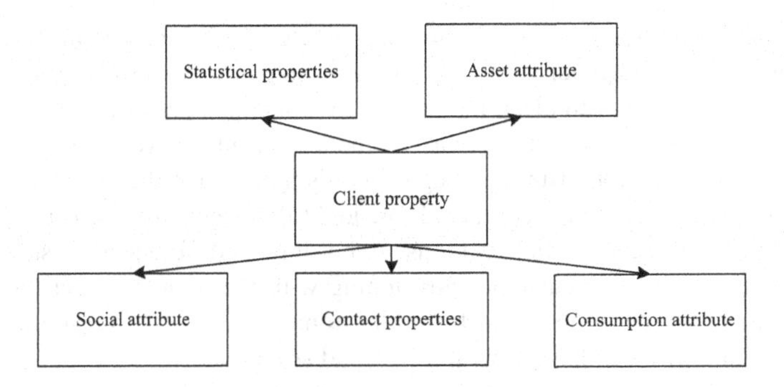

Fig. 3. Customer attributes

information, channel sources, areas of concern, real estate of interest, area of interest, and recommended time. According to the user's microblog ID in the user collection, the original text-based blog posts of these users in the database are retrieved and formed into a large document. When the user scale exceeds the critical value, it will strengthen the advantage of the number of products on the user's utility, thereby affecting consumers' purchase of products, and this has become the cause of network externalities that people are more concerned about. To achieve the fusion of the three user interest lists, the most important thing is to merge synonyms, that is, to calculate the similarity between two words and combine the two words whose similarity is greater than a certain threshold into the same word [9]. Big data analysis can filter out valuable information from massive, complex and unrelated customer data. The conceptual semantics of a content word consists of four parts: similarity of basic semantic descriptions, similarity of other basic semantic descriptions, similarity of relational semantic descriptions, and similarity of symbolic semantic descriptions. The formula for calculating the similarity of two words is as follows:

$$\gamma(\beta_1, \beta_2) = \sum_{4} \lambda_j \prod_{4} \eta_j \tag{5}$$

In formula (5), $\gamma(\beta_1, \beta_2)$ represents the similarity between the two words β_1 and β_2; λ_j represents the preset parameter; η_j represents the conceptual semantic similarity of the content word; j represents each part of the semantic composition. Accurate analysis and judgment of customer behavior patterns and value through precise segmentation technology is the mainstream method in the field of customer segmentation. The formula for calculating the similarity between semes is as follows:

$$\vartheta(\mu_1, \mu_2) = \frac{\nu}{\nu + D} \tag{6}$$

In formula (6), $\vartheta(\mu_1, \mu_2)$ represents the similarity between the two semes μ_1 and μ_2; ν represents the adjustment parameter, which is 1.5 in this paper; D represents the distance between the two semes. Mining and vectorizing user interests from user behavior records using big data analysis. According to the evaluation result of customer value, carry out value positioning to customers, and determine customer level classification. The variables of customer value positioning include: customer's current living situation, family demographics, occupational characteristics, family annual income, purchasing needs, source channels, etc. [10]. The third step is to use the Chinese text mining preprocessing technology to clean the text of the document and use the Chinese word segmentation technology to perform word segmentation, and then convert the segmented text data into word vectors using the bag of words model, and then use the LDA topic model to mine the document's content. Topic and topic keywords, so as to obtain a set of user interests based on social relations. According to different real estate projects, match customer value and customer positioning with the project, select the customer group that matches the project, and form a customer system. The customer hierarchy generated in this way will have a pyramid-shaped structure.

2.4 Design a Social Network Real Estate Advertising Push Model

The influence of social network users has a significant impact on the dissemination of advertisements. Understanding the influence dissemination mode among social network users can better understand the advertising recommendation process. The dissemination of rumors and the replacement of information in social networks are time-sensitive, especially when a crisis occurs, the news media must report quickly and timely, and advertising recommendations should also give full play to the timeliness, and push advertisements to the entire social network in the shortest time. At the consumer level, because the socialization of online advertisements has a positive impact on consumer utility, consumers should make more efforts to socialize online advertisements, such as online advertisements in the form of Weibo or WeChat Moments. Social interaction behaviors of online advertising, such as likes, comments, and retweets, to improve consumer utility and welfare, especially when consumer social preferences are high. The centrality and similarity of nodes in a social network have a profound impact on the dissemination of information in the network and the importance of nodes in the network. This section defines centrality as follows:

$$\chi(o) = \frac{|\omega(o)|}{\max(|\omega|)} \tag{7}$$

In formula (7), $\chi(o)$ represents the centrality of node o; $\omega(o)$ represents the set of all nodes adjacent to node o (including in-degree nodes and out-degree nodes); ω is the length of all nodes in the statistical network. The influence of a node is not only related to the relationship between the node and neighbor nodes, but also the relationship between neighbor nodes. Personalized advertisement recommendation is to recommend advertisements that may be of interest to users based on their interests. In this paper, user interests are mainly manifested in two aspects. One is a user interest model constructed based on user-related blog posts, which extracts the user's interest features. One is the user's interaction record with real estate advertisements, which reflects the user's interest through the user-advertising score. Among the three social behaviors of liking, commenting and forwarding online advertisements, consumers should also focus on the social behavior of online advertisements. Especially for Weibo advertisements, users' likes on advertisements can best improve their consumer welfare. Relationships in social networks are complex and ever-changing, and two unrelated neighbor nodes may become friends in the future. Advertising recommendation in social networks is based on the word-of-mouth effect of marketing. If there are more nodes with friend relationships among neighbor nodes around a node, it means that its neighbor nodes are more closely connected, and the probability of successful advertising recommendation is higher. In the three-dimensional advertising precision perception of browsing habits, positioning information and chat records, advertising precision based on user browsing habits has a positive impact on consumer utility, while advertising precision based on user positioning information has a negative impact on consumption. Utility. Calculate the similarity of users and obtain a set of similar users. The similarity of users is mainly calculated by the similarity of interest features between users and the similarity of scores between users. In a social network, the number of out-degrees of a node represents the number of neighbor nodes that believe in the node. The more out-degree nodes of a node, the more nodes that follow the node, and the greater the influence in the social network. When a node recommends advertisements to neighboring nodes, the higher the number of recommendable nodes, the larger the influence scope. According to the user interest feature vector of each user, the cosine distance formula is used to calculate the interest similarity between the candidate set user and the target user. Calculated as follows:

$$\varsigma(q_1, q_2) = \frac{\sum_k q_1 q_2}{\sqrt{\sum_k q_1^2}\sqrt{\sum_k q_2^2}} \tag{8}$$

In formula (8), $\varsigma(q_1, q_2)$ represents the similarity between user feature q_1 and item feature q_2; k represents the spatial dimension of the feature vector. Finally, the 10 users with the largest similarity are obtained according to the reverse ranking of users to form a set of similar users. On the basis of obtaining user characteristics and item characteristics, the similarity between the two is calculated by the similarity calculation formula, and finally the calculation results are sorted in reverse order of similarity to generate a recommendation list for users. Through the analysis of the existing data, after obtaining the characteristics of different customer groups, it is necessary to formulate targeted marketing strategies for each customer group based on the needs of the enterprise. In the process of formulating marketing strategies, staged marketing objectives are required. Such as cross-selling, increasing the number of customers, etc. So far, the design of social network real estate advertisement push method based on big data analysis has been completed.

3 Experimental Study

3.1 Experiment Preparation

In this paper, we evaluate the advertising push results of the designed method through experiments. The experiment tested the social network real estate advertising push method based on big data analysis on the Epinions data set. Epinions is a consumer review website established in 1999. In the website, users can read old and new reviews on various items to decide whether to buy items. The website uses a scoring mechanism to determine the reliability of each review. Users can also selectively establish interactive relationships with other users. The development environment of this paper is as follows: the operating system is Windows10 64 bit operating system, the memory is 8g, the algorithm development language is python, Pycharm is used as the development environment, and the Scikitlearn toolkit based on Python is used for experiments. Because of the huge amount of data and the long running time, this experiment excludes the isolated nodes in the Epinions data set, and selects 10000 nodes out of 49290 nodes as the number of users receiving advertisements for the experiment.

3.2 Results and Analysis

Choosing a reasonable standard is of great significance to the scientificity of the experimental results. This paper evaluates the performance of the social network real estate advertising push method based on big data analysis from three aspects: recommendation accuracy, running time and user satisfaction.

The experimental results of this method are compared with those of reference [2] and reference [3]. The recommended accuracy results of 10000 user nodes are shown in Table 1.

Table 1. Comparison results of recommendation accuracy of 10000 user nodes

Testing frequency	Social network real estate advertising push method based on big data analysis/%	Reference [2] method/%	Reference [3] method/%
1	99.716	70.626	77.587
2	98.723	71.639	78.574
3	97.718	74.648	76.568
4	99.665	73.615	75.586
5	98.680	72.623	76.575
6	98.691	72.626	74.597
7	99.674	71.608	76.588
8	98.657	70.635	77.554
9	97.692	71.611	78.567
10	98.718	73.624	76.582
Mean value	98.79	72.32	76.87

In the test with 10,000 advertising user nodes, the average recommendation accuracy of the social network real estate advertisement push method based on big data analysis is 98.79/%, which is 26.47% higher than the average recommendation accuracy of the reference [2] method, and is higher than the reference [2]. The mean recommendation accuracy of the method [3] is improved by 21.92%. Therefore, it shows that the method in this paper can improve the push accuracy of real estate advertisements.

The running time test results of 10000 user nodes are shown in Table 2.

Table 2. Comparison results of running time of 10000 user nodes (ms)

Testing frequency	Social network real estate advertising push method based on big data analysis	Reference [2] method	Reference [3] method
1	2506	2946	3174
2	2679	2988	3268
3	2785	3064	3350
4	2663	2926	3246
5	2620	2955	3387
6	2855	2977	3221
7	2787	3012	3102
8	2611	3120	3368
9	2642	3253	3297
10	2778	3179	3392
Mean value	2692.6	3042	3280.5

In the test with 10000 advertisement user nodes, the running time of the social network real estate advertisement push method based on big data analysis is 2693 ms, which is 349.4 ms and 587.9 ms shorter than the reference [2] method and the reference [3] method. With the increase of network scale, the running time of each advertising push method increases significantly, and the running time of the method designed in this paper is lower than that of the two comparison methods. The efficiency of the social network real estate advertisement push method based on big data analysis is verified on the real network dataset. The results show that the design method in this paper improves the push accuracy while ensuring the operation efficiency.

From the comparison results of user satisfaction shown in Table 3, it can be seen that the user satisfaction of the method in this paper is significantly higher than that of the two literature comparison methods. The highest user satisfaction of the method in this paper is 99.5%, the highest user satisfaction of the method in reference [2] is 59.3%, and the highest user satisfaction of the method in reference [3] is 62.4%.

Table 3. User satisfaction result (%)

Testing frequency	Social network real estate advertising push method based on big data analysis	Reference [2] method	Reference [3] method
1	95.0	58.6	59.6
2	95.6	58.7	59.3
3	97.2	57.5	57.5
4	96.2	53.6	58.0
5	98.8	59.3	58.1
6	95.4	56.0	59.9
7	97.7	58.1	61.7
8	99.5	55.2	62.4
9	98.6	57.4	60.5
10	96.3	58.6	60.8

4 Concluding Remarks

The real estate platform faces the characteristics of complex real estate transaction market and relatively long transaction cycle, which makes it necessary to deal with the storage, analysis and matching of massive housing listings, hundreds of millions of buyers and numerous sales information. How to quickly and accurately complete the above information Processing requires the use of big data analysis technology. This paper proposes a social network real estate advertisement push method based on big data analysis, which can improve the recommendation accuracy and reduce the overall running time. The user's interest is constantly changing with time, and this paper will construct a dynamic user interest model to extract user interest features in the follow-up research.

References

1. Zhang, Y.-j., Dong, Z., Meng, X.-W.: Research on personalized advertising recommendation systems and their applications. Chin. J. Comput. **44**(3), 531–563 (2021)
2. Hu, J.: Information intelligent push method based on user profile. Wuxian Hulian Keji **17**(19), 161–162 (2020)
3. Li, X.: Design of pushing mechanism of art advertising media based on user behavior information. Mod. Electronics Techn. **43**(1), 143–147 (2020)
4. Wang, C.: Design of web advertising intelligent push system based on visual information transmission. Mod. Electronics Tech. **43**(20), 160–163 (2020)
5. Zhihua, X., Yuying, L.: Video recognition and adaptive recommendation algorithm based on machine learning. J. Shenyang Univ. Technol. **44**(03), 336–340 (2022)
6. Chen, J.-x, Zhang, T.: Intelligent push simulation of interactive design based on data sparse feature. Comput. Simulation **37**(12), 166–170 (2020)
7. Wang, C.: Detection model of fake commercial advertisements on social network based on block chain technology. Economic Res. Guide (09), 44–47+95 (2021)

8. Tang, H., Zeng, J., Li, F., et al.: Point of interest recommendation based on location category and social network. J. Chongqing Univ. (Nat. Sci. Ed.), **43**(7), 42–50 (2020)
9. Su, B., Liang, D.: Face face detection in advertising recommendation system based on face-eyes co-detector. Comput. Technol, Developm. **31**(7), 134–139 (2021)
10. Xun, Y.-l, Bi, H.-m, Zhang, J.-f: Heterogeneous social network recommendation based on weak ties. Comput. Eng. Design **42**(6), 1526–1534 (2021)

Prediction Method of Crack Depth of Concrete Building Components Based on Ultrasonic Signal

Kangyan Zeng[1(✉)], Yan Zheng[1(✉)], Jiayuan Xie[2], and Caixia Zuo[1]

[1] Department of Architectural Engineering, Chongqing College of Architecture and Technology, 401331 Chongqing, China
zheny1546@163.com

[2] Chongqing Architectural Technician College, Chongqing 400041, China

Abstract. Cracks in concrete components have a serious impact on the safety of building structures. On the one hand, with the increase of service life, cracks will reduce the safety of building structures, and with the deepening of cracks, the service life of building structures will be reduced. Therefore, it is very necessary to predict cracks in concrete components. In the crack prediction of concrete building components, the predicted results deviate from the actual value due to the deviation of the measured strain value of concrete. Based on ultrasonic signal, a method for predicting crack depth of concrete building components is proposed. In the process of concrete ultrasonic transmission, the number and length of micro cracks will increase and expand due to stress concentration. According to this phenomenon, the finite element simulation of concrete building components is carried out to obtain the damage model. The characteristics of concrete cracks are extracted based on ultrasonic signals, and the law of acoustic frequency changing with time can also reflect the state of medium stress. The extracted crack signal features of concrete building components are input into CNN model for prediction and recognition. The test results show that the prediction method of crack depth of concrete building components based on ultrasonic signal can improve the accuracy of prediction results and has high engineering application value.

Keywords: Ultrasonic signal · Concrete · Building components · Crack analysis · Crack depth · Depth prediction

1 Introduction

With the rapid growth of concrete infrastructure construction in China, more and more attention has been paid to the health prediction and maintenance of later facilities. Cracks will inevitably appear in the construction or long-term use of concrete structures. If cracks are not found and repaired in time, it will not only affect the appearance, but also affect the normal use of concrete structures. Therefore, it is very important to predict concrete cracks.

W. Fu and L. Yun (Eds.): ADHIP 2022, LNICST 468, pp. 634–647, 2023.
https://doi.org/10.1007/978-3-031-28787-9_47

Concrete is one of the most widely used building materials in infrastructure construction, which is widely used in infrastructure construction such as roads, bridges, tunnels, dams and houses. Cracks are inevitable in the construction process or long-term use of concrete structures. The main reasons for cracks are as follows: cracks are caused by uneven stress due to low tensile strength of concrete structures and excessive load for a long time; Because the concrete structure is often exposed to the outside, it is easy to deform under the influence of external temperature changes, and it is often eroded by rain to cause cracks; The quality of construction materials and the level of construction technology directly affect the quality and service life of concrete structures [1].

If the cracks in the concrete structure can not be found and repaired in time, the damage to the structure will be aggravated. When the depth and width of the crack exceed the critical value borne by the concrete structure, it will not only affect the external beauty, but also affect the normal use of the concrete structure, and may even lead to major safety accidents and unnecessary losses. Therefore, the prediction and prevention of concrete cracks are of great significance for maintaining the stability of concrete structures and extending the service life.

The method in reference [2] proposes a concrete crack prediction method based on the improved convolutional neural network, which uses the entropy threshold method to process the image, improves the traditional convolutional neural network from three aspects, and uses the improved convolutional neural network to train the sample image, so as to complete the prediction of concrete cracks. Reference [3] proposes a concrete crack prediction method based on laser ultrasonic technology, establishes a plane strain finite element model, and studies the interaction law between laser-excited surface waves and concrete surface cracks. Analyze the scattering echo characteristics of surface wave and crack front (the longitudinal edge where the wave and crack first act), discuss the influence of crack depth on the time difference of scattering echo characteristic points, and complete the prediction of concrete cracks. Reference [4] proposes a concrete crack prediction method based on the array ultrasonic imaging method. According to the principle and calculation method of the array ultrasonic imaging method to predict the crack depth, the imaging prediction is carried out across a single crack and the whole specimen, and the prediction results of concrete cracks are obtained. Although the above method can complete the prediction of concrete cracks, there are problems such as insufficient prediction accuracy and low data recall.

Ultrasonic nondestructive testing technology is widely used in the field of civil engineering in recent years because of its convenient, fast and accurate operation. The frequency of the acoustic wave method used in the ultrasonic method is far lower than that of the electromagnetic wave, which can achieve nondestructive testing in a large depth range in concrete. Moreover, the price of ultrasonic equipment is low and the detection algorithm is flexible. Therefore, this paper proposes a method for predicting the crack depth of concrete building components based on ultrasonic signals. The overall research technical route of this method is as follows: through ultrasonic technology, the number and length data of concrete cracks are collected, and the finite element simulation of concrete components is carried out according to the collected data. Based on the finite simulation results, the damage model of components is established. The ultrasonic frequency is calculated to extract the characteristics of concrete cracks and the stress state of concrete structures. The extracted crack features are input into the CNN model to complete the prediction of concrete cracks.

2 Prediction Method of Crack Depth of Concrete Building Components Based on Ultrasonic Signal

2.1 Ultrasonic Conduction Mode of Concrete

At the meso material level, concrete is a multiphase composite composed of mortar, aggregate, interface transition zone and various defects such as pores and inclusions. The failure process of concrete structure is a progressive process of damage, damage accumulation, macro crack occurrence and macro crack propagation. Under the action of no external load, there are a certain amount of micro cracks in it. When the crack length and stress concentration of concrete are low, the number of cracks will increase. The conduction mode of ultrasonic wave in concrete building components is shown in Fig. 1.

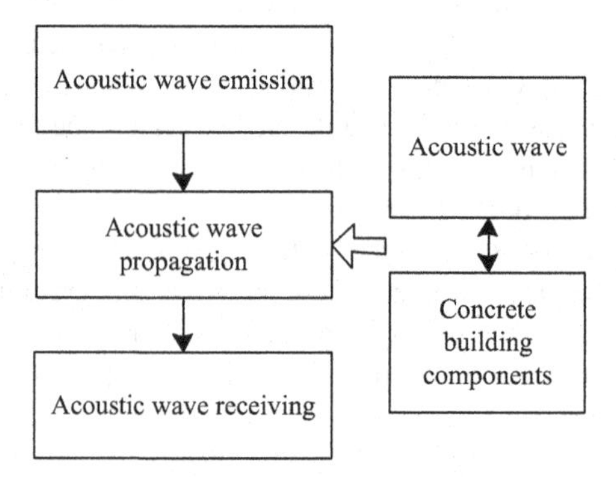

Fig. 1. Ultrasonic conduction mode

Assuming that the crack inside the concrete is a narrow cut, the inhomogeneity of the medium will cause slight differences in the sound pressure and density at each particle, which is expressed as additional sound pressure and additional density. When the displacement vector of a particle caused by a wave such as an elastic wave is small, any product of higher-order vectors can be ignored. Therefore, for any point in the inhomogeneous medium, its sound pressure and density are expressed as:

$$\begin{cases} A = A_0 + A_1(t) \\ B = B_0 + B_1(t) \end{cases} \tag{1}$$

In formula (1), A_0 represents the average sound pressure, B_0 represents the average density, A and B respectively represent the sound pressure and density at any point in the concrete building component; A_1 and B_1 represent additional sound pressure and additional density; t represents the propagation time of ultrasonic wave.

According to the law of conservation of mass and momentum, it can be proved that the spatial and temporal distribution of sound pressure in medium is closely related to the spatial distribution of density. For the damage equation of concrete crack, it is considered that when the concrete is subjected to stress, the crack volume changes accordingly, resulting in the displacement of particles in the material. For any point in the non-uniform subspace, the sound field propagating to the particle satisfies the free boundary Green's function, and the calculation formula is as follows:

$$W(r_1) = \frac{\alpha t \beta}{4\pi \|r - r_1\|} \tag{2}$$

In formula (2), W represents the sound field at the particle; α represents scattering operator; β represents wave velocity; r and r_1 represent any point and particle respectively.

The second-order amplitude generated by ultrasonic propagation is directly proportional to the square of wave propagation distance, frequency and intensity. The total additional scattered sound field can be obtained by integrating the fields of all points in space at the particle. When the ultrasonic ranging is certain, the higher the intensity and frequency of the input wave, the more obvious the second-order amplitude. However, the amplitude of the input wave is limited. The second-order harmonic of ultrasonic is often detected by inputting a higher frequency. Therefore, concrete has randomness at the meso material level, and its density, elastic modulus and other mechanical parameters are random quantities related to spatial location. However, the higher the ultrasonic frequency, the corresponding signal attenuation will increase. Because the wavelength of ultrasonic wave in concrete is equivalent to the size of coarse aggregate, the propagation of ultrasonic wave will be affected by the non-uniformity of concrete medium (interference), resulting in the randomness of the measured ultrasonic signal.

2.2 Finite Element Simulation of Concrete Building Components

In the process of using ultrasonic method to detect concrete structure, the received signal often carries complex information inside the structure. Through fast Fourier transform, we can explore the law of ultrasonic propagation inside the structure. Concrete, rock and other materials are structural analysis based on the assumption of continuous medium and the concepts of stress and strain. ABAQUS is a set of highly functional finite element software, which can solve all engineering linear problems to nonlinear problems. The specific flow chart of finite element analysis is as follows (Fig. 2):

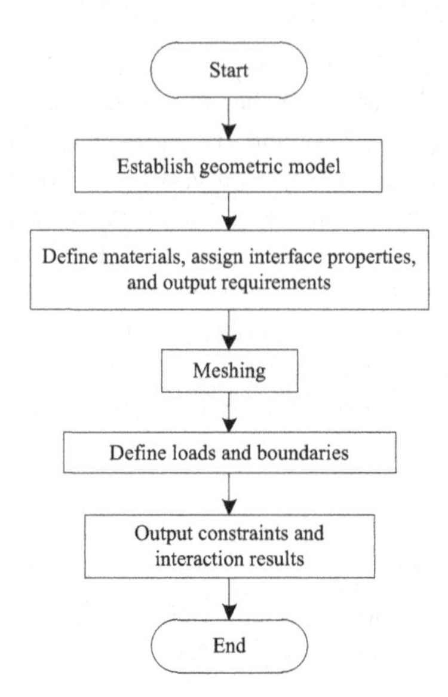

Fig. 2. Finite element analysis process

When establishing the constitutive model of damage, the study of damage variables is often the most important content. For the definition of damage variables, we can study a representative volume element in solid materials from a macro point of view and use the method of continuum mechanics to investigate the changes of macro mechanical property parameters caused by damage. In the step of establishing the geometric model, the aggregate of concrete building components is randomly generated according to the two-dimensional circle, and the coordinates of the corresponding aggregate are read in MATLAB. The corresponding geometric model is directly established in ABAQUS, and three material properties are defined, including aggregate, mortar and interface. As long as the external load does not continue to increase, the cracks in the concrete will not continue to increase and expand, and new cracks will appear, and a few cracks will be closed during unloading. When defining the analysis step, the time length needs to meet the following constraints:

$$\Delta\tau = \frac{1}{20\varphi_{\max}} \tag{3}$$

In formula (3), $\Delta\tau$ represents the length of time; $\varphi_{\max}$ represents the maximum frequency of time domain excitation.

The integrity of concrete materials has not changed, and the constitutive relationship of concrete at this stage can also be approximately regarded as linear. Since the shear wave velocity is smaller than the longitudinal wave velocity, that is, the transverse wave length is smaller than the longitudinal wave wavelength, the minimum wavelength of the shear wave (generally 10 nodes) is used to control the grid size. In fact, at this stage of crack formation, the number and length of mortar cracks and interface cracks are very small and can be ignored. In grid division, except that the grid size close to aggregate and mortar is 0.8 mm, the other grid sizes are LMM. If the aggregate in concrete is not easy to be damaged, its density is defined as high. From the physical level, the ultrasonic nonlinear coefficient is a quantitative index that can characterize the degree of waveform distortion when ultrasonic waves pass through concrete materials. The strength of cement slurry and interface materials is relatively low and easy to be damaged, so the concrete damage model is used to describe the nonlinearity of loaded concrete in the finite element software simulation. Ultrasonic nonlinear coefficient can not only reflect the damage degree of material, but also reflect the growth rate of damage variable with strain. In traditional linear ultrasonic testing, the value of ultrasonic wave velocity will be directly affected by the elastic modulus of materials. In the stress field, referring to the concrete damage plasticity (CDP) model in ABAQUS, it mainly uses the damage factor to express the stiffness degradation of materials. The elastic modulus at any time can be expressed as:

$$P = P_0(1 - \chi)^2 \tag{4}$$

In formula (4), P and P_0 represent the elastic modulus and initial modulus of any node; χ represents the damage factor.

Under the action of long-term continuous load, while the concrete material is creep, the hydration reaction also continues. The hydration reaction makes the concrete material more dense and the elastic modulus increases gradually. In the sound field, in order to simulate the propagation of ultrasonic wave in concrete, the signal modulated by harm window function is used as the excitation signal in this paper. Considering the frequency resolution and spectrum leakage, Hanning window is adopted this time [5]. The corresponding function form is:

$$\gamma(t) = \frac{1}{2}\left[1 - \cos\left(\frac{2\pi\eta t}{m}\right)\right] \tag{5}$$

In formula (5), $\gamma(t)$ represents Hanning window function; η represents frequency; m indicates the number of windows.

Ultrasonic signal is constrained by subjective factors (human operation) and external conditions (instrument and environment), so it is very necessary to verify or predict with the help of relevant finite element calculation software.

2.3 Extraction of Concrete Crack Characteristics Based on Ultrasonic Signal

As a measurement signal, ultrasonic itself has a certain degree of uncertainty. The measurement error caused by the accuracy of the acquisition instrument and the standard deviation of the amplitude are analyzed. In addition, the measurement error of the amplitude of the waveform will also be caused by the ambient temperature and humidity, ambient noise and coupling system. The crack propagation and extension will occur when the concrete is under the stress that is not higher than the critical stress of failure for a long time, which is about 70% - 90% of the stress level. At this time, the cracks will expand to the mortar, resulting in the increase of mortar cracks, and the cracks will be connected with each other to form relatively large cracks. Therefore, the interior of the concrete will be affected and the integrity will be weakened, and the stress–strain relationship is also nonlinear [6]. However, as long as the load remains unchanged or unloaded, the crack propagation will tend to stop. The primary goal of ultrasonic coda test is to improve the stability of waveform and control the quality of coda data, which is conducive to analyzing the load relationship between stress amplitude matrix and characteristic vector. For an acoustic acquisition signal, from the perspective of the field of digital signal processing, the signal is the equal spacing sampling of the real analog signal, and the sampling frequency determines the accuracy of the digital signal [7]. In each frequency position combination, it should also be noted that several pairs of identical signals will be generated during the transmission and reception of ultrasonic signals. This is determined by acoustic reciprocity, so only one signal needs to be taken for analysis. Using MATLAB to carry out fast Fourier transform FFT on the discrete acoustic signal data, the complex set of the acoustic signal about the number of sampling points can be obtained.

The calculation formula of amplitude $\mu(s)$ corresponding to the s-th frequency is as follows:

$$\mu(s) = \vartheta(t)e^{-\frac{2\pi st}{n}} \tag{6}$$

In formula (6), $\vartheta(t)$ represents the amplitude sequence corresponding to the discrete time of the continuous signal $\phi(t)$, n represents the number of sampling points; e is the natural constant. In order to facilitate observation, in the subgraph corresponding to each frequency position combination, the balance position of each signal is shifted upward by a certain value relative to the balance position of the previous signal to avoid overlap [8], and the complex part is expressed by the triangular function through the Euler formula.

$\mu(s)$ is a conjugate complex set symmetrical about s, that is, the digital signals at these sampling points can obtain $\frac{n}{2} + 1$ frequency amplitude information including zero frequency. The arrival time of ultrasonic signals collected at the same location is almost the same even if the frequency is different; Even at the same distance from the excitation source and the same frequency, the amplitude of the ultrasonic signal has great variability; Within the measurement frequency range, the waveform consistency is better with the increase of frequency [9]. At this time, the expression of the continuous signal with respect to the time variable t can be deduced from the complex set obtained by the FFT processing of any waveform $\vartheta(t)$ with respect to the discrete time series. The calculation formula is as follows:

$$\vartheta(t) = |\mu(s)| \cos\left(\frac{2\pi st}{n} + \lambda_0\right) \tag{7}$$

In Eq. (7), λ_0 represents the initial phase, which is the angle between $\vartheta(t)$ and the real axis on the complex plane.

The discrete Fourier transform of ultrasonic sampling data can fully display the frequency domain information. On the premise that the total number of samples remains unchanged, its relationship with the time-domain signal is that the higher the sampling frequency is, the higher the resolution of the time-domain signal is and the lower the resolution of the frequency-domain signal is. The speed of sound is equal to the ratio of the propagation distance to the arrival time of the wave. The propagation distance of wave is known information, that is, the distance between the transducer collecting the signal and the excitation transducer. In order to determine the arrival time of the wave, this paper uses the threshold method, that is, the maximum value of time-domain noise is defined as the threshold, and the time corresponding to the first sampling point exceeding the threshold is the arrival time of the wave. In practical engineering, the receiving frequency of ultrasonic detection signal is often concentrated in one frequency band. When analyzing the signal in frequency domain, it is necessary to conditionally amplify the resolution of useful frequency band and reduce the resolution of useless frequency band (such as noise) [10]. Finally, the variation law of acoustic frequency with time can also reflect the stress state of the medium, highlighting the time-domain characteristics of ultrasonic signal at the cracks of concrete building components.

2.4 The Prediction Model of Crack Depth of Building Components is Established

Based on the signal characteristics of cracks in concrete building components, a crack depth prediction model is established. The extracted crack features of concrete building components are input into CNN model for classification and recognition. In this paper, inception V3 neural network is used for training and recognition. Inception V3 is a CNN model proposed by Google. The biggest feature of the network model is the mixed layer, which is a multi-layer network and network nested structure, that is, the original node is also a network. Compared with V2 version, the biggest change of V3 version is to divide the convolution kernel of 7*7 into two one-dimensional convolution kernels of 7*1 and 1*7. Splitting one convolution kernel into two convolution kernels can not only accelerate the calculation speed, but also further deepen the network, so as to strengthen the nonlinear characteristics of the network. When training CNN model, it is necessary to select an appropriate target loss function to evaluate the consistency between the crack prediction results and the real label. Cross entropy loss function is one of the most widely used loss functions in deep learning. The cross entropy loss function is still used in this paper. The formula of cross entropy function is shown in Eq. (8).

$$F = z \log z' + (1 - z) \log\left(1 - z'\right) \tag{8}$$

In formula (8), F represents the cross entropy function; z represents the crack value of the real mark; z' is the predicted value of the model.

The high-order harmonic signal is sensitive to the cracks in the damage area, and the cracks are randomly generated and uncontrollable in the signal propagation path, resulting in faster energy attenuation of the high-order harmonic signal. In addition, the human operation error is easy to cause the dispersion of the test result data. This can also explain the so-called "filtering function of concrete", that is, high-order signal waves are filtered with the increase of structural damage. After determining the loss function, the function is optimized by small batch gradient descent. Each gradient update selects a small number of samples randomly from the training samples for learning, which reduces the oscillation of the convergence process. If the learning rate is set too small, the convergence speed of the model will be very slow. If the learning rate is set too large, the model will oscillate around the minimum or even deviate. In order to solve the problem that the model oscillates back and forth at the local minimum, the gradient descent method with momentum is generated. Suppose that after a neural network model is trained on the data set, the full connection layer at the top of the model is removed, that is, the layer that outputs the set category probability is removed, and the rest is fixed as a feature extractor, Then train a linear classifier on a new data set according to the classification set by yourself. The commonly used linear classifier is softmax classifier, which is connected with the feature extractor to customize its own task. So far, the design of crack depth prediction method of concrete building components based on ultrasonic signal has been completed.

3 Experimental Study

3.1 Experiment Preparation

The section of the concrete cube specimen in this experiment is designed to be 1000 mm * 1000 mm * 2000 mm, and a small amount of reinforcement is arranged to prevent temperature cracks. The experimental system is composed of signal generator, signal amplifier, transducer, oscilloscope and computer. After the electrical signal emitted by the signal generator is amplified, it is transmitted into the concrete through the transmitting transducer (the electrical signal is converted into acoustic signal), and then the receiving transducer (the acoustic signal is converted into electrical signal) feeds back the signal carrying the damage information of concrete material to the oscillograph, and then Fourier transform the time-domain signal to obtain the spectral characteristics of the signal. The standard value of the compressive strength of the concrete used in the experiment is 30 MPa. When pouring the concrete, it shall be fully vibrated to discharge the bubbles. The concrete pouring is in summer, and the ambient humidity and temperature are relatively stable. Therefore, under natural conditions, the test piece is covered with gunny bags and watered for 28 days to ensure that the test piece has no visible cracks and hidden cracks as far as possible. The particle size of coarse aggregate for concrete is 5 mm to 25 mm. The size of smart aggregate (SA) is equivalent to that of coarse aggregate, so it will not interfere with the mechanical properties of concrete. The water cement ratio of concrete material is 0.47. Ubuntu16.04 LTS 64 bit operating system is used as the working platform, Python is used as the main programming language, and the whole network model is built by using Pytorch version 1.4.0 deep learning framework.

3.2 Results and Analysis

The cracks of concrete building components are selected as the research objects, and the methods of this paper, reference [2] and reference [3] are used to carry out the experiment of crack prediction effect. The cracks of concrete building components are shown in Fig. 3.

Fig. 3. Cracks in concrete building components

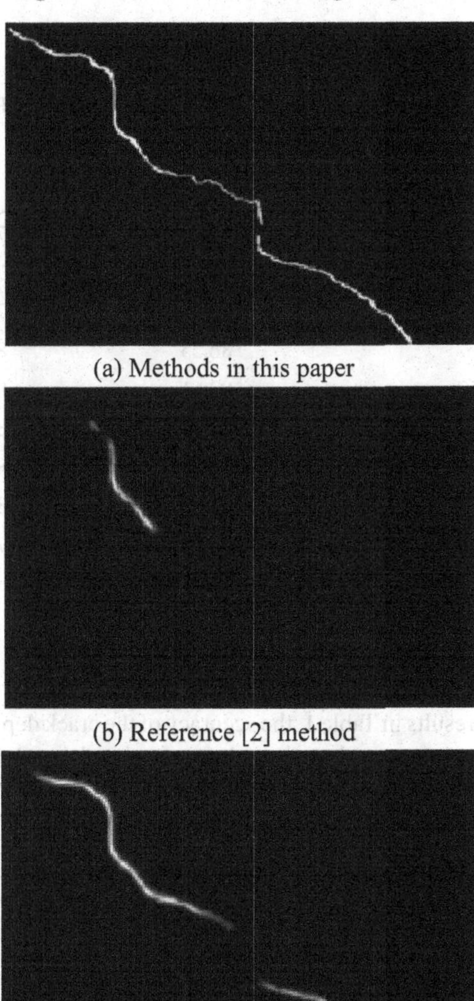

(a) Methods in this paper

(b) Reference [2] method

(c) Reference [3] method

Fig. 4. Crack prediction results of concrete building components with different methods

The fracture prediction results of the three methods are shown in Fig. 4.

From the crack prediction results of concrete buildings shown in Fig. 4, it can be seen that the crack prediction results of the method in this paper are closest to the real crack results, while the crack prediction results of the methods in reference [2] and reference [3] are poor, intermittent and incoherent, which can not guarantee the safety of concrete buildings.

In order to further verify the performance of the crack depth prediction method of concrete building components based on ultrasonic signals proposed in this paper, the method in this paper is compared with the method in reference [2] and the method in reference [3]. The accuracy, precision, recall and F1 value are selected as the evaluation indicators, and the test results are shown in Table 1–4.

Table 1. Comparison results of accuracy (%)

Number of tests	Prediction method of crack depth of concrete building components based on ultrasonic signal	Reference [2] method	Reference [3] method
1	92.43	85.47	78.46
2	94.85	84.79	72.57
3	93.56	85.56	79.58
4	95.62	86.25	81.25
5	94.57	82.61	82.66
6	92.19	83.24	83.32
7	91.58	84.38	82.03
8	92.26	82.56	82.22
9	93.35	83.23	79.55
10	94.12	86.02	80.92

According to the results in Table 1, the accuracy of the crack depth prediction method of concrete building components based on ultrasonic signals is 93.45%, which is 9.04% and 13.19% higher than the methods in reference [2] and reference [3].

Table 2. Comparison results of precision rate (%)

Number of tests	Prediction method of crack depth of concrete building components based on ultrasonic signal	Reference [2] method	Reference [3] method
1	96.43	77.46	81.71
2	95.87	79.83	82.94
3	96.56	78.69	80.58
4	94.21	79.25	82.85
5	95.55	72.52	86.66
6	96.32	76.91	85.22
7	95.56	75.34	85.33
8	94.29	78.58	82.52
9	93.60	76.25	84.01
10	95.26	77.62	83.42

According to the results in Table 2, the accuracy rate of the crack depth prediction method of concrete building components based on ultrasonic signals is 95.37%, which is 18.12% and 11.85% higher than the methods in reference [2] and reference [3].

Table 3. Comparison results of recall rate (%)

Number of tests	Prediction method of crack depth of concrete building components based on ultrasonic signal	Reference [2] method	Reference [3] method
1	85.09	77.49	72.43
2	91.56	75.88	76.87
3	90.62	74.51	75.94
4	88.23	76.62	73.65
5	87.07	75.25	72.26
6	85.16	75.36	75.52
7	86.52	75.60	76.33
8	84.25	74.23	71.22
9	85.84	78.18	72.51
10	86.41	76.51	70.18

According to the results in Table 3, the recall rate of the crack depth prediction method of concrete building components based on ultrasonic signals is 87.08%, which is 11.12% and 13.39% higher than the methods in reference [2] and reference [3].

Table 4. Comparison results of F1 value (%)

Number of tests	Prediction method of crack depth of concrete building components based on ultrasonic signal	Reference [2] method	Reference [3] method
1	90.41	77.48	76.79
2	93.67	77.80	79.79
3	93.50	76.54	78.19
4	91.12	77.91	77.98
5	91.11	73.86	78.81
6	90.40	76.13	80.08
7	90.82	75.47	80.58
8	88.99	76.34	76.45
9	89.55	77.20	77.84
10	90.62	77.06	76.23

According to the results in Table 4, the F1 value of the method for predicting the crack depth of concrete building components based on ultrasonic signals is 91.02%, which is 14.44% and 12.75% higher than the methods in reference [2] and reference [3].

Based on the above results, the method proposed in this paper can obtain the signal characteristics of cracks. The samples used for training and testing can well show the difference between positive and negative samples, and show the signal characteristics of cracks in the way of training. Therefore, it shows better prediction performance of crack depth, so as to provide support for crack detection and health assessment of concrete building components.

4 Concluding Remarks

Cracks are one of the main diseases of concrete structures. Timely and effective detection of these cracks is of great significance to maintain the sustainable use of concrete structures. In this paper, a method for predicting the crack depth of concrete building components based on ultrasonic signal is proposed. This method can improve the accuracy of prediction, is convenient and reliable, and has great engineering practical value. In the actual concrete ultrasonic testing work, the changes inside the structure may be more complex. If you want to improve the reliability of acoustic testing technology, more improvements are needed. In terms of signal processing, there will be occasional abnormalities in the waveform measured in the test, but if the error is not obvious, it can not be found in time, which will have a certain impact on the follow-up analysis work.

References

1. Jin, H., Chun, Q., Hua, Y.: Life prediction method for corrosion-induced crack of historical reinforced concrete buildings built in the Republic of China. J. Southeast Univ. (Nat. Sci. Edn.) **50**(5), 797–802 (2020)
2. Zhang, Z.-h., Lu, J.-g.: Concrete bridge crack detection based on improved convolution neural network. Comput. Simulation **38**(11), 490–494 (2021)
3. Liu, X., Duan, J., Sang, Y., et al.: Laser ultrasonic detection method for concrete crack depth. J. Central South Univ. (Sci. Technol.) (3), 839–847 (2021)
4. Zhang, J., Gu, S., Pan, Y., et al.: Concrete crack depth detection based on array ultrasonic imaging method. Nondestructive Testing **42**(03), 32–37 (2020)
5. Liu, P., Zhang, D., Xu, Z., et al.: Research on application of crack control technology for mass concrete wall. New Building Mater. **49**(05), 84–87+109 (2022)
6. Fan, Q., Liu, D.: Fracture characteristics of FRP reinforced precast cracked concrete. J. Building Mater. **23**(02), 328–333+371 (2020)
7. Zhang, G.-t., Chen, Y., Lu, H.-b., et al.: Fractal characteristics of fiber lithium slag concrete cracks under sulfate attack. Chin. J. Eng. **44**(2), 208–216 (2022)
8. Qian, K., Cheng, P., Zhang, L., et al.: Analysis of cracking of concrete floor slabs in steel-tube-bundle structure. J. Architecture Civil Eng. **38**(01), 107–116 (2021)
9. Yang, X., Ji, Q., Chen, Q., et al.: Study on cracks control method in concrete structures in European and American codes and their comparisons with Chinese code. Building Structure **50**(7), 99–106 (2020)
10. Li, G., Yan, B.: Modeling and simulation of building component location distribution based on genetic algorithm. Comput. Simulation **38**(07), 266–270 (2021)

Virtual Reconstruction of Museum Spatial Information Based on Unity3D

Zi Yang[1(✉)] and He Wang[2]

[1] Hubei Institute of Fine Arts, Wuhan 430060, China
yangzi1232022@163.com
[2] Changchun Humanities and Sciences College, Changchun 130051, China

Abstract. Museum space scene structure is complex, color information contains more, it is difficult to obtain complete and accurate three-dimensional scene information. Based on Unity3D, a virtual reconstruction method of museum spatial information is designed. Aiming at the notable structured information in museums, the three-dimensional space coordinates of museums are calculated. After the scale space is established, the extreme points in the scale space are detected, and the image sequences are detected and matched. The surface information of the point cloud model is obtained by solving Poisson's equation, and the surface with geometric entity information is obtained by fusing the point cloud. The museum 3D virtual reconstruction model is built in Unity3D to realize the transformation from 2D image to 3D scene. The test results show that the proposed method improves the similarity of museum structure to a certain extent, and can restore the spatial information of museum more truly.

Keywords: Unity3D · Museum · Spatial reconstruction · Spatial information · Virtual reconstruction · 3D reconstruction

1 Introduction

It is well known that humans perceive their surroundings mainly through vision. Through the eyes to obtain, transmission, memory and understanding of the environment around the external information, thus understanding the world, so the importance of vision for human beings is self-evident. But in some high-risk or precision and other special places, human incompetence. Therefore, with the development of computer technology, the computer vision technology has emerged, which simulates the human vision digitally, acquires and interprets the information needed in the work. With the development of artificial intelligence industry, the ability to perceive the space and process the information of the surrounding environment has been greatly improved. Nowadays, we can often see the scene of applying the 3D reconstruction technology of indoor scene [1]. For example, interior architects can measure the whole room size in advance, do layout planning and furniture customization in advance by using 3D virtual reconstruction technology. The museum can use the indoor scene 3D virtual reconstruction technology to

W. Fu and L. Yun (Eds.): ADHIP 2022, LNICST 468, pp. 648–659, 2023.
https://doi.org/10.1007/978-3-031-28787-9_48

build an online museum to meet the needs of more people. Mobile phone manufacturers can reconstruct the 3D model of indoor scene by using the 3D virtual reconstruction technology, generate various annotated images more conveniently and train convolution neural network.

A museum is an institution, building, place or public institution that collects, collects, displays and studies objects representing the natural and human cultural heritage, classifies those objects of scientific, historical or artistic value, and provides the public with knowledge, education and appreciation of culture and education. Aiming at indoor scene 3D reconstruction is one of the most valuable directions in 3D reconstruction. How to make 3D reconstruction more effective, and how to make machine obtain all kinds of 3D information by image is of great significance. Virtual museum is a kind of digital exhibition hall constructed by computer 3D modeling technology and virtual reality technology. It is a kind of 3D interactive experience mode. Based on the traditional museum, virtual museum and its exhibits are transplanted onto the Internet by using virtual technology. It breaks through the limitation of time and space, enriches the exhibition mode of museum and enhances the publicity and education functions of museum. Three-dimensional data, the computer to its acquisition, cognition, transformation, there is still a lot of room for improvement. It can be seen that the computer has the same powerful function as the human eye, that is, it can get the information from the three dimensional world at anytime and anywhere, and directly process, describe and judge the obtained information. The process of obtaining 3D information by means of computer or digital sensor is called 3D virtual reconstruction. The research goal of 3D virtual reconstruction is to input one or several images to computer for analysis and processing, so as to restore the 3D spatial information of reconstructed objects.

Unity3D, a multi-platform integrated development tool for building visualizations, real-time 3D animation and other types of interactive content developed by Unity3D Technologies, is a fully integrated professional engine that provides a fully fledged interface, one of its greatest features being support for cross-platform development. The Unity3D engine has a rich library of tutorials and descriptions of physics engines, animations, sounds, timelines, scripts, and sample code for developers to use to get a faster understanding of the interface. Based on Unity3D, a virtual reconstruction method of museum spatial information is proposed in this paper, and the information from different angles can be seen directly and concretely.

2 Virtual Reconstruction of Museum Spatial Information Based on Unity3D

2.1 Computing Museum Three-Dimensional Space Coordinates

Virtual museum is the use of virtual reality technology, simulation models or scenes, used to show the museum as a whole, or the history of real existence, has now disappeared or endangered scene. Virtual museums are generally presented as a separate roaming system for visitors to browse, learn to use, can also be embedded in the web page, as part of the digital museum functions to display to visitors. Three-dimensional virtual reconstruction is the process of obtaining 3D scene from the matching disparity map. The image taken by binocular camera is processed, and the 3D information of spatial objects is calculated by parallax map. The information that the camera simulates as the imaging model is the camera parameters, including the internal parameters and external parameters. Only when the specific parameters of the camera are determined, the spatial geometry of the target can be further reconstructed. Among them, internal parameters include the optical characteristics, scale factor, focal length and lens distortion of the camera itself, while external parameters mainly include the camera's position in the world coordinate system. Projecting a 3-D point onto a normalized image plane, then calculating the radial and tangential distortions of the points on the normalized plane, and projecting the de-distorted points onto the pixel plane through the internal parameter matrix, the correct position of the point on the image is obtained [2]. In the following discussion, we assume that all images have been de-distorted and that there is no need to include distortion parameters in the calculation. Firstly, the museum image taken by binocular camera is transformed to coordinate, and the three-dimensional space coordinate is obtained. Any spatial object point in a real-world scene is projected simultaneously in two images. According to the principle of binocular imaging system, the coordinate of image points in 3D space is solved by triangular geometry. Assuming that the parallax of a matching point A_1 and A_2 on the left and right image is h, the following relationships can be deduced according to the imaging scale relationship:

$$\begin{cases} p_1 - p_0 = g_a \dfrac{a_1}{c_1} \\ q_1 - q_0 = g_b \dfrac{b_1}{c_1} \\ p_2 - p_0 = g_a \dfrac{a_1 - h}{c_1} \\ q_2 - q_0 = g_b \dfrac{b_1}{c_1} \end{cases} \tag{1}$$

In Formula (1), (p_0, q_0) represents the central coordinate value of the left and right images; (p_1, q_1) and (p_2, q_2) represent the pixel coordinates of a matching point on the left and right images respectively; (a_1, b_1, c_1) represents the three-dimensional space coordinates of the pixel in the left camera coordinate system; and g_a and g_b represent the focal length of the left and right cameras on the horizontal and vertical axes. The expression of orientation in the position and pose of a camera is often described indirectly by the expression of rotation. The pointing of the camera to the Z axis in the current camera coordinate system (generally default to camera orientation) is expressed through a rotation transformation that completes the conversion of the camera from the Z axis in the world coordinate system to the Z axis in the current camera coordinate system [3]. Epipolar alignment of the image, the matching points are on the same line, so the pixels are the same vertical. Therefore, the spatial coordinate data of any point on the image is obtained, and the calculation formula is as follows:

$$\begin{cases} a_1 = \dfrac{h(p_1 - p_0)}{p_1 - p_2} \\ b_1 = \dfrac{hg_a(q_1 - q_0)}{g_b(p_1 - p_2)} \\ c_1 = \dfrac{hg_a}{p_1 - p_2} \end{cases} \tag{2}$$

According to the location of two points, the camera calibration result is obtained. The 3D coordinates of the point in the camera coordinate system in space can be solved by the inner and outer parameter matrices.

2.2 Detection of Spatial Information Characteristics of Museums

After calibrating and correcting the museum 3D space coordinates, it is necessary to detect and match the features of the images taken by the museum scenes. In the work of 3D reconstruction of image sequence, a key part is to find out the relationship between images in image sequence, and then get the 3D information according to the mutual information of these images. Although there is not much noise and weather in the indoor scene, the scene structure is complex and the color information contains more. Direct grayscale algorithm will lose part of the original color information of the image. This paper proposes a grayscale compensation algorithm to extract more effective feature points in feature detection. The resulting gray scale value is based on the original gray scale value to increase color compensation, the formula is as follows:

$$d = \gamma + \gamma' + \varphi \tag{3}$$

In Formula (3), d represents the gray value after compensation; γ and γ' represent the gray value and color compensation amount of the original image respectively; and φ represents illumination compensation amount. The formula for the amount of color compensation is as follows:

$$\gamma' = \delta_1 \mathrm{sgn}(200\beta_1)|200\beta_1|^{\delta_2} + \delta_1 \mathrm{sgn}(50\beta_2)|50\beta_2|^{\delta_2} \tag{4}$$

In formula (4), δ_1 is the color contrast parameter, the value of which is 2.3; δ_2 is the range parameter, the value of which is 0.6; sgn is the sign function; β_1 is the chroma signal; and β_2 is the saturation signal. The extraction equation for the amount of light compensation is as follows:

$$\varphi = \varepsilon\gamma' e\left|-\frac{\gamma}{2\eta^2}\right| \tag{5}$$

In formula (5), the value of the contrast parameter of ε illumination is 1.8; the value of e is the natural constant; and the value of η is the standard deviation, which is 0.25. Because there are many pixels involved in the 3D virtual reconstruction of spatial features of museums, and because of occlusion and other problems, it is easy to track the end points of line segments in images because of different viewpoints, the end points of line segments do not belong to the corresponding situation, but deteriorate the performance [4]. So the descriptor is constructed here, and then the descriptor matching method is used to extract the line feature descriptor. Feature points are extracted using Harris corner extraction algorithm. After extraction, it is necessary to check the quality of its checkpoints, judge whether it contains the required information, and it can not be the edge of the image.

Image points are different from other points in the image, that is, some special image points can provide more useful information, such points are called feature points. Obtaining the 2-D coordinates of such feature points by using the relevant feature point detection technology is the foundation and key step of 3-D reconstruction. Generally speaking, the size of the image entropy value is proportional to the size of the information contained in the image. After the scale space is established, the extreme points in the space are detected. Then, the resulting point is chosen as the key candidate point. Specifically, in the detection, each pixel is compared with a certain point. Taking a pixel as an example, this paper compares the pixel with 8 neighboring pixels in the same scale and 18 neighboring pixels in the corresponding position of upper and lower neighboring scale. At the same time, because the feature corners are reduced, the matching time can be reduced, which improves the real-time of the algorithm. The feature corners extracted by Harris replace the feature points extracted by Gaussian convolution algorithm in SIFT algorithm. SIFT algorithm has a certain stability, its extraction operator in the brightness, scale changes, such as interference, can still detect a large number of accurate feature points. Markov distance is used to judge the similarity between the two eigenvectors. At the same time, the correlation between feature vectors is eliminated, and the matching precision is improved.

2.3 Point Cloud Fusion of Museum Spatial Information

Point cloud fusion of museum spatial information is to eliminate redundant observation points of the same name. By point cloud fusion, data redundancy can be reduced and dense sampling points on the surface of objects can be obtained. The flow of the point cloud fusion is shown in Fig. 1.

It can be seen from Fig. 1 that first, it is necessary to collect point cloud data from multiple viewpoints, and then perform point cloud filtering. On this basis, multi-view point cloud splicing and matching are performed, and the integration of point cloud data is completed. Keep the equipment in the structured light system fixed and set the number of shots in advance according to the complexity of the target model. The scanned object is placed on the rotating table, and the rotating table drives the target model to rotate at a certain speed. The camera collects 8 point cloud data with different viewing angles. The key point of point cloud fusion is how to distinguish the same name from multiple observations. According to the spatial resolution of the reference image under the light beam of the same name in the image to be matched, determine whether the reference image is a point of the same name. The formula is as follows:

$$w(n) = \frac{|\kappa_1(n) - \kappa_2(n)|}{\kappa_2(n)} \tag{6}$$

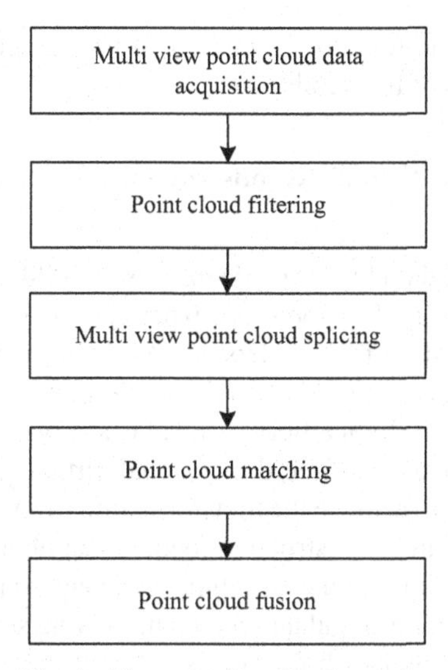

Fig. 1. Flow of point cloud merging

In Type 6 (6), n represents spatial resolution; $w(n)$ represents judgment function; $\kappa_1(n)$ represents the corresponding depth value in the image to be matched; and $\kappa_2(n)$ represents the depth value of the reference image guided by object points in the image to be matched. In 3D space, the distribution of point cloud is scattered and disordered, and it is difficult to establish a connection between them, but there is a topological relationship between points and points in space. At this point, all the points calculated by the search are Gaussian normal distribution, the rest are removed as outliers by filtering algorithm. If the point points in the space of the current processing pixel have the same names in the fused point cloud, they are directly ignored, otherwise, the fused point cloud data are added, and the same names in the reference image and the image to be matched are marked as processed [6]. The plane information is detected from the depth map, and then sparsely sampled in the plane to reduce the data representation redundancy. The surface information of point cloud model is obtained by solving Poisson's equation, and the surface model with geometrical entity information is obtained by reconstructing the target model. According to local information to evaluate the smoothness of object points, the potential plane position of object is given, and then sparse sampling is carried out in object plane to reduce the redundancy of expression in dense point cloud [7]. The formula for calculating the mean value of the difference between objects is:

$$\vartheta = \frac{\sum_{N=1}^{n} N(\tau_1 - \tau_2)}{|\tau_2|} \tag{7}$$

In Formula (7), ϑ indicates the mean difference of the points; N indicates the total number of points; τ_1 indicates the neighborhood depth value in the depth chart; and τ_2 indicates the depth value of the current point. After obtaining the probability of each point in the object plane in the depth map, the object plane can be changed into a connected region by the way of region growth. Then we can use the grid in the image plane to sample the connected area of the depth map.

2.4 Establishment of 3D Virtual Reconstruction Model of Museum Based on Unity3D

It has become a reliable method to obtain 3D models of building level objects by 3D virtual reconstruction technology. The good user experience in the 3D virtual reconstruction of museum spatial information mainly depends on the immersion, and the immersion mainly depends on the reality of the 3D scene. Therefore, the fineness of the 3D model in the scene, the mapping and the light efficiency all play a very important role. In this paper, Unity3D is used to complete the three-dimensional virtual reconstruction of museum spatial information, that is, to achieve the transformation from two-dimensional image to three-dimensional scene. The reconstruction method can obtain the 3D structure of the object quickly without the limitation of equipment and environment, and the point cloud is rich in information and the triangulated model has strong stereo. The 3D model is too rough and crude, the reality of the whole scene will be reduced, the user's immersion will be reduced or even all disappeared. However, if the scene is too complex, too refined, the higher the requirements for computer performance, the average user's computer system will run faster, real-time will also be greatly reduced, resulting in a direct impact on the user's experience. Therefore, it is necessary to consider the trade-off in modeling, so as to enhance the authenticity of the model. In Unity3D, the rotation of the camera enables the rotation of the 3D spatial field of view for viewing museum models in 3D space. To achieve the rotation function, the following scheme is set. The camera position shall remain unchanged, and the camera angle shall be changed correspondingly only by moving the mouse (PC end) or sliding gestures (mobile end). The implementation steps for this scenario are shown in Fig. 2.

The code is simple and easy to maintain, only dynamic change the camera angle, rotation speed and angle easy to control, the user experience is better. The 3D virtual reconstruction model of museum spatial information is the most suitable choice because of the limitation of space occupied and the actual operation and interaction. So the interface design is also based on the PC. In Unity3D, the translation of 3D field of vision is realized by the translation of the camera, mainly by the direction vector of the camera left and right, and the direction vector of camera up and down. The translation of the viewport in this way requires special attention to the speed control of the translation, which should be related to the size of the viewport to achieve a comfortable translation effect. According to the rules, location, layout, quantity, visual rendering, associated business items, constraints and other structured processing of node attributes. Then it realizes visualization display and 3D engine rendering. Compared with other prototyping software, Unity3D is more convenient and easy to modify and export. It can also set up connections and actions and interactive demo. Low-fidelity prototypes can efficiently test the usability of a system, improve the user experience, and reduce the cost of program development. Three-dimensional zoom is achieved by controlling the viewport size of the camera in Unity3D and dynamically changing the viewport size by sliding the mouse wheel or two fingers. Increase your camera viewport by a multiple of 1.1 as the mouse wheel scrolls forward or slides outwards, and decrease your camera viewport by a

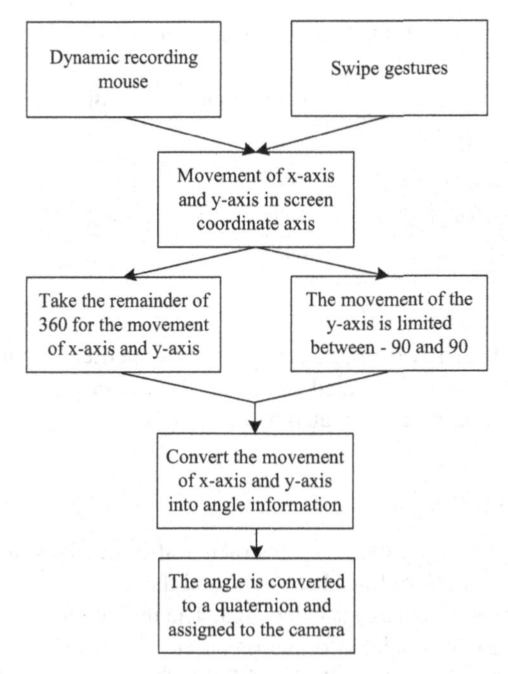

Fig. 2. Realization steps of 3D visual field rotation

multiple of 1.1 as the mouse wheel scrolls backward or slides inward. By encapsulating the JSON format string data, the node attributes in the JSON string are filtered, and the complex multi-layer topology is parsed and filtered. The pre-processing data (multi-layer, processing, filtering) is input, and then the algorithm is invoked to calculate the layout of the network topology. According to the constraints of the node attributes, the layout strategy, the layout algorithm, the layout rules are associated with nodes, and the nodes are displayed, which makes the visualization model of museum spatial information clear and flexible. Thus, the design of 3D virtual reconstruction method of museum spatial information based on Unity3D is completed.

3 Experimental Study

3.1 Experimental Preparation

In order to evaluate the performance of 3D virtual reconstruction method more accurately, the performance of the method is tested on the open dataset. The dataset used is ICLNUIM dataset, it is the dataset that reconstructs the indoor scene as the target. It has the true value of the point cloud reconstruction model, which is mainly used for the evaluation of 3D reconstruction. The test images selected for this article are the same scene shot from the front, left, right, and above four directions, corresponding to four sequences, each with the number of frames and sequence length shown in Table 1.

In order to quantitatively analyze the results of spatial 3D virtual reconstruction, structural similarity is used to evaluate the results. Structural similarity is a measure

Table 1. Scene sequence features

Sequence	Direction	Frame number	Binocular image sequence
1	Front face	1623	32
2	Left lateral surface	1027	34
3	Facies lateralis	1236	42
4	Above	982	38

of the similarity between two images, which is symmetrical, bounded and uniquely maximal. The brightness is estimated by mean, the contrast is estimated by standard deviation, and the similarity is estimated by covariance.

3.2 Results and Analysis

A 3D virtual reconstruction model is constructed after the same angle is obtained in the front, left, right and above four directions, and the structure similarity between the output image and the scene image is calculated. The measurement results of 3D virtual reconstruction of museum spatial information based on Unity3D proposed in this paper are compared with those based on SFM and BIM. The experimental results are shown in Tables 2–5.

Table 2. Structural similarity comparison of sequence 1

Number of tests	Virtual reconstruction of museum spatial information based on unity3D	Virtual reconstruction of museum spatial information based on SFM	Virtual reconstruction of museum spatial information based on BIM
1	0.753	0.683	0.674
2	0.762	0.685	0.678
3	0.775	**0.698**	0.665
4	0.786	0.675	0.662
5	**0.788**	0.682	0.673
6	0.775	0.686	0.675
7	0.782	0.693	**0.682**
8	0.766	0.682	0.681
9	0.752	0.685	0.664
10	0.781	0.674	0.671

In the front modeling of indoor scene, the similarity of 3D virtual reconstruction method based on Unity3D is 0.772, which is higher than that based on SFM and BIM.

Table 3. Structural similarity comparison of sequence 2

Number of tests	Virtual reconstruction of museum spatial information based on unity3D	Virtual reconstruction of museum spatial information based on SFM	Virtual reconstruction of museum spatial information based on BIM
1	0.747	0.644	0.644
2	0.744	0.657	0.638
3	0.738	0.668	0.646
4	0.746	**0.676**	0.655
5	0.745	0.652	0.662
6	0.732	0.665	0.663
7	0.753	0.658	0.656
8	**0.759**	0.666	0.652
9	0.745	0.652	0.641
10	0.742	0.643	**0.667**

Table 4. Structural similarity comparison of sequence 3

Number of tests	Virtual reconstruction of museum spatial information based on unity3D	Virtual reconstruction of museum spatial information based on SFM	Virtual reconstruction of museum spatial information based on BIM
1	0.747	0.650	**0.672**
2	0.738	0.647	0.642
3	0.746	**0.678**	0.666
4	0.759	0.666	0.655
5	0.733	0.652	0.668
6	0.732	0.640	0.655
7	0.745	0.673	0.642
8	0.749	0.636	0.653
9	**0.763**	0.665	0.660
10	0.751	0.652	0.631

In the left side modeling of indoor scene, the similarity of 3D virtual reconstruction method based on Unity3D is 0.745, which is 0.087 and 0.093 higher than that based on SFM and BIM.

In the right side modeling of indoor scene, the similarity of 3D virtual reconstruction method based on Unity3D is 0.746, which is 0.090 and 0.092 higher than that based on SFM and BIM.

Table 5. Structural similarity comparison of sequence 4

Number of tests	Virtual reconstruction of museum spatial information based on unity3D	Virtual reconstruction of museum spatial information based on SFM	Virtual reconstruction of museum spatial information based on BIM
1	0.713	0.604	0.590
2	0.704	0.609	0.597
3	0.718	0.622	0.588
4	**0.726**	**0.623**	0.616
5	0.702	0.615	**0.625**
6	0.703	0.612	0.593
7	0.715	0.598	0.592
8	0.702	0.583	0.611
9	0.701	0.591	0.624
10	0.714	0.612	0.602

In the modeling of indoor scene, the similarity of 3D virtual reconstruction method based on Unity3D is 0.710, which is 0.103 and 0.106 higher than that based on SFM and BIM. From the above results, we can see that the structure similarity of the proposed method is improved to a certain extent, which shows that the proposed method has higher precision and can more accurately restore museum 3D virtual scene information. Because the method in this paper detects the extreme points of the space on the basis of the scale space. On this basis, the image sequence is matched, and the surface with geometric entity information is obtained by fusing point clouds. This can further improve the virtual matching accuracy. Finally Unity3D built a 3D virtual reconstruction model of the museum.

4 Conclusion

Museum virtual space display is a new experience learning method for visiting learners, which can greatly improve learners' immersion and experience. Virtual museums provide learners with rich and real information and sensory senses, enable more people to receive cultural knowledge anytime, anywhere, enrich the methods of experiential learning, and make the educational function of museums play the greatest role and effect. In this paper, the three-dimensional spatial coordinates of the museum are calculated based on the significant structured information in the museum. The extreme points in the scale space are detected, and the image sequence is detected and matched, which improves the accuracy of 3D virtual reconstruction. The built-in 3D virtual reconstruction model of the museum in Unity3D realizes the conversion from 2D images to 3D scenes. In the process of feature extraction and matching, there are few feature points on the surface of the cabinet with smooth surface and weak texture representation, which will cause hollows in the subsequent reconstruction. Experiments show that the method in this paper improves the similarity of the museum structure and can restore the spatial information of the museum more realistically. How to extract features from smooth and weak texture surfaces and fill in the voids smoothly and without distortion is the future research direction.

References

1. Yu, Z., Peng, X., Qiu, C., et al.: An improved approach of indoor space three-dimensional reconstruction with RGB-D SLAM. Power Syst. Big Data **23**(5), 30–37 (2020)
2. Chen, L., Yu, L.: Visual rationality design of interior three-dimensional space based on virtual optics. Laser J. **41**(11), 183–187 (2020)
3. Wang, Z., Peng, M., Xu, H.: Research on Information Visualization design of bronze ware in museums. HuNan BaoZhuang **35**(5), 57–61 (2020)
4. Cao, J., Ye, L.-Q.: Multi view 3D reconstruction method of virtual scene in building interior space. Comput. Simul. **37**(9), 303–306, 381 (2020)
5. Zhang, S., Zhao, W., Peng, J., et al.: Augmented reality museum display system based on object 6D pose estimation. J. Northwest Univ. (Natural Science Edition), **51**(5), 816–823 (2021)
6. Yang, K.: Design of digital 3D panoramic super-resolution reconstruction system based on virtual reality. Modern Electron. Technique **43**(10), 145–147, 152 (2020)
7. Chao, Y.: Design of museum space multi?element interactive system based on three?dimensional representation. Modern Electron. Technique **44**(12), 155–158 (2021)

Research on Encrypted Transmission Method of Survey Data of Offshore Engineering Buoy

Wenyan Wang[1,2,3](✉) and Huanyu Zhao[1,2,3]

[1] Institute of Oceanographic Instrumentation, Qilu University of Technology (Shandong Academy of Sciences), Qingdao 266061, China
wangwenyan086@163.com

[2] Shandong Provincial Key Laboratory of Marine Monitoring Instrument Equipment Technology, Qingdao 266061, China

[3] National Engineering and Technological Research Center of Marine Monitoring Equipment, Qingdao 266061, China

Abstract. The measurement data of offshore engineering buoy includes vector data and grid data. Due to the large difference in structure and characteristics, the encrypted transmission time is long. Therefore, an encrypted transmission method of offshore engineering buoy measurement data is designed. On the basis of analyzing the motion performance of the buoy structure, the measurement data of the marine engineering buoy are obtained, and the characteristics of the marine environmental load are extracted. Unpack and sort the uplink data, sort the data that needs to be transmitted to other nodes into frames and output them to the downlink, and transfer the data that needs to be sent to the external network to the external interface, so as to improve the transmission efficiency of marine engineering buoy measurement data. The improved Logistic chaotic mapping algorithm is adopted as the main algorithm of data encryption transmission mode to ensure the security of buoy measurement data. The test results show that the encryption transmission method of marine engineering buoy measurement data designed this time can effectively reduce the encryption time and improve the encryption efficiency while ensuring the data security.

Keywords: Ocean engineering · Buoy survey · Load characteristics · Data encryption · Data transmission · Encrypted transmission

1 Introduction

Marine engineering buoy measurement data encryption takes geospatial cognition as a bridge, abstracts the real geographical world step by step, and obtains spatial concepts of different abstract levels, thus realizing the simulation of geographical systems. Ocean spatial data model is composed of three different levels: conceptual model, logical data model and physical data model. As one of the main ways of marine monitoring, marine data buoy is a modern marine observation facility. It is a surface floating automatic monitoring platform used to obtain physical and biochemical parameters such as marine

W. Fu and L. Yun (Eds.): ADHIP 2022, LNICST 468, pp. 660–672, 2023.
https://doi.org/10.1007/978-3-031-28787-9_49

hydrology, meteorology and water quality. It has the characteristics of long-term, all-weather, all-time, continuous, synchronous, comprehensive and low cost, and can realize automatic collection and automatic transmission of data [1]. Vector data and grid data are commonly used in offshore engineering buoy measurement. Due to the different structures and characteristics of these two data models, it is more difficult to encrypt and transmit data.

Reference [2] proposes a data encryption transmission method based on DES algorithm, which is a block encryption system with 64 bit cipher and a symmetric encryption algorithm with the same encryption key and decryption key. Through the entry parameters of DES algorithm, the replacement method of 64 bit block cipher is analyzed, and the data encryption steps are determined to complete the encrypted transmission. Reference [3] proposes a data encryption transmission method based on AES algorithm, introduces AES algorithm to encrypt sensitive data, designs a data encryption transmission scheme, introduces key expansion, data encryption and decryption optimization algorithm, and analyzes the flow of data encryption transmission method and the realization of data encryption. Reference [4] proposes a data encryption transmission method based on chaotic algorithm to obtain high-end equipment command sensitive data to be encrypted, so as to narrow the encryption range and remove redundant data. The encryption method based on chaotic mapping is used to realize the encrypted transmission of data.

Although the above method can complete the encrypted transmission of data, it has the problems of long time consumption and poor security. In order to solve the problems existing in the above encryption transmission methods, a new encryption transmission method of marine engineering buoy measurement data is proposed.

2 Research on Encrypted Transmission Method of Survey Data of Offshore Engineering Buoy

2.1 Obtain Survey Data of Offshore Engineering Buoy

Offshore engineering buoy is a kind of drifting buoy floating on the ocean surface. It has the characteristics of small volume, low cost and easy delivery. It can be used in large-scale marine environment monitoring, oil spill tracking, maritime search and rescue, military and other fields. Once the ocean engineering buoy came out, it has become an important means to obtain marine data. With the rapid development of satellite positioning and communication technology, its role in ocean observation can not be replaced. As a kind of offshore floating body, the basic design and theoretical research of offshore engineering buoy refer to the general theoretical research methods of offshore floating body structure. In order to better analyze the structure of offshore engineering buoy, this paper is based on three-dimensional potential flow theory. The complete system of offshore engineering buoy is usually composed of three parts: buoy body, water sail and data acquisition unit. Flexible connection is adopted between water sail and buoy body. The forms of water sail include umbrella type, curtain type and cylindrical type. The selection of water sail and its connecting length with buoy shall be appropriate to ensure the wave induction of buoy. Based on the basic analysis method of panel integration, aiming at the basic structural parameters (shape, size, draft, center of gravity position, moment of inertia, etc.) affecting the motion response of buoy, the effects of these basic parameters on the force, motion performance and hydrodynamic parameters of buoy are analyzed in time domain and frequency domain. Natural period is an important index to measure the motion performance of floating body. Since the main floating body of the circular buoy is a symmetrical structure, when arranging the carrying instruments, the carrying instruments are generally symmetrically arranged. Therefore, the natural period of the buoy in the rolling and pitching directions is roughly the same. In the calculation of the natural period of the buoy, only the natural period in the rolling direction of the buoy needs to be considered. According to the theory of ship seakeeping, the approximate natural frequency of free rolling of floating body is:

$$Q = \frac{1}{2\pi}\sqrt{\frac{G_{\delta}}{\Delta V_{\delta-1}}} \tag{1}$$

In formula (1), G represents the displacement, V represents the moment of inertia of the rolling motion of the floating body, and δ represents the additional moment of inertia of the rolling motion of the floating body. At first, the ocean engineering buoy was limited to the measurement of ocean current. With the increasing maturity of ocean engineering buoy technology, the newly developed ocean engineering buoy covers the measurement of precipitation, ice drift, humidity, sea conditions and other data. It plays an irreplaceable role in business, civil and military, such as marine oil spill accident tracking, disaster prediction and so on. Since the motion response of the buoy is mainly affected by the external environmental load (mainly wave load), in order to better study the influence of the basic parameters of the buoy on the motion performance of the buoy, based on the analysis of the motion performance of the buoy structure, this paper will focus on the further detailed analysis of the hydrodynamic parameters of the buoy in the wave environment. The calculation formula of buoy heave natural frequency is as follows:

$$L = \frac{1}{2\pi}\sqrt{\frac{V}{\frac{E}{\eta} + d}} \tag{2}$$

In formula (2), E represents the additional mass of the free heave of the floating body, η represents the displacement of the floating body, and d represents the gravitational acceleration. Usually, ocean engineering buoys can continuously monitor ocean current data, meteorological data and marine hydrological information for a long time without interference and poor sea conditions. It can be put in manually or in large quantities by machinery. Small offshore engineering buoys can be evaluated for recovery cost to decide whether to recover or not. Offshore engineering buoys equipped with solar charging panels can work on the sea for a long time. During this period, a large amount of marine data can be obtained, and the smaller volume is less likely to be found and damaged. Therefore, they have been widely used in a short time. When analyzing the motion performance and hydrodynamic parameters of floating body at sea, the floating body at sea is usually considered as a rigid body structure. As a kind of rigid body, the buoy body can generally move freely in space. Therefore, when analyzing the motion performance and hydrodynamic force of the ocean buoy, we should first establish an appropriate coordinate system for the studied ocean buoy structure, otherwise there will be a lack of reference. The numerical simulation method based on potential flow theory has the advantage of fast simulation speed, but because it ignores the viscosity of the fluid itself, it can not accurately describe the impact of wave climbing, diffraction, eddy current and other phenomena after the interaction between wave and buoy. It has a particularly obvious impact on the calculation results of small floating body, and finally causes irreversible error to the results. For the floating body on the sea, when it is only affected by gravity and buoyancy, the buoy is stationary in the vertical direction. If there is an external couple acting on the object, the object will tilt. When the external couple is withdrawn, the object rotates under the action of moment H, where:

$$H = \gamma \times \frac{\sqrt{\frac{G}{d}}}{r \times \sin \theta} \tag{3}$$

In formula (3), γ represents the distance from the center of gravity of the object to r, r represents the intersection of the vertical line of the floating center position of the object and the central axis of the object, and θ represents the inclination angle of the object. In addition, the draft is an important factor affecting the carrying performance of the buoy. On the basis of meeting the stability and movement performance, more environmental monitoring instruments should be carried as much as possible to maximize the economic benefits of the buoy. Therefore, it is necessary to analyze the stability and movement performance of the buoy based on the draft under the premise of determining the shape of the buoy, in order to provide a basis for the carrying performance design of the buoy. With the improvement of computer computing power, numerical wave flume based on incompressible viscous fluid shows its advantages in the field of ship and ocean engineering and is gradually widely used. This chapter will introduce the mathematical model, basic theory and numerical simulation method of numerical flume based on incompressible viscous fluid in detail, so as to provide a theoretical basis for the establishment and simulation calculation of subsequent numerical wave flume. For the shape of the hemispherical buoy selected above, keep the center of gravity position and mass moment of inertia unchanged in the analysis process, and analyze the motion of several degrees of freedom by changing the draft, that is, changing the carrying and counterweight, so as to provide reference for the selection and design of offshore small buoy. Hydrodynamic parameters (such as additional mass, additional damping, etc.) are important parameters to measure the motion performance of buoy. Therefore, before analyzing the influence of draft on the motion performance of buoy, it is necessary to analyze the hydrodynamic parameters of buoy according to different draft.

2.2 Extracting the Characteristics of Marine Environmental Load

As a kind of floating body on the sea, the loads borne by the buoy in the basic design process of its buoy structure mainly include the permanent load of the buoy's own weight and environmental loads such as wind, wave and current. These two parts of loads together constitute the main external environmental loads in the movement process of the buoy structure. The earth's surface is composed of the atmosphere. Due to the influence of the earth's gravity, the atmosphere is unevenly distributed in the horizontal direction, which leads to the movement of air from the high-pressure area to the low-pressure area. This kind of air flow is wind. The wind has the following characteristics: the wind speed increases with the increase of height. As a kind of floating body on the sea, the upper instrument rack of offshore small ocean buoy is equipped with solar panels, which makes its wind receiving area account for a large proportion compared with the overall surface area, and the wind speed on the ocean is generally very high, which makes the movement of buoy more sensitive to the environmental factor of wind. Due to the influence of friction, the flow velocity of the air near the ground is small. With the increase of the ground, the flow velocity becomes larger, and the final velocity tends to be stable. Due to the Coriolis force, compared with the wind direction of the atmospheric boundary layer, the wind direction at the surface will deflect at a certain angle, and the deflection angles at different times and in different regions are also different, up to tens of degrees. When the offshore structure is subjected to the force of wind load, it will produce a large overturning moment. Therefore, the force generated by wind load needs to be considered when calculating the external load. Assuming that the wind field is two-dimensional and there is only wind parallel to the sea level, the instantaneous wind can be regarded as the superposition of average wind speed and fluctuating wind speed, which will produce average load force and fluctuating wind load force on the main body of the drifting buoy. The calculation formula is:

$$W_{\varphi} = \frac{1}{2}\sqrt{\mu + s^2} \times \frac{1}{\sqrt{(\varphi + 1)^2}} \tag{4}$$

In formula (4), μ represents the leeward area of the buoy body on the waterline, s represents the wind coefficient, and φ represents the windward area of the buoy body on the waterline. Under the action of the environmental factor of wind, the small floating body such as buoy will produce large overturning moment and anchoring force, and cause dynamic load with large frequency range, which will have a great impact on the motion response of small floating body such as buoy. Wind is one of the environmental factors in the design of offshore structures. The air in the atmospheric boundary layer has random turbulent flow. Its structure can be expressed by turbulence degree, Reynolds stress, correlation function and spectrum. The turbulence degree can reach 20%. There are three kinds of waves in the marine environment: wind wave, swell and mixed wave. Wind and waves are caused by sea breeze, which is the focus of this paper. Swells are waves caused by sea winds from other areas to this area. Ocean current is mainly composed of wind-driven current, tidal current, ocean circulation, internal wave current, etc. The current velocity varies with water depth. Under normal circumstances, in the ocean surface, the velocity is generally constant in direction, stable in magnitude and decreasing with water depth, so the calculation formula of velocity is:

$$R(\varpi) = T_{(\varpi)} + Z_{(\varpi - 1)} + \sqrt{\frac{1}{\mu^2}} \tag{5}$$

In formula (5), ϖ represents the velocity of wind-driven current, T represents the velocity of tidal current, and Z represents the velocity of ocean circulation. Because the wind load will have a large overturning moment on the floating buoy on the sea surface and have a great impact on the buoy anchoring system to a great extent, it is necessary to select an appropriate wind load calculation formula when analyzing the movement of offshore small marine buoy in the marine environment. Due to the small water entry depth of the floating buoy, it can be approximately considered that the speed and direction of the current remain unchanged within a certain period of time, so the calculation formula of the current load force is:

$$M_{\sigma} = \frac{1}{2} \times N_{\sigma} + \left\| \frac{Z}{s} \right\|^{2} \tag{6}$$

In formula (6), N represents the drag force coefficient and σ represents the density of seawater. Ocean current factor is also one of the important environmental factors in the design of offshore structures. The ocean current has a great drag force on offshore structures, which puts forward great requirements for the stability of small offshore floating bodies such as buoys and the design of mooring system. Mixed waves are waves with both wind and swell. The drifting buoy is affected by the wave load in the marine environment. The wave load is mainly divided into three parts: wave frequency force. The load force is obtained by deriving the first-order velocity potential. The average drift force, the load action is similar to the current load, and the wave height is positively correlated. The load force is obtained by solving the second-order velocity potential. Compared with the deep-water buoy, in the offshore sea area, due to the shallow water depth, the drag effect of ocean current on the mooring system is produced on most of the anchor cables. Therefore, ocean current is an environmental factor that can not be ignored in the process of buoy structure design. The characteristics of waves in different sea areas are different, and the motion law of wave particles in sea areas is also different. If the wavelength, relative water depth and relative wave height are different, the motion law of wave water quality points is also different. Linear wave theory, Stokes wave theory and stream function wave theory are often used in the field of marine engineering research. Although the differential equations and boundary conditions of traditional wave theory are different, the results are similar. The fundamental difference lies in the satisfaction of nonlinear motion and wave surface dynamic boundary conditions. Because the average drift force accounts for a relatively small proportion of wave load, it is ignored. The slowly varying drift force is neglected again because it accounts for a small proportion of the wave load. When the drifting buoy flows with the current at sea, it will not only be affected by the wind, waves and currents, but also be completely submerged by the sea due to some uncertain factors. Therefore, it is necessary to analyze the stability of the main body of the drifting buoy and judge its instability conditions. In the natural environment, the formation and movement of ocean waves is a random phenomenon, and their eigenvalues such as wave height, period and water depth are also changing randomly. In order to analyze the changes of waves on the ocean, some experts put forward the application of random process theory to waves, put forward the concept of wave spectrum, and put forward the means of wave spectrum analysis on this basis.

2.3 Build Data Frame Demand Model

In the transmission process, the concentrator multiplexes the uplink data into downlink frames with an interval of 125us in two ways. One is to receive the uplink data of each node independently; The other is to use time division multiplexing (TDMA) technology to centrally receive the uplink data of each node, and the concentrator has different processing methods for the uplink data received by these two methods. The security technology strategy of data transmission is an integrated and comprehensive solution. Mobile security access should adopt a number of security measures including terminal management and control, terminal reinforcement, channel encryption, authentication access, access control, gateway isolation and security management to build a security environment for mobile applications [5]. Unpack and sort the uplink data, sort the data to be transmitted to other nodes into frames and output to the downlink, and transmit the data to be sent to the external network to the external interface. The advantage of this method is that it improves the utilization of system channel to a great extent. At the same time, the technical means of intelligent perception and hybrid encryption are adopted in practical applications, such as adopting security technology at the transmission layer, session layer and application layer of the network, doing a good job in internal and external network data isolation, authentication access and access control, and implementing data preprocessing before data transmission. Such a comprehensive solution can play a more comprehensive security protection effect [6]. However, the concentrator is required to complete the functions of router, gateway and network management, which makes the structure of the concentrator complex and greatly increases the difficulty of implementation. The uplink data is forwarded in two ways: full forwarding and dynamic forwarding. The data forwarding mode is simple, but it is not easy to realize the data forwarding system, regardless of whether the uplink structure is simple or not [7]. The dynamic forwarding method is to monitor the node data and control the forwarding according to the data flow. The measurement data frame structure of offshore engineering buoy is shown in Fig. 1:

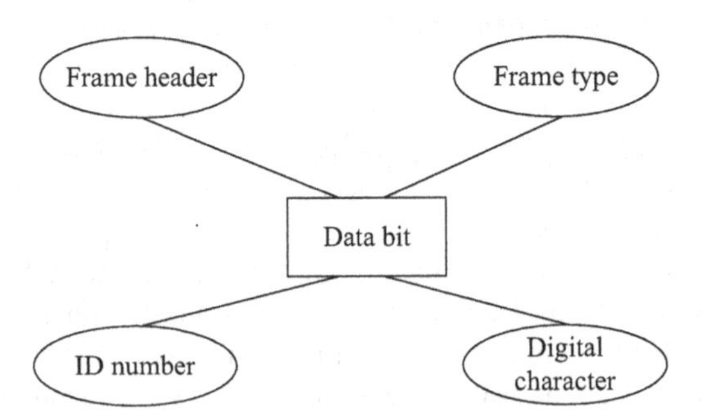

Fig. 1. Frame structure of measurement data of offshore engineering buoy

As can be seen from Fig. 1, the concept of data frame demand model is produced with the development of computer technology. Data frame demand model is a tool to describe the relationship between data content and data. It is one of the main symbols to measure the strength of database capacity. The concept and idea of computer data modeling soon extend to the modeling of spatial data. People transform their cognition of the real geographical world into an implementation model that is convenient for mutual communication and understanding and suitable for computer interpretation and processing. This is the abstract process of geospatial. When the data flow demand is large, the transmission rate can be expanded to Zn times of the basic rate. When there is no data, the forwarding will be stopped. The complexity and efficiency of the system are between the processing forwarding mode and the full forwarding mode. In short, the efficiency of forwarding mode is lower than that of processing forwarding mode. Its advantage is that the connected equipment can be independent, the design of the equipment is relatively simple, and the system function can be increased by adding equipment. Geospatial cognitive process can be summarized into three abstract levels to realize modeling, that is, people first recognize, abstract and describe the real world, gradually obtain the conceptual model, then convert it into logical data frame demand model through coding, establishing spatial relationship and expression, and finally establish physical data frame demand model through data organization and structure. The time-sharing upload of uplink data is divided into fixed time slot upload and dynamic upload. Fixed time slot upload divides the upload channel into different time slots. Nodes upload data according to the fixed time slot. The concentrator generates time positioning data by ranging each node and transmits the time positioning data to each node to ensure time slot synchronization [8]. Dynamic upload is to send channel application when each node has data to upload, and the concentrator dynamically allocates time slots as needed. Dynamic upload mode is more efficient than fixed slot upload mode, but it is more complex to deal with.

2.4 Design Data Encryption Transmission Mode

The previous selective encryption technology of measurement data takes a specific format data as the research object, starting with the process of data coding and compression. There are many parameters involved in the coding process of multimedia data, such as I-frame data, P-frame data, B-frame data, macroblocks in I-frame, DCT coefficients or symbols, wavelet transform related parameters and entropy coding process related parameters, etc. Symmetric key cryptosystem means that the keys used to encrypt and decrypt data are the same or similar (that is, it is easy to push from one to another). Almost all mature symmetric key cryptosystems are characterized by fast encryption speed, high encryption strength (generally only strong cracking, also known as exhaustive cracking), easy implementation of software and hardware, especially suitable for encrypting large amounts of data. It is the first step of almost all data confidentiality work, With the computing power of today's computers, without symmetric key cryptography algorithm, there will be no data security and encryption. The fundamental purpose of data encryption transmission is to ensure the security of data. No matter what encryption algorithm and encryption method are used to encrypt and protect data, the effect of data encryption is the primary problem that needs to be seriously considered. Through the analysis and summary of previous studies, it can be found that in terms of encryption algorithms, most of them choose mature encryption algorithms that have been widely used and verified, such as DES, AES, RC4, RSA and so on. Through the analysis of these parameters acting on the encoding process, some parameters that have a great impact on the encoding and decoding process are selected for encryption, so as to reduce the total amount of data encryption and ensure better encryption effect [9]. The research premise of dynamic data encryption technology is similar to the previous research on data selective encryption, which is also based on the analysis of encrypted data. Symmetric key cryptography can be divided into block cipher and sequence cipher according to different methods of data encryption. The encryption idea of block cipher is to divide the plaintext data into groups of equal size, and then encrypt each plaintext group with the key to obtain the ciphertext group. The size and quantity of ciphertext group and plaintext group are equal, that is, the encrypted data usually has no size expansion and compression. With the development of hardware technology, the bandwidth of both wired and wireless communication is greatly improved in the network, and the capacity of storage devices is also increasing in the storage aspect. Therefore, the lossy compression coding process, which aims to improve the compression ratio, is no longer particularly necessary. In this paper, the improved Logistic chaotic mapping algorithm is used as the main algorithm of data encryption transmission mode. The encryption scheme is shown in Fig. 2.

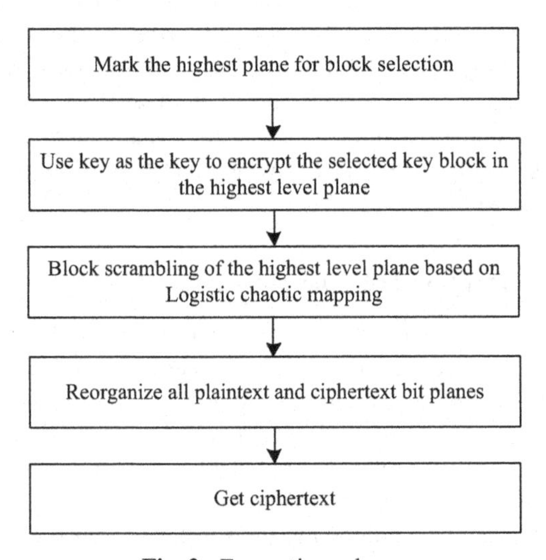

Fig. 2. Encryption scheme

Lyapunov exponent represents the numerical characteristics of the average exponential divergence rate of adjacent trajectories in phase space. It is an important numerical feature for identifying chaotic motion. It is the logarithm of geometric mean. The Lyapunov exponent of Logistic mapping is calculated as follows:

$$\alpha=\frac{1}{m}\sum_{m=0}^{m-1}\ln\left|\frac{df(\varpi,\vartheta)}{d\varpi}\right| \tag{7}$$

In formula (7), α represents Lyapunov exponent; m represents the number of iterations; f represents the mapping relationship; ϖ represents a state of the function; ϑ

is the control parameter. In terms of encryption methods, we should focus on encrypting the parts with high importance in the plaintext based on the characteristics of the encrypted data plaintext. The method of selecting the original text of fixed proportion data according to the equal spacing or random spacing has very limited practical significance, because this method does not distinguish the importance of different parts in the original text of data. Adopting this method will reduce the overall data encryption strength in equal proportion while reducing the encryption operation cost [10]. In addition, by referring to the two complete encryption methods introduced above, it is a desirable way to encrypt the less important part of the data with lightweight encryption algorithm. A great feature of most cryptographic algorithms based on this cryptosystem is iterative encryption. Iterative encryption refers to generating multiple key packets using the key and performing multiple operations with the plaintext data. The steps of each operation are often the same, but the key packets are different. After multiple operations, the ciphertext is generated, which fully realizes the confusion and diffusion of plaintext, making the relationship between plaintext and ciphertext very complex, Thus, the ciphertext has good anti statistical analysis ability and anti differential attack ability. In addition to ensuring data security and improving encryption efficiency, the dynamics of dynamic encryption technology is also reflected in meeting the needs of different security levels. Through the research and analysis of the plaintext of the encrypted data, it is not only necessary to select the most important part and encrypt it with a cryptographic algorithm with high encryption strength to ensure the data security. We should also further subdivide the importance of different parts of the data plaintext, so as to provide the possibility of encryption with different encryption algorithms to meet the needs of different security levels. The number and value range of parameters in chaotic mapping are related to the size of key space and directly affect the security of encryption algorithm. The improved logistic algorithm divides the chaotic mapping into two stages, and the number of pre iterations is significantly reduced compared with the classical logistic algorithm. The algorithm formula is:

$$\varpi_{m+1} = \begin{cases} 4\vartheta\varpi_m(0.5 - \varpi_m) \\ 1 - 4\vartheta(0.5 - \varpi_m)(1 - \varpi_m) \end{cases} \tag{8}$$

In formula (8), ϖ_m and ϖ_{m+1} represent the mapping states before and after iteration, respectively. After the vector data undergoes integer transformation, difference transformation and mean transformation, Gzip encoding is used for secondary data compression, and the compressed binary stream is encrypted. The encryption process adopts the symmetric encryption method, which is based on the improved Logistic chaotic map encryption to enhance the confidentiality of the data. In many applications, it is necessary to use or transmit lossless raw data without compression encoding. This kind of raw data generally does not have special compression coding and does not have a complex coding format, so it is impossible to select specific important parameters and apply the previous selective encryption method for multimedia data to encrypt such data. And because this kind of data is not compressed and encoded, compared with the general multimedia data that has been compressed and encoded, the data length is often several times that of the latter. So far, the design of the encryption transmission method of marine engineering buoy measurement data has been completed.

3 System Test

3.1 Test Preparation

In order to verify the security and effectiveness of the dynamic data encryption method proposed in this paper, the encryption transmission method of marine engineering buoy survey data designed in this paper is programmed. The motion performance and stability of the buoy meet the general specification requirements of the micro-buoy body. The swing period is between 1–3 s, and the maximum swing angle is less than 30° under normal conditions. For buoys with a main scale of 1.2–1.8 m, high initial stability is required between 0.3–0.4 m, and buoys with a main scale of less than 1.1 m. According to the environmental load of the working sea conditions and the functional requirements of the buoy design, and referring to the design parameters of the existing small drifting buoys, the basic parameters of the buoy body, such as the height of the center of gravity, the moment of inertia, etc. The arrangement provides the basis for the buoy to achieve optimum performance. Collect marine engineering buoy measurement data for encrypted transmission. The experimental server uses Microsoft Visual Studio as the development tool, the development language is C#, the supporting environment is Miscrsoft.Net Framework 4.6, and the database uses SQLite. The client development environment is Xamarin for Visual Studio, and the mobile GIS functions are developed based on the ArcGIS Runtime for Xamarin framework environment. The online mode directly calls the network map service published by the server, and uses the SQLite database to store the local data of the mobile terminal.

3.2 Test Results

The data volume of ocean engineering buoy measurement information is set to four Set the data volume of offshore engineering buoy measurement information to 10000 bits. Under the same experimental environment, take data encryption time and security of data encryption transmission as experimental comparison indicators, and compare and verify the method in this paper with the methods in reference [2], reference [3] and reference [4].

Four data encryption transmission methods are used to encrypt and transmit the buoy measurement information, and the time spent in encryption by different methods is counted. The comparison results of encryption time of different methods are shown in Table 1.

As shown in Table 1, the average data encryption time of the method in this paper is 2865.6 ms. Compared with the three literature comparison methods, the data encryption time of the method in this paper is reduced.

In order to further verify the performance of this method, the security of data encryption transmission is taken as the experimental comparison index, and the four methods are also tested. The comparison results of encrypted transmission security of the four methods are shown in Fig. 3.

From the comparison results of encrypted transmission security shown in Fig. 3, it can be seen that among the four methods, the method in this paper has the lowest intrusion rate, and the highest intrusion rate of the method in this paper does not exceed 0.5%. The

Table 1. Comparison of data encryption time (ms)

Number of experiments	Methods in this paper	Reference [2] method	Reference [3] method	Reference [4] method
10	2819	3401	3418	3214
20	2704	3533	3587	3395
30	2858	3459	3665	3363
40	2986	3586	3636	3225
50	2865	3648	3523	3438
60	2837	3574	3502	3406
70	3021	3418	3525	3382
80	2954	3426	3458	3201
90	2876	3535	3482	3319
100	2847	3544	3323	3460
110	2915	3558	3309	3327
120	2922	3421	3264	3448
130	2788	3430	3551	3291
140	2762	3356	3585	3138
150	2830	3491	3626	3264
Mean value	2865.6	3492	3496.9	3324.7

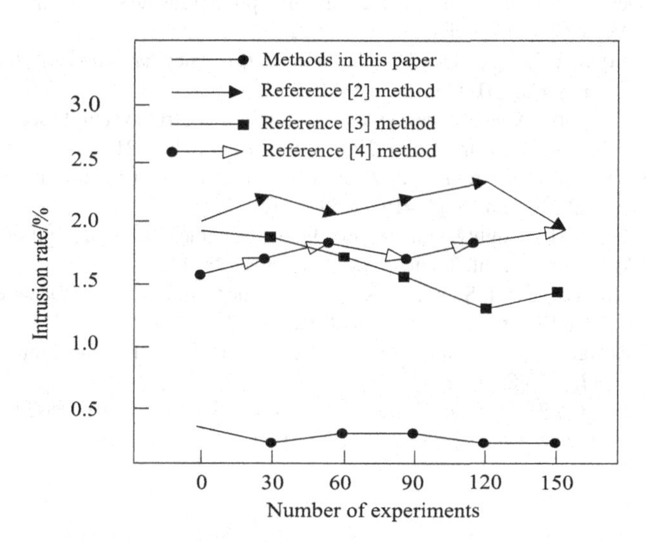

Fig. 3. Comparison results of encrypted transmission security

three literature comparison methods not only have strong volatility of intrusion rate, but also have high intrusion rate, which cannot fully ensure the security of data encryption transmission. Therefore, the above experimental results show that the method in this paper can effectively improve the security of the marine engineering buoy measurement data.

4 Concluding Remarks

In this paper, an encryption transmission method of marine engineering buoy survey data is designed, which can shorten the encryption time and ensure the security of data encryption and the availability of decrypted data at the same time. In order to improve the efficiency of real-time encryption, the raster data encryption algorithm used in this paper, the sequence generated by the chaotic map is directly operated with the plaintext in the way of stream cipher, and no plaintext feedback or ciphertext feedback is used to affect the iteration of the chaotic map. Therefore, the anti-attack performance needs to be further improved.

References

1. Fengxia, Z.: Development and design of traffic identification system based on DPI. Electron. Design Eng. **28**(03), 93–97 (2020)
2. Dandan, G., Fei, W., Dan, L.: Mobile communication data encryption method based on DES algorithm. Telecom Power Technol. **37**(20), 136–137140 (2020)
3. Chai, C.: Design of network information data encryption transmission method based on AES algorithm. Modern Transmission **2**, 39–42 2022
4. Huo, Y.: Encryption method of high-end equipment instruction data based on chaos algorithm. J. Ordnance Equip. Eng. **41**(11), 190–193 2020
5. Yang, L.: Design of robot data encrypted transmission control system based on blockchain technology. Comput. Measure. Contr. **29**(06), 119–122+163 (2021)
6. Zhenfeng, Z.: Embedded remote data encryption transmission method based on discrete chaotic map. Tech. Autom. Appl. **41**(07), 58–61 (2022)
7. L, W., S, Y., Q, Li, I.: A lightweight implementation scheme of data encryption standard with cyclic mask. J. Electron. Inf. Technol. **42**(8), 1828–1835 (2020)
8. Wang, X., Xin, G.Q., Hui, S., et al.: Design of an encryption protocol for wireless network data transmission. Electron. Design Eng. **28**(02), 73–77 (2020)
9. Chen, X.: Design of access data encryption scheme in ship communication network. Ship Sci. Technol. **42**(20), 88–90 (2020)
10. Fang, W., Lei, G., Xu, C.: Encryption method for smi sensitive data based on datasocket technology. Comput. Simul. **38**(8), 217–221 (2021)

Tracking Method of Ocean Drifting Buoy Based on Spectrum Analysis

Huanyu Zhao[1,2,3](✉) and Wenyan Wang[1,2,3]

[1] Institute of Oceanographic Instrumentation, Qilu University of Technology (Shandong Academy of Sciences), Qingdao 266061, China
huanyuzhao2021@163.com
[2] Shandong Provincial Key Laboratory of Marine Monitoring Instrument Equipment Technology, Qingdao 266061, China
[3] National Engineering and Technological Research Center of Marine Monitoring Equipment, Qingdao 266061, China

Abstract. In view of the poor accuracy of buoy estimation and tracking results caused by ignoring the influence of the marine environment on buoys in the current tracking process of ocean drifting buoys, a tracking method of ocean drifting buoys based on spectrum analysis is proposed. By calculating the disturbing force and moment of the ocean current on the ship, the relative operating characteristics between the drifting buoy and the ocean current are obtained, and the ocean current motion model is constructed; According to the Newton Eulerian dynamic equation, the acceleration and external force of the floating body are calculated, and the motion spectrum characteristics of the floating buoy are extracted; According to the result of feature extraction, the trajectory of ocean drifting buoy is located and tracked by Kalman filter. So far, the tracking of the ocean drifting buoy is completed. The experimental results show that the tracking accuracy of this method is up to 100% without interference, and 99.81% with interference.

Keywords: Spectrum analysis · Underwater recovery · Trajectory tracking · Ocean drifting buoys

1 Introduction

Humanity has entered the 21st century, and the resources on land have been exhausted. Before human beings have found renewable energy sources, the ocean, as an untapped virgin land, has become the focus of national strategic competition, and thus has become an important field of high-tech research. my country is also a big country with 3 million square kilometers of "blue land", rich in marine resources, especially in the South China Sea, where considerable oil and gas resources have been proven, but many countries near the South China Sea are competing to exploit and plunder resources [1]. At present, many countries have clearly realized the importance of the ocean to the sustainable development of the country, society, economy and national security. In order to achieve the development goal of a maritime power, it is necessary to actively explore and develop

W. Fu and L. Yun (Eds.): ADHIP 2022, LNICST 468, pp. 673–687, 2023.
https://doi.org/10.1007/978-3-031-28787-9_50

the ocean. The ocean contains rich mineral resources and biological resources, and the ocean, which accounts for 49% of the earth's area, is a public area. The resources in this part of the area do not belong to any country. If a country has the technical strength, it can enjoy this resource exclusively. Independently develop this area. While developing renewable resources, people also turn their attention to the ocean. The boundless ocean not only provides human beings with shipping, aquatic products and rich minerals, but also contains huge energy. Ocean energy refers to the renewable energy attached to seawater. The ocean receives, stores and releases energy through various physical processes. These energy sources exist in the ocean in the form of tides, waves, temperature differences, salinity gradients, and ocean currents.

Only ordinary observations can only analyze the basic conditions of coastal areas and islands, and cannot play a role in ocean navigation. In order to deepen the research of marine resources, the marine buoy system has been established in the past few decades. The ocean drift buoy is an unmanned automatic ocean observation station, which is fixed in the designated sea area and fluctuates with the waves, just like the navigation marks on both sides of the channel [2]. It is not easy to affect the passing ships in the ocean, and can carry out long-term, continuous, all-weather work in any harsh environment, and regularly measure and report various hydrology, water quality and meteorological elements every day. It plays a vital role in the development and utilization of marine resources. The ultimate purpose of monitoring the marine environment is to benefit mankind. At present, there is a general shortage of energy worldwide, and the current energy shortage is increasingly becoming a "bottleneck" restricting the economic development of many countries. The development of renewable energy that can replace coal, oil and natural gas has become the focus of widespread attention. At the same time, the environmental impact brought by fossil fuels Pollution seriously affects the living environment of human beings.

There are existing literatures on the track tracking of ocean drifting buoys at home and abroad. At present, the research on track tracking of ocean drifting buoys mainly focuses on the track tracking and path following of ocean drifting buoys. For trajectory tracking problems, if conventional nonlinear control methods are used, at present, local feedback linearization and system model decoupling are mostly used. Linearization methods to solve the trajectory tracking problem [3].

The method in document [4] proposes a passive location and tracking method of short range moving target based on combined linear array, which adopts mutually perpendicular structural layout to solve the problem of single array azimuth positioning. The Kalman filter algorithm is used to predict and estimate the target trajectory, and the target positioning information is matched with the tracking trajectory information to complete the target tracking process. The method in document [5] proposes the design and research of the operation monitoring system of the marine meteorological drifting buoy, and uses the C/S framework to design the client software based on the net platform. All modules of the system are componentized, and the buoy operation monitoring system is designed by combining OpenGL, GDI+, GIS, SQL and other technologies. However, the above method has a problem of low tracking accuracy. In this regard, the tracking method of ocean driving buoy based on spectrum analysis is proposed to obtain the relative operating characteristics between the drifting buoy and the ocean current by calculating the disturbing force and moment of the ocean current on the hull, and to build the ocean current motion model; According to the Newton Eulerian dynamic equation, the acceleration and external force of the floating body are calculated, and the motion spectrum characteristics of the floating buoy are extracted; According to the result of feature extraction, the trajectory of ocean drifting buoy is located and tracked by Kalman filter. So far, the tracking of the ocean drifting buoy is completed. According to the experimental results, the tracking accuracy of the method in this paper can reach 100% at the highest under the condition of no interference, and 99.81% under the condition of interference. It can be seen that the tracking accuracy of the method in this paper is high, which can effectively improve the tracking accuracy of ocean drifting buoys, accelerate people's understanding and management of ocean resources, and have certain practicality.

2 Design of Track Tracking Method for Ocean Drifting Buoy Based on Spectrum Analysis

After a detailed analysis of the current track tracking process of ocean drifting buoys, the spectrum analysis technology is used as the core technology in this study to apply it to the track tracking process of buoys. In order to ensure the integrity of the ocean drifting buoy trajectory tracking method based on spectrum analysis, the trajectory tracking process is set as shown in Fig. 1.

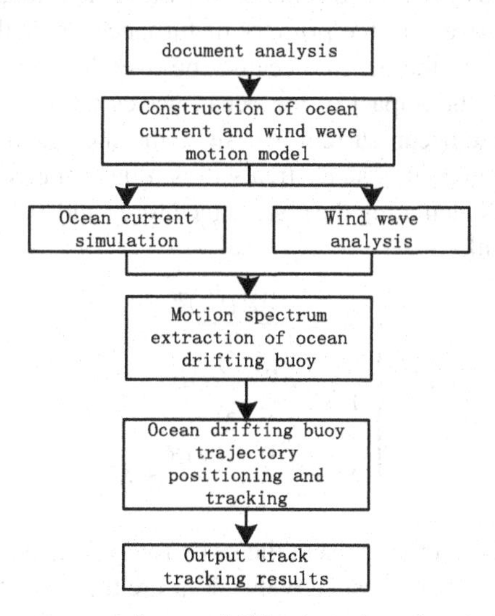

Fig. 1. Tracking process of ocean drifting buoy based on spectrum analysis

As shown in the figure above, this research will mainly address three questions:

(1) Analyze the impact of ocean currents on the trajectory of ocean drifting buoys, and overcome the impact of the external environment on the accuracy of trajectory tracking results.

(2) Analysis of the motion characteristics of ocean drifting buoys. Firstly, based on the force analysis of ocean drifting buoys, based on series expansion, it is explained that there is strong coupling and nonlinearity in the movement process of drifting buoys, and the physical constraints of the buoys and the controllability analysis are analyzed.
(3) Obtain the motion spectrum of the drifting buoy, combine it with the motion characteristics of the ocean drifting buoy, use appropriate techniques to analyze the overall motion process, and obtain the drifting trajectory of the ocean drifting buoy, so as to achieve the goal of this research.

2.1 Constructing Ocean Current Motion Model

At present, the moving target tracking technology has been widely used in various tasks on land, sea and air. The purpose of moving target tracking is to make the subject successfully track the last moving object, mainly including moving target modeling, estimation and tracking. How to extract sea surface state information from radar sea surface echo Doppler spectrum, it is necessary to understand the mechanism of the interaction between high frequency electromagnetic waves and the ocean surface. Therefore, it is necessary for us to understand the characteristics of high-frequency radar waves propagating on the ocean surface and their scattering. Since the object of this study is the ocean drifting buoy, when tracking its trajectory, it is necessary to refer to the marine environment issues and the influence of ocean currents on the movement range of the buoy. On the surface of the ocean, waves are the primary disturbance to anything. When the ocean drifting buoy is moving in the ocean, because the overall weight of the ocean drifting buoy is relatively light, the impact of the ocean waves on it is small, and it is mainly strongly disturbed by the ocean current [6]. Since the speed and direction of the ocean current change very slowly, this study treats it as a quantitative constant disturbance, and gives a formula for calculating the disturbance force and disturbance moment of the ocean current on the hull:

$$\begin{cases} X = \dfrac{\rho W A_x(\alpha) B}{2} \\ Y = \dfrac{\rho W A_y(\alpha) B'}{2} \\ N = \dfrac{\rho W A_n(\alpha) B}{2} * K \end{cases} \tag{1}$$

Among them, X, Y, and N represent the sway force, sway force and roll moment of the current on the drifting buoy, respectively; W, α are the speed and encounter angle of the current relative to the drifting buoy, respectively; B, B' are the water of the drifting buoy, respectively The transverse section and the longitudinal section of the lower part; K represents the resistance of the ocean water body; A_x, A_y and A_n are the current's sway force coefficient, sway force coefficient and roll force coefficient, respectively. In order to ensure that the calculation effect of this model is relatively stable, it is supplemented in this study. Assuming that the ocean current has a constant velocity $Q = (u_i, v_i, 0)$ in the earth coordinate system, and the velocity in the water coordinate system is $Q' = (u_i', v_i', r)$, the relationship between the ocean current velocity in these two coordinate systems can be expressed as:

$$Q = \Delta Q' \tag{2}$$

At this time, the relative velocity between the drifting buoy and the ocean current can be expressed as $V_I = (V_{Ix}, V_{Iy}, V_{Iz})$, and its relative operating characteristics can be expressed as:

$$\begin{cases} V_{Ix} = u_i - u_i' \\ V_{Iy} = v_i - v_i' \\ V_{Iz} = 0 - r \end{cases} \tag{3}$$

In the calculation, the size of the current in the vertical direction is not considered, so that the component of the current in the vertical direction is zero. The above calculation process is summarized and the ocean current motion model is constructed:

$$\kappa(t) = \kappa_0 + \sum_{j=1}^{n} e_j \cos(\varpi_i t + \zeta_i) \tag{4}$$

Where, κ represents the sea surface height, κ_0 represents the average sea surface height, κ_0 represents the amplitude of the j wave train, ϖ_i represents the angular frequency of the i wave train, and ζ_i represents the phase angle of the i wave train. Thus, simultaneous interpreting rough sea surface by waves with different wavelengths and different directions of propagation. The formula is combined with the above ocean current model as the calculation basis of the impact of external environment on buoy movement in this study. The specific construction process of ocean current motion model is as follows:

2.2 Extraction and Analysis of Motion Spectrum of Ocean Drifting Buoy

According to the ocean current motion model constructed above, the motion spectrum of the ocean drift buoy is extracted. Through literature analysis, it can be seen that the good conductor characteristics of sea water make the high-frequency electromagnetic wave spread further on the sea surface. At the same time, the propagation of electromagnetic wave on the sea surface is affected by the roughness of the sea surface, and the additional attenuation of electromagnetic wave is also different under different sea states [7]. With the increase of frequency, the additional attenuation value increases. In general, the higher the marine environmental interference, the greater the additional attenuation caused. The radar system used in this study is a fully coherent radar system, and the synchronous controller provides all timing control signals required by the system. It provides timing and control signals and interrupt request signals to each module of the system to control the coordinated work of frequency synthesizer, receiver switch, transmitter switch, sampling signal and sampling trigger signal. Through this radar system, the motion spectrum of ocean drifting buoy is obtained, sorted, extracted and analyzed.

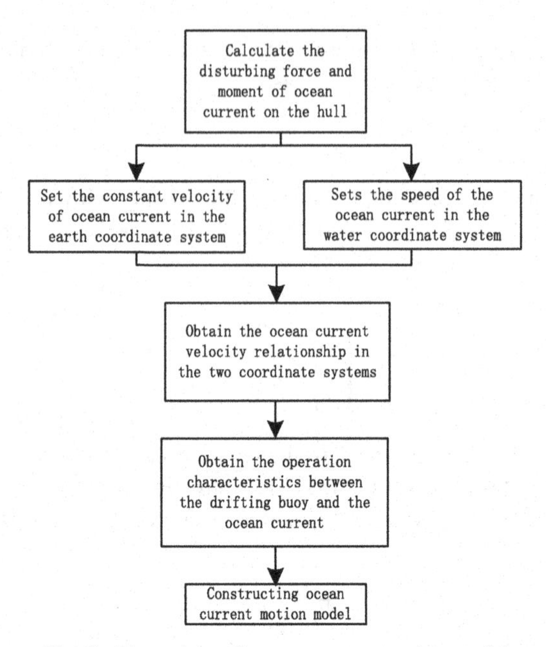

Fig. 2. Flow chart of ocean current motion model

n this study, based on the three-dimensional potential flow theory and Morrison equation, the hydrodynamic performance of the drifting buoy is analyzed, especially considering the influence of hydrodynamic interaction on the hydrodynamic performance of the buoy. The frequency domain analysis is carried out under the condition of free floating without considering the mooring system. For the ocean drifting buoy, according to the Newton Euler dynamic equation, the acceleration and external force of the floating body can be known as follows:

$$kg = F \tag{5}$$

$$[U]\overline{\psi} + \psi[U\psi] = D \tag{6}$$

Here, k is the mass of the floating body, g is the coordinate of the center of gravity of the floating body, U represents the moment of inertia, ψ represents the angular velocity of the floating body, and F and D represent the external force and bending moment on the floating body. Assuming that the rotational motion of the floating body is a small amount, formula (5) and formula (6) are simplified and combined as follows:

$$E(t) = FD \tag{7}$$

The degrees of freedom of the mass matrix, motion vector and force vector in formula (7) will be extended to $12n * 12n$, $12n$ and $12n$, where n is the number of floating bodies. The motion of multiple floating bodies is to consider the interaction of the flow field between floating bodies. First, two floating bodies are considered. By extracting the motion spectrum characteristics of ocean drifting buoys, the mass matrix, stiffness matrix and moment matrix of buoys are obtained [8]:

$$K = \begin{bmatrix} L^i & 0 \\ 1 & L^{ii} \end{bmatrix} \tag{8}$$

$$E = \begin{bmatrix} E^i & E^{i,ii} \\ E^{ii,i} & E^{ii,ii} \end{bmatrix} \tag{9}$$

$$\begin{cases} R_i = \begin{bmatrix} R_i^i \\ R_i^{ii} \end{bmatrix} \\ R_w = \begin{bmatrix} R_w^i \\ R_w^{ii} \end{bmatrix} \\ R_m = \begin{bmatrix} R_m^i \\ R_m^{ii} \end{bmatrix} \end{cases} \tag{10}$$

Superscript i and ii represent floating body 1 and floating body 2, and each floating body has 6 degrees of freedom. Where K is the mass submatrix of $12\,\mathrm{n} \times 12\,\mathrm{n}$, E is the additional mass submatrix, R_i is the delay function, [C] is the stiffness matrix, R_w is the displacement vector of the floating body, and R_m is the combined external force vector, where the force vector includes wave excitation force, drift force, wind force, current force and mooring force. The subscript N indicates the number of the object. When the subscripts are the same, it indicates the existence of a single floating body, and when the subscripts are different, it indicates the role from other floating bodies. In this study, based on the three-dimensional potential flow theory and starting from the boundary value problem, the first-order linear wave theory and the second-order nonlinear wave theory are derived by using the stoke method, and the diffraction potential theory of the drifting buoy is further analyzed. At the same time, the first-order and second-order velocity potential problems are solved by using the Green's function method (boundary element method). At the same time, the hydrodynamic interaction of the floating buoy system is determined, and the frequency domain and time domain equations of the dynamic response of single floating body and multi floating body are obtained. It provides a theoretical and analytical basis for the follow-up ocean drift buoy trajectory tracking.

2.3 Ocean Drifting Buoy Trajectory Positioning and Tracking

According to the above extracted motion spectrum characteristics of the ocean drift buoy, the trajectory of the ocean drift buoy is located and tracked by Kalman filter. The track location and tracking of ocean drifting buoy is an important link in this research. In order to track the ocean drifting buoy, we must first estimate the motion state of the ocean drifting buoy. The state estimation and prediction of ocean drifting buoy mainly includes: formation and processing of measurement data, target motion modeling, target identification and maneuver detection, filtering and prediction, etc. Motion estimation and target state estimation are very important. Common filtering methods include Kalman filter, particle filter and grey prediction [9, 10]. This chapter mainly introduces the Kalman filter, which is the most commonly used filter theory, and estimates the motion of the target based on the Kalman filter method. Before motion estimation, the buoy motion model is constructed according to the above research results. Due to the large lateral damping, the rolling motion of the buoy can be ignored in the analysis. Therefore, the following assumptions can be made:

$$B = \widetilde{B} = \delta = \overline{\delta} = 0 \tag{11}$$

Thus, the six degree of freedom spatial motion equation can be simplified to the motion in the direction of five degrees of freedom, so as to simplify the control design. Since the rolling angle and angular velocity are not included, the previous buoy motion model is rewritten assuming that the center of gravity and floating center of the buoy coincide, the size of gravity and buoyancy are equal, and the interference factors of external ocean currents are not considered. It can be seen from the optimized formula that when the longitudinal velocity $v1$ is stable, the vertical plane motion model is independent, while the horizontal plane motion model is affected by δ and $\overline{\delta}$. Therefore, it can be divided into three steps and analyzed by cascade system. The three-dimensional linear trajectory tracking of the buoy can be defined as finding the appropriate control inputs h and p on the basis of obtaining the stable longitudinal velocity $v1$ for the buoy at any initial position, so that the transverse error and vertical error tend to zero. The problems faced by trajectory tracking are the same as path following. It is difficult to obtain the model and ensure the robustness, adaptability and control performance of the system under complex environmental interference and model perturbation. Therefore, it is difficult to be applied in engineering practice. Aiming at these problems and considering the application of practical engineering, this paper proposes a trajectory tracking method of ocean drifting buoy based on spectrum analysis under the condition of ocean current interference. Referring to the pure tracking method, the motion relationship between buoys is displayed as shown in Fig. 3.

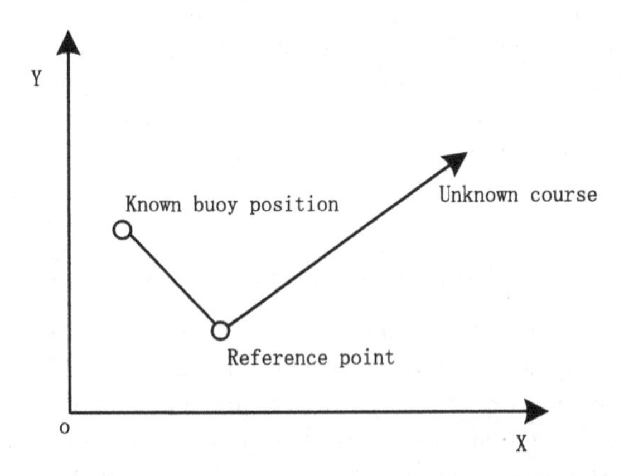

Fig. 3. Schematic diagram of buoy movement relationship

Combined with the content in the above figure, considering the original position information of the buoy, in the inertial coordinate system, the relative motion geometric relationship between the known buoy position and the target point is shown in Fig. 2, the reference line MN is selected, the drifting speed of the buoy is v'', and the speed of the target point is v''', τ and τ' are the angle between the speed of the buoy and the target point and the positive direction of MN, the distance from the buoy to the target is K, $\Im$ is the angle between the target line of sight and the reference line MN, and the movement direction of the buoy is along the line of sight direction, and finally achieve K reduced to a safe distance d, speed, $v'' = v'''$, $\Im = \Im'$. According to the above settings, the movement speed and basic movement distance of the buoy are obtained. According to this data, combined with the current trajectory tracking algorithm, the reference frequency analysis results and the marine environment simulation results, the buoy trajectory tracking in the marine environment is realized. Then there are:

$$\overline{V} = V_t + \frac{(V_{\max} + \lg(Z_{tt}) - V) * (Z_{tt}^2 - \overline{Z_{tt}})}{\sqrt{Z_{tt}^2 + \Delta_{tt}^2}} \tag{12}$$

$$\Re = a\tan(y_i - y(t), x_i - x(t)) \tag{13}$$

Among them, V_t represents the moving speed of the target point, $\overline{V}$ and $\Re$ represent the expected tracking speed and heading of the buoy; $V_{\max}$ represents the maximum drifting speed of the buoy; Z_{tt} represents the distance from the known buoy position to the unknown buoy, and Δ_{tt} represents the parameters for adjusting the calculation convergence speed. Arrange the above calculation contents to ensure the integrity and coherence of the calculation process. The calculation link set in this paper is combined with the current trajectory tracking method. So far, the design of the ocean drifting buoy trajectory tracking method based on spectrum analysis is completed.

3 Analysis of Experimental Demonstration

In this study, a method for tracking the trajectory of ocean drifting buoys based on spectrum analysis is proposed. Before this method is applied to practical work, an experimental link is first constructed to analyze its application effect to ensure that this method meets the current buoy trajectory tracking. Related requirements.

3.1 Experiment Preparation

This experiment mainly carries out simulation tests on the above-mentioned buoy trajectory tracking method. Under the conditions of no interference and water flow interference, the application effects of the method in this paper, the method in literature [3] and the method in literature [4] in buoy trajectory tracking are compared. The buoy model in the simulation experiment is the same as the path following, and the calculation parameters remain unchanged. The maximum drifting speed of the buoy is set to $V_{\max} = 2.5$, and the initial drifting speed, drifting direction, position, and rudder angle are all zero in the simulation.

Assuming that the initial position of the buoy is (100, 200), it moves in a uniform straight line along the y axis at a speed of 10 m/s. at $t = 500$–700 s, the buoy makes a 90° slow turn to the x axis with an acceleration of $q_x = q_y = 0.12\,\text{m/s}^2$. After the slow turn is completed, the acceleration drops to zero. At $t = 750\,\text{s}$, the target makes a 90° fast turn to the y axis with an acceleration of $0.3\,\text{m/s}^2$ at $t = 800\,\text{s}$, end the turn, reduce the acceleration to zero, and then move at a uniform speed for a period of time. The parameters used by all methods in the simulation are as follows: weighted attenuation factor $\alpha = 0.90$, maneuver detection threshold, and exit maneuver detection threshold $T_J = 9.50$. Take the above contents as the basic calculation parameters of this experiment, and simulate the motion process of the buoy on the basis of this simulation parameters, so as to provide the basis for the subsequent experimental process.

In order to ensure the authenticity of this experiment, set the target buoy in the form shown in Fig. 4, and draw the ideal buoy trajectory according to the above simulation parameters. The specific contents are as follows.

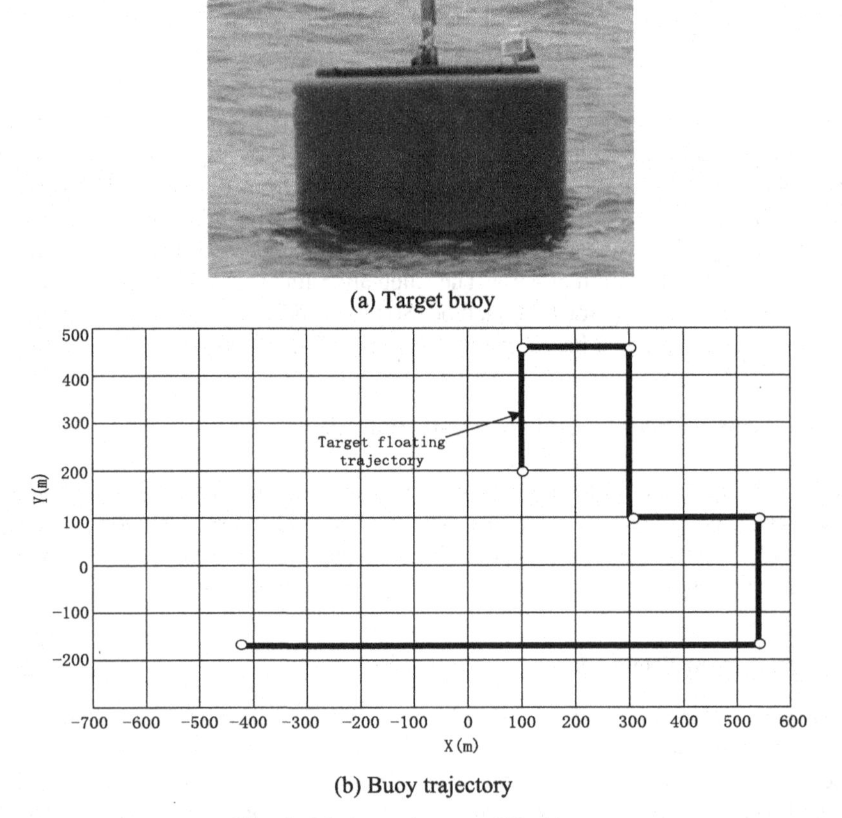

(b) Buoy trajectory

Fig. 4. Motion trajectory of ideal buoy

As shown in the figure above, in this study, the six segments of the motion trajectory and the distance between each two nodes are taken as an experimental segment, and the three selected methods are used to track it and compare the tracking accuracy between different methods. According to the calculation formula of accuracy index, it is set as follows:

$$A = \frac{s}{M} * 100\% \tag{14}$$

The obtained tracking result is s; M represents the ideal buoy trajectory; A indicates tracking accuracy. Use this index to analyze the use effect of different methods.

3.2 Analysis of Experimental Results

3.2.1 Analysis of Test Results Without Interference

According to the above settings, the experimental results of non-interference conditions are obtained, as shown in Fig. 5.

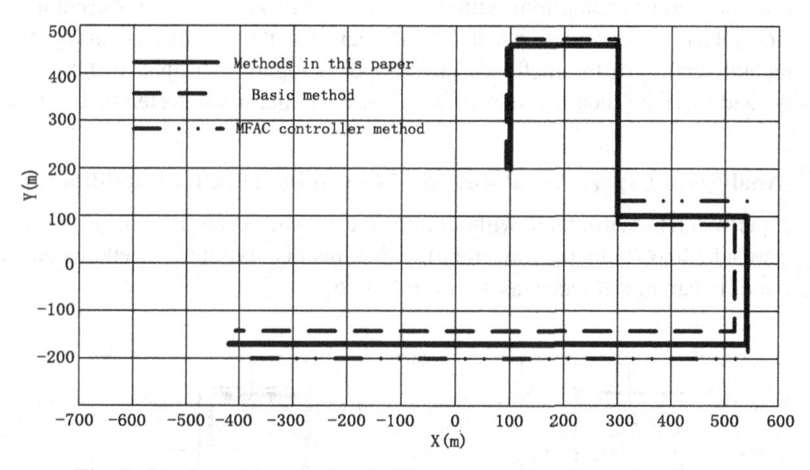

Fig. 5. Track tracking results of different methods without interference

Table 1. Track tracking accuracy of different methods without interference (unit /%)

Experimental section	The tracking accuracy of the proposed method	Tracking accuracy of method in reference [3]	Tracking accuracy of method in reference [4]
1	99.15	95.14	97.15
2	99.75	96.15	98.62
3	99.81	94.25	97.66
4	99.45	94.62	98.78
5	100.0	90.15	96.15
6	100.0	89.65	97.80

It can be seen from Fig. 5 that the trajectory tracking results of different methods are different. It can be preliminarily determined that the use effect of the method in this paper is better only through image observation. In order to make a more systematic analysis, the trajectory tracking accuracy of different methods is calculated, and the calculation results are shown in Table 1.

Through the analysis of the above experimental results, it can be seen that the tracking accuracy of the methods in this paper is above 99%, and the highest is 100%; The tracking accuracy of the method in literature [3] is 96.15% at the highest and 89.65% at the lowest; The tracking accuracy of the method in reference [4] is 98.78% at the highest and 96.15% at the lowest. The tracking accuracy of the method in this paper is obviously better than that of the method in reference [3] and that of the method in reference [4], but the overall difference between these three methods is small under the condition of no interference. By fusing the track tracking results of different methods under non-interference conditions with the track tracking accuracy of different methods under non-interference conditions, it can be seen that the tracking accuracy of each experimental section of the method in this paper is higher than that of the other two methods, and it can be seen that it has a better effect under non-interference conditions.

3.2.2 Analysis of Experimental Results of Ocean Interference Conditions

In the experimental environment without interference, the ocean current and wave interference are added to obtain the trajectory tracking results of different methods under the condition of ocean interference, as shown in Fig. 6.

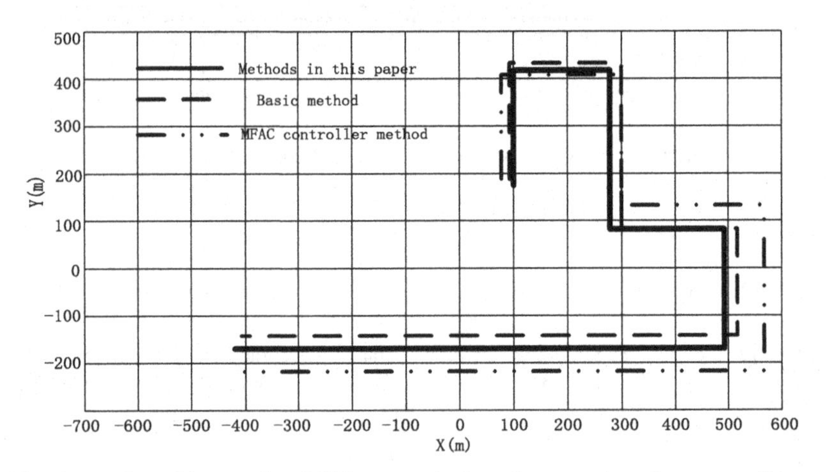

Fig. 6. Track tracking results of different methods under ocean interference conditions

It can be seen from Fig. 6 that after adding interference items in the experimental environment, the track tracking results of different methods are quite different. From image observation, we can see the price difference of the tracking effects of the two methods currently used. In order to make a more systematic analysis, the trajectory tracking accuracy of different methods is calculated, and the calculation results are shown in Table 2.

Table 2. Track tracking accuracy of different methods under ocean interference conditions (unit /%)

Experimental section	The tracking accuracy of the proposed method	Tracking accuracy of method in reference [3]	Tracking accuracy of method in reference [4]
1	99.15	89.17	91.52
2	99.75	89.68	93.51
3	99.81	88.62	93.54
4	99.45	87.94	93.44
5	98.62	89.15	92.11
6	98.75	87.05	92.77

Through the analysis of the above experimental results, it can be seen that the tracking accuracy of the method in this paper is as high as 99.81% when the ocean interference is added to the experimental environment; The tracking accuracy of the method in reference [3] is 89.68%; The tracking accuracy of the method in reference [4] is 93.51%. The tracking accuracy of the method in this paper is obviously better than that of the method in reference [3] and that of the method in reference [4]. The tracking accuracy of the method in document [3] is less than 90%. The tracking accuracy of the method in document [4] is relatively high, but not as good as the tracking accuracy of the algorithm in this paper. Combining the track tracking results of different methods under ocean interference conditions and the track tracking accuracy experimental results of different methods under ocean interference conditions, it can be seen that the method has a good effect in this experimental environment.

4 Discussion and Analysis

In this study, an ocean drifting buoy trajectory tracking method based on spectrum analysis is proposed. Through the experiments of non-interference conditions and ocean interference conditions, it is confirmed that the application effect of the method in this paper is better than the current method. For the buoy trajectory tracking problem, the spectrum analysis, water dynamics and trajectory tracking methods are applied to this study to solve the influence of water flow interference and increase the robustness of the tracking method. The simulation results show that the buoy tracking path can be obtained effectively and quickly under the influence of uncertainty.

In the process of spectrum analysis, combined with the characteristics of multi working state of radar, the design idea and specific scheme of time-sharing multi frequency/dual frequency are put forward. Multifrequency radar system is a system engineering, which needs to cooperate with multi band or broadband radar antenna and corresponding transmitting system. In the follow-up research, it is necessary to give relevant design and corresponding synchronous control sequence test results for radar control system and receiving and processing system. Buoy target tracking is a very complex problem and a research direction of great value. Due to the limitation of knowledge level and ability, this research is done under some simplified models and assumptions, so there are many places to be improved. The research on the following aspects has practical significance:

(1) This study only estimates the position of the target, does not estimate the speed and acceleration of the target, and only studies the motion estimation of the target in the horizontal plane, not the motion estimation of the space target;
(2) Although the simplified model used in the simulation verification of this study can reflect the motion characteristics of the buoy, it is not accurate enough, and in order to facilitate the design of the controller, the off diagonal elements of some model parameters are ignored. Therefore, in the future research, we should consider the actual motion characteristics of the buoy and establish a more comprehensive buoy mathematical model.

5 Concluding Remarks

In view of the shortcomings of the current methods of tracking the trajectory of drifting buoys in the application, a new method of tracking the trajectory of drifting buoys is proposed in this study. By calculating the disturbing force and moment of the ocean current on the ship, the relative operating characteristics between the drifting buoy and the ocean current are obtained, and the ocean current motion model is constructed; According to the Newton Eulerian dynamic equation, the acceleration and external force of the floating body are calculated, and the motion spectrum characteristics of the floating buoy are extracted; According to the result of feature extraction, the trajectory of ocean drifting buoy is located and tracked by Kalman filter. Complete the tracking of ocean drifting buoys. According to the experimental results, the tracking accuracy of this method can reach 100% at the highest under the condition of no interference, and 99.81% under the condition of interference, which proves that this method effectively improves the tracking accuracy of the trajectory of ocean drifting buoys, and the use effect also meets the current research requirements. This method still has some shortcomings in some links, and it needs to be optimized in future research.

References

1. Carlson, D.F., Pavalko, W.J., Petersen, D., et al.: Maker buoy variants for water level monitoring and tracking drifting objects in remote areas of greenland. Sensors **20**(5), 1254 (2020)
2. Hu, M., Yu, S., Li, Y., et al.: Finite-time trajectory tracking control of full state constrained marine surface vessel based on command filter. J. Nanjing Univ. Sci. Technol. **45**(3), 271–280 (2021)
3. Hu, Y., Hua, T., Chen, M.Z.Q., et al.: Application of semi-active inerter in a two-body point absorber via force tracking: Trans. Inst. Meas. Control. **43**(12), 2809–2817 (2021)

4. Xu, G., An, Y., Yang, F., et al.: Passive location and tracking method of short range moving target based on combined linear array. Appl. Acoust. **39**(1), 149–156 (2020)
5. Zhang, D., Cao, X., Wang, Z., et al.: Design and research of marine meteorological drifting buoy operation monitoring system. J. Zhejiang Meteorol. **41**(4), 34–37+48 (2020)
6. Chen, X.: Research and performance analysis of combined mathematical model for ship trajectory tracking. Ship Sci. Technol. **42**(22), 22–24,66 (2020)
7. Jiang, H., Wang, F., Dong, F., et al.: A novel trajectory tracking control of collaborative robot based on udwadia-kalaba theory. Modular Mach. Tool Autom. Manufact. Technique **1**, 78–83 (2021)
8. Li, C., Li, S., Ma, W., et al.: Research on automatic identification of air traffic flow based on flight path clustering. Comput. Simul. **38**(10), 73–77 (2021)
9. Kourani, A., Daher, N.: Marine locomotion: a tethered UAV-Buoy system with surge velocity control. Robot. Auton. Syst. **2**, 103858 (2021)
10. Leng, Y., Zhao, S.: Explicit model predictive control for intelligent vehicle lateral trajectory tracking. J. Syst. Simul. **33**(5), 1177–1187 (2021)

Compensation Method of Infrared Body Temperature Measurement Accuracy Under Mobile Monitoring Technology

Jun Li[1,2], Jun Xing[1,2], Yi Zhou[3], Qingyao Pan[2](✉), and Miaomiao Xu[4]

[1] Shenzhen Academy of Inspection and Quarantine, Shenzhen 518033, China
[2] Shenzhen Customs Information Center, Shenzhen 518033, China
[3] Testing and Technology Center for Industrial Products of Shenzhen Customs, Shenzhen 518067, China
[4] Anhui Xinhua University, Hefei 230088, China

Abstract. Infrared human body temperature measurement is widely used in medicine because of its high safety and flexibility. But in the process of application, there is a big error with the actual temperature. Identify the environmental characteristics of infrared temperature field, judge the high and low temperature area, construct the nonuniformity correction model, convert the electrical signal to the temperature value of the target object, optimize the human body temperature measurement process, introduce the concept of brightness temperature, and use the moving monitoring technology to design the precision compensation mode. The experimental results show that the error mean of the infrared human body temperature compensation method and the other three methods is smaller, which shows that the performance of the method is better.

Keywords: Mobile monitoring technology · Temperature field · Non-contact temperature measurement · Contact temperature measurement · Environmental radiation · Precision compensation

1 Introduction

In the past hundred years, with the progress of human society, human beings have higher requirements for the accuracy, range, sensitivity, real-time and convenience of temperature measurement. The infrared radiation of the detection target is transferred to the infrared detector through the optical imaging objective lens of the infrared thermal imager, and the distribution of the infrared radiation energy is displayed by amplification, shaping and digital mode conversion. Scientists have developed thermocouples, infrared thermometers and infrared thermal imagers and other temperature measuring tools are becoming increasingly mature technology, these tools are not yet perfect because of their limitations. The image produced by the infrared thermal imager reflects the heat distribution of the object surface, and can show the temperature value of each point. It is a very difficult task to use these tools to measure the temperature of the target

W. Fu and L. Yun (Eds.): ADHIP 2022, LNICST 468, pp. 688–701, 2023.
https://doi.org/10.1007/978-3-031-28787-9_51

objects in high temperature, high risk and other harsh environments. As a non-contact thermometer, infrared thermometer has been widely used, such as infrared thermometer in public places such as airports and railway stations, etc. [1, 2]. In many applications, there is an urgent need for a high precision, large dynamic range thermometer to achieve reconnaissance, navigation, fire control, NDT and other important applications. Along with people's demand for production and life, more and more applications began to use non-contact temperature measurement technology. Especially in recent years, with the rapid development of infrared focal plane technology and embedded processing technology, the intelligent degree of infrared temperature measurement technology is getting higher and higher. Non-contact temperature measurement uses the detector to detect the radiant energy of the object to achieve temperature measurement. The non-contact temperature measurement method has no limitation of temperature range and can be applied to both medium and low temperature, high temperature and ultra-high temperature. With the development of uncooled focal plane refrigeration technology, the cost of infrared thermal imaging instrument is getting lower and lower, which lays a foundation for the wide application of infrared temperature measurement technology. According to different principles, non-contact temperature measurement methods can be divided into acoustic temperature measurement, spectral analysis, laser interference temperature measurement and radiation temperature measurement. Infrared point thermometer is the earliest infrared thermometer, but its measuring range is too narrow to show the outline of the object. Therefore, the accuracy of the application is very limited. Therefore, the compensation of infrared human body temperature measurement accuracy is deeply studied. Identify the environmental characteristics of the infrared temperature field, judge the high and low temperature areas, build a non-uniformity correction model, convert the electrical signal into the temperature value of the target object, optimize the human body temperature measurement process, and design a precision compensation mode using mobile monitoring technology. The results show that the mean error of temperature measurement data is smaller and the performance is better.

2 Compensation Method for Infrared Human Body Temperature Measurement Accuracy Under Mobile Monitoring Technology

2.1 Identify Environmental Characteristics of Infrared Temperature Field

The electrons, atoms and molecules inside matter are in constant motion, and energy is released in the process of downward transition in the form of electromagnetic waves. Radiation is energy emitted from the interior of matter. The measurement system of infrared temperature field is mainly composed of infrared optical lens, infrared thermal imager, image acquisition equipment and data processing software. When the temperature of the object is higher than the thermodynamic temperature of 0 K, it will continuously radiate electromagnetic waves to the surroundings. Different temperatures of the object will radiate different energy and different wavelengths of electromagnetic waves. Infrared optical lenses are used to focus the infrared energy radiated by the infrared scene to be measured on the photosensitive surface of the infrared detector for imaging. Usually called infrared radiation or infrared, infrared is also a kind of electromagnetic wave, its wavelength range from 0.78 μm to 1000 μm. In the actual radiation model, in addition to its own radiation, an object will reflect and transmit radiation from the environment. Therefore, the energy that reaches infrared optics is mainly composed of three parts, namely:

$$G = \frac{G_1 + G_2 + G_3}{2} \tag{1}$$

In formula (1), G_1 represents the radiant energy of the environment reflected by the target surface, G_2 represents the radiant energy of the target's own surface, and G_3 represents the radiant energy of the environment transmitted from the target surface. If the target emissivity is e, the reflectivity is l, the transmittance is δ, the ambient radiant energy is η (related to ambient temperature), and the radiant energy of the blackbody at the same temperature as the target surface is R, then the formula (2) can be written as follows:

$$H = \sum (el + \delta\eta)^2 - \frac{\delta}{R} \tag{2}$$

The infrared thermal imaging system can convert the received infrared radiation energy into the gray-scale image that can be displayed on the image display device for human eyes to watch. The gray-scale value of the image target corresponds to the radiation energy value of the target point. We can visually judge the regions with high temperature and the regions with low temperature, and can also judge the temperature change of the target by video, but the temperature is the radiation temperature of the target surface, not the real temperature. The infrared thermal imager converts the infrared image light signal of the object to be observed gathered by infrared lens into electrical signal and outputs it after image processing. Temperature determines the thermal radiation characteristics of objects, so radiation can be called temperature radiation. The temperature radiation characteristics of objects are the basis of optical temperature sensing and photoelectric sensing. At room temperature, the object is mainly infrared radiation, the human eye is invisible. The image acquisition device is used to collect and store the image output by the infrared thermal imager for subsequent processing and analysis. This topic uses the standard equipment to carry on the collection. Only when the temperature is raised to about 500 °C will some dark red visible light be emitted. The temperature continues to rise and white light emits. The intensity of thermal radiation is mainly determined by the temperature of the object, the higher the temperature, the stronger the radiation energy, the more infrared radiation. Temperature measurement data processing software is used to process the infrared image collected by the image acquisition equipment, and convert the image data into temperature data for analysis. Infrared radiation, like electromagnetic wave, will be absorbed and reflected in the process of propagation. An object absorbs infrared radiation and converts it into heat energy. As the distance it travels increases, infrared radiation decays. The accuracy of infrared temperature measurement is related to many factors, such as emissivity of object, ambient radiation, tube radiation, nonuniformity of response, gray drift, calibration and data processing algorithm, lens nonuniformity, and distance from target, etc. Because the object at a thermodynamic temperature higher than 0 K, can always spontaneously continue to emit infrared to the surrounding space, so any object under certain conditions can be regarded as infrared radiation source. Blackbody is an ideal source of infrared radiation, which absorbs all electromagnetic radiation at any wavelength. Emissivity is a function of wavelength, temperature, material and radiant surface conditions. It is very difficult to obtain reliable data. The emissivity of the metal is low, but it increases with temperature and can increase tenfold or more when the oxide layer is formed on the surface. The absorbed radiant energy is equal to the emitted radiant energy, which is the biggest characteristic of blackbody. Its absorptivity and emissivity are defined as 1. The temperature value L corresponding to the target with the same radiant energy shall correspond to the radiant energy received by the detector as follows:

$$Q = \frac{(1-\eta)}{\sum |L-e|} \tag{3}$$

In addition, as a result of careless operation caused by fingerprints, dust, dirt and surface scratches, etc., can cause emissivity measurement changes. The emissivity of nonmetals is higher, generally greater than 0.8, and decreases with increasing temperature. Radiation from metals or other opaque materials occurs within a few microns of the surface, so the emissivity is a function of the surface state of the material, independent of size. The gray body as an absorber is not able to absorb all the radiation incident on it, the emissivity is a constant less than 1. Infrared radiation in addition to some of the properties of visible light, but also follow some of the inherent special laws. The emissivity of a coated or painted surface is a property of the coating itself, not of the substrate surface. For the same material, the emissivity may vary from one technical manual to another. The difference is caused by the change of the surface condition of the sample.

2.2 Establishment of a Nonuniformity Correction Model

Inhomogeneity has two definitions, one is the inhomogeneity of infrared detector, the other is the inhomogeneity of infrared imaging, including the inhomogeneity of optical system. The inhomogeneity of infrared imaging refers to the inconsistency of output between different pixels when the input radiation of infrared imaging is uniform. It is well known that the basic theory of infrared thermometry is the principle of infrared radiation and the signal reception and conversion of related electronic devices [3, 4]. Nowadays, infrared thermometer system appears frequently in our daily life. Infrared thermometer is generally composed of optical focusing device, photoelectric detector, operational amplifier, filter circuit, microprocessor and display module. The nonuniformity of infrared imaging system will seriously affect the quality of infrared image, and the accuracy of temperature measurement will be affected. There are many reasons for the nonuniformity of IR imaging response, including inhomogeneity of IR focal plane itself, inhomogeneity of external input and inhomogeneity of working state. The function of the optical focusing device is to gather the infrared radiation energy of the object. The focusing ability is determined by the optical element of the temperature measuring system and its position. When infrared light is focused on a photoelectric sensor through an optical system, the photoelectric sensor converts the light signal into an electrical signal, usually a voltage or inductance signal, that an ADC converter can collect. Ideally, the output signal of the focal plane of the infrared detector should be exactly the same under the uniform infrared radiation. But because of the unevenness of semiconductor material, mask error, defect, technology and other factors, the output of IR detector will appear unevenness, which is the inherent characteristics of infrared detector. The inhomogeneity of the device itself is mainly caused by three aspects: the inhomogeneity of pixel response rate, the inhomogeneity of signal transmission and the inhomogeneity of dark current. In order to obtain the correction parameters, adjust the blackbody temperature to p (low temperature), collect the output q of the infrared focal plane array, and calculate the average value of the whole image as U:

In order to obtain the correction parameters, adjust the blackbody temperature to (low temperature), collect the output of the infrared focal plane array, and calculate the average value of the whole image as follows:

$$U = \frac{1}{p \times q} \sum \frac{p^2}{|q - p|} \tag{4}$$

According to the calculation result of the formula (4), the electric signal can be converted into the temperature value of the target object through amplification, filtering and the internal algorithm of signal processing to achieve the purpose of temperature measurement. In addition, the natural conditions such as ambient temperature, background disturbances and atmospheric conditions should also be taken into account. The emissivity of an object greatly affects the accuracy of the measurement. The inhomogeneity introduced by external input includes: the inhomogeneity of the output of infrared imaging system caused by the change of ambient temperature, external interference, radiation change of optical system and shell. The output signal of focal plane array of infrared detector is closely related to its working state. The change of working state will directly affect the injection efficiency of detector, the gain of readout circuit and dark current, and thus affect the uniformity of the whole infrared imaging system. Emissivity is a key physical parameter for highlighting the surface radiation characteristics of materials. Its value is related to many factors, such as surface temperature, wavelength, incident angle, polarization direction, etc. This close relationship is also affected by surface conditions, including roughness, coating thickness, surface color and physical impurities and other factors. The response curve of the detector is divided into two intervals. Each interval is corrected by the two-point correction method. The correction formula is as follows:

$$T = \frac{q}{\varphi} + E_p + \sum \frac{1}{D} \tag{5}$$

In formula (5), φ represents the gain parameter, E represents the offset parameter and D represents the correction temperature range. It is difficult to achieve the same transmittance in the whole field of view of the lens. Generally, the intermediate transmittance is higher than the edge transmittance. The nonuniformity is mainly shown as multiplication. Because there are many factors affecting the emission frequency, it is not easy to obtain the exact emission frequency of an object. The emissivity of the blackbody radiation source used in the daily calibration work can be said to be the biggest uncertain factor that affects the accuracy of infrared thermometer temperature measurement. Due to the existence of the factors affecting the nonuniformity, the image is characterized by fixed space noise and crosstalk, which seriously affects the quality of the image. Therefore, in order to obtain high quality images and ensure the accuracy of temperature measurement, the nonuniformity of infrared imaging response must be corrected.

2.3 Optimize Human Body Temperature Measurement Process

When we use infrared thermometer to measure human body temperature, the measured object is human skin, but most of the skin is often wrapped by clothes, only the head is exposed, so the measured part is selected for forehead and eardrum. The eardrum is measured because its temperature is similar to the body's internal temperature. But the ear warm gun needs to contact with the human body, the use is not convenient. In the process of temperature measurement, infrared ray is a kind of electromagnetic wave, which is located at the right end of the visible spectrum of light. Forehead temperature is the highest point of the skin temperature of the head, most of the infrared thermometers choose the forehead as the body's temperature measurement target. However, the temperature of the forehead is not equal to the temperature of the human body, the temperature of the two is affected by many factors, the relationship between the two is not constant, so the forehead temperature needs to be corrected. Infrared radiation is a natural phenomenon on all objects, any object above absolute zero temperature is the source of infrared radiation, which mainly includes ultraviolet, visible light, infrared and so on. Therefore, people should be measured in a resting state, so that their body temperature and ambient temperature to achieve a balance. Only when the heat exchange of the object is stable, can the temperature instrument be used to measure the result. The time response is slow, so it is not suitable for monitoring small target and instantaneous temperature. With the development of science and technology, it has been studied to use contact temperature measurement technology and high-speed data acquisition technology in industrial temperature measurement. Non-contact temperature measurement plays an important role in infrared technology because it has many advantages, such as non-contact, high sensitivity, high precision, wide application, simple operation and so on. In addition, the use of infrared detection of human body temperature, people's age, gender will also affect the detection results. Older people and women had the weakest correlation with body temperature compared to other people, with young people having the best correlation. Because of the reason of old people's physical quality, the influence of cold on body temperature is not as obvious as that of young people. Infrared thermometer is firstly applied to measure temperature in clinic. At the initial stage of infrared technology, its promotion is hindered by the application of detector. Infrared detectors can convert infrared radiation emitted by an object to electrical energy, which can be converted to voltage or current or to detect changes in physical properties of materials. The comparison between infrared thermometer and mercury thermometer shows that infrared thermometer can also ensure high accuracy and correlation, and can measure temperature in 2 s, and let children feel comfortable. The main wavelength of radiant energy is functionally related to temperature, surface state and emissivity. Common temperature measurements are shown in Fig. 1:

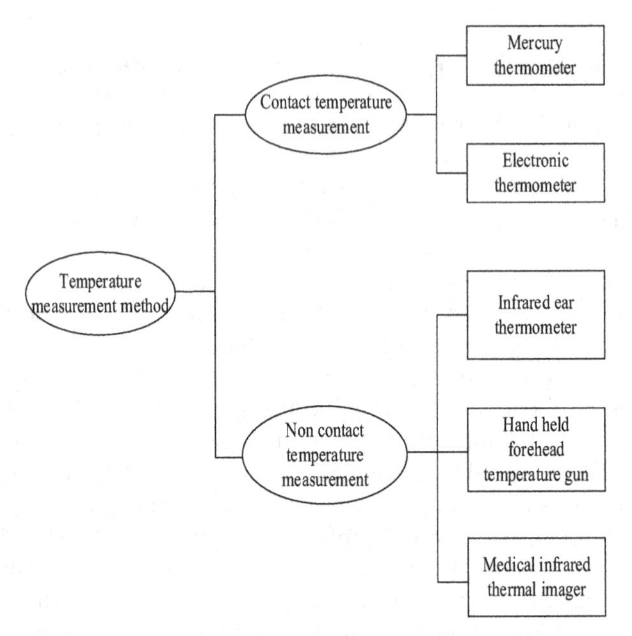

Fig. 1. Diagram of common temperature measurement methods

As can be seen from Fig. 1, the common temperature measurement methods include: contact and non-contact temperature measurement. Among them, contact thermometer includes: mercury thermometer and electronic thermometer. Non-contact thermometers include infrared ear thermometers, hand-held forehead thermometers, and medical infrared thermal imagers. A large number of field and clinical trials show that there are obvious differences between human body surface temperature and internal body temperature. The skin temperature varies according to ambient temperature, physiology, and movement. Only know the skin temperature is unable to determine the real temperature of the human body. The skin temperature is determined by the internal temperature of the human body and is affected by external factors. The infrared band is divided into many different names according to the electromagnetic wave spectrum. The division method is based on the dissemination method, the production way as well as the use situation different and so on principle divides. The infrared band can be divided into four types, as shown in Fig. 2:

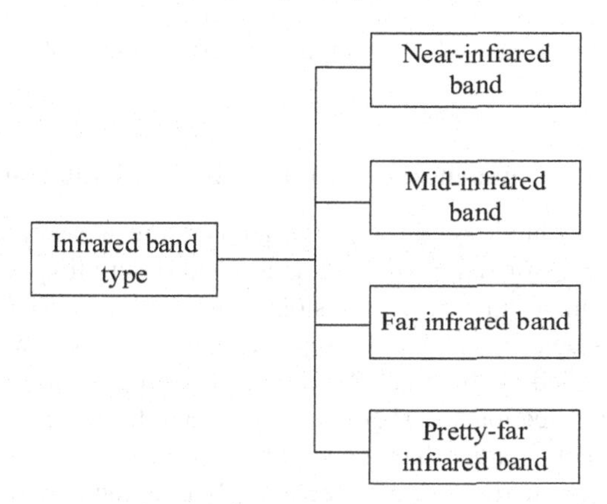

Fig. 2. Infrared band types

As can be seen from Fig. 2, infrared band types include: near infrared, middle infrared, far infrared and far infrared, the specific way of classification may vary according to the subject area. Forehead temperature is determined by brain temperature. Many researchers have established a brain heat transfer model, studied the theory of brain heat transfer, and discussed the effect of heat conduction, convection and metabolism on brain temperature. The conclusion that ambient temperature and convective heat transfer coefficient have a great effect on the temperature of the head surface is obtained by theoretical analysis and practical verification. The skin temperature of the head changed linearly with the environment temperature. The skin temperature of the head increased by 2.06 °C for every 10 °C increase of the environment temperature. The theoretical basis of monochromatic radiation thermometry is Wien's formula. The radiation power of a blackbody is a function of wavelength and temperature, and when the wavelength is fixed it is only related to temperature. When the measurement is received by the measured object in a certain direction of the radiation emittance, called spectral radiation brightness. The monochromatic radiant thermometer is calibrated by the spectral radiance of the blackbody, and it will be deviated if the measured object is not blackbody. The skin temperature was influenced by the ambient temperature in different sites and seasons, which affected the accurate measurement of the forehead temperature. Convective heat transfer coefficient is determined by wind speed, and the change of convective heat transfer can significantly affect the surface temperature of the head. Therefore, when we use infrared thermometer to measure human body temperature, we should choose the place with stable ambient temperature and low wind speed, or modify it according to the different environment. Because the spectral radiance of the blackbody is higher than that of the non-blackbody at the same temperature, the temperature of the non-blackbody measured by the instrument is lower than the true temperature. In order to calibrate the temperature deviation, the concept of brightness temperature is introduced. Generally speaking, there are two types of thermometer: contact and non-contact. The physiological parameters and activities of human body also affect the temperature of brain, such as blood perfusion rate, metabolism rate and skin thermal conductivity coefficient. Among them metabolic rate is bigger to skin temperature effect, metabolic taller, the heat that produces more.

2.4 Design Precision Compensation Mode of Mobile Monitoring Technology

Mobile monitoring technology, through the temperature measurement of the target behavior characteristics of sampling, calculation and description of the target behavior parameters after the formation of precision compensation [5–7]. Compared with the normal behavior model, the results of temperature measurement with large statistical deviation were identified as abnormal. Through statistical analysis or transformation of the data, the normal detection accuracy can be separated from the temperature data and noise. When the target is larger than the field of view of the infrared thermometer and the influence of atmospheric transmittance is negligible, the radiation intensity to the detector is independent of the distance between the object and the system. Generally, thermal imager is used in outdoor environment. Because of the change of atmospheric absorptivity, atmospheric background temperature, ambient temperature and the influence of nearby objects, the influence of thermal imager on infrared instrument is more complex and changeable, so that it affects the difficulty of using infrared instrument to measure temperature accurately in reality. Although filters filter out most of the infrared wavelengths, there are still a lot of infrared wavelengths of objects in the human body and similar to the 9 to 14 μm band. For example, the infrared wavelength of the Earth is 10 μm, so when using infrared thermometer to measure, we should try to reduce the infrared interference of other targets. There is an eyepiece above the infrared sensor, which determines the angle of view of the infrared detector to detect the infrared radiation. To reduce the interference of other objects to detect the target, the measured target should be full of the whole angle of view. However, for the infrared ear thermometer, the target is close to the infrared ear thermometer, and is in the relatively airtight ear canal environment, so the influence of measurement distance on the infrared ear thermometer is almost negligible. If the temperature is only determined according to a certain data of the input emissivity, the accuracy of the temperature measurement will be greatly affected. Moreover, the influence of emissivity is different with different temperature measuring instruments. If the monochromatic thermometer is used, the error of the luminance temperature is:

$$\left(\frac{\sigma}{Y}\right)^{-Y} = \frac{(1)}{\sigma^2} \tag{6}$$

In formula (6), σ represents ambient temperature and Y represents radiant energy error. To sum up, the factors affecting infrared temperature measurement are complex, such as measuring distance, emissivity, ambient temperature and atmospheric factors, etc. The infrared radiation emitted by the temperature target radiates in all directions, and the infrared radiation received by the sensor is only a small part of it. The rest of the energy is released into the surrounding space. But the angle coefficient of the surrounding environment to the lens is very small, so the influence of the environment to the direct thermal radiation of the lens is very small. For infrared ear thermometers, the ear thermometer needs to be inserted into the ear canal, and the thermometer is located in a relatively airtight environment. It can be seen from the analysis that the infrared thermometer has a great influence on the temperature when the distance of the object is more than 5 cm from the receiver, while the distance of the infrared thermometer from the tympanic membrane is small. The corrected value of an infrared ear thermometer is mathematically expressed as follows:

$$\Delta = \psi - k \frac{\psi}{\sum_{x=1} k_{|x-d|}} \tag{7}$$

In formula (7), ψ stands for the temperature of the black body, k stands for the average temperature of the black body measured by the infrared ear temperature system, and d stands for the d temperature reading. The infrared ear thermometer is located in a relatively closed space, and the atmospheric factors have little influence on the measurement results. According to the infrared radiation theory, the target emissivity is always less than 1, and the emissivity of different materials is different, so we need to set the target emissivity before detecting the target. According to the nonblackbody radiation theory, the ratio of the radiant energy of the gray body to that of the black body is the emissivity, so the different emissivity will cause the error of the measurement. The emissivity of human skin is fixed, but for different people, the material and surface roughness of earwax will directly affect the emissivity, which will affect the measurement accuracy of infrared ear thermometer. To sum up, the radiant energy received by the lens mainly depends on the temperature and ambient temperature, the emissivity of the temperature target, the wavelength and area of the temperature lens, the distance between the lens and the target and the alignment of the target. However, a closed space creates a dark environment that can significantly reduce individual differences in emissivity, which can affect measurements, although the impact of emissivity cannot be ignored [8]. And the drift caused by the external environment temperature is compensated. Based on the above analysis, the influence of target emissivity, atmosphere factor and measurement distance on the error of infrared ear thermometer is limited, and the ambient temperature has great influence.

3 Experimental Analysis

3.1 Test Readiness

Prepare the test according to the test requirement. When the human body temperature measurement system began to carry out human body temperature measurement, initialization of infrared temperature sensor MLX90614, ranging sensor VL53L1. When the button is pressed, the MLX90614 infrared temperature sensor begins to collect the forehead temperature and ambient temperature. The driver is selected under the big option Device Drivers, in which, select Samsung S3C2440/S3C2442 Serial port support under Character Devices I > Serial driver to add the serial port driver. Select support for Video for Linux under Multimediadevices, and select USB SPCASXX Sunplus/Vimicro/Sonix Jupeg Cameras to add a driver for the USB camera. VL53L1 rangefinder begins to measure distance. The temperature and distance data collected by the sensor are preprocessed and filtered by MCU controller. When the pretreatment is finished, the forehead temperature is judged. The DSP for the USB camera selected in the experiment is owned by SOnix, so be sure to choose this option. In Graphics support, select Support for frame buffer devices, and select the appropriate branding and size of the LCD screen. If the forehead temperature value is within the normal forehead temperature range (32–36 °C), the human body infrared temperature measurement system calculates the collected data by multiple linear regression algorithm to fit the human body sublingual temperature value, and Linux abstracts FrameBuffe: This device can be used for user mode process to realize direct writing screen. By imitating the function of video card, the Framebuffer mechanism abstracts out the hardware structure of video card, and can directly operate on video memory through the Framebuffer. If the collected forehead temperature is not in the range of normal forehead temperature, the collected data can not be corrected by multivariate regression algorithm, and the forehead temperature can be directly output and buzzer can give an alarm. Users can think of a Framebuffer as an image of display memory that, once mapped into a process address space, can be directly read and written, which can be immediately reflected on the screen.

3.2 Test Results

When the human body is chosen as the object of temperature measurement, the armpit temperature is firstly measured by medical mercury thermometer, and the forehead temperature is measured by DS18B20 before measurement. Because the normal body temperature change is not big, basically in 37 °C, change range in ±1 °C or so, but not always constant. Therefore, it is necessary to ensure that the temperature measurement environment is indoor without wind and the ambient temperature remains unchanged. The armpit temperature and forehead temperature of human body should be measured before the infrared measurement, and then measured again after the infrared measurement. In order to verify the effectiveness of the compensation method of infrared human body temperature measurement precision designed in this paper, the compensation method of infrared human body temperature measurement precision based on data mining, the compensation method of infrared human body temperature measurement precision based on association rules and the compensation method of infrared human body temperature measurement precision based on genetic algorithm are selected respectively, and the experimental comparison is made with the compensation method of infrared human body temperature measurement precision proposed in this paper. Under different distance conditions, the error of temperature measurement data of four compensation methods is tested respectively. The experimental results are shown in Tables 1, 2, and 3:

Table 1. Error of 5 cm distance temperature data (°C)

Number of experiments	Compensation Method for Infrared Human Body Temperature Measurement Accuracy Based on Data Mining	Compensation Method for Infrared Human Body Temperature Measurement Accuracy Based on Association Rules	Compensation Method for Infrared Human Body Temperature Measurement Accuracy Based on Genetic Algorithm	Compensation Method of Infrared Human Body Temperature Measurement Accuracy
1	0.631	0.552	0.487	0.312
2	0.584	0.547	0.569	0.254
3	0.612	0.569	0.612	0.331
4	0.538	0.499	0.613	0.214
5	0.622	0.523	0.647	0.206
6	0.529	0.511	0.685	0.319
7	0.642	0.602	0.652	0.256
8	0.566	0.542	0.636	0.311
9	0.608	0.536	0.548	0.256
10	0.577	0.629	0.602	0.229

From the Table 1, we can see that the average error of the infrared human body temperature compensation method is 0.269 °C, 0.591 °C, 0.551 °C, 0.605 °C.

Table 2. Distance 20 cm Temperature Data Error (°C)

Number of experiments	Compensation Method for Infrared Human Body Temperature Measurement Accuracy Based on Data Mining	Compensation Method for Infrared Human Body Temperature Measurement Accuracy Based on Association Rules	Compensation Method for Infrared Human Body Temperature Measurement Accuracy Based on Genetic Algorithm	Compensation Method of Infrared Human Body Temperature Measurement Accuracy
1	2.361	2.216	2.364	1.023
2	2.225	3.066	2.358	1.114
3	2.198	2.108	3.169	1.028
4	2.647	3.669	2.087	1.136
5	3.105	2.874	3.647	1.087
6	2.697	3.692	2.848	1.154
7	2.684	2.588	2.154	1.364
8	3.005	3.615	2.199	1.154
9	2.647	2.347	3.647	1.255
10	2.712	2.165	2.088	1.346

From the Table 2, the mean error of the infrared human body temperature compensation method is 1.166 °C, 2.628 °C, 2.834 °C, 2.656 °C.

Table 3. Error of 35 cm temperature data (°C)

Number of experiments	Compensation Method for Infrared Human Body Temperature Measurement Accuracy Based on Data Mining	Compensation Method for Infrared Human Body Temperature Measurement Accuracy Based on Association Rules	Compensation Method for Infrared Human Body Temperature Measurement Accuracy Based on Genetic Algorithm	Compensation Method of Infrared Human Body Temperature Measurement Accuracy
1	5.647	5.648	5.697	2.647
2	5.649	6.021	5.879	3.316
3	5.558	6.213	6.112	2.226
4	5.648	5.879	5.948	2.314
5	6.115	6.337	6.377	3.051
6	6.233	5.874	5.869	3.099
7	6.487	6.599	6.451	2.874
8	6.255	5.788	5.846	2.694
9	6.369	6.347	6.464	3.008
10	5.998	6.124	6.997	3.154

From the Table 3, the mean error of the temperature compensation method and other three methods is 2.838 °C, 5.996 °C, 6.083 °C and 6.164 °C.

To sum up, the infrared human body temperature measurement accuracy compensation method under mobile monitoring technology has low error mean value of temperature measurement data and good performance. The reason is that this method carries out statistical analysis or transformation on the data to separate the normal detection accuracy from the temperature measurement data and noise. When the target is larger than the field of view of the infrared thermometer and the influence of atmospheric transmittance can be almost ignored, the radiation intensity incident on the detector has nothing to do with the distance between the object and the system, which is conducive to reducing the temperature measurement error to a certain extent.

4 Conclusion

A large number of temperature data are measured in the absence of external environment interference, and the precision compensation formula is fitted by moving monitoring technology and least-squares method. The method solves the problem that the traditional infrared body temperature measurement is easily affected by the ambient temperature and measurement distance, and improves the accuracy of infrared body temperature measurement.

Due to the limited time and energy, the samples measured in the temperature measurement experiment only have the measured data values at normal body temperature. For the high fever data with fever, no experimental measurement is carried out. It needs to be further improved in the future. The specific contents are as follows:

(1) In order to better facilitate the service of patients, the size of the data collection node needs to be further reduced, and the shape needs to be designed for different age groups to bring more comfortable experience to patients.
(2) The data in the data acquisition system is finally collected in the coordinator module. In actual use, when the number of data nodes reaches thousands, the coordinator module may not be able to carry such high-intensity data traffic. The method of transmitting data to the PC client through serial port is no longer applicable, and other communication methods need to be used to transmit data to the database.
(3) Although the PC client software can realize the functions of real-time display and query of data, it can not diagnose the measured body temperature. It can be considered to realize the early warning and diagnosis of some common diseases on the basis of the system in combination with the specific application of physiological parameters of body temperature in clinic.

Fund Project. Shenzhen Science and Technology Plan Project (JSGG20210901145534003).

References

1. Guan, H.: An infrared thermometer developed with touch-screen technology. J. Liaodong Univ. (Nat. Sci.) **27**(4), 238–242 (2020)
2. Yu, W., Ma, F., Li, P., et al.: Analysis of the factors affecting the accuracy of infrared thermometer and the countermeasures. Popul. Sci. Technol. **22**(5), 99–101 (2020)
3. Chandrasekar, B., Rao, A.P., Murugesan, M., et al.: Ocular surface temperature measurement in diabetic retinopathy. Exp. Eye Res. **211**(2), 108–119 (2021)
4. Bai, X.: Remote monitoring and simulation of infrared radiation intensity of space target. Comput. Simulat. **37**(5), 317–321 (2020)
5. Liang, Y., Li, H., Luo, Y., et al.: Development of intelligent mobile monitoring system of facility agriculture based on WSN. Comput. Technol. Developm. **30**(7), 164–168 (2020)
6. Sullivan, S.J., Seay, N., Zhu, L., et al.: Performance characterization of non-contact infrared thermometers (NCITs) for forehead temperature measurement. Med. Eng. Phys. **93**(7), 93–99 (2021)
7. Zhang, H.: Mobile monitoring platform based on coal mine safety monitoring system. Saf. Coal Mines **51**(11), 133–136 (2020)
8. Liu, S., Wang, S., Liu, X., et al.: Fuzzy detection aided real-time and robust visual tracking under complex environments. IEEE Trans. Fuzzy Syst. **29**(1), 90–102 (2021)

Risk Identification Model of Enterprise Strategic Financing Based on Online Learning

Xiang Zou[1](✉) and Huishu Yuan[2]

[1] School of Accounting and Finance, Wuxi Vocational Institute of Commerce, Wuxi 214153, China
sln13526@163.com

[2] Chengdu College of University of Electronic Science and Technology of China, Chengdu 611731, China

Abstract. The current financing risk identification model only identifies according to the financial situation of enterprises. The limitations of risk indicators affect the accuracy of identification, and have great limitations for enterprises in different industries. In view of the above problems, this paper will study and build an enterprise strategic financing risk identification model based on online learning. After analyzing the causes of enterprise strategic financing risk, the financing risk identification index system is designed. Based on the structure of impulse neural network, the risk identification model is constructed by using online learning algorithm. The model test results show that the minimum identification accuracy of the model is 92.4%, which has a good risk identification effect for enterprises in different industries.

Keywords: Online learning · Enterprise strategic financing · Financing risk · Risk identification · Risk matrix · Impulse neural network

1 Introduction

The competitive pressure of domestic enterprises from all aspects has further increased. In order to remain invincible in the fierce competition, enterprises must have strong financial management ability. The core of financial management is fund management. The important link of fund management is how to finance and how to control the financing risk. In recent years, the financing scale and number of events of enterprises have shown an explosive growth trend. Due to different economic environments, different development stages and different industry development characteristics, enterprises have a variety of financing needs and financing purposes, and they also face different financing risks [1]. In the early stage of development, enterprises will generally face considerable capital pressure. In the early stage of development, their profitability is weak, their investment is large, and their R & D cycle is long, which will lead to a considerable degree of financing risk. In the context of the new normal of economic growth slowdown and development structure transformation, the capital market, as an important support for

W. Fu and L. Yun (Eds.): ADHIP 2022, LNICST 468, pp. 702–714, 2023.
https://doi.org/10.1007/978-3-031-28787-9_52

economic transformation and innovation, provides a huge amount of financial support for the sustainable development of enterprises, and the active financing activities provide a strong driving force for the development and growth of enterprises.

Scholars at home and abroad have not studied the strategic financing risk of enterprises for a long time. Most developed countries mainly control and solve the financing problems of cultural enterprises through the appropriate guidance and support of the government and the self-regulation of market mechanism. Risk identification is a process in which economic units judge, classify and sort out the faced and potential risks, and identify the nature of risks. Japan has also set up special departments for small and medium-sized enterprises, set up branches everywhere, and created a more perfect environment for enterprise development through the introduction of a number of policies, regulations and financial schemes [2]. In Europe, where the world economy is very active, it is generally direct government subsidies or direct financial support, which is simpler and more direct than other countries and regions. Most scholars believe that the enterprise financing risk is caused by the increase of liabilities. In the process of enterprise financing, the capital structure is unbalanced. For example, if the proportion of debt financing is too large, the enterprise's ability to pay will be weakened and the financing risk will increase. Scholars have concluded that the factors affecting financing risk include asset liability ratio, total asset profit ratio, profitability, solvency, liquidity and growth ability. Scholars also explored the interaction between corporate value and enterprise risk management. According to different research methods, foreign financing risk identification methods are mainly divided into univariate analysis, multivariate statistical analysis and artificial intelligence analysis. Risk identification methods are mainly divided into qualitative identification and quantitative identification. Qualitative identification methods are widely used in matrix analysis, ANP network analysis, Delphi method, etc., but qualitative identification methods have certain subjectivity and obvious limitations in the identification of risk matters [3, 4].

Online learning can adjust the model in real time and quickly according to the online feedback data, so that the model can reflect the online changes in time and improve the accuracy of online prediction [5]. The process of online learning includes: presenting the prediction results of the model to users, then collecting the feedback data of users, and then training the model to form a closed-loop system. The enterprise's business strategy is constantly adjusted to adapt to the changes of the market, and the enterprise financing risk is also changing with the business situation, resulting in a certain timeliness of the enterprise's strategic financing risk, and higher requirements for the speed and accuracy of risk identification. Therefore, based on the above analysis, in order to reduce the losses caused by financing risks in enterprise operation, this paper will study and build an enterprise strategic financing risk identification model based on online learning, which is of great significance for improving enterprise financing risk control, optimizing enterprise financing structure and mode, and exploring enterprise sustainable development.

2 Construction of Enterprise Strategic Financing Risk Identification Model Based on Online Learning

Enterprise strategic financing has effectively alleviated the bottleneck problems such as lack of financial resources and shortage of funds in different periods of enterprise development, promoted the flow of capital, activated the capital market, effectively promoted the improvement of enterprise capital operation efficiency and the expansion and strengthening of scale, and played an important role in ensuring steady economic growth and smooth adjustment of economic structure. The risk of enterprise strategic financing may be caused by policy change, interest rate change, low management level, lack of integrity and so on. The problems existing in enterprise financing risk management include: backward risk management technology and means, imperfect legal basis for risk management, poor implementation of foreign exchange control policies, increased foreign exchange risk, objective conditions of risk management hinder its development, cumbersome financing procedures and poor financing channels, which increase financing risk [6]. Enterprise financing risk identification should establish a comprehensive risk management system that covers all business processes and operation links and can continuously monitor, regularly evaluate and accurately warn risks. In the process of enterprise strategic financing risk identification, we should adhere to the four principles of comprehensive identification, key identification, dynamic identification and objective identification. Therefore, this paper will build a financing risk identification model after analyzing the causes of enterprise strategic financing risk.

2.1 Cause Analysis of Enterprise Strategic Financing Risk

The financing risk cannot be completely eliminated, and because of the instability of the value determination of intangible assets and income acquisition of enterprises, the changing factors affecting financing activities will only increase, and the financing risk of enterprises must exist objectively. It is precisely because of the objective existence of risk that determines the objectivity and necessity of financing risk. This paper analyzes the causes of enterprise strategic financing risk from the external and internal environment of enterprise operation. Figure 1 is the schematic diagram of financing risk classification of enterprise business strategy.

Enterprise external environmental factors refer to the analysis of the impact from the outside of the enterprise from the perspective of politics, economy and society. From the development of enterprises in different industries in China, there are still deficiencies in China's market. As for the financing mode of enterprises, there are few financing tools available. As a traditional financing institution, commercial banks can play a limited role in the current situation, and some emerging financing methods still lack a relatively mature capital market to give full play to their greatest advantages. This kind of financing means is more restricted by enterprises. With the rapid development of China's economic market, although the financial financing channels in the market are constantly enriched, the relevant laws are not perfect. In the process of implementation, there will still be some problems such as weak operability of laws and regulations, which will affect the process of enterprise strategic financing. In addition, the overall credit level of some enterprises is low, and all kinds of fraud are prohibited repeatedly. The reasons for

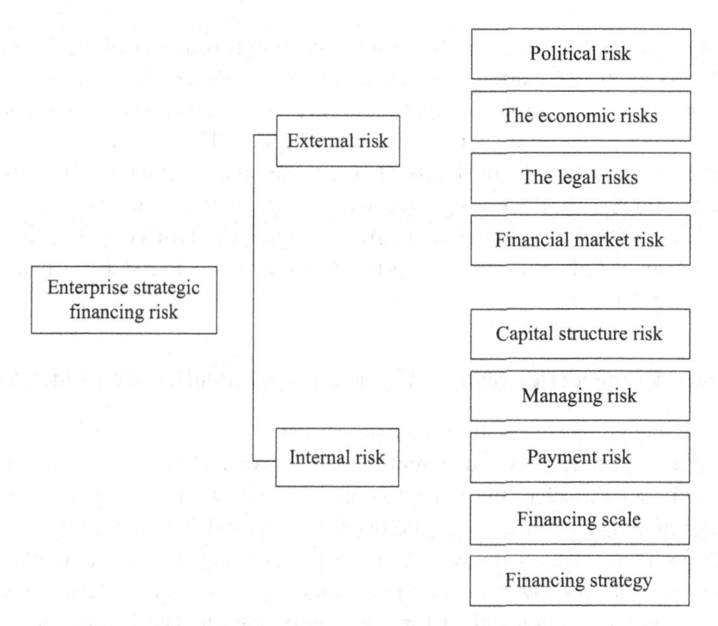

Fig. 1. Classification of enterprise strategic financing risks

the formation of enterprise financing risk are the confusion of internal operation and management and the unreasonable formulation of strategy. There is a large gap in the strength of enterprises in different industries in China, and there are a large number of enterprises. These enterprises usually lack strong operation capacity and efficient financing strength, so they can not form enterprises with economies of scale [7]. These enterprises are still very backward in the concept of operation and management, lack of Zhuoyuan development strategic planning, and can not gather excellent management talents to plan for them, which makes them more blind in the financing process, and do not clearly identify and evaluate the potential financing risks of enterprises, laying hidden dangers for the subsequent sustainable development of enterprises.

The main factors causing financing risks within enterprises are enterprise financing scale, financing structure, financing strategy and so on. The excessive debt financing scale of enterprises will also increase the financing risk. It greatly increases the possibility of insufficient solvency or eventual bankruptcy of enterprises due to the lack of guarantee of income. At the same time, the interest rate of debt financing is too high, so that enterprises will pay more and more interest, which further increases the risk of enterprise strategic financing. The debt financing structure is unreasonable. The short-term or long-term financing under different business strategies will have varying degrees of impact on the business situation of enterprises, which will lead to the change of enterprise financing risk. In addition, although the financing methods and Strategies of enterprises in different periods of strategic financing need to be adjusted, the financing opportunity is also very important. If the enterprise misses the financing opportunity and the financing cost. Moreover, under different debt financing methods, enterprises are also different in the difficulty of obtaining funds and restrictive terms for enterprises, and their corresponding debt financing costs and potential financing risks are also quite different. In addition, there is also a certain correlation between the enterprise's own capital and the financing anti risk ability. The strong accumulation ability of the enterprise's own capital will enhance the enterprise's ability to resist the financing risk. The imperfection of enterprise capital structure increases the difficulty of enterprise financing; At the same time, if the enterprise cannot recover the sales money in time, it will cause trouble to the enterprise's capital chain, resulting in a greater crisis. According to the causes of enterprise strategic financing risk analyzed above, the enterprise strategic financing risk identification index system is designed.

2.2 Design of Enterprise Strategic Financing Risk Identification Index System

Enterprise financing risk evaluation is the process of determining the risk level of financing risk factors of enterprises in different industries. According to the above analysis of financing risk causes, select the indicators that comply with the principles of authenticity, integrity, independence, logic and operability, and design the enterprise strategic financing risk identification index system. In the comprehensive evaluation of enterprise financing risk, too few evaluation indicators can not reflect the actual risk, and too many indicators are not convenient for practical operation. The key lies in the role of the selected evaluation indicators in risk identification. Table 1 shows the financing risk identification index system designed according to the causes of strategic financing risk analyzed above.

After designing the enterprise strategic financing risk identification index system shown in Table 1, according to its identification and judgment criteria, use the risk matrix to analyze and determine the level of each risk factor. It is an evaluation method to judge the degree of risk by constructing the risk rating table and risk index and arranging experts to score the risk factors. Experts estimate based on reasonable judgment, information available at that time and their own accumulated experience. Therefore, the experience and knowledge of experts determine the quality of risk assessment. The occurrence probability and risk impact of enterprise strategic financing risk are integrated to obtain the financing risk comparison table shown in Table 2.

In the constructed risk identification indicators, some indicators may belong to "very large", "very small", "medium" and "interval". The larger the index value, the better the index, which is called very large index; The smaller the index value, the better the index, which is called very small index; For some indicators, it is expected that the more centered the indicator value is, the better, which is called the centered indicator; The expected value of some indicators falls within a certain interval, which is called interval indicators. In order to facilitate the data processing of indicators, various types of indicators are transformed into very large indicators j. For very large indexes, linear transformation is carried out according to the following formula:

$$q_{ij} = \frac{p_{ij}}{p_j^{\max}} \tag{1}$$

Table 1. Enterprise strategic financing risk identification index system

Risk type	Risk causes	Risk factor
Market risk	Market competition	Low barriers to entry
		Market competition intensity
		The degree of difficulty of enterprise elimination
		Enterprises exit high threshold
	Level of enterprise competitiveness	Companies have difficulty meeting consumer demand for their products
		The market share of the enterprise is unstable and there are no fixed customers
		Product marketing ability
Production and operation risk	Technology research and development risk	The difficulty and complexity of technology are high
		Uncertainty of technology life cycle
		Conformity of production capacity with market demand
	Manage risk	Imperfect internal structure of the enterprise
		Weak management ability of managers
Environmental risk	Macroeconomic environment	Changes in interest rate, exchange rate and inflation rate
	Government policy	Government support and policy changes
Financial risk	Financial market risk	Enterprises are punished for trading in the market
		The enterprise industry is impacted by the international market
	The credit risk	Decline in corporate credit rating
		Insufficient repayment ability of enterprises
	Operational risk of flow	Inactive market transactions make financing limited

(*continued*)

Table 1. (*continued*)

Risk type	Risk causes	Risk factor
		Problems in financing operation calculation

Table 2. Comparison of risk levels

Risk probability range (%)/risk impact	Negligible	Small	Moderate	Serious	Crux
0–10	Low	Low	Low	Intermediate	In
11–35	Low	Low	Intermediate	Intermediate	High
36–55	Low	Intermediate	Intermediate	Intermediate	High
56–85	Intermediate	Intermediate	Intermediate	Intermediate	High
85–100	Intermediate	High	High	High	High

For the very small index j, carry out linear transformation according to the following formula:

$$q_{ij} = \frac{1 - p_{ij}}{p_j^{\max}} \tag{2}$$

Then, the index value is transformed from 0 to 1:

$$q_{ij} = \frac{p_{ij} - p_j^{\min}}{p_j^{\max} - p_j^{\min}} \tag{3}$$

After the preliminary processing of the indicators in the enterprise strategic financing risk identification index system, based on the pulse neural network structure, the online learning algorithm is used to improve and optimize, and the enterprise strategic financing risk online identification model is constructed.

2.3 Realize the Construction of Online Identification Model of Enterprise Strategic Financing Risk

The financing risk of enterprises is positively correlated with financing cost, financing structure, financing scale and the proportion of intangible assets in total assets. According to the analysis of MM capital structure theory, the higher the financing cost, the smaller the proportion of financing income relative to financing cost, the greater the possibility of financing interruption or failure and the greater the financing risk. In addition, the higher the debt financing ratio of enterprises, that is, the higher the asset liability ratio, the greater the financing risk. Before using impulse neural network to identify financing risk, establish the correlation coefficient matrix between each point through the relevant data of each risk index, find out the factors with large relative connection degree, and then connect them with edges according to the relevant principles.

The enterprise strategic financing risk network can be represented by $G = (V, E)$. Let the network have N nodes, M edges, $V = \{v_1, v_2, \cdots, v_n\}$ represents the set of nodes, nodes represent the risk identification index, $E = \{e_1, e_2, \cdots, e_m\}$ represents the set of edges, edges represent the relationship between nodes, and the small weight represents the relationship between nodes. The degree of correlation between each node indicates the degree of correlation between the enterprise's strategic financing risk indicators. The higher the degree of node aggregation, the closer the correlation between the enterprise and its risk factors. After determining the weight of each risk index according to the expert evaluation results, the network structure of enterprise strategic financing risk is determined. According to the correlation between enterprise financing risk indicators reflected in enterprise strategic financing risk network, a pulse neural network is established. In order to facilitate the subsequent derivation of spatiotemporal back propagation learning algorithm based on discrete time steps, the explicit iterative representation of pulse neurons with known sampling time will be used to ensure the traceability of pulse calculation and facilitate the calculation of chain rules. The combination of equations describing the specific dynamic characteristics of neurons is as follows:

$$I_i^n[t] = \sum_{j=1}^{l(n-1)} w_{ij} s_j^{n-1}[t] + \beta \tag{4}$$

$$K_i^n[t] = K_i^n[t-1] + \frac{I_i^n[t] - K_i^n[t-1]}{\tau} + \alpha s_j^{n-1}[t] \tag{5}$$

$$s_j^n[t] = g\left[K_i^n[t] - \alpha\right] \tag{6}$$

$$g\left[K_i^n[t] - \alpha\right] = \begin{cases} 1, K_i^n[t] \geq \alpha \\ 0, K_i^n[t] < \alpha \end{cases} \tag{7}$$

In the above equation, the superscript n, subscript i and square brackets t of each variable represent various internal states of the FF pulse neuron located in the n layer of the network at time n. K, I and s respectively represent the network parameters, total input vector and its own pulse issuing state of pulsed neurons (1 indicates issuing pulses and 0 indicates keeping silent). τ Value represents the membrane time constant, $l(n)$ represents the total number of neurons in layer n, w_{ij} is the weight value between presynaptic pulse neuron j and postsynaptic pulse neuron i, and β represents the constant input of pulse neuron i at each time.- once the K value of pulse neuron i reaches or exceeds the threshold 0, the neuron will generate and send out pulses and reset the K value at the same time. The output pulse activity of pulse neurons is controlled by gate function $g[]$ to generate or not generate pulses (1 means to generate pulses and 0 means not to generate pulses).

The input of the pulse neuron i located in the hidden layer of layer n will also affect its own output pulses in time t and $t+1$, and then affect the final number of pulses by affecting the total input of the pulse neuron in the postsynaptic layer $t+1$. The parameter update formula of pulse neural network output bias is as follows:

$$\delta_j^n[t] = c\sum \delta_j^{n+1}[t]w_{ij}^{n+1}[t+1]\frac{\partial s_j^n[t+1]}{\partial K_i^n[t+1]} \quad (8)$$

The above formula is a component of the weight update value in the hidden layer. It is necessary to understand the accurate symmetrical weight information to complete the calculation. However, this demand is considered impossible in the brain. At the same time, the back-propagation process in the time dimension requires the neuron to understand the input data information of time $t+1$ at time t, which makes the multi-layer SNN unable to learn the time characteristics in the data in real time. Due to the internal state of the pulse neuron i at time t, the idle update rule needs to know the accurate symmetrical weight information at time t and $t+1$. However, it is unlikely to meet such strong architectural constraints in the actual network connection. In order to avoid this problem, this paper uses the fixed weight of random initialization in the network to replace the weight of connecting the same two neurons at different times, so as to realize faster and simpler learning rules. After the input pulse neuron sends out a pulse, the neuron transmits the pulse to the postsynaptic pulse nerve; At time t, if the state of postsynaptic neurons issuing pulses is inconsistent with their expected value, the value of δ will change, which will affect the state of. In addition, because it is also related to time, its change curve will be smoother relative to the change curve of δ value. The updated value of the weight will finally be determined by the value of c and the state of the input pulse. In the process of implementation, the online learning algorithm only needs the pulse neuron to update its internal variables iteratively, without storing all the intermediate data generated by the input data in a period of time. Taking the index of financing risk identification system as the node vector of neural network, the relevant enterprise data are input to the neural network. After the iterative processing of neural network, the enterprise financing risk identification results are obtained.

According to the above process, the construction of enterprise strategic financing risk identification model based on online learning is completed. Using this model can help enterprises identify financing risks timely and effectively in the process of strategic financing, facilitate enterprises to adjust strategic financing strategies in time, take corresponding risk prevention measures, and reduce the losses caused by enterprise financing risks.

3 Model Test

3.1 Test Preparation

Test Scheme Design

The performance of the above-mentioned enterprise strategic financing risk identification model based on online learning will be compared and tested. Before the model test, actual enterprise data will be extracted to form a sample data set for model identification accuracy and efficiency. Compare the enterprise strategic financing risk identification model constructed in this paper with the financing risk identification model based on fuzzy fmea-vikor and the financing risk identification model based on sample weighted support vector machine. The identification model deals with the financing risks of enterprises in different industries respectively. Compare the model identification results with the results of professional institutions, and get the corresponding test data. The performance of the risk identification model is comprehensively evaluated by comparing the risk identification accuracy and risk identification time of the identification model.

Selection of Test Data

This paper selects 36 companies in different cities and industries as the research objects. These enterprises can be divided into large-scale manufacturing, Internet and FMCG enterprises. Among them, the risks of each enterprise in strategic financing are known, and all enterprise financing risks are measured by professional institutions. The samples are divided into two groups: training samples and test samples. There are 12 enterprises in the modeling samples and 24 enterprises in the test samples. This test collected the financial data of the first 20 quarters of each company.

3.2 Test Results

Tables 3, 4, and 5 respectively show the test data of strategic financing risk identification of three risk identification models for large manufacturing, Internet and FMCG enterprises.

Table 3. Comparison of strategic financing risk identification test data of large manufacturing enterprises

Number	Risk identification model based on online learning		Risk identification model based on fuzzy fmea-vikor		Risk identification model based on sample weighted support vector machine	
	Accuracy/%	Recognitiontime/s	Accuracy/%	Recognition time/s	Accuracy/%	Recognition time/s
1	96.9	14.25	92.4	18.11	89.4	26.36
2	95.8	11.37	92.2	19.18	89.7	25.80
3	97.6	12.56	94.7	18.17	87.4	26.18
4	95.1	14.38	92.5	18.47	89.7	26.27
5	95.3	13.83	91.9	19.58	86.8	26.11
6	97.4	14.83	94.8	18.32	87.1	26.14
7	96.7	13.84	94.6	18.51	88.6	26.22
8	96.8	14.68	91.1	19.38	89.2	26.34
9	97.2	13.61	94.6	18.36	89.5	25.73
10	97.5	10.82	91.8	18.62	87.7	26.56

By analyzing the data in Table 3, it can be seen that when identifying the strategic financing risk of large manufacturing enterprises, the identification accuracy of the identification model in this paper is higher than 95%, while the highest identification accuracy of the risk identification model based on fuzzy fmea-vikor is only 94.8%, and the highest identification accuracy of the risk identification model based on sample weighted support vector machine is only 89.7%. The recognition time of the recognition model in this paper is significantly less than that of the other two recognition models.

Table 4. Comparison of test data of strategic financing risk identification of Internet enterprises

Number	Risk identification model based on online learning		Risk identification model based on fuzzy fmea-vikor		Risk identification model based on sample weighted support vector machine	
	Accuracy/%	Recognition time/s	Accuracy/%	Recognition time/s	Accuracy/%	Recognition time/s
1	93.9	15.76	88.8	23.48	83.2	29.66
2	92.6	14.25	89.4	23.51	82.4	29.14
3	93.3	13.83	87.2	25.64	83.6	29.52
4	93.9	12.27	87.7	23.80	80.5	28.97
5	95.2	14.03	86.9	23.97	82.1	30.38
6	92.4	15.41	88.5	23.61	81.4	29.26
7	92.7	13.32	89.9	23.78	83.9	29.46
8	94.2	14.54	89.1	24.73	82.3	29.53
9	94.3	15.39	88.6	25.24	81.6	29.82
10	94.8	13.82	89.3	24.41	83.7	29.50

By analyzing the data in Table 4, we can see that when identifying the strategic financing risk of Internet enterprises, the lowest identification accuracy of this model is 92.4%, which is higher than the highest identification accuracy of the other two models. The recognition time is also much shorter than the other two recognition models.

By analyzing the data in Table 5, it can be seen that the recognition accuracy of the model in this paper is higher than 93.9%, and the recognition accuracy is higher than 91.7% of the risk recognition model based on fuzzy fmea-vikor and 89.3% of the risk recognition model based on sample weighted support vector machine.

To sum up, for the financing risk identification of enterprises in highly variable industries, the identification accuracy of this model is significantly higher than the other two models, and the identification time of this method is far less than the other two identification models. It shows that this model has better recognition effect and reliability for different industries. Summarizing the analysis of the test data of the above three tables, the identification accuracy and efficiency of the enterprise strategic financing risk identification model based on online learning constructed in this paper are significantly improved when identifying the enterprise financing risk, and the practical application effect is better.

Table 5. Comparison of strategic financing risk identification test data of FMCG enterprises

Number	Risk identification model based on online learning		Risk identification model based on fuzzy fmea-vikor		Risk identification model based on sample weighted support vector machine	
	Accuracy/%	Recognition time/s	Accuracy/%	Recognition time/s	Accuracy/%	Recognition time/s
1	94.6	15.68	89.2	20.92	88.4	27.66
2	93.9	15.36	91.3	21.58	88.9	27.53
3	94.2	13.78	89.8	20.07	88.7	27.42
4	94.6	14.93	90.5	22.78	88.2	27.03
5	94.7	14.77	91.7	23.78	87.6	26.64
6	95.4	15.02	90.9	20.49	89.3	27.38
7	95.2	15.21	89.1	21.32	87.1	27.45
8	95.1	14.48	90.0	22.94	88	27.53
9	95.2	13.79	89.6	23.07	87.4	26.71
10	94.8	13.76	89.4	20.36	86.8	27.16

4 Conclusion

After the financial crisis, with the development of Internet technology, Internet of things technology, blockchain technology and the improvement of risk management level, the research on enterprise financing business transformation has also become mature. Enterprises carry out strategic financing related operations according to the development strategy and capital management needs of enterprises at different operating stages. In this process, many organizational elements forming strategic financing present a systematic change. However, there are usually different degrees of risks in enterprise financing operations. Although they cannot be eliminated in the financing process, risk management can be strengthened through various ways and means such as risk transfer and dispersion, so that the high risks of financing activities can be controlled within the scope that all stakeholders are willing to bear. This is also the key to enterprise strategic financing. Therefore, accurately identifying the strategic financing risks of enterprises is very important for actively building a security system for effectively managing financing risks and forming the unique competitiveness of enterprises. The risk response of financing risk is an indispensable part of risk management. This paper constructs an enterprise strategic financing risk identification model based on online learning. Through research, the following conclusions are reached:

(1) By virtue of the good training characteristics of online learning, this model makes the risk identification model adjust in time according to the changes of the enterprise's business strategy, and improves the accuracy of financing risk identification.

(2) The risk identification model constructed in this paper can help business decision-makers to make rational induction and summary of financing risk management methods and measures, reveal the mechanism of financing risk formation and change, guide financial practice, and implement more effective management of enterprise financing risk after it is actually applied to enterprise operation.

Fund Project. Jiangsu University "Qinglan Project" Outstanding Young Backbone Teacher Training Project; Wuxi Commercial Vocational and Technical College provincial and ministerial-level and above subject cultivation project (KJXJ21602); Wuxi Commercial Vocational and Technical College doctoral research start-up project (RS18BS03); Wuxi Commercial Vocational and Technical College "Eagle Program" training project (RS21CY04).

References

1. Song, R., Dong, L.: Early warning of financing risks of intellectual property pledge in artificial intelligence enterprises based on fuzzy FMEA-VIKOR. Fuzzy Syst. Math. **35**(05), 106–117 (2021)
2. Geng, C., Li, X.: Financing risk prediction of science and technology enterprises based on sample weighted SVM model. Indust. Technol. Econ. **39**(07), 56–64 (2020)
3. Xing, M., Dong, X.: Research on risk assessment of intellectual property pledge financing for small and medium-sized technological enterprises: Based on the perspective of supply chain finance. Sci. Technol. Manag. Res. **40**(18), 196–202 (2020)
4. Liu, Z., Li, H., Wang, L.: A study on the early warning mode of financing risk of pledge of intellectual property based on FMEA method. On Econ. Prob. **02**, 58–66 (2020)
5. Wen, C., Yang, X., Zhang, J., et al.: Deep learning optimization algorithm based on momentum term separation. Comput. Simulat. **39**(02), 337–342 (2022)
6. Gu, L., Zhang, C., Wu, T.: Financing risk assessment of major water conservancy PPP projects based on cloud model. Yellow River **43**(11), 116–121 (2021)
7. Xu, K., Zhang, C., Bao, X.: Analysis on the risk formation mechanism of supply chain financing based on the self-owned funds of E-commerce platform. Chin. J. Syst. Sci. **29**(02), 100–104+115 (2021)

Research on Autonomous Learning Management Software Based on Mobile Terminal

Han Yin[1](✉) and Qinglong Liao[2]

[1] School of Traffic and Transportation, Xi'an Traffic Engineering Institute, Xi'an 710300, China
xinqiba001@sina.com

[2] State Grid Chongqing Electric Power Research Institute, Chongqing 401121, China

Abstract. There is a large amount of data calculation in the data transmission from the mobile terminal to the server edge. Psychologically assisted autonomous learning management software provides users with learning resources and learning activities to stimulate and maintain learning motivation. According to the software requirement, the whole development architecture is established based on the mobile terminal device, and the client mobile terminal device responds to the user operation and sends the data request to the Web server. The server adopts module entity to reflect the system database concept, divides the task into different modules according to the decision result, and reduces the delay of data transmission. The client is mainly composed of three interfaces: system login module, online learning module and support service module, which correspond to different activities respectively. The test results show that the psychologically aided autonomous learning management software based on mobile terminal device can reduce CPU usage and memory usage and improve performance.

Keywords: Mobile terminal equipment · Psychological assistance · Autonomous learning · Management software · Software development · Learning management

1 Introduction

The progress of science and technology, the improvement of people's living standards, promote the popularity of mobile phones. In the era of mobile phone popularization, people get more and more information through handheld mobile devices and wireless communication networks, and with the further development of Internet technology, mobile development technology has developed to make the performance of developed apps better. In the 21st century, wireless communication technology is gradually replacing wired communication technology to become the mainstream of the industry, which also prompted the innovation and progress of distance learning. With the development of mobile technology more and more mature, there are a variety of mobile software on the market such as mobile hitchhiking software, mobile ordering software to provide our lives with greater convenience. The development trend of mobile learning is

W. Fu and L. Yun (Eds.): ADHIP 2022, LNICST 468, pp. 715–727, 2023.
https://doi.org/10.1007/978-3-031-28787-9_53

also advancing rapidly, so the research on the implementation technology of mobile online education is of great practical significance. Mobile terminal devices with mobile phone and other media can effectively overcome the disadvantages of book learning and computer learning. As a new type of learning resource, the learning resource based on mobile device shows strong interactivity and interest by virtue of powerful multimedia technology, various forms of touch and simple and easy-to-learn operation habits, which is an important difference between mobile learning resources and traditional learning resources [1, 2]. Today's management software is facing the impact of the Internet and cloud. In particular, the development of mobile Internet challenges the timeliness of management software. Developers need to master important information and data at the first time. Therefore, it is of great significance to develop and study learning software.

Reference [3] proposes the development and Application Research of the data evaluation management software for primary school students' physical health test. The data evaluation and management software of primary school students' physical health test is divided into three important functional modules: score scoring, score warning and sports advice, and students' personal score sheet. It has the characteristics of strong operability and clear scoring interface. Setting up color scoring early warning can dynamically understand the physical health development of students, provide sports suggestions for early warning scores, and effectively assist teachers in teaching and students in autonomous learning. This method can reduce CPU utilization, but the data transmission rate is low. Reference [4] proposes the development and application of teaching management software for nursing interns. The software system is composed of three modules: Intern information entry, internship quality assessment and rotation management. After referring to relevant literature, the software quality evaluation questionnaire is compiled based on the theoretical framework of the international standard ISO/IEC9126 quality model. SPSS17.0 software was used for data entry and statistics. This method is feasible, but the CPU occupation is high.

Therefore, the research on the development of self-regulated learning management software based on mobile terminal equipment is proposed. After fully understanding the user's needs, formulate corresponding functional modules according to the needs. Use the web server, streaming media server and database server to build the overall development architecture of the mobile terminal device and reduce the memory consumption. Develop the server-side program of the psychological assistance independent learning management software through MyEclipse program to improve the overall performance of the system.

2 Development of Psychologically Assisted Autonomous Learning Management Software Based on Mobile Terminal Devices

2.1 Requirement Analysis of Psychologically Assisted Autonomous Learning Management Software

Before you can design software, there must be clear requirements. Only when the requirements are clear, can the R&D personnel develop the corresponding functional software according to the requirements and ensure the smooth development and testing of the system. The software system designed in this way can meet the needs of more people. On the other hand, it can also ensure the feasibility of the project research practice. Psychological Aided Autonomous Learning can make use of the learning resources and activities provided by the network to choose the learning content and learning goal, stimulate and maintain the learning motivation, make the learning plan, determine the learning time, adjust the learning strategy, take the initiative to create learning situations for meaning construction, and promote the activities of problem solving and social interaction. Different user roles have different needs, from the perspective of learners, we hope that learning forms can replace the PCs through mobile terminal device learning, to obtain better learning experience and improve learning efficiency. It is hoped that the mobile terminal will be more user-friendly, beautiful, smooth in performance and can watch the learning videos you want to learn. Psychologically assisted autonomous learning management software diagnoses, evaluates and gives feedback to self-learning activities, continuously monitors and adjusts learning, perfects the learning process of self-knowledge system, and enables learners to constantly enrich and improve themselves, thereby improving learning and work efficiency and laying a good foundation for lifelong learning. From the point of view of software function, this paper analyzes the requirements of psychologically assisted autonomous learning management software. The specific functionality you should have for your software is shown in Fig. 1.

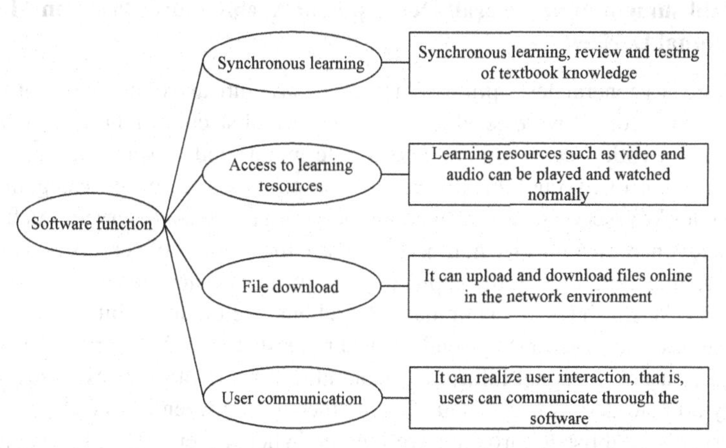

Fig. 1. Functions of psychologically assisted autonomous learning management software

Software for learners to provide learning resources to assist classroom learning, to help learners expand the classroom learning content, through the test to help learners master the learning content. Simultaneously these study content and the test content also can carry on the renewal along with the classroom content renewal. Learners should make self-planning and self-management according to their actual situation, and advocate respecting, trusting and exerting learners' initiative in the process of learning. In the process of autonomous learning, learners must actively reflect on their own learning process and learning results, and adjust the learning plan and learning objectives according to the results so as to learn effectively. Simple text and image content will make learners feel bored in the learning process, which requires the software to present video and audio resources to help learners easily grasp the knowledge to learn, so the software must be able to support the normal play of audio and video materials. On the one hand, autonomous learning advocates self-planning, self-monitoring, self-reflection and self-evaluation, which are all open in the whole process of autonomous learning, and on the other hand, abundant resources in the network can be shared freely by learners according to their needs.

With the software running, the software will occupy more and more devices, so it needs software to support the uploading and downloading of files, so as to reduce the memory occupied by the software. Uploading important files from mobile devices to the server or obtaining the latest learning resources from the server can help learners to share the resources and ensure the normal operation of the software on mobile devices [5]. Through network or multimedia teaching resources and modern information technology, network autonomous learning communicates with students and teachers to solve difficult problems in the process of autonomous learning. Help learners to establish contact through communication, so that learners can get timely help in the learning process. For developers, the reliability and implementation of software products and the stability and maintainability of the software are the key requirements in the process of software development, hoping to develop a popular and stable software products.

2.2 Establishment of an Overall Development Architecture Based on Mobile Terminal Devices

Architecture is a general description of the overall structure and components of software, and a framework for software development. Users establish contact through mobile communication networks between mobile devices terminals and servers. The server side of the development of psychologically assisted autonomous learning management software includes Web servers, streaming media servers and database servers. As functional requirements increase and user numbers increase, the number of servers increases with the maintenance iterations of the application, requiring cluster management of the various servers. When the user manipulates the client, the client mobile terminal device responds to the user's actions by sending data requests to the Web server. The database server stores users' personal information and all kinds of learning resources to ensure the safety and storage of users' data. The connection between a user and a streaming media server is established through a web server. When a user applies for playing video and audio, the user first makes an application to the streaming media server through the web server, and the streaming media server responds to the content requested, and then feedback the content of the streaming media response to the user through the web server [6]. The server receives the response from the client and responds to the connection request or data request according to the logic rules in the corresponding interface. Finally, the queried data is returned to the client in a JSON string format. The database is used to store user information, video information and all kinds of self-learning resource information. The overall development architecture based on mobile terminal devices is shown in Fig. 2.

The presentation layer on the server side directly contacts the user, receives and transmits user data and displays data fed back from the business logic layer. The cloud server, which is the bridge between the user side and the edge service side. On the one hand,

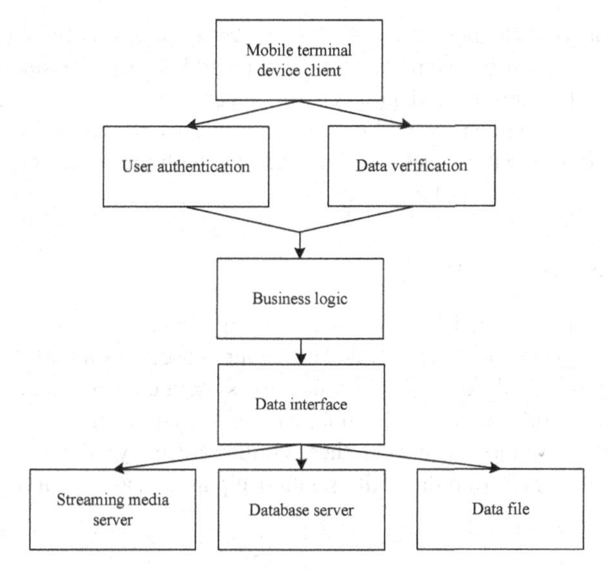

Fig. 2. Overall development architecture based on mobile devices

the cloud needs to communicate with the end user to achieve user login, personal information management and analysis; on the other hand, the cloud needs to communicate with the edge server to regularly update the local database resources of psycho-assisted autonomous learning resources server and to detect and monitor the running status of the whole software. Business logic handles logic, such as accessing the data layer according to the business logic and then receiving feedback from the data access layer to the presentation layer according to the business logic to implement software functions. Software through the network course information stored in the database, the database will have a corresponding record, and through the relevant attributes of the entity to describe the meaning of each field expression. The edge server and the cloud server establish a long TCP-connected communication link, through which the cloud and the edge can interact. When the user clicks the video playback, the data of the video in the server is read by the link address, and the video is played at the mobile terminal for the user to watch and learn. The related attributes of network curriculum entity include curriculum name, curriculum category, curriculum description and so on. The main purpose of this information is to record and manage the course data. The main function of the data access layer is to operate the database, receive the instructions from the business logic layer to add, delete and check the database, and feed back the results to the business logic layer.

The architectural pattern of the MVC used by the client. The mobile terminal operating system is open source and easy to operate, and has an optimized graphics library and powerful multimedia functions. The built-in database can meet the developer's requirements for data. For example, the Android operating system open source, developers can get the source code for free and use the platform to develop Android applications. In the process of coding, the common class and the same logic code are extracted for all the network requests, which makes the logic code of the whole project clear and avoids the bloatiness of the whole project [7]. At the same time, the Android system provides developers with good services, allowing developers to develop applications in the process to solve their problems in a timely manner. Part of the business logic can be realized through the centralized data processing mechanism on the cloud, for example, monitoring and management of mobile terminal devices can be realized through the cloud server, and terminal devices can communicate with the cloud server through the public network bandwidth HTTP.

2.3 Software Server-Side Design

The Psychologically Assisted Autonomous Learning Management Software server-side program is developed using MyEclipse. The server is used to support data interaction. The server side uses module entities to reflect the system database concept. According to the functional business points of the mobile learning system, the users and courses are designed, and the correlations between these entities are analyzed in detail. The system E-R diagram is obtained from the entity relationship model, as shown in Fig. 3.

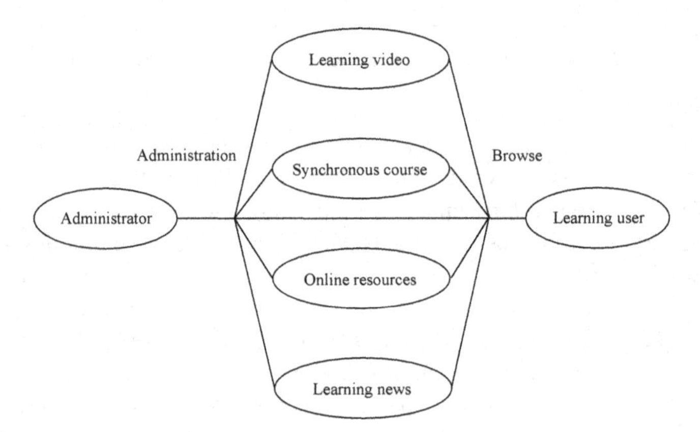

Fig. 3. Database entity relationships

According to the business requirements of the application software system, the edge server needs to send the edge data generated by mobile client to the cloud server through TCP long connection. In order to solve the problem, this paper proposes a method to calculate the network delay threshold, which is suitable for different application environments according to the requirements of different systems. In the cloud server, the edge server database resources should be updated regularly and mobile terminal devices should be monitored. We ran multiple script files in the Worker Man architecture to enable real-time communication between the edge server and the cloud server. The learning video entity is the storage of information such as the address of the video, the ID of the video, etc. Each video played has an associated record in the video table. The video resources of the mobile platform are saved on the video server by dynamically generating the video playback address, and then saved in the video table by linking. Through the analysis, we know that the choice of network threshold will mainly affect the following three factors: the accuracy of recognition, network transmission delay, mobile computing. Therefore, the optimization goal can be summarized as the following three objectives: to improve the accuracy of system identification, reduce the network transmission delay and reduce the computational complexity. The migration decision engine determines whether a task is executed locally, in the cloud, or on an edge server, and divides the task into modules to be executed on different devices according to the decision. There is a transmission delay in the data transmitted from the mobile end to the edge end, and the recognition accuracy of the transmission task can be described as follows:

$$p = \sum_{m=1} q_m \times \alpha \tag{1}$$

In formula (1), p represents the overall identification accuracy of the software; m represents the number of mobile terminal devices; q_m represents the identification accuracy of the inference task of the checker; and α represents the network latency threshold to be solved. If the system determines that the current network delay is high, the task requires the mobile side to carry out an additional reasoning, which increases the amount of mobile computing. When a terminal application sends a task execution request, the monitor detects the availability of cloud and edge server resources, including computing load, network transmission capacity, response aging, and so on. Each additional computation on the mobile device can be seen as a loss of overall system performance, so we use a loss function to express the effect of increased computational pressure on software performance on the mobile side. The formula is:

$$g(\alpha) = \sum_{m=1} h_m k(\alpha)\beta \tag{2}$$

In formula (2), $g(\alpha)$ represents the increased calculated pressure loss at the mobile end; h_m represents network transmission latency; $k(\alpha)$ represents the time taken by the mobile device to perform a single inference; and β represents the checker score. When the network is congested, it will increase the computational pressure of the mobile terminal, reduce the communication delay between the mobile terminal and the edge terminal, and reduce the overall software response time. On the one hand, the edge server sends packets to the cloud periodically, including user's personal information, user's behavior data, edge server's CPU load, memory usage and so on. Edge servers take on the computing and storage capabilities of part of the cloud servers, helping to complete tasks with high real-time requirements. The decrease in response time is an improvement in the overall performance of the software, so we use the gain function to represent the increase in the performance of the software due to the decrease in response time. The formula is:

$$f(\alpha) = \sum_{m=1} h_m k(\alpha)(\gamma - \alpha) \tag{3}$$

In formula (3), $f(\alpha)$ represents the gain in software performance; γ represents the delay in communication for this task. In the process of migration, it is necessary to consider not only the energy consumption for data uploading from the terminal to the edge, but also the energy consumption for receiving the results.

2.4 Software Client Design

The client of psychologically assisted autonomous learning management software is coded in Java language, and the SDK is selected as the development environment. The client of psycho-assisted autonomous learning management software is mainly composed of system login module, online learning module and support service module, which corresponds to different activities. After the client program starts, the system first runs the welcome interface, then enters the system login module, starts the entire program flow. The system login module is implemented by LoginActivity. User download apk installation successfully enter the user name and password, click on the login, the client java code through the post request to the user name, password, mobile phone IMEI value and other information sent to the back end. The backend uses a set of validation logic to determine whether the current user has access to the mobile learning system, and if so, enters the main interface or prompts the user to contact the administrator for device binding [8]. The user login process is shown in Fig. 4.

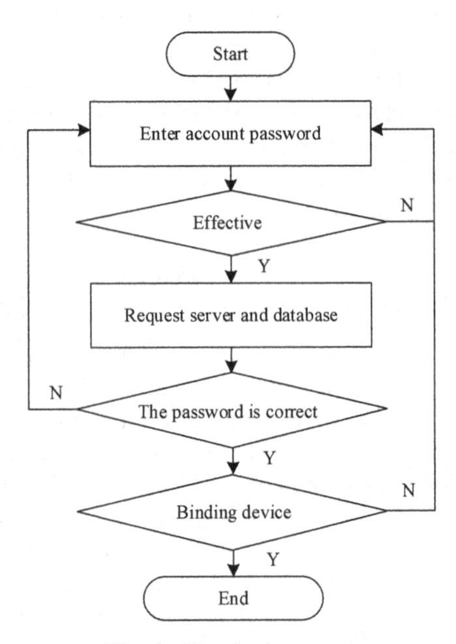

Fig. 4. User login process

After the user login interface is used to input the username and password, the system sends an HTTP request to verify the validity of the learner's identity, and returns the verification result to the customer, if the verification is valid, the user enters the system; otherwise, the system gives an error message. Learners can also register a new account through the system login module. Use the getText ().toString () method to get the account number and password entered by the user. If the current user does not enter any information, clicking Login will prompt for the account number or password via Toast. The account and password entered by the user are then sent to the server via a post request. The client needs to interact with the server side of the mobile learning system at runtime, such as user login validation, requesting course information data, sending feedback data, etc. [9]. The server will check the database according to the transmitted information. If the account password is wrong, the feedback will be given to the client if the account password is wrong; if the account password is correct, the server will further determine whether the current mobile device is bound. If the device has been bound, the server will be allowed to log in otherwise, please contact the administrator to bind the device. After the server responds to the information, it judges what is returned. If the server returns 2, it indicates that the user name has been registered, prompting the user to "User Name Already Exists"; if the server returns 1, it indicates that the user name has not been registered, then the user can enter the password and nickname in the text box, save the user name, password and nickname entered by the user in the User table in the server database, and return the message "Registration Succeeded". When the registration is successful, the software will automatically close the current page and jump to the software homepage, and display the user's nickname in the homepage. Data requests are sent to the server when the client program is implemented using the Apache Http Client project built into Android. When the server side completes the data operation, it returns the results as JSON data to the client. The course of this topic is mainly through the mobile phone player defined on the end to play video, and through the mobile phone to watch learning.

The online learning module is implemented by LoginActivity. After the learner enters the main interface of the system, all the functions of the system are displayed through the main form page, MainActivity. When the user clicks on the course list, the click event response from the RecycleView jumps to the video playback interface through Intent and passes the url of the video playback to the playback interface through the intent putExtra () method. Online learning module includes course browsing, resource downloading, communication and discussion, and system notification. Learners can choose according to their needs. Courseware learning is mainly composed of mobile learning resources such as text, pictures and audio. Through the presentation of detailed text and pictures, the learners can understand the learning content systematically. Video learning provides users with short videos to help them learn better on the move. In order to help learners to broaden their knowledge field and have a more comprehensive and objective understanding of the new technology or practice research, the latest cutting-edge technology and practice research are provided to the learners through diversified media presentations in extracurricular extension. The playback interface gets the object of the Intent through the method of getIntent (), which then calls the getStringExra () method to get the playback link. Finally, the playback address is set through the myvideo.setVideo Path () method for video playback.

Upload and download module helps users download resources to their own smart terminal through the client, but also can upload their own resources to the server, and other users can share resources through download. Support service modules are implemented through the SupportActivity. The module includes four functions: modifying the name, modifying the password, logging off and logging on, and relevant information, which are displayed by switching between main form pages. When the user jumps to the upload and download interface, they will see the download list in the download area. Click the download button next to the download list to download the files they want. When a user wants to upload a file, clicking the upload button will bring up a dialog asking the user to select the file to upload. The user can upload the file to the server by clicking the upload button. Based on the above process, the software of autonomous learning management based on mobile terminal device is developed.

3 Experimental Study

3.1 Experimental Preparation

The aim of software testing is to ensure the final quality of software, and the testing of software is carried out with the whole process of software development. The software client side test selects the handset which the user occupies in the market to use many carries on the real machine test. Through the real machine test, the overall functions of the system are evaluated and analyzed. The server system adopts the common hardware equipment as the test platform. The hardware test environment is as follows: CPU: Core i78 Series; Memory capacity: 64 GB; Graphics card: Intel supercore graphics card 630. The test environment for the database server is as follows: operating system: Windows 10; database system: MySQL; database administration tools: Navicat Premium in Chinese. In the process of development, we can use USB connection and APK to run on Android smartphone, which can realize the content of ALM software design. In unit testing, the function of each module must be tested. Only the function of each module is normal can the software run normally in the process of integration testing. Then, the compatibility and performance of mobile application are tested to test the dependence of software on mobile hardware and whether the portability is satisfied. Select 50 Android mobile terminals with high usage rate for performance testing to verify the performance of the designed software.

3.2 Results and Analysis

Through unit test, the user's basic information, including user's username, login password and nickname, can be uploaded to the database on server side. Click the menu button of each learning interface can normally jump to the corresponding interface, click the text button of each interface can display the corresponding text. Click on the video and audio buttons, video and audio can play. In the Mobile Application Compatibility Test, all kinds of mobile terminals have passed the adaptability test, which shows that the psycho-assisted autonomous learning management software has a high adaptability. The CPU usage and memory usage were selected as evaluation items, and the tests were carried out under the conditions of 10 and 50 mobile terminals. The test results of the psychological assisted autonomous learning management software designed this time are compared with the method of reference [3] and the method of reference [4]. The test results are shown in Tables 1, 2, 3, and 4.

Table 1. Number of mobile terminals with 10 CPU usage (%)

Number of tests	The proposed method	The method of reference [3]	The method of reference [4]
1	4.56	13.92	18.47
2	4.62	15.47	22.84
3	4.64	16.84	**19.68**
4	7.57	18.64	15.56
5	**8.88**	17.28	17.22
6	6.26	15.59	19.35
7	5.63	14.32	18.93
8	7.22	**19.64**	16.01
9	8.35	13.28	17.24
10	5.51	16.52	18.52

When the number of mobile terminal devices is 10, the CPU utilization of the proposed method is 6.32%, which is 9.83% and 12.06% lower than that of the comparative method.

Table 2. Number of mobile terminals with 50 CPU usage (%)

Number of tests	The proposed method	The method of reference [3]	The method of reference [4]
1	8.47	23.40	19.49
2	9.88	24.87	18.07
3	10.04	**25.58**	17.25
4	8.67	22.29	20.66
5	11.21	21.66	**28.33**
6	12.35	20.39	21.52
7	**13.52**	22.21	19.05
8	9.96	21.54	22.84
9	8.23	23.13	20.22
10	10.12	20.50	16.43

When the number of mobile terminal devices is 50, the CPU utilization of the proposed method is 10.25%, which is 12.31% and 10.14% lower than that of the comparative method.

Table 3. Number of mobile terminals 10 Memory usage (MB)

Number of tests	The proposed method	The method of reference [3]	The method of reference [4]
1	69.41	109.67	115.24
2	71.28	110.48	106.48
3	60.07	108.84	118.75
4	65.64	102.51	104.86
5	71.35	103.20	109.23
6	72.52	105.32	112.02
7	**73.16**	112.66	118.65
8	64.83	103.93	114.51
9	65.29	115.59	**121.94**
10	68.92	**116.25**	103.27

When the number of mobile terminal devices is 10, the memory occupation of the proposed method is 68.25 MB, which is 40.60 MB and 44.25 MB lower than that of the comparative method.

Table 4. Number of mobile terminals 50 Memory usage (MB)

Number of tests	The proposed method	The method of reference [3]	The method of reference [4]
1	79.42	154.48	182.09
2	78.85	**162.89**	175.64
3	72.58	150.56	172.31
4	82.66	153.63	146.34
5	79.33	151.30	163.27
6	78.55	149.24	169.85
7	64.22	148.57	175.66
8	**88.14**	146.11	**182.63**
9	81.45	145.72	163.45
10	80.73	142.58	167.72

When the number of mobile terminal devices is 50, the memory occupation of the proposed method is 78.59 MB, which is 71.92 MB and 91.31 MB lower than that of the comparative method. Because the method in this paper combines mobile terminal equipment and wireless network communication technology, it can reduce the CPU occupation and memory occupation of autonomous learning management software, and effectively improve the performance of management software. Therefore, the psychologically aided autonomous learning management software based on mobile terminal device has passed the performance test, which meets the basic requirements of online.

4 Conclusion

With the rapid development of Internet and mobile devices, smart mobile devices have become more and more popular. Smart mobile devices have become an irreplaceable part of people's life. In this paper, the management software of psychologically assisted autonomous learning is developed based on mobile terminal devices. After analyzing the requirements of the psychological assisted self-learning management software, the overall development architecture is established through the web server, streaming media server and database server. According to the decision-making results, the tasks are divided, and the development and research of the management software for psychological assisted autonomous learning are completed. Through the real machine test, the software has good compatibility and performance test results, to meet the development needs. In the design of mobile learning project, how to combine mobile devices and learning situation effectively, strengthen the impact of mobile technology on the social relations of learners, and better enable learners to enjoy the advantages of mobile technology in supporting situational cognition, social learning and informal learning.

References

1. Xu, S., Zhang, Y., Wang, J., et al.: Research on students online autonomous learning behavior and psychological characteristics under the Internet+Classroom background. J. Jilin Teach. Inst. Eng. Technol. **37**(1), 33–35 (2021)
2. Li, C.: Practical application of "activity-guided" teaching mode in college students' mental health education—Based on the exploration of life skills development project. Heilongjiang Sci. **12**(17), 61–62+65 (2021)
3. Xiao, H.: Development and application of data evaluation management software for pupils' physical health test. Bull. Sport Sci. Technol. **29**(11), 70–72 (2021)
4. Zhang, J., Xie, X., Zhang, X., et al.: Development and application of teaching management software for nursing interns. Chin. Nurs. Res. **34**(7), 1242–1245 (2020)
5. Liu, Y., Chen, Q.: Personalized information mining of mobile terminal users based on tag mapping. Comput. Simulat. **39**(1), 177–180+208 (2022)
6. Xu, M.-W., Liu, Y.-Q., Huang, K., et al.: Autonomous learning system towards mobile intelligence. J. Softw. **31**(10), 3004–3018 (2020)
7. Chen, Z., Zhang, Y., Yang, X.: Analysis of resource construction mode based on mobile terminal learning. Shaanxi Radio TV Univ. J. **22**(2), 20–24 (2021)
8. Zhang, L.: Discussion on computer software development and database management. Digit. Technol. Appl. **38**(8), 60–62 (2020)
9. Wan, M., Cao, L.: Software defined networking security based on reinforcement learning. Comput. Eng. Design **41**(8), 2128–2134 (2020)

Research on Online Monitoring of Power Supply Reliability of Distribution Network Based on Mobile Communication Technology

Qinglong Liao[1(✉)], Qunying Yang[2], Dongsheng Zhang[3], Xiao Tan[1], Xiaodong Wu[1], and Han Yin[4]

[1] State Grid Chongqing Electric Power Research Institute, Chongqing 401121, China
hch852421@126.com
[2] State Grid Chongqing Electric Power Company, Chongqing 400014, China
[3] State Grid Liaoning Electric Power Company Anshan Power Supply Company, Anshan 114000, China
[4] School of Traffic and Transportation, Xi'an Traffic Engineering Institute, Xi'an 710300, China

Abstract. The distribution network plays an important role in distributing electric energy in the power network. Because of the huge and complex structure of the distribution network, the power supply effect is affected. Therefore, an on-line monitoring method for power supply reliability of distribution network based on mobile communication technology is proposed. Use mobile communication technology to collect fault information in distribution network. Through the reliability evaluation algorithm, the reliability management system is constructed and the monitoring process is designed. The overall detection system is divided into four modules, and online detection is realized through communication interconnection. The experimental analysis results show that the online monitoring method of power supply reliability of distribution network based on mobile communication technology has high accuracy and practicability, and fully meets the research requirements.

Keywords: Mobile communication technology · Distribution network · Power supply reliability · Online monitoring

1 Introduction

With the rapid development of modern society and the popularization of high technology, the application of electric power has been fully popularized in the society. However, the improvement of social benefits also makes social development and people's life have higher requirements for the reliability of power supply. According to the incomplete statistical data of power grid monitoring and management, most of the power outages currently occur in the distribution network, and the power loss generated by the distribution network accounts for nearly half of the power loss of the power grid. Relatively speaking, the research on the reliability of distribution network has not attracted the

W. Fu and L. Yun (Eds.): ADHIP 2022, LNICST 468, pp. 728–743, 2023.
https://doi.org/10.1007/978-3-031-28787-9_54

strong attention of scholars. The biggest reason for this phenomenon is that the facilities of the power generation system are concentrated compared with the distribution network, the original investment in the facilities is large, the establishment takes a long time, and the degree and scope of the impact of power failure caused by insufficient power supply on society and ecology are easy to attract people's attention. Compared with many foreign countries, although the research on the reliability of distribution network in China started late, after a long time of hard exploration, it also pays more and more attention to the research on the power supply reliability of distribution network. A series of reliability evaluation algorithms are proposed, the corresponding evaluation and calculation software is developed, a relatively perfect database of power grid basic parameters and accident records is established, and a system for efficient monitoring and management of power grid power supply capacity and power quality is established.

Reference [1] proposes a study on power supply reliability monitoring of active distribution network based on load characteristics. Select the collection equipment with load characteristics, and transmit the collected data using optical fiber network; Combining the power consumption in different time periods, a load probability model is established to obtain the time series characteristics of all load points in the distribution network; Analyze the reliability indexes of the load points of the series and parallel structures respectively, and evaluate the reliability of the distribution system by using the outage frequency, outage time and other indexes; Take these indicators as the monitoring basis, obtain the probability formula of normal operation of the distribution network, input basic information into the network, set the iteration times and sampling density, and judge whether the power supply can meet the power supply demand of all loads during the power outage The algorithm has high accuracy, but the calculation process is complicated. Reference [2] proposes a study on reliability of distribution network with DG based on load optimality. First, a load importance measurement index system is established. Secondly, each index in the load is subjectively weighted based on AHP. Then, each index is objectively weighted using entropy weight method. Finally, kender harmony coefficient is introduced to obtain the comprehensive weight of the index. This algorithm has high detection efficiency, but the accuracy of detection needs to be improved.

At present, there are two kinds of quantitative calculation methods of reliability indexes: analytical method and simulation method. The analytical method is based on the structural characteristics of the network and carries out reliability calculation through mathematical modeling [3]. Simulation method is an experimental method. On the premise that the original parameters of equipment reliability in the distribution network are known, the possible operating states of the system are determined by sampling from the probability distribution function of the equipment, and then the operating states of various systems are analyzed and calculated, and the reliability indexes are obtained according to the statistics of the results of the simulation experiment. The method of reliability analysis mainly adopts the minimum path method, which is to seek the shortest power supply path for all user points in the network, and equivalent the equipment with non shortest power supply path to the shortest power supply path. Therefore, only the influence of nodes and load nodes on the shortest power supply path on network reliability needs to be considered, and the methods and evaluation indexes suitable for power supply reliability statistics, calculation and analysis of distribution network users are formulated. The shortest path algorithm is applied to the theoretical analysis of the actual calculation example, and the reliability improvement measures are analyzed one by one by calculating the reliability index and using the income increment/cost increment evaluation method, so as to find the reliability improvement measures suitable for the actual power network. Therefore, an on-line monitoring method of power supply reliability of distribution network based on mobile communication technology is proposed.

2 Research on Online Monitoring of Power Supply Reliability of Distribution Network

2.1 Identification of Power Supply Fault Information of Distribution Network

Due to the complex and changeable operation of the actual distribution network system, there are still many problems in the production and operation process, and the practical value of these algorithms is limited. Using mobile communication technology to collect fault information, construct fault judgment matrix or locate the distribution network through hot arc search is also affected by the collected information to a great extent, and the fault tolerance is poor. In case of distorted information or mistransmission of information, it will cause misjudgment and missing judgment. Mobile communication technology has good interpolation results, while extrapolation may have large error and slow convergence speed. The monitoring method based on mobile communication technology can effectively simulate the process of fault diagnosis, but there are many defects in practical application. With the development of hardware technology and integrated intelligent technology, as well as the in-depth research on the principle of fault location by more experts and scholars, and the comprehensive use of various advanced and optimized intelligent methods, the electrical quantity after fault can be collected more efficiently and accurately, which will make the fault location of distribution network achieve better development. The advantages and disadvantages of various fault location algorithms are shown in Table 1.

Reliability mainly refers to the ability of a certain element, a certain equipment or a certain system to realize the specified functions under normal specified conditions during the predetermined operation period. Its measure is called reliability or reliability. Reliability represents the ratio or probability of equipment and system success. The reliability research of power system involves all links, coupled with the rapid progress of computer information technology, at present, many power grids have realized the supervision and management under the computer system. Therefore, it is the general trend to carry out the system reliability analysis through the computer. Due to the wide range of power system, comprehensive reliability evaluation is difficult to achieve. It is usually discussed and studied after the subsystem is divided. According to the composition of power system, it can include the reliability evaluation of power generation, transmission, distribution and other systems. The reliability index of distribution network can be divided into load point index and system index, in which the system index can be further divided into frequency time index and load electricity index [4]. The load point index describes the reliability of

Table 1. Comparison of advantages and disadvantages of various fault location algorithms

Fault location method	Advantage	Shortcoming
Matrix algorithm	Strong adaptability, high system storage efficiency and good engineering practicability	It takes a long time to judge the fault, and it is easy to miss and misjudge
Overheated arc search algorithm	The principle is simple and the detailed fault degree can be obtained	It has high requirements for the completeness of fault information and complex processing of special switches contained in distribution network
Expert system	Avoid limitations and be more careful and accurate; Strong applicability, not affected by environment, time and space and human factors; Analyze the problem objectively, comprehensively and logically	It is difficult to acquire knowledge and verify its completeness; The diagnosis speed of large system is very slow; Difficult maintenance; Lack of learning ability; Poor fault tolerance
Wavelet neural network	Have certain generalization ability; Strong fault tolerance; Fast execution speed	Learning convergence speed is generally slow; The error of extrapolation is large

a single load point, and the system index is used to describe the reliability of the whole system. The system reliability index can be calculated through the load point index. The commonly used load point reliability index mainly includes the average failure rate of load point, the average annual outage time of load point and the average outage duration of each fault of load point. Its specific meaning is:

$$r = \frac{U}{s\lambda} \tag{1}$$

In formula (1), the average failure rate of load point λ the average failure rate of load point is the expected value of the number of power outages at load point within the statistical time (usually one year), and the unit is generally the next year (Times). Average annual outage time of load point U average annual outage time of load point refers to the expected value of outage duration of load point within the statistical time (usually one year), and the unit is generally hour/year (h/a3). The average power outage duration s per fault at the load point, the average power outage duration of each fault at the load point can be calculated from the first two indicators by the formula. Point-of-load reliability metrics do not always fully characterize the system. In order to reflect the severity and importance of system outage, its reliability needs to be evaluated from the perspective of the whole system. All system reliability indexes can be calculated from the average failure rate λ_i and average outage time U_i of load point. The reliability indexes of load point can be calculated from the following formula:

$$\begin{cases} \lambda_i = r\sum_k \lambda_{i,k} \\ U_i = t\sum_k U_{i,k} \end{cases} \tag{2}$$

In formula (2), $\lambda_{i,k}$ and $U_{i,k}$ represent the average failure rate and outage time caused by the failure of component k at the load point, respectively.

2.2 Power Supply Reliability Evaluation Algorithm of Distribution Network

When studying the reliability of power system, the research object is generally divided into two levels: component and system. Component is the basic unit of the system, and it cannot be divided in the system. The system is made up of components, and it is the whole of components. The main components of the distribution network are: overhead lines, buried cables, air switches, voltage regulators, distribution transformers, cables, isolating switches, fuses, etc. [5]. Their reliability is directly related to the reliability of distribution system and even the reliability of power system. From the perspective of reliability theory, the electrical equipment or components used in power system can be divided into two types: Repairable components and non repairable components. If the component is put into use, once it fails, it cannot be repaired, or although it can be repaired but it is not economical, this kind of equipment is called non-repairable component. A component fails after a period of use. Reliability and unreliability Reliability refers to the probability that a component can perform a specified function under specified conditions and within a predetermined period of time. It is a function of time and is recorded as R(). It can also be said that reliability is the probability that the lifetime T is greater than the time t, that is, the probability that the element can still work reliably at the time t:

$$R(t) = P[T\lambda_i > t], \quad t \geq 0 \tag{3}$$

The probability of failure (failure) of the component from the start of use to time t, or the probability that the life T of the component is less than or equal to time t, is called the unreliability (or failure function), which is also a function of time, denoted as F(t), that is:

$$F(t) = P[U_i T > t], \quad t \geq 0 \tag{4}$$

The data exchange of the mobile communication architecture is shown in Fig. 1.

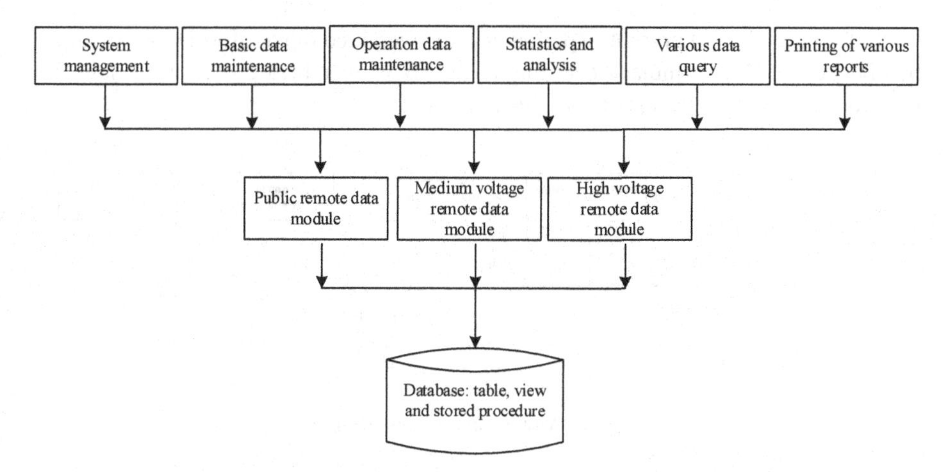

Fig. 1. Data exchange of mobile communication architecture

The system consists of client, middle layer and database server. The client interacts with the database through the remote data module of the middle layer. The remote data module not only provides an intermediary for clients to access the database, but also encapsulates a large number of business logic. Among them, the public remote data module includes parts other than the data table and business logic for medium voltage data and high voltage data, such as various coding data, system management part and application server initialization. The medium voltage remote data module and the high voltage remote data module respectively encapsulate the medium/high voltage basic data, operation data, statistical analysis data and corresponding business rules. The work of the database server includes the maintenance of the database structure, that is, tables, indexes, views, stored procedures, etc. [6]. The collected information is preprocessed, and the relationship between the information output and the input components is analyzed according to the processing results. The specific calculation formula is as follows:

$$f(x_n, y_m) = \frac{R(t)l(x_n, y_m)}{F(t)\sqrt{(x_n)(y_m)}} \tag{5}$$

In formula (5), $l(x_n, y_m)$ represents the covariance between the input information x_n and the output information y_m; k represents the total information quantity. According to the formula, the normalized calculation formula of input information x_n and output information y_m can be obtained:

$$x_n' = \frac{x_n - x_{min}}{x_{max} - x_{min}}, \quad y_n' = \frac{y_n - y_{min}}{y_{max} - y_{min}} \tag{6}$$

The formula is weighted to obtain the input data vector expression, as shown in the formula:

$$x = \left(f_1(x_1', y')x_i', f_2(x_2', y')x_{i-1}', \ldots, f_1\left(x_j', y'\right)x'\right)^T \tag{7}$$

In formula (7), T represents a period, and the weighted input information vector can be obtained by this formula. According to the above weighted processing results, the monitoring process is designed, as shown in Fig. 2.

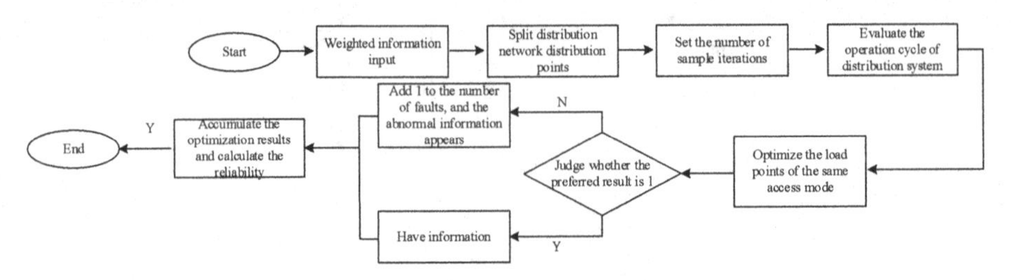

Fig. 2. Monitoring process design

The specific monitoring process is as follows: input the weighted information, divide the distribution points of the distribution network in the set, count the number of power sources, and read the corresponding loads. Set the initial number of sample iterations and evaluate the distribution network operation. In normal operation, judge whether the power supply is connected correctly. If the access is correct, the load points of different access modes are optimized, and then it is judged whether the preferred result is 1. If it is 1, it means that there is correct access information; otherwise, abnormal information occurs. If the access is wrong, the same access mode load point needs to be optimized, and then judge whether the preferred result is 1, if it is 1, it means that there is correct access information; otherwise, abnormal information occurs. Count the optimization results twice, and use this as an index to determine whether the number of optimizations is greater than the preset number of times. If so, output the correct access information, and the process ends; if not, return to step (2). According to the above monitoring process, a monitoring implementation scheme is designed.

2.3 Realization of Online Reliability Detection of Distribution Network

The overall structure of the online monitoring system is divided into system data acquisition module, system data storage module, system operation analysis module and graphical interface display module. In order to realize the communication interconnection of hardware equipment between each module and within the module, different interfaces need to be designed to realize the online collection of system power data, the monitoring and analysis of reliability information and the storage and management of data [7]. The sample data of the power supply bureau and its subordinate district and county-level power supply bureaus in a two-year power grid are collected. Through the analysis of the impact indicators of power supply reliability, the indicators affecting power supply reliability are summarized as shown in Table 2.

Table 2. Impact indicators of power supply reliability

Indicator type	Specific indicators
Grid structure	Ring network rate, switchable rate, average length of each line, network connection standardization rate, inter station connection rate and average number of sections of the line
Equipment quality	Average life value of scrapped distribution transformer, average life value of scrapped switch cabinet, failure rate of bare conductor medium voltage line tread, medium voltage failure rate of insulated wire and medium voltage cable failure rate
Technical equipment level	Insulation rate, wire rate, number of timely data connected to EMS system by master station, number of quasi real-time data connected to metering front by master station, number of equipment accounts connected to GIS system
Cause of failure	Failure times caused by natural factors, failure times caused by external factors, failure times caused by operation and maintenance construction factors
Operation and maintenance capability	Average continuous duration of medium voltage fault outage, fault emergency repair in place, positioning, average duration of power restoration, number of live work, households during power outage, number of live work optimization, and total number of pre trial and regular inspection

In order to establish the JDBC connection mode between the master station system and the database, read the power data information collected by the terminal and the power failure event information actively reported, and establish the original data table of power failure event, the combined power failure event data table and the relationship table between the original event and power failure event in the database [8]. Secondly, through the collected voltage information of metering points, the shutdown and power supply status of users are discriminated and analyzed according to the calculation model of medium voltage users, so as to realize the calculation of RS-1, aihc-1 and aitc-1 indicators. For the outage event information, the outage event merging algorithm is used to judge whether to merge the outage event. If the outage event can be merged, the last outage event information is written into the merged outage event data table. At the same time, the last power failure event information is overwritten with the original power failure event information in the relationship table between the original event and power failure event. If the power failure event cannot be merged, the last power failure event information is directly written into the relationship table between the original event and power failure event, and the data information in the relationship table between the original event and power failure event is statistically calculated to obtain the regional power failure frequency index [9]. Finally, the JDBC connection mode is established again, and the calculation results of power supply reliability index are written into the database for storage and management. The calculation function flow of system power supply reliability index is shown in Fig. 3.

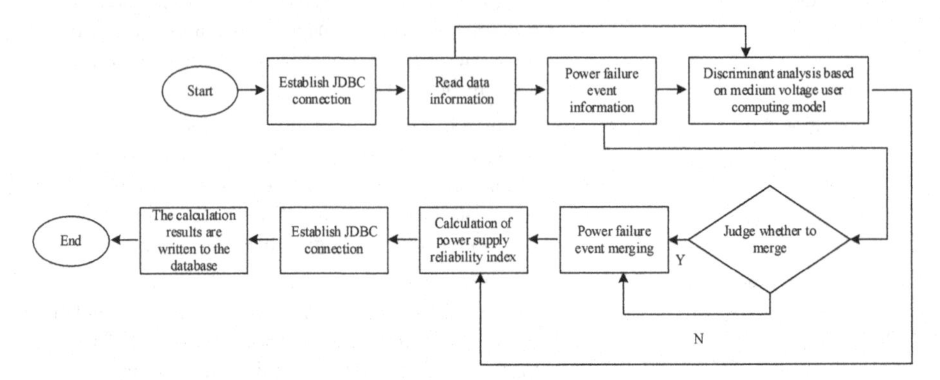

Fig. 3. Function flow chart of system power supply reliability index calculation

The query date and query type combo boxes in the power supply reliability index calculation interface are combobox controls based on Ajax technology to realize the data display of the page without refresh. Users can select different query conditions in different combo boxes according to their needs to realize the online calculation and query of power supply reliability indexes with different monthly values and cumulative values. The evaluation index can reflect whether the distribution system is reliable or not, and provide data basis for reference decision-making. In terms of reliability data statistics of distribution network, it is mainly divided into two categories: reliability data statistics of components and operation reliability data statistics of power supply system. The power supply reliability index for distribution network has a long history. Its earliest proponents are Edison Electric Power Research Institute (EEI), American Public Power Association (APA) and Canadian Electric Power Association (CEA). Among them, the most important index is the reliability index of each load point, ENS, SAF1, CAIF1, CAID and SAIDI. Their specific definition is 2nsaifi, which represents the average outage frequency of the power supply system. It mainly refers to the average outage times of power customers in the power supply system within a year. Its calculation formula is as follows:

$$SAIFI - \frac{\sum_{i} \lambda i Ni}{f(x_n, y_m) - \sum_{\mathrm{i}} N_i} \tag{8}$$

In formula (8), A and N respectively represent the user outage rate and the number of users at the load point, and ID represents the duration index of the average outage of the power supply system. In its main system, the monitoring master station and terminal for the average duration of power outage within the time range of one year constitute an intelligent condition monitoring system. Its terminal is installed on the distribution network site and is mainly responsible for the transmission of line operation status information. The monitoring master station can receive and upload terminal information and provide visual display information for the system. Therefore, it is also necessary to monitor the main station and terminal line. The principle of short circuit fault monitoring is shown in Fig. 4.

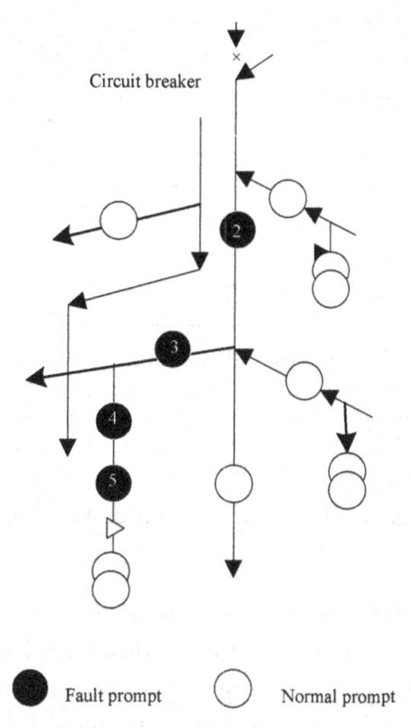

Fig. 4. Principle of short-circuit fault monitoring

The principle of short-circuit fault is shown in Fig. 4. There are three forms of fault judgment for phase-to-phase short-circuit faults in the distribution network, namely self-positioning, local-positioning and re-positioning fault judgment. When phase to phase short-circuit fault occurs in the monitoring system line, the intelligent distribution network monitoring system collects the fault waveform. Through this waveform, we can see the sudden change of short-circuit fault current, accurately monitor the fault location and the working state of electrical equipment (components), which can generally be divided into operation state and shutdown state. The running state is also known as the available state, that is, the element is in the state in which it can perform its specified function. Shutdown state is also called unavailable state, that is, it is unable to perform its specified function due to component failure. The whole life of an element is in the alternation of "operation" and "shutdown", which is a cycle. Both continuous working time t and continuous shutdown time t are random variable reliability (R), which refers to the probability of no failure in the time interval [0, n] under normal operation conditions at the starting time. During the operation of the transmission line, the reliability calculation can be expressed by the formula:

$$R(t) = \frac{T_U}{NT_S} = 1 - \frac{T_D}{NT_S} \tag{9}$$

In formula (9), T is the operation cycle and N is the annual average number of transmission lines. The reliability evaluation of distribution network under complex conditions is carried out on the basis of single load. Count the abnormal values monitored by the distribution grid within 1 min, and design the sliding window processing model, so as to reduce the memory occupied by big data in the sliding window. The design of the sliding window processing model is shown in Fig. 5.

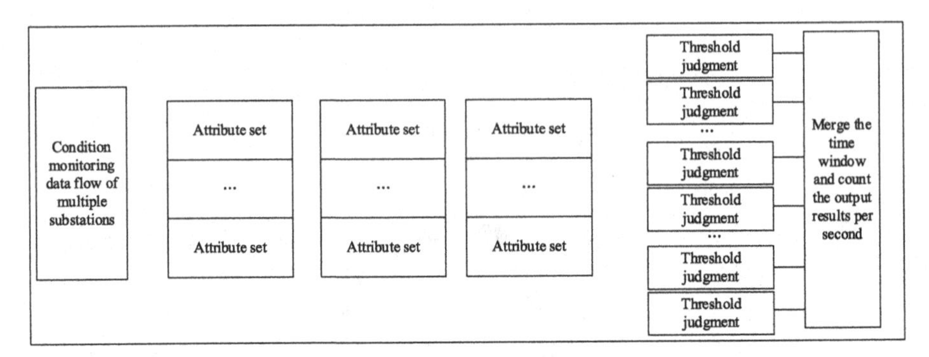

Fig. 5. Sliding window processing model

According to the processing model, the sliding window topology is task programmed and cluster processed, the window structure is designed by using storm programming model, the program is developed by using programming language, and the data flow is processed in batches within the specified time to ensure data continuity. The specific implementation process is shown in Fig. 6.

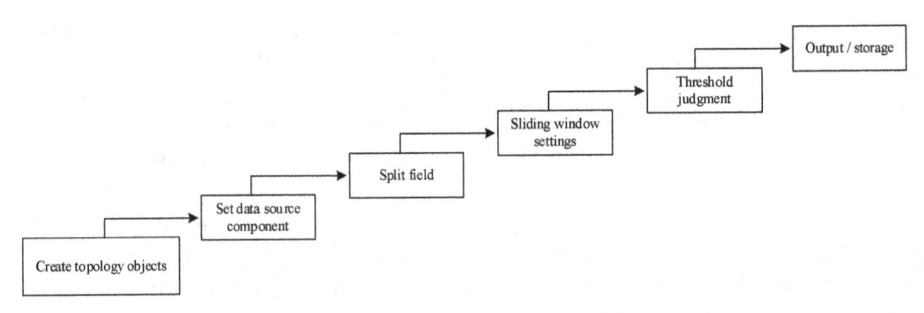

Fig. 6. Implementation process of sliding window topology

In the model, the data source component and logic processing component are topologies formed by data flow. According to the needs of distribution business, the distribution system is directly connected with users, so its voltage quality and power supply reliability are easy to change due to load changes. Especially when the difference between peak and valley of load curve is large, it will restrict the reliability evaluation under constant load. Therefore, in the process of reliability research of most distribution systems, when calculating the power consumption and outage cost of power customers, it is usually assumed that the load of power customers is constant, and then these reliability indexes are calculated through average load or peak load. However, the load of the power supply system is actually a function of changing time and is affected by factors such as the type of power customers and seasons. Therefore, the application of average load or peak load as parameters is bound to affect the accuracy of reliability calculation. If the variation range of load is large, the error interval of the index will also increase, which will affect the accuracy of reliability evaluation. Therefore, it is necessary to analyze the reliability of complex distribution network under load curve.

3 Analysis of Experimental Results

In order to fully verify the accuracy of the algorithm established in this paper, the model is used to verify the actual data of a municipal power supply bureau respectively, and to detect whether the algorithm has certain applicability and effectiveness for the reliability evaluation and prediction effect of the actual situation in this area. The statistics of the verification results of power supply reliability are shown in Table 3.

Table 3. Statistical results of power supply reliability

Power Supply Bureau No	Power supply reliability (%)	BP network prediction results			Paper method		
		Estimate	Absolute error	Relative error	Estimate	Absolute error	Relative error
A	99.1468	106.3256	7.1788	7.241%	96.6589	2.4879	2.509%
B	99.9868	93.6523	6.3345	6.335%	95.6582	4.3286	4.329%
C	99.1625	105.6528	6.4903	6.545%	95.6259	3.5366	3.566%
D	99.2685	93.6526	5.6159	5.657%	95.6523	3.6162	3.643%
E	99.6985	92.6523	7.0462	7.068%	**102.3256**	2.6271	2.635%
F	99.1685	**108.6528**	**9.4843**	**9.564%**	95.6523	3.5162	3.646%
G	**99.9951**	93.6528	6.3426	6.343%	94.9856	**5.0095**	**5.100%**
H	99.3656	92.3658	6.9998	7.044%	96.0568	3.3088	3.330%
I	99.9863	92.9875	6.9988	7.000%	95.6482	4.3381	4.339%
J	99.7156	90.4518	9.2638	9.290%	94.9649	4.7507	4.764%

The statistics of the verification results of the average outage time of users are shown in Table 4.

Table 4. Statistics of user average outage time verification results

Power Supply Bureau No	Power supply reliability (%)	BP network prediction results			Paper method		
		Estimate	Absolute error	Relative error	Estimate	Absolute error	Relative error
A	11.49	12.11	0.62	5.396%	11.98	0.49	4.426%
B	10.49	11.85	1.36	12.965%	9.86	**0.63**	6.001%
C	11.49	10.03	1.46	12.707%	11.87	0.38	3.307%
D	12.93	12.18	0.75	5.800%	13.42	0.49	3.790%
E	6.48	5.82	0.66	10.185%	6.89	0.41	6.327%
F	10.18	12.05	1.87	**18.369%**	10.65	0.47	4.617%
G	9.43	10.02	0.59	6.257%	9.59	0.16	1.697%
H	14.99	15.58	0.59	3.936%	15.52	0.53	3.536%
I	**15.25**	**17.03**	1.78	11.672%	**15.66**	0.41	2.689%
J	8.18	10.49	**2.31**	18.240%	8.85	0.67	**8.191%**

The comparison between Table 3 and Table 4 shows that the training error of this method is lower and the number of iterations is less. In order to further verify the fitting degree and efficiency of the two evaluation algorithms, this paper continues to use the determination coefficient R2 in the neural network algorithm for comparative analysis, which can analyze the fitting effect between independent variables and dependent variables. Generally, the calculation result of R2 is between 0 and 1, and the closer it is to 1, the better the fitting degree of the algorithm can be obtained. The statistical results of the determination coefficient R2 of BP algorithm and the method in this paper are shown in Table 5.

Table 5. Statistical results of determination coefficient R2

Decision coefficient	BP algorithm		Paper method	
	Training sample set	Test sample set	Training sample set	Test sample set
R2	0.763	0.866	0.905	0.958

The training time of monitoring data is shown in Table 6.

Table 6. Training time of monitoring data

	BP algorithm	Paper method
Training time (s)	426.28	199.68

When the working process remains unchanged, change the data source component and logic processing component to test the system throughput. The data source components are set in two cases. The first group is: keep the number of data source components unchanged, add logical components, and set them to (4,4), (4,5), (4,6); The second group is: the number of logical processing components remains unchanged, and the data source components are added, which are set to (4,4), (5,4), (6,4). The six groups of topologies are tested experimentally, and the system throughput under different conditions is counted. The results are shown in Table 7.

Table 7. System throughput of different components

Number of components	System throughput/(10000 pieces/s)		
	50000 pieces of data	150000 data	250000 data
(4,4)	1.02	1.07	1.13
(4,5)	1.05	1.11	1.10
(4,6)	**1.13**	**1.16**	**1.14**
(5,4)	0.89	0.91	1.01
(6,4)	0.83	0.86	0.87

It can be seen from Table 7 that the system throughput shows different characteristics by changing the number of concurrent logical processing components. The experimental group in the comparison table can fix the data source components. With the increase of logical components, the processing of condition monitoring data components can be accelerated, so as to improve the processing throughput of the system. Under the condition of fixed throughput, the monitoring efficiency of traditional system and automatic monitoring system is compared and analyzed, and the results are shown in Fig. 7.

According to Fig. 7, when the weight value and threshold parameters achieve the best effect, this method will further increase the training efficiency, reduce the number and degree of blind training, ensure the training efficiency, reduce the correlation between various variables to a certain extent, fully improve the effectiveness and adaptability of data, and ensure the reliability and stability in the training process. Therefore, as a new, reliable and stable prediction model, this method can be fully applied to the reliability evaluation of distribution network, ensure the safe and reliable operation of distribution network, and even provide an effective technical support for the safe and reliable development of distribution network in various cities across the country.

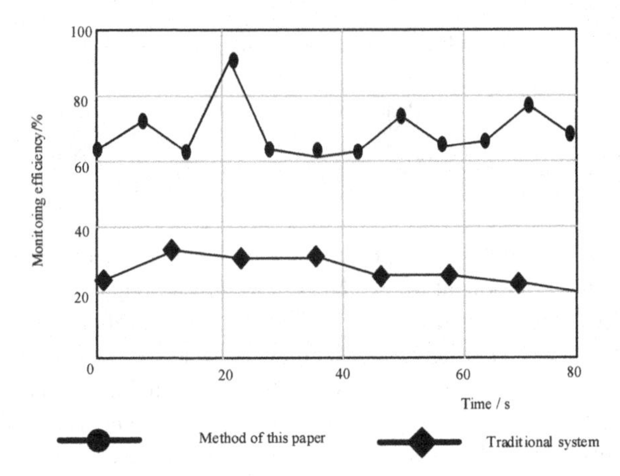

Fig. 7. Comparative analysis of monitoring efficiency of two systems

4 Conclusion

In order to improve the reliability of power supply network, the research on on-line monitoring of power supply reliability of distribution network based on mobile communication technology is proposed. Using mobile communication technology and reliability evaluation algorithm to identify fault information in power supply network. According to the processing model, the sliding window topology is subject to task programming and cluster processing to ensure data continuity. By expounding the relationship between reliability and cost, this paper puts forward the evaluation method of income increment and cost increment to evaluate the reliability improvement measures. The relationship between cost and reliability improvement is comprehensively considered, and expressed by drawing Pareto curve. Finally, some basic grid structures of a city power supply bureau are selected as research examples for reliability improvement and quantitative calculation and analysis. The historical reliability statistics and basic parameters of the city are selected, and the best reliability improvement scheme is obtained by using this method, which verifies the effectiveness of the algorithm.

References

1. Chen, Z., Wang, P., Tian, H., et al.: A study on power supply reliability monitoring of active distribution networks based on load characteristics. Power Syst. Clean Energy **38**(5), 65–70, 78 (2022)
2. Liu, X., Zang, J., Liu, C., et al.: Reliability research of distribution network with dg based on load superiority degree. J. Shanghai Univ. Eng. Sci. **35**(1), 67–74 (2021)
3. Lian, J., Fang, S., Zhou, Y.: Model predictive control of the fuel cell cathode system based on state quantity estimation. Comput. Simul. **37**(07), 119–122 (2020)
4. Wei, J., Xu, W., Liu, W., et al.: Design and implementation of low voltage power network monitoring system. Foreign Electron. Meas. Technol. **39**(01), 118–122 (2020)
5. Wang, W., Song, J., Wang, D., et al.: Intelligent fault diagnosis and disposal platform based on monitoring data of power system. Jilin Electric Power **48**(02), 31–35 (2020)

6. Xu, Y., Han, X., Yang, M., et al.: Decision-making model of condition-based maintenance for power grid with equipment on-line monitoring. Autom. Electr. Power Syst. **44**(23), 72–81 (2020)
7. Zhao, H., Feng, J., Ma, L.: Infrared image monitoring data compression of power distribution network via tensor tucker decomposition. Power Syst. Technol. **45**(04), 1632–1639 (2021)
8. Wang, X., Huang, X., Li, T.: Optimal design of winding of permanent magnet motor based on high frequency PCB stator. Proc. CSEE **41**(06), 1937–1946 (2021)
9. Zhang, H., Wang, H., Li, Q., et al.: Power supply scheme of traction cable co-phase connected power supply system for heavy-haul railway powered by AT. Electr. Power Autom. Equip. **41**(01), 204–213 (2021)

Multi Node Water Quality Monitoring System of Fish Pond Based on Unmanned Ship Technology

Shaoyong Cao(✉), Xinyan Yin, Huanqing Han, and Dongqin Li

School of Industrial Automation, Zhuhai College of Beijing Institute of Technology, Zhuhai 519085, China
suinkln45222@126.com

Abstract. In order to better improve the multi node water quality of fish pond, a design method of multi node water quality monitoring system of fish pond based on unmanned ship technology is proposed, the hardware configuration of the system is optimized, the operation process of system software is simplified, and the multi node water quality monitoring algorithm of fish pond is constructed, so as to realize the effective management of UAV ship and the real-time monitoring goal of water quality state of fish pond. Finally, it is confirmed by experiments, The multi node water quality monitoring system of fish pond based on unmanned ship technology has high practicability in the process of practical application.

Keywords: Unmanned ship technology · Fish pond · Multi node water quality monitoring · Monitoring system

1 Introduction

The water quality of pond water is very important for aquaculture. The water quality not only affects the growth rate of aquatic products, but also affects the quality of aquatic products [1]. Pond farmers' management of fish ponds is mostly based on experience and on-site observation. The renewal of water bodies is blind, either causing waste or damage during intervention. If the development trend of water quality can be predicted, the breeding environment can be intervened in time before water quality deterioration to prevent waste or loss caused by water quality deterioration. Therefore, it is necessary to monitor the water quality of breeding pond [2]. Whether the water body is suitable for fish growth has a great relationship with the substances contained in the water, among which the dissolved oxygen content, water pH, temperature and ammonia nitrogen content have the greatest impact on fish growth. Reference [3] proposes an intuitive remote monitoring framework for fish pond water quality based on cloud computing. In aquaculture, due to the openness of the surrounding environment, the changes of water quality parameters are nonlinear, dynamic, unstable and complex. The multi-sensor and private cloud integration framework is used to collect data in real time and enhance remote

W. Fu and L. Yun (Eds.): ADHIP 2022, LNICST 468, pp. 744–759, 2023.
https://doi.org/10.1007/978-3-031-28787-9_55

monitoring capabilities. Reference [4] proposes to detect abnormal fish behavior using motion trajectories in a ubiquitous environment. First, an image enhancement algorithm is used to enhance the color of water images and enhance fish detection. Then a target detection algorithm is used to identify fish. Finally, the classification algorithm is used to detect the abnormal behavior of fish. Establish an automated system to monitor the fish farm, so as to reduce the cost and time of fish farmers, and provide a more efficient and simple operation mode. At present, most fish pond farmers breed based on personal experience. They judge whether the water environment of fish pond is conducive to the survival of fish through the phenomena of water temperature, water color and floating head of fish. This kind of judgment based on experience is not accurate or intelligent. There are also a few fish pond farmers who introduce monitoring systems, but their systems generally adopt 3G4G mobile communication network or ZigBee network communication technology. In this way, the monitoring system has problems such as inflexible expansion of detection nodes and large long-distance transmission interference, and the cost is also very high when the number of detection sensors is large. In view of the above situation, this paper presents a design of fish pond water quality monitoring system based on unmanned ship technology.

2 Multi Node Water Quality Monitoring System of Fish Pond

2.1 System Hardware Structure Design

It is designed according to the project requirements of the fish pond monitoring system, involving the sensor collection technology of dissolved oxygen, pH value, conductivity and water temperature, the oxygen enrichment device and the fish pond area intrusion detection device. The system uses the corresponding sensors to detect these data of the water quality of the fish pond, and automatically starts the oxygen increasing device remotely when the dissolved oxygen is insufficient; Intrusion detection is carried out in the fish pond area by using infrared reflection sensors and video monitoring devices. In case of fish theft at night, the system can automatically start the alarm device. Each detection device in the fish pond contains relevant sensor modules and unmanned ship modules. The data collected by all detection devices in the fish pond is forwarded to the unmanned ship gateway through the unmanned ship node. The unmanned ship gateway uploads the data to the system's cloud platform in accordance with MQT protocol. The general processing program of the cloud platform part of the system provides relevant interfaces for PC clients and Android clients to call. The cloud platform of the system includes an MQT server, which establishes an MQT connection with each gateway device. The gateway device uploads data to the cloud platform in real time through the corresponding MQT connection. The cloud platform can also issue some control instructions to the oxygen enrichment device, alarm device Gate equipment, water pump and other equipment [5]. The lightweight data interaction technology of JSON data format is adopted between the system cloud platform and the PC client. The PC client of the system includes functions such as fish pond information overview, data analysis, fish pond monitoring and intelligent gate. In the fish pond overview, users can see various water quality data of the fish pond environment; The data analysis part can dynamically display the oxygen content and pH value of the collected fish pond through a broken line diagram [6]; The fish pond monitoring part can help you understand the current fish pond environment in real time; There is also an intelligent gate function, which can be set to automatically open the gate and manually open the gate. When the water depth exceeds 3M, the gate will automatically open to discharge water, so that the water depth in the fish pond is always at a suitable position [7]. The system can be operated in the cloud in terms of data acquisition, remote control, unmanned ship, data analysis, early warning information release, decision support and so on, making the fish pond supervision easy and simple. The whole water quality monitoring system is composed of upper computer control terminal, coordinator, routing node, monitoring node and so on. The ZigBee module that can communicate with each other and automatically network is used for data transmission of monitoring nodes. The monitoring node is both a terminal node and a routing node. The network structure of unmanned ship equipment is shown in Fig. 1:

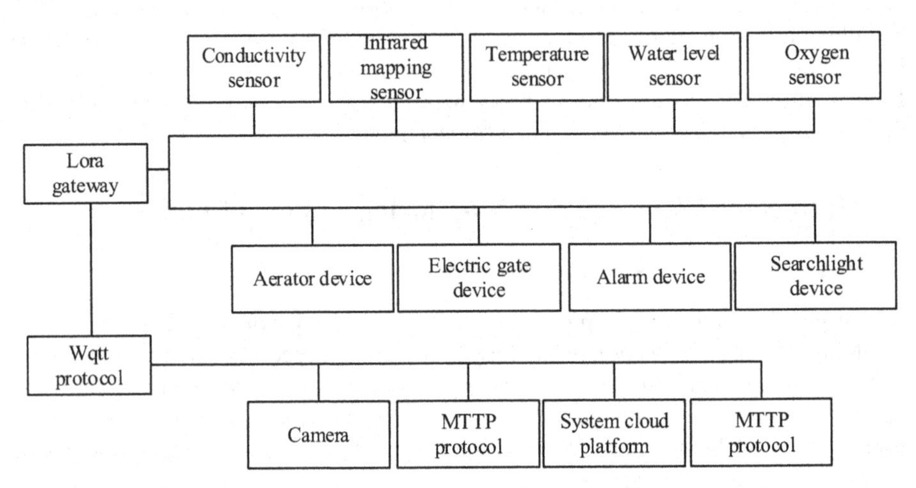

Fig. 1. Structure of unmanned ship equipment monitoring network

The monitoring terminal takes the personal computer as the hardware platform, provides a friendly interactive interface, and serves as the center of data centralization, analysis and processing of the system. The system monitoring terminal software receives and processes the data of wireless nodes, and can also start the response mechanism to deal with abnormal water quality. The monitoring software is VB6 The man-machine interactive control interface written by 0 can complete the functions of data acquisition, storage, management and analysis and historical data query. The coordinator is connected with the serial port of the monitoring terminal through RS-485 bus to 232 to monitor the water quality of the pond in real time [8]. In order to collect and upload the data of dissolved oxygen, pH value, water temperature and conductivity of fish pond water quality, the system needs to be equipped with nodes and gateways. Dora node first forwards the data to the gateway through the network, and then the gateway uploads the data to the system cloud platform through TCPP network. Each node collects values every 2 min or so and sends data to the gateway according to the following data frame format. The communication baud rate is 115200 (network data sending and receiving has been realized in the project) (Fig. 2).

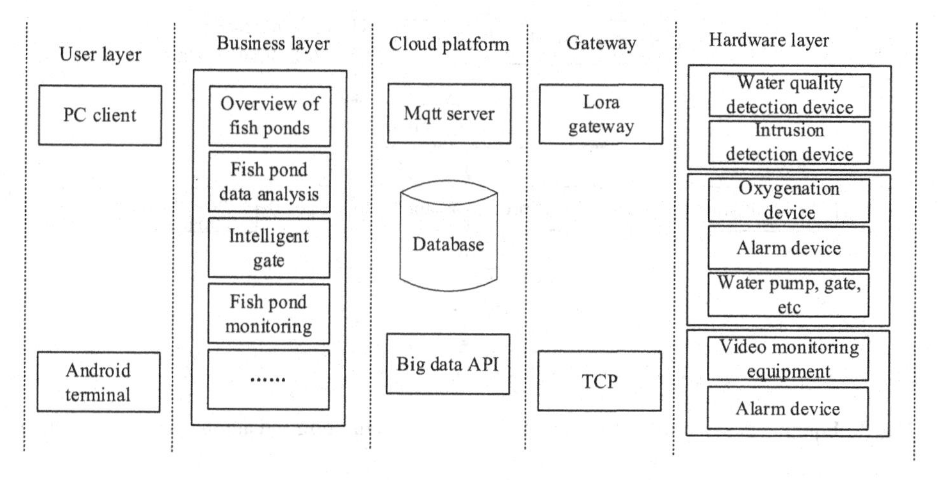

Fig. 2. System structure framework optimization

The main interface of the fishpond environment monitoring client of the system is shown in the figure. In this interface, you can overview the real-time environment of the fishpond. Click the "intelligent gate" navigation button in the main interface to enter the water level control setting of the fish pond gate. The fish pond intelligent gate interface can remotely control the fish pond gate through the "open gate" and "close gate" buttons in the fish pond gate, or select automatic control after setting the fish pond water level value. The system will automatically open the gate and discharge water when it is higher than the set value according to the comparison between the water level value collected by the fish pond water level detection device and the set value [9]. The "data analysis" function of the system is mainly to periodically display the data collected by the water quality monitoring sensors deployed in each area of the fish pond. In order to reduce the network flow of the system platform and ensure the dormancy of the acquisition terminal as much as possible, the system sets the acquisition cycle as 10 min, gives an alarm pop-up when the data is not within the set normal range, and pushes a reminder message to the system administrator. According to the different functions of each module, the intelligent monitoring and automatic oxygenation system is divided into: power supply module, data processing module, voice alarm module, wireless communication module and water quality identification relay module [10]. As an energy supply module, the power module plays a decisive role in the whole system and determines the normal working time of the monitoring system. The working time of the module will affect the promotion and use of the monitoring system. The system selects lithium battery as the power supply, and the output voltage is 5 V through the voltage transformation module to provide energy for the system (Fig. 3).

The fish pond information acquisition module is composed of dissolved oxygen temperature integrated sensor AD59O, pH sensor HH5-WQ201, methane gas sensor, multiplexer 74HC4051 and AD converter MA $\times$ 1240. The dissolved oxygen temperature integrated sensor AD590 can be widely used to measure the dissolved oxygen content in various occasions, especially in aquaculture water, photosynthesis, respiration and

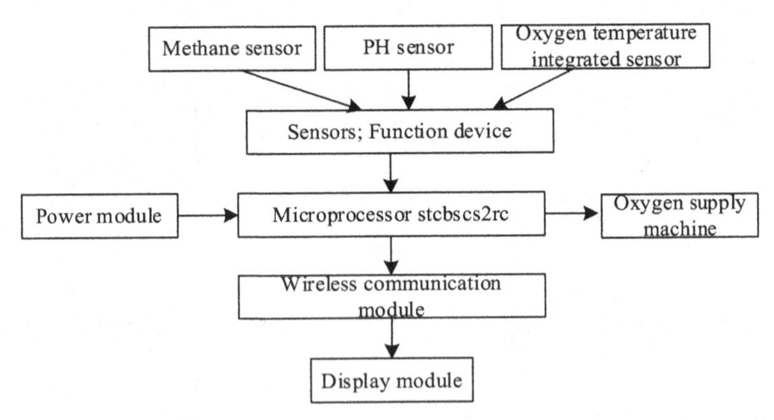

Fig. 3. Equipment structure of fish pond water quality monitoring and automatic control system

field measurement; Methane gas sensor is a general intelligent sensor. It is a high-precision methane sensor designed based on the principle that the appropriate micro power laser can be absorbed by different concentrations of methane. The sensor has the characteristics of small volume, low power consumption, stable performance, long service life of sensor elements and high sensitivity. It continuously detects the methane concentration in the air and transmits relevant data in real time. At the same time, the sensor also has the function of internal temperature and air pressure detection; The amplifier of pH sensor hhvq201 inside the sensor is a circuit with standard pH through data monitoring, and a plug is connected with the data collector at the end of the sensor connecting line. Specific voltage will be generated in neutral solution, and the voltage will increase by 0.25 for each increase of pH value; When the pH value decreases by 1, the voltage decreases by 0.25 (Fig. 4).

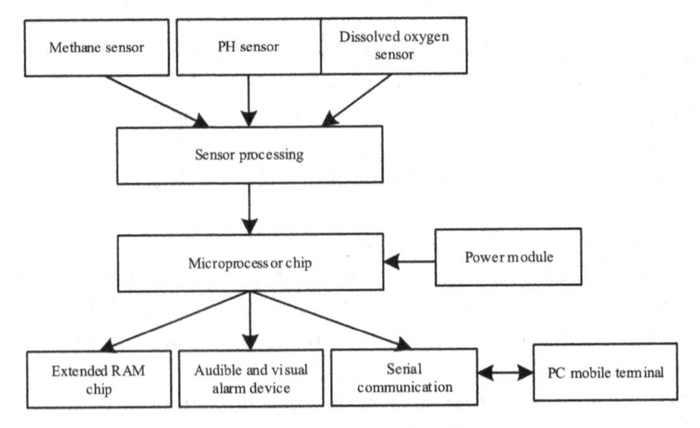

Fig. 4. System hardware configuration

The equipment control terminal controls the on-off of the relay by sending RS485 communication command, so as to control the operation of on-site execution equipment (water quality identification and circulating water pump). The system selects SR-311A485 control relay. The PCB of this type of control relay is made of FR-4 plate. It uses industrial chip and has lightning protection circuit. Its module address can be configured through RS485 communication, and the configuration set by jumper lock module can be used to prevent misoperation. The product has the advantages of power saving, energy saving, long service life and stable and reliable performance (Table 1).

Table 1. Equipment parameters of unmanned ship monitoring system

Technical attributes	Specific parameters
Electric shock capacity	10 A/30 V DC
Temperature range	−35 °C–+85 °C
Durability	110000 times
Rated voltage	DC10 ~ 25 V
Communication interface	RS498
Baud rate	9700 (default)
Default communication format	9700,

When designing the hardware circuit of the monitoring system, the modular design principle is adopted. Each module of the system hardware circuit is designed independently, and each module is connected by interface. This design method not only facilitates the hardware circuit to quickly find and solve problems in the debugging process, and improves the efficiency of hardware development, but also brings great convenience to the later maintenance and upgrading of the system.

2.2 System Software Function Optimization

The design purpose of the intelligent monitoring system for fishery water quality is to dynamically monitor and regulate the water environment in the process of fishery breeding. Through the analysis of the overall functional requirements of the monitoring system, its client platform should have the following six functions. The functional structure diagram of the unmanned ship client of the intelligent monitoring system for fishery water quality is shown in Fig. 5.

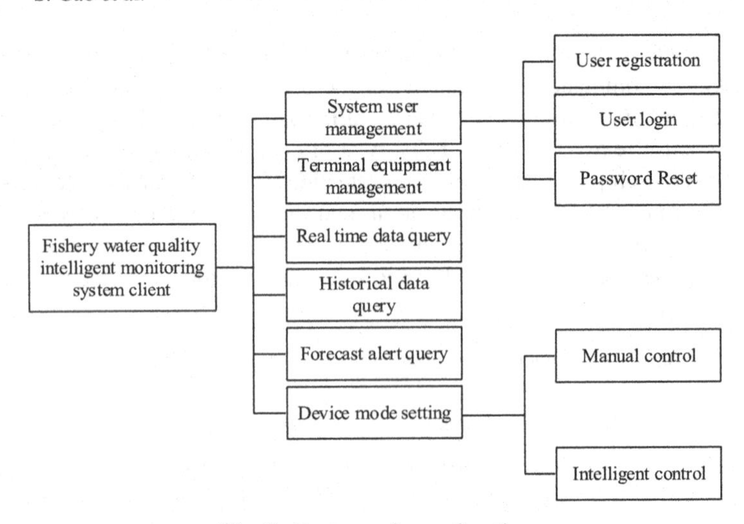

Fig. 5. System software function

The ZigBee module that can communicate with each other and automatically network is used for data transmission of monitoring nodes. The monitoring node is both a terminal node and a routing node. The monitoring terminal software takes the personal computer as the hardware platform, provides a friendly interactive interface, and serves as the center of data centralization, analysis and processing of the system. The system monitoring terminal software receives and processes the data of wireless nodes, and can also start the response mechanism to deal with abnormal water quality. The monitoring software is VB6 The man-machine interactive control interface written by 0 can complete the functions of data acquisition, storage, management and analysis and historical data query. The coordinator is connected with the serial port of the monitoring terminal through RS-485 bus to 232 to monitor the water quality of the pond in real time.Using wireless communication GSM requires short message protocol. The short message from the application layer sends the short message to the MSc through the transmission layer, relay layer, connection management sublayer cm and the following mobile management sublayer mm and wireless management sublayer RRM. It is divided into the following points: (1) Short message transmission layer (SMTL) service; (2) Information relay layer (Smrl) service; (3) Information (SML control management, mobile management, radio resource management sublayer) services. Denoising the collected aquaculture water quality is a crucial step in the process of prediction and early warning. Its working principle is to use signal processing to remove the interference factors hidden in the water quality data, avoid data deviation and ensure the accuracy of subsequent calculation intelligent prediction. When using the unmanned ship client, the breeding user needs to register as the system user first, and log in to the client according to the registered user name and password. The user can bind the field terminal equipment independently according to the demand, and the system will upload the user information and the bound equipment information to the database. After the binding operation is completed, the user can log in to the control interface for relevant monitoring operations. The process flow of unmanned ship monitoring operation is shown in Fig. 6:

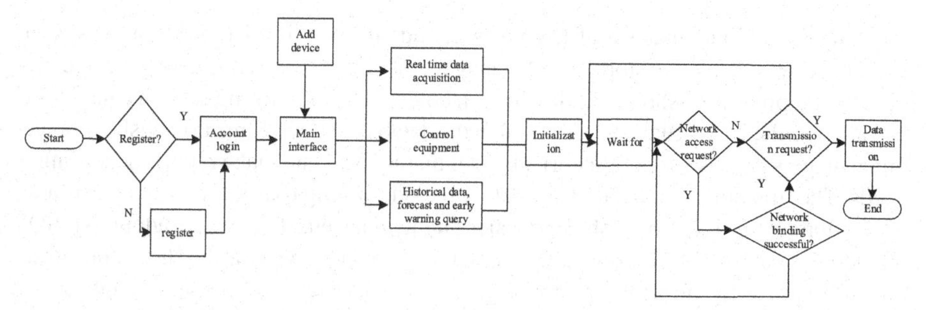

Fig. 6. Processing flow of unmanned ship monitoring operation

The monitoring system monitors the water environment parameters of the fish pond, opens or closes the on-site execution equipment according to the changes of the fish pond water environment, adjusts the water quality, and ensures the safety and stability of the fish pond water environment. The design of embedded control module is mainly aimed at the implementation of field control terminal. The control terminal is responsible for controlling the execution equipment on site, including water quality identification, circulating water pump, etc. The microprocessor of the control terminal connects the relay through the RS485 interface, and controls the on or off of the relay by sending the RS485 control signal, so as to realize the on or off of the on-site execution equipment. The automatic control flow chart of on-site water quality identification is shown in the figure. The control terminal adopts threshold control. The operation of water quality identification is automatically controlled according to the preset dissolved oxygen parameter standard, that is, the upper and lower limits of dissolved oxygen. When the dissolved oxygen concentration in the fish pond is less than the lower limit, the water quality identification is automatically turned on, and when it is greater than the upper limit, the water quality identification is automatically turned off. The automatic control flow of water quality identification is shown in Fig. 7:

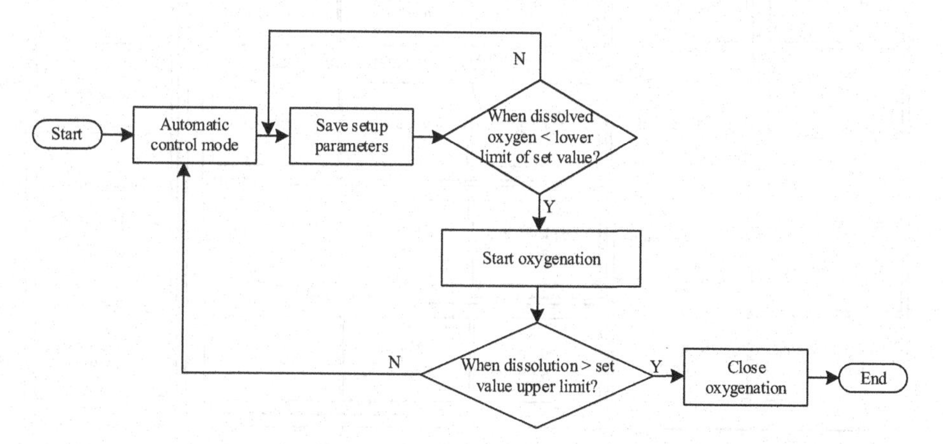

Fig. 7. Automatic control flow chart of water quality identification

Establishing the database of fishery water quality intelligent monitoring system is conducive to the effective storage and management of fishery water quality data and provide data support for fishery breeding management. The monitoring system can process and analyze the water quality data stored in the database. At the same time, fishery breeding managers can also export historical water quality data and make reports according to needs. The functions of the database mainly include storing user information, user binding equipment information, fish pond water quality data, etc. The system adopts MySQL relational database. According to the actual needs of the system, the data table structure is designed for information storage, mainly including user information table, user binding equipment information table, fish pond water quality data table and relationship correspondence table.

2.3 Realization of Multi Node Water Quality Monitoring in Fishpond

For the data interface of server, it mainly provides API for embedded terminal equipment and client. In the system, the data interface of the server is realized by post request. It responds to the data upload of the on-site water quality collection terminal and the data query of the unmanned ship client in real time. As soon as the cloud server receives the request information, it processes it immediately and returns the request result to the client or the on-site terminal device. The operation object of the data interface is MYSQL in the server, which relies on the database to respond to various request information. The persistence layer architecture selected in the data interface design is mybatis, and its specific structure is shown in Fig. 8.

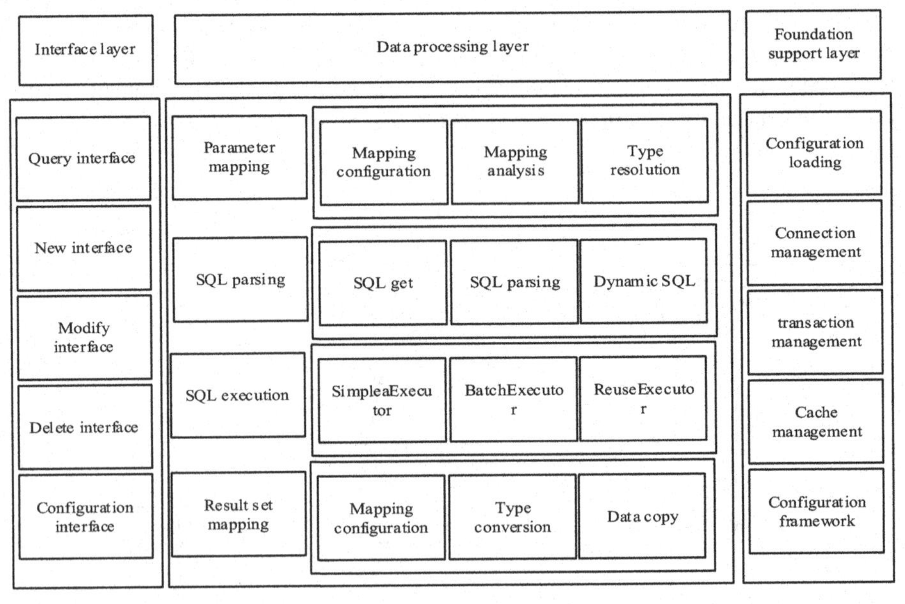

Fig. 8. Mybatis water quality monitoring application model

The application steps of mybatis are as follows: first write the configuration file of mybatis, create the database connection session, then configure the relationship mapping, complete the operation on the database, and finally end the session. Aiming at the problem of low accuracy in the prediction of fishery water quality parameters by support vector regression machine, this paper proposes and adopts the eemd-gwo-svr combined model prediction method based on set empirical mode decomposition method, gray wolf optimization algorithm and support vector regression machine. Firstly, EEMD is used to decompose the collected water quality data into a series of relatively stable characteristic modulus function components, and then the mixed noise sequence is eliminated. Then GWO optimization algorithm is used to optimize the C (penalty factor) and γ (kernel function parameters) parameters for Intelligent Optimization: use the optimized SVR to establish a prediction model for the fishery water quality data after noise reduction. According to the water quality environment of on-site fishery breeding in Jiangxinzhou, combined with the case analysis of breeding samples and relevant literature reference, the water quality early warning standard is constructed, and the fishery water quality early warning model is established based on the prediction of fishery breeding water quality data. The logic flow chart of fishery water quality prediction and early warning model based on unmanned ship is shown in the figure. Through dynamic prediction and early warning of environmental parameters of fish pond in Jiang Xin Zhou, the water quality in advance is adjusted according to the dynamic forecast and early warning information, and the water quality of aquaculture is refined.

The early warning model of fishery water quality is based on the data of water temperature, dissolved oxygen DO, pH, ammonia nitrogen and other parameters predicted by the water quality prediction model, so as to ensure the stability of aquaculture water quality in a suitable environment. By mastering the suitable water quality environment of cultured fish, and determining the alarm situation and dividing the alarm level according to the specific range of water quality parameters, we can carry out fishery water quality early warning. According to the actual needs of application, the alarm degree of fishery water quality early warning model is divided into four levels, and the emergency degree is no alarm, light alarm, medium alarm and heavy alarm from low to high. The specific description is shown in Table 2:

Table 2. Alarm level of water quality early warning model

Alarm level	Describe
No police	The water quality of the fish pond has not been damaged and is very suitable for fishery and aquaculture activities
Light police	Water quality parameters reach the appropriate upper or lower limit and last for a certain period of time, but are not enough to affect the normal survival of water organisms, which should be paid attention to
Central Police	The water quality parameters exceed the appropriate value of fishery breeding, and corresponding treatment measures need to be taken
Heavy police	The water quality parameters exceed the limit value of fishery breeding, and corresponding treatment measures need to be taken immediately

The water environment of fishery breeding is a comprehensive reflection of the changes of various water quality parameters. The quality of fishery breeding water is directly related to the output of fishery breeding. Therefore, it is necessary to monitor and analyze the important parameters in the fishery aquaculture water environment, and formulate fishery water quality standards in combination with examples, so as to provide reference for the early warning control of the monitoring system and ensure the safety of fishery aquaculture water quality. According to the fishery water quality standards, surface water environmental quality standards and other relevant regulations, and in strict compliance with the principles of scientificity, economy and timeliness, this system constructs the fishery water quality early warning standards in combination with the case analysis of fish pond culture samples in Jiangxinzhou and relevant literature references, and uses the key water quality parameters such as water temperature, dissolved oxygen, pH and ammonia nitrogen of fishery culture as the basis for judging the water quality alarm level, The standards of water environment parameters are shown in Table 3:

Table 3. Standard table of water environment parameters

Police level	Grade	Water temperature/°C	Dissolved oxygen (do)	pH value	Ammonia nitrogen
No police	Good	21–26	6–9	6.6–7.7	0–0.16
Light police	Middle	16–21 26–29	4–6 9–13	6.1–6.6 8.7–9	0.15–0.3
Central Police	Difference	5–16 29–31	3–5 12–16	4–7 9–11	0.2–0.4
Heavy police	Very bad	−11–5 31–51	0–4 16–21	0–5 11–15	0.4–10

In the system data manipulation interface, users can select real-time monitoring, remote control, system setting, historical data, data analysis, system management and other data manipulation functions according to their own manipulation needs. Among them, the real-time monitoring function is to monitor the water temperature information, pH value information and dissolved oxygen information on the water quality monitoring site, and the remote control function is to issue the water quality identification control command and conduct the oxygenation operation according to the water quality information monitored by the water quality on-site collection terminal; The system setting function is to safely set the upper and lower limits of various parameters of water quality, so as to create a safe growth environment for aquaculture organisms. The historical data function is to view the water quality monitoring information at a specified time by calling the system database according to the user's needs: the data analysis function is to display the collected data information in the form of a graph in front of the user to help the user classify the changes of water quality; The system management function is used to manage the user's personal account and number.

3 Analysis of Experimental Results

Taking a pond as the object to test the system, the node in the system automatically transmits the collected data to the upper computer every hour, and the upper computer stores and analyzes the received data. The node selection button can display the data at this node. In order to verify the stability and reliability of remote data and information transmission of the system, the GPRS DTU unmanned ship module is tested in this paper. Firstly, build a test platform and connect it according to the connection scheme between the unmanned ship module and the on-site acquisition terminal and the server of the monitoring center. Secondly, write the program, generate HEX file, and use the burning software to burn the program into the single chip microcomputer through the USB port. Thirdly, start the SMS card GPRS traffic service and install it in the SIM card slot to connect the line. Finally, turn on the system hardware equipment and server software, collect data by using temperature sensor, pH sensor and dissolved oxygen sensor, transmit data information through GPRS DTU module, and check whether the data received by the server of the monitoring center is different from the data at the sending end. The construction of the experimental network environment is a typical star network. The wireless access point is located at the core of the star network. The data communication between all stations in the network can only be realized by forwarding data packets through the wireless access point in the center. It can be seen that the operation security of wireless WiFi AP is very important in this network topology, and general wireless routers can meet the needs of experimental testing. The specific flow of the experiment is shown in Fig. 9.

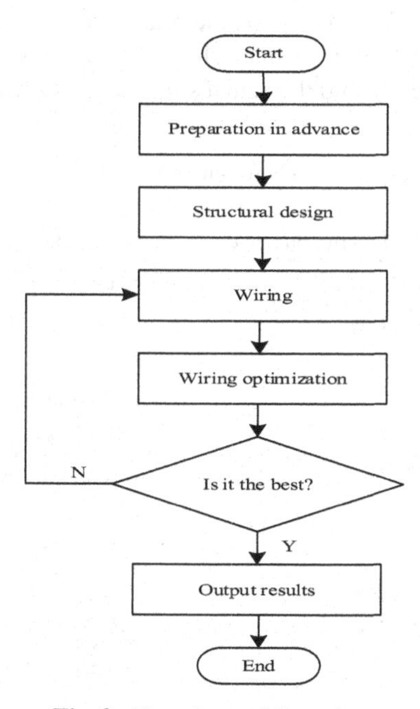

Fig. 9. Experimental flow chart

Table 4 shows the test sample data:

Table 4. Test sample data

Test group No	Test value	Field data	Server data	Accuracy (%)
A	Water temperature/°C	11.9	11.9	100
	pH value	7.6	7.6	100
	Dissolved oxygen (mg/L)	8.5	8.5	100
B	Water temperature/°C	18.7	18.7	100
	pH value	7.6	7.6	100
	Dissolved oxygen (mg/L)	8.1	8.1	100
C	Water temperature/°C	23.8	23.8	100
	pH value	7.5	7.5	100
	Dissolved oxygen (mg/L)	7.5	7.5	100

NB-lot signal quality directly reflects the communication situation of unmanned ship. By understanding the signal quality of on-site NB-lot, we can give better play to the advantages of unmanned ship monitoring communication. By using the "AT+CSQ" instruction on the NB-lot chip SIM7020C, the signal quality of its NB lot can be obtained. The instruction can detect the strength index and channel bit error rate of the received signal. Use the serial port tool to send the "AT+CSQ" command to SIM7020C, and the chip replies "+CSQ:30,0". Where "30" is the return value of signal strength and "0" is the return value of channel bit error rate. The corresponding relationship between the return value of the instruction and the actual situation is shown in Table 5:

Table 5. Correspondence between return value and actual situation

Signal quality	Return value (integer)	Actual value
Signal strength (RSSI)	0	Rssi < = −111 dBm
	1	−109 dBm < = rssi < −106 dBm
	2	−106 dBm < -rssi < −104 dBm
	3–31	−104 dBm < -rssi < −48 dBm
	32	−48dBm <= rssi
	99	Unknown or undetectable
Channel bit error rate (BER)	0	ber < 0.2%
	1	0.2% <= ber < 0.5%
	2–4	0.5% <= ber < 1.7%
	5–6	1.7% <= ber < 6.4%
	7–8	Basically unable to communicate normally

By comparing the test results of various test methods under the same test conditions and conditions, it is found that pH is a main reference index for monitoring the health status of water bodies. The test results of pH value of water quality are shown in Fig. 10:

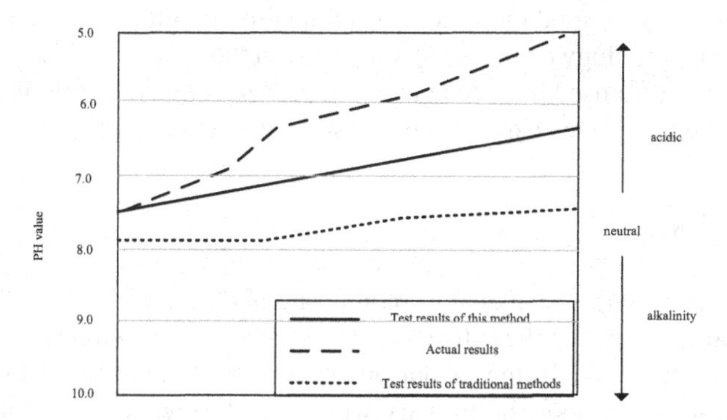

Fig. 10. Test results of pH value of water quality

Through the comparison of the above tests in Fig. 9, it can be seen that the pH value of the water quality pH test conducted by the rotating grid method described in this paper is 6.5–4.5, which belongs to weak acid. Compared with the measured pH value, the conventional detection method is 7.0–6.5, which belongs to neutral. There is a great difference between the model and the measured value, which verifies the accuracy of the algorithm in the actual test. Because the samples in the same sea area contain different impurities and pollution components, the impact on the water body is also different (mentioned earlier), so there may be some subtle differences during the test, which can be ignored here. By comparing with pH value, the acquisition accuracy of data is compared, and the results are shown in Fig. 11:

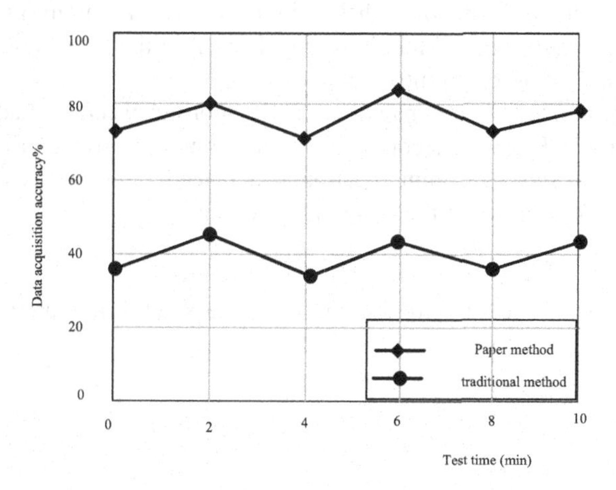

Fig. 11. Comparison of test results

As can be seen from the above Fig. 11, compared with the conventional water sample test, the accuracy of the water sample test data used has been greatly improved compared with the conventional method. The test results show that this method is feasible and can solve the practical problems.

To sum up, the fish pond multi node water quality monitoring system based on the unmanned ship technology can verify the accuracy in the actual test. The accuracy of the water sample test data used has been greatly improved compared with the conventional method. The method in this paper can better solve the actual problem.

4 Conclusion

The designed fish pond water quality monitoring and automatic oxygenation system arranges sensors, power supplies, displays, data processors and wireless communication modules beside the pool. In the layout process, a small container is used to replace the environment of the pool, an ordinary wire is used to replace the waterproof wire, and a lithium battery is used to replace the large energy storage battery, ignoring the small factors of the system. The test results show that the water quality monitoring and automatic oxygenation system of the fish pond can accurately monitor the water quality of the pond and achieve the expected results. At the same time, the monitoring personnel can timely understand the water quality environment of the fish pond through the display. The test shows that the system is stable and has a good market scene.

Although the above research has made some progress, due to the short time, further research can be carried out in the following aspects:

(1) In the future, we can check the dissolved oxygen concentration and temperature of the fish pond water body, pay attention to other impact factors that need to be monitored, and increase the diversity of the monitoring system. Therefore, we need to gradually increase the monitoring capacity of the system in the later stage.
(2) Considering the cost in depth, with the addition of pH value monitoring, temperature monitoring, video security monitoring and so on in the later system, it is easy to increase the cost of the monitoring system.
(3) In terms of wireless communication, the monitoring system can adopt the technically mature WiFi wireless communication technology, giving priority to external interference to wireless communication, which needs to be studied in the later stage to increase the security of the monitoring system.

Fund Project. The Special projects in Key Areas of Guangdong Province(Grant 2021ZDZX4050).

References

1. Rand, J.M., Nanko, M.O., Lykkegaard, M.B., et al.: The human factor: weather bias in manual lake water quality monitoring. Limnol. Oceanogr. Meth. **20**(5), 288–303 (2022)
2. Qiu, Q., Dai, L., Rijswick, H., et al.: Improving the water quality monitoring system in the Yangtze River basin—legal suggestions to the implementation of the Yangtze river protection law. Laws **10**(25), 1–13 (2021)
3. Sivakumar, S., Ramya, V.: An intuitive remote monitoring framework for water quality in fish pond using cloud computing. IOP Conf. Ser. Mater. Sci. Eng. **1085**(1), 012037 (2021)
4. Anas, O., Wageeh, Y., Mohamed, E.D., et al.: Detecting abnormal fish behavior using motion trajectories in ubiquitous environments. Procedia Comput. Sci. **175**, 141–148 (2020)
5. Jing, L., Siyu, F., Yafu, Z.: Model predictive control of the fuel cell cathode system based on state quantity estimation. Comput. Simul. **37**(07), 119–122 (2020)
6. Huang, Y., Wang, X., Xiang, W., et al.: Forward-looking roadmaps for long-term continuous water quality monitoring: bottlenecks, innovations, and prospects in a critical review. Environ. Sci. Technol. ES&T **56**(9), 5334–5354 (2022)
7. Tsafack, J., Tchuenchieu, D., Mouafo, H.T., et al.: Microbial assessment and antibiotic susceptibility profile of bacterial fish isolates in an aquaculture production site in Mefou Afamba division of Cameroon. Environ. Sci. Eng. B **10**(1), 20–30 (2021)
8. Jia, J., Qiu, S., Bai, R., et al.: Design and application of intelligent river water quality monitoring and early warning platform based on soa. Comput. Appl. Softw. **38**(02), 13–18+26 (2021)
9. Kuang, L., Shi, P., Ji, Y., et al.: Data fusion method for water quality monitoring using WSN based on improved support function. Trans. Chin. Soc. Agric. Eng.**36**(16), 192–200 (2020).
10. Zhang, J., Sheng, Y., Chen, W., et al.: Design and analysis of a water quality monitoring data service platform. Comput. Mater. Continuum **1**, 389–405 (2021)

Evaluation Model of Teaching Quality of College English Integrated into Ideological and Political Course Under Social Network

Yang Gui[1](✉) and Jiang Jiang[2]

[1] Anhui Province Anhui International Studies University, Hefei 231201, China
guiyang0510@163.com

[2] Digital Department of Guangdong Power Grid Corporation, Guangzhou 510800, China

Abstract. In the process of application, the teaching quality evaluation model of college English integrated ideological and political courses has the problem of low accuracy. Therefore, under the social network, a teaching quality evaluation model of college English integrated ideological and political courses is designed. The consistency characteristics of college English and ideological and political courses are extracted, the teaching quality evaluation index is selected, the role of teachers in a certain learning stage is quantified, and the teaching quality evaluation model is constructed by using social networks. The experimental results show that the evaluation accuracy of the proposed model is higher than that of the other two quality evaluation models.

Keywords: Social network · College English · Ideological and political courses · Teaching quality evaluation · Teaching level · Teaching links

1 Introduction

The fundamental task of colleges and universities is to cultivate talents to meet the needs of the society, and teaching is the most important means of talent training. In order to better cultivate talents and meet the needs of the increasingly competitive talent market, we must strengthen the management of teaching quality in colleges and universities. The evaluation of teaching quality is an indispensable and effective means in the process of teaching quality management, and is the key to improve teaching quality, The quality of teaching evaluation directly affects the learning efficiency of students and directly reflects the teaching level of teachers. Therefore, how to evaluate teachers' teaching quality objectively and reasonably so as to promote the improvement of teaching quality is one of the important contents of university management. The evaluation of teaching quality is of great significance for improving teaching quality and promoting the standardization and scientization of university management [1]. School leaders and administrators need to adjust teaching quality objectives and make decisions in time based on objective information. Find out the deficiencies in the teaching process, teaching quality evaluation

W. Fu and L. Yun (Eds.): ADHIP 2022, LNICST 468, pp. 760–770, 2023.
https://doi.org/10.1007/978-3-031-28787-9_56

is a very strict implementation procedure, which can judge the objectivity of information comprehensively and scientifically, so as to help timely adjust the teaching objectives.

The objectives, methods and standards of teaching quality evaluation point out the direction of specific goals for teachers, and can stimulate teachers to actively invest in education reform to the maximum extent [2]. Teachers can exchange teaching experience with each other through the evaluation process, learn the strengths of others and make up for their own shortcomings [3]. The information obtained through the feedback of teaching quality evaluation can help teachers find what they ignore in the teaching process and find their shortcomings. In addition, the evaluation results can also make teachers themselves recognized to a certain extent, and their teaching achievements are affirmed by everyone, which plays an incentive role for teachers.

Most universities only do simple average processing for the feedback information of evaluation, the consideration of problems is not comprehensive, so that the evaluation results deviate from the real value. If there may be some irresponsible evaluation in the evaluation of teaching, which will affect the final evaluation results, managers should try to avoid such mistakes in the statistical results. After the evaluation, many schools only tell the teachers the final results of the evaluation, and the specific evaluation results are not timely fed back to the teachers. The teachers cannot understand the shortcomings in their teaching, nor can they improve the teaching level. Therefore, it is necessary to carry out in-depth research on the teaching quality evaluation model of college English integrated into ideological and political courses.

2 Construction of Quality Evaluation Model

2.1 Extract the Consistency Characteristics of English and Ideological and Political Courses

The humanistic and ideological nature of College English provides a necessary condition for universities to integrate ideological and political courses. Language can carry culture, ideology and thought. These factors are only expressed in a certain language. College English teaching should pay attention to cultivating students' awareness of industry cultural knowledge and professional quality contained in professional English knowledge. First of all, both the design of teaching links and the selection of subject textbooks always focus on the topic of moral education. In addition to the five explicit compulsory courses in Colleges and universities, there are also implicit ideological and political courses in the course of Ideological and political education, but the ultimate goal of the course is to cultivate the ideological and moral quality of the educated, so as to promote their all-round development. We should not only focus on imparting College English professional knowledge, but also focus on the goal of knowledge. English teachers have active thoughts and flexible teaching methods. They are also one of the teachers loved by students. Students are more willing to kiss their teachers and believe in their way. Therefore, these factors determine that college English classroom is also an effective place for ideological and political education in addition to the "Two Courses", which provides the possibility for the realization of Ideological and political curriculum. The improvement of the ideological and moral quality of the educated benefits from many factors, including the ideological and moral character of the educator, the infiltration of teaching methods, the integration of discipline and professional knowledge and ideological and political elements in teaching materials, and so on. However, because higher education takes curriculum as the carrier, only educators flexibly use the penetration function of curriculum can make the educated improve their moral cultivation in the general environment of colleges and universities. Hidden curriculum refers to the general name of various educational elements in school education that can affect students' thought and behavior in addition to explicit curriculum. In the previous traditional "Two Courses", students are not willing to accept the theoretical knowledge blindly preached by ideological and political teachers who are not interested in them. This direct and explicit teaching method can not only cause students' interest and enthusiasm in learning ideological and political courses, but also lead to students' disgust, which is more unfavorable to students' Ideological and political education. At the same time, English courses have more educational contents and themes than other science and engineering courses. Although these themes are expressed in English, the main idea of the content is closely related to the ideological and political content. Therefore, it can be said that college English and ideological and political education are highly consistent in terms of educational objectives. Through the unique function of hint, imitation and assimilation, hidden curriculum can make students unconsciously improve their ideological and moral quality, unknowingly improve their character and establish correct three views. College English is a good carrier of the hidden curriculum of Ideological and political education. Compared with ideological and political courses, College English courses can help students quietly change their thoughts and behaviors without defense, help them establish a correct world outlook, outlook on life and values, and cultivate their professional ethics. Even if the phenomenon of emphasizing knowledge over morality and the marginalization of school moral education is widespread, and even if most college educatees are under the pressure of employment and further study, these difficulties can not become a stumbling block to the ideological and political work of college education. The ultimate goal of higher education is still to grasp both "professional knowledge" and "Ideological and moral character", and the ultimate goal is to promote the educatees to move forward in the direction of all-round development. From this, we can get the consistency characteristics of College English and ideological and political courses, as shown in Fig. 1:

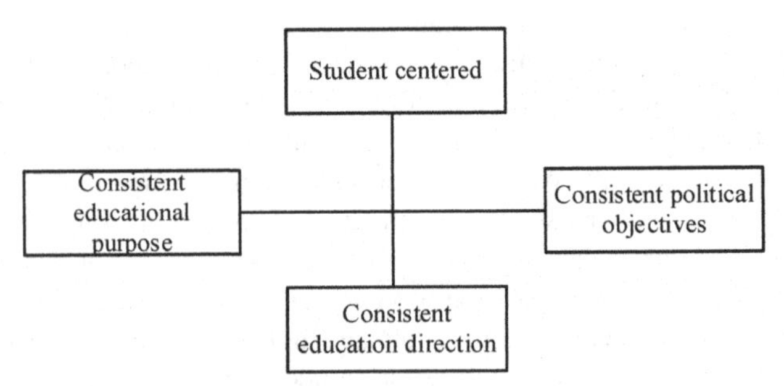

Fig. 1. Consistency characteristics of College English and ideological and Political Courses

As can be seen from Fig. 1, the consistency characteristics of College English and ideological and political courses include: student-centered, consistent political objectives, consistent educational direction and consistent educational objectives. College English takes English learning as the medium to help learners understand relevant industry knowledge at home and abroad, be familiar with industry terms and expressions, and master the ability to effectively communicate and solve problems in English under specific scenes. It is not only an extension and supplement of the basic English learning stage, but also lays a foundation for the professional English learning stage. As a compulsory course for college students, ideological and political theory course also has more unique advantages than other basic courses to cultivate the international vision of the educated. In the information age of rapid development, students will inevitably be impacted by bad thoughts. In addition, their own weak ability to identify good and bad cultures can easily be influenced by these bad thoughts [4]. In order to improve such a cultural environment, eliminating bad culture from the source can not solve the fundamental problem. Only by cultivating the correct ability of international vision can we guide us not to be controlled by decadent ideas. At the same time, it is also to meet the actual work needs and personal learning needs in the future. College English courses are humanistic and ideological. English teaching can not be separated from discourse and teach words and grammar directly. There are many humanistic materials in College English teaching content. Therefore, college English, as a basic course, should pay attention to cultivating students' ability to continue learning at work after graduation. Attention should be paid to cultivating students' interest in learning and helping them form good study habits. For college students in the new era, the easiest and most convenient way is through the correct guidance of ideological and political education. Only by comprehensively cultivating college students' correct values and thoroughly understanding China can we see the world from the perspective of China, thus improving the educated's international understanding ability. Ideological and political education always adheres to the principle of basing itself on the national conditions in terms of teaching objectives, teaching features and teaching contents, so that college students can have a profound understanding of the history and present situation of China, thus helping students to accurately understand the value orientation in line with the national conditions of China. Language learning should follow certain rules, starting with simple words and grammar points, and then learning complex texts. We should go from simple to deep. College English teaching should also follow this rule, from simple to complex, and finally help students achieve the ability to solve problems in English in the workplace. This is consistent with the process of ideological and political education.

2.2 Selection of Teaching Quality Evaluation Indicators

Scientific and reasonable evaluation index is the basis and premise to ensure the quality of students' evaluation of teaching. The indicators of student evaluation should reflect the internal law of the formation of teaching quality, and should be representative, comprehensive, formative, operable and quantitative. The core of the evaluation indicators should highlight all the emphasis on enhancing the learning effect of students, and fully reflect the principle of student-oriented teaching and teacher-oriented teaching [5, 6]. There should be clear guidance instructions when formulating the evaluation form, and the part of students' self-evaluation should be added. More qualitative indicators can be set that are conducive to teachers' personality and creativity, and the weight of qualitative indicators and quantitative indicators should be allocated reasonably. The content of the index should be specific and clear, and the degree of differentiation should be high, so that students can better grasp, and according to the type of courses, the evaluation form should be personalized, and try to make the evaluation index simple [7, 8].

From the perspective of the existing teaching level evaluation system, the design of indicators is mainly reflected in the following aspects:

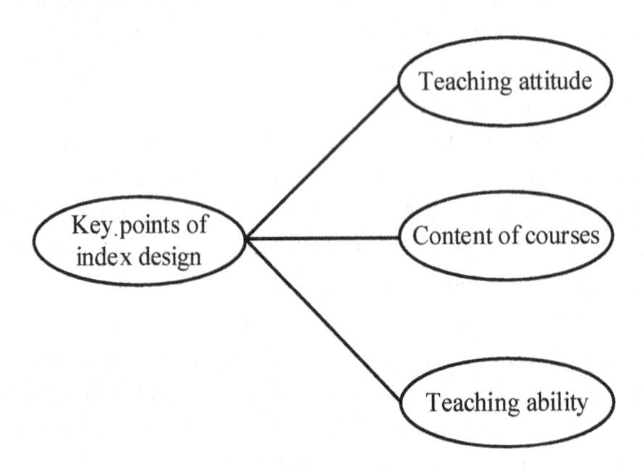

Fig. 2. Key points of index design

As can be seen from Fig. 2, this model combines the evaluation results of these different evaluation subjects and obtains the comprehensive evaluation results for their empowerment. To do the evaluation work well, we must solve two major problems, one is to establish the evaluation index system, the other is to deal with the evaluation results. The following is a specific analysis of these two problems:

(1) Analysis of the evaluation index system

The determination of the evaluation index system is the first step of teaching quality evaluation, and whether the evaluation system is reasonable directly relates to the rationality of the evaluation model. Generally the better evaluation system is determined through the analytic hierarchy process. Firstly, an evaluation table divided into three levels of indicators was obtained through literature review, in which evaluation indicators at all levels were given.

(2) Processing of evaluation results

Effective processing of the evaluation results is the most important step in the evaluation model. Many current models simply calculate the average value of the evaluation results, which will ignore a lot of useful evaluation information. In order to make full use of the evaluation information, this model adopts the interval value fuzzy judgment method to replace the average value method. Firstly, the three-level indicators are evaluated to find out the effective interval of each three-level indicator, and then the effective interval of each second-level indicator is calculated according to its corresponding weight coefficient, and then the effective interval is obtained by further and further. In addition, because different evaluation subjects have different starting points and angles of evaluation, the effective interval should be calculated for them respectively. Then the effective interval of comprehensive evaluation is obtained according to its corresponding weight, and the final comprehensive evaluation result is determined. Whether it can promote students' positive thinking, whether students' grades are improved, and whether students' mastery of knowledge points is comprehensive. To evaluate teachers fairly and objectively, the key is to establish a standardized and reasonable evaluation index system of curriculum teaching quality. To establish the evaluation index system of curriculum teaching quality, we must adhere to the principles of objectivity, independence, operability and integrity. At the same time, they should have clear learning motivation, full learning motivation and low participation in the classroom. Therefore, students should clarify their career planning and ensure their sufficient learning motivation. Teachers should actively introduce discussion in the classroom to improve students' classroom participation.

In order to evaluate the fairness, objectivity and accuracy of the results, different content index systems are formulated for different evaluation subjects, namely supervision experts, peer teachers and students.

2.3 Social Network Building Model

Integrating technical principles of social networks into the process of model construction, first of all, clarifying the purpose of evaluation, which is the guideline of evaluation [9, 10]. Clarify why to evaluate, what aspects of things to evaluate, and what accuracy requirements to evaluate. Define the object system, that is, determine the evaluation object and evaluator. It directly determines the content, mode and method of evaluation. And the factors to be included in the evaluation activities and the relationship between various factors should be determined. Under the condition of known weight vector and input sample, Suppose ϕ represents the weight vector, l represents the learning rate, the weight adjustment formula is obtained:

$$\phi(l+1) = \frac{\|\phi - \varepsilon\|^2}{\Delta l} \tag{1}$$

In formula (1), ε represents the input sample. Assuming that the goal of neuron weight correction is to minimize the scalar function, γ represents the correction amount at the current moment, then each correction is expected to have:

$$D = \sum \frac{\gamma^2 - \eta}{|l+1|} \tag{2}$$

In formula (2), η represents the gradient vector of l. The output error square of neuron is obtained:

$$Q(R) = \frac{1}{2}\left(F - R^T\right) \times \eta \tag{3}$$

In formula (3), F represents the original weight of neuron, R represents the current weight of sample, and T represents the weight adjustment amount. The teaching design

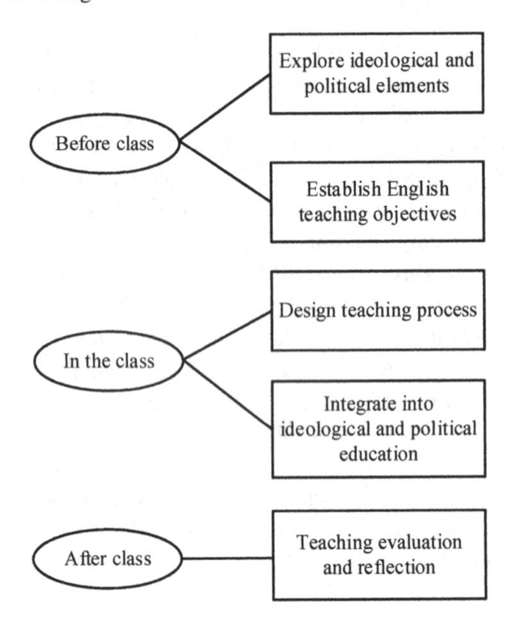

Fig. 3. Schematic diagram of teaching design process

process of the teaching quality evaluation model is described according to the three stages before, during and after class, as shown in Fig. 3:

According to Fig. 3 that the teaching design process of the model:

(1) Before class: excavate ideological and political elements, establish the goal of English teaching;
(2) Class: design the teaching process, into the ideological and political education;
(3) After class: teaching evaluation, reflection.

Reasonably determining the weight is of great significance to evaluation or decision-making, that is, multiple indicators are "synthesized" into a comprehensive evaluation value through a certain mathematical model. Selecting a more appropriate synthesis method is the key link of comprehensive evaluation. Output the evaluation results and explain their meaning, and make decisions according to the evaluation results. The teachers will attend classes in the classroom, and the director of the teaching and Research Office will organize teachers to listen to and communicate with each other. Peer teachers' evaluation of teaching includes six first-class indicators, the specific contents are: teaching management, teaching content. The different paths of the central node and the long tail node in the communication process of mobile Internet provide a reference for social regulators to effectively control the dissemination of negative information on mobile social networks. The educated are forced by the pressure of entering a higher school and employment. They need to work hard to learn basic knowledge to obtain more certificates and improve the passing rate of English level, which leads to the strengthening of their own learning purpose, which means that the student group has little energy to think about other problems, including how to improve their comprehensive quality.

3 Simulation Experiment

In order to verify the validity of the model, a simulation experiment was carried out. The basic reference model [2] and the reference model [3] were selected as the control group, and the model in this paper was selected as the experimental group for the comparative experiment. The evaluation performance of the three models was tested with the same number of samples, as shown below.

Table 1. Number of samples 100 model evaluation accuracy (%)

Number of experiments	Reference [2] model	Reference [3] model	Model in this paper
1	55.312	52.131	59.336
2	53.774	54.126	62.005
3	52.017	53.998	61.748
4	51.466	54.812	63.007
5	52.853	53.714	62.555
6	53.847	52.815	63.718
7	52.117	53.606	62.554
8	53.646	52.749	61.949
9	51.842	53.812	62.505
10	52.694	52.123	63.717
11	53.220	54.918	64.285
12	54.916	53.007	65.231
13	52.154	52.114	64.718
14	50.140	53.466	65.912
15	51.206	52.948	66.311

According to Table 1 that the average evaluation accuracy of the College English teaching quality evaluation model and the other two quality evaluation models are 63.303%, 52.747% and 53.356% respectively.

According to Table 2 that the average evaluation accuracy of the College English teaching quality evaluation model and the other two quality evaluation models are 54.118%, 44.184% and 43.983% respectively.

According to Table 3 that the average evaluation accuracy of the College English teaching quality evaluation model and the other two quality evaluation models are 46.453%, 40.127% and 38.099% respectively.

Further verify the above conclusion, that the evaluation accuracy of this model is high, analysis of the model in the implementation process of the similarity calculation indicators, the accuracy of similarity was evaluated by using the model in this paper, the model in reference [2] and the model in reference [3]. The result is shown in Fig. 4:

Table 2. Number of samples 200 model evaluation accuracy (%)

Number of experiments	Reference [2] model	Reference [3] model	Model in this paper
1	45.166	42.080	53.644
2	43.718	41.566	52.911
3	45.815	43.788	53.485
4	46.310	45.915	52.977
5	44.619	44.613	56.548
6	42.117	45.718	57.331
7	43.695	44.663	56.748
8	44.506	42.519	57.916
9	43.129	43.878	52.144
10	44.706	42.516	53.646
11	43.819	43.757	52.818
12	45.602	44.909	51.404
13	44.337	45.612	52.619
14	42.005	43.221	53.215
15	43.221	44.997	54.371

Table 3. Sample size 300 model evaluation accuracy (%)

Number of experiments	Reference [2] model	Reference [3] model	Model in this paper
1	36.394	35.007	46.123
2	38.515	36.949	45.157
3	39.214	35.212	47.109
4	41.007	39.458	48.209
5	40.649	41.006	44.774
6	42.336	38.575	45.338
7	41.815	39.402	46.157
8	36.878	38.466	45.310
9	38.144	39.157	44.718
10	37.466	36.228	45.122
11	39.979	37.991	46.371
12	41.164	38.231	45.111
13	42.377	37.455	49.122
14	43.845	38.902	48.517
15	42.122	39.452	49.662

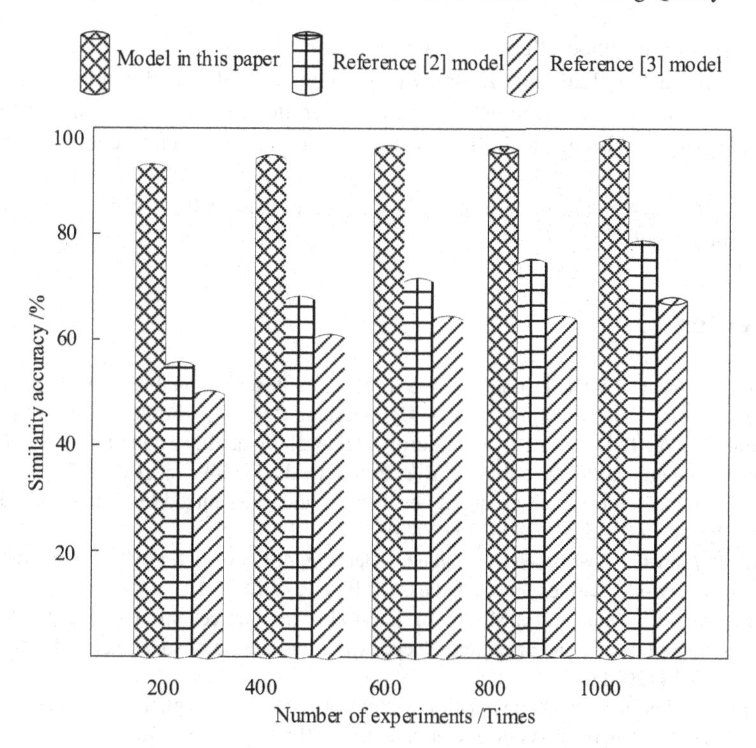

Fig. 4. Test results of similarity calculation accuracy

According to the analysis results in Fig. 4, the evaluation accuracy of the model in this paper is higher than that of the model in reference [2] and the model in reference [3], both of which are above 95%, which improves the accuracy of the evaluation results, lays the foundation for the completion of the evaluation of the teaching quality of College English integration into Ideological and political courses, and increases the evaluation accuracy.

To sum up, the model studied in this paper is a kind of evaluation model, which can be applied to the determination of all kinds of weights, the model is perfect in the determination of the index system and the processing of the evaluation results, and has been widely used in various fields of real life, such as the evaluation of service quality, the evaluation of excellent students and so on.

4 Conclusion

The model constructed in this paper improves the teaching quality evaluation system of theoretical and experimental courses in colleges and universities from the aspects of teachers' quality, teaching contents, attitudes, methods and effects, and makes it more in line with the teaching development of colleges and universities. In addition, on the premise of retaining a large amount of original information, the dimension of the evaluation index is reduced to avoid the influence of the complexity of the network model on the prediction effect. Through the research, the following conclusions are drawn:

(1) The model has good effect and high precision;
(2) The similarity calculation accuracy of the constructed model is good, and it has a good evaluation effect, which is of great significance for improving the teaching quality of schools and promoting the communication between teachers and students.

In the future, we can make a more in-depth analysis of the integrity and available value of the model.

References

1. Xiao, Y., Wang, J.: The evaluation system of online teaching quality in universities: value orientation and construction strategies. Heilongjiang Res. High. Educ. **10**, 141–144 (2020)
2. Wang, B., Wu, T., Duan, Y., et al.: Experimental teaching quality evaluation system based on engineering education accreditation. Res. Exp. Lab. **39**(5), 149–152,181 (2020)
3. Wu, N.: Research on teaching quality evaluation under the mixed teaching mode. J. Hubei Open Vocat. Coll. **33**(24), 144–145 (2020)
4. Choi, J., Jeon, C.: Cost-based heterogeneous learning framework for real-time spam detection in social networks with expert decisions. IEEE Access **29**(7), 103–110 (2021)
5. Derksen, M.E., Kunst, A.E., Murugesu, L., et al.: Smoking cessation among disadvantaged young women during and after pregnancy: exploring the role of social networks. Midwifery **98**(98), 7–14 (2021)
6. Song, Q.: Research on the evaluation system of the teaching quality of entrepreneurship-oriented entrepreneurship course. Continue Educ. Res. **6**, 82–84 (2020)
7. Xiang, S., Lu, L.A., Aa, A., et al.: ED-SWE: event detection based on scoring and word embedding in online social networks for the internet of people. Digit. Commun. Netw. **7**(4), 559–569 (2021)
8. Roozbahani, Z., Rezaeenour, J., Katanforoush, A., et al.: Personalization of the collaborator recommendation system in multi-layer scientific social networks: a case study of ResearchGate. Exp. Syst. **39**(5), 1–18 (2021)
9. Wang, Z., Li, P.: Simulation of parallel validation for heterogeneous social network attributes based on string space. Comput. Simul. **37**(8), 409–413 (2020)
10. Tang, H., Zeng, J., LI, F., et al.: Point of interest recommendation based on location category and social network. J. Chongqing Univ. (Nat. Sci. Edn.) **43**(7), 42–50 (2020)

Application of Cloud Security Terminal in Information Management of Power Industry

Jiang Jiang[1](✉), Hanyang Xie[2], Yuqing Li[3], and Jinman Luo[4]

[1] Digital Department of Guangdong Power Grid Corporation, Guangzhou 510800, China
ljhh201@163.com
[2] Information Center of Guangdong Power Grid Corporation, Guangzhou 510000, China
[3] Logistics Service Center of Dongguan Power Supply Bureau of Guangdong Power Grid Corporation, Dongguan 523900, China
[4] Information Center of Dongguan Power Supply Bureau of Guangdong Power Grid Corporation, Dongguan 523900, China

Abstract. With the powerful computing power of cloud computing, cloud security terminals are widely used in many industries. The information construction of the power industry will generate a large amount of data. In order to improve the efficiency of information management in the power industry and reduce the success rate of information cracking, this paper applies the cloud security terminal to the power information management. After deploying the power industry information management cloud security terminal architecture, build a terminal access control model. Cloud node load is predicted through Bayesian model to ensure smooth power information management. The improved Wu-Manber algorithm is used to protect the power information layer and ensure the safety of power circuit information. In the example verification, after the cloud security terminal is applied, the success rate of information being cracked is less than 10%, the response speed of this method is significantly improved, and the level of power information management is improved.

Keywords: Cloud security terminal · Power industry · Information management · Application research · Access control · Bayesian

1 Introduction

The electric power industry is a technology- intensive and asset-intensive basic industry of the national economy. The informationization of the industry started early, but its complexity and difficulty are greater than those of modern service industry and manufacturing industry. The development and maturity of network, computer and communication technologies have basically eliminated the technical bottleneck in the process of enterprise informatization, especially the informatization application technologies. While surrounding the development of Web technologies, the informatization technologies, the optimization, upgrading and integration of equipment have reached a considerable

W. Fu and L. Yun (Eds.): ADHIP 2022, LNICST 468, pp. 771–783, 2023.
https://doi.org/10.1007/978-3-031-28787-9_57

level, the mainstream products and technologies have basically taken shape, and the technical risks for enterprises to implement informatization have been effectively reduced. The construction of information engineering projects in the power industry covers a wide range, including the digital construction of substations, the construction of power transmission and distribution networks, the construction of communication networks, and the construction of more information systems supporting the daily maintenance and management work [1]. At present, the electric power industry has basically realized the informationization management, and has promoted the electric power industry informationization construction ability. The construction of informatization project is a very complicated system project, which has the characteristics of high technical content, fuzzy task boundary, imprecise objectives, strong pertinence, frequent change of business requirements and tight schedule [2]. Because of the lag of the understanding of management personnel and the construction of management mechanism, the management level of informationization construction project is far from the same proficiency as traditional management project. Therefore, to improve the ability of organization informatization project management and make use of informatization project construction to ensure the realization of informatization strategic goal have become the most important and frontier subject.

Cloud security terminal comes into being under the era background of cloud computing and big data, and has advantages in many aspects. People can use Cloud security terminal system to integrate information resources, and also can realize the real improvement of information management level, truly keep up with the trend of development of the times, and not be eliminated by the market. Especially, the contradiction between terminal system and information security will be aggravated when the scale of enterprise is enlarged. Therefore, the managers of enterprise should realize the seriousness of the problem and take some effective measures to avoid the information leakage, which will be beneficial to the electric power industry of our country moving towards the direction of intelligence.

Some scholars have proposed a method of hierarchical management of power information, taking into account factors such as distributed power sources, distribution lines, and equipment controllable levels, to solve the problem of efficient and intelligent management of diverse load equipment under its jurisdiction in the park. The park managers realize the overall collaborative management of the diverse loads to achieve the purpose of improving energy efficiency. However, this method needs to be further improved in the balance of information load. Some scholars have proposed an information management method based on edge computing. This method introduces edge computing as a data processing model, disperses the computing load pressure of the data processing center to the edge side of the device, thereby improving the data response speed, and can still process local data in an offline state. However, there is room for further improvement in data loss prevention. In order to enhance the capability of data disaster tolerance and prevention, strengthen data sharing and business integration, enhance the business support capacity, and ensure the operation security of the power industry, the information management method based on cloud security terminals in the power industry will be studied to explore the practical application of cloud security. This paper deploys the electric power industry information management cloud security terminal architecture, and constructs the terminal access control model. The Bayesian model predicts the cloud node load passing to ensure the smooth power information management. The improved Wu-Manber algorithm is used to protect the power information layer and manage the university of power information.

2 Application Research of Cloud Security Terminal in Information Management of Power Industry

2.1 Deployment of Cloud Security Terminal Architecture for Information Management in Power Industry

By integrating resources and assigning to different users according to their needs, cloud terminals maximize the utilization of scheduling management resources, and can uniformly carry out the standardized configuration of software, realize the application system online, version update, unified repair of bugs, active and passive offline server maintenance and other processes, as well as flexible expansion and easy management. Using cloud security terminals can realize unified management and scheduling of office resources, actively control illegal outreach channels, uniformly carry out computer registration and installation of anti-virus software, and eliminate file leaks, software vulnerabilities and hacker attacks. In complex network environments, software deployment platforms usually run in independent trust domains. In this open environment, trust assessment and evidence collection between entities are carried out between untrusted entities. Especially the trusted evidence collection and software credibility assessment in software runtime [3].

The cloud security terminal adopts the virtualization architecture as the bottom design, provides a highly scalable, reliable and stable resource platform for the basic server layer of the cloud terminal system, has built-in business continuity and disaster recovery functions, can protect desktop data and its availability, and provides a powerful backstage guarantee for desktop virtualization. The logical structure of the cloud security terminal is shown in Fig. 1. From bottom to top, the logical structure is: hardware resource layer, third-party cloud monitoring management control layer, virtualization and cloud platform layer, regional central node and terminal access layer. The hardware resource layer integrates and pools existing resources, divides them according to the needs of users, and stores all processing data in fixed data storage devices to improve data security; the third party cloud monitoring management control layer comprises five parts, namely, user access to information parsing interface, parser, filtering and aggregation engine, configuration management module and monitoring rule base [4]. Virtualization and cloud platform layer realizes the redirection of peripherals by adopting Xen virtualization technology and SPICE Desktop Transfer Protocol; Domain center node layer is the core layer of monitoring execution, responsible for subscribing and publishing data. It mainly includes command interface, control components SubAgent _ Manager, TopicAgent _ Manager and PubAgent _ Manager, monitoring agent components SubAgent, TopicAgent and PubAgent. Terminal access layer uses the smaller client to access, and only configures the interface including embedded processor, local flash memory and various peripherals. In the process of transmission, it provides a more secure environment by only transmitting high-strength encryption transform values of terminal signals and images. Different types of monitoring agents can be deployed, including application monitoring agents that monitor application information, virtual machine monitoring agents that monitor the running information of VMs, and physical infrastructure agents that monitor the running information of the infrastructure.

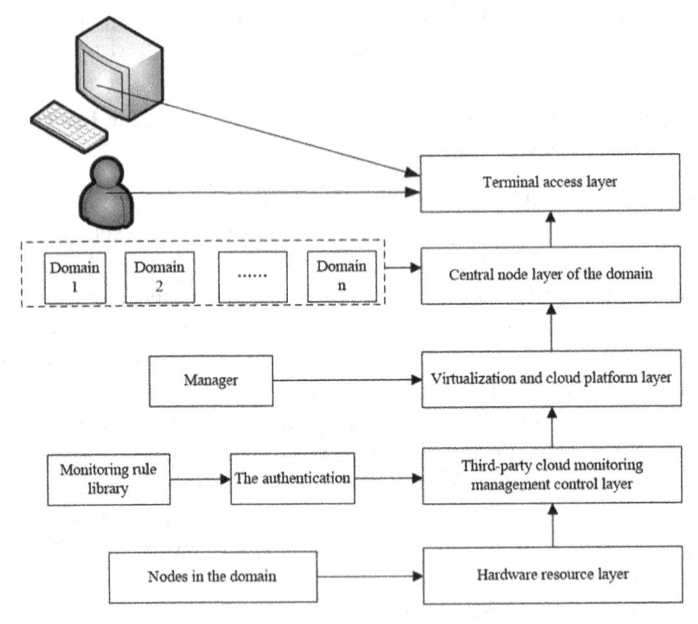

Fig. 1. Cloud security terminal deployment architecture

Under the cloud security terminal deployment architecture, the process of obtaining data related to the information management of the power industry is as follows:

Step 1: The user submits service request information and performs identity authentication;

Step 2: The user access information parsing interface receives the access request from the user, and parses the access request information into a triple, (user ID, cloud service type), monitoring category, monitoring content);

Step 3: The parser receives monitoring information from the cloud user and is responsible for parsing the monitoring information triple (user ID, cloud service type, monitoring category, monitoring content) into domain, topic, content);

Step 4: The configuration management module receives parsing information from the parser, accesses the monitoring rule base, and configures the monitoring file according to the monitoring content;

Step 5: The domain adptor command interface receives the parsed monitoring configuration file from the monitoring configuration management module and, based on the match result of the domain information, determines whether the domain needs to be monitored by the user. If the match succeeds, execute. Otherwise, the domain does not perform monitoring requests.

Step 6: TopicAgent _ Manager enables related TopicAgent; SubAgent _ Manager enables SubAgent; PubAgent _ Manager enables PubAgent based on the topic information of the monitoring configuration file.

Step 7: SubAgent sends a subscription request to the TopicAgent based on the monitoring profile topic information;

Step 8: Agent writes monitoring data to PubAgent in the domain according to data type and organization form;

Step 9: PubAgent publishes data to the TopicAgent by topic number;

Step 10: SubAgent extracts subscription data from the TopicAgent based on topic;

Step 11: SubAgent delivers monitoring data to the filtering and aggregation engines.

Step 12: Filtering and aggregation engines filter and reduce data streams to minimize the additional overhead associated with monitoring data streams;

Step 13: Generates a specific monitor view to the user and displays a warning message based on the refresh of the monitor.

For systems with higher security requirements, it is difficult to meet the requirements only by using autonomous access control mechanism. DPS-CMS can increase or decrease the number of agents flexibly according to the scale of monitoring facilities, thus effectively reduce the storage and computing pressure of monitoring nodes, and is more suitable for cloud computing system composed of large-scale virtual organizations than centralized agent deployment. DPS-CMS adopts P/S communication model, takes data as center, network delay is small, real-time is strong; four-layer agent mode, each layer has corresponding data preprocessing method, which reduces the amount of data transmission and improves the ability of data transmission. From this point of view, DPS-CMS can meet the adaptive requirements of cloud monitoring. DPS-CMS can be implemented by adding or enabling a corresponding PubAgent if there is a new monitoring entity that needs to be monitored, in this sense DPS-CMS meets the elastic requirements of cloud monitoring. The DPS-CMS publish-subscribe app is modular, and PubAgent and SubAgent can join or leave dynamically to meet flexible application requirements [5].

When the amount of tasks processed by the user cloud node increases significantly, the load of the node will also increase accordingly. The user needs to pre-allocate more resources in time to meet the needs of task processing and prevent the excessive load from affecting the normal use of the service.. On the contrary, in order to save expenses, users can reduce some resources without affecting the normal provision of services, so as to prevent the waste of idle resources caused by excessively low load. If we can predict the change of node load in advance and grasp the trend of load change, it will have important guiding significance for the timely allocation and recovery of node resources. This approach can improve the optimization of node service performance and utilization.

2.2 Cloud Security Terminal Access Control Model

The distributed nature of cloud computing means that storage or computing in the cloud may involve resources in different autonomous domains, requiring multiple entities in different domains to coordinate with each other to complete the entire operation. Resource sharing is becoming more and more important, and the security issue of cross-domain resource access that accompanies cross-domain access cannot be ignored. From the point of view of practical security issues, users upload their important data to the cloud, and it is necessary to consider the ability of cloud computing to ensure data security and integrity in the process of cloud services, and it is also necessary to establish trust in cloud service providers. In the cloud computing environment, each cloud application belongs to different security management domains, each security domain manages local resources and users, and different domains correspond to different security management policies and rules. When users access resources across domains, each domain has its own access control strategy. When sharing and protecting resources, a common and mutually agreed access control strategy must be formulated for the shared resources [6].

First, the X domain Y and the domain start preparations at the same time, and the domain X defines the risk function $\Gamma\left(x_i, x_j, r\right) \in [0, 1]$ of this domain. And calculate the initial risk value Γ_i of the nodes in the domain, the larger the risk value, the higher the risk. Where m is the total number of nodes in domain X. At the same time, the initial value of the risk cursor R^X is set subjectively by the domain management node. In addition to the two initial work of the above-mentioned domain, the preparation work of domain Y also includes defining the shared function related to it. The risk cursor set by the domain Y is R_i^X. In this example, i takes Y, which represents the risk threshold value in the domain Y for the visit from the Y domain.

Then, the user u of the domain Y initiates an access request r. At this time, the management node of domain Y compares the risk value Γ_i of this visit with the risk cursor R^X of domain Y. And compare the counter value c with the set maximum access times F_i. The decision of the management node of domain Y for this visit is defined as follows [7].

$$D(r)^X = \begin{cases} A, \Gamma_i(r) < R_i^X \wedge (c < F_i) \\ D, else \end{cases} \quad (1)$$

If the result is A, go to the next step; otherwise, the access request of u is rejected, and the process is terminated. After that, domain Y will calculate user u's mapping role set $f_Y(X)$ in domain Y according to u's request r and sharing function $S(Y)(X)$. And calculate the risk value of this access, compare the risk value and the size of the risk cursor, and decide whether to allow this access request. If access is allowed, continue to step 5. If access is denied, the cross-domain access process is terminated. The detailed process is shown in Fig. 2.

If access is allowed. Domain Y verifies the identity of user u. If the verification passes, assign the mapping role set $Y(X)$ to u. And allow it to access resources in domain Y with this role.

Based on the role-based access control model, the concept of "trust" is introduced, which is the so-called trust-based access control technology. That is, on the basis of granting permissions to roles, the user's trust degree must be verified. Only when the user's identity verification and trust degree both meet the requirements set by the system, can the user access rights be assigned. The cloud computing environment has potential vulnerabilities in the application layer and the basic service layer. The vulnerability identification refers to the correlation analysis, which may be attacked and exploited. The weak point of the system is an important link in the security situation assessment of static indicators. Organizational and logical errors in software, inconsistency in coordination

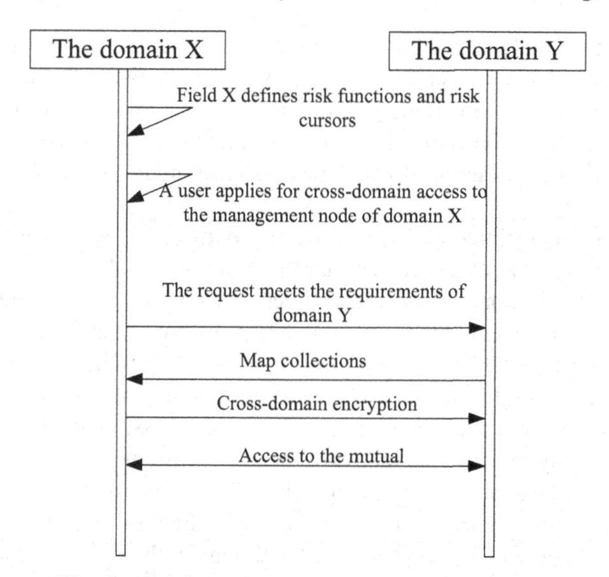

Fig. 2. Cloud security terminal cross-domain access

between different devices and other factors inevitably lead to security policy defects, identify weaknesses that may be exploited by threats, identify technical vulnerabilities and identify objects, such as physical environmental factors, internal and external network structure Access policy, apply the security protocol type in the middleware, and then perform qualitative or quantitative assignment.

Initialization of trust relationship between nodes The establishment of trust relationship between nodes needs to go through two stages: service discovery stage and trust evaluation stage, monitoring the impact of all interactions between nodes on the cloud security management system, evaluation, update and decision-making of trust values. The justification is largely dependent on the observer. The main task to observe is how the trust context changes when the nodes interact, and another task is to store, manage and trigger the dynamic update of the trust value. When the observation system observes that the behavior of a node exceeds the "allowed" range, that is, it determines that the behavior of this node is an aggressive behavior. At this time, it is necessary to trigger the re-evaluation of the relevant trust degree. During cross-domain access, there are generally two access policies for cloud security terminals: Boolean function or LSSS matrix. Boolean functions can be represented by an access structure tree, the attributes are leaf nodes, and "and" and "or" represent other nodes except leaf nodes. In general, the form of the access policy is represented by a matrix, so we need to convert the form represented by the Boolean function to the LSSS matrix representation. The algorithm is roughly as follows: Assume that the root node is the vector l, and the child node is the vector of the parent node. If node i is "AND", then its left and right child nodes are $(v_i|l)$ and $(0, -1)$, respectively. if node i is "OR". Then its left child node and right child node are consistent with their own vectors.

In the attribute encryption algorithm, it is necessary to judge whether the user attribute set that sends the request to the cloud security terminal matches the access policy tree of the file. The general judgment method is: set the root node of the access policy tree T as G. T' is a subtree of T, and its root node is g. 5If a set ξ satisfies the access policy tree T' that is $T'(g) = 1$. And the node is a leaf node, then if and only if $\xi(g) \in G$. If this node is a non-leaf node, the value of $T'(g)$ of all child nodes of g needs to be calculated. $T'(g) = 1$ can only be established when the number of nodes whose $T'(g)$ value is 1 is more than k.

When using the attribute encryption access control system to protect the security of the data stored in the cloud security terminal, the key ciphertext encrypted by the CP-ABE algorithm is stored in the cloud together with the data ciphertext, which can reduce the key management of the system To a large extent, it can improve the efficiency of key distribution and the flexibility of users to access data. When attacks from outside the cloud continue to intensify, in addition to causing damage to public cloud tenant applications, for example, customers cannot log in to access. In addition, the attacker's ability will gradually increase the malicious behavior with the gradual accumulation of attack resources. The attack ability threat value TH is a measure of the threat to cloud security caused by the attack ability possessed by malicious attacks at a certain time. Affected by the degree of control over the target tenant, the sensitive data obtained by the attack. The level of attack capability assessment is attack capability threat value → component threat value → security data threat value → threat interval.

$$TH = \sum_{i=0}^{n} th_{next}[i]w[i] \tag{2}$$

Malicious attacks transform read permissions into remote login permissions by acquiring permissions layer by layer. Therefore, by setting the access control of the cloud security terminal, the operation security of the cloud security terminal can be guaranteed.

2.3 Cloud Security Endpoint Access Load Prediction

When the cloud security terminal protects the information management of the power industry, the difference in the volume of information and data exchange will cause the overload of the security node. Therefore, this paper will predict the load of cloud security terminal nodes to ensure the smoothness of power information management.

Due to the high correlation between cloud node loads, it is usually manifested as adjacent short time gaps. A series of load situations with time intervals of different lengths are then predicted. Each time interval starts from t_0, and the average load of each time period is recorded as $L_1, L_2, \ldots, L_n$. Assuming that the current moment is t_0, two load averages have now been predicted. L_i and L_{i-1} correspond to two different time intervals $[t_0, t_i]$ and $\left[t_0, t_{i-1}\right]$, respectively. The load value $\overline{L}_i$ corresponding to the $\left[t_{i-1}, t_i\right]$ time can be calculated according to the following formula.

$$\overline{L}_i = L_i + \frac{t_{i-1} - t_0}{t_i - t_{i-1}}(L_i - L_{i-1}) \tag{3}$$

Applying Bayesian theory to load prediction of cloud security endpoints should be implemented in the following six steps:

1) Determine the load state vector of the cloud security node, that is, the vector $Z = \{z_1, z_2, \ldots, z_m\}^T$ of the class. Where m is the number of states associated with the security endpoint.
2) An attribute vector that determines the cloud node's load, $AT = \{at_1, at_2, \cdots, at_n\}^T$, where n is the number of attributes.
3) Calculate the prior probability $P_B(Z_i)$ of each state using the sample data set (historical data).
4) Calculate the joint probability $P(at_i|Z_i)$ for each state separately.
5) The posterior probability is calculated based on the historical data and the following formula.

$$P(Z_i|at_i) = \frac{P(at_i|Z_i)P(Z_i)}{\sum_m P(at_i|Z_k)P(Z_k)} \tag{4}$$

6) The predicted value of the state is selected according to the following formula

$$\overline{L_i} = E(Z_i|at_i) = \sum_{i=1}^{m} Z_i P(Z_i|at_i) \tag{5}$$

If the predicted node load of the cloud security terminal meets the node migration condition, the running task will be migrated to other nodes. The node migration condition includes two restrictions, namely the upper limit trigger condition and the lower limit trigger condition. The main function of the upper limit trigger condition is to reduce the load of nodes running under high load. If the cloud security terminal node used by the user is often under high load, in order to ensure the quality of service, the user can select a suitable cloud node on the running node that meets the upper limit trigger condition and migrate it to other running nodes. This ensures that the computing power running on this node meets the needs of users.

2.4 Power Information Level Protection of Cloud Security Terminal

When using the cloud security terminal for power information transmission, in order to ensure the information security of the power industry, it is necessary to carry out transmission authentication according to the dynamic protocol. The dynamic protocol structure in this study is used to describe the dynamic behavior of a certain moment or a certain state, focusing on the description of the control logic, examining the state and connection between objects in the protocol at any moment. In UML, sequence diagrams and activity diagrams are used to represent the dynamic structure model of the protocol: the sequence diagram depicts the interaction model of the protocol and describes the interaction of some elements of the protocol in time. The sequence diagram is oriented to the process of protocol execution; the activity diagram describes the The state model of the protocol, the dynamic behavior of a class, can also describe concurrent activities, and the activity diagram is oriented to specific objects. In this section, the dynamic structure model of the protocol is described, mainly from the initialization of the protocol and the execution phase of the protocol.

The initialization of the protocol is initiated by the back-end server, the back-end server completes all initialization work, and the information receiving end passively accepts the identification and writes it into the memory unit. At the beginning of the protocol execution, the message sender initiates a session and transmits the message to the server. The server receives the information, performs corresponding calculations, and feeds back a message to the information receiver. The receiving end receives the message, and performs preliminary verification on the identity of the receiving end through calculation and comparison. The back-end server receives the message and calculates it, searches the database to determine whether there is a calculated value, and continues to calculate and transmits the message to the access controller if it exists. The access controller receives the message for calculation, and transmits the message to the authentication server. The authentication server receives the message for verification. If it passes, the authentication is realized. If it fails, the authentication fails.

This study uses the improved Wu-Manber algorithm to protect different information levels in the power industry. The Wu-Manber algorithm uses three tables, SHIFT, HASH, and PREFIX, to store the information that the algorithm needs to run. The lookup of the two tables, SHIFT and HASH, is realized by hashing the suffix of the pattern string, and there are certain repetitions. For example, in the preprocessing stage, each pattern string suffix will calculate the hash twice, once to initialize the SHIFT table and once to initialize the HASH table. In the matching process, the algorithm calculates the hash value h according to the last B characters of the current window according to the HASH function. First use h to find the SHIFT table to determine whether the window slides backwards. If SHIFT[h] $= 0$, then use h to look up the HASH table. In this way, the same hash value h is searched twice. If SHIFT $= 0$, then look up the HASH table by hash calculation to get all possible keywords that match the suffix of the current matching window. Then look up the PREFIX table to determine whether the prefixes of these keywords match. This process involves the lookup operation of two tables, and the efficiency is relatively low. Merge the SHIFT and PREFIX tables into the HASH table, and modify the data structure of the HASH linked list. Add a data field to store the shift value in the main table node. The prefix data field used to store the prefix hash value of the pattern string is added to the child linked list node. It can be seen that only the main entry with the shift value of 0 has a sub-linked list, and the sub-linked list pointer fields of the main entry with the shift value of not 0 are all NULL. The maximum shift value in the algorithm is $m - 1$, which is generally not very large in practical situations. When the cloud security terminal creates the HASH table, the memory space of the sub-linked list is dynamically applied. The process of the improved algorithm to protect the power information level is as follows:

1) According to the last B characters of the current window, the hash value h is calculated according to the HASH function.
2) Check the main table of the HASH table to determine the size of the data field value d. If $d < m$, the data field value represents the shift value, slide the window data positions to the right, and return to step (1). If it is $d \geq m$, the value of the data field represents the address of the head node of the sub-chain list, and the step (3) is entered.
3) Calculate the current window prefix hash value text prefix.

4) Find all the pattern strings with the suffix hash value *h* according to the HASH table, and compare text_prefLx with the prefix hash value prefix of these pattern string nodes one by one in the order of the pattern strings in the HASH table sub-linked list. If the two are not equal, go to the next pattern string. If the two are equal, complete comparison is performed until it is judged whether the match is successful.
5) Slide the current window back one character, turn step (1).

According to the above process, the research on the information management method of the power industry based on the cloud security terminal is completed. In order to clarify the application of cloud security terminal in the actual information management of the power industry, the next section will carry out example verification.

3 Example Verification

3.1 Validate the Design

This experiment will verify the information management method of the power industry based on the cloud security terminal studied above from the two aspects of the security and efficiency of the management method. In order to make the verification results of the example more authentic and reliable, in this verification, the information management method based on hierarchy and the information management method based on edge computing are used as comparison groups. This verification is completed by comparing the success rate of the power industry information being deciphered and the response time of the method under the application of the method.

3.2 Validation Results

The following Table 1 is for the verification of this example, when different management methods are applied to the information management of the power industry. The comparison of the success rate of power information transmission and the response time of the method to the attack.

Analysis of the data in Table 1 shows that when the management method proposed in this paper is used in the information management of the power industry, the probability of information being successfully deciphered is less than 10%, which is better than other management methods. From the perspective of the response time of the method, after using the method in this paper, it can respond quickly to attacks on power information transmission, reduce the risk of data damage and loss, and effectively manage power information. That is to say, the application of cloud security terminal in the information management of the power industry can fully protect the security of power information data and ensure the normal operation of the power industry.

Table 1. Instance validation data

Group	Management method based on cloud security terminal		Hierarchical Information Management Method		Information management method based on edge computing	
	Cracked success rate/%	Response time/ms	Cracked success rate/%	Response time/ms	Cracked success rate/%	Response time/ms
1	6.3	4.03	22.9	7.96	15.9	5.80
2	7.7	3.76	21.5	7.84	13.2	5.86
3	8.8	3.53	22.4	7.37	16.8	6.55
4	9.2	3.54	19.3	7.90	13.1	6.17
5	6.5	3.76	21.1	7.21	16.5	6.46
6	7.4	3.95	18.0	6.92	16.6	6.43
7	9.3	3.80	19.2	8.15	15.4	6.22
8	9.6	3.87	17.6	7.73	14.2	6.84
9	8.1	3.93	22.7	7.64	13.3	6.63
10	6.5	4.09	20.8	7.06	15.7	6.25

4 Conclusion

With the deepening of electricity market reform, electricity informatization has developed by leaps and bounds. In recent years, the level of informatization in the electric power industry has been getting higher and higher, all of which are inseparable from the support of information technology, and the corresponding requirements for the security of information systems have also increased. The cloud security terminal is an innovative office system based on cloud computing technology, which can realize the push of the desktop system on the remote server to the terminal and the redirection of the terminal and terminal peripherals to the desktop system on the remote server. It will release applications including computing resources, storage resources, and management services to end users through a variety of terminal types. At the same time, the storage and management of data is centralized in the cloud, and the client does not participate in any calculation and application. It is efficient and green, and can effectively solve problems such as office terminal operation and maintenance and security management. With the help of the good characteristics of cloud security terminal, this paper explores the application of cloud security terminal in the information management of power industry. It has been verified by examples that the use of cloud security terminals can effectively deal with the security protection problems in the information management of the power industry. Due to the limited time, this paper still needs to improve in the resource-domain interaction and resource scheduling. The extensive and in-depth study of these problems will undoubtedly have a profound impact on the development of intelligent cloud computing technology.

References

1. Wang, H., Liu, G.: Security audit strategy of e-government cloud in big data age. Audit Econ. Res. **36**(04), 1–9 (2021)
2. Gan, Y., Chen, X.: Trusted client identity authentication simulation for power security control process. Comput. Simul. **38**(08), 181–184+189 (2021)
3. Wang, J.: Implementation of cloud security terminal in power information management. Appl. IC **38**(10), 118–119 (2021)
4. Yang, Q.: Research and construction of data quality management application in electric power enterprises based on HAWQ. Process Autom. Instr. **41**(12), 67–71 (2020)
5. Guan, G., Song, Q., Liu, H., et al.: Research on distribution network management and operation and maintenance system based on edge computing. Adv. Power Syst. Hydroelectr. Eng. **36**(10), 90–96 (2020)
6. Yang, R.: Research on regional big data analysis method based on cloud security and deep learning. Electron. Design Eng. **29**(10), 15–18+23 (2021)
7. Zhao, G., Chao, M., Xie, B., et al.: Application of deep belief network in cloud security situation prediction. J. Chin. Comput. Syst. **41**(06), 1195–1202 (2020)

Author Index

W. Fu and L. Yun (Eds.): ADHIP 2022, LNICST 468, pp. 785–787, 2023.
https://doi.org/10.1007/978-3-031-28787-9

Printed in the United States
by Baker & Taylor Publisher Services